AF658615

Basics of Marine and Estuarine Ecology

Abhijit Mitra • Sufia Zaman

Basics of Marine and Estuarine Ecology

Springer

Abhijit Mitra
Department of Marine Science
University of Calcutta
Kolkata, West Bengal
India

Sufia Zaman
Department of Oceanography
Techno India University
Kolkata, West Bengal
India

ISBN 978-81-322-2705-2 ISBN 978-81-322-2707-6 (eBook)
DOI 10.1007/978-81-322-2707-6

Library of Congress Control Number: 2015957433

Springer New Delhi Heidelberg New York Dordrecht London

Printed on acid-free paper

Springer (India) Pvt. Ltd. is part of Springer Science+Business Media (www.springer.com)

Acknowledgement

The purpose of this book is to focus the marine and estuarine ecosystem, which is one of the *vital* resource reservoirs of the planet Earth. These resources are not only the abiotic resources like oil, natural gas or minerals, but also encompass living resources like coastal vegetation, fishes, molluscs, arthropods and representatives of almost all the animal phyla. We hope this book will be a useful source of information for those working for a world that is ecologically sustainable.

This book is not a final work but should be viewed as part of an ongoing process, of a continuing effort to understand a set of complex interrelated global issues. It is part of a continuous flow of ecosystem related research papers and books. Indeed, some of the material in this book have been taken from our own earlier papers, where research scholars like Dr. Ananda Gupta, Dr. Amitava Aich, Dr. Kiran Lal Das, Dr. Aftab Alam, Dr. Harekrishna Jana, Dr. Rajrupa Ghosh, Mr. Prosenjit Pramanick, Mr. Shankhadeep Chakraborty, Mr. Atanu Roy, Ms. Bulti Nayak, Ms. Suhana Datta, Mr. Deepta Chakravartty, Ms. Kasturi Sengupta, Mr. Kunal Mondal, Mr. Subhasmita Sinha, Ms. Mahua Roychowdhury, Mr. Saumya Kanti Ray, Mr. Rahul Bose and Mr. Saurav Sett contributed their findings through rigorous field work.

Our debt to Dr. Kakoli Banerjee, Subhdra Devi Gadi and Dr. Rajrupa Ghosh is uncommonly large. Their contribution went far beyond that of editors as they added texture and colour to our prose. At times it was difficult to tell when editing ended and writing began.

Each of our colleagues, Dr. Subhra Bikash Bhattacharya and Mr. Tanmay Ray Chaudhuri, put great effort in updating the book. We are thankful to Dr. Pardis Fazli of the Department of Biological and Agricultural Engineering, University Putra, Selangor, Malaysia, for her effort in representing our data in graphical forms.

We are also indebted to the Ministry of Environment and Forest, Ministry of Earth Science, Department of Science and Technology (DST), Government of India, and IUCN for funding our venture through various projects. Several innovative programmes that constitute the annexure sections of the present book may serve as a road map in climate change mitigation and adaptation process. Few examples of such innovations are mangrove based fruit products by Mr. Prosenjit Pramanick, carbon content in gastropods by

Ms. Bulti Nayak etc., which are valuable assets of the present knowledge reservoir.

Finally, Dr. Abhijit Mitra expresses his gratefulness to his wife Shampa, daughter Ankita and mother Manjulika whose inspirations and encouragements acted as boosters to complete the manuscript. In many rigorous field works in Sundarbans, the assistance provided by Ankita has helped the authors to add case studies of various dimensions. The sacrifice and inspiration that Dr. Mitra received from his father Late Dhanesh Chandra Mitra can be considered as the foundation pillar of his effort in learning ecology and grasping the subject.

Dr. Sufia Zaman expresses her deepest gratitude to her mother Mrs. Ayesha Zaman for her unconditional love and practical day-to-day support, and to her father Mr. Salim-uz-Zaman who gave her immense moral support. Dr. Zaman also acknowledges the support of her beloved husband Dr. Sahid Imam Mallick. Dr. Zaman wishes to accord her deep sense of gratitude to her family members including her uncle (Mr. Pradip Kumar Mitra) and aunt (Late Mrs. Kanika Mitra), younger sister (Ms. Sharmilee Zaman), her in-laws and beloved grandmother (Mrs. Shibani Dhar) for their encouragement and inspiration throughout the strenuous period of manuscript preparation.

Contents

About the Authors

Abhijit Mitra Associate Professor and former Head, Department of Marine Science, University of Calcutta, India, has been active in the sphere of Oceanography since 1985. He obtained his Ph.D. as a NET qualified scholar in 1994. Since then he joined Calcutta Port Trust and WWF (World Wide Fund for Nature-India), in various capacities to carry out research programmes on environmental science, biodiversity conservation, climate change and carbon sequestration. Dr. Mitra is also serving as the advisor of Oceanography Division of Techno India University, Kolkata. He has to his credit about *325* scientific publications in various national and international journals, and *28* books of postgraduate standards. Dr. Mitra is presently the member of several committees like PACON International, IUCN, SIOS, Mangrove Society of India etc. and has successfully completed about *16* projects on biodiversity loss in fishery sector, coastal pollution, aquaculture, alternative livelihood, climate change and carbon sequestration. Dr. Mitra also visited as faculty member and invited speaker in several universities of Singapore, Kenya, Oman and USA. In 2008 Dr. Mitra was invited as visiting fellow at University of Massachusetts at Dartmouth, USA, to deliver a series of lecture on climate change. Dr. Mitra also successfully guided *24 Ph.D. students*. Presently his research areas include environmental science, mangrove ecology, sustainable aquaculture, alternative livelihood, climate change and carbon sequestration.

Sufia Zaman presently serving as Adjunct Assistant Professor in the Department of Oceanography in Techno India University (Kolkata, India) started her career in the field of Marine Science since 2001. She worked in the rigorous region of Indian Sundarbans and has a wide range of experience in exploring the floral and faunal diversity of Sundarbans. She has published *3* books on carbon sequestration, *75* scientific papers and contributed chapters in several books on biodiversity, environmental science, aquaculture and livelihood development. Dr. Zaman is presently a member of Fisheries Society of India. She is also running projects on carbon sequestration by mangroves of Indian Sundarbans. She is the recipient of DST Women Scientist and Jawaharlal Memorial Doctoral fellowship awards. Her areas of research include aquaculture, fish nutrition, phytoplankton diversity, climate change mangrove ecology and alternative livelihood. Dr. Zaman is also the first researcher in the maritime state of West Bengal (India), who initiated trial experiments on iron fertilization and subsequent enhancement of primary (phytoplankton) and secondary (fish) productions in the brackish water ponds of Indian Sundarbans with the financial assistance of the Department of Science and Technology, Government of India. Dr. Zaman is also providing consultancy on green technology to several industries, NGOs and corporate sectors.

1 Marine Ecosystem: An Overview

Contents

About 71 % of the surface of the planet Earth is covered by saline water. The water depth averages 3.8 km, a volume of 1370×10^6 km^3. This is the marine ecosystem and is the reservoir of food, oil, natural gas, minerals and several bioactive substances that have immense importance in the pharmaceutical industries. The ecosystem sustains the livelihood of millions of people. The oceans serve as the main highway for international trade as well as the main stabilizer of the world's climate. The oceanic waters and sediments are also the dwelling place of a large variety of flora and fauna. According to the Global Biodiversity Assessment, produced by the United Nations Environment Programme (UNEP), there are 178,000 marine species in 34 phyla.

The oceans are basically huge mass of saline water mainly separated from each other by the continents. On the other hand, the seas are varying extents separated from the oceans by island chains or by submarine ridges rising from the seafloor, subjected to considerable changes in the course of geological history. The ocean is divided into five major areas, namely, Antarctic Ocean, Arctic Ocean, Atlantic Ocean, Indian Ocean and Pacific Ocean, while the nomenclature of sea expands to the North Sea, Mediterranean Sea, Red Sea, Black Sea, Arabian Sea, Caribbean Sea, Baltic Sea (Table 1.1), etc. Unlike oceans, the seas are more prone to changes by climatic conditions.

A. Mitra, S. Zaman, *Basics of Marine and Estuarine Ecology*,
DOI 10.1007/978-81-322-2707-6_1

Table 1.1 Mean area and volume of oceans and seas

Name	Area (km^2)	Volume (km^3)
Atlantic Ocean	82,441,500	323,613,000
Indian Ocean	73,442,700	291,030,000
Pacific Ocean	165,246,200	707,555,000
Oceans	**321,130,400 (88.94 %)**	**1322,198,000 (96.50 %)**
Arctic Sea	14,090,100	16,980,000
Malay Sea	8,143,100	9,873,000
Central American Sea	4,319,500	9,573,000
Mediterranean Sea	2,965,900	4,238,000
Inter-continental seas	**29,518,600 (8.20 %)**	**40,664,000 (2.96 %)**
Baltic Sea	422,300	23,000
Hudson Bay	1,232,300	158,000
Red Sea	437,900	215,000
Persian Gulf	238,800	6000
Smaller enclosed seas	**2,331,300 (0.64 %)**	**402,000 (0.03 %)**
Bering Sea	2,268,200	3,259,000
Okhotsk Sea	1,527,600	1,279,000
Japan Sea	1,007,700	1,361,000
East China Sea	1,249,200	235,000
Andaman Sea	797,600	694,000
California Sea	162,200	132,000
North Sea	575,300	54,000
English Channel and Irish Seas	178,500	10,000
Laurentian Sea	237,800	30,000
Bass Sea	74,800	5000
Fringing seas	**8,078,900 (2.22 %)**	**7,059,000 (0.51 %)**
Hydrosphere	**361,059,200**	**137,032,500**

Both ocean and seawater are saline in nature. The salinity exhibits significant spatio-temporal variations. The salt in the marine ecosystem originates from various sources. It is stated that from the very beginning, the crust of the Earth is subject to slow dissolving action of water. Ever since seawater has been constantly washing the shores and rains falling on the land have been draining debris into rivers and seas. Thus, seawater contains all the elements of the Earth's crust.

The rivers bring about 5.4×10^8 of the total solids of the ocean. However, it is surprising that there exists great similarity between river water and seawater in terms of chemical composition. The latter has chloride in abundance (about 75 %), while the former has mostly calcium (60 %). There is only 2 % of sodium chloride in river waters. The conclusion is naturally that either rivers do not convey enough salts to the sea which makes it saline or whatever they pour in the sea is continually lost. The quantity of salts brought by the rivers is too small and estimated to be two millionth part of the ocean. Some of it is absorbed by animals and plants to form skeletons of lime or silica. A certain proportion called cyclic salts is swept off the sea towards the land by windblown spays and picked up by water vapour and again brought into the sea. Thus, they further reduce the gross contribution of salt which the rivers make to the ocean. Moreover, a great amount of calcium brought by rivers is constantly being withdrawn by marine animals like coral polyps, foraminifera, gastropods, oysters, etc. Inspite of the above facts, the small amount of salt brought annually by rivers increases the salinity of the sea.

Salts are also added either due to erosion of the earth crust or by some agencies which add minerals from the obscure sources buried deep within the earth, e.g. volcanic ash brings some material—especially chlorine, boron, sulphur and iodine.

The marine ecosystem can be divided into benthic (substratum based) and pelagic (aquatic) zones.

1.1 Benthic Compartment

The benthic compartment of the ocean starts from the shoreline and extends up to the hadal zone (Fig. 1.1). The zone is vast and complex and is characterized by the presence of several interesting sculptures and features. For the convenience of the readers, the benthic zone of the marine ecosystem is discussed here on the basis of subdivisions like supralittoral zone, eulittoral zone, sublittoral zone, continental shelf, continental slope, bathyal zone, abyssal plains and hadal zone.

1.1.1 Supralittoral Zone

The supralittoral zone also known as the splash zone or spray zone is the area above which the highest tidal water reaches. The organisms inhabiting this zone are adapted to minimum availability of water and include species of gastropods and isopods. Some thorny plants (mangrove variety) like *Acanthus* sp. are common in the supralittoral zone. The thorns are the features of xerophytes that reduce the loss of water from the plant body.

Sand dune is a common feature of tropical supralittoral zone. A sand dune is formed by a mound of sand deposited by winds, which rises to a summit and possesses a slip face. In the Sundarbans, the dunes run parallel to the southern coastline and have no definite shapes during the initial stages of formation. The eolian sand deposits make up an area known as the dune field. Wind sand ripples also characterize many areas of the dune field. Sand dunes in the Sundarbans are generally oriented parallel to the relevant wind and are barchan or transverse dunes. Barchan dunes are crescent-shaped formed by unidirectional wind and advance by avalanching of sand on the slop face. In contrast, transverse dunes are elongated, nearly straight and regularly spaced sand ridges.

The windward sides of dunes contain superimposed eolian ripples, wherever they are devoid of vegetation. These slopes, under permissible angle of repose, also show development of eolian ripples and wind ripples. The formation of dunes follows a basic sequence of dune formation stages.

During the initial stage of dune formation, the perennial grass *Paspalum vaginatum* of the family *Poaceae* forms an arc-shaped train in the sand with the convexity of the arc pointed along the direction of wind. This grass helps trap sand and initiates the growth of mounds of sand. The growth of the vegetation continues, which further accelerates the accumulation of sand and which, in turn, results in the growth of the dunes.

Hummocks of shrubs like *Clerodendrum inerme* may coalesce sometimes to give rise to complexities in dune morphology. A typical dune configuration is finally attained through the upwind deposition of more sand, away from the convex front of the arc-shaped hummock covered with vegetation. As a result, sand grains are deposited on the upper part of the slip face and may eventually avalanche. Further drifting of sands along the windward slope creates a steep face to the dune structure.

Plants play an important role in dune formation. The root acts as sand binders and helps stabilize the dunes while the leaves and stems help trap airborne sand. The vegetation protects the dune from destructive winds and arrests dune migration to a considerable extent. In many desert areas, vegetation cover has been found to be the most effective and permanent mechanism to arrest dune migration.

The plant communities associated with dunes may be separated into two distinct groups, namely, those which help in the establishment and growth of dunes and these which arrest dune migration. Some plant species such as grass *Paspalum* and the herbs *Aeluropus lagopoides*, *Cynodon dactylon*, *Opuntia* sp., *Salicornia* sp. and *Suaeda maritima* are associated with embryonic dunes (pro-dunes) and new dunes.

Ipomoea pes-caprae, *Launea* sp. and *Sesuvium portulacastrum* are generally associated with post-embryonic to mature dunes and arrest dune migration, to a great extent. The herb *S. portulacastrum* has extremely long, penetrating roots which may reach 2–3 m in length. *Ipomoea*

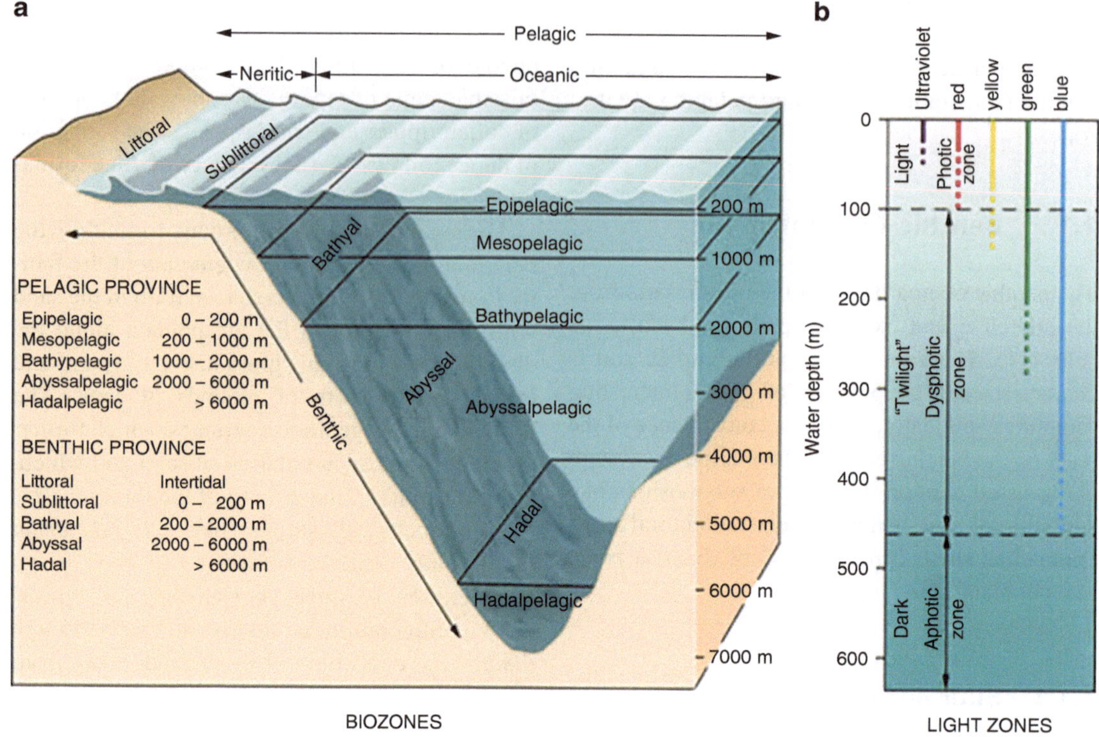

Fig. 1.1 Benthic compartment of the marine ecosystem

Fig. 1.2 *Ipomoea pes-caprae* serves as the sand binder in the intertidal zone

Fig. 1.3 View of *Launea* sp. that forms a dense cover on the dune surface and imparts stability

pes-caprae (Fig. 1.2) has good network of roots that often penetrate to depth of 1–2 m. The species, together with *Launea* sp (Fig. 1.3), can form a dense cover on the dune surface.

Investigations of the internal structure of the dunes revealed that wedge, lenticular and planar tabular type of cross-laminations in sections cut both parallel and normal to the lee directions of dunes. The boundary surfaces of the cross-laminations had, in many cases, been eroded. These surfaces were either straight or were concave upward. The sets varied in thickness from a few cm to 40 cm. The bounding surfaces showed variable degrees of inclination (5°–40°) in sections parallel to the lee-slope of the dunes.

Dune sections cut normal to the lee-slope also revealed bounding surfaces with variable inclinations. Sliding of the sands down the lip face caused high dip of the bounding surfaces, while erosion by sourcing on the stoss side was responsible for creating upward concave or straight low dipping bounding surfaces. The cross-laminations in between the upper and lower bounding surfaces within a set showed dips ranging between 20° and 35°. Laminar with dips of more than 30° were not uncommon. High moisture enabled the sand to remain cohesive at high angles of repose.

In some cases, the azimuthal directions of the cross-laminations in the same dune showed 180° reversals. The variability of inclinations of cross-laminations reflected avalanche faces, variations in wind directions and vegetation hummocks occurring in different pats and levels of the same dune. The internal laminations of the dunes were often disrupted by the roots of various types of vegetation.

1.1.2 Eulittoral Zone

It is the zone between high tide and low tide and sustains a wide spectrum of organisms. The zone is also referred to as midlittoral zone and is the dwelling ground of wide variety of organisms (Fig. 1.4).

The eulittoral zone sustains a wide spectrum of flora and fauna. In tropical regions, this zone is the survival ground of mangroves, which are specialized halophytes with pneumatophores, supporting root system and possess features like viviparous germination (Fig. 1.5).

The eulittoral zone is the matrix for ecological succession, and the pH of the substratum (Fig. 1.6) associated with the mangrove forest is slightly acidic in nature, with values ranging between 5.5 and 6.8 (Fig. 1.7).

Fig. 1.4 Intertidal mudflat in the estuarine ecosystem falls under the eulittoral zone; *HTL* high tide level, *LTL* low tide level

Fig. 1.5 Eulittoral zone: survival ground of coastal vegetation

Fig. 1.6 Eulittoral zone adjacent to mangrove forest

Fig. 1.7 Slightly acidic pH of the eulittoral substratum

1.1.3 Sublittoral Zone

This is the zone below eulittoral zone and is always exposed to saline water. The lower limit of sublittoral zone extends up to the edge of the continental shelf. Several species of seaweeds are found in the sublittoral zone, and their growth is luxuriant due to presence of sun light.

1.1.4 Continental Shelf

Continental shelves are generally flat areas, averaging 68 km (40 miles) in width and 130 m (430 ft) in depth, which slope gently towards the bottom of the ocean basin. The width of a continental shelf is frequently related to the slope of the land it borders. Mountainous coasts, like the West Coast of the United States, usually have a narrow continental shelf, whereas low-lying land, like the East Coast of the United States, usually has a wide one. Continental shelves are the extensions of the continents to which they are attached (Fig. 1.8). The criterion for defining the seaward edge of the continental shelf is a marked change in slope. The continental shelf has a gradient of about 1:1000.

The continental shelves are basically part of the landmasses, and during geologic past, they witnessed rise and fall of the sea level. When the sea level was low during, the ice ages (periods of increased ice on land), erosion deepened valleys, waves eroded previously submerged land and rivers left their sediments far out on the shelf. When the ice melted and sea level rose, these areas were flooded and sediments built up in areas close to the new shore. Although presently submerged, these shelf areas bear the signatures of old riverbeds and glaciers, features they acquired when exposed as part of the continent. Some continental shelves are covered with thick deposits of silt, sand and mud derived from the land; for example, the Mississippi and Amazon Rivers deposit large amounts of sediments at their mouths. Other shelves are bare of sediments, such as where the Florida Current sweeps the tip of Florida, carrying the shelf sediments northwards to the deeper water of the Atlantic Ocean.

The repeated emergence and submergence of coast have been instrumental in shaping the morphology of the continental shelves. A study conducted on this dynamicity in the Indian subcontinent (East Coast) is highlighted as Annexure 1A.

The continental shelves (Fig. 1.9) act as the reservoir of several commercially important items like sand, petroleum, natural gas and

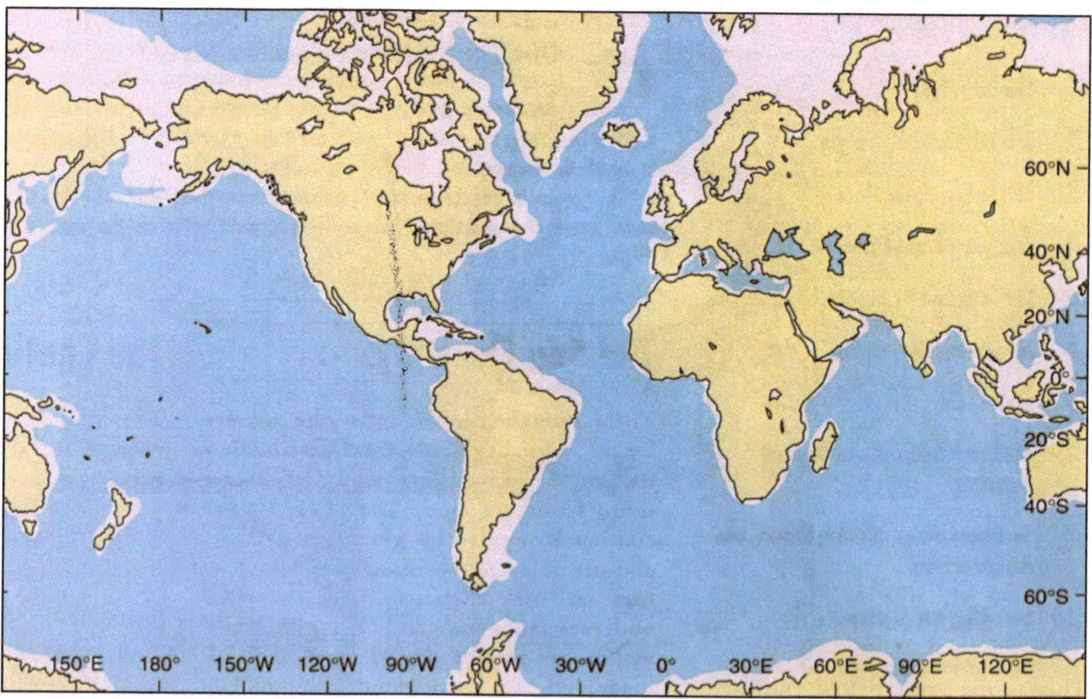

Fig. 1.8 Pinkish colour denoting the continental shelves

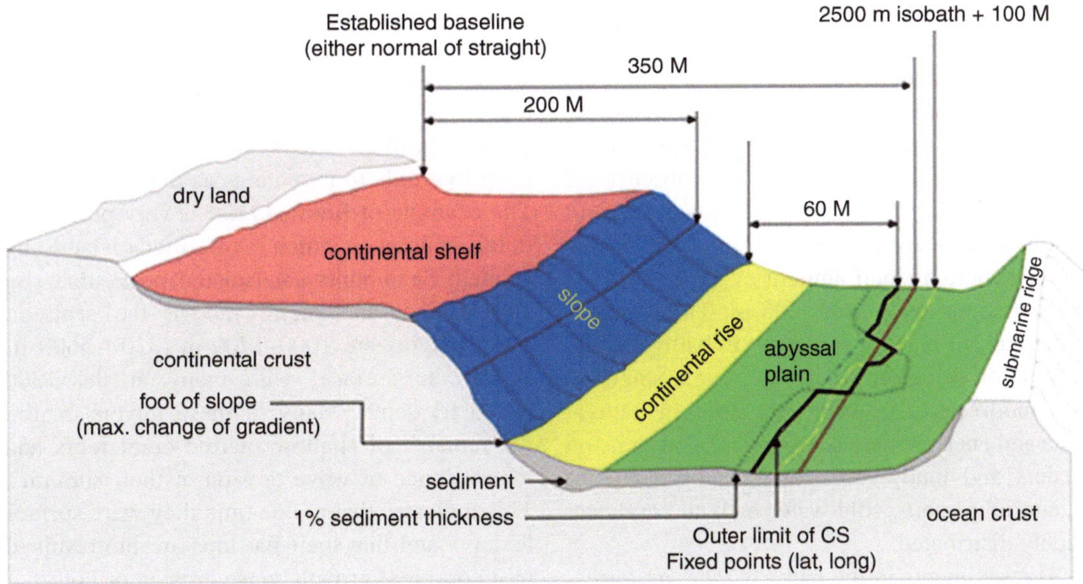

Fig. 1.9 Continental shelf: a zone for resource exploration and utilization

different categories of shellfish, finfish, seaweeds, etc. These areas are also treated as the bin of civilization, as several wastes of complex nature are dumped on the continental shelves. As the continental shelves are extremely important in terms of commercial valuation and ecosystem services, therefore some regulations have been imposed on the issue of commercial exploration and valuation. In the domain of laws and policies, continental shelves as mentioned in Article 76 encompasses prolongation of the land territory of all the maritime states that remains in submerged state including the seabed and subsoil and extends up to a distance of some 200 nautical miles.

1.1.5 Continental Slope

Seawards from the continental shelf is the continental slope, which extends to a depth of 1.6–3.2 km. The continental slope has a gradient between 1:2 and 1:40. The angle and extent of the slope may be of varying nature. In some oceans, the angles may be low, whereas in some other cases, the angles may be high that causes steepness. The depth may increase rapidly from 200 m (600 ft) to 3000 m (10,000 ft) or it may drop as far as 8000 m (26,000 ft) into a deep-sea trench, as it does off the West Coast of South America. The continental slope may show rocky outcroppings, and it is often relatively bare of sediments because of its steepness. Some continental slopes possess submarine canyons that are similar to canyons found on land. Many of these submarine canyons are aligned with river systems on land and were probably formed by these rivers during periods of low sea level. The Hudson River canyon on the East Coast of the United States is an example of this. Other submarine canyons have ripple marks on the floor, and at the ends of the canyons sediments fan out, suggesting that they were formed by moving sediments and water. Oceanographers believe that these canyons were formed by turbidity currents. Turbidity currents are swift avalanches of sediment and water that erode a slope as they sweep down and pick up speed. At the end of the slope, the current slows and the sediments fan out. Turbidity currents can be caused by earthquakes or the accumulation of large amounts of sediments on steep slopes that overload the slope's capacity to hold them.

1.1.6 Bathyal Zone

The bathyal zone lying between 200 m and 2000 m is characterized by absolute darkness and low temperature, making the survival of organisms highly difficult. In the absence of solar radiation, photosynthesis is practically nil in the bathyal zone except in some regions of tropics where a small amount of solar radiation can penetrate up to some 600 m. The speed of water current is extremely slow in bathyal zone, and in some regions, water is almost stagnant. The biodiversity is extremely poor in bathyal zone and encompasses few species of suspension feeders and mud scavengers. In sub-Arctic to equatorial regions, cold water bathyal corals are widely distributed.

The sediments in the bathyal zone are terrestrial, pelagic or authigenic (formed in place) in nature. Terrestrial (or land-derived) sediments are predominantly clays and silts and are commonly coloured blue because of accumulated organic debris as well as bacterially produced ferrous iron sulphides. Coarser terrigenous sediments are also brought to the bathyal seafloor by sporadic turbidity currents originating in shallower areas. Where supplies of terrigenous materials are scarce, microscopic shells of phytoplankton (coccolithophorids) and zooplankton (foraminiferous and pteropods) fall through the water grain by grain, accumulating as white calcareous ooze deposits. Authigenic sediments result from the interaction of clay, feldspar, and volcanic-glass particles with seawater, forming the minerals glauconite, chlorite, phillipsite and palagonite. These sediments are characteristically green because of their chlorite and glauconite contents.

1.1.7 Abyssal Plains

The term abyssal plain indicates the flatness of the surface. Abyssal plains are found at the base of the continental rise and are relatively flat plains having a gradient less than 1:1000. They are formed due to even deposition of sediments from the continental rise carried down by the turbidity currents. Abyssal hills and seamounts are scattered across the seafloor. Submerged flat-topped seamounts, known as guyots, are found most often in the Pacific Ocean. They may vary from low hills to mountains as high as 1525 m. The example of Bermuda rise is very prominent in this context on which the Bermuda Islands are formed. Seamounts are isolated peaks that rise several thousand metres above the seafloor. These guyots are 1000–1700 m (3300–5600 ft) below the surface, with many at the 1300 (4300 ft) depth. Many of these guyots exhibit the remains of shallow marine coral reefs and the evidence of wave erosion at their summits. This indicates that at one time they were surface features and that their flat tops are the result of past coral reef growth, wave erosion or both.

1.1.8 Hadal Zone

The hadal zone represents the deepest part of the ocean. This zone is found from a depth of around 6000 m (20,000 ft) to the bottom of the ocean. The hadal zone has extremely poor biodiversity due to stressful environmental conditions in terms of temperature, pressure and light availability. In the absence of any solar radiation, life in the hadal zone is triggered by the process of chemosynthesis.

Trenches are long narrow depressions in the hadal zone that are over 6100 m deep. The deepest known trench in the ocean compartment is the Mariana Trench of the western North Pacific that is about 11,000 m deep. The Peru–Chile Trench extends for over 6120 km (3600 miles) along the coast of South America and is the longest of the ocean trenches. The Java Trench extends for a distance of almost 4760 km (2956 miles) along the coast of the islands of Indonesia. By comparison, there are only two relatively short trenches in the Atlantic, namely, the South Sandwich Trench and the Puerto Rico–Cayman Trench. Trenches are invariably associated with the systems of active volcanoes and are believed to be caused by down wrapping of the oceanic crust beneath the continental crust.

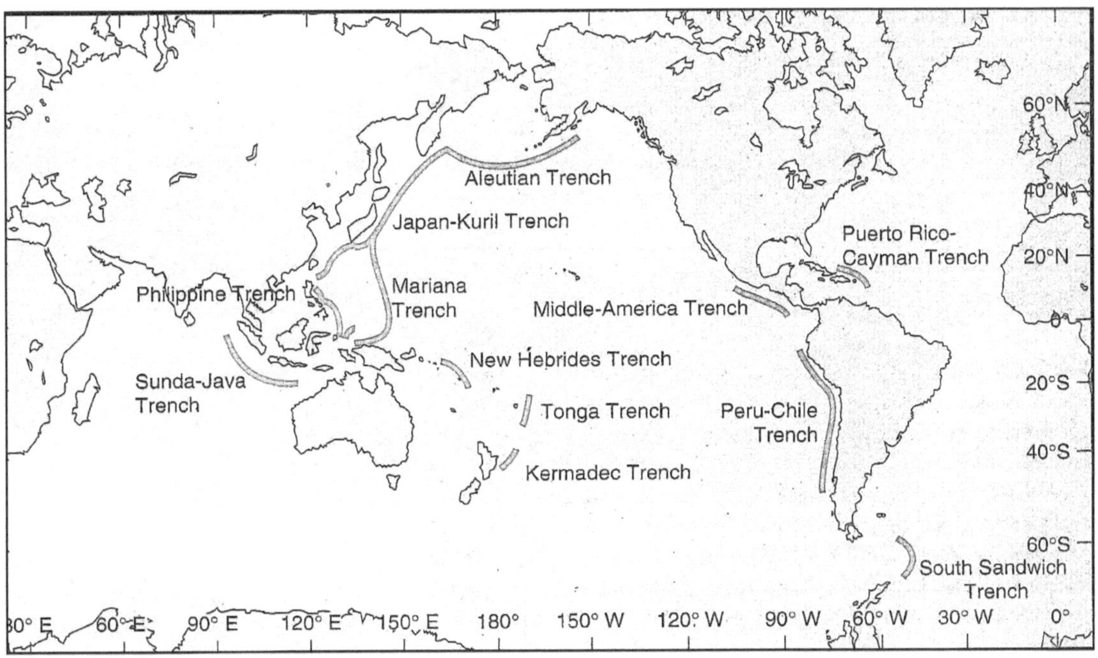

Fig. 1.10 Major trenches of the world

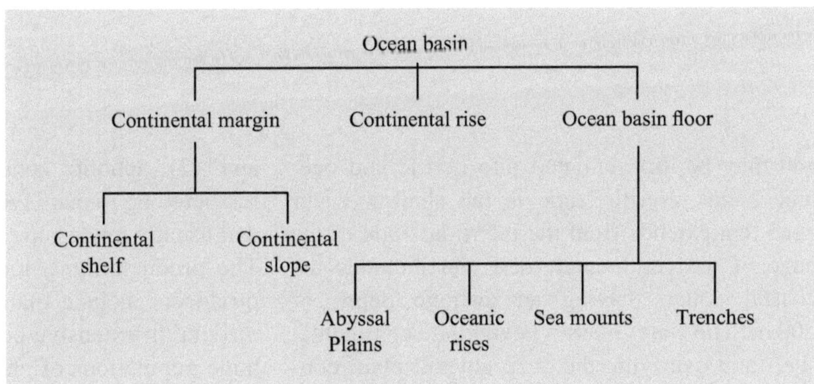

Fig. 1.11 Components of the ocean basin (after Mitra 2000)

Figure 1.10 highlights some major trenches of the world.

It is interesting to note from the above discussion that the ocean basin has unique designs and sculptures similar to the land features above it (Fig. 1.11). Mountain ranges, canyons, valleys and vast plains are all the important components of the underwater landscape. These physical features of the ocean bottom are called bathygraphic features, and unlike their counterpart topographic features on land, they change at relatively slow pace. Erosion is slow in the relatively calm recesses of the ocean, and changes are mainly attributed to sedimentation, uplifting and subsidence.

The structures and features of ocean basin are summarized in Fig. 1.11.

1.2 Aquatic Compartment

The entire water mass above the ocean floor (benthic substratum) with all the diverse types of organisms is referred to as the pelagic zone

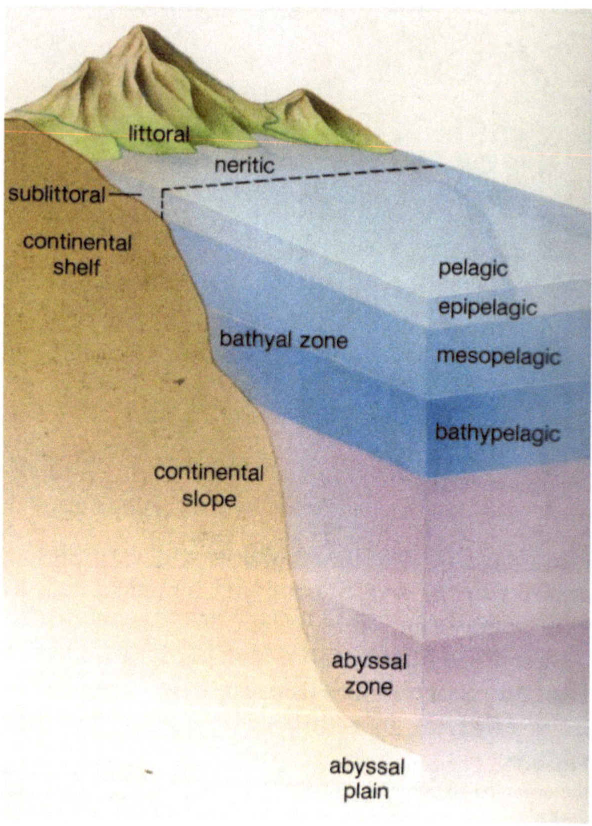

Fig. 1.12 Division of pelagic zone of the ocean

and may be differentiated into neritic and oceanic zones. Neritic zone is the shallow-water zone that extends from the intertidal zone to the edge of the continental shelf and includes the coastal waters having an average depth of 200 m. The water masses beyond the continental shelf and overlying the deep abyssal plain constitute the oceanic zone. The depth of this zone is more than 6 Km. The pelagic zone is divided vertically into sub zones like epipelagic zone (up to 200 m), mesopelagic zone (200–1000 m), bathypelagic zone (1000–4000 m) and abyssopelagic zone (4000–6000 m) (Fig. 1.12).

1.2.1 Classification on the Basis of Light Penetration

On the basis of distribution/penetration of light, the oceanic compartment may be divided into two distinct zones, namely, the (1) photic zone and (2) aphotic zone. The **photic zone** is restricted up to that layer where light is available and usually ranges up to 200 m from the surface. The productivity is more in this zone and the producers include many species of large algae thriving in extensive beds (Fig. 1.13), as well as huge populations of phytoplankton that serve as nutrient base for the rest of the consumers.

The **aphotic zone** is the deeper lightless zone of the ocean usually below 200 m. Apart from the above two zones, there is another zone known as the **dysphotic zone**, sometimes referred to as the twilight zone. This zone has very low levels of illumination because the water above absorbs more than 95 % of the sunlight. Photosynthesis under these dim conditions is not possible.

In the euphotic zone, there is plenty of light to perform the process of photosynthesis by macrophytes and macrophytes due to which organic material is produced. This production exceeds the loss of organic matter through the

Fig. 1.13 Large algae on hard substrata in the photic zone

process of plant respiration. The amount of light required for photosynthetic productions to just balance the respiratory losses in plants is referred to as the 'compensation light intensity', and the depth at which the photosynthetic production is balanced by plant respiration is known as 'compensation depth'. Basically compensation depth defines the lower boundary of the euphotic zone.

Light has immense role in regulating the behaviour of marine organisms. Light has regulatory role on the behaviour of marine and estuarine organisms particularly on zooplankton (Clarke 1934; Duval and Geen 1976; Pagano et al. 1993; Atkinson et al. 1996). Ultraviolet (UV) radiation, especially with a wavelength of 280–315 nm, has adverse impact on many zooplankton species (Hunter et al. 1981; Kouwenberg et al. 1999). Some animals, for example, larvae of Coregonidae fishes, use skin pigmentation and avoidance behaviour to protect themselves against UV radiation (Ylognen et al. 2005). Larvae of several benthic organisms and deep-dwelling shrimp species escape the adverse impact of UV radiation by changing their dwelling place (Frank and Widder 1994; Adams 2001). Copepod (Martin et al. 2000; Rhode et al. 2001) and cladoceran (Johnsen and Widder 2001) species avoid UV stress by vertical migrations. Some crustaceans have evolved biochemical methods of avoiding the UV-induced stress, including pigmentation (Rhode et al. 2001); more coloured shrimps tend to occur in deeper water layers (Vestheim and Kaartvedt 2009). Different responses of cladoceran species to UV exposure in various freshwater lakes have been documented (Leech and Williamson 2000; Leech et al. 2005). Light-dependent behaviour of the abundant zooplankton species inhabiting the White Sea (situated adjacent to North Polar circle) was studied experimentally by Daria et al. (2010).

1.2.2 How to Estimate the Age of Ocean Water?

The age of ocean water refers to the time or period since the water mass was last at the surface and in contact with the atmosphere. The computation of age of ocean water has great relevance. It throws light on the rate of overturn of ocean water. This has important implications in terms of dumping of conservative pollutants like radioactive wastes and also replenishment of nutrients. If the average time of overturn is much less than the half-life of such materials, it would be dangerous to dump them in the ocean because they would be brought to the surface while still active and might be picked up by fish and so conveyed back to man.

There are several methods to estimate the age of ocean water, which are discussed here.

1.2.3 Dissolved Oxygen (DO) Consumption Method

This method was devised by Worthington in 1954. He observed in the deep water of North Atlantic Ocean (at depths of 2500 m) the average dissolved oxygen content decreased by 0.3 ppm between 1930 and 1950. Considering the consumption of oxygen to be constant, the rate of decrease of DO is 0.3 ppm/20 = 0.015 ppm/year. Now this data can be used to calculate the age of the ocean in the following way:

Suppose, initial DO in the surface water of Bay of Bengal = 7.6 ppm (in 2014).
Assume final DO (at a certain depth) in Bay of Bengal = 5.8 ppm
Change in DO = 1.8 ppm
Considering the rate of decrease of DO = 0.015 ppm/year, the age of the water is then 120 years.

This implies that the present parcel of water was last at the surface of Bay of Bengal during 1894.

This calculation is based on several assumptions as, for example, (i) the consumption of oxygen occurs by chemical combination with detritus, (ii) the consumption rate of oxygen is constant through seasons and years and (iii) physical processes like wave action, turbulence, upwelling, etc., have been ignored in this calculation.

1.2.4 Decay Rate of ^{14}C Method

This method was first reported by Kulp et al. (1952). According to this estimation, the atmosphere at the sea surface was only source of ^{14}C to ocean waters. Away from the surface, the ^{14}C content would not be replenished and would decay with its half-life of about 5600 years. The early measurements suggested an age of the order of 2000 years for water at 2000–5000 m in the North Atlantic, but this was subsequently shown to be too high on account of contamination in the chemical processing.

1.2.5 ^{90}Sr Content Method

Measurement of the ^{90}Sr content of ocean water has revealed significant amounts at depths to 1000 m. As the only source of this isotope is presumed to be the residue from atom bombs, starting in 1954, this indicates that the rate of vertical mixing in the upper waters may be quite rapid.

1.2.6 Hydrogen Isotope Method

Radioactive tracer like tritium (^3H) with a half-life of about 12 years is also used to estimate the age of ocean water. It occurs in the upper layers of the ocean at concentrations of the order of only one tritium atom for 10^{17} or 10^{18} ordinary hydrogen atoms (^1H), but with new techniques, it can be measured quantitatively for age determinations of seawater which have been routinely measured in the systematic redetermination of the distribution of the main elements in the world oceans in the GEOSECS (Geochemical Ocean Sections Study). This was a multinational, multi-institutional study of the main oceans and whose objective was the study of the geochemical properties of the oceans with respect to large-scale circulation problems.

Brain Churners

1. What causes the water salty in the ocean?
2. Why the environment is stressful in supralittoral zone?
3. How the salts of the ocean water are used by the organisms?
4. Why continental shelves are highly productive compared to continental slopes and abyssal plains?
5. Why the length of the food chain and the number of trophic levels are more in the continental shelves and low in the hadal zone?
6. Why photic zone of the ocean has greater fishery production compared to dysphotic or aphotic zone?
7. Why seamounts are flat topped?
8. How submarine canyons are formed on the continental slopes?
9. How trenches are formed at the bottom of the ocean?
10. Why the potential fishing zones mostly coincide with the zone of upwelling?

Annexure 1A: Oscillation of Seafloor in the Gulf of Mannar

1. Introduction

The floor below the ocean compartment is never stable. Several researches have been carried out on the changes in seafloor, which may be attributed to alteration of land surface elevation. Changes in absolute water surface levels occur worldwide as the oceans are interlinked. Such alterations are referred to as eustatic changes. In many cases, it has been observed that the extent of continental shelf undergoes changes with the passage of time. In case of Indian subcontinent, a pronounced variation in the extent of continental shelf is observed between the East Coast and the West Coast. The shelf on the West Coast is broad with thin layer of sediment, while the shelf in the East Coast is narrow with thick layer of sediment. The case studies of the area of Gulf of Mannar, extending from Tuticorin to Rameswaram Island, are very pertinent in this context. The area lies between 8°47′ and 9°15′ N latitudes to 78°5′ and 79°30′ E longitudes and sustains several types of habitats like seagrasses, seaweeds and coral reefs (Fig. 1A.1).

2. Materials and Methods

The main components used for this study are Naval Hydrographic Chart (1975), Topographic Sheet of Survey of India (SOI), Eco-Sounder and Global Positioning System (GPS). The bathymetry map was prepared on the basis of Naval Hydrographic (1975) chart data and was subsequently digitized into ARC-INFO to prepare a digital elevation model using ERDAS imagine software. A survey on the bathymetry of the area was carried out during April 1999 using eco-sounder and GPS along Mandapam and Tuticorin coastal area in the Gulf of Mannar. The depth values were recorded with respect to chart datum. The measured depths were tide corrected with respect to time and then converted with respect to chart datum.

3. Results and Discussion

In the shore between Tuticorin and Vaippar River, the seafloor is sloping gently down to 5 m depth, while in Sippikkulam coast where the seafloor is very steep down to 2 m depth, it extends up to 0.129 km from the coast. In between 4 m and 5 m depths, some elevated rises and islands are observed. At a distance of 4.30 km (8°49′ N–78°12′16″ E) from Tuticorin coast, a 3.9 m elevated rise was observed. Just north-east of this rise a 6 m elevated island (8°49′35″ N–78°12′28″ E) called Van Island, which is situated 5 km away from Tuticorin coast. Around this island, the seafloor is sloping very gently down to 2 m depth in southeast, east,

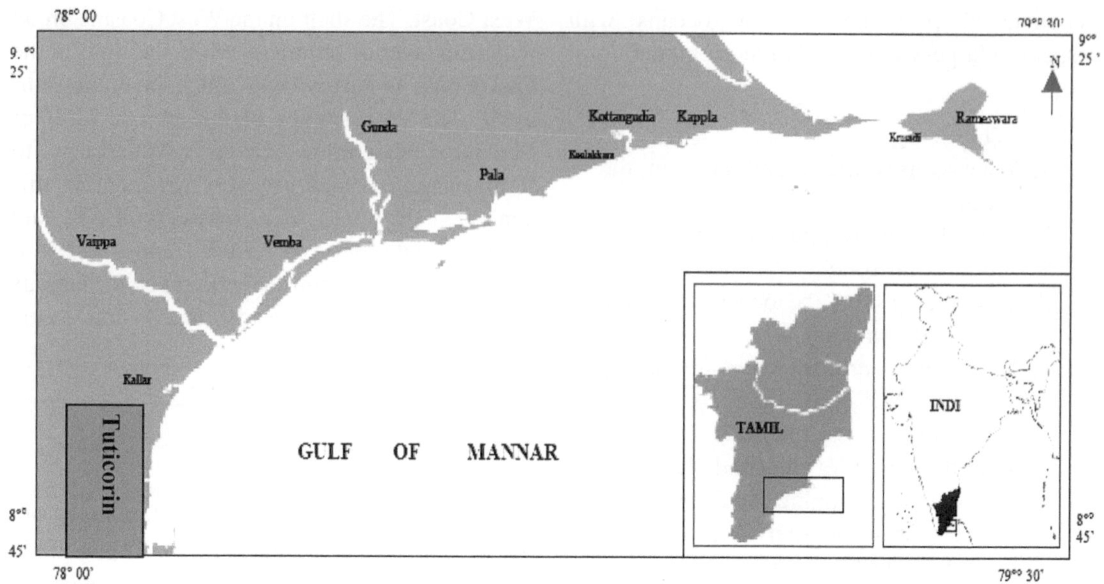

Fig. 1A.1 Location of Gulf of Mannar

north-east and north directions. Whereas in the west, southwest and northwestern parts of the island, the seafloor slope is very steep. On the north-eastern, eastern and southern sides of the island, the seafloor is covered by fringing coral reef, extending down to 2 m depth with an approximate distance of 1.50 km from the coast of the island. Between the depths of 3 m and 4 m, there is another island called Koswari Island, which is located at 8°52′2″ N–78°13′22″ E, with a distance of 6.09 km from Tharuvaikulam coast. In between 5 m and 10 m depths, the seafloor slopes moderately. From the depth of 10 m to 20 m, the seafloor slopes gently, having a distance of 16.65 km between them. At a depth of 20 m, the seafloor falls suddenly with a very steep slope (continental slope) extending till 30 m depth. The width of this slope has been calculated as 7.27 km. The total width of continental shelf in Tuticorin region has been calculated as 26.75 km. In the coast from Veppalodai to Sippikkulam, between 4 m and 6 m depths, elevated islands, namely, Kariya Shuli and Vilangu Shuli Islands, are situated at a distance of 4.77 km and 6.56 km from Sippikulam coast respectively. The seafloor is found to be sloping gently towards north, east and south from the shore of Vilangu Shuli Island, whereas west of the seafloor tends to slope very steeply. An extensive well-developed fringing reef has been identified around Vilangu Shuli and Karia Shuli Islands extending to 3 m depth.

In the shore between Vaippar and Muttiapuram, the seafloor is at a depth of 3 m, which extends to an approximate distance of 3.51 km from the shore. In the area near the mouth of Vembar, the seafloor topography is plain with a depth of 2 m. Between 2 m and 20 m, the seafloor slopes very gently and extends to a distance of 19.75 km. A sudden steep slope is encountered at 20 m depth; this area is the starting point of the continental slope and it extends to the depth of 30 m. The width of the slope has been calculated as 5.45 km, and the width of the continental shelf has been calculated as 19.75 km. In the area from Terkku Narippeyur to Gundar River, the shelf is found at a depth of 3 m, extending to 0.36 km south and southeast of Narippeyur coast and 0.88 km south of Gundar River-mouth. Between the depths of 7 and 8 km, the seafloor is considerably broad and the slope is very gentle. An elevated island is found exposed above sea level, viz., Uppu Tanni Island situated at 6.72 km from Gundar River-mouth. The seafloor is plain up to 1 m depth in all directions around the island except in the north and north-eastern directions where the slope is very steep. In the shore between Gundar and Palar River, the seafloor is found at a depth of 3 m, and this deep extends to 1.19 km from the coast. At the south of Mel Mendal coast, the seafloor slopes very steeply to 7 m depth and extends to 1.44 km from the coast. Between the depths of 10 and 20 m, the seafloor slopes moderately with a width of 7.52 km. At an average depth of 20–30 m, there is a continental slope with 16.56 km width. In between 7 and 8 m depths, the seafloor is broad and has a very gentle slope. In this region particularly at latitude of 9°6′5″ N and longitude of 78°32′10″ E, an island, namely, Shalli Island, having 9 m elevation from seafloor is encountered. On the northern and southern sides of this island, the seafloor slopes very steep, while on the eastern and western sides, the slope is gentle. Towards the east from Shalli Island, another island, namely, Nalla Tanni Island, is encountered (9°6′11″ N–78°34′29″ E). Around this island, the seafloor slopes very gently, particularly in the north-east, east and southeastern sides of the island, which extends to a depth of 4 m from the shore of the island. Whereas in the northern and southwestern sides of this island, the seafloor have a very steep slope, extending to a depth of 4 m at a distance of about 0.36 m from the coast of the island. In the region between the Palar River-mouth and Dhanushkodi shelf, the topography has some irregularities. In the area opposite to the Palar River-mouth, there is a gentle depressed channel, having an approximate length of 4.41 km towards south. In the coast between the Palar River-mouth and Kalachimundal, the seafloor is found at a depth of 1 m, and it extends to a distance of 0.28 km from the shore. Between these areas, the seafloor gradient is very steep to a depth of 7 m. The width of this area is 1.60 km. After reaching 7 m depth, the seafloor rises up to 3 m depth; this depressed channel runs to a distance of 21.27 km towards the north-east and south and lies between 9°9′15″ N–78°40′4′ E and –9°13′8″ N–78°47′37″ E. The average width of this channel has been measured as 0.75 km. After reaching 3 m depth, the seafloor has a sudden fall to 10 m depth. In between the two depressions, a flat-topped continental rise has been observed. The average width of the continental rise is 1.58 km and is located 5.7 km from the shore. On this rise, there are two elevated islands rising 4 m from the seafloor and are located between 9°9′3″ N–78°41′28″ E and 9°9′4″ N–78°43′32″ E. These islands are called as Anaipar and Pilliyarmunai Islands. Around these islands, the seafloor slopes very gently, and extensively developed fringing corals are found within 2 m depth. Between 4 and 5 m depth, at a lat long of 9°9′12″ N–78°45′8′ E, an island, namely, Puvarasanpatti Island, having an elevation of 4.50 m from the seafloor, is encountered. At a depth between 10 m and 20, the seafloor slopes gently. The width of this continental shelf is 18.17 m. In the area between Keelakarai and Pudumadam, the seafloor is almost plain and has a depth of 3–4 m. This plain extends to a distance of 9.72 km from the coast. In this plain, some low elevated rise and Islands are observed. Located at a lat long of 78°49′10″ E 9°9′31″ N, an island, namely, Appa Island, having an elevation of 5 m from seafloor, has been noticed. Around this island, the seafloor slopes gently to 2 m depth. It extends to an average distance of 1.50 km. At 9°14′2″N–78°51′25″ E and 9°14′32″ N–78°52′31″ E, two continental rise have been observed having an elevation of 1 m from the seafloor. In between 4 and 5 m depths, there are some low elevated islands (5.50 m from seafloor), namely, Talairi, Valai and Muli Islands. They are located at an average distance of 8.45 km from Kaplar River-mouth. In this region, the seafloor gradient is very steep extending from 5 m depth to 10 m depth and then it slopes gently up to 30 m depth.

In the coast between Pudumadam and Thoniturai, the seafloor is almost plain having a depth of 2-3 m. This plain extends 6.95 km from the coast. Near Seeniappa Dargah and Thoniturai, this plain is encountered at 2 m depth, and it extends to an average distance of 2.17 km from the coast of Seeniappa Dargah and 0.50 km from Thoniturai coast. Along this plain, some low elevated continental rise (9°12′14″

N–79°5′19″ E) and chain of islands have been observed. From Seeniappa Dargah to 6.65 km towards south, an island, namely, Musal Island, having an elevation of 3.5 m from seafloor is encountered. Around this island, fringing corals have developed very extensively to a depth of 2 m; they extend 1.45 km towards the north, 1.75 km towards the northwest, 1.69 km towards the west and 1.42 km towards the south from the coast of the island. Around this island, the seafloor slopes gently up to a depth of 2 m. On the seaward side of this island, the seafloor slope is very steep to 10 m depth and to an average distance of 2 m. About 6.18 km away from Maraikayar Pattinam, there are two islands, namely, Manalli and Manalliputti Islands (9°12′23″ N–79°7′26″ E and 9°12′23″ N–79°8′16″E). Around these islands, the seafloor is encountered at 1 m depth, and the topography of the seafloor around this island has a moderate gradient. On the seaward side of this island, the seafloor slope is very steep, between 3 m and 6 m depths. The width of this slope is 0.73 km. At a depth of 6–7 m, a plain having a width of 3.52 km is encountered. The seafloor slopes gently between 7 and 30 m depths. In the area between Thoniturai and Pamban canal, the seafloor is encountered at a depth of 1 m and extends to an average distance of 1.07 km from the coast of Thoniturai and Velupilliyarkovil. In this area, the seafloor slopes very gently up to 2 m depth. In between 2 and 3 m depths, there is a vast plain seafloor extending 3.80 km from north to south and 23.3 km from east to west. Along this plain, there are low elevated chains of islands (9°14′28″ N–79°10′28″ E and 9°13′28″ N–79°14′16″ E), namely, Pumurichan, Kovi, Kursadi and Shingle Islands. Around these islands, the seafloor is encountered at a depth of 0.5 m and extends to an average distance of 0.49 km towards north, 1 km towards south, 0.19 km towards east and 0.27 km towards west. In the north of these islands, the seafloor slope is steep (till a depth of 1 m), whereas it is very gentle in the south (till a depth of 3 m). In between 3 m and 10 m, the seafloor slope is steep having a width of 1.83 km. From 10 to 30 m, the seafloor slopes moderately. In the area between Dhanushkodi and Kundugal, the seafloor is at a depth of 6 m and slopes moderately up to 30 m depth. The width of the continental shelf in this area has been measured approximately as 26.25 km.

A comparative study was conducted using the contour map of 1999 and bathymetry map of 1975. The study exhibited a decrease in the seafloor level in the study area. This may be attributed to emerging of land due to tectonic activity. Several marine scientists have also confirmed that the coast of Gulf of Mannar is on an emerging phase due to tectonic movement (Foot 1888; Ahmad 1982; Stoddart and Pillai 1972; Loveson and Rajamanickam 1988; Ramasamy 1989). Ramasamy (1997) has build up a post collision tectonic model for the southern part of Indian in which he has observed a series of geo-environmental problems being caused due to such ongoing tectonic movement. In some areas (particularly at river-mouths and some islands), the seafloor level increased as a result of erosional activities. In and around the Tuticorin group of islands, the illegal removal of corals resulted in an increase in depth (Thanikackalam and Ramachandran 2002). The average depth reduction of seafloor along the coast of the study area has been estimated as 0.51 m over a period of 24 years, which is equivalent to 0.021 m/year along the coast and 0.023 m/year around the island. This estimation is based on the assumption that the rate of change of seafloor is uniform throughout the study period. Thus, we conclude from this result and discussion that tectonic movement coupled with anthropogenic activities (like coral mining) play a crucial role in regulating the anatomy, physiology and morphology of seafloor.

References

Adams, N. L. (2001). UV radiation evokes negative phototaxis and covering behavior in the sea urchin *Strongylocentrotus droebachiensis*. *Marine Ecology Programme Series, 213*, 87–95.

Atkinson, A., Ward, P., & Murphy, E. J. (1996). Diel periodicity of sub-Antarctic copepods: Relationships between vertical migration, gut fullness and gut evacuation rate. *Journal of Plankton Research, 18*, 1387–1405.

Clarke, G. L. (1934). Factors affecting the vertical distribution of the copepods. *Ecology Monograph, 4*, 67–90.

Duval, V. S., & Geen, G. H. (1976). Diel feeding and respiration rhythms in zooplankton. *Limnological Oceanography, 21*, 823–829.

Frank, T. M., & Widder, E. A. (1994). Evidence for behavioral sensitivity to near-UV light in the deep sea crustacean *Systellaspis debilis. Marine Biology, 118*, 279–284.

Hunter, J. R., Sandor, S. E., & Taylor, J. H. (1981). Effects of solar and artificial ultraviolet-B radiation on larval northern anchovy, *Engraulis mordax. Photochemical Photobiology, 34*, 477–486.

Johnsen, S., & Widder, E. A. (2001). Ultraviolet absorption in transparent zooplankton and its implications for depth distribution and visual predation. *Marine Biology, 138*, 717–730.

Kouwenberg, J. H. M., Browman, H. I., & Cullen, J. J. (1999). Biological weighting of ultraviolet (280–400 nm) induced mortality in marine zooplankton and fish. I. Atlantic cod (*Gadus morhua*) eggs. *Marine Biology, 134*, 269–284.

Kulp, J. L., Tyron, L. E., Eckelmann, W. R., & Snell, W. A. (1952). Lamont natural radiocarbon measurements. *Science, 116*, 409–414.

Leech, D. M., & Williamson, C. E. (2000). Is tolerance to UV radiation in zooplankton related to body size, taxon, or lake transparency? *Ecological Application, 10*, 1530–1540.

Leech, D. M., Padeletti, A., & Williamson, C. E. (2005). Zooplankton behavioral responses to solar UV radiation vary within and among lakes. *Journal of Plankton Research, 27*, 461–471.

Martin, G. G., Speekmann, C., & Beidler, S. (2000). Photobehavior of the harpacticoid copepod *Tigriopus californicus* and the fine structure of its nauplius eye. *Invertebrate Biology, 119*, 110–124.

Martynova, D. M., & Gordeeva, A. V. (2010). Light dependant behaviour of abundant zooplankton species in the White Sea. *Journal of Plankton Research*, 1–16. http://plankt.oxfordjournals.org

Mitra, A. (2000). *Foundation of environmental science*. New Delhi: Narendra Publishing House.

Pagano, M., Gaudy, R., & Thibault, D. (1993). Vertical migrations and feeding rhythms of mesozooplanktonic organisms in the Rhone River plume area (North-West Mediterranean Sea). *Estuary Coast Shelf Science, 37*, 251–269.

Rhode, S. C., Pawlowski, M., & Tollrian, R. (2001). The impact of ultraviolet radiation on the vertical distribution of zooplankton of the genus Daphnia. *Nature, 412*, 69–72.

Vestheim, H., & Kaartvedt, S. (2009). Vertical migration, feeding and colouration in the mesopelagic shrimp *Sergestes arcticus. Journal of Plankton Research.* doi:10.1093/plankt/fbp077.

Worthington, L. V. (1954). Preliminary note on the timescale in the North Atlantic circulation. *Deep-Sea Research, 1*, 244–251.

Ylognen, O., Huuskonen, H., & Karjalainen, J. (2005). Effects of UV radiation on the vertical distribution of vendace [*Coregonus albula* (L.)] larvae in Finnish lakes. *Ecology of Freshwater Fish, 14*, 161–167.

Annexure References

Ahmad, E. (1982). 'India, coastal morphology', The encyclopedia of earth sciences. In M. L. Schwartz (Ed.), *The encyclopedia of beach and coastal environments.* Stroudberg: Hutchinson Ross Co. (Vol. 15, pp. 481–484).

Foot, R. B. (1888). Notes on Rameswaram Island. *Madras Christian College Magazine* (July 1998) (pp. 828–840).

Loveson, V. J., & Rajamanickam, G. V. (1988). Progradation as evidenced around as evidenced around a submerged ancient port, Periyapattinam, Tamilnadu, India. *International Journal of Land System Ecological Studies, 12*, 94–98.

Ramasamy, S. M. (1989). Morpho-tectonic evolution of east and west coast of Indian Peninsula. In *Geological survey of India special publish* (Arabian Sea Seminar, No. 24, pp. 333–339).

Ramasamy, S. M. (1997). Remote sensing and creation challenging coastal engineering geological problems of Tamilnadu coast, India. In *Proceedings of international symposium on engineering geology and the environment*, Greece (pp. 345–348).

Stoddart, D. R., & Pillai C. S. G. (1972). Raised reefs of Ramanathapuram district, south India. *Transactions, Institute, British Geographical Society* (Vol. 56, pp. 111–125).

Thanikachalam, M., & Ramachandran, S. (2002). Management of coral reefs in Gulf of Mannar using remote sensing and GIS techniques-with reference to coastal geomorphology and land-use. *Map Asia 2002, Asian conference on GIS, GPS, aerial photography and remote sensing*, Bangkok, August 7–9, 2002. http://www.gisdevelopment.net

Estuarine Ecosystem: An Overview 2

Contents

Estuaries are unique spots on our planet, which are situated and sandwiched between the continents and the seas. Penetrated by the sea through the recurring tides and flushed by the freshwater outflows of the lotic system, an estuary is a dynamic system where the freshwater meets seawater. Blessed by the fertile flows of both the seas and the rivers; these fascinating biotopes are by far the most productive ecosystem on our planet; the abode of unique species of plants and animals; the cradle of several species of finfish; the nursery of commercially important shellfish; the reservoir of food, chemicals, mineral, oils and natural gas; and the ideal spots for tourism and aquaculture.

2.1 Definition and Ecosystem Services

An estuary may be defined as the zone of intersection of fresh and marine waters with unique physico-chemical and biological characteristics. Estuaries generally occupy those areas of the coasts, which are least subject to marine features/activities and wave actions and thus are major sites for development of harbours, recreational activities and aquacultural farms. In estuaries, freshwater collected over vast regions of the land pours into the ocean, which sends saltwater upstream far beyond the river-mouth. According to Odum (1971), estuaries belong to different class of 'fluctuating water level ecosystems'.

A. Mitra, S. Zaman, *Basics of Marine and Estuarine Ecology*,
DOI 10.1007/978-81-322-2707-6_2

Each estuary has its own physical features that influence its ecology. These features primarily encompass the amount of river discharge, depth and general topography, specific circulation pattern, climatic regime and vertical tide range. Although the physico-chemical condition in estuaries is extremely stressful (particularly in terms of salinity and turbidity), yet the food availability is so favourable that the ecosystem is packed with life. Organisms living in this habitat exhibit wide range of tolerances due to tidal and seasonal variations of hydrological parameters (i.e. they are mainly euryhaline). Estuaries rank first among the most productive regions of marine ecosystems as they typically contain diverse group and high biomass of benthic algae, seagrasses, salt marsh grasses and phytoplankton that support large number of fishes and birds. This is mainly because estuaries are enriched by nutrients that are contributed through land drainage. Also rivers contribute substantial load of nutrients in the estuarine system. The estuarine system possesses considerable nutrient retention capacity due to which the flora and fauna thriving within such system never faces the scarcity of nutrients.

The astronomical tide at the estuarine mouth determines the nature of flow in an estuary near its mouth. The tide along the coast of India is mixed, i.e. it consists of a mixture of oscillations with a period of about 12.5 h (semidiurnal oscillation) and with a period of about a day (diurnal oscillation). The range is particularly high in gulfs such as the Gulf of Kutch or the Gulf of Khambhat or a funnel-shaped estuary like the Hooghly estuary of West Bengal, all of which get narrower away from their sea end (Fig. 2.1).

Estuaries provide unique ecosystem services. Only a few types of fish species leave permanently in an estuary, but many species develop there; thus, both larval and juvenile fishes are present in large numbers. It has been estimated that well over half of all marine fishes develop in the protective environment of an estuary, and because of this reason estuaries are often referred to as the nurseries of the sea. Apart from commercially important finfish species, shellfishes like shrimps, molluscs and crabs also use the estuary as their nursery ground (Fig. 2.2).

Several definitions have been forwarded to depict the features of estuarine compartment, some of which are highlighted here.

According to Cameron and Pritchard (1963), 'an estuary is a semi-enclosed coastal body of water which has a free connection with the open sea and within which seawater is measurably diluted with freshwater derived from land drainage'.

Pritchard (1967) stated estuary as 'a semi-enclosed coastal body of water, which has a free connection with the open sea; it is thus strongly affected by tidal action and within it seawater is mixed with freshwater from land drainage'.

Perillo (1995) defined estuary as 'a semi-enclosed coastal body of water that extends to the effective limit of tidal influence, within which seawater entering from one or more free connections with the open sea or any other saline coastal bodies of water is significantly diluted with freshwater derived from land drainage and can sustain euryhaline biological species, either a part or whole of their life cycle'.

Estuaries described by these definitions do not include the narrowness of the system, neither these definitions address the situation where evaporation exceeds the freshwater supply from rivers and from local rain. Considering all these gaps, in recent times estuary has been defined as 'a narrow semi-enclosed coastal body of water which has a free connection with the open sea at least intermittently and within which salinity of the water is measurably different from the salinity in the open ocean' (www.es.flinders.edu.au/~mattom/ShelfCoast/chapterl1.html).

In an estuarine system, salinity and other hydrological parameters are controlled by the tidal action. It is observed that exactly in the same location (that can be fixed by GPS), an environmental variable alters significantly with the change in tidal phase. A case study on the effect of tidal influence on major hydrological parameters is presented Chap. 3 of this book.

The time required for an estuary to exchange its water with the ocean is its **flushing time**. If the net circulation is fast and the total volume of an estuary is small, flushing is rapid. A rapidly flushing estuary has a high carrying capacity for wastes, because the dissolved or suspended

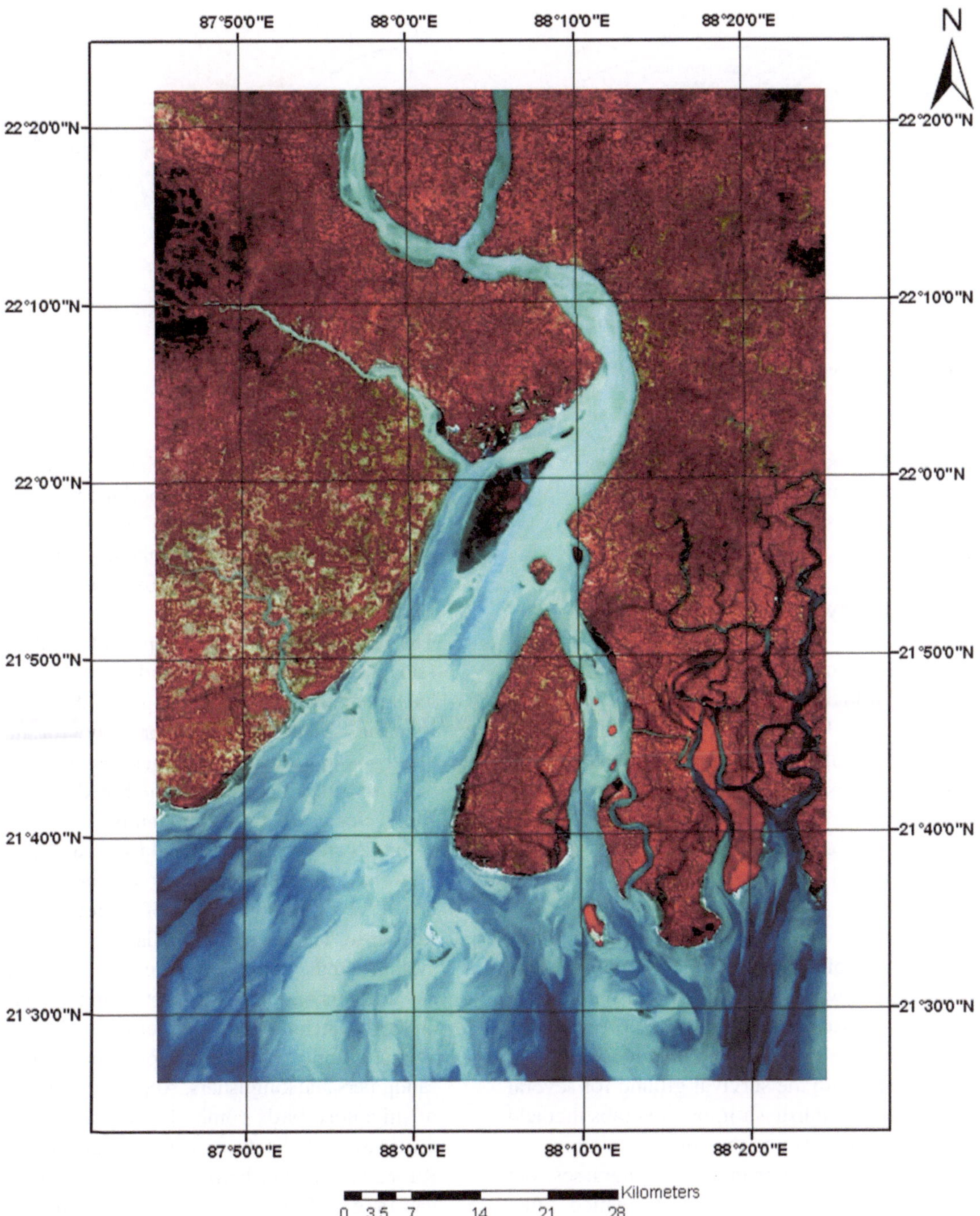

Fig. 2.1 Hooghly estuary with *funnel-shaped mouth* and gradually becomes narrower towards upstream

wastes are moved quickly out to sea and diluted. The flushing time has great relevance in maintaining the aquatic health of an estuary. If the flushing time is slow, there is high probability of accumulation of waste materials in the estuarine systems that are derived from the adjacent landmasses or aquacultural units. The flushing time of estuaries is determined by dividing their volume by the rate of net seaward flow. Understanding the circulation and flushing time

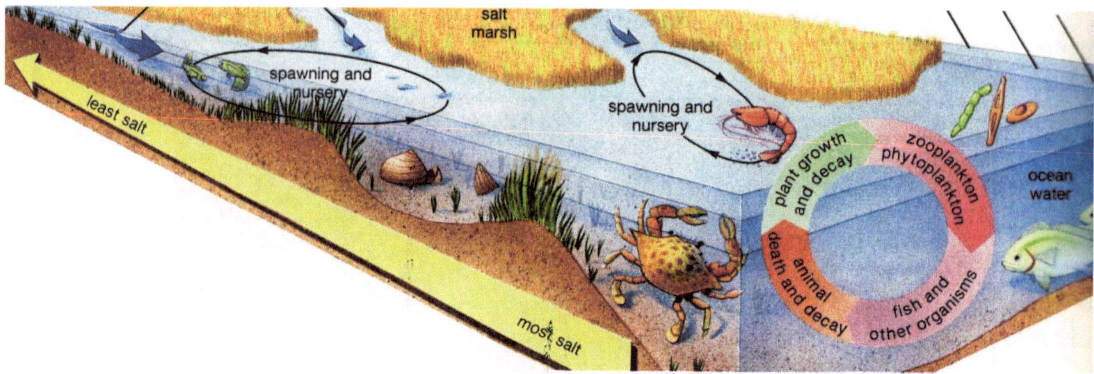

Fig. 2.2 Estuaries are breeding grounds for finfish and shellfish species

of the estuarine system is, therefore, extremely important for managing and accelerating the ecosystem services of the system.

2.1.1 Ecosystem Services

Estuaries are noted for taxonomic diversity of species, biological productivity and unique aesthetic value. The water, sediment and biotic communities of the estuaries offer several ecosystem services that provide both direct and indirect benefits to run the wheel of human civilization. Few common ecosystem services are highlighted here:

1. Estuaries are the breeding and nursery grounds of several species of finfish and shellfish, which are commercially important and are also included in the export basket of the nation. The estuaries of Sundarbans, for example, are the survival ground for several species of finfish, shrimps and crabs that add foreign currency to the country economy.
2. Estuaries sustain mangroves, seagrasses, salt marsh grasses and seaweeds, which often serve as agents of bioremediation. Reports of heavy metal accumulation by macrophytes of estuaries are plenty (Mitra et al. 1994; Trivedi et al. 1994), which show the adaptive capability of estuarine flora to thrive in contaminated media (both water and sediment). Such biological treatments of conservative pollutants have immense economic and ecological importance.
3. Estuaries are the sites for recycling of nutrients where microbes are the key players.
4. Estuaries offer sustainable fisheries to the local population.
5. The mangroves of estuaries provide timber, fuel, wax, honey to the island dwellers and local people of the fringe villages (Table 2.1).
6. Estuaries serve as the migratory route for the anadromous and catadromous fishes, and because of such migration, the breeding and life cycles of the fishes are completed.
7. Estuaries are the home ground for a variety of migratory birds and endangered species that can serve as the foundation of ecotourism in real sense. The intertidal mudflats, the aquatic system and floral communities of the estuaries may attract the tourist with gorgeous and eye catching array of herons, pelicans, storks, eagles, ospreys, plovers, gulls, sandpipers and kingfishers. About 95 varieties of migratory birds come all the way in the estuaries of India from the Caspian Sea in Russia and the northern parts of the Himalayan ranges. The Sundarban estuaries are the home ground of Royal Bengal tiger (*Panthera tigris tigris*), which can swim across the estuary over a distance of 15 km.
8. The aquatic system of the estuary is extensively used for aquaculture, although in many cases the adverse impacts of shrimp culture have been identified (Mitra 1998).

Table 2.1 Traditional uses of mangrove species

Mangrove species	Use
Aegiceras corniculatum	Bark used as fish poison; also contains tannin
Avicennia alba	Used for fodder and fuel
Avicennia officinalis	Used for firewood; bark contains tannin
Bruguiera caryophylloides	Wood used for firewood and timber; bark used as tannin
Bruguiera gymnorrhiza	Wood used for house posts; excellent fuel; bark contains tannin
B. parviflora	Timber used as fuel; leaves and bark contain tannin
B. sexangula	Timber used in house building
Ceriops tagal	Used for keels of boats and house posts. Provides good fuel charcoal. Bark rich in tannin used for dyeing fishing nets
Rhizophora mucronata	Bark used for tannin and cattle fodder
Sonneratia apetala	Wood used in house building, packing cases and yields excellent fuel
Xylocarpus granatum	Yields gum, resin which is used in local medicine; bark contains tannin

9. Estuaries are treated as the final receptacle of all wastes. Deterioration of the water quality and commercially important fauna has been reported from many estuaries of the world.
10. The biotic communities of the estuarine system (particularly the mangroves, salt marsh grass, seaweeds, etc.) and the underlying soil act as the store house of carbon. This ecosystem service is of great relevance in context to carbon dioxide rise and subsequent climate change. A study conducted by the present authors in 24 stations of Indian Sundarbans (Fig. 2.3) exhibits considerable percentage of soil organic carbon in the intertidal mudflats ranging from 0.66 % to 1.41 % (Fig. 2.4).
11. The salt marsh grass found abundantly in the intertidal mudflats of estuaries is used for fish feed preparation (*Vide* Annexure 2A for detailed experimental approach). Research findings also reveal that fish feed incorporated with estuarine seaweed has resulted in improved performance, better feed efficiency, better pellet stability and improved animal (fish) product quality, which may be attributed to the enormous genetic potential of these lower group of plants with a genome that is more than twice the size of yeast. This quality has brought the seaweeds in serving many industries such as food, animal feed, cosmetics, pharmaceuticals and biofuels.
12. Flora and fauna of the estuaries are today used as sources of bioactive substances, which have immense importance in the pharmaceutical industries. Examples of **CAL** (*Carcinoscorpius* amoebocyte lysate) and **TAL** (*Tachypleus* amoebocyte lysate) are very relevant in this context. They are derived from the blue blood of horseshoe crabs (*Carcinoscorpius rotundifolia* and *Tachypleus gigas*) found abundantly in the estuaries of West Bengal and Odisha, which are two maritime states in the North-east coast of the Indian subcontinent. CAL and TAL are used for bacterial endotoxin test.
13. Estuarine fishes are the sources of omega-3 fatty acids, which are good for the heart. Coronary heart disease occurs due to atherosclerosis and thrombosis. Atherosclerosis is initiated by endothelial damage that may result from stress-related events and may finally lead to thrombosis and arrhythmia, the main cause of sudden death. The omega-3 fatty acid synthesizes with the help of enzymes, eicosanoids. The antithrombotic effect of

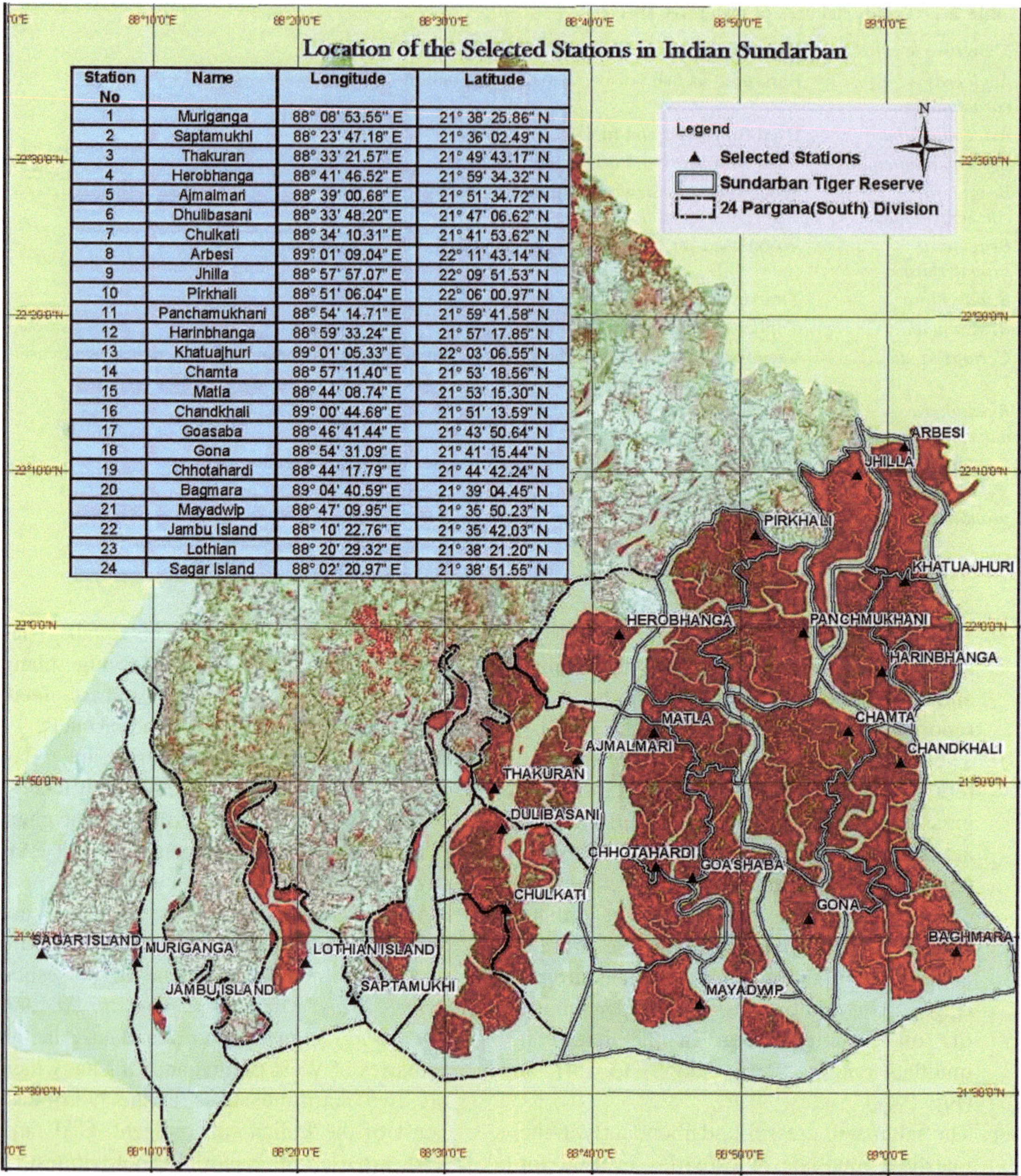

Station No	Name	Longitude	Latitude
1	Muriganga	88° 08' 53.55" E	21° 38' 25.86" N
2	Saptamukhi	88° 23' 47.18" E	21° 36' 02.49" N
3	Thakuran	88° 33' 21.57" E	21° 49' 43.17" N
4	Herobhanga	88° 41' 46.52" E	21° 59' 34.32" N
5	Ajmalmari	88° 39' 00.68" E	21° 51' 34.72" N
6	Dhulibasani	88° 33' 48.20" E	21° 47' 06.62" N
7	Chulkati	88° 34' 10.31" E	21° 41' 53.62" N
8	Arbesi	89° 01' 09.04" E	22° 11' 43.14" N
9	Jhilla	88° 57' 57.07" E	22° 09' 51.53" N
10	Pirkhali	88° 51' 06.04" E	22° 06' 00.97" N
11	Panchamukhani	88° 54' 14.71" E	21° 59' 41.58" N
12	Harinbhanga	88° 59' 33.24" E	21° 57' 17.85" N
13	Khatuajhuri	89° 01' 05.33" E	22° 03' 06.55" N
14	Chamta	88° 57' 11.40" E	21° 53' 18.56" N
15	Matla	88° 44' 08.74" E	21° 53' 15.30" N
16	Chandkhali	89° 00' 44.68" E	21° 51' 13.59" N
17	Goasaba	88° 46' 41.44" E	21° 43' 50.64" N
18	Gona	88° 54' 31.09" E	21° 41' 15.44" N
19	Chhotahardi	88° 44' 17.79" E	21° 44' 42.24" N
20	Bagmara	89° 04' 40.59" E	21° 39' 04.45" N
21	Mayadwip	88° 47' 09.95" E	21° 35' 50.23" N
22	Jambu Island	88° 10' 22.76" E	21° 35' 42.03" N
23	Lothian	88° 20' 29.32" E	21° 38' 21.20" N
24	Sagar Island	88° 02' 20.97" E	21° 38' 51.55" N

Fig. 2.3 Stations in Indian Sundarbans selected for organic carbon study

omega-3 fatty acid is due to these eicosanoids, which inhibit platelet aggregation, vasoconstriction and adhesion, and induces vasodilation. Omega-3 fatty acid is also good for diabetic patients as it improves insulin resistance syndrome that leads to heart disease. It is also reported to be good for arthritis.

14. Several medicinal plants widely distributed in the coastal and estuarine regions were tested for antiviral, antifungal and mosquito larvicidal activities. In the Parangipettai coast of India (Southeast coast), 113 medicinal plants have been recorded. Out of 36 plants, 3 plants showed high anti-HIV activity. The anti-HIV

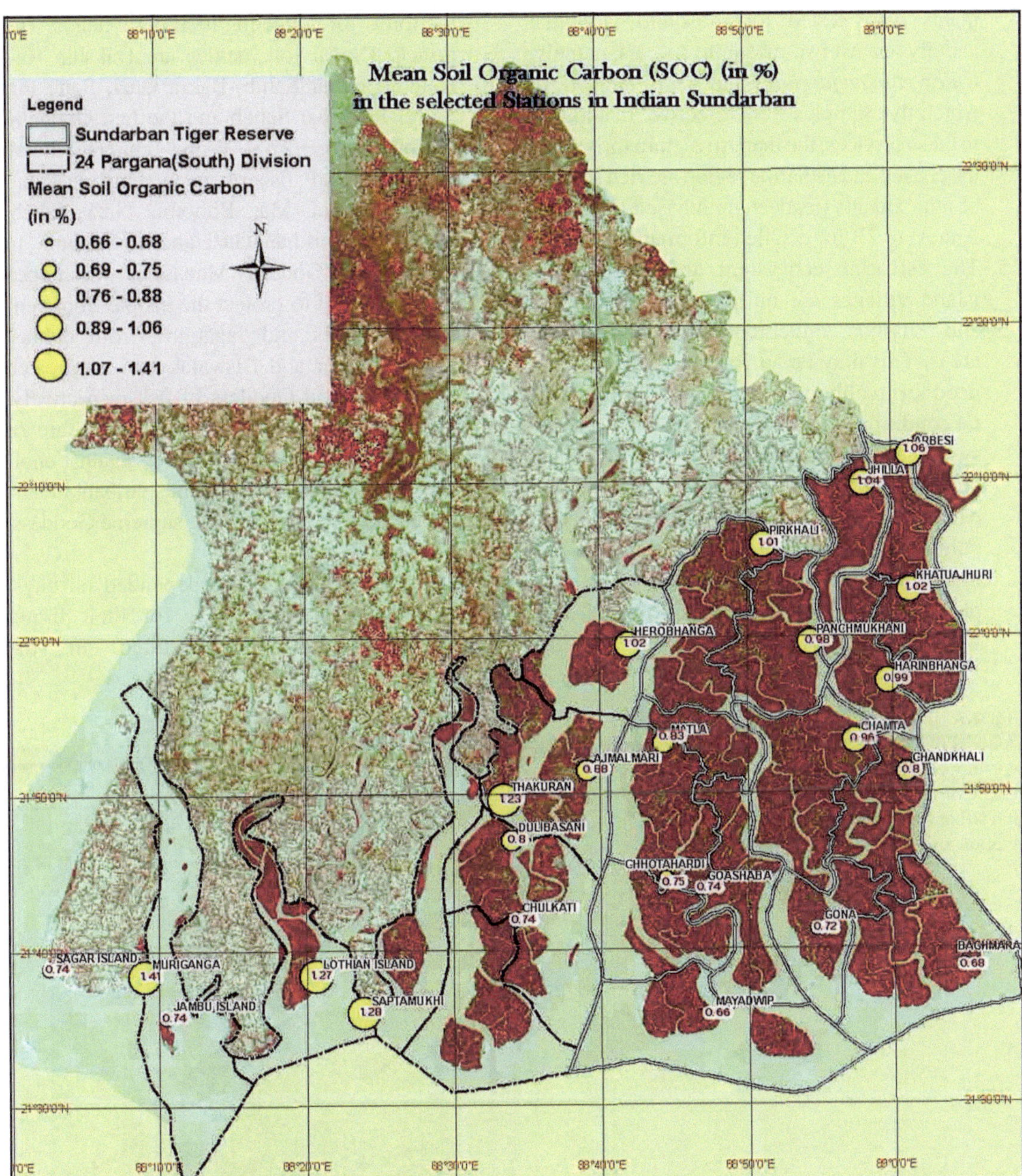

Fig. 2.4 Percentage of organic carbon in the intertidal mudflats of Indian Sundarbans

activity of coastal population was also tested in comparison with their terrestrial counterparts. In general, coastal populations exhibited better anti-HIV activity than terrestrial ones. 23 plants were screened for antibacterial activity against 13 human pathogenic bacterial strains, of which 23 extracts showed antibacterial activity against one or more bacteria. Twenty-five plants were screened for antifungal activity against six fungi. All the plant extracts examined showed activity against one or more fungi. Two plant species exhibited high antifungal activity. In general, the coastal samples showed higher activity than terrestrial ones. Eleven plants were tested for mosquito larvicidal activity against two mosquito species, namely, *Culex quinquefasciatus* and *Aedes aegypti*, of which five samples were effective. In addition to these services, the mangrove flora thriving in the coastal and estuarine region are rich sources of antioxidants (readers are advised to consult Annexure 2B for detailed information.

15. The estuarine ecosystem and surrounding island villages are unique sites of cultural and religious convergence unlike the urban areas. This may be attributed to the extreme need for livelihood and adverse environmental conditions prevailing in the estuarine ecosystem that often break the barrier of caste and religion among the people. A relevant example in this context is the religious and cultural canvas of Indian Sundarbans. In this mangrove-dominated deltaic complex, people of both religions—Islam and Hinduism—worship the same deities before entering into the forest for collecting their daily needs. The principal local folk deities are Dakshin Rai, Kalurai, Badar Sahib, Bakra Gazi, Sona Pir, Sawal Pir, Gazi Saheb and the first Goddess, Bonbibi (Fig. 2.5). Wood cutters, honey collectors and fishermen worship Dakshin Rai, Narayani Ma, Bonbibi, Gazi Saheb, Kalu Rai, Barkhan Gazi and Sa Janguli. In addition, the Goddess Manasa is worshipped with the belief to protect the people from venomous snakes and Jagatguru from cobras. Makal Thakur and Biswalakshmi are treated as the God and Goddess of fish respectively. Manik Pir is worshipped for the welfare of cows and Olabibi for protection against cholera. For protection of the embankments, people refer to Ganga as the supreme Goddess.

Historically, people have depended on bays, estuaries and coastal regions for their living requirements; this has resulted in the

Fig. 2.5 Bonbibi: 'the Goddess of forest' worshipped by the local people of Indian Sundarbans irrespective of religion and caste

establishment of major population centres along the shores of estuaries and coastal seas. In the United States today, over one-half of the population lives within 80 km (50 mi) of the coasts (including the Great Lakes). This population, with its necessary industries, energy-generating facilities, recreational activities and waste-treatment plants, has created a significant negative impact on the coastal and estuarine zones.

The ecosystem services of the estuaries are presently under threat due to rapid industrialization, urbanization and unplanned tourism. Added to the uncontrolled fishing and extraction of natural resources, estuaries and brackish waters are subjected to constant changes in their physico-chemical condition due to various anthropogenic activities. The widespread reclamation of water areas, construction of barrages and other saltwater extraction structure, dredging and aquatic pollution, etc., cause substantial negative impact on the positive health of the estuaries. Days are not too far when the ecosystem services of the estuaries will be greatly squeezed.

In general, the total economic valuation of an estuarine system can be calculated as the sum of four components. These are (direct) use value, indirect use value, option value and non-use value (Table 2.2).

For monetary valuation, many of these components are easier to measure, as, for example, use values are easy to measure because they are observable quantities of products consumed as well as their standard market prices can be utilized to determine the economic value. Recreational use can also be measured from the log book entries maintained in the ecotourism camps, hotels, resorts, sanctuaries, eco parks or reserve forest. The charges for visiting and staying in the heart of Indian Sundarbans are highlighted in Fig. 2.6 and Table 2.3.

Certain ecosystem services are extremely difficult to measure as, for example, the pollution control, erosion control and flood control potential of an estuarine system or the oxygen generation and carbon dioxide sequestration by the estuarine microphytes and macrophytes, etc. The mangroves of estuarine system, for example, are noted for erosion control and combat sea level rise (Fig. 2.7). In such cases 'shadow value' must be estimated in order to 'price' the produced services.

It is also difficult to measure the option values and non-use values of the estuary as they are not reflected as physical commodities or in any form of observable/measurable items. These values are estimated by using surveys that ask people a series of questions about their willingness to pay for ecosystem services they do not use (Table 2.4).

2.2 Classification

In standard text books on oceanography, four basic types of estuaries are described (Table 2.5).

Pritchard (1967) classified estuaries on the basis of the following four characteristics. They are (a) salinity, (b) geomorphology, (c) water circulation and stratification and (d) systems energetics.

Table 2.2 Components of economic valuation of an estuarine ecosystem

Type	Description	Example
Direct use value	Resources that provide direct benefits to the stakeholders	Fishery resources, irrigation facilities, aquaculture activities, existing tourism units, etc.
Indirect use value	Users get indirect benefits and often some distance away from where they originate	Pollution filtering capacity of estuarine system, flood control, etc.
Option value	Users may be willing to pay for acquiring benefits from estuarine resources in future	May be a future source of drinking water or introduction of mangrove tourism
Non-use value	Users may be willing to pay for conservation of resources, which will never be used directly	Preservation or protection of endangered species arising out of a sense of environmental stewardship that is unrelated to the indirect and direct uses in the current or in the future

Fig. 2.6 Price for visiting and staying in a forest camp in the heart of Indian Sundarbans

Table 2.3 Approximate price (in $) for travelling and staying in the heart of Indian Sundarbans

Items	Price ($)
Staying charge per head per day	1
Air condition launch (per day)	33.34
General launch (per day)	10
Boat	5.83
Still camera	Free
Video camera	3.33
Cinema, documentary (per day)	250
Ethnic cottage	20
Round cottage (per day)	16.67
Govt. employee on duty (excluding forest department staffs) (per day)	6.67
Dormitory (of 10 beds)	2.5
Educational trip per student (per day)	0.17
Children under 5 years	Free

Note: Conversion to $ has been done on the rate of $1 = INR.60.00

I. **On the basis of salinity, estuaries can be grouped into the following types:**

(a) **Oligohaline:** In this type of estuary, the freshwater mixes with the saline water in such proportion that the water ultimately becomes uniformly saline. Such estuaries are common where small rivers meet the seas, e.g. Haldi River of West Bengal (India).

Fig. 2.7 Mangroves are potential controlling agents of erosion

Table 2.4 Economic valuation methods for water

Method	Approach	Water service appropriate for method	Data needs	Limitations
Contingent valuation method	Ask people directly their willingness to pay (WTP)	All use values and non-use value (e.g. drinking water, fishing, protecting species)	Survey with scenario description and questions about WTP for specific services	Potential biases due to hypothetical nature of scenarios
Travel cost method	Estimate demand curve from data on travel expenditures	Recreation: boating, fishing, swimming	Survey on expenditures of time and money to travel to specific sites	Only captures recreational benefits; difficult to apply for multiple destination trips
Hedonic property value method	Identify contribution of environmental quality to land values	Water quality, wetland services	Property values and characteristics including environmental quality	Requires extensive information about ecosystem services of hundreds of specific sites
Change in productivity method	Assess impact of change in water service on produced goods	Commercial fisheries, agricultural uses	Impact of change in water service on production; net values of produced goods	Information on biological impacts of changes in ecosystem services often unavailable

Source: Pagiola et al. (2004)

Table 2.5 Types of estuaries and its characteristics

Name	Figure	Description
Salt-wedge estuary	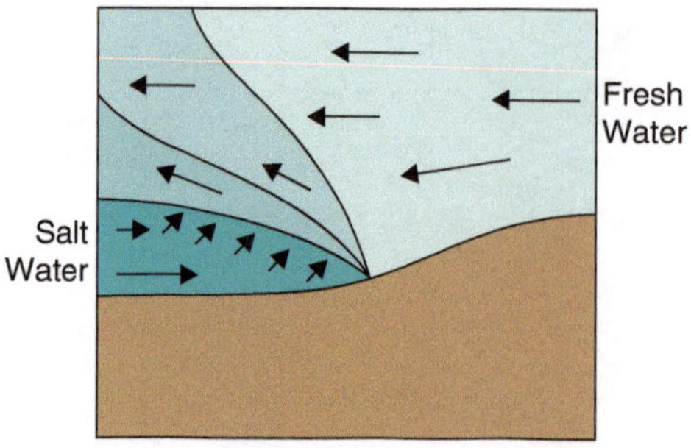 Salt Water; Fresh Water	The high flow rate of the river holds back the lesser flow of saltwater. The saltwater is drawn upwards into the fast-moving river flow
Well-mixed estuary	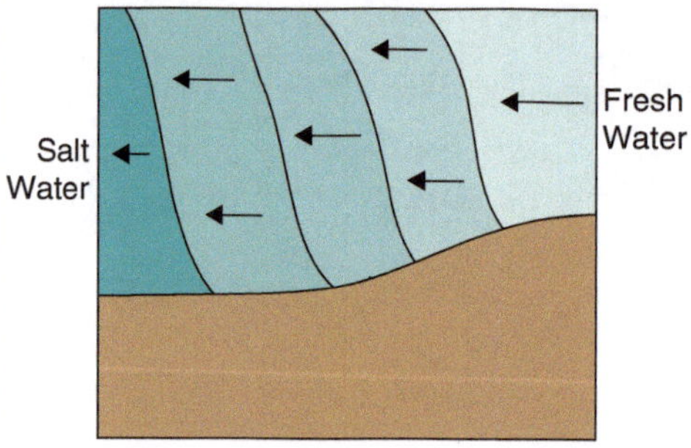 Salt Water; Fresh Water	Strong tidal currents distribute and mix the seawater throughout the shallow estuary. The net flow is weak and seawards at all depths
Partially mixed estuary	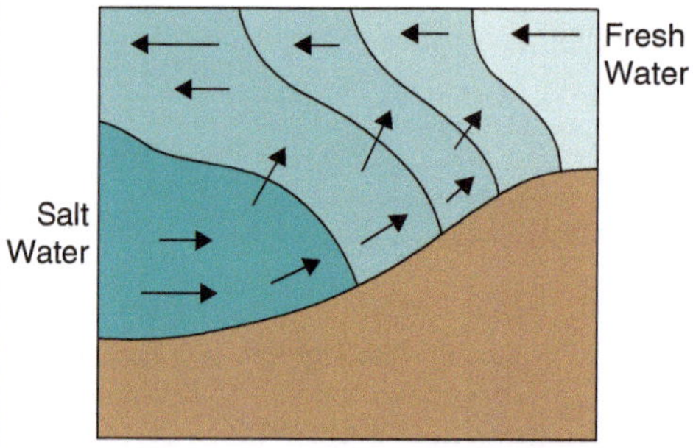 Salt Water; Fresh Water	Seawater enters below the mixed water that is flowing seawards at the surface. Seaward surface net flow is larger than river flow alone
Fjord-type estuary	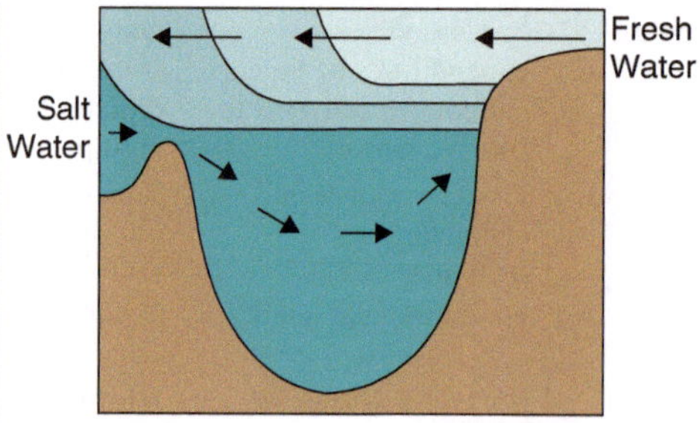 Salt Water; Fresh Water	River water flows seawards over the surface of the deeper seawater and gains salt slowly. The deeper layers may become stagnant due to the slow inflow rate of saltwater

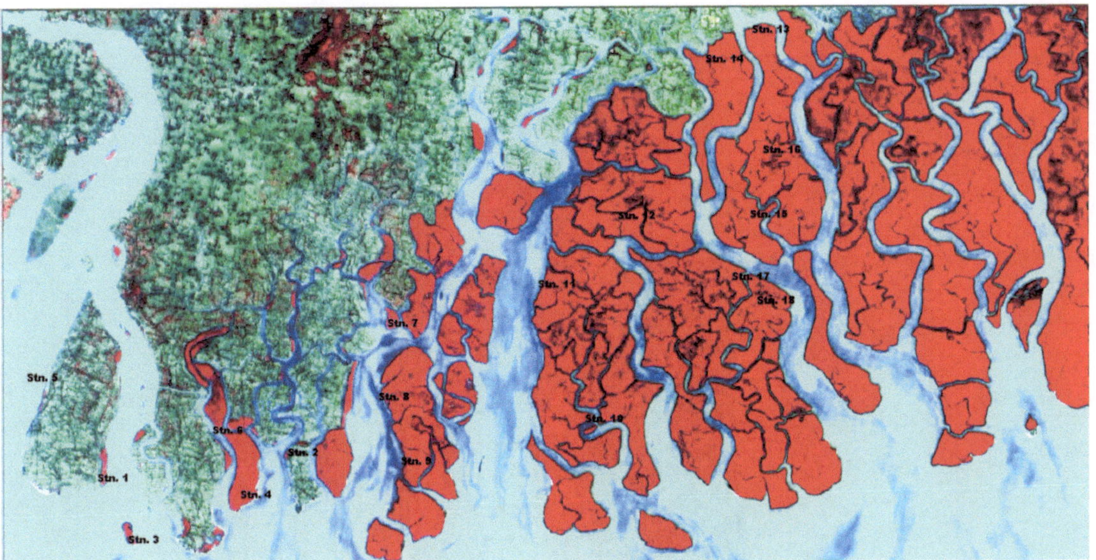

Fig. 2.8 Location of sector-wise sampling stations in Indian Sundarbans; the red colour indicates the mangrove vegetation

(b) **Mesohaline:** It is that type of estuary where the estuarine water has medium salinity due to proportional mixing of the fresh and saline water, e.g. Mahanadi river of Odisha (in the East coast of India).

(c) **Polyhaline:** In this type of estuary, there are areas where there is distinct variation of salinity owing to the variation in tidal water intrusion into the river-mouth, e.g. estuaries adjacent to the Bay of Bengal and the Indian Sundarbans. The estuarine complex in this mangrove-dominated deltaic complex can be subdivided into three sectors, namely, western, central and eastern sectors (Fig. 2.8), each with different signatures of salinity (Fig. 2.9a–c). More than two decades of data (1984–2013) were compiled from the archives of the Department of Marine Science, University of Calcutta, for this study. A number of studies on different aspects of the Sundarban complex have been published over the years, which include description of the data (and methods) at different times over the past three decades (Mitra et al. 1987, 1992, 2009; Chakraborty and Choudhury 1985; Mitra and Choudhury 1994; Saha et al. 1999; Banerjee et al. 2002, 2003, 2013; Mondal et al. 2006; Sengupta et al. 2013). Real-time data (through field sampling by the authors) were also collected simultaneously since 1998 from 18 sampling stations (Table 2.6) in the lower Gangetic region during high tide condition to assure quality and continuity to the data bank. For each observational station, at least 30 samples were collected within 500 m of each other, and the mean value of 30 observations was considered for statistical interpretations.

In the western sector, the salinity decrease ranged from 0.58 psu/year (at Jambu Island) to 1.46 psu/year (at Harinbari) (Fig. 2.9a). Considering all the six stations in the western sector, the decadal decrease of salinity is 7.50 psu/decade. In the western sector,

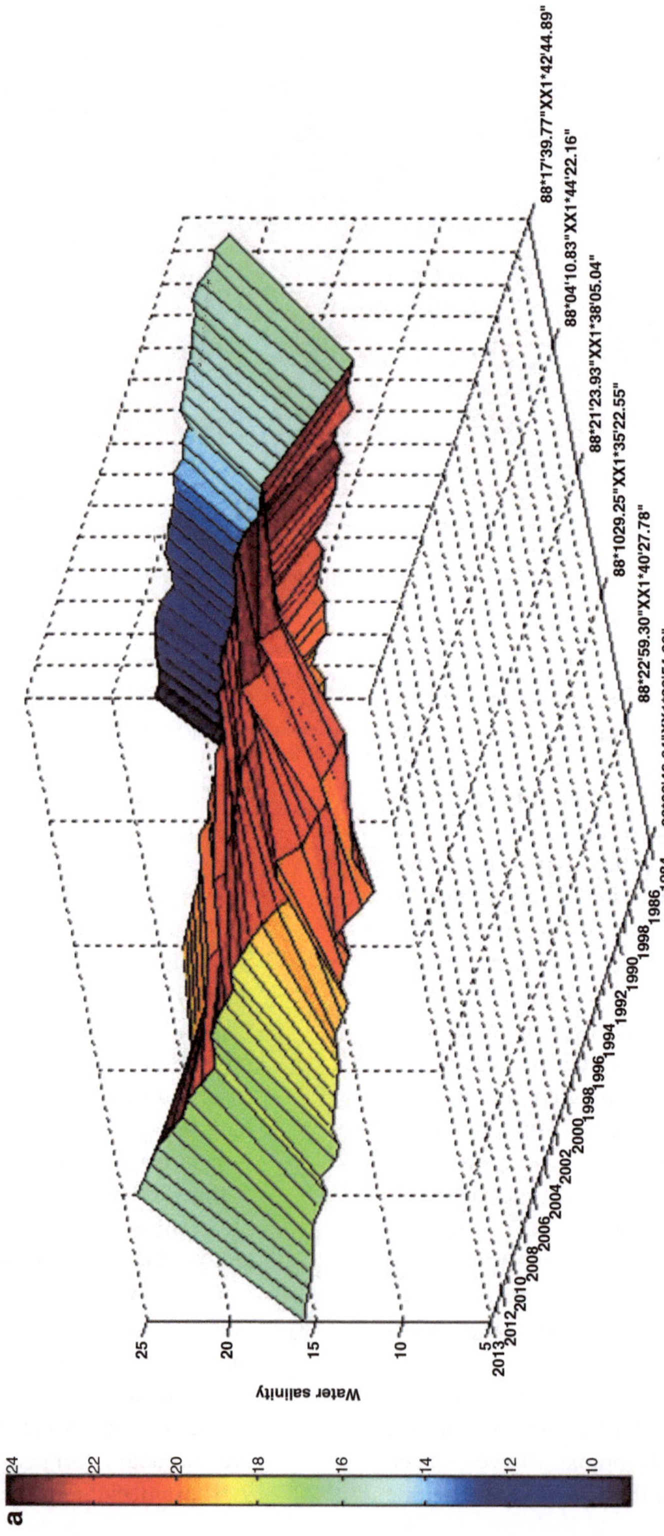

Fig. 2.9 (**a**) Surface water salinity in six stations of western Indian Sundarbans. (**b**) Surface water salinity in six stations of central Indian Sundarbans. (**c**) Surface water salinity in six stations of eastern Indian Sundarbans

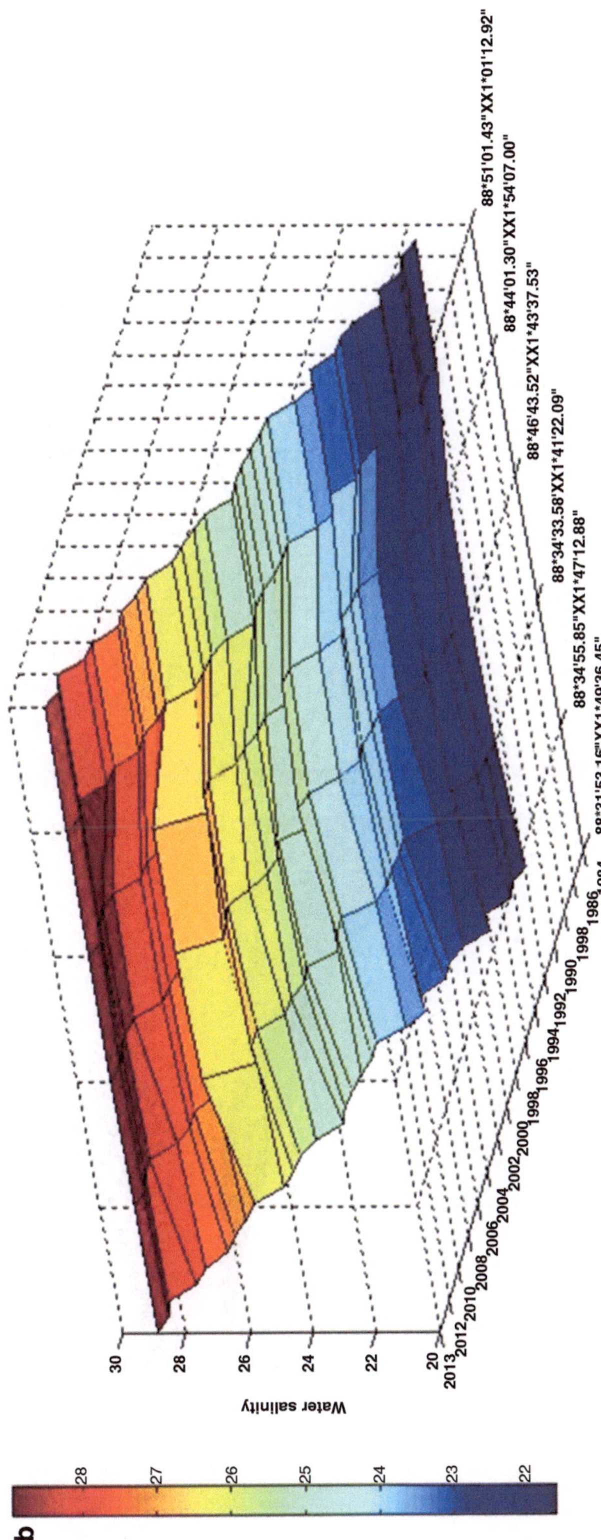

Fig. 2.9 (continued)

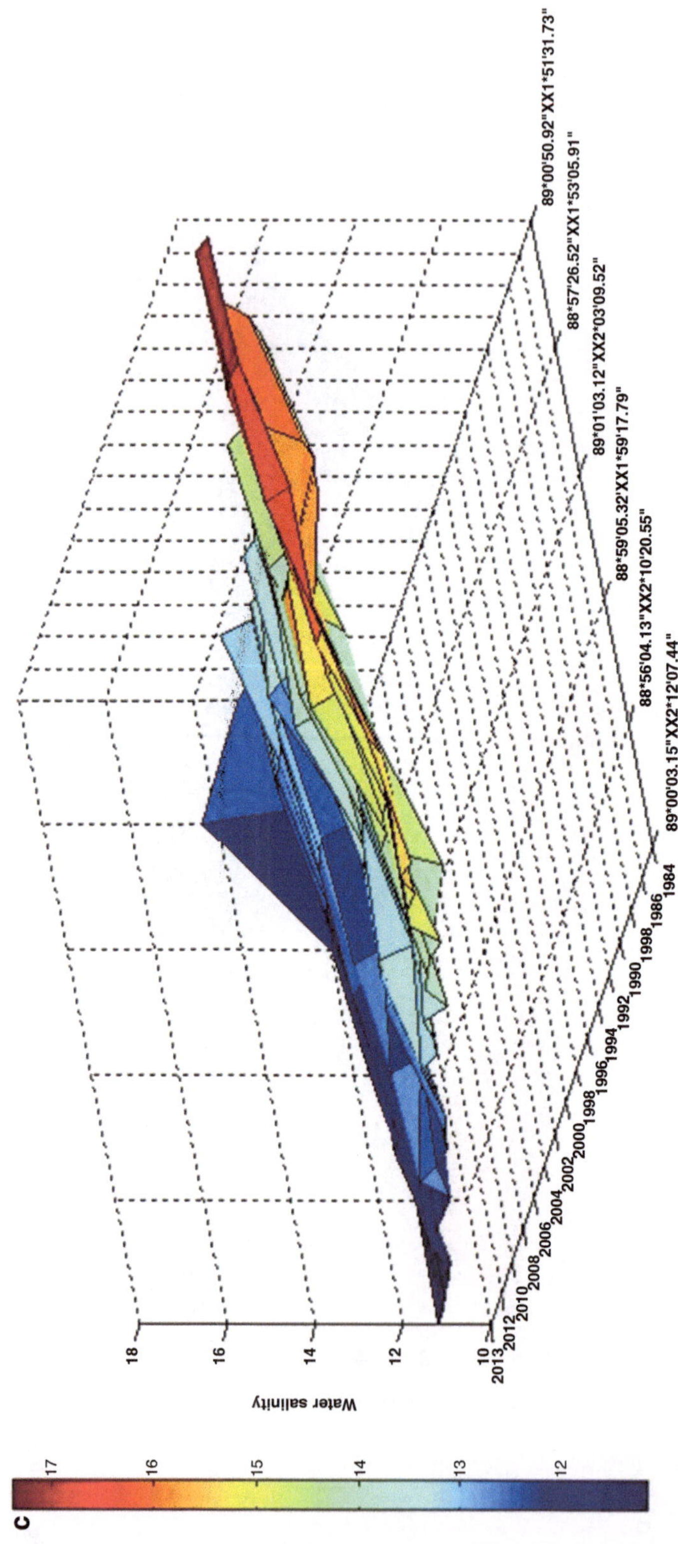

Fig. 2.9 (continued)

Table 2.6 Sampling stations in the western, central and eastern sectors of Indian Sundarbans in the lower Gangetic delta region

Sectors		Sampling station	Latitude	Longitude
Western sector	**Stn. 1**	Chemaguri (W_1)	21°38′25.86″N	88°08′53.55″E
	Stn. 2	Saptamukhi (W_2)	21°40′02.33″N	88°23′27.18″E
	Stn. 3	Jambu Island (W_3)	21°35′42.03″N	88°10′22.76″E
	Stn. 4	Lothian (W_4)	21°38′21.20″N	88°20′29.32″E
	Stn. 5	Harinbari (W_5)	21°44′22.55″N	88°04′32.97″E
	Stn. 6	Prentice Island (W_6)	21°42′47.88″N	88°17′55.05″E
Central sector	**Stn. 7**	Thakuran Char (C_1)	21°49′53.17″N	88°31′25.57″E
	Stn. 8	Dhulibasani (C_2)	21°47′06.62″N	88°33′48.20″E
	Stn. 9	Chulkathi (C_3)	21°41′53.62″N	88°34′10.31″E
	Stn. 10	Goashaba (C_4)	21°43′50.64″N	88°46′41.44″E
	Stn. 11	Matla (C_5)	21°53′15.30″N	88°44′08.74″E
	Stn. 12	Pirkhali (C_6)	22°06′00.97″N	88°51′06.04″E
Eastern sector	**Stn. 13**	Arbesi (E_1)	22°11′43.14″N	89°01′09.04″E
	Stn. 14	Jhilla (E_2)	22°09′51.53″N	88°57′57.07″E
	Stn. 15	Harinbhanga (E_3)	21°57′17.85″N	88°59′33.24″E
	Stn. 16	Khatuajhuri (E_4)	22°03′06.55″N	89°01′05.33″E
	Stn. 17	Chamta (E_5)	21°53′18.56″N	88°57′11.40″E
	Stn. 18	Chandkhali (E_6)	21°51′13.59″N	89°00′44.68″E

the salinity intrusion is confined to 70 km from the mouth even during the dry season. The tidal variation at the mouth is from 6.1 m at springs to 0.22 m at neaps. The freshwater discharge ranges from a peak value of 4250 $m^3\ s^{-1}$ to almost zero in the dry season. The average values of freshwater discharge are 3000 $m^3\ s^{-1}$ during SW monsoon season (June–September) and 1000 $m^3\ s^{-1}$ during a dry season (November–May). Normally, the freshwater discharges are regulated from Farakka barrage to maintain water levels at Calcutta (Biswas 1985; Sadhuram et al. 2005).

The central sector presents a completely reverse picture in terms of aquatic salinity. Irrespective of stations, salinity has increased (Fig. 2.9b) between the range 1.05 psu/year (in Chulkathi) and 1.12 psu/year (in Matla and Pirkhali). The average decadal increase of salinity is 13.04 psu/decade.

In the eastern sector, salinity has decreased (Fig. 2.9c) which ranges from 0.54 psu/year (in Chamta) to 0.98 psu/year (in Jhilla). The average decadal decrease of the salinity in this sector is 10.28 psu.

II. **On the basis of geomorphology, estuaries are of the following types:**

(a) **Drowned river valleys:** The coastlines with relatively low and extensively wide coastal plains, e.g. Chesapeake Bay on the mid-Atlantic coast of the United States.

(b) **Fjord-type estuaries:** The deep U-shaped indentures due to glacial erosion with a shallow sill at their mouth formed by terminal glacial deposits, e.g. Norway, British Columbia and Alaska.

(c) **Bar-built estuaries:** These are shallow basins often partly exposed at low tide enclosed by a chain of offshore bars or barrier islands, broken at intervals by inlets. These bars are sometimes deposited offshore or are remnants of former coastal dunes, e.g. North Carolina and Georgia.

(d) **Estuaries produced by tectonic processes:** Geological faulting or local subsidence results in the formation of coastal indented structures, e.g. San Francisco Bay.

III. **On the basis of water circulation and stratification:**

(a) **Highly stratified or 'salt-wedge' estuary:** When the freshwater discharge from the rivers dominate over the tidal action, the freshwater tends to overflow the heavier saltwater which forms a 'wedge' extending along the bottom for a considerable distance upstream. This flow of freshwater is again governed by the Coriolis force, which forces the freshwater to move strongly along the right shore (if the observer faces the sea in the Northern Hemisphere and vice versa in the Southern Hemisphere). Such 'stratified' or 'bilayered' estuary will exhibit a salinity profile with a 'halocline' or zone of sharp change in salinity from top to bottom, e.g. Mississippi River.

(b) **Partially mixed or moderately stratified estuary:** When the freshwater and the tidal water mix in equal proportion due to turbulence caused by periodicity of waves, such estuary is formed. Due to this, the energy is dissipated in vertical mixing, thus creating a complex pattern of layers and water masses, e.g. Chesapeake Bay.

(c) **Completely mixed or vertically homogenous estuary:** When the tidal action is more than the freshwater discharge, the water tends to mix well from top to bottom and the salinity is relatively high. When there is wide variation in salinity and temperature, then horizontal estuaries are formed, e.g. bar-built estuaries.

IV. **On the basis of systems energetics:** H. T. Odum (1971) have described estuaries on the basis of systems energetics as stated here:

(a) **Physically stressed systems of wide latitudinal range:** These include mainly the high energy breaking waves, strong tidal currents, severe temperature and salinity shocks, low dissolved oxygen level during night or high rate of sedimentation. Due to severe environmental stress in this system, few species (opportunistic species) are able to thrive in this fluctuating condition. It is because of this fact such systems exhibit extremely poor species diversity. The index of dominance is high in these systems owing to the ability of fewer numbers of species to cope with adverse environmental conditions. However, in areas of intertidal zone, sharp zonation of species and seasonal replacement of communities is very distinct because adaptation is more efficiently accomplished by species replacement along a gradient than by adaptation within the species. Rocky sea fronts, intertidal rocks, sand beaches, high velocity tidal channels, sedimentary deltas and hypersaline lagoons are included in this category.

(b) **Natural arctic ecosystems with ice stress:** Extreme cold conditions (in Arctic and Antarctic coasts) exert considerable stress on the intertidal zones of the areas in which light is limiting due to very short summer season. Productivity is extremely low in these systems due to poor diversity of phytoplankton. In most of the times of the year, the phytoplankton species remain in encysted condition to get rid of the chilled water.

(c) **Natural temperate coastal ecosystems with seasonal programming and reproduction:** Seasonal changes in primary productivity of the ocean and the behaviour of animals are the common characteristics in estuarine biology. The more subdued tides, waves and currents in the semi-enclosed basins provide energy subsidies rather than stresses in comparison to deeper sounds and offshore waters which often get charged with nutrients and organic matter from fertile shallow zones. Temperate estuaries, though very fertile, are vulnerable to damages caused by pollution, dredging, filling and diking. Some important features of the temperate estuaries include tide pools, salt marshes, eel-grass (*Zostera* sp.) beds, seaweed bottoms, kelp beds, oyster reefs and mudflats, which harbour dense population of clams and sea worms.

(d) **Natural tropical coastal ecosystems of high diversity:** These systems are characterized by minimum environmental stress, and therefore negligible energy is spent by the species for adaptive modifications. Due to congenial environment, the species diversity is very high in these systems. Examples of such systems are mangroves, seagrass meadows, salt marsh grass and seaweed community and coral reefs in shallow-water zones. These biotic forms contribute nutrients due to which a wide spectrum of planktonic diversity exists in this system. The fishery potential is extremely high, and hence the modern trend of monoculture (e.g. culture of *Penaeus monodon*) is perhaps a wrong choice in this system. These productive systems with varied ecological niches are suitable for polyculture practice.

(e) **Emerging new systems associated with man:** In modern societies, estuaries are treated as the ultimate sink of different categories of waste. Estuaries have varying capacities to handle 'degradable' material depending on the size of the system, flow patterns, hydrological parameters, microbial load, type of estuary and climatic zone. Materials such as treated sewage and pulp mill wastes, seafood and other food processing wastes, petroleum wastes and dredging spoil can be decomposed and dispersed provided that (i) the system is not stressed by poison (insecticides, acids, etc.) and that (ii) the rate of input is controlled at low to moderate levels and not subjected to sudden shocks produced by periodic massive dumping. Of all the man-made changes, impounding estuarine water has perhaps the greatest effect. It must be recognized that impounded waters are a completely different type of ecosystem that have nearly no natural capacity for waste treatment. In countries like India and Bangladesh, huge numbers of impounded waterbodies have been created to promote shrimp or tiger prawn (*Penaeus monodon*) culture in the coastal areas, due to which the ecology and environment of the surrounding areas have been greatly damaged. The rate of reclamation of marshes and particularly mangrove swamps has accelerated in recent years in some parts of the tropics due to the rapid expansion of pond farming of shrimps for export (Fig. 2.10). About 50 % of the mangrove forests in the Philippines have been developed into brackishwater fish ponds (Saclauso 1989). The area converted in Thailand is estimated to be about 27 % and in Ecuador about 13–14 %. Such large-scale conversions have aroused considerable environmental concern among the public and development agencies.

Tabuchi (2003) estimated that on a global scale, the area under mangroves is shrinking by 100,000 ha annually due to clear cutting of timber and conversion into aquaculture projects.

Fig. 2.10 Shrimp culture ponds in mangrove patches: a common scene in tropical mangrove ecosystem (Photograph taken on 04.12.2013 by Mr. Tanmay Ray Chaudhuri (Environmentalist and a researcher of Techno India University, Kolkata))

In many countries, the coastal area is exploited for procuring salt from the seawater. About 30 % of the world's supply of salt comes from seawater. In order to keep the cost of production low, natural evaporation is used for extracting salt from the seawater. In the south of France, Puerto Rico, central California, the Bahamas, Hawaii and the Netherland Antilles, sea salt is extracted and reined to produce table salt. The process begins by allowing seawater to enter shallow ponds. Evaporation of the water produces a concentrated salt solution to which more seawater is added. Finally, the water is allowed to totally evaporate, leaving a thick salt layer behind, which can be processed commercially. In the Indian subcontinent, salt pans (Fig. 2.11) are developed in mangrove patches located along the coastlines of maritime states which cause vertical and horizontal migration of salts in the adjacent areas. This has profound negative impact on the survival and growth of coastal vegetation. High salinity reduces the growth of mangroves and even results in complete extinction of certain species like *Heritiera fomes*.

The marshy areas and mangrove patches are also destroyed in few areas of lower Gangetic delta through development of brick kilns (Fig. 2.12). The soil excavated for developing shrimp farms are used to manufacture bricks, which is a spin-off product of shrimp industry. The construction of brick kilns in swampy areas not only adds carbon dioxide to the atmosphere (a local level effect) but also clears the vegetation and modify the marsh soil, which otherwise are rich reservoir of carbon.

Estuaries may also be classified as tide-dominated and wave-dominated (Fig. 2.13) estuary on the basis of several important (positive and negative) hydrological processes (Table 2.7).

Fig. 2.11 Coastal waters stocked in shallow ponds for procuring salt through evaporation (Photograph taken on 04.12.2013 by Ms. Ankita Mitra (Environmentalist, Asutosh College, Kolkata))

Fig. 2.12 Brick kiln in the marshy land near estuary in the lower Gangetic delta region (Photograph taken on 04.12.2013 by Mr. Tanmay Ray Chaudhuri (Environmentalist and a researcher of Techno India University, Kolkata))

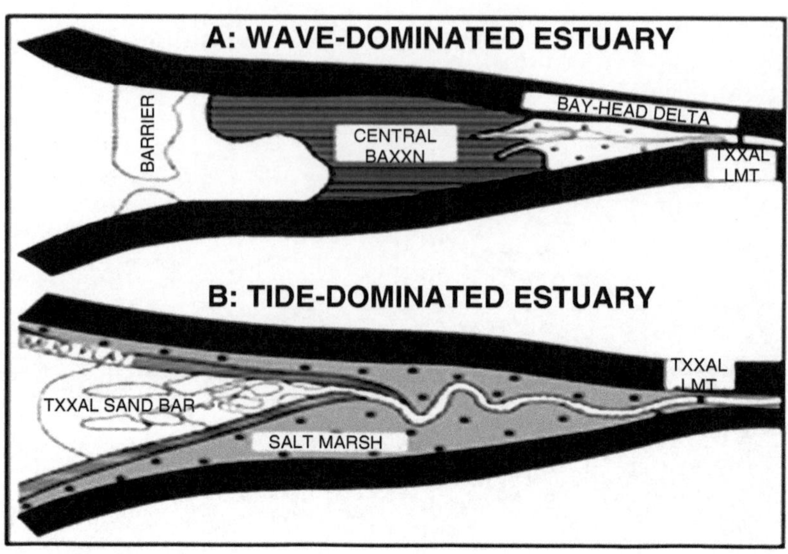

Fig. 2.13 Wave and tide-dominated estuary

Brain Churners

1. Why are the tidal amplitude and range higher in a funnel-shaped estuary?
2. How is flushing time important in maintaining the health of the estuary?
3. What is the source of soil organic carbon in the intertidal mudflats of mangrove forests?
4. How can the estuarine resources be linked with pharmaceutical industries?
5. Why is religious convergence observed in highly stressful environmental condition?
6. Will you consider CAL and TAL in the domain of direct use value of estuarine resources?
7. How the ecosystem services of the estuary may be hampered?
8. How are estuaries affected by climate change and subsequent sea level rise?
9. Why is productivity extremely low in Arctic and Antarctic coasts?
10. Can the estuarine system be considered as a store house of carbon? Explain your answer with case studies.

Annexure 2A: Blending Mangroves and Livelihood: A March Towards a New Dimension

The phenomenon of climate change was triggered since the dawn of human civilization, but it took an extraordinary long period to reveal this naked, but uncertain truth that anthropogenic factors are the raw materials for changing the climate of the planet Earth. The Intergovernmental Panel on Climate Change (IPCC) stated the fact with high confidence in their report of 2007. And this real but indigestible issue has converged two US presidential candidates of opposite polarity into one principal focus. Obama says that the United States is responsible for leaving the planet in better shape for future generations, while Clinton points towards moral issue in context to global warming. She thinks to replace the Kyoto Protocol, which is set to expire in 2012. The heat of these statements hardly reaches the poor villagers of Amalamethi, or Kalidaspur, which are located in the remotest corner of this planet at the apex of Bay of Bengal. The places in the deltaic complex of Indian Sundarbans are known for magnificent mangrove vegetation (Fig. 2A.1), tiger habitat (Fig. 2A.2), man–animal conflict, tiger prawn seed collection (Fig. 2A.3) and frequent natural calamities due to

Table 2.7 Comparative study of tide and wave-dominated estuaries

Point	Tide-dominated estuary	Wave-dominated estuary
Catchment's input	Freshwater entering from the catchments is relatively lower. In negative estuaries, the net inflow of marine water exceeds the outflow of catchment-derived (fresh) water. In such cases, the hypersaline water is usually exported to the ocean	Freshwater enters from the catchments. Although the volume of freshwater input varies spatially and temporally (depending on local catchments and climate conditions), it is often relatively high in positive estuaries
Freshwater input	The volume of freshwater entering the estuary is too low to cause significant level of stratification. High tidal ranges may tend to accelerate mixing of any freshwater inputs and marine water	Water circulation in wave-dominated estuaries generally ranges from well mixed to salinity-stratified, depending on the degree of wave mixing, volume of freshwater input and climate. 'Positive' wave-dominated estuaries have lower salinity water towards their head. The volume of freshwater causes stratification (or layering) in the water column, which varies with seasonal flow. Buoyant low-salinity freshwater floats above the denser, high-salinity ocean water
Salinity	High rates of evaporation cause increases in salinity within the estuary. The resulting high-density hypersaline water sinks beneath the buoyant marine water which penetrates through the estuary mouth and flows out of the estuarine entrance into the coastal ocean through a process known as reverse stratification. A large degree of mixing occurs between the two layers	A 'salt-wedge' or intrusion of denser saline marine water penetrates through the entrance along the bed of the estuary. Some mixing occurs at the interface between the fresh and marine water. The distance that the salt-wedge penetrates is dependent on the tidal range and the amount of fluvial flow received by the estuary. During high fluvial flow events (which may be seasonal), fresh floodwater may push the saltwater beyond the mouth. However, the large volume of central basins typical of wave-dominated estuaries tends to reduce this effect
Evaporation	Evaporation is the dominant process in negative tide-dominated estuaries due to arid climatic conditions and the extensive area of shallow intertidal environments. Aridity and the degree of evaporation may vary seasonally; however by definition, evaporation in 'negative' estuaries is much larger than freshwater input. Consequently, negative estuaries tend to have longer residence times than positive estuaries	While significant evaporation can occur in wave-dominated estuaries characterized by positive circulation, evaporation (by definition) does not exceed the amount of freshwater input
Water exchange	Exchange of seawater and estuarine water occurs through the wide entrance of the estuary. Flood and ebb tides may follow different routes into and out of the estuary, and the tidal prism tends to be large. In negative estuaries, the net inflow of marine water exceeds the outflow of freshwater derived from the catchments. In such cases, the hypersaline water is usually exported to the ocean	Exchange of ocean water and estuarine water occurs through the entrance of the estuary, although the magnitude of exchange depends on the size and length of the entrance channel. In positive wave-dominated estuaries, the outflow of freshwater exceeds the inflow of marine water. During dry conditions, the entrance of the estuary may be intermittently closed

cyclonic depressions and equinoxus tides. Nitai Gayen, a son of the mangrove soil, was a poacher, but after witnessing the repeated disasters due to tidal surges now thinks to shift to a new livelihood pattern. Like Netai, a large fraction of the people of these islands are thinking to accept new livelihood schemes like oyster culture, seaweed culture or fish feed preparation from mangrove flora. It is in this context a study was undertaken at Kalidaspur village in the eastern part of Indian Sundarbans during 2007 to find the role of specially formulated feed prepared from *Porteresia coarctata* (commonly known as salt marsh grass) to boost up the growth of freshwater prawn,

Fig. 2A.1 Mangrove-dominated Indian Sundarbans at the apex of Bay of Bengal

Fig. 2A.2 Sundarbans, the home of Royal Bengal tiger (*Panthera tigris tigris*) (Photo credit: Mr. Biswajit Roy Chowdhury, NEWS)

Fig. 2A.3 Prawn seed collection by island dwellers: a major threat to the ecosystem

Macrobrachium rosenbergii. This mangrove associate species is abundantly available in the mudflats of Indian Sundarbans (Fig. 2A.4) and is noted for its rich protein content. Dried powder of this floral species was mixed with the prawn feed as a source of protein. The Department of Marine Science, University of Calcutta, became a partner to this project in context to technology transfer. A significant profit was obtained after the final harvest (Fig. 2A.5) through this innovative scientific venture, and the beneficiaries witnessed the differences in terms of success indicators (Table 2A.1).

The Sundarbans mangrove region is a threatened ecosystem due to multitude of factors like prolonged over-exploitation of its natural resources, its use as a sink of anthropogenic wastes, industrial and maritime wastes generated in the upstream of the rivers flowing through the region, high population pressure around the region and the resultant shrinkage of the area brought about by clearing of forest land for agriculture and tiger prawn culture and lack of proper ecological management. Continuous erosive actions due to tidal currents and waves along with changing salinity have aggravated the magnitude of threat. The present programme has immense ecological and economic relevance in connection to these issues in the following ways:

- Utilization of only the freshwater system (ponds, ditches, rainwater harvested canals, etc.) and therefore clearance of mangrove areas for the culture of *Penaeus monodon* (shrimp) may be totally avoided.
- Involvement of the local people in three livelihood tiers: preparation of fish feed, ecofriendly prawn culture practice and nursery development of *Porteresia coarctata* for raw material backup to sustain the floral-based fish feed industry. It is expected that such involvement will restrict a sizable fraction of the people from illegal poaching and several anti-conservation activities.
- Economic upliftment of the local people.

Fig. 2A.4 Salt marsh grass *Porteresia coarctata*: a natural protein reservoir of deltaic Sundarbans

Fig. 2A.5 Final harvest of freshwater prawn grown with proteins of salt marsh grass

Annexure 2B: Mangrove Vegetation: A Natural Source of Antioxidant

1. Introduction

Oxygen is an indispensable molecule for the growth and survival of aerobic organisms in the planet earth. The entire mechanism of aerobic respiration resulting in the liberation of ATP is triggered by oxygen. However, this gaseous lifeline of the planet has an important demerit as it poses oxidative stress. Oxidative stress has been defined as a disturbance in the cell or organism related to pro-oxidant–antioxidant balance in favour of the former (Sies 1991). It differs from

Table 2A.1 Cost benefit analysis of the project

	Items	Pond (area = 500 m^2) maintained by traditional method (control pond)	Pond (area = 500 m^2) maintained by new technology (experimental pond)
Cost	Number of seeds	2500	3250
	Seed cost (in Rs.)	1000	1300
	Feed quantity (in kg)	336.9	571.35
	Feed cost (in Rs.)	13,476.00	14,283.75
	Experimental cost	6000	6000
	Labour/ management cost	2000	2000
	Total cost (in Rs.)	25,312.90	27,405.10
	Total unit cost (in Rs./m^2)	50.63	42.16
Benefit	Production return (in kg)	104	156
	Economic return (at Rs. 350/kg)	36,400	54,600
	Expenditure (in Rs.)	25,312.90	27,405.10
	Total profit/pond (in Rs.)	11,087.10	27,194.90
	Profit/unit area (in Rs./m^2)	22.17	41.84

any other stresses in that its primary effectors, the reactive oxygen species (ROS), can arise largely in the course of normal cell metabolism (Marova et al. 2005). Oxidative stress is involved in several pathological problems, especially in chronic degenerative diseases as diabetes, atherosclerosis, cancer, Alzheimer disease, etc. A question arises on how the organisms get rid of the oxidative stress. Till date the answer is related to the antioxidant defence mechanism of aerobic organisms. This defence mechanism is provided by an integrated antioxidant system, which has three distinct components (Scheme 2B.1), each equipped to reduce oxidative stress and the resultant adverse effects.

Preventative antioxidants suppress the formation of free radicals. Radical-scavenging antioxidants, such as the flavonoid compounds and vitamin C, serve to 'mop up' excess free radicals. Thus, scavenging antioxidants remove the ROS once formed, thereby preventing radical chain reaction (Marova et al. 2005). Repair enzymes play an important role in repairing and removing ROS-damaged molecules. Vitamin E and the carotenoids are very important biological antioxidants that have both preventative and radical-scavenging roles.

Astaxanthin is a carotenoid. It belongs to a larger class of phytochemicals known as terpenes. It is classified as a xanthophyll, which means 'yellow leaves'. Like many carotenoids, it is a colourful, fat-/oil-soluble pigment. Astaxanthin can be found in microalgae, yeast, salmon, trout, krill, shrimp, crayfish, crustaceans and the feathers of some birds. Professor Basil Weedon was the first to map the structures of astaxanthin (Fig. 2B.1).

Astaxanthin, unlike some carotenoids, does not convert to vitamin A (retinol) in the human body. Too much vitamin A is toxic for a human, but astaxanthin is not. However, it is a powerful antioxidant; it is ten times more capable than other carotenoids, which is due to its pure antioxidant nature, unlike other pigments which have pro-oxidant features (Fig. 2B.2).

In nature, a typical xanthophylls-producing unicellular microalgae is *Haematococcus pluvialis*, well known for its massive accumulation of ketocarotenoids, mainly astaxanthin up to

Scheme 2B.1 Structure of astaxanthin

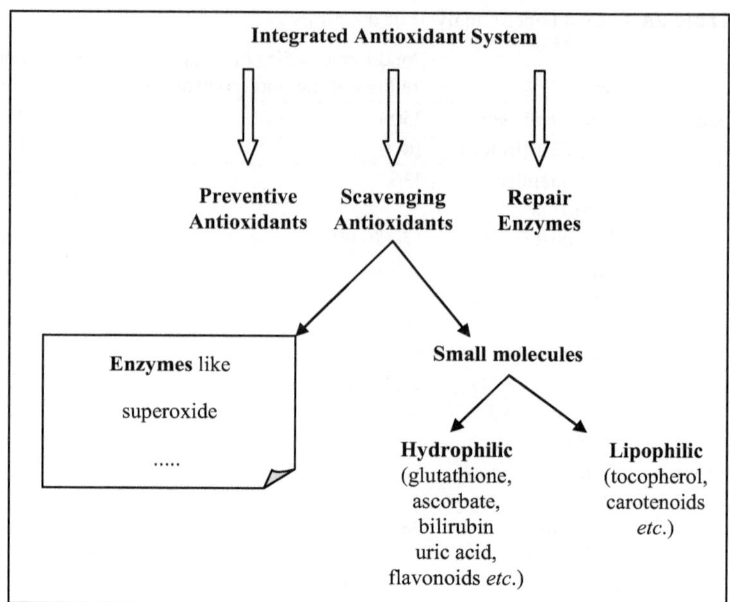

Fig. 2B.1 Astaxanthin structure

Fig. 2B.2 Pro-oxidants

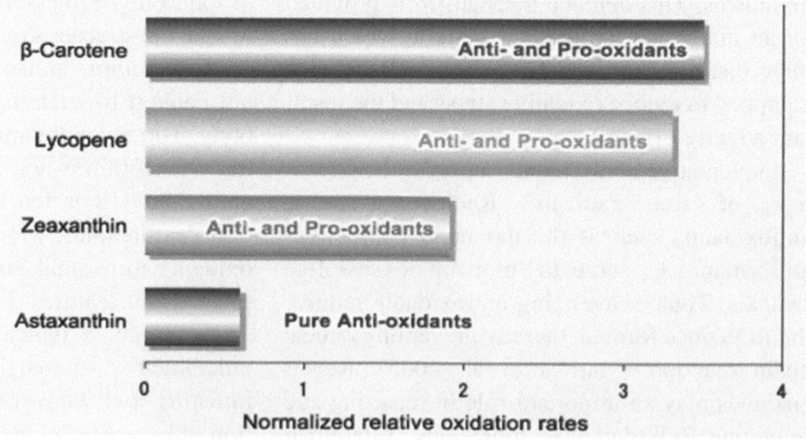

4 % of its dry mass and its acyl esters, in response to various stress conditions, e.g. nutrient deprivation or high irradiation. Also, the yeast *Phaffia rhodozyma* has been widely used for astaxanthin production in fed-batch fermentation processes using low cost materials as substrates (An et al. 2001; Chociai et al. 2002; Vazquez et al. 1998). Because of antioxidative properties and the increasing amount of astaxanthin needed as a supplement in the aquaculture of salmonoids and other seafood, there is growing interest in finding out the natural reservoir of astaxanthin. The present paper is the outcome of a research endeavour undertaken during March 2007 to screen the mangrove vegetation for astaxanthin at Jharkhali Island in the eastern sector of Indian Sundarbans region. This deltaic lobe is situated at the apex of the Bay of Bengal and has been designated as World Heritage Site for its marvellous genetic diversity with respect to mangroves and its associated flora and fauna. There are 34 species of true mangroves in the present geographical locale with several ecological, pharmaceutical and economic utilities (Mitra and Banerjee 2005), but the leaves of only ten dominant species collected from the Jharkhali island were selected for the present study.

2. Materials and Methods

The entire network of the present programme consists of the sampling of the leaves of ten dominant mangrove species during the low tide period from the Jharkhali Island during March 2007. Leaves of the selected species were collected from two different portions (submerged lower zone and exposed upper zone) of the same plant. The lower region of the tree gets inundated during the high tide condition and upper region of the same plant remains unexposed to tidal water. Salinity, pH, temperature, dissolved oxygen and nutrient load of the ambient water were analysed simultaneously to pinpoint the hydrological parameters to which the plant species are exposed in natural condition. The collected leaves were thoroughly washed with ambient water followed with deionized water and oven dried at 110 °C overnight. The extraction of astaxanthin was done in organic solvent as per the standard method and analysed spectrophotometrically.

3. Results and Discussion

The astaxanthin level in the selected mangrove species exhibits significant variations. It is of the order *Heritiera fomes* > *Bruguiera gymnorrhiza* > *Avicennia alba* > *Avicennia marina* > *Avicennia officinalis* > *Sonneratia apetala* > *Aegiceras corniculatum* > *Aegialitis rotundifolia* > *Ceriops decandra* > *Rhizophora apiculata* (Table 2B.1). The relatively greater astaxanthin content in the submerged leaves of mangroves confirms the synthesis of astaxanthin content under stressful condition. However, more studies are needed to confirm the influence of tidal influx and subsequent salinity fluctuation on astaxanthin level in the mangrove floral parts. The present data may serve as baseline information on the regulatory role of tidal submergence on astaxanthin level in the estuarine and coastal vegetation. The enhancement of astaxanthin production under stressed condition of organisms is a matter of interest, and several researches are still being undertaken to pinpoint the reaction pathway of astaxanthin production by inducing stress of varied nature. Many types of yeast have been described with an increase ability to produce carotenoids when they grow under unfavourable environment (Certik et al. 2005). Several workers have reported both in the dark and light the enhancement of the accumulation of astaxanthin in cysts of *Haematococcus pluvialis* under salt stress conditions. The present study has pointed higher astaxanthin level in those leaves of the mangroves that are inundated for 10–12 h by tidal waters (Table 2A.1) having typical estuarine water characteristics (salinity = 25.85 %0, pH = 7.97, temperature = 31.5 °C, dissolved oxygen = 6.10 mg/l, NO_3 = 21.04 μgat/l, PO_4 = 1.39 μgat/l and SiO_3 = 83.16 μgat/l). The steep enhancement of astaxanthin level in the inundated Sundari leaves (*Heritiera fomes*)

Table 2B.1 Astaxanthin content in ten mangrove species collected from Jharkhali Island of Indian Sundarbans during March 2007

Mangrove species	Astaxanthin content (mg/kg)		Percentage increase (%)
	Submerged	Exposed	
Avicennia officinalis	396.56	297.75	33.19
Avicennia alba	441.33	328.02	34.54
Avicennia marina	435.59	319.10	36.51
Sonneratia apetala	162.80	103.49	57.31
Aegiceras corniculatum	120.86	98.15	23.14
Aegialitis rotundifolia	105.92	82.41	28.53
Ceriops decandra	91.09	67.44	35.07
Heritiera fomes	761.00	398.54	90.95
Rhizophora apiculata	84.90	56.22	51.01
Bruguiera gymnorrhiza	461.38	397.11	16.18

clearly reflects the highest degree of stress posed by water salinity on this species. *Heritiera fomes*, being freshwater-loving mangrove species, cannot tolerate high salinity (Mitra and Pal 2002), and thus acceleration of astaxanthin production may probably be a part of its adaptation to cope with the stenohaline condition of coastal and estuarine environment that becomes acute during high tide. The astaxanthin level of mangrove flora is thus a function of its physiological system, which is extremely species specific.

The level of astaxanthin estimated in the mangrove floral species of Indian Sundarbans is less than the existing natural mega-reservoir of astaxanthin like *Phaffia rhodozyma* and *Haematococcus pluvialis*. However, considering the huge stock of mangrove vegetation in and around the area of Sundarban Biosphere Reserve (SBR), the reservoir of the natural antioxidant may reach the point of compensation, provided few backup nurseries in the region for extraction of bioactive substances are maintained.

4. Looking Forward

The mangrove ecosystem of Indian Sundarbans is one of the most biologically productive and taxonomically diverse ecotone of Indian subcontinent with a unique reservoir of bioactive substances. The detection of antioxidant astaxanthin in the floral parts of these typical estuarine and coastal vegetations adds a new dimension to these halophytes. Astaxanthin is an important feed ingredient with wide application both in pisciculture and animal husbandry sector owing to its antioxidant nature. Since the animals cannot synthesize carotenoids within their system, pigments must be supplemented to their feeds, allowing the assimilation and providing the characteristic pigmentation of the cultured aquacultural species, egg yolk, etc., for increasing the quality and consumer acceptance in the market place (Johnson and An 1991). This will not only upgrade the nutrition sector of animal husbandry and aquaculture but will also increase the immunity power of the cultured fish species and domesticated livestock. An ecologically fragile ecosystem sustaining a large fraction of poverty-stricken population like Sundarbans needs an alternative livelihood programme not only for upgrading their economic profile but also to realize the utility and application of their surrounding vegetation as a part of strengthening the root of conservation. The present programme may open an avenue of preparing fish feed, poultry and cattle feed by utilizing the antioxidant base of mangroves through involvement of the local people. This will defray the people from illegal entry into the forest and will also improve the animal and fish nutrition sector of the area through setting up of small-scale feed units. The antioxidant reservoir of Sundarbans mangrove ecosystem has several future applications, and hence proper policy is needed to blend the biotechnological approach

Table 2B.2 Astaxanthin level in different organisms

Natural astaxanthin sources	Astaxanthin concentration (ppm)
Salmonoids	~5
Plankton	~60
Krill	~120
Arctic shrimp	~1200
Phaffia yeast	~8000
Haematococcus pluvialis	~40,000

Source: http://algatech.com/astax.htm

Table 2B.3 Application mangrove antioxidant property

Application field	Avenues
Conservation of biodiversity	Diversion of local inhabitants from illegal intrusion into the forest through introduction of nonconventional livelihood programmes
	Maintaining carotenoid enriched floral nursery or floral park, grinding and mixing of carotenoid rich mangrove leaf with poultry feed, cattle feed, fish feed, etc.
Livelihood schemes	Setting up of fish, poultry and animal-feed units based on mangrove originated astaxanthin
	Initiation of ecofriendly aquaculture practice through fish feed rich in astaxanthin or any other natural antioxidant Production of export quality aquacultural items
Ecology and environment	Upgradation of water and pond bottom health
	Minimization of microbial load in the aquaculture sector

with the livelihood components of the island dwellers (Tables 2B.2 and 2B.3).

References

Banerjee, K., Mitra, A., Bhattacharyya, D. P., & Choudhury, A. (2002). Role of nutrients on phytoplankton diversity in the north–east coast of the Bay of Bengal. In A. Kumar (Ed.), *Ecology and ethology of aquatic biota* (pp. 102–109). New Delhi: Daya Publishing House.

Banerjee, K., Mitra, A., & Bhattacharyya, D. P. (2003). Phytopigment level of the aquatic subsystem of Indian Sundarbans at the apex of Bay of Bengal. *Sea Explorers, 6*, 39–46.

Banerjee, K., Sengupta, K., Raha, A., & Mitra, A. (2013). Salinity based allometric equations for biomass estimation of Sundarban mangroves. *Biomass and Bioenergy, 56*, 382–391.

Biswas, A. N. (1985). Geohydro-morphometry of Hooghly estuary. *Journal of Institute of Engeneering (India), 66*, 61–73.

Cameron, W. M., & Pritchard, D. W. (1963). Estuaries. In M. N. Hill (Ed.), *The sea* (Vol. 2, pp. 306–324). New York: Wiley.

Chakraborty, S. K., & Choudhury, A. (1985). Distribution of fiddler crabs in Sundarbans mangrove estuarine complex, India. In *Proceedings of national symposium on biology, utilization and conservation of Mangroves* (pp. 467–472).

Mitra, A. (1998). Status of coastal pollution in West Bengal with special reference to heavy metals. *Journal of Indian Ocean Studies, 5*(2), 135–138.

Mitra, A., & Choudhury, A. (1994). Heavy metal concentrations in oyster *Crassostrea cucullata* of Henry's Island, India. *Journal of Ecobiology, 6*(2), 157–159.

Mitra, A., Ghosh, P. B., & Choudhury, A. A. (1987). Marine bivalve *Crassostrea cucullata* can be used as an indicator species of marine pollution. In *Proceedings of national seminar on estuarine Management* (pp. 177–180). Department of Ocean Development, India.

Mitra, A., Choudhury, A., & Zamaddar, Y. A. (1992). Effects of heavy metals on benthic molluscan communities in Hooghly estuary. *Proceedings of the Zoological Society, 45*, 481–496.

Mitra, A., Trivedi, S., & Choudhury, A. (1994). Interrelationship between trace metal pollution and physico-chemical variables in the frame work of Hooghly estuary. *Indian Ports, 2*, 27–35.

Mitra, A., Banerjee, K., Sengupta, K., & Gangopadhyay, A. (2009). Pulse of climate change in Indian sundarbans: A myth or reality. *National Academy of Science Letters, 32*, 1–7.

Mondal, K., Mukhopadhyay, S. K., Biswas, H., De, T. K., & Jana, T. K. (2006). Fluxes of nutrients from the tropical River Hooghly at the land–ocean boundary of Sundarbans, NE Coast of Bay of Bengal, India. *Journal of Marine Systems, 62*, 9–21.

Odum, H. T. (1971). *Environment, power and society* (p. 331). New York: Wiley.

Pagiola, S., Konrad, V. R., & Joshua, B. (2004). *Assessing the value of ecosystem conservation*. World Bank Environmental Department paper No 101. The World Bank Environmental Department, in Collaboration with the Nature Conservancy and IUCN – The World Conservation Union.

Perillo, G. M. E. (1995). Definitions and geomorphological classifications of estuaries. In G. M. E. Perillo (Ed.), *Geomorphology and sedimentology of estuaries* (Developments in sedimentology 53, pp. 17–47). Amsterdam: Elsevier.

Pritchard, D. W. (1967). What is an estuary: Physical stand point? In G. H. Lauff (Ed.), *Estuaries* (American association for the advancement of science publication 83, pp. 3–5). Washington, DC: AAAS.

Saclauso, C. A. (1989). Brackishwater aquaculture: Threat to the environment? *Naga, 12*, 6–8.

Sadhuram, Y., Sarma, V. V., Murthy, R. T. V., & Rao, P. B. (2005). Seasonal variability of physico–chemical characteristics of the Haldia channel of Hooghly estuary. *Journal of Earth System Science, 114*(1), 37–49.

Saha, S. B., Mitra, A., Bhattacharyya, S. B., & Choudhury, A. (1999). Heavy metal pollution in Jagannath canal, an important tidal water body of the north Sundarbans aquatic ecosystem of West Bengal. *Indian Journal of Environmental Protection, 19*, 801–804.

Sengupta, K., Roy Chowdhury, M., Bhattacharya, S. B., Raha, A., Zaman, S., & Mitra, A. (2013). Spatial variation of stored carbon in *Avicennia alba* of Indian Sundarbans. *Discovery Nature, 23*(8), 19–24.

Tabuchi, R. (2003). *The rehabilitation of mangroves in Southeast Asia*. Retrieved June 1, 2007 from http://www.unu.edu/env/plec/marginal/proceedings/TabuchiCH12.pdf

Trivedi, S., Mitra, A., Gupta, A., Chaudhuri, A., Neogi, S., Ghosh, I., et al. (1994). Inter-relationship between physico-chemical parameters and uptake of pollutants by estuarine plants *Ipomea pescarpes*. In *Proceedings of the seminar: On our environment: Its challenges to development projects* (pp. SC-1–SC-6). American Society of Civil Engineers – India Section, Kolkata (India)

Internet Reference

www.es.flinders.edu.au/~mattom/ShelfCoast/chapterl1.html

Annexure References

An, G. H., Jang, B. G., & Cho, M. H. (2001). Cultivation of the carotenoid hyperproducing mutant 2A2N of the red yeast *Xanthophyllomyces dendrorhous* (*Phaffia rhodozyma*) with molasses. *Journal of Bioscience and Bioengineering, 92*, 121–125.

Certik, M., Masrnova, S., Sitkey, V., Minarik, M., & Emilia, B. (2005). Biotechnological production of astaxanthin. *Chemicke Listy, 99*, s237–s240.

Chociai, M. B., Machado, I. M. P., Fontana, J. D., Chociai, J. G., Busta, S. B., & Bonfim, T. M. B. (2002). Cultivo da levedura *Phaffia rhodozyma* (*Xanthophyllomyces dendrorhous*) em processo descontinuo alimentado para producao astaxantina. *Brazillian Journal of Pharmaceutical Science, 38*, 457–462.

Johnson, E. A., & An, G. H. (1991). Astaxanthin from microbial sources. *Critical Review in Biotechnology, 11*, 297–326.

Marova, I., Macuchova, S., Kotrla, R., & Jiri, H. (2005). How antioxidant intake influences lipid metabolism and antioxidant status in patients with type 2 diabetes and/or hypelipidemia. *Chemicke Listy, 99*, s251–s253.

Mitra, A., & Banerjee, K. (2005). *Living resources of the sea: focus Indian Sundarbans* (p. 120). New Delhi: WWF-India Secretariat, Sunderbans Landscape Project.

Mitra, A., & Pal, S. (2002). *The oscillating mangrove ecosystem and the Indian Sundarbans*. In S. Banerjee, & F. Tampal (Eds.). WWF-India, WBSO, Kolkata (India)

Sies, H. (1991). *Oxidative stress II. Oxidants and antioxidants*. London: Academic.

Vazquez, M., Santos, V., & Parajo, J. C. (1998). Fed-batch cultures of *Phaffia rhodozyma* in xylose-containing media made from wood hydrolysates. *Food Biotechnology, 12*, 43–55.

3 Physical Processes in the Marine and Estuarine Ecosystems

Contents

The major physical processes that configure the coastal zone, estuarine mudflats and shorelines of islands are waves, tides and currents. These processes may be broadly divided into destructive and constructive processes. The destructive processes include shoreline weathering, coastal erosion, etc., and the constructive processes encompass accretion or deposition of the sediment. The coastal zone is constantly exposed to waves, currents and tides of varying degree and types. Each of these processes is discussed in brief.

3.1 Waves and Tides

The most identifiable signature of the restlessness of the sea is the form and type of waves. Waves are of various categories due to their different mode of origin. There are tidal waves, seismic sea waves and wind or storm waves.

Waves are usually generated by wind passing over the surface of the sea. As long as the waves propagate slower than the wind speed just above the waves, there is an energy transfer from the wind to the waves. Both air pressure differences between the upwind and the lee side of a wave crest, as well as friction on the water surface by the wind, making the water to go into the shear stress, causes the growth of the waves.

Wave height is determined by wind speed, the duration of time the wind has been blowing, fetch (the distance over which the wind excites the

A. Mitra, S. Zaman, *Basics of Marine and Estuarine Ecology*,
DOI 10.1007/978-81-322-2707-6_3

waves) and by the depth and topography of the seafloor (which can focus or disperse the energy of the waves). A given wind speed has a matching practical limit over which time or distance will not produce larger waves. When this limit has been reached, the sea is said to be 'fully developed'. In general, larger waves are more powerful, but wave power is also determined by wave speed, wavelength and water density.

Oscillatory motion is highest at the surface and diminishes exponentially with depth. However, for standing waves (clapotis) near a reflecting coast, wave energy is also present as pressure oscillations at great depth, producing microseisms. These pressure fluctuations at greater depth are too small to be interesting from the point of view of wave power.

The waves propagate on the ocean surface, and the wave energy is also transported horizontally with the group velocity. The mean transport rate of the wave energy through a vertical plane of unit width, parallel to a wave crest, is called the wave energy flux.

Wind acts in several ways to help the formation of waves. Waves are generally produced by the action of wind blowing across the surface of the ocean and are formed as a result of the transfer of energy from the wind to the water. Winds can generate waves as small as a ripple or as high as 30 m. The relation between wind direction and the energy of the wave may be elucidated in the following cases:

(a) When waves advance in the direction of the wind, with a speed lesser than of the wind, there is an increase of pressure on windward side of each crest and decrease on leeward side. Hence, wind pressure is positive in order to increase the energy of waves.
(b) When waves advance in the direction opposite to the wind, the atmospheric pressure is negative in its action for wave energy.
(c) If speed of wave is greater than that of wind and waves are moving in the same direction, then the wave energy will have negative effect due to atmospheric pressure.

In addition to these actions, wind also works on the water surface by exerting a tangential frictional stress. The fetch of the wind acting upon the water is also responsible for the increased energy and consequent increase in the height of waves.

In a wave train, there is a regular succession of crests and troughs (Fig. 3.1). The horizontal distance between successive crests is the wavelength (L), while the wave height (H) is the vertical distance from a crest to a trough.

The period (T) is the time for passage of two successive wave crests passed through a fixed point and is usually expressed in seconds. Wave speed (V) can be calculated by $V = L/T$.

At the surface, water moves in circular vertical orbits with a diameter equal to the wave height. Away from the surface, orbital motion decreases, and orbits become smaller. At a depth of half a wavelength ($L/2$), orbital motion essentially vanishes. Thus, wave-generated water movements only occur near the surface. Where the water depth is less than half a wavelength, the

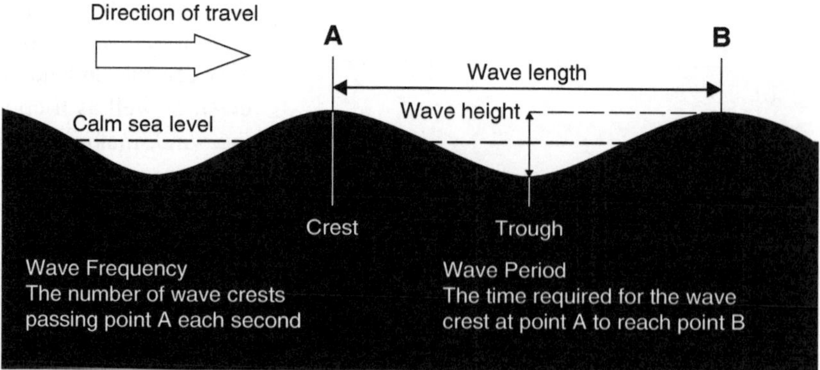

Fig. 3.1 Schematic representation of a typical wave

bottom interferes with the water motion. Water particles near the bottom cannot move vertically. This causes elliptical-shaped orbits. Wave size and speed depend on wind speed, wind duration and distance of water over which the wind blows. Theoretically, a wave's height cannot exceed 0.14 of its wavelength. If this is exceeded, white caps and breaking waves are formed (Fig. 3.2a, b, c).

Ocean surface waves can be classified in several ways. The most common way is by the period of the waves (Table 3.1).

(a) The shortest-period waves under this classification are the capillary waves. When the wind starts to blow on the ocean surface, capillary waves are generated. They seem just like a fine structure of small ripples of nearly capillary dimension. They basically contribute the largest amount of energy from the wind to the ocean water. Their wavelength is up to 1.74 cm.
(b) Ultra-gravity wave is another type of wave with periods between 0.1 and 1.0 s.
(c) The third division is ordinary gravity waves with periods between 1 and 30 s. They are composed of two types of waves namely 'sea waves' and 'swell waves'.
(d) The infra-gravity waves are generally long-period waves having periods between 5 min and 12 h. 'Tsunamis' and 'storm surges' belong to this division.
(e) The ordinary tidal waves are astronomical waves with fixed time period of about 12 h (semidiurnal) and 24 h (diurnal).
(f) The trans-tidal waves are those having periods greater than 24 h. Under this the longer-period components of astronomical tides are also included.

Waves can also be classified on the basis of the relationship of the waveform to the depth of water.

(a) A 'deep-water wave' is one in which the ratio depth (d)/wave/length (L) is more than 0.5. Capillary waves belong to this category.
(b) A 'shallow-water wave' is one in which D/L is less than 1/20. 'Tsunamis' are shallow-water waves even in the deepest portion of the ocean.

Another mode of wave classification is by the nature of controlling force.

(a) The shortest of the waves, the capillary waves, have a restoring force due to the surface tension of water. When water has been displaced, the force brings it back to normal position due to surface tension.
(b) In case of ultra-gravity waves and gravity waves, gravity acts as the major restoring force.
(c) Gravity together with Coriolis force acts as the restoring force in case of long-period waves.

Waves can also be classified on the basis of their generating force:

(a) Winds are responsible for the generation of capillary waves, ultra-gravity waves gravity waves and infra-gravity waves.
(b) Storms and earthquakes or movement of the crustal plates are responsible for long-period waves like tsunamis. Basically seismic sea waves (or tsunami) are generated by displacement of the ocean floor. Volcanic eruptions and catastrophic submarine mass movements can produce them, but by far, the most significant cause is earthquake. The displacement of a large mass of ocean water, often at a great depth, generates waves of small amplitude (usually less than 1 m), considerable length (up to 200 km or more) and high velocity. As with translatory waves, the wave velocity is related to ocean depth, and waves can travel at a velocity of about 600 km/h over water 3000 m deep. Seismic waves can therefore cross the entire Pacific Ocean within a span of few hours.

Fig. 3.2 (**a**) Initiation of breaking waves. (**b**) Approaching of breaking waves. (**c**) Formation of white caps and breaking waves

Table 3.1 Classification of ocean surface waves

Classification	Period range	Depth classification	Restoring force	Generating force
Capillary waves	<0.1 s	Deep-water wave	Surface tension	Winds
Ultra-gravity waves	0.1–1 s	Deep-water wave	Surface tension and gravity	Winds
Gravity waves	1–30 s	Deep-water wave to shallow-water wave	Gravity	Winds
Infra-gravity waves	30 s–5 min	Deep-water wave to shallow-water wave	Gravity and Coriolis force	Winds and gravity
Long-period waves	5 min–12 h	Transitional to shallow-water wave	Gravity and Coriolis force	Storms and earthquake
Ordinary tidal waves	(Semidiurnal and diurnal) 12–24 h	Shallow-water wave	Gravity and Coriolis force	Gravitational attraction of sun and moon
Trans-tidal waves	>24 h	Shallow-water wave	Gravity and Coriolis force	Storms, sun and moon

(c) The gravitational attraction of the sun and the moon acts as generating force for ordinary tidal waves.
(d) Storms, the sun and the moon are effective causes for trans-tidal waves.

On the basis of motion of waveform, wave can be classified into two categories.

(a) **Progressive waves** are those, which propagate laterally.
(b) **Standing waves** are those, which does not propagate laterally, but moves in a vertical direction. A standing wave may form when two progressive waves move in opposite direction in such a way that interference results. 'Seiches' are standing waves.

In addition to the above categories of waves, several researchers have stated about **coastal trapped waves** in the Sydney region of Australia. These are not waves in true sense, since they have little observable effects at the sea surface. These are characterized by timescales of 7–20 days and scales of longshore variability of many hundreds of kilometres (Church et al. 1986). They may be thought of as a current oscillation which moves equatorwards from its generation source in Bass Strait affecting Sydney waters some 2 days after being generated by strong winds in the Strait (Griffin and Middleton 1991). They are, therefore, predictable to some extent and can have the effect of producing strong currents in the nearshore region, which may have either northward flowing or southward flowing.

Very little information is available on the classification of waves in the Bay of Bengal region adjacent to the Indian part of Sundarbans. Surface waves in the Bay of Bengal are caused due to the wind system whose direction and velocity are mainly controlled by northeast and southwest monsoons. The wind from north and northeast starts blowing from the beginning of October to the end of March. The months of January and February are relatively calm with an average speed of around 3.5 km/h. It again commences to blow violently from southwest around the middle of March and continues till September. During this period, several low-pressure systems are formed in this region, among which few take the shape of depressions and cyclonic storms of varying intensity. The wind speed during this time rises above 100 km/h and is usually accompanied by gigantic tidal waves causing much loss of lives, damage to properties, forests and habitats of animals and plants thriving in this area. Sea waves in this region rarely become destructive except during cyclonic storms. When the cyclonic incidences

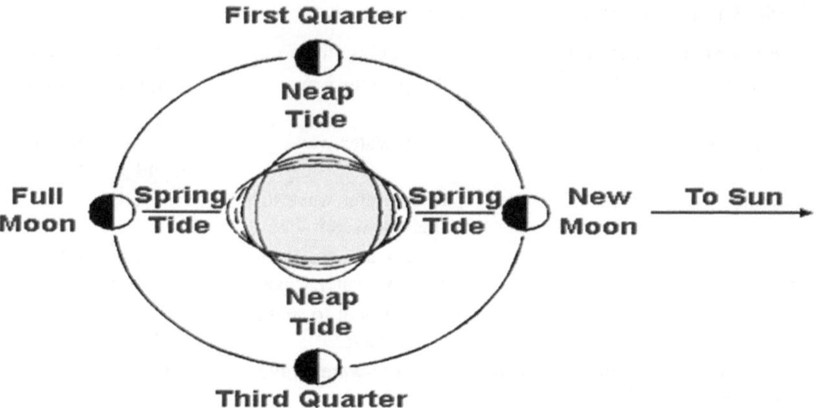

Fig. 3.3 Spring and neap tides

coincide with the spring tides, wave height sometimes exceeds over 5 m above the mean tide level. Ripple waves appear in the months of October, November and December, when wind-generated wave height varies approximately from 0.20 to 0.35 m. In the months of April to August, when the normal wind speed ranges between 15 and 40 km/h, large wavelets are formed in the shelf region, which starts breaking as they approach towards the coastal margin. In this condition, the wave height rises up to 2 m that causes maximum scouring action to adjoining of landmasses.

3.1.1 Tides

Tide is a phenomenon that occurs in the marine and estuarine systems in response to the pull of the moon and the more distant sun. Its character, pattern and variation are dynamic due to changing position of the attracting bodies. Hence, to explain the tide, it is necessary to account for these forces separately as well as in coordination. Newton was the first to put forth that the gravitational attraction exists even between the minutest drop of ocean and the outermost star of the universe—which may be negligent in amount. In general, the theory of gravitation states that the force of attraction is directly proportional to the product of masses and inversely proportional to the square of the distance between them—implying that the bigger mass will exert more gravitational pull as compared to the smaller mass and the farthest situated mass will exert lesser force than those situated nearby. Based on this background, the gravitational attraction of the moon felt on the Earth is more powerful than that of other planets and stars, because of its proximity to the planet Earth.

In oceanographic terminology, tides may be defined as the periodic rise and fall of the sea level and occur as a result of the gravitational pull of the sun and the moon. The magnitude of this attraction varies directly as the product of the mass of attracting bodies and inversely as the square of the distance between them. Though the sun's mass is 26×10^6 times bigger than that of the moon, the ratio of the square of the distance from the sun to the Earth and that from the moon to the Earth is roughly 58×10^6. It is because of this, the moon's pull is about 2.2 times stronger than that of the sun. Depending on the position of the heavenly bodies, the tides may be of two types: (i) spring tides (tides that are witnessed during the new moon condition), when the sun, the moon and the Earth are in a line (linear alignment) leading to maximum tidal range and (ii) neap tides (tides that occur when the sun and the moon are at right angles to each other during the first or third quarter of the lunar cycle). Figure 3.3 represents the positions of the sun and the moon in relation to the Earth during the two types of tides.

It may so happen that, while the sun, the moon and the Earth are in conjunction, the moon is at perigee, i.e. shortest distance in its orbit from the Earth. The tidal forces are mightiest under such

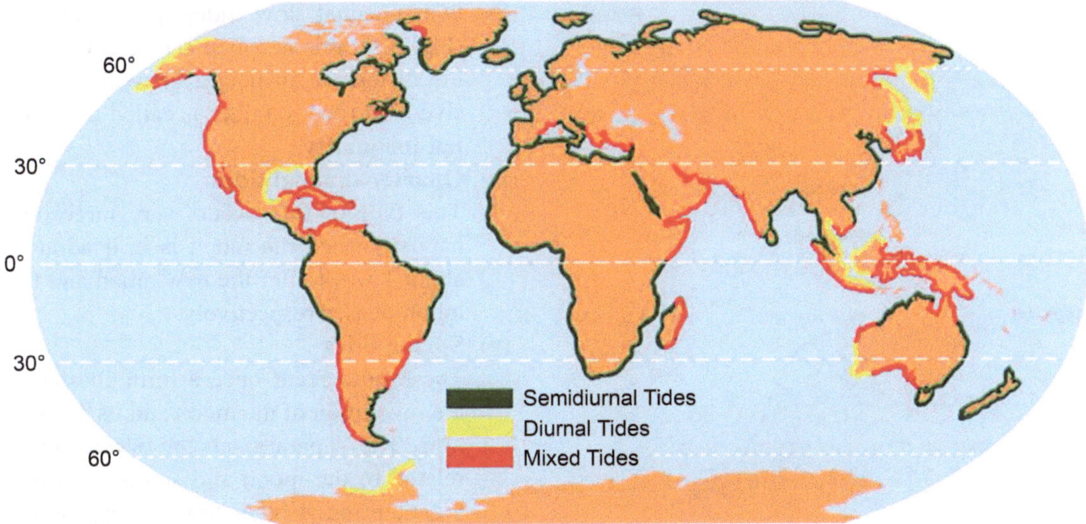

Fig. 3.4 Global distribution of the three major tidal types. Most of the world's coastlines have semidiurnal tides

circumstances, and the tides are higher than the usual. If in addition, the sun is on the equator, the perigee spring tides are the strongest. This happens in the **equinoxes** in March and September.

Tides are normally semidiurnal in nature. However, diurnal and mixed tides are also observed in some places in the ocean and bays. Hence, tides are classified as one of three types, semidiurnal, diurnal or mixed according to the characteristics of the tidal pattern (Fig. 3.4).

In **semidiurnal tide**, two high waters and two low waters occur in each day with relatively small differences in the respective highs and lows. Examples of such tides are the tides on the Atlantic coast of United States. In **diurnal tides**, only a single high and a single low water occur each tidal day. Tides of diurnal type occur along the northern shore of the Gulf of Mexico, in the Java Sea and in the Gulf of Tonkin. **Mixed tides** are characterized by large diurnal inequalities in heights and/or times of successive high and/or low waters. In general, this type of tide intermediates between predominantly semidiurnal and a predominantly diurnal. Such tides are prevalent along the Pacific coast of the United States and in many other parts of the world.

Tides may therefore be classified from various angles as stated here in brief:

(i) **Semidiurnal tides**

These tides occur at intervals of 12 h 30 min. When the moon is directly over the Earth's equator, its associated tidal bulges are centred on the equator. In theory, all locations on the planet except at the highest latitudes would rotate through the two tidal bulges and experience two equal high tides and two equal low tides per tidal day; this is known as a **semidiurnal tide**. Semidiurnal tides have an approximate period of 12 h and 25 min (Fig. 3.5) and theoretically have a wavelength of more than half the circumference of the Earth.

(ii) **Diurnal tides**

In this case, a single high and a single low tide occur each day. They tend to occur in certain areas when the moon is at its furthest from the equator. When the moon is above the Tropic of Cancer or Tropic of Capricorn, the diurnal inequality is at its maximum and the tides are called tropic tides. When the moon is above or nearly above the equator, the diurnal inequality is

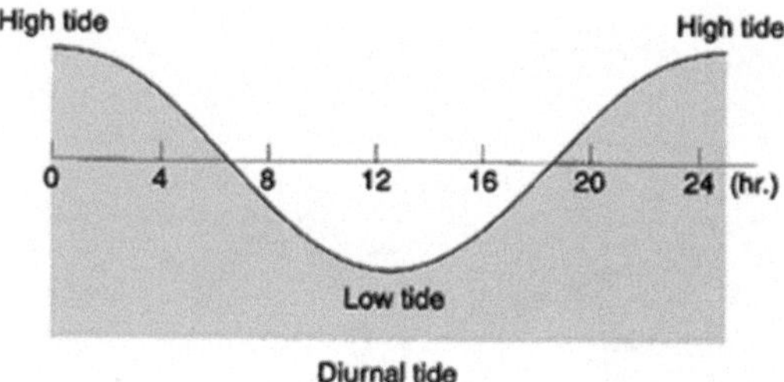

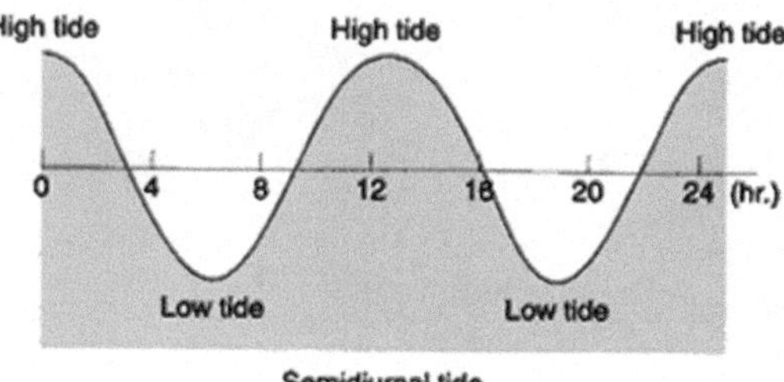

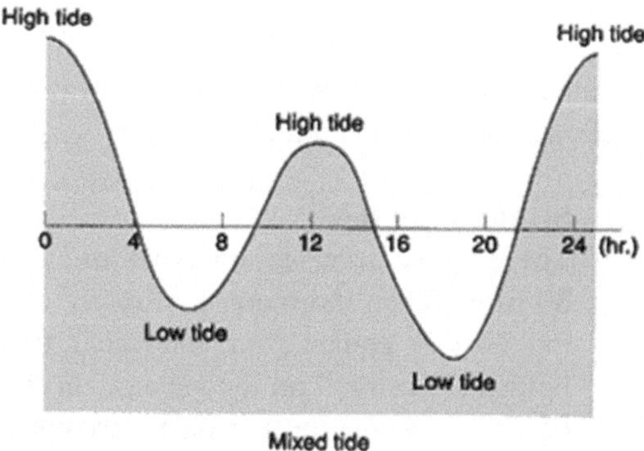

Fig. 3.5 Types of tides

minimum and the tides are known as equatorial tides. When the moon and its associated tidal bulges are either north or south of the equator, most points at high latitudes in theory would be impacted by one tidal bulge and would experience one high tide and one low tide per tidal day. This so-called **diurnal tide** has a period of 24 h and 50 min (Fig. 3.5).

(iii) **Mixed tide**

Different types of tides occur when the moon is either north or south of the equator. Whereas semidiurnal tides are observed at the equator at all times, most locations north or south of the equator experience two unequal high tides and two unequal low tides per tidal day (Fig. 3.5); this is called a **mixed tide** and the difference in height between successive high (or low) tides is called the **diurnal inequality**.

(iv) **Quarter-diurnal tides**

This type of tide occurs very rarely and happens when the moon is in quadrature about 1 week after the new moon and full moon phases respectively.

(v) **Spring tides**

These tides recur once a fortnight due to the revolution of the moon and its declination. The separate sets of ocean bulges related to the moon and sun act at times together and at other times in opposition. About every 2 weeks, the positions of the sun, moon and Earth form a straight line (Fig. 3.3). At these times of new and full moon phases as viewed from the Earth, the lunar- and solar-related ocean bulges also line up (and add up) to produce tides having the greatest monthly tidal range (i.e. the highest high tide and lowest low tide).

(vi) **Neap tides**

These tides recur once a fortnight due to the revolution and declination of the moon. At the first and third quarter phases of the moon, the sun's pull on the Earth is at right angles to the pull of the moon (Fig. 3.3). At this time, tides have their minimum monthly tidal range (i.e. unusually low high tide and unusually high low tide); these are **called neap** tides or 'fortnightly tides'. Furthermore, the moon orbits the Earth in an ellipse (rather than a circle) so that the moon is closest to the Earth (stronger tide-generating force) at perigee and farthest from the Earth (weaker tide-generating force) at apogee. The moon completes one perigee–apogee–perigee cycle once in every 25.5 days.

(vii) **Monthly tides**

The tides that occur due to the revolution of the moon and its position at perigee and apogee

(viii) **Equinoctial spring tides**
These tides recur at an interval of 6 months, due to the revolution of the Earth round the sun and sun's varying declination.

(ix) **Yearly tides**
Tides that occur due to revolution of the Earth and the position at perihelion and aphelion

The tidal currents, produced by the ebb and flow of tides, have little significance in open sea, but in the nearshore zones (depending on the size, configuration and topography of the marine basins), they control the coastal morphology to a great extent. The coasts can be categorized into macrotidal (<2 m), mesotidal (2–4 m) and microtidal (>4 m) depending on the tidal range.

3.1.2 Theories Related to Tides

Numerous theories have been put forwards to explain the origin of the tides. The tidal theories involve difficult mathematical deductions and analysis. The first mathematical theory which states the response of the ocean to the tidal forces was given by Newton in 1687. Newton's approach made the foundation on which subsequently the later works are based. Some of the important theories are given here.

3.1.2.1 The Equilibrium Theory

Isaac Newton in 1687 in his *Principia* developed the theory of equilibrium in which he assumed a mass of equilibrium under the combined forces of gravitation of celestial bodies. According to the equilibrium theory, the tidal bulge is caused by the gravitational attraction, and the moon is the major component regulating the ocean tide. The tide here is exaggerated. The actual bulge is on the order of 1 m. An equilibrium tide is a theoretical tide.

The effect of the sun's and moon's gravity and the rotation of the Earth on tide is most easily explained by studying equilibrium tides on a smooth water-covered sphere. Let us consider the Earth and the moon as a single unit that orbits the sun (Fig. 3.6). The moon orbits the Earth, held by the Earth's gravitational force acting on the moon (B), which must be balanced by a force called centrifugal force (B́′) acting to pull the moon away from the Earth and send it spinning out into space, and this balance maintains the moon in its orbit. Similarly, the moon's gravitational force acting on the Earth (C) must be balanced by the centrifugal force (C′). These centrifugal forces are caused by the Earth–moon system rotating about an axis at the centre of the system's mass, which is 4640 km. away from the Earth's centre. Again the Earth–moon system is held by the sun's gravitational attraction (A). A centrifugal force again acts to pull the Earth–moon system away from the sun (A′). To remain in this orbit, the Earth–moon system requires that the gravitational forces equal the centrifugal forces.

The moon's gravitational force is stronger on particles on the side of the Earth closest to the moon, while the centrifugal force is stronger on particles on the side of the Earth furthest from the centre of mass of the Earth–moon system. The distribution of forces tends to pull surface particles away from the centre of the Earth and creates the tide-generating force field on the Earth.

The stronger gravitational force of the moon acting on a unit mass of water at the Earth's surface closest to the moon is proportional to Mr/R^3, where 'M' is the mass of the moon, 'r' is the radius of the Earth and 'R' is the distance between the centres of the Earth and the moon. Because the water covering the Earth is liquid and deformable, the moon's gravity moves water towards a point under the moon, producing a bulge in the water covering. At the same time, the centrifugal force acting on the surface opposite the moon creates a similar bulge of water. The Earth model develops two bulges with two depressions in between the bulges or two crests and two troughs or two high tide levels and two

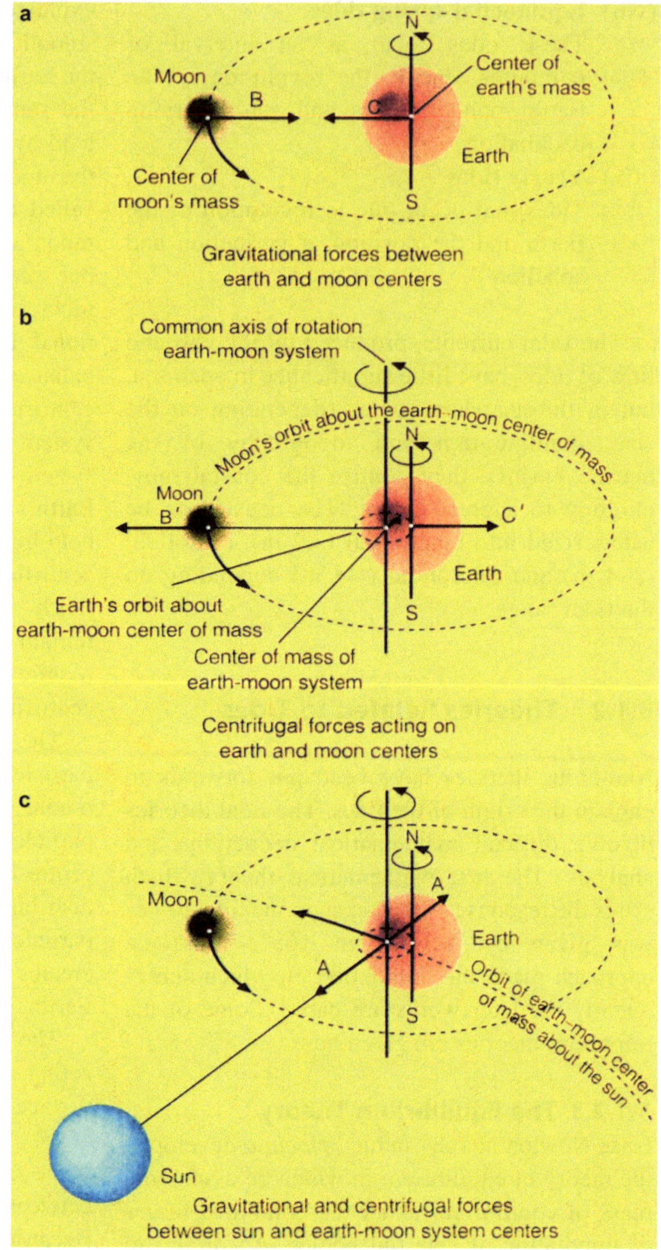

Fig. 3.6 Gravitational and centrifugal forces associated with tides (**a**) Gravitational forces between earth and moon centers; (**b**) Centrifugal forces acting on earth and moon centers; (**c**) Gravitational forces between sun and earth-moon system centers

low tide levels. The wavelength is equal to half the circumference of the Earth.

Limitations of the Theory

The equilibrium theory has several limitations. Firstly, the bulge of seawater as proposed by this theory cannot take place unless water masses actually change their positions, that is, a horizontal movement of tide is essential for this. Secondly, high tide should occur only when the moon is exactly on its meridian, and all the places on the same longitude should have high tides at the same time. But in practical this does not happen. In addition to this, the bottom relief/ topography and its complicated features (which have been ignored in this theory) obstruct the passage of the tidal flow from the sea to the estuary and vice versa.

3.1.2.2 The Dynamical Theory of Laplace

In 1775, the famous mathematical astronomer, Laplace approached the phenomenon of tides from the viewpoint of dynamics as the water in the sea is always in motion. In his work, *Mecanique Celeste*, he proved that the horizontal tide-producing forces are more important than the vertical forces, because in the latter the periodical variation of the acceleration of gravity is absent. Under the impact of oscillation, there is a horizontal movement of water towards the moon's meridian where the bulge is created. As the Earth rotates, different places automatically come beneath the moon, and changes in the horizontal movements of water follow. Hence, the water runs sometimes in one direction and sometimes in another in order to raise the bulge at points opposite the moon occupying different positions at different times.

3.1.2.3 Progressive Wave Theory

William Whemwell in his paper entitled 'Essay towards a First Approximation to a Map of Co-Tidal line' in 1833 and G.B. Airy in his article 'Waves and Tides' in 1842 presented the concept of progressive waves while explaining the varying intensity and varying amplitude of tides at different places. Whemwell also prepared a map of the world on which were drawn the so-called co-tidal lines, that is, the lines joining points at which high water occurs at the same hour. He called his theory the 'progressive wave theory', whereas G.B. Airy's approach is known as the 'canal theory'. The progressive wave theory for the first time took a different approach, considering the Earth as a heterogeneous body and not a perfect fluid. Secondly, tide is seen on the same longitude at different times, while simultaneously a lagging in the time of tide is observed on the same meridian away from the source. Thirdly, on a globe surrounded by water, a tide appears to take the form of a tidal wave travelling from east to west.

The progressive wave theory at the outset observes the tides as the movement of a wave in the sea from the place of its origin under the moon's tidal influence. In a globe completely surrounded by water, the tides will have the form of a wave having crests (tide) and troughs (ebb). The length is measured from crest to crest, which will be half of the circumference of that latitude on which it is proceeding. The rate of the progressing wave is dependent on the depth of the water bodies in which it is moving. On a rotating Earth, the progress of the wave would be from east to west. Due to the location of the continents, the movement of the waves and the generation of the progressive waves are hampered.

3.1.2.4 Stationary Wave Theory

The stationary wave theory probably seems to offer the best explanation for local differences in tides, their varieties and their age. Dr. R.-A. Harris of the US Coast and Geodetic Survey first propounded this theory, which is almost opposed to the ideas advanced by the progressive wave theory. The idea of this theory is to insist upon the regional occurrence of tidal phenomenon in the water bodies such as the Atlantic, the Indian and the Pacific Oceans. Hence, the solid Earth with its water hemisphere is made to oscillate under the lunar force. The resulting waves are known as the stationary waves.

3.1.3 Tidal Bore and Tide in Rivers

The tidal bore occupies a significant position in a tidal phenomenon. It is created when the tide enters the river mouth, and there is a forced entry of tidal wave against the running water of river or the channel. The forces are directly or indirectly connected with the tides of the open sea. Therefore, the nature and magnitude of the tides in the rivers and estuaries are affected by a variety of factors such as slope of river, flow of water, range of tide in the adjacent seas and the factor of friction. Once the tide originates by tidal forces in the open sea, it takes the form of the tidal wave moving inside the river or the estuary. Hence, the feature of river tide is considered as the wave motion in progress. According to mathematical calculations, the progress of the tidal

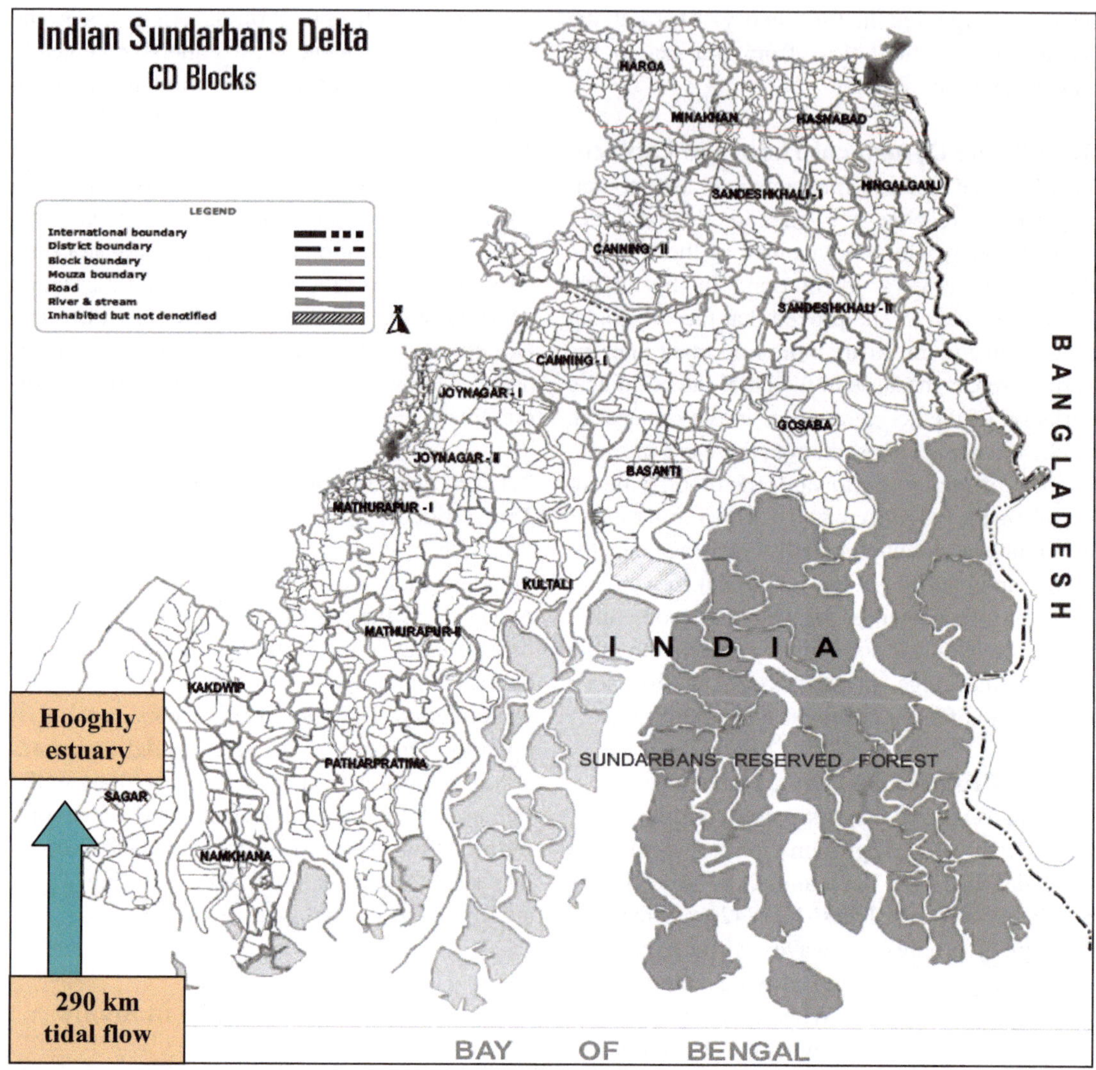

Fig. 3.7 Tidal pulse in the western Indian Sundarban estuary (Hooghly estuary) reaches 290 km upstream

wave should vary with the depth, the formula being $r = gh$, where, r stands for the rate of progress of the tide, g for the acceleration of gravity and h for the depth of the water way. The average value of the acceleration of gravity is 3217 ft/s, so r in terms of nautical miles per hour is $3.36 \sqrt{h}$.

The extent of tidal movement, upwards in the river channel, is determined by the slope of the riverbed, the velocity of the river current and the force of the tide. Under normal conditions, the tidal waves go up the stream until they are stopped by fall, rapids or dams. In the Amazon River, the upward movement of tides in the river channel goes to the extent of 45 nautical miles. In the Hudson River, the tide reaches up to 131 miles inside, but in the Mississippi River, it is limited up to the delta because of tides of lesser magnitude and intensity in the Gulf of Mexico. The tidal water in the Hooghly estuary (the western most estuaries of Indian Sundarbans) reaches about 290 km upstream from the estuarine mouth (Fig. 3.7). The tide penetrates up to 170 km in the southwest and 0–50 km in the southeast area of

Fig. 3.8 Tidal pulse in the western and eastern Bangladesh Sundarban estuary reaches maximum up to 170 and 50 km, respectively

Bangladesh during lean period (April–May) depending on the topography and channels in the area (Fig. 3.8).

The tidal activity is an important mechanism for the movement of water and nutrients especially in the estuarine areas, and this is probably a reason for the wide range of biodiversity in the estuarine water. It is also interesting to observe that the hydrological parameters in an estuarine system significantly changes with tidal phase. Same sampling station/site (fixed with GPS) exhibits different values of environmental/hydrological parameters during high and low tidal phases. For further information on the variation of environmental parameters with tide, the readers are advised to go through Annexure 3A.

3.2 Currents

Ocean current is the general movement of a mass of water in a fairly defined direction under the influences of various forces, internal as well as external. It is one of the most important characteristics of the seawater. Like the planetary wind system, surface-ocean currents play a significant role in helping equalize the energy imbalance between the tropical and polar regions. In addition, surface-ocean currents greatly influence the climate and shoreline configuration of coastal locations. The Earth's surface-wind system is the primary controlling component of the major surface currents and drifts. Other controls are the Coriolis effect and the size, shape and depth of the sea or ocean basin. Currents may also be caused by differences in density due to variations in temperature and salinity, tides and wave action. The major surface currents move in broad circulatory patterns, called **gyres,** around the subtropical highs. Because of the Coriolis effect, the gyres flow clockwise in the Northern Hemisphere and counterclockwise in the Southern Hemisphere (Fig. 3.9). As a general rule, the surface currents do not cross the equator.

Waters near the equator in both hemispheres are driven west by the tropical easterlies or the trade winds. The current thus produced is called the equatorial current. At the western margin of the ocean, its warm tropical waters are deflected polewards along the coastline. As these warm waters move into higher latitudes, they move through waters cooler than themselves and are identified as 'warm currents' (Fig. 3.9). In the Northern Hemisphere, warm currents, such as the Gulf Stream and the Kuroshio Current, are deflected more and more to the right (or east) because of the Coriolis effect. At about 40°N,

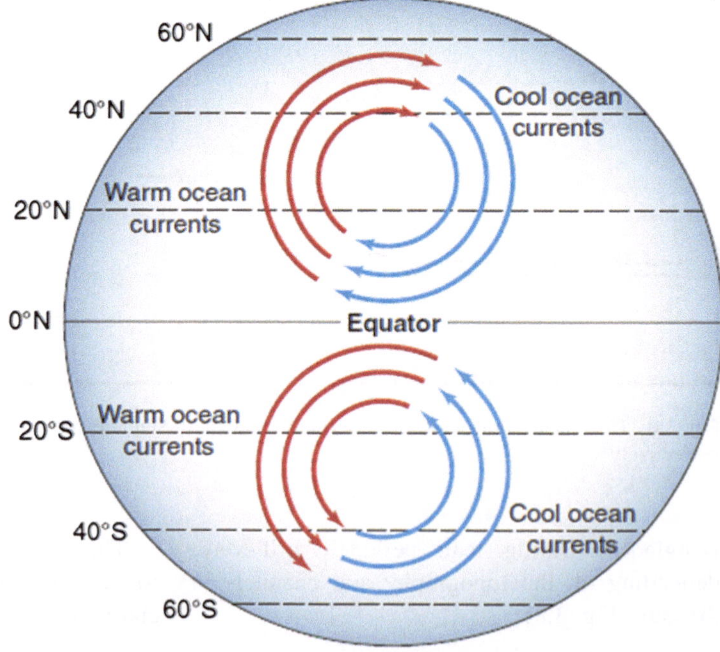

Fig. 3.9 The major ocean currents flow in broad gyres in opposite directions in the Northern and Southern Hemispheres

the westerlies begin to drive these warm waters eastwards across the ocean, as in the North Atlantic Drift and the North Pacific Drift. Eventually, these currents run into the land at the eastern margin of the ocean, and most of the waters are deflected towards the equator. By this time, these waters have lost much of their warmth, and as they move equatorwards into the subtropical latitudes, they are cooler than the adjacent waters. They have become cool, or cold, currents (Fig. 3.9).

In general the warm currents move polewards as they carry tropical waters into the cooler waters of higher latitudes, as in the case of the Gulf Stream or the Brazil Current. Cool currents deflect water equatorwards, as in the California Current and the Humboldt Current. Warm currents tend to have a humidifying and warming effect on the East Coasts of continents along which they flow, whereas cool currents tend to have a drying and cooling effect on the West Coasts of the landmasses. The contact between the atmosphere and ocean currents is one reason why subtropical highs have a strong side and a weak side. Subtropical highs on the West Coast of continents are in contact with cold ocean currents, which cool the air and make the eastern side of a subtropical high more stable and stronger. On the East Coasts of continents, contacts with warm ocean currents cause the western sides of subtropical highs to be less stable and weaker. The general circulation just described is consistent throughout the year, although the position of the currents follows seasonal shifts in atmospheric circulation. In addition, in the North Indian Ocean, the direction of circulation reverses seasonally according to the monsoon winds. The cold currents along the West Coasts in subtropical latitudes are frequently reinforced by **upwelling.** As the trade winds in these latitudes drive the surface waters offshore, the wind's frictional drag on the ocean surface displaces the water to the west. As surface waters are dragged away, deeper, colder water rises to the surface to replace them. This upwelling of cold waters adds to the strength and effect of the California, Humboldt (Peru), Canary and Benguela Currents.

About 10 % of the water in the world ocean is involved in surface currents, which are water flowing horizontally in the uppermost 400 m (1300 ft) of the ocean surface, driven mainly by wind friction. Winds drive currents that are at or near the ocean's surface. These currents are generally measured in metres per second or in knots (1 knot = 1.85 kilometres per hour or 1.15 miles per hour). Surface currents move water above the pycnocline, the zone of rapid density change with depth. The primary force responsible for surface currents is the wind. Most of the Earth's surface-wind energy is concentrated in each hemisphere's trade winds (easterlies) and westerlies. Waves on the sea surface transfer some of the energy from the moving air to the water by friction. This tug of wind on the ocean surface begins a mass flow of water. The water flowing beneath the wind forms a surface current. The moving water 'piles up' in the direction the wind is blowing. Water pressure is higher on the 'piled up' side, and the force of gravity pulls the water down the slope—against the pressure gradient—in the direction from which it came. In this case, the Coriolis effect intervenes. Northern Hemisphere surface currents flow to the right of the wind direction, because of the Coriolis effect. Southern Hemisphere currents flow to the left. Continents and basin topography often block continuous flow and help deflect the moving water into a circular pattern. This flow around the periphery of an ocean basin is called a gyre.

Gyres in balance between the pressure gradient and the Coriolis effect are called geostrophic gyres, and their currents are called geostrophic currents. The geostrophic gyres are largely independent of one another in each hemisphere, because of the patterns of driving winds and the present positions of continents. There are six great current circuits in the world ocean: two in the Northern Hemisphere and four in the Southern Hemisphere. Five are geostrophic gyres: the North Atlantic gyre, the South Atlantic gyre, the North Pacific gyre, the South Pacific gyre and the Indian Ocean gyre. Though it is a closed circuit, the sixth and largest current, known as the West Wind Drift (or Antarctic Circumpolar Current), is

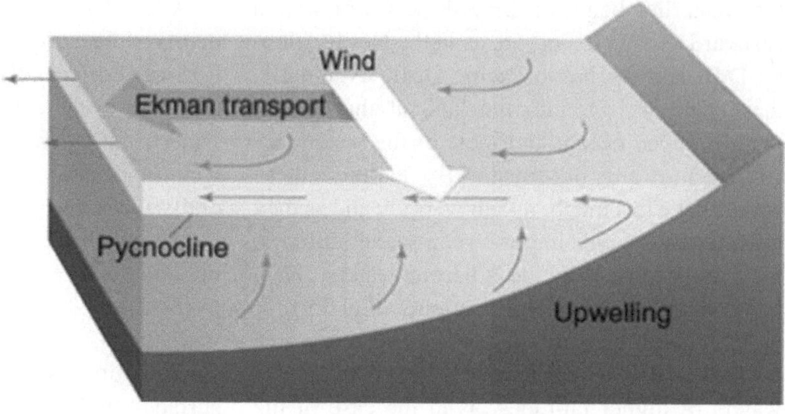

Fig. 3.10 Ekman transport moves surface waters away from the coast, and surface waters are replaced by water that wells up from below in the process known as upwelling

technically not a gyre since it does not flow around the periphery of an ocean basin. The West Wind Drift flows endlessly eastwards (i.e. never deflected by a continent) around Antarctica, driven by powerful, nearly ceaseless westerly winds. While it might be assumed that the two gyres in the North and South Pacific (and the two gyres in the North and South Atlantic) converge exactly at the geographical equator, instead the junction of equatorial currents (referred to as the meteorological equator) lies a few degrees north of the geographical equator. The meteorological equator and the Intertropical Convergence Zone (the band at which the trade winds converge) are displaced 5° to 8° northwards primarily because of the heat accumulated in the Northern Hemisphere's greater tropical land surface area. Ocean circulation like the atmospheric circulation is balanced around the meteorological equator.

3.2.1 Upwelling and Downwelling: Causes and Significance

In some coastal areas of the ocean (and large lakes such as the North American Great Lakes), the combination of persistent winds, the Earth's rotation (the Coriolis effect) and restrictions on lateral movements of water caused by shorelines and shallow bottoms induces upward and downward water movements. The Coriolis effect along with the frictional coupling of wind and water (Ekman transport) causes net movement of surface water at about 90° to the right of the wind direction in the Northern Hemisphere and to the left of the wind direction in the Southern Hemisphere. Coastal **upwelling** (Fig. 3.10) occurs where Ekman transport moves surface waters away from the coast; surface waters are replaced by water mass from the bottom of the ocean.

Where Ekman transport moves surface waters towards the coast, the water piles up and sinks in the process known as coastal **downwelling** (Fig. 3.11). Upwelling and downwelling illustrate mass continuity in the ocean; that is, water is a continuous fluid so that a change in distribution of water in one area is accompanied by a compensating change in water distribution in another area.

Upwelling and downwelling influence sea-surface temperature and biological productivity. Upwelling waters may originate below the pycnocline and are therefore colder than the surface waters they replace. Where the thermocline is shallow, the upwelling waters are usually rich in the dissolved nutrients (e.g. nitrogen and phosphate compounds) required for phytoplankton growth. This nutrient transport into the surface waters where sunlight, also required for phytoplankton growth (photic zone), is present results in the rapid growth of phytoplankton populations. Since phytoplankton form the base of marine food webs, the world's most productive fisheries are located in areas of coastal upwelling that bring cold nutrient-rich waters to the surface (especially in the eastern boundary regions of the subtropical gyres); about half the world's total fish catch comes from upwelling zones.

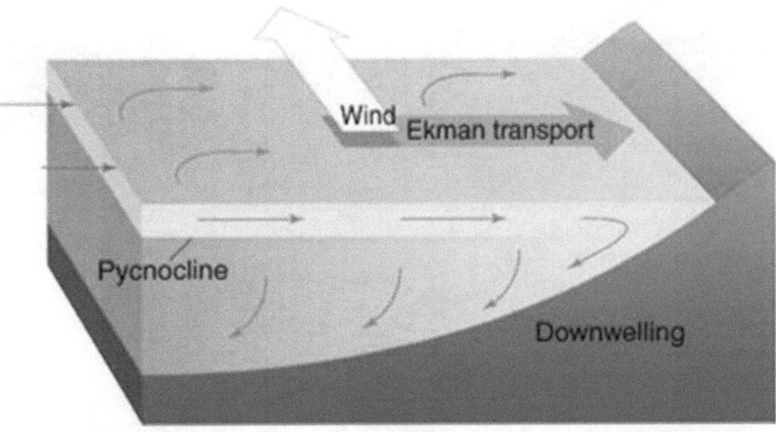

Fig. 3.11 Ekman transport moves surface waters towards the coast, where the waters pile up and sink in the process known as downwelling

Downwelling reduces biological productivity and transports heat, dissolved materials and surface waters rich in dissolved oxygen to greater depths. This occurs along the West Coast of Alaska in the eastern boundary region of the Gulf of Alaska gyre (driven by winds in the Aleutian low).

3.2.2 Factors Regulating the Pattern and Distribution of Currents

Most of the currents of the oceans follow certain rules as listed here (Fig. 3.12).

1. Warm currents move towards the cold seas, and cool currents move towards the warm oceans.
2. Cold dense upwelled waters are found in the middle latitude (trade belt) on the western shores of the continents.
3. From the colder seas, the cold waters of lesser density move into warmer oceans along the eastern coasts of higher latitudes.
4. In the lower latitudes, warm currents are found on the eastern shores and the cold on the western shores, whereas in the higher latitudes, warm currents flow along the western shores and cold currents on the eastern shores.
5. Convergences and divergences in the oceans also control the currents. In each hemisphere, there are three different water masses and two convergences, namely, subtropical and subpolar convergence.
6. The currents and their nature vary according to the seasonal changes in the year.

The pattern and distribution of currents in the world ocean is regulated by several factors, which may arise within or outside the planet Earth. For convenience of the readers, these factors are classified here in brief (Tables 3.2, 3.3 and 3.4).

3.2.3 What Happens Practically?

The shape of the planet Earth is not exactly round; it is ellipsoidal, and hence, there is a deviation in the current pattern of the ocean. In ellipsoidal structure, the force of gravity is not directed towards the centre of the equator. In Fig. 3.13 at surface (P), the gravitation AB (Fig. 3.14) may be resolved into two components, namely, AD at right angles to the surface and AE, which is tangential. The former presses the particle against the surface, while the latter tends to move towards the pole. In Fig. 3.14, the centrifugal force in AC direction is at right angles to the axis of rotation, which may be resolved again into two components—AF at right angle to the surface and AG, which is tangential. The former makes the particle relatively lighter, and the latter tends to move it towards the equator. When the particle is at rest on the rotating Earth, the forces AE and AG balance each other. Otherwise, if AE becomes more than AG, it will drive the mass towards poles and vice versa. This deflecting force is calculated

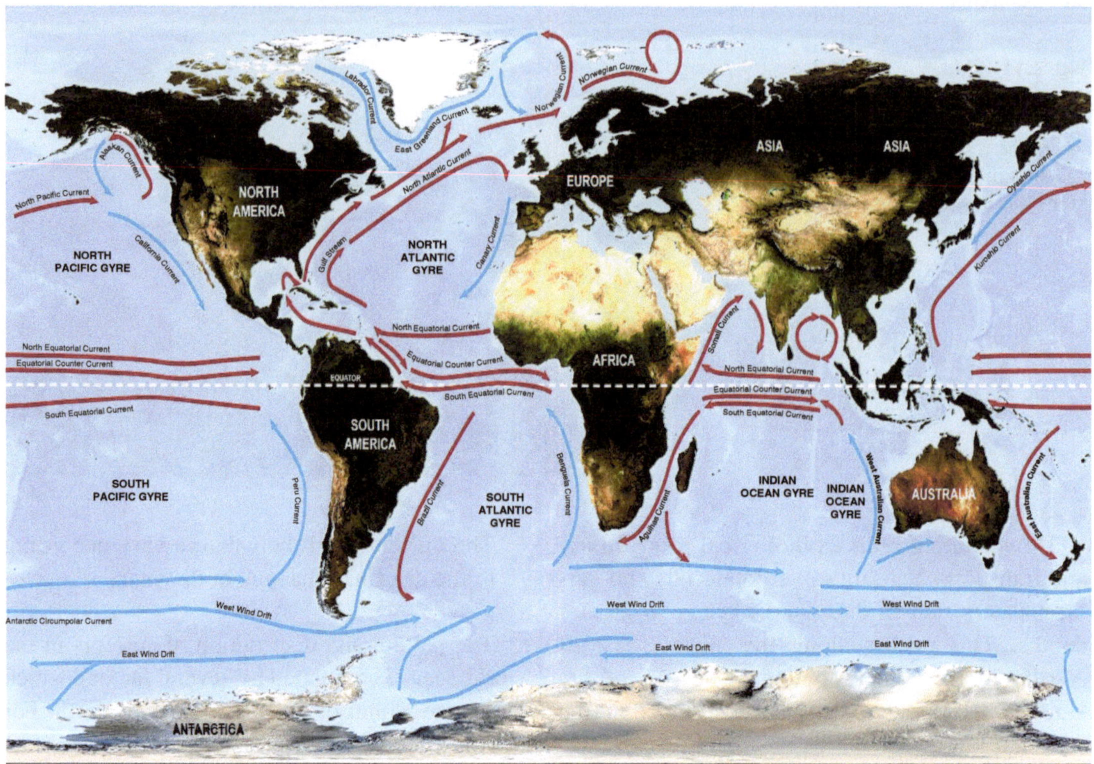

Fig. 3.12 Map of the major world ocean currents, showing warm and cool currents

Table 3.2 Factors in relation to the Earth's nature

Factor	Description
Gravitational force	A tiny particle of the ocean mass is also subject to a pure gravitational force directed towards the central region of the Earth. The magnitude and the dimensions of the acceleration of gravity of a moving mass vary with latitude and depth depending upon the position near to the centre of the Earth. Polar diameter is lesser than the equatorial diameter; hence, acceleration is greater at poles. The normal value at the pole is 9.83205, and at the equator, it is 9.78027. In other words, the gravity increases polewards; similarly deeper places in the ocean are near to the centre; hence, gravitational pull varies. In general the resultant effect on the currents would be a pull directed towards the centre. This force is also known as centrifugal force
Deflective force by the Earth's rotation	The rotation of the Earth from west to east results into the deflective Coriolis force which has a tendency to throw away particles from the centre of the Earth, opposing to the gravitational pull. Otherwise weaker, this force acts as an important controlling factor of the currents. The general tendency of the water at the surface in equatorial region is to reach the polar seas. But due to the rotational force, the currents are deflected as observed in the Baltic and the Mediterranean Sea. There it was found that the wind current existed up to 5 m only, and on many occasions, its speed was registered below 3 on Beauford Scale, thus unable to develop strong currents. Similarly, observations were made at the time when strong wind blew; then the speed of the current was 3.7 miles per 24 h. By computing the actual direction of the current and the wind, the deviation on the right was made up and was attributed to the rotation of the Earth. This deflection of the current, as computed and summarized by Ferrel in his Law, is the result of the movement of any particle on the rotating Earth which is acted upon by two forces, the force of gravity in the direction of AD towards the centre at right angles to the surface and the centrifugal force in the direction of DB at right angles to the axis of rotation as shown in Fig. 3.13. The former is directed downslope, as against the upslope direction of the latter. AD will be at right angles to the surface and hence can have no tendency to make the particle move in any direction. As the force DB is oblique to the surface, it will not only lift the particle but will cause it to slide towards the equator or the pole

Table 3.3 Ex-oceanic factors

Factor	Description
Atmospheric pressure and its variations	Atmospheric pressure is of primary significance in the fundamental equation of motion
Wind and the frictional force	Wind blowing regularly on the surface of the sea exercises a very considerable force in the movement of seawater. It glides with friction over the sea surface and drags onwards the pile of water with a considerable modification of the theoretical directions of the current to a remarkable depth
Precipitation	The excess of precipitation on oceans disturbs the level and results in the slope currents to equalize it
Nature of evaporation and insolation	Differences in the distribution of insolation and amount of evaporation over the oceans leads to current formation

Table 3.4 Physico-chemical factors

Factor	Description
Pressure gradient	The variation in the pressure gradient within the ocean may result in an internal disturbance. The surface of the sea over which the pressure is uniform or all points on which same pressure is found is known as 'isobaric surface'. Within the sea, pressure from all sides acts over a particle, hence the amount of pressure on any small element can be computed from the depth at which it is found, and on the coefficient of compressibility of seawater. Generally, the pressure increases downwards in the sea, and isobaric surfaces are parallel to sea level
Temperature difference	Turbulent mixing produced by winds and waves transfers heat downwards from the surface. In low and middle latitudes, this creates a surface **mixed layer** of water of almost uniform temperature which may be a few metres deep to several hundred metres deep. Below this mixed layer, at depths of 200–300 m in the open ocean, the temperature begins to decrease rapidly down to about 1000 m. The water layer within which the temperature gradient is steepest is known as the **permanent thermocline** The temperature difference through this layer may be as large as 20 °C. The permanent thermocline coincides with a change in water density between the warmer low-density surface waters and the underlying cold dense bottom waters. The region of rapid density change is known as the **pycnocline,** and it acts as a barrier to vertical water circulation; thus, it also affects the vertical distribution of certain chemicals which play a role in the biology of the seas
Salinity	Variability in salinity is closely linked to ocean currents. Salinity in surface waters is increased by the removal of water through evaporation, and it is decreased primarily through the addition of freshwater via precipitation and continental run-off, either in the form of rain or snow or from river inflow. At higher latitudes, salinity also is decreased by ice and snowmelt The average salinity of the oceans is about 35 psu, and variability in the global distribution of surface salinity greatly influences the movement of water mass. It has been observed that the highest salinity values are found at about 20°–30° latitude in both hemispheres, in areas having high evaporation and low precipitation. Low salinities are found in polar areas, which have high precipitation as well as melting ice, and in areas influenced by polar water
Density	In the marine system, the densest and deepest water masses originate primarily around Antarctica or in the vicinity of Greenland and Iceland. During winter in high latitudes, surface waters become colder, and because seawater density continues to increase to the freezing point, there is a continual sinking of water until that point is reached. As sea ice forms, it is less saline than the seawater so the salinity of the water is elevated, and the density further increased. This very dense polar water sinks and flows towards the equator at intermediate depths (Antarctic Intermediate Water and North Atlantic Deep Water) or along the seafloor (Antarctic Bottom Water). Antarctic Bottom Water in particular penetrates far into the northern parts of the Atlantic and Pacific Oceans. Deep water will eventually be returned to the surface by wind-driven mixing, and thus, there is a continuous, but very slow (on the order of several hundred to a thousand years), cycling between surface and deep waters

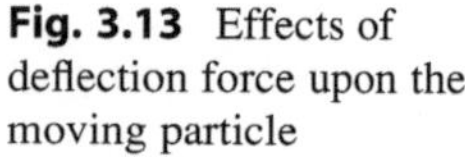

Fig. 3.13 Effects of deflection force upon the moving particle

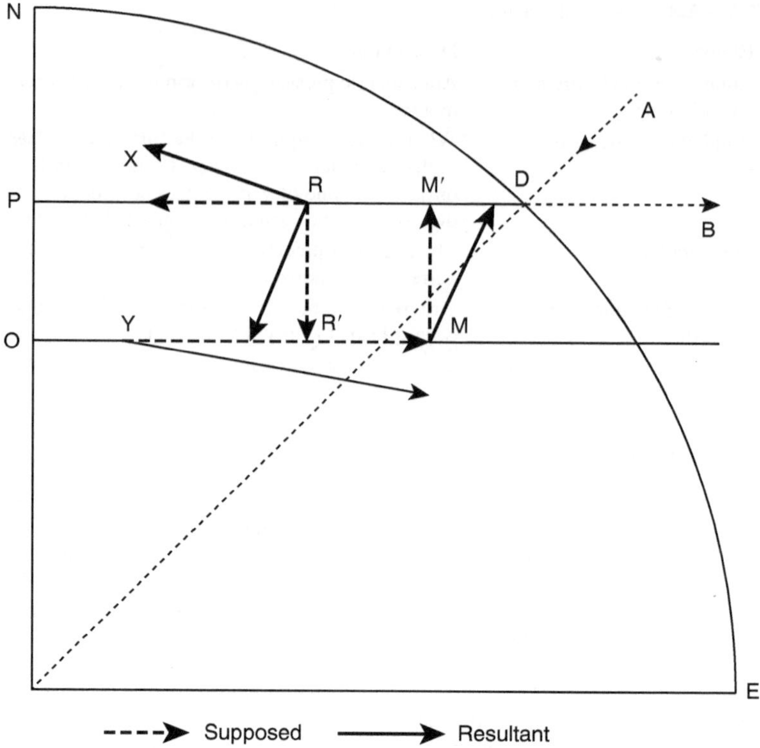

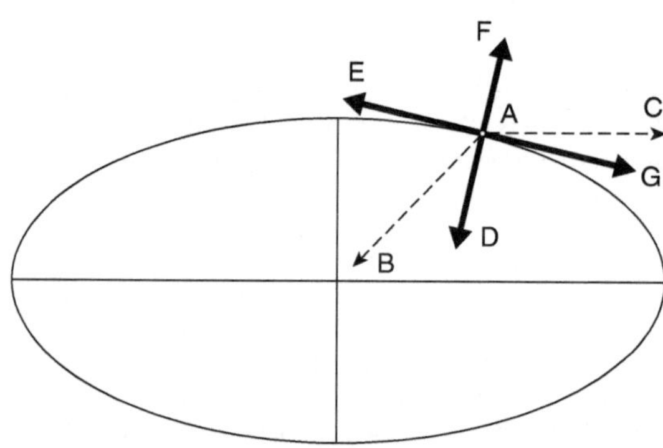

Fig. 3.14 Effects of deflection force upon the moving particle on the ellipsoidal Earth

as a function of the latitude, and it increases from the equatorial region to the pole.

Salinity, temperature and pressure establish the density of seawater. Changes at the sea surface, which result in higher density, will result in the downwelling of that water. Very dense water formed at high latitudes sinks to form the bottom water masses of the oceans, and this process is important in maintaining oxygen levels at all depths.

The ocean currents are also influenced and modified by factors like shape of the coastline, seasons and bottom topography. There role in modifying the direction and pattern of currents is discussed here in brief.

3.2.3.1 Direction and the Shape of the Coastline

Direction and shape of the coastline—coastal form modifies the direction of the currents to some extent. The westward movement of the South Equatorial Current is obstructed by the presence of the South American continent, and due to its projecting nose at Cape Sao Roque, this current is bifurcated into two: one part enters into the Caribbean Sea moving along northern coast of South America and mixes with the North Equatorial Current, whereas the other moves to the south following the direction of the coast of Brazil. Similar deflections are found on the western parts of equatorial oceanic regions where piling up of water takes place. In the Indian Ocean also, drift closely follows the form of the Indian coastline. In the Pacific Ocean, the islands obstruct the westward flow of the equatorial current and govern their directions.

3.2.3.2 Seasonal Variations

Seasonal variations in direction and in the volume of the currents are found in specified areas of wind shift belt or in special climatic regions. The currents of the Indian Ocean show a marked change in their direction. During northeast monsoons in winter, the direction of the monsoon drift is from east to west along the coast. But reversed direction is noted in summers when southwest monsoon drags the currents in the northeast direction. Consequently, changes in direction are also noted in the equatorial current of the Indian Ocean. The former appears only during winter months, and the latter only in summers.

3.2.3.3 Bottom Topography

The nature of the bottom topography to some extent modifies the currents on the surface as well as on the bottom of the sea. Ekman has examined the above fact by taking into account the rotation of the Earth. According to him, in low latitudes, the currents tend to flow east to west, independent of the slope of the bottom, but in middle and high latitudes, the currents tend to be influenced by the bottom contours.

The curvature or the deflection observed in the surface current as a result of a bottom barrier would be greatest where the change in the slope is greatest. In a homogenous layer, the presence of the slope current within which the velocities are independent of depth, except within a layer, are influenced by bottom friction. But in a nonhomogeneous water, the isosteric surfaces would be horizontal at bottom and sloping above in such a manner that looking towards the direction of flow the water of greatest specific volume (lowest density) is to the right and of east specific volume is to the left. When there is a submarine ridge under the water mass, the surface of water rises and the distribution of mass is altered in such a way that the surfaces slope upwards when approaching the ridge and downwards after passing the ridge.

3.2.4 El Niño

Figure 3.14 represents the cold Humboldt Current that flows equatorwards along the coasts of Ecuador and Peru. When the current approaches the equator, the westward-flowing trade winds cause upwelling of nutrient-rich cold water along the coast. This process triggers the phytoplankton bloom and results in the increase of fish population. Fishing, especially for anchovies, is a major local industry. Every year usually during the months of November and December, a weak warm counter current replaces the normally cold coastal waters. Without the upwelling of nutrients from below to feed the fish, fishing comes to a standstill. Fishermen in this region are acquainted with this phenomenon for hundreds of years. In fact, this is the time of year they traditionally set aside to tend to their equipment and await the return of cold water. The residents of the region have given this phenomenon the name **El Niño,** which is Spanish for 'The Child', because it occurs about the time of the celebration of the birth of the Christ Child. The warm-water current usually lasts for 2 months or less, but occasionally the disruption to the normal flow lasts for many months. In these situations, water temperatures are raised not just along the coast but for thousands of kilometres offshore (Fig. 3.15). Over the past decade, the term *El Niño* has come to describe these exceptionally strong episodes and not the annual event. During the past 50 years, approximately 18 years qualify as having El Niño conditions (with sea-surface temperatures 0.5 °C higher, or warmer, than normal for 6 consecutive months).

Brain Churners

1. How does the wind passing over the surface of the sea generate waves?
2. What is clapotis?
3. How are the seismic waves generated?
4. What are coastal trapped waves?
5. How is the tidal bulge influenced by the sun and moon?

(continued)

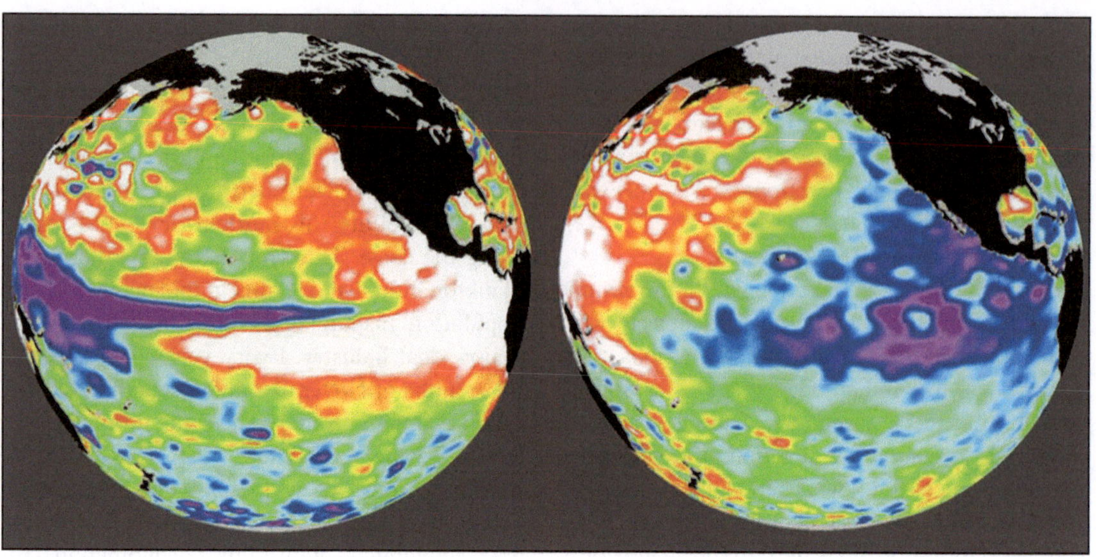

Fig. 3.15 These enhanced satellite images show a significant El Niño (*left*) and La Niña (*right*) episodes in the Tropical Pacific. The *red* and *white* shades display the warmer sea-surface temperatures, while the *blues* and *purples* mark areas of cooler temperatures

6. Why do gyres flow clockwise in the Northern Hemisphere and counterclockwise in the Southern Hemisphere?
7. What are upwelling and downwelling? How is the primary and secondary productivity influenced by these processes?
8. What are the factors regulating the salinity of seawater?
9. Why does the polar water from the Antarctic Region sink down and flow towards the equator?
10. How is the South Equatorial Current modified by the shape of the coast line?

Annexure 3A: Oscillation of Hydrological Parameters in and Around Indian Sundarbans

1. Introduction

The deltaic complex of the mighty River Ganga can be said to start from the extreme upstream region of Farakka in the maritime state of West Bengal. The River Ganga divides into two arms about 40 km southeast below Farakka at Khejurtala village in Murshidabad district. The right arm of the river (which was the original course of Ganga) continues to flow south in West Bengal in the name of the Bhagirathi (called Hooghly in its downstream stretch), which crosses 500 km to the sea (Bay of Bengal). The left arm flows into Bangladesh after flowing by the border of Murshidabad for 60 km in the name of Padma and joined by the Brahmaputra and the Meghna, these rivers form the huge deltaic lobe before meeting the Bay of Bengal. The Hooghly estuary is the westernmost estuary in the Gangetic delta and serves as the lifeline of millions of people inhabiting the mangrove-dominated Sundarbans and highly urbanized city of Kolkata, Howrah and the newly emerging Haldia port–cum–industrial complex. Multifarious industries are situated on the banks of the Hooghly River, namely, paper, textiles, chemicals, pharmaceuticals, plastics, shellac, food, leather, jute, pesticides, etc. (UNEP 1982). A considerable quantity of toxic and hazardous substance is being released into this important aquatic system through these industrial effluents along with huge organic load emanating from agricultural and shrimp culture activities and several non-point sources. The present study was therefore undertaken during 15–30 May 2014 to scan the stretch of the Hooghly estuarine system, the major arm of the River Ganga in terms of some important physico-chemical variables.

2. Study Area

The River Ganga emerges from the glacier at Gangotri, about 7010 m above mean sea level in the Himalayas, and flows down to the Bay of Bengal covering a distance of 2525 km. In this length, Ganga passes along 29 class I cities (population over 100,000), 23 class II cities (population between 50,000 and 100,000) and 48 towns having less than 50,000 populations. Stakeholders of several tiers are associated with this mighty river. About 50 % of Indian populations live in the Ganga basin. 43 % of the total irrigated area in the country also falls within the Ganga basin, and there are about 100 urban settlements with a total population of about 120 million on its banks.

The Hooghly estuary, the first deltaic offshoot of the River Ganga, is a coastal plain estuary and lies approximately between 21°31′–23°20′N and 87°45′–88°45′E. The southern part of the estuary flows through the marshy deltaic complex covered with thick mangrove forest called Sundarbans (declared as Biosphere Reserve of the country with an area of 9630 km^2). The estuary has a funnel-shaped mouth. The area of cross section can be related to distance by the expression $A_x = A_o e^{-kx}$, where A_x is the area of cross section of distance x, x is the distance measured landwards and A_o is the area of cross section at the mouth and is a constant (0.0241). The tide in the estuary is predominantly semidiurnal. The vertical tide range at the mouth varies from 5.2 m during spring to 1.8 m at the neap period. Due to the funnel-shaped mouth configuration of the estuary, a near constancy of ranges is maintained over a long stretch of the estuary. Twelve stations were selected in this estuarine stretch (Table 3A.1 and Fig. 3A.1) to monitor the spatial variations of physico-chemical characteristics during two tidal phases in the premonsoon period (May, 2014). The coordinates of the stations were measured by GPS (Model Garmin eTrex H).

Table 3A.1 Local name and coordinates of sampling stations

S. No.	Sampling station	Coordinates
1	Raichak (Stn.1)	22°12′12.00″N and 88°07′42.09″E
2	Diamond Harbour (Stn.2)	22°11′04.02″N and 88°10′50.52″E
3	Kulpi (Stn.3)	22°36′28.86″N and 88°23′28.32″E
4	Balari (Stn.4)	22°07′02.16″N and 88°11′35.34″E
5	Haldi River mouth (Stn.5)	22°00′26.07″N and 88°03′29.64″E
6	Nayachar (Stn.6)	22°00′30.42″N and 88°03′32.52″E
7	Khejuri Reserve Forest (Stn.7)	21°54′51.66″N and 88°00′56.52″E
8	Ghoramara Island (Stn.8)	21°56′15.24″N and 88°07′33.06″E
9	Harwood point (Stn.9)	21°56′15.24″N and 88°07′33.06″E
10	Harinbari (Stn.10)	21°46′54.12″N and 88°04′02.64″E
11	Chemaguri (Stn.11)	21°39′49.32″N and 88°09′11.88″E
12	Sagar South (Stn.12)	21°39′04.68″N and 88°01′47.28″E

3. Materials and Methods

The entire network of the present programme consists of the analysis of physico-chemical characteristics of Hooghly estuarine water during May 2014 with respect to selective variables like surface water temperature, salinity, pH, alkalinity, DO, BOD, COD, NO_3-N, PO_4-P, SiO_3-Si, extinction coefficient, chloride, sodium, potassium and total nitrogen. In order to evaluate the impact of tides, water samples were collected both during high tide and low tide periods. For each observational station, triplicate water samples were collected from the surface during two tidal conditions at a distance of 100 m of each other and analyzed for the selected parameters.

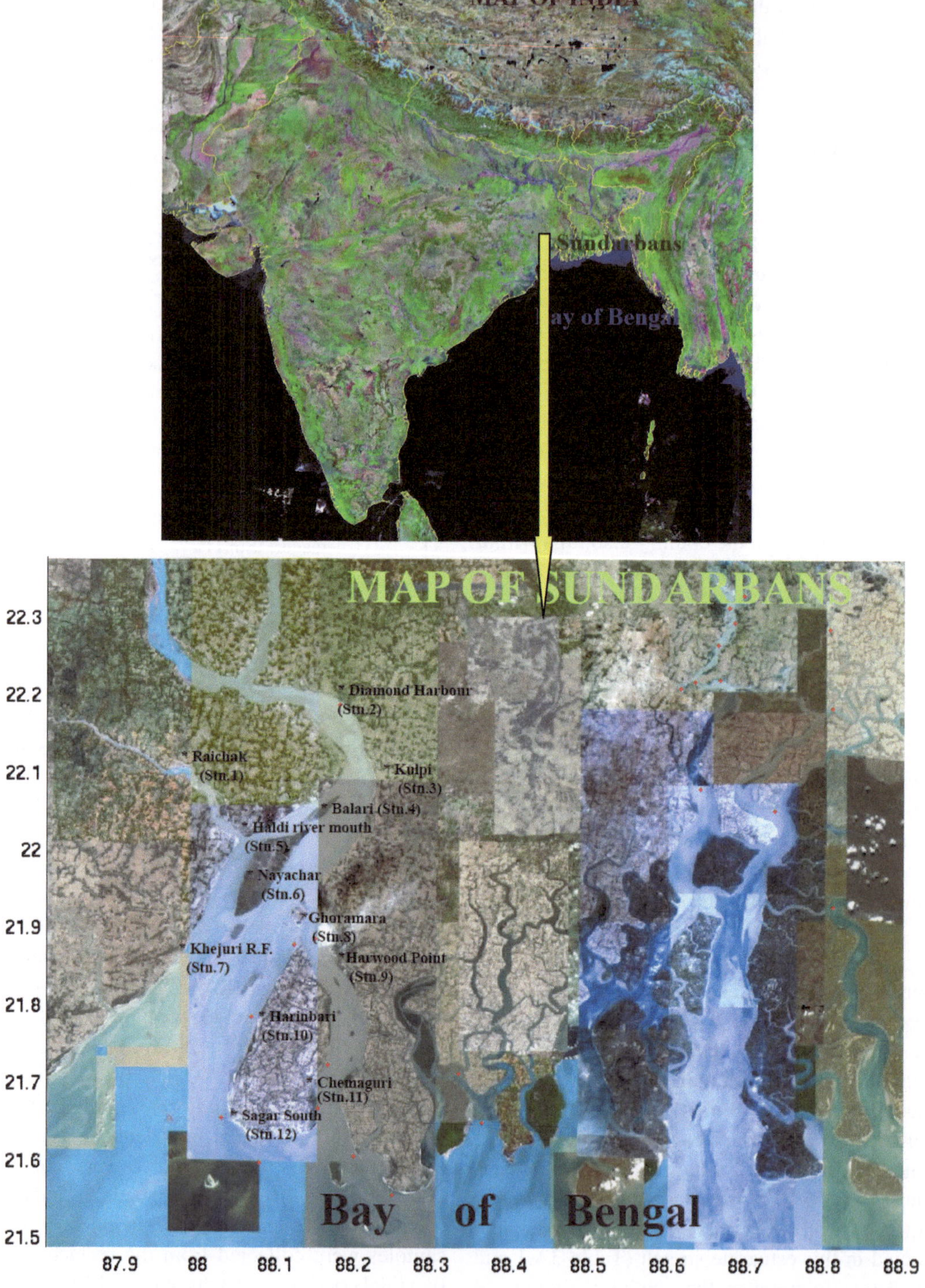

Fig. 3A.1 Map showing the lower stretch of Gangetic delta and location of sampling stations

Table 3A.2 Precision level of hydrological parameters

Parameter	Precision	Detection limit
Water temperature (°C)	±0.05 °C	0–100 °C
Salinity (‰)	±0.5 ‰	0–100 ‰
pH	±0.02	1–14
Alkalinity (mg/l)	±5 mg/l	10–2000 mg/l
DO (mg/l)	0.3 mg/l (±0.2 °C)	2–20 mg/l
BOD (mg/l)	±0.2 mg/l	1–20 mg/l
COD (mg/l)	±0.5 mg/l	10–300 mg/l
NO_3 (μg at/l)	±0. 2 μg at/l (1 μg at/l at 1.7 cm Cell)	0.1–45 μg at/l
PO_4 (μg at/l)	±0. 3 μg at/l (3 μg at/l at 1.7 cm Cell)	0.05–5 μg at/l
SiO_3 (μg at/l)	±1.5 μg at/l	20–300 mg at/l
Extinction coefficient (m^{-1})	±0.2 m^{-1}	2–10 m^{-1}
SO_4 (mg/l)	±3 mg/l	5–1500 mg/l
Na (mg/l)	±5 mg/l	10–15,000 mg/l
K (mg/l)	±3 mg/l	10–500 mg/l
Cl (mg/l)	±10 mg/l	100–20,000 mg/l
Total N (μg at/l)	±0.5 μg at/l	5–75 mg/l

A Celsius thermometer was used to measure the surface water temperature; pH and alkalinity were measured in the field using a portable pH meter (sensitivity = ±0.02) and micropipette titration method, respectively. Surface water salinity was measured by refractometer and cross-checked in the laboratory by argentometric method. The salinity of the standard seawater procured from NIO was analyzed by the same method, and a deviation of 0.02 % was obtained. Transparency was measured in the field by using a Secchi disc of 30 cm in diameter and converted into extinction coefficient. D.O., BOD, COD, NO_3-N, PO_4-P, SiO_3-Si, chloride, sodium, potassium and total nitrogen were measured as per the procedure stated in Strickland and Parsons and APHA 20th Edition (Strickland and Parsons 1972; APHA 2001). The precision of analysis is highlighted in Table 3A.2.

Finally, the tidal and spatial variations of the selected physico-chemical variables were evaluated through Duncan's Multiple Range Test (Table 3A.3). The alphabetical notations were used to mark the similarity and differences at significant level of an alpha 0.05 (Gomeez and Gomez 1984).

4. Results and Discussion

Estuaries are important segment of biogeochemical cycle as they regulate the amount of riverborne major and minor elements entering the coastal environment and ultimately the deep ocean. Estuarine ecosystems are complex and dynamic due to strong gradients in chemical composition of water, variable suspended matter concentration and complex hydrodynamic processes. When river water mixes with seawater, different types of physical and chemical processes take place that may affect the partitioning of trace metals between particulate and dissolved phases and hence the composition of the deposited sediments (Forstner 1983). Recently, the importance of estuarine processes in modifying the chemistry of the materials accumulating and passing through this interface has been realized. Several geochemical processes, such as precipitation and flocculation of the dissolved and colloidal substances (Coonley et al. 1971; Sholkovitz 1976; Gobeil et al. 1981), desorption–adsorption phenomenon, chemical diagenesis and exchange with the bottom sediments (Yeats and Bewers 1982), have been studied within the mixing zone.

In the present *era*, because of rapid industrialization and urbanization, wastes of complex nature have posed an adverse impact on the coastal and estuarine waters. The waters, rich in nutrients and heavy toxic metals, damage the living aquatic organisms (Bryan 1971, 1976, 1984; Hobbie 1976; Andrew et al. 1983). Our study area is no exception to this trend. The chain of factories and industries situated on the western bank of the Hooghly estuary is a prominent cause behind the gradual transformation of this beautiful ecotone into stinking cesspools of the megapolis. Our observation on the tidal and spatial variations of important physico-chemical variables in the major

Table 3A.3 Duncan's Multiple Test Range to determine the spatial and tidal variation of hydrological parameters

Station No.	Water temperature (ºC)		Salinity (‰)		pH		Alkalinity (mg/l)		DO (mg/l)		BOD (mg/l)		COD (mg/l)		NO_3 (μg at/l)		PO_4 (μg at/l)		SiO_3 (μg at/l)		Extinction coefficient (m^{-1})		SO_4 (mg/l)		Na (mg/l)		K (mg/l)		Cl (mg/l)		Total N (μg at/l)	
	HT	**LT**	**HT**	**LT**	**HT**	**LT**	**HT**	**LT**	**HT**	**LT**	**HT**	**LT**	**HT**	**LT**	**HT**	**LT**	**HT**	**LT**	**HT**	**LT**	**HT**	**LT**	**HT**	**LT**	**HT**	**LT**	**HT**	**LT**	**HT**	**LT**	**HT**	**LT**
1	a	a	A	b	a	b	a	b	a	b	a	b	a	b	a	B	a	b	a	b	a	b	a	b	a	b	a	a	a	b	a	b
	c	c	C	d	c	d	c	d	c	d	c	d	c	d	c	d	c	d	c	d	c	d	c	d	c	d	c	c	c	d	c	d
2	a	a	A	b	a	b	a	b	a	b	a	b	a	b	a	B	a	b	a	b	a	b	a	b	a	b	a	a	a	b	a	b
	c	c	C	d	c	d	c	d	c	d	c	d	c	d	c	d	c	d	c	d	c	d	c	d	c	d	c	c	c	d	c	d
3	a	a	A	b	a	b	a	b	a	b	a	b	a	b	a	B	a	b	a	b	a	b	a	b	a	b	a	a	a	b	a	b
	c	c	C	d	c	d	c	d	c	d	c	d	c	d	c	d	c	d	c	d	c	d	c	d	c	d	c	c	c	d	c	d
4	a	a	A	b	a	b	a	b	a	b	a	b	a	b	a	B	a	b	a	b	a	b	a	b	a	b	a	a	a	b	a	b
	c	c	C	d	c	d	c	d	c	d	c	d	c	d	c	d	c	d	c	d	c	d	c	d	c	d	c	c	c	d	c	d
5	a	a	A	b	a	b	a	b	a	b	a	b	a	b	a	B	a	b	a	b	a	b	a	b	a	b	a	a	a	b	a	b
	c	c	C	d	c	d	c	d	c	d	c	d	c	d	c	d	c	d	c	d	c	d	c	d	c	d	c	c	c	d	c	d
6	a	a	A	b	a	b	a	b	a	b	a	b	a	b	a	B	a	b	a	b	a	b	a	b	a	b	a	a	a	b	a	b
	c	c	C	d	c	d	c	d	c	d	c	d	c	d	c	d	c	d	c	d	c	d	c	d	c	d	c	c	c	d	c	d
7	a	a	A	b	a	b	a	b	a	b	a	b	a	b	a	B	a	b	a	b	a	b	a	b	a	b	a	a	a	b	a	b
	c	c	C	d	c	d	c	d	c	d	c	d	c	d	c	d	c	d	c	d	c	d	c	d	c	d	c	c	c	d	c	d
8	a	a	A	b	a	b	a	b	a	b	a	b	a	b	a	B	a	b	a	b	a	b	a	b	a	b	a	a	a	b	a	b
	c	c	C	d	c	d	c	d	c	d	c	d	c	d	c	d	c	d	c	d	c	d	c	d	c	d	c	c	c	d	c	d
9	a	a	A	b	a	b	a	b	a	b	a	b	a	b	a	B	a	b	a	b	a	b	a	b	a	b	a	a	a	b	a	b
	c	c	C	d	c	d	c	d	c	d	c	d	c	d	c	d	c	d	c	d	c	d	c	d	c	d	c	c	c	d	c	d
10	a	a	A	b	a	b	a	b	a	b	a	b	a	b	A	B	a	b	a	b	a	b	a	b	a	b	a	a	a	b	a	b
	c	c	C	d	c	d	c	d	c	d	c	d	c	d	C	d	c	d	c	d	c	d	c	d	c	d	c	c	c	d	c	d
11	a	a	A	b	a	b	a	b	a	b	a	b	a	b	A	B	a	b	a	b	a	b	a	b	a	b	a	a	a	b	a	b
	c	c	C	d	c	d	c	d	c	d	c	d	c	d	C	d	c	d	c	d	c	d	c	d	c	d	c	c	c	d	c	d
12	a	a	A	b	a	b	a	b	a	b	a	b	a	b	A	B	a	b	a	b	a	b	a	b	a	b	a	a	a	b	a	b
	c	c	C	d	c	d	c	d	c	d	c	d	c	d	C	d	c	d	c	d	c	d	c	d	c	d	c	c	c	d	c	d

a-a combination = tidal similarity, a-b combination = tidal variation, c-c combination = spatial uniformity, c-d combination = spatial variation; the similarity and differences are at significant level of an alpha 0.05, *HT* high tide, *LT* low tide

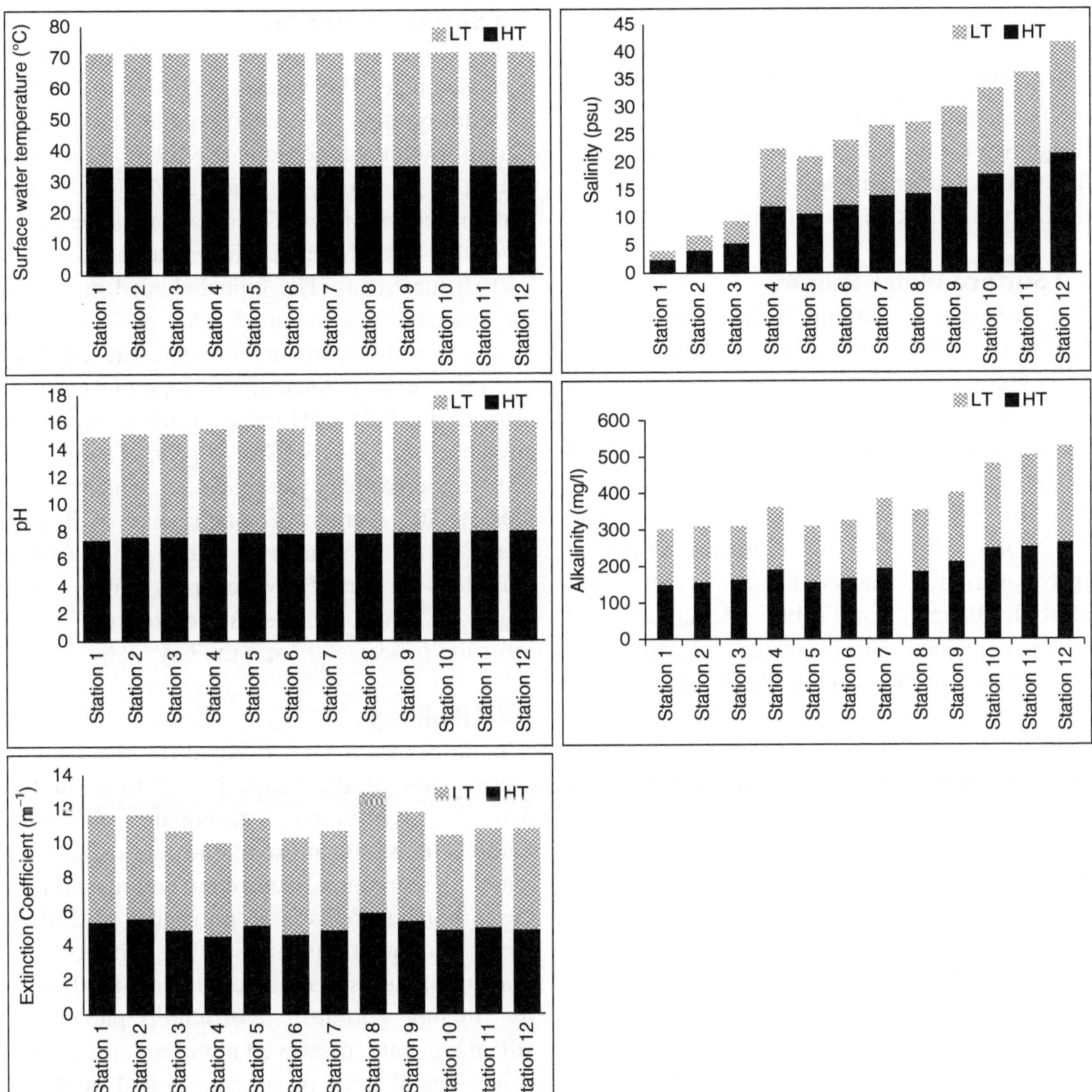

Fig. 3A.2 Tidal variations of surface water temperature, salinity, pH, alkalinity and extinction coefficient in the sampling stations

estuarine stretch of the Gangetic delta region is summarized in Table 3A.3

4.1 Surface Water Temperature

The surface temperature varied between 36.1 °C and 36.8 °C during high tide and 36.2 °C to 36.9 °C during low tide. The estuarine stretch did not exhibit significant tidal variation of surface water temperature; neither the spatial variation was prominent (Table 3A.3, Fig. 3A.2). The less fluctuation of water temperature is due to high specific heat of the aquatic phase, which enables water to resist much fluctuation of temperature than the adjacent landmasses. The aquatic subsystem in the present geographical locale therefore acts as a stabilizing factor upon the temperature profile of the Gangetic delta protecting the deltaic biodiversity from drastic thermal shock. The surface water temperature has considerable effect on phytoplankton population density by influencing the process of cyst germination (Ishikawa and Taniguchi 1994; Blanco 1995). The fluctuation of this variable has profound influence on estuarine food chain. Insignificant tidal and spatial variations of surface water temperature seem to have no impact on the phytoplankton community of the estuary.

4.2 Surface Water Salinity

The surface water salinity values ranged from 2.02‰ to 21.59‰ during high tide and 1.28‰ to 19.66‰ during low tide. The salinity values (mean of high tide and low tide) decreased from the downstream to the upstream zone as per the order Stn. 12 > Stn. 11 > Stn. 10 > Stn. 9 > Stn. 8 > Stn. 7 > Stn. 6 > Stn. 5 > Stn. 4 > Stn. 3 > Stn. 2 > Stn. 1, and the significant spatial variation was confirmed by Duncan's Multiple Range Test (Table 3A.3). La Fond (1954) explained that the decline of salinity of the surface waters is mainly due to the riverine contribution, which is much higher in the upstream stations (Stns. 1–4). The discharge from the Farakka barrage has got a significant influence on salinity in the present study area (Mitra et al. 2009). The barrage was constructed during 1975 to augment water flow in the Hooghly channel for the purpose of navigation, and during our study period, the average discharge was 34,195 cusec of freshwater per day. Five-year surveys (1999–2003) on water discharge from Farakka barrage revealed an average discharge of $(3.4 \pm 1.2) \times 10^3$ m^3/s. Higher discharge values were observed during the monsoon with an average of $(3.2 \pm 1.2) \times 10^3$ m^3/s and the maximum of the order 4200 m^3/s during freshet (September). Considerably lower discharge values were recorded during premonsoon with an average of $(1.2 \pm 0.09) \times 10^3$ m^3/s and the minimum of the order 860 m^3/s during May. During postmonsoon discharge values were moderate with an average of $(2.1 \pm 0.98) \times 10^3$ m^3/s. Significant tidal variation of surface water salinity (Fig. 3A.2) in all the sampling stations is being regulated by the discharge of freshwater from the barrage.

4.3 Surface Water pH

The pH of the seawater showed a variation within a small range. The values ranged from 7.64 to 8.19 during high tide condition and 7.50 to 8.10 during low tide condition. The relatively higher values of pH during high tide in all the selected stations are the effect of intrusion of saline water from the Bay of Bengal that penetrates almost 250 km upstream. The funnel-shaped mouth of the estuarine system forces more penetration of seawater in the upstream zone that caused alkaline effect even in the extreme uppermost station around Stn. 1. The pH values in the downstream stretch of the estuary (Stn. 10, Stn. 11 and Stn. 12) are approximately around 8.15, which is very close to the average pH of the Bay of Bengal water (8.28). Significant tidal and spatial variations of surface water pH (Table 3A.3, Fig. 3A.2) may thus be linked strongly to the intrusion of seawater from the bay.

4.4 Alkalinity

Alkalinity of seawater is equal to the stoichiometric sum of the bases in solution. In the natural environment, carbonate alkalinity tends to make up most of the total alkalinity due to the common occurrence and dissolution of carbonate rocks and presence of carbon dioxide in the atmosphere. Other common natural components that can contribute to alkalinity include borate, hydroxide, phosphate, silicate, nitrate, dissolved ammonia, the conjugate bases of some organic acids and sulphide. The major components contributing to alkalinity in the present geographical locale are carbonate rocks and other substances like nitrate, phosphate, ammonia, etc., that originate from sewage, municipal wastes (from the city of Kolkata, Howrah and Haldia) and large number of shrimp culture units in the Gangetic delta region. The alkalinity values ranged from 160 to 240 mg/l during high tide condition and 131 to 250 mg/l during low tide condition. Significant tidal and spatial variations of alkalinity (Table 3A.3, Fig. 3A.2) were confirmed in the study stretch. The relatively higher alkalinity values in the downstream stations (Stns.

9 to 12) may be related to the proximity of the stations to the Bay of Bengal and the presence of luxuriant mangrove vegetation in these zones. The relatively higher pH values in the downstream stations are due to the mixing of seawater with estuarine waters and by the mangrove photosynthetic activity, which utilize CO_2, thereby shifting the equilibrium towards highly alkaline (Ruttner 1953). Mangroves in the present study area are restricted in and around Stns. 9, 10, 11 and 12.

Alkalinity is important for fish and aquatic life because it protects or buffers against rapid pH changes. Living organisms, especially aquatic life, function best in a pH range of 6.0–9.0. Higher alkalinity levels in surface waters of downstream stations will buffer acid rain and other acid wastes and prevent pH changes that are harmful to aquatic life. The downstream stations of Gangetic delta are therefore relatively less stressful in comparison to upstream zones.

4.5 Dissolved Oxygen

Dissolved oxygen values ranged from 4.43 to 6.60 mg/l during high tide and 4.95 to 6.91 mg/l during low tide. The higher values of DO in the upstream stations (Stn. 1 to Stn. 4) may be apparently due to DO-rich freshwater conveyed through rivers and mere dilution of the zone with Farakka barrage discharge. Significant tidal variation of DO with relatively higher value during low tide period is the result of more freshwater during this phase of the tidal cycle. The contributory role of freshwater to increase dissolved oxygen was confirmed by Nair (Nair 1985) while working in the Kalapakkam waters. Significant variation of DO through space in the estuarine stretch (Table 3A.3, Fig. 3A.3) may be attributed to dilution factor (high in the upstream stations) and anthropogenic pressure (maximum at Stns. 4 and 9) due to the presence of Haldia port–cum–industrial complex (at Stn. 4), passenger jetties, fish landing units and busy markets draining untreated sewage (at Stn. 9) into the estuary.

4.6 BOD and COD

In the present study, the BOD values ranged from 2.8 to 6.5 mg/l during high tide condition and 2.7 to 8.1 mg/l during low tide condition. The COD values ranged from 67 to 145 mg/l during high tide condition and 71 to 152 mg/l during low tide condition. High BOD, COD and low DO levels observed at Stn. 2, Stn. 5 and Stn. 9 may be attributed to discharge of untreated municipal sewage and effluents from industries. The COD is used as a measure of the oxygen equivalent of the organic matter content of a sample that is susceptible to oxidation by a strong chemical oxidant. The mushroom growth of hotels and resorts and the presence of fish landing centres in and around Stn. 2 and Stn. 9 may be another prominent cause of high BOD and COD values. Most pristine rivers have a 5-day BOD below 1 mg/L, and moderately polluted rivers have a BOD value in the range of 2–8 mg/L. The present estuarine stretch is therefore not congenial in terms of organic load and nutrients. The anthropogenic activities of diverse nature have caused significant spatial variation of BOD and COD values (Fig. 3A.3). The relatively higher dilution of the system during low tide brings more sewage from the upstream stations centering the thickly populated city of Kolkata that increased BOD and COD values significantly (Table 3A.3).

4.7 Nutrients

Nitrate represents the highest oxidized state of nitrogen. The most important source of nitrate is biological oxidation of organic nitrogenous substances, which come through sewage and industrial wastes. Nitrate and phosphate usually exhibited higher values towards the upstream stations, while a reverse picture was noticed for silicate. The nitrate values ranged from 15.02 to 30.45 μg at/l during high tide and 16.29 to 35.14 μg at/l during low tide. Phosphate values ranged from 1.42 to 3.98 μg at/l during high tide and 1.81 to 6.01 μg at/l during low tide. The silicate values showed an increasing trend while proceeding from the coastal zone to riverine zone, although the values were significantly low at Stns. 9, 10, 11 and 12 near the bay. The values

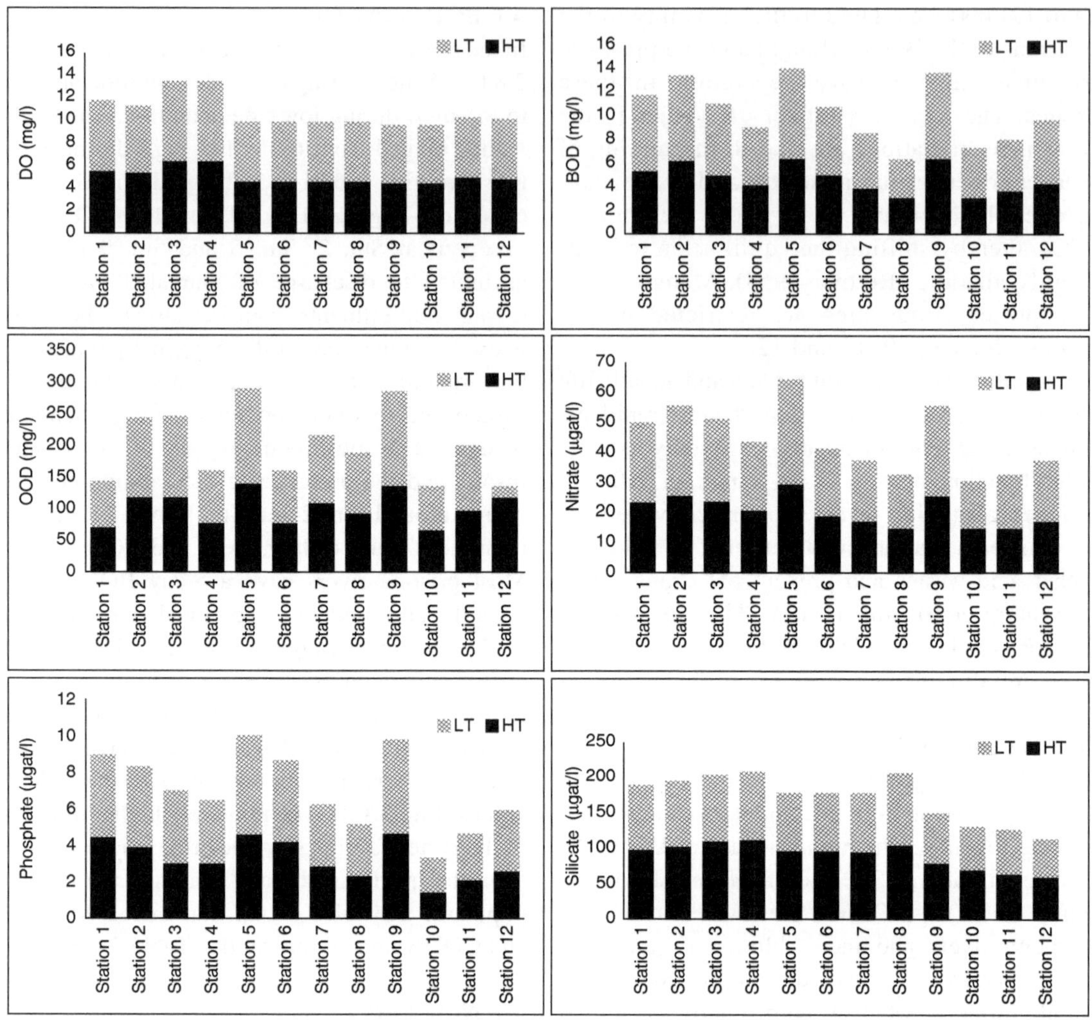

Fig. 3A.3 Tidal variations of surface water DO, BOD, COD, mitrate, phosphate and silicate in the sampling stations

ranged from 61.80 to 110.22 µg at/l during high tide and 49.78 to 95.00 µg at/l during low tide (Table 3A.3, Fig. 3A.3). Observations of increase in nutrients with decrease in salinity have been reported in the Indian estuaries by various workers (Sankaranarayanan and Qasim 1969; Solarzano and Ehrilich 1975). The increase of nutrient load in the present study area may be attributed to (i) increased industrialization and urbanization, (ii) unplanned expansion of shrimp culture units, (iii) large-scale use of fertilizers (urea, superphosphate, etc.) for boosting crop production in monocropping areas of the islands of Sundarbans, (iv) mushrooming of tourism units, (v) considerable number of unorganized fish landing sites with no provision for proper sewage and garbage disposal, (vi) increased number of fishing vessels and trawlers (Mukherjee et al. 2007) (Table 3A.4), (vii) erosion of embankments and mudflats due to wave action and (viii) contribution of litters and mangrove detritus from the adjacent landmasses.

The variations of nutrients were significant between two tidal conditions and also with space. The higher NO_3 and PO_4 levels in and around Stn. 2, 5 and 9 are the results of anthropogenic activities. The considerable nutrient load around Stn. 8 is the result of erosion activities

Table 3A.4 Official records of fishing vessels/trawlers in and around Indian Sundarbans

Year	No. of mechanized boat/trawlers
1998–1999	3362
1999–2000	3585
2000–2001	3622
2001–2002	3763
2002–2003	4175
2003–2004	4481
2004–2005	4521
2005–2006	4575
2006–2007	4630

due to sea level rise (Hazra et al. 2002). The spatial oscillation of nutrients (NO_3, PO_4 and SiO_3) is therefore the resultant of several causes, and it is extremely difficult to isolate the exclusive contribution of each factor in enrichment of nutrients.

4.8 Extinction Coefficient

The values of extinction coefficient ranged from 4.65 to 6.12 during high tide and 5.33 to 6.91 during low tide (Table 3A.3, Fig. 3A.2). The high value of extinction coefficient around Stn. 5 may be related to the location of the station near Haldia port–cum–industrial complex that discharge considerable amount of domestic sewage and industrial effluents. Stn. 9 is a busy market place with fish landing units and passenger vessel jetties, which generate considerable amount of sewage. The domestic sewage contains colloidal and finely divided suspended matter that impact turbidity in the water column thereby raising the value of extinction coefficient. Similar lowering of water transparency was confirmed by Satyanarayana et al. (1990) while working in the other Indian estuaries. Extinction coefficient was also higher around Stn. 8, which is an erosion-prone zone contributing appreciable load of silt particles to ambient water. Relatively higher extinction coefficient values during low tide indicate more riverine contribution of suspended particulate matter from the populated cities and towns of the upstream zone. Significant tidal and spatial variations of the parameter were confirmed through Duncan's Multiple Test Range (Table 3A.3).

4.9 Sulphate

Sulphate is a major constituent of seawater and is used as an electron acceptor for the oxidation of organic matter in the absence of oxygen. The site of sulphate reduction is more often on the surface of sediments than the water column. The reaction kinetics of sulphate reduction and associated sulphur cycling depend on the interplay between sulphate concentration and availability of organic matter (Goldhaber and Kaplan 1974). In coastal sediments, the reduction of sulphate by the respiratory metabolism of sulphate by the sulphate-reducing bacteria is reported to be a common phenomenon (Jorgensen 1977). A principal controlling factor of sulphate reduction is the abundance and type of organic matter present (Berner 1972).

In the present study, the sulphate values ranged from 492 to 1293 mg/l during high tide condition and 354 to 1123 mg/l during low tide condition. Significant spatial and tidal variations of sulphate were observed in the estuarine stretch of the delta complex (Table 3A.3, Fig. 3A.4). Sulphate, like Cl, exhibited a higher concentration in the downstream stations under the influence of seawater mixing. The relatively lower concentration of sulphate in the upper estuary may be due to consumption of the anion by paddy fields along the upstream stations, whereas removal is compensated in the lower part of the estuary by the contribution of seawater and decay of organic matter from mangroves.

4.10 Chloride

Chloride is the major ion in seawater, which regulates the estuarine salinity. Chloride level exerts considerable influence on the biotic community of marine and estuarine system. It has got biological effects at different levels such as at cellular level and in animal tissues and is also responsible for mortality. Studies on fish have shown that the first detectable damage occurs in the blood cells. Osmoregulatory dysfunction and disturbance to calcium and magnesium levels precede haemolysis. These effects have been observed in the number of species (Hose et al. 1983; Cohen and Valenzuela 1977;

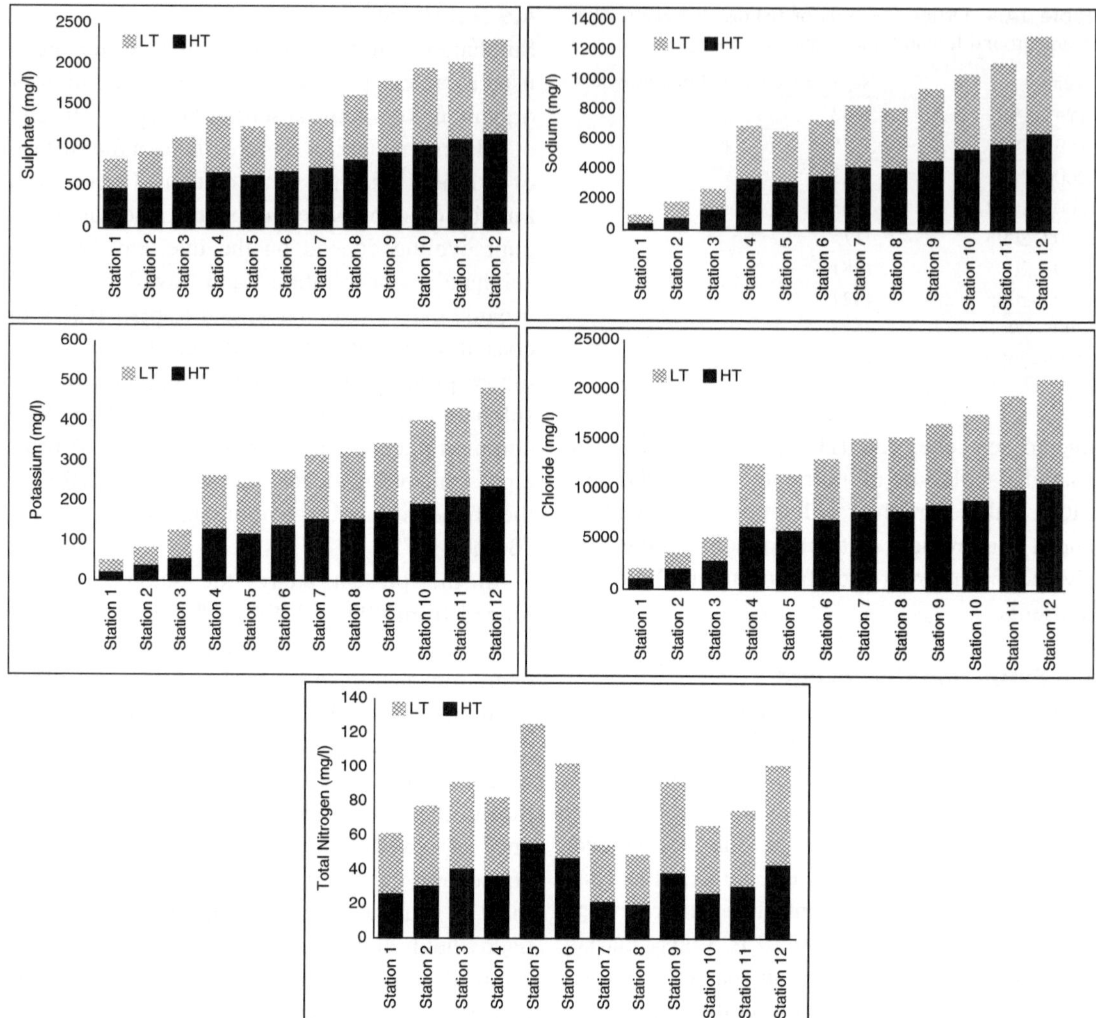

Fig. 3A.4 Tidal variations of surface water sulphate, sodium, potassium, chloride and total nitrogen in the sampling stations

Middaugh et al. 1977; Buckley et al. 1976). The degree of damage has been reported to occur at a concentration of 0.1 mg/l in 30 min (Hose et al. 1983). Kidney damage has been observed in a chlorine concentration as low as 0.06–0.3 mg/l (Bass and Heath 1977). Other than fishes, several marine invertebrate species are also affected by lethal concentration of chlorine in ambient waters. In the present study, the Cl values ranged from 1412 to 12,104 mg/l during high tide condition and from 799 to 11,180 mg/l during low tide condition. The higher values of Cl in the downstream stations are due to the proximity of the stations to the Bay of Bengal. Significant spatial and tidal variations of chloride ion in the estuarine stretch (Table 3A.3, Fig. 3A.4) confirm the influence of seawater intrusion in the upstream zone of the Gangetic delta region. The discharge from Farakka barrage, however, acts as a shield against seawater penetration in the upper estuarine stretch due to which the values were lower.

4.11 Sodium

The Na values ranged from 711 to 685 mg/l during high tide and 439 to 6099 mg/l during low tide condition and represent the dominant cation in the system contributing to 28 % of surface water salinity (Fig. 3A.4). These values are comparable to Na level in the water of Pichavaram mangrove, Vellar estuary and Coleroon estuary in the southeast coast of India (Ramanathan et al. 1999). The progressive increase of Na from Stn. 1 to Stn. 12 is due to the proximity of the downstream stations to the Bay of Bengal, and the higher values during high tide are the result of intrusion of seawater from the Bay. Significant tidal and spatial variations of the cation are confirmed by Duncan's Multiple Range Test (Table 3A.3).

4.12 Potassium

Seawater contains about 400 ppm potassium. It tends to settle and consequently ends up in sediment mostly. Rivers generally contain about 2–3 ppm potassium. This difference is mainly caused by a large potassium concentration in oceanic basalts. Calcium-rich granite contains up to 2.5 % potassium. In water, this element is mainly present as K^+ (aq) ions. Potassium is a naturally abundant radioactive potassium isotope. Seawater contains a natural concentration of about 4.5×10^{-5} g/L. Potassium occurs in various minerals, from which it may be dissolved through weathering processes. Some clay minerals also contain potassium, which ends up in seawater through natural processes and settles in sediments.

In the present study, the K values ranged from 29 to 246 mg/l during high tide condition and 23 to 241 mg/l during low tide condition. Unlike other variables, the tidal difference of K is not significant, although the spatial variation of the cation is highly significant (Table 3A.3, Fig. 3A.4). The higher values of K in the lower estuarine stretch are due to the deposition of clay particles in the high-saline downstream stations at the apex of the Bay of Bengal. The extreme erosion in the lower estuarine stretch is another prominent cause of origin of clay particles enriched with K. Earlier study in this estuarine system recorded lower K level in the upper estuary in comparison to downstream zone (Chaudhuri and Choudhury 1994).

4.13 Total Nitrogen

Water samples from the study area were significantly enriched in nitrogen (NH_4^+ NO_2^- and NO_3^-) indicating that inputs from sewage, agricultural run-off and benthic fluxes are not fully assimilated within the system. In the present study, the total nitrogen values ranged from 20.6 to 56.8 μg at/l during high tide condition and from 26.5 to 67.9 μg at/l during low tide condition. The significant tidal and spatial variations of total nitrogen (Table 3A.3) reflect the anthropogenic source of the variable. The present study stretch acts as sink of the highly urbanized city of Kolkata. The city is one of the major megacities in India where the population has increased from 9,194,018 (in 1981) to 11,021,918 (in 1991) and finally 13,216,546 (in 2001). This means an exponential growth rate of 1.72 during 1981–1991 and 1.82 during 1991–2001 (http://www.india-seminar.com/2007/59/579/_k_c_sivaramakrishnan.htm). The wastes generated from such quantum of population load reaches the Hooghly estuarine stretch. Apart from these, the area supports large numbers of shrimp (*Penaeus monodon*) culture farms often coexisting with brick manufacturing units that are potential sources of nitrogen in the Gangetic delta complex (Table 3A.3, Fig. 3A.4).

5. Conclusion

In the present study, some environmentally important hydrological parameters and nutrients were measured during the premonsoon period (May, 2014) at different sampling stations in the Hooghly of the lower Gangetic delta. Few core findings from this study are listed here: (i) There is pronounced impact of tidal cycle on the hydrological parameters of the Hooghly estuary as confirmed through Duncan's Multiple Range Test. (ii) The presence of highly populated and industrialized cities like Kolkata, Howrah and Haldia port–cum–industrial complex has posed an adverse impact on the health of the ecosystem. (iii) Ecorestoration of this fragile ecosystem is of utmost importance as it is a hot spot of biodiversity with some 34 true mangrove floral species and 69 mangrove associate floral species and the home ground of Royal Bengal tiger (*Panthera tigris tigris*).

References

Airy, G. B. (1842). On the laws of the rise and fall of the tide in the River Thames. *Philosophical Transactions of the Royal Society of London, 132*, 1–8. doi:10.1098/rstl.1842.0002.

Church, J. A., White, N. J., Clarke, A. J., Freeland, H. J., & Smith, R. L. (1986). Coastal trapped waves on the east Australian continental shelf. Part II: Model verification. *Journal of Physical Oceanography, 16*, 1945–1958.

Griffin, D. A., & Middleton, J. H. (1991). Local and remote wind forcing of New South Wales inner shelf currents and sea level. *Journal of Physical Oceanography, 16*, 1652–1664.

Laplace, P. S. (1775). Recherches sur plusieurs points du systeme du monde. *Memoires of the Academy of Royal Science, 88*, 91–92.

Newton, I. (1687). *Philosophiae naturalis principia mathematica* (in Latin). Swem Library: Jussu Societatis Regiae ac Typis Josephi Streater. New-York: Daniel Adee, 1846.

Whewell, W. (1833). Essay towards a first approximation to a map of cotidal lines. *Philosophical Transactions of the Royal Society, 123*, 147–236.

Annexure References

Andrew, O. M., Grasshoff, F., Koroloff, K., Kremling, K., & Olaffason, J. (1983). *Methods of sea water analysis*. New York: Verlag Chemie.

APHA (American Public Health Association). (2001). *Standard methods for the examination of water and waste water* (20th ed., p. 380). Washington, DC: APHA.

Bass, M. L., & Heath, A. G. (1977). Toxicity of intermittent chlorination to Bluegill *Lepomis macrochirus*: Interaction with temperature. *Bulletin of Environmental Contamination and Toxicology, 17*(4), 416–423.

Berner, R. A. (1972). Sulphate reduction, pyrite formation and the oceanic sulphur budget. In D. Dyrssen & D. Jnagner (Eds.), *Novel symposium 20: The changing chemistry of oceans, Almqvist and Wiskel* (p. 347). Stockholm: Wiley.

Blanco, J. (1995). Cyst production in four species of neritic dinoflagellates. *Journal of Plankton Research, 17*, 165–182.

Bryan, G. W. (1971). The effects of heavy metals (other than mercury) on marine and estuarine organisms. *Proceedings of the Royal Society of London B, 177*, 389.

Bryan, G. W. (1976). Some effects of heavy metal tolerance in aquatic organisms. In A. P. M. Lockwood (Ed.), *Effects of pollutants on aquatic organisms* (p. 7). Cambridge: Cambridge University Press.

Bryan, G. W. (1984). Pollution due to heavy metals and their compounds. In O. Kinne (Ed.), *Marine zoology* (Vol. V, Part 3, p. 1289). New York: John Wiley & Sons.

Buckley, J. A., Whitemore, C. M., & Mastuda, R. T. (1976). Changes in blood chemistry and blood cell morphology in coho salmon following exposure to sublethal levels of chlorine in municipal wastewaters. *Journal of the Fisheries Research Board of Canada, 33*(4), 776–782.

Chaudhuri, A. B., & Choudhury, A. (1994). *Mangroves of the Sundarbans* (Vol. I, p. 165). New Delhi: IUCN—The World Conservation Union.

Cohen, G. M., & Valenzuela, M. (1977). Gill damage in the mosquito fish *Gambusia affinis* caused by chlorine. *Institute of Freshwater Science and Biology Journal, 3*(4), 361–365.

Coonley, L. S., Baker, E. B., & Holland, H. D. (1971). Iron in the Mullica River and in Great Bay. *Chemical Geology, 7*, 51–63.

Forstner, U. (1983). Assessment of metal pollution in rivers and estuaries. In I. Thornton (Ed.), *Applied environmental geochemistry* (Vol. I, pp. 395–419). New York: Academic Press.

Gobeil, C. B., Sundby, B., & Silverberg, N. (1981). Factors influencing particulate matter geochemistry in the St. Lawrence Estuary turbidity maximum. *Marine Chemistry, 10*, 123–140.

Goldhaber, M. B., & Kaplan, I. R. (1974). The sulphur cycle. In E. D. Goldhaber (Ed.), *The sea* (Vol. 5, p. 569). New York: Wiley.

Gomeez, K. A., & Gomez, A. A. (1984). *Statistical procedures for agricultural research* (2nd ed., p. 680). New York: Willey.

Hazra, S., Ghosh, T., Dasgupta, R., & Sen, G. (2002). Sea level and associated changes in the Sundarbans. *Science and Culture, 68*, 309–321.

Hobbie, J. E. (1976). Nutrients in estuaries. *Oceanus, 19*, 41–45.

Hose, J. E., King, T. D., Zebra, K. E., Stoffel, R. J., Stephans, J. S., & Dickson, J. A. (1983). Does avoidance of chlorinated seawater protect fish against toxicity? Laboratory and Field observations. In R. L. Jolley, W. A. Brungs, J. A. Cotruvo, R. B. Cummins, J. S. Mattice, & V. A. Jacobs (Eds.), *Water chlorination: Environmental impact and health effects* (Vol. 4, Book 2, pp. 967–982). Ann Arbor: Ann Arbor Science.

Ishikawa, A., & Taniguchi, A. (1994). The role of cysts on population dynamics of Scrippsiella spp. (Dinophyceae) in Onagawa Bay, Northeast Japan. *Marine Biology, 119*, 39–44.

Jorgensen, B. B. (1977). The sulphur cycle of coastal marine sediment (Limfjorden, Denmark). *Limnology Oceanography, 22*, 815.

La Fond, E. C. (1954). On the upwelling and sinking off the east coast of India. *Andhra University Memoirs in Oceanography Waltair, 29*, 117.

Middaugh, D. P., Couch, J. A., & Crane, A. M. (1977). Responses of early life-history stages of the Striped Bass *Morone saxatilis* to chlorination. *Chesapeake Science, 18*(1), 141–153.

Mitra, A., Banerjee, K., Sengupta, K., & Gangopadhyay, A. (2009). Pulse of climate change in Indian Sundarbans: A myth or reality? *National Academy Science Letters, 32*, 1–7.

Mukherjee, M., Roy Choudhury, M., & Tripathi, R. (2007). Fishers, fishery and gears in Sundarban wetlands. In M. Mukherjee (Ed.), *Sundarbans wetlands* (pp. 115–131). West Bengal: Department of Fisheries, Aquaculture, Aquatic Resources, and Fishing Harbours, Govt. of West Bengal.

Nair, K. V. K. (1985). *Impact of nuclear power station on the hydrobiological characteristics of Kalpakkam waters*. Proceedings of the symposium on seawater quality demands, NCML, Bombay, pp. 13.1–13.10.

Ramanathan, A. L., Subramanian, V., Ramesh, R., Chidambaram, S., & James, A. (1999). Environmental geochemistry of the Pichavaram mangrove ecosystem (tropical), southeast coast of India. *Environmental Geology, 37*(3), 223–233.

Ruttner, F. (1953). *Fundamentals of limnology* (p. 242) (trans: Frey, D. G., Frey, F. E.). Toronto: Toronto University Press.

Sankaranarayanan, V. N., & Qasim, S. Z. (1969). Nutrients of Cochin backwaters in relation of environmental characteristics. *Journal of Marine Biology, 2*, 236.

Satyanarayana, D., Rao, I. M., & Prasada Reddy, B. R. (1990). Primary productivity, plants pigments and particular organic carbon of Visakhapatnam Harbour – A seasonal study. Proceeding of International Symposium of Marine Pollution, Department of Ocean Development, New Delhi, pp. 287–300.

Sholkovitz, E. R. (1976). Flocculation of dissolved organic and inorganic matter during the mixing of river water and seawater. *Geochimica et Cosmochimica Acta, 37*, 851–880.

Solarzano, L., & Ehrilich, B. J. (1975). Chemical investigations of Loch Etive Scotland: I. Inorganic nutrients and pigments Exch. *Marine Biology and Ecology, 29*, 45.

Strickland, J. D. H., & Parsons, T. R. (1972). *A practical handbook of seawater analysis* (Fisheries Research Board of Canada Bulletin, Vol. 167, p. 310). Ottawa: Queen's Printer.

UNEP. (1982). *Pollution and the marine environment in the Indian Ocean* (UNEP regional seas reports and studies, Vol. 13). Geneva: UNEP Regional Seas Programme Activity Centre.

Yeats, P. A., & Bewers, J. M. (1982). Discharge of metals from the St. Lawrence River. *Canadian Journal of Earth Sciences, 19*, 982–992.

http://www.india-seminar.com/2007/579/579/_k_c_siva ramakrishnan.htm

4 Abiotic Variables of the Marine and Estuarine Ecosystems

Contents

4.1 Chemical Components

Aquatic phase of the marine and estuarine ecosystems is an ideal solvent, and because of this reason seawater is a well-mixed solution of several salts and gases. The churning of seawater due to wave, currents and tidal actions accelerates the process of solubility of substances in the solvent. It is mainly because of thorough mixing the ionic composition of the major ions of the seawater (except bicarbonate and fluoride) exhibits uniform spatial variation, i.e. the composition is almost the same from place to place as well as from depth to depth. Thus, the ratio of one major ion to the other remains almost constant.

The total salinity may change as freshwater is removed or gained, but the major ions exist in the same proportions. This principle of constant proportions may not apply along the shores, where rivers may bring in large quantities of dissolved substances or reduce the salinity to very low values. Some anthropogenic factors like discharge of freshwater from dams and barrages also lower the salinity of estuarine system. The case study of the estuaries of Indian Sundarbans is very relevant in this context. The discharge from Farakka barrage situated in the upstream of the Hooghly estuary (westernmost estuaries of Indian Sundarbans) is the primary cause of freshening of the water, whereas factors like massive siltation (leading to total death of

A. Mitra, S. Zaman, *Basics of Marine and Estuarine Ecology*,
DOI 10.1007/978-81-322-2707-6_4

the Bidyadhari channel that used to carry freshwater to Sundarbans) has made the central zone hypersaline by cutting off the freshwater supply from the upstream region. The factors controlling the salinity of nearshore region or estuaries are explained through a case study highlighted as Annexure 4A.

The salinity of the surface water is basically zonal in distribution although not as clearly so as the temperature. The average surface salinity distribution is different from that for temperature in that it has a minimum just north of the equator and maximum values in the subtropics at about 25°N and S of the equator. The minimum and maxima are evident in the individual oceans. Values decrease towards high latitudes. Observations make it clear that surface salinity is determined by the opposing effects of evaporation increasing it and precipitation decreasing the value. The salinity maxima of the trade wind regions exist where the annual evaporation (E) exceeds precipitation (P), so that (E − P) is positive. Examples of the areal distribution of (E − P), as distinct from global average, were given as annual and seasonal averages for the North Atlantic Ocean by Schmitt et al. (1989) together with references to other sources of such data.

The range of surface salinity values in the open oceans is from 33 to 37 psu. Lower values occur locally near coasts where large rivers empty and in the Polar Regions where the ice melts. Higher values occur in regions of high evaporation such as the eastern Mediterranean (39 psu) and the Red Sea (41 psu). On the average the North Atlantic is the most saline ocean at the surface (35.5 psu), the South Atlantic and South Pacific less so (about 35.2 psu) and the North Pacific the least saline (34.2 psu).

4.1.1 Spatio-temporal Variations of Salinity

Information on temporal variations of salinity is much less than for temperature which is more easily measured. Annual variations of surface salinity in the open ocean are probably less than 0.5. In regions of marked annual variation in precipitation, such as the eastern North Pacific and the Bay of Bengal, and near ice, there are large annual variations. These variations are confined to the surface layers because in such regions the effect of reduced salinity may override the effect of temperature in reducing the seawater density. This keeps the low-salinity water in the surface layer. Diurnal variations of salinity appear to be very small.

Temporal salinity variations may be large at water mass boundaries where advective and diffusive changes occur. These water mass boundaries can be found both at the surface and at depth. Again, these boundaries are usually marked by fluctuations in salinity maxima or minima.

The spatio-temporal variation of salinity has become very vivid as impact of climate change induced sea level rise and several anthropogenic factors cropped up in the present *era* (Vide Annexure 4A). The alteration of salinity is an important topic in marine and estuarine ecology as this variable poses a significant impact on the coastal vegetation (Vide Annexure 4B), on fishery (Vide Annexure 4C) and even on carbon sequestration potential of mangroves (Vide Annexure 4D).

4.1.2 Deep-Water Salinity

In deep waters, 4000 m or deeper, the salinity is relatively uniform at 34.6–34.9 psu throughout the world ocean. Remembering that the deep-water temperature also has a small range (−0.9 °C to 2 °C), this means that the deep-water environment is very uniform in character.

4.1.3 Conservative and Non-conservative Ions

The major ions present in seawater are also called **conservative** ions, because they do not change their ratios to each other with changes in salinity and because they are not generally removed or added by living organisms. Certain ions present in much smaller quantities, and some dissolved gases and assorted organic molecules change their concentrations in seawater with biological and chemical processes; these are called **non-conservative** ions (Table 4.1).

4.1.4 Dissolved Gases in Seawater

The most abundant gases in the atmosphere and the oceans are nitrogen (N_2), oxygen (O_2) and carbon dioxide (CO_2). The percentages of each of these gases in the atmosphere and in seawater are given in Table 4.2. Oxygen and carbon dioxide play important roles in the ocean because they are necessary to life, and biological activities modify their concentrations at various depths. Nitrogen is not used directly by living organisms except for certain bacteria. Gases such as argon, helium and neon are present in small amounts, but they neither interact with the ocean water nor are used by its inhabitants.

The amount of any gas that can be held in solution without causing the solution either to gain or to lose gas is the saturation value. The **saturation value** changes because it depends on the temperature, salinity and pressure of the water. Colder water holds more gas than warmer water; less salty water holds more gas than more salty water, and water under more pressure holds more gas than water under less pressure.

In the process known as **photosynthesis**, plants use carbon dioxide to form organic molecules and produce oxygen as a by-product. Since plants require sunlight for photosynthesis, plant-like marine organisms are confined, on the

Table 4.1 Major constituents of seawater

Constituent	Symbol	g/kg in seawater[1]	Percentage by weight	
Chloride	Cl^-	19.35	55.07	
Sodium	Na^+	10.76	30.62	
Sulfate	SO_4^{2-}	2.71	7.72	
Magnesium	Mg^{2+}	1.29	3.68	**99. 36**
Calcium	Ca^{2+}	0.41	1.17	
Potassium	K^+	0.39	1.10	
Bicarbonate	HCO_3^-	0.14	0.40	
Bromide	Br^-	0.067	0.19	
Strontium	Sr^{2+}	0.008	0.02	
Boron	B^{3+}	0.004	0.01	
Fluoride	F^-	0.001	0.01	
Total		**~35.00**	**99.99**	

Source: Riley and Skirrow (1975), Chemical Oceanography, Vol. 1, 2nd Edition., p. 366

Table 4.2 Dissolved gases in seawater

Gas molecule	% in atmosphere	% in surface seawater	ml/l of seawater	Molecular wt.	μmol/kg
Nitrogen N_2	78.0	47.5	10	28.014	0.446
Oxygen O_2	21.0	36.0	5	31.998	0.219
Carbon dioxide CO_2	0.03	15.1	40	42.009	2.142
Argon	1	1.4	–	39.948	0.01

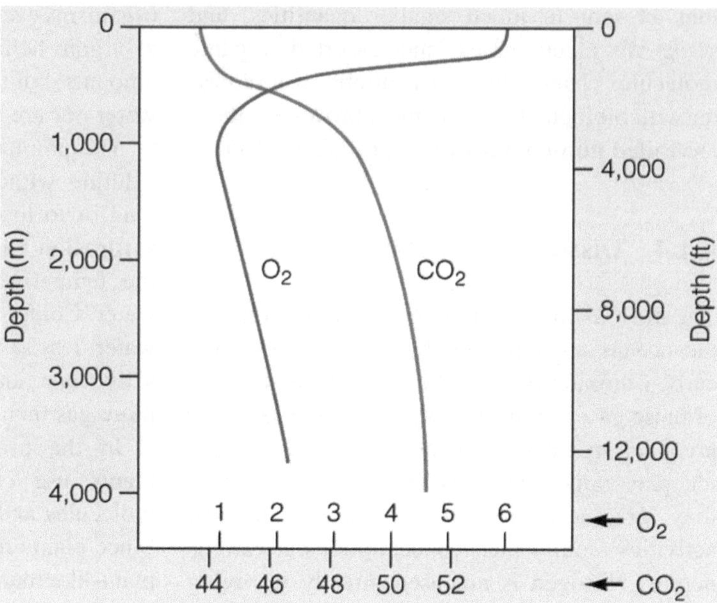

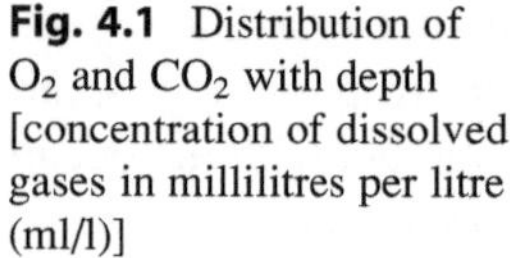
Fig. 4.1 Distribution of O_2 and CO_2 with depth [concentration of dissolved gases in millilitres per litre (ml/l)]

average, to the top 100 m (330 ft) of the sea, where sufficient light is available. Therefore, oxygen is produced in surface water, and carbon dioxide is consumed there. By contrast, **respiration**, which breaks down organic substances to provide energy, requires oxygen and produces carbon dioxide. All living organisms respire in order to produce energy in living cells, and respiration occurs at all depths in the oceans. **Decomposition**, the bacterial breakdown of non-living organic material, also requires oxygen and releases carbon dioxide.

Oxygen can be added to the oceans only at the surface, from exchange with the atmosphere or as a waste product of photosynthesis. Carbon dioxide also enters from the atmosphere at the sea surface, but it is available at all depths from respiration and decomposition (Fig. 4.1).

If the water is quiet, the nutrients and sunlight are abundant, and a large population of plants (preferably phytoplankton) is present; oxygen values at the surface can rise above the saturation value to 150 % or more. This water is **supersaturated**. Wave action tends to liberate oxygen to the atmosphere and return the condition to the saturated state.

Much below the surface, the **oxygen minimum** occurs at about 800 m (2600 ft) depth due to absence of photosynthetic process, but presence of respiratory events by organisms dwelling at this aphotic zone. At depths greater than this, the rates of removal of oxygen and production of carbon dioxide both fall because the population density of animals and the abundance of decaying organic matter have decreased. The slow supply of oxygen to depth by cold water sinking from the surface gradually increases the oxygen concentration again at depth. Very low or zero concentrations of oxygen occur in the bottom waters of isolated deep basins, which have little or no exchange or replacement of water. These include the bottoms of trenches, deep basins behind a shallow entrance sill (e.g. the Black Sea) and the bottoms of deep fjords. The water is trapped and becomes stagnant; respiration and decomposition use up the oxygen faster than the slow circulation of water to this depth can replace it. The bottom water thus becomes anoxic, or stripped of dissolved oxygen.

The common features for dissolved oxygen profiles for the Atlantic and the Pacific are (1) the high values close to the surface, (2) the oxygen minimum in the upper 1000 m between the tropics, (3) the relatively high values below 1000 m in the Atlantic (North Atlantic Deep Water), (4) low values in the North Pacific and

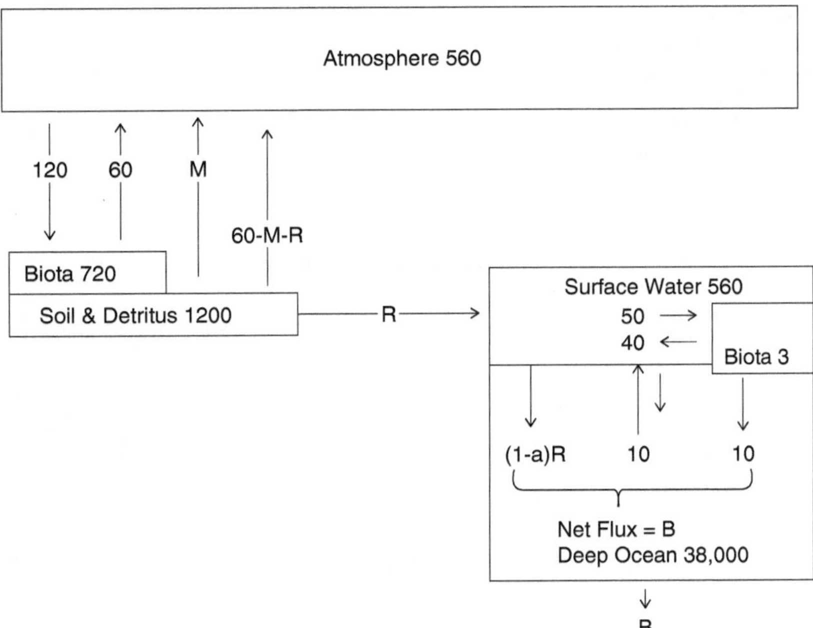

Fig. 4.2 Carbon reservoir of the planet Earth. Numbering in boxes is the amounts of carbon in each reservoir (unit Gt). Numbers between *arrows* are the annual rates of carbon transfer (in Gt); *M* methane flux, *R* riverine flux of particulate carbon, *B* burial flux

(5) similar distributions in the southern latitudes in both oceans. Distributions in the Indian Ocean are similar to this in the Pacific (south and tropics). The lower values in the deep water of the pacific compared with the Atlantic indicate that this water has been away from the surface for a longer period of time and that the deep-water circulation may be slower in the Pacific. In certain regions, such as the Black Sea and the Cariaco Trench (off Venezuela in the Caribbean), there is no oxygen, but hydrogen sulphide is present instead (from the reduction of sulphate ion by bacteria). This indicates that the water has been stagnant there for a long time.

Carbon dioxide is a major component of seawater. Figure 4.2 illustrates the major carbon reservoirs of the planet Earth as a series of interconnected boxes. The number in each box is the estimated amount of the carbon in the reservoir whose unit is gigaton or Gt or billions of tons. All these data are devoid of anthropogenic noise.

Carbon occurs in the atmosphere primarily as CO_2 and to a much smaller extent as CH_4, in the living biota and soils as organic matter and in the oceans primarily as dissolved CO_2, HCO_3^- (bicarbonate ion) and CO_3^{2-} (carbonate ion). The CO_2, HCO_3^- and CO_3^{2-} are collectively referred to as dissolved inorganic carbon (DIC).

When CO_2 enters seawater, the following chemical reactions take place:

$$CO_2(\text{gas}) + H_2O(\text{liquid}) \Rightarrow H_2CO_{3(aq)}(\text{carbonic acid})$$

$$H_2CO_3 \Rightarrow H^+ + HCO_3^-$$

$$CO_3^{2-} + H^+ \Rightarrow HCO_3^-$$

giving the net reaction:

$$H_2O + CO_2 + CO_3^{2-} \Rightarrow 2HCO_3^-$$

About 90 % of the inorganic carbon in the oceans is in the form of HCO_3^-, about 10 % is in the form of CO_3^{2-} and less than 1 % is in the form of CO_2. The transfer of CO_2 between the atmosphere and ocean is driven by the difference in CO_2 partial pressures (pCO_2). If, for example, the pCO_2 in the atmosphere is greater than of the ocean surface, there will be a net flow of CO_2 into the ocean. The fact that only 1 % of oceanic carbon is in the form of CO_2 means that it would be otherwise. This in turn allows the ocean to hold a large amount of carbon without creating a large back pressure that would drive CO_2 back into the atmosphere.

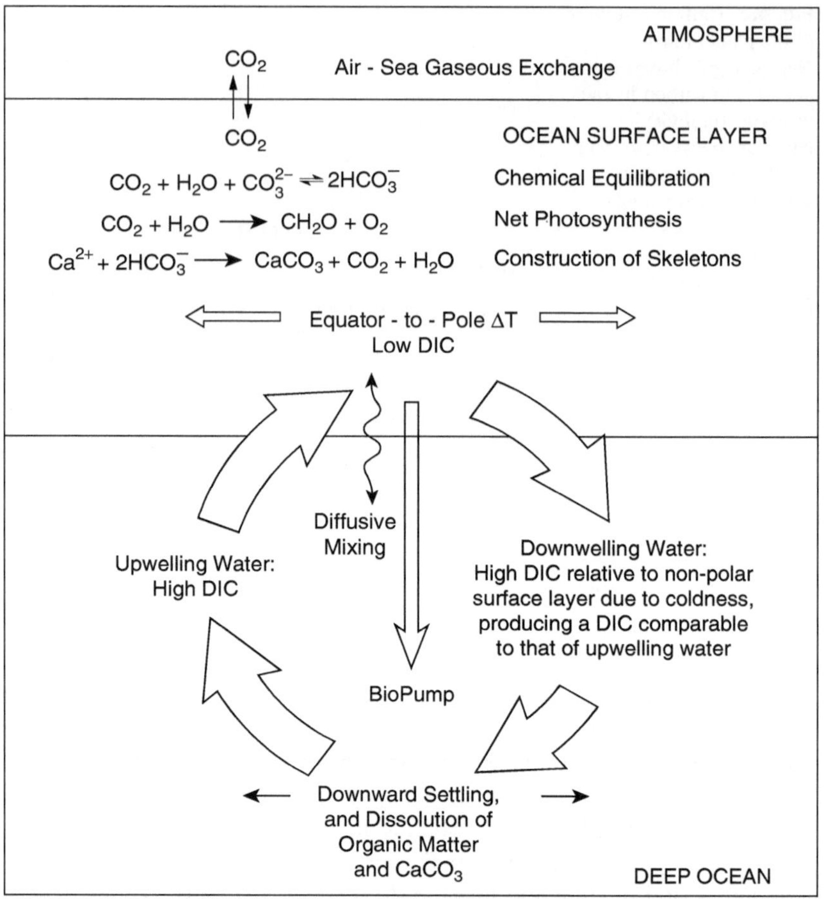

Fig. 4.3 Schematic diagram of interaction and transference of carbon between atmosphere, surface layer of ocean and deep layer of ocean

In the surface layer of the ocean, two biologically driven processes of importance occur: the construction of soft organic tissues by photosynthesis, which can be represented by the reaction:

$$CO_2 + H_2O \Rightarrow CH_2O + O_2{}^+$$

and the construction of calcareous skeletons, through the net reaction:

$$Ca^{2+} + 2HCO_3{}^- \Rightarrow CaCO_3 + H_2O + CO_2$$

Photosynthesis in the oceans is estimated to take up about 50 Gt C/year. Most of this goes into short-lived microorganisms and is rapidly returned to ocean surface water when these organisms die or are eaten and 'burned' by larger organisms. However, some of the soft tissue and skeletal material produced in the mixed layer ends up in the deep ocean through sinking dead microorganisms. The combined flux is crudely estimated to be about 10 Gt C/year and is referred to as the biological pump. About 60–80 % of this flux is due to falling soft tissue, and the remainder is due to falling $CaCO_3$ particles. These and other processes by which carbon is transferred between the surface layer and deep ocean are illustrated in Fig. 4.3. Most of the organic tissue decomposes in the upper 1 km of the ocean, while most of the $CaCO_3$ dissolves below a depth of 3–4 km. The net result is to deplete the surface water of nutrients (primarily phosphate and nitrate) and DIC, to create maximum nutrient concentrations around a depth of 1 km and to create a maximum DIC concentration at a somewhat greater depth in the ocean.

Biological activity transfers DIC from areas of low DIC (surface water) to areas of large DIC (the deep water), which is against the concentration gradient. This transfer is therefore referred to as the 'biological pump', since pumps are devices that transfer air or liquid against a pressure gradient.

Because of the sharp increase in the DIC concentration with increasing depth in the upper ocean, turbulent mixing (diffusion) tends to transfer DIC (and nutrients) upwards. This is the main process that balances the downward transfer of DIC by the biological pump. Convection also causes an upward flow of DIC, since the water that sinks tends to have a lower DIC concentration than the water that rises. One might expect that advection would also transfer DIC upwards, since the concentration of DIC in the surface layer which feeds the downwelling water in polar regions is much less than in the water that upwells from below. However, the solubility of DIC in seawater increases with decreasing temperature, and as warm surface waters flow polewards, they are able to absorb more CO_2 from the atmosphere as they cool. As a result, the DIC concentration in sinking water is about 8 % greater than the global average surface concentration and, it turns out, very close to that in the water that upwells at lower latitudes. Consequently, the net vertical transfer of DIC due to advective overturning is rather small. However, the effect of the solubility variation alone is to cause a downward transfer of DIC, and this downward transfer is referred to as the 'solubility pump'.

The average overall concentration of carbon dioxide throughout the oceans is not substantially affected by biological processes, but tends to remain almost constant, controlled by temperature, salinity and pressure. The general profile of carbon dioxide exhibits relatively lower values at the surface (effect of photosynthesis by producers of the aquatic phase) and a higher value at depth (this may be attributed to only respiration and decomposition and no photosynthesis in the aphotic zone coupled with high column pressure).

An interesting feature in connection to variation in carbon dioxide level in seawater is the alteration (decrease) of pH, which is referred to as acidification. A case study on the acidification of estuarine water due to carbon dioxide rise in the atmosphere and seawater (an effect of climate change) is presented as Annexure 4D.

4.1.5 Nutrients

Nitrate, phosphate and silicate are the three important nutrients in seawater. For all three nutrients, the values are low in the upper few hundred metres with higher values in the deeper water. In the Pacific, these deeper distributions are in the form of mid-depth maximum value tongues extending from north to south with cores at 1000/2000 m for phosphate and nitrate and at 2000/3000 m for silicate and values decreasing from north to south. In addition there are maxima at the south in the Antarctic Bottom Water.

In the Atlantic, the mid-depth tongues extending from north to south are minimum value cores associated with the North Atlantic Deep Water and again there are maximum values in the south in the Antarctic Bottom Water.

The actual values in the mid-depth cores are higher by a factor of about 2 for phosphate and nitrate in the Pacific than in the Atlantic and by a factor of 3–10 for silicate. This is attributed to the slower and mid-depth and deep circulations in the Pacific than in the North Atlantic. The lower dissolved oxygen values in the Pacific than in the Atlantic are attributed to the same cause. A conspicuous feature of the phosphate and nitrate concentrations is that their ratio is everywhere close to 1 phosphorous atom to 15 nitrogen atoms, although the reason for this is not known.

The low values of the nutrients in the upper layers is because of their utilization by phytoplankton in the euphotic zone, while the increase in deeper waters is because of their release back to solution, by biological processes during the decay of detrital material sinking from the upper layers. It is assumed that replenishment in the surface layers is chiefly by physical processes of vertical diffusion, overturn and upwelling. The upwelling processes active along the east sides of the oceans are the reasons for the high biological productivity observed there. In addition, in the upper 200–300 m, physical processes of up- and downwelling associated with upper-layer divergences and convergences modify nutrient distributions locally. For example, in the equatorial Pacific, higher surface values of phosphate (and silicate) occur in the upwelling regions at and north of the equator associated with the divergence of the South Equatorial Current at the equator and between the North Equatorial Counter Current and the North Equatorial Current at about 10°N, and these are regions of high biological productivity.

Excessive nutrient load is currently the largest problem in the coastal rivers and bays throughout the world. Human activity has an enormous influence on the global cycling of nutrients, especially on the movement of nutrients to estuaries and other coastal areas. For phosphorous (P), global fluxes are dominated by the essentially one-way flow of phosphorous carried in eroded materials and wastewater from the land to the oceans, where it is ultimately buried in ocean sediments (Hedley and Sharpley 1998). The size of this flux is currently estimated at 22 Tg P/year (Howarth et al. 1995), suggesting that current human activities cause an extra 14 Tg of P to flow into the ocean sediment sink each year or approximately the same as the amount of P fertilizer (16 Tg P) applied to agricultural land annually (NRC 2000). The effect of human activity on the global cycling of nitrogen (N) is equally immense, and the rate of change of pattern of uses is much greater (Galloway et al. 1995). The single largest global change in the N cycle comes from increased reliance on synthetic inorganic fertilizers, which accounts for more than half of the human alteration of the N cycle (Vitousek et al. 1997). The process for making inorganic N fertilizer was invented early in the twentieth century (Smil 2001), but was not widely used until the 1950s. The rate of use increased steadily until the late 1980s, when the collapse of the former Soviet Union led to great disruptions in agriculture and fertilizer use in Russia and much of Eastern Europe. These disruptions resulted in a slight decline in global N fertilizer use for a few years (Matson et al. 1997). By 1995 the global use of inorganic N fertilizer was again growing rapidly with much of the growth driven by increased use in China (NRC 2000). Approximately half of the inorganic N fertilizer that was ever used on Earth has been applied during the last 15 years. Production of N fertilizer is the largest process whereby human activity mobilizes N globally. Others human control processes, such as combustion of fossil fuels and production of N-fixing cropping agriculture, convert atmospheric N into biologically available forms of N. Overall, human fixation of N (including production of fertilizer, combustion of fossil fuel and production of N-fixing agricultural crops) increased globally twofold or threefold between 1960 and 1990 and continues to grow (Galloway et al. 1995). By the mid-1990s human activities made new N available at a rate of 140 Tg N/year (Vitousek et al. 1997), matching the natural rate of biological N fixation on all the land surfaces of the world (Vitousek et al. 1997; Cleveland et al. 1999). The rate at which humans have increased the supply of reactive, biologically available N, leading to global eutrophication, far exceeds the rate at which humans have increased carbon dioxide in the atmosphere, leading to global warming (Vitousek et al. 1997).

4.2 Sediment Compartment

Sediment is the term given to any natural particle that accumulates in a loose form. The sand, gravel and even the dust floating off are all examples of sediment. The continental crust and the oceanic crust that constitute the floor of the planet Earth are vast blankets of material called 'sediment'. This material accumulates over time as it drifts down through the water from the surface. The material can originate from various sources—it may come from the land and from animals and plants living in the oceans and estuaries and may even come from the outer space. A large portion of the sea bottom is covered with sediments, which contains rock fragments, animal debris, sand, silt, clay, etc.

4.2.1 Transportation of Marine Sediments

Marine sediments present at the bottom of the sea are deposited by the transporting agents like rivers, winds, ice, waves, etc. According to Twenhofel (1932) these agents are of different categories.

4.2.1.1 Rivers

Rivers are considered to be the main agents, which help to bring the disintegrated rock particles up to the sea and deposit them on the shelf. A variety of terrigenous material, ranging from sedimentary to igneous, from red to black, and from big to small, is brought into the seawater by them. They carry down 33 times as much sediment as worn off by the waves along the coasts.

4.2.1.2 River Wash

It causes most of the landslides, occurring along the river banks as well as slidings along the coasts. Landslides occurring on the heads of submarine canyons also help to widen their valleys on the shelves and carry the sediment into the deep sea. Thus, a vast amount of sediment is being carried away.

4.2.1.3 Wave Action

The mighty waves erode the land along the shores and break it into fragments and boulders which are further reduced to sand and mud. They carry these particles, far away from the shores, and deposit it along waveless coasts or rather quiet areas. Researchers have observed that eroding zones of the islands coincide with low organic carbon (expressed in %) in the sediment.

4.2.1.4 Work of Ice

The work of glaciers is of great importance in higher latitude. They carry with them whatever debris cones the way either by erosion, plucking or disintegration. According to Twenhofel (1932), in the Arctic Ocean, shallow-water sea ice is formed during winter, capturing within it in the sea bottom material present there. Due to melting of ice, the surface of the icebergs rises and it floats away; thus, the material is carried far away from the source. This material, as deposited ordinarily or during glacial age, is characterized by a variety of debris, ranging in size and in character depending on the rocks. Murray and Hjort (1912) and Bramlette and Bradley (1940) have also noted icebergs carrying land debris far away from the source. After melting, the glacial material is deposited along with the pelagic deposits.

4.2.1.5 Organic Rafting

Sometimes trees, vegetation, etc., are transported during the floods which carry them to a great distance. Thus, terrigenous plants are found deposited in the deep-sea bottom. Man also becomes an agent, who helps the deposition of cinders and ash from vessels which are found in samples of mud.

4.2.1.6 Volcanic Activity

Apart from the transportation of volcanic material by winds, volcanic activity itself has great amount of force which throws the material far into the sea, later to be deposited on the ocean floor. The volcanic material ranges from big lava fragments, lapilli, dust cinder and ash to various gases. The thickness of the material decreases with the distance from the source. Near the origin, the thickness of the sediment is greatest. Assortation of sediment is not found in volcanic deposit of the ocean floor.

4.2.1.7 Wind

Wind is a great force in transporting the finer particles of the land into the sea. It is found that the products of volcanic eruptions either on the land or on the sea are blown thousands of miles away from the origin. The volcanic dust of Krakatoa in 1883 is deposited into the sea. Hence, the volcanic dust is found in almost each part of the ocean. The lack of vegetation in the regions of wind erosion and lack of agricultural activity on a large scale also helps the wind to carry away the material to distant regions.

Table 4.3 Classification of sediments on the basis of particle size

Division	Subdivision	Diameter (mm)	Phi (Ø) size	Sinking rate (cm/s)
Gravel	Boulder	>256	<−8	>4.29 × 10^6
	Cobble	64–256	−6 to −8	2.68 × 10^5 to 4.29 × 10^6
	Pebble	4–64	−2 to −6	1.05 × 10^3 to 2.68 × 10^5
	Granule	2–4	−1 to −2	2.62 × 10^2 to 1.05 × 10^3
Sand	Very coarse	1–2	0 to −1	65.6–262
	Coarse	0.5–1	+1 to 0	16.4–65.5
	Medium	0.25–0.5	+2 to +1	4.09–16.4
	Fine	0.125–0.25	+3 to +2	1.02–4.09
	Very fine	0.0625–0.125	+4 to +3	0.256–1.02
Mud	Silt	0.0039–0.0625	+8 to +4	<9.96 × 10^{-4}
	Clay	<0.0039	>8	

Radezewski (1939) noted that in the west of the African coast, wind carries a great amount of desert sediment in the form of dust clouds which falls into sea. In general the size of the particles is between 5 and 15 μ. Smaller particles than this are not easily distinguished from the waterborne particles. But typical desert minerals could be recognized such as quartz, feldspar, mica, organic siliceous remains, calcite, aggregates of small particles, etc. Quartz grains coated with reddish iron oxide known as desert 'quartz' are most easily recognized among the sea deposits.

4.2.1.8 Transportation Within the Sea

Other forces inside oceans also work to assort or carry away the material brought by different agencies. The material that is brought into the sea is not deposited on the floor at once but is continuously transported by waves and currents, in suspension. According to the specific gravity, size and shape of the particle and the specific gravity and viscosity of the water, the material begins to settle. Various observations regarding this have also been carried out. Wadell proves that turbulence reduces the rate of fall; and currents and eddies also affect the rate of deposition. Gripenberg (1934) found that fine-grained material when mixed with the seawater flocculates into units of greater size and settles with similar velocity of quartz sphere (5–15 μ in diameter), that is, at the rate of 1–20 m/day. The coarser material brought near the shore by icebergs is immediately deposited, but finer material of lesser settling velocity is carried far away from the coast by currents. The distance to which the particle is carried away depends on the settling velocity. Some of the very fine sediments in the seawater move in suspension, but in the course of time they also settle down.

4.2.2 Classification of Marine Sediments

Sediments may be categorized as sand, if the diameter of the individual particle exceeds 62 μm, or mud, if the diameter of the individual particle is less than 62 μm. Geologists use a phi or Ø scale, based on the **negative of the power of 2** to differentiate between the sizes of the sediment particles (Table 4.3).

Marine scientists have reported the presence of three classes of mud based on the colour of the sediments. These are blue mud, red mud and green mud. A brief account of each of these mud types is discussed here.

(a) **Blue mud:** It is the most common and widespread deposit in the deeper areas surrounding continental lands and in partially enclosed seas. This fine mud, mostly bluish-black, is fine detritus resulting from the disintegration of the rock containing iron sulphide and organic matter. Their upper layer has a reddish tint due to the presence of ferric oxide or ferric hydrate. Blue mud is considered to be mainly formed of land detritus; hence, carbonate of lime

Table 4.4 Types of sediment on the basis of origin

Type	Source	Favourable deposition site	Examples
Biogenous	Biota	Seabed under neritic and oceanic zone	Coral, siliceous and calcareous oozes
Cosmogenous	Outer space	Seabed under pelagic zone	Meteorites
Hydrogenous	Chemical precipitation from seawater	Seabed under neritic and oceanic zone	Phosphates, manganese nodules
Lithogenous	Eroded rock, volcanoes and airborne dust	Seabed under neritic and oceanic zone	Boulder–sand, sand–silt, silt–clay

ranges up to 35 %. The process of formation is helped by sulphur, which reduces the iron into ferrous state containing 1 % or 2 % of amorphous black organic substances. Most of the deposits of blue mud are found along the Atlantic, the Mediterranean, the Arctic and the Banda Sea.

(b) **Red mud:** It is a variety of mud differentiated from others by the presence of iron oxide and ochreous matter. The percentage of calcium carbonate ranges from 6 to 61, whereas the average remains 32. The siliceous organisms such as radiolarian and diatom are rare, while mineral particles of neighbouring land form 10–25 % of the red mud. As compared to blue mud, this variety is rare. Typical localities of its occurrence are the Yellow Sea, the coasts of Brazil and large areas of the floor of the Atlantic Ocean.

(c) **Green mud:** It is mostly seen off high coasts free from large rivers and their deposits, such as the Pacific and the Atlantic coasts of North America (south of Cape Hatteras and California specially), and off the coasts of Japan, Australia and South Africa. Due to this, terrigenous particles are thus longer exposed to the dissolving action of seawater and are not soon overlain by sediments accumulating on their tops; hence, gradually they are converted into green mud, containing green silicate of potassium and iron known as glauconite. As regards the mineral percentage, glauconite ranges up to 7–8 % and carbonate of lime 0.56 %. Green mud is found between 100 and 900 fathoms deep. In fine green sand composed of angular glauconite, particles are found off the Cape of Good Hope where cold and warm currents meet.

Apart from describing the marine sediments as mud or sands, the particles may be categorized by origin like biogenous, hydrogenous, lithogenous, etc. The biogenous sediment is derived from the marine biota, particularly the debris of the shell. If the sediment contains more than 30 % of the biogenous particle, then only the sediment may be referred to as biogenous type. Essentially, biogenous particles may be rich in calcium carbonate (calcareous), silicate (siliceous) or phosphate (phosphatic) depending upon whether they are derived from the shells of foraminifera or diatoms or from the bones/scales of marine vertebrates. The hydrogenous particles are derived from inorganic chemical reactions in water, which comprises the manganese nodules of the ocean floor in the Pacific Ocean. The lithogenous particles originate from the erosion of rocks, the particles of which are transported in the seas or rivers. A summary of the major types of sediment is given in Table 4.4. On the continental slope, the particles are sorted by size according to the action of waves and currents. The largest particles are deposited, whereas the smallest grains move seawards in deep and are eventually deposited in deep water.

Fig. 4.4 Patterns created by sediment fauna

4.2.3 Ooze: Definition and Type

Sediments which contain more than 30 % of material of organic origin are known as oozes. Oozes may of various types as stated here.

Globigerina ooze	in which the calcium carbonate is in the tests of pelagic foraminifera
Pteropod ooze	which contains conspicuous shells of pelagic molluscs
Diatom ooze	which contains frustules of diatoms and is primarily restricted to a virtually continuous belt around Antarctica and a band across the North Pacific Ocean
Radiolarian ooze	which is almost entirely limited to the Pacific Ocean, where it covers a wide band in the equatorial region

In coastal Bay of Bengal, the intertidal zone registers a gradual change in the textural characteristics from high water level to low water level, indicating a sediment change from sandy to silty nature. The sand flats of the mixed and open-sea intertidal zone consist of 90–95 % of fine to very fine sand. Sorting of sediments is very systematic due to natural agents in the present geographical locale. The surface of the sand flat is covered with crescentic ripples, asymmetrical ripples, backwash ripples, ladder back or cross ripples, swash marks, rhomboid marks, rill marks and current crescents. Bioturbation structures are special attractions to nature lovers that are witnessed in the sediment bed of muddy and sandy beaches (Fig. 4.4). These structures result from the activities of nearshore benthic organisms and reflect a unique fabric work of the biotic community on the abiotic matrix. The main agents behind these marvellous fabric works and sculptures are the anemones, gastropods, crustaceans, decapods, bivalves, holothuroids and hemichordates (Fig. 4.5). In some pockets of Indian Sundarbans, animals like *Macoma birmanica*, *Telescopium telescopium*, *Uca* sp. and *Cerithidea* sp. prepare a muddy substratum and produce bioturbation (Fig. 4.6) in the form of mounts, trails and burrows.

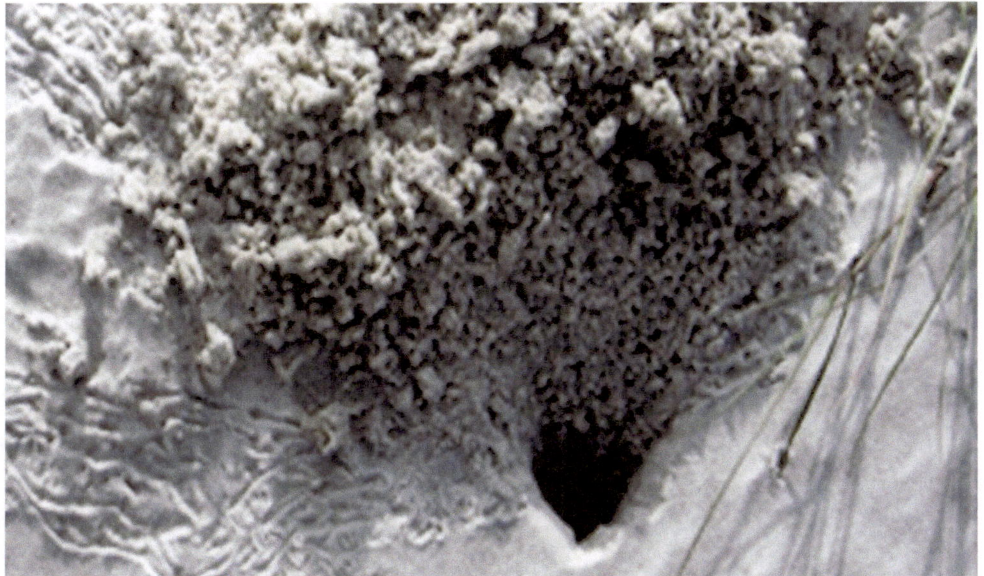

Fig. 4.5 Holes of crabs on the sea beach

Fig. 4.6 Ripple marks and bioturbation

4.2.4 Sediment Deposition: A Boon or a Curse?

Deposition of sediment in the coastal zone and estuarine bed creates a suitable habitat for coastal vegetation, particularly mangroves. The intertidal mudflats, which are caused by gradual deposition of sediments sustains a rich genetic resource (Fig. 4.7). Deposition of sediment also increases the area of islands as seen in case of Thakuran Island in central Indian Sundarbans (Fig. 4.8).

Fig. 4.7 Intertidal mudflats formed due to deposition of sediments

Excessive sedimentation reduces the draft of the navigational channel leading to an adverse effect on the economic growth of the region. The case study of Gangetic delta can be a relevant example in this context. The major portion of the Bengal Basins and the Ganges delta is floored with quaternary sediments eroded from the high lands on three sides and deposited by the Ganges, Brahmaputra and Meghna rivers and their tributaries and distributaries (Chowdhury 2010). The Ganges river originated in Gangotri glacier in the southern slopes of the Himalayas and carries discharge from catchments of about 865,000 km^2 in India to Bangladesh (Islam and Gnauck 2008; Joseph 2006). Erosions and accretion are common phenomena and threats in the coastal zone of the Bangladesh Sundarbans, where nearly a billion tonnes of sediment is brought by the rivers. About two thirds of these sediments are discharged in the Bay of Bengal, while the rest contributes to the formation of new land and islands. However, widespread loss of mature land through river erosion is a severe problem in the estuaries areas (Islam 2004).

The major rivers of Bangladesh Sundarbans transport 2.4 billion tons of sediment annually. This situation has an adverse impact on the coastal vegetation. For example, silt deposition in the north-eastern part of the Bangladesh Sundarbans hinders the vigorous growth of mangrove vegetation. In many cases it has been observed that mangroves are not properly regenerated under condition of excessive sedimentation. Deposition of sediment also possesses adverse impact on the benthic communities, particularly oysters. Enhanced siltation of the riverbed and subsequent drainage congestion are some other problems associated with sediment deposition.

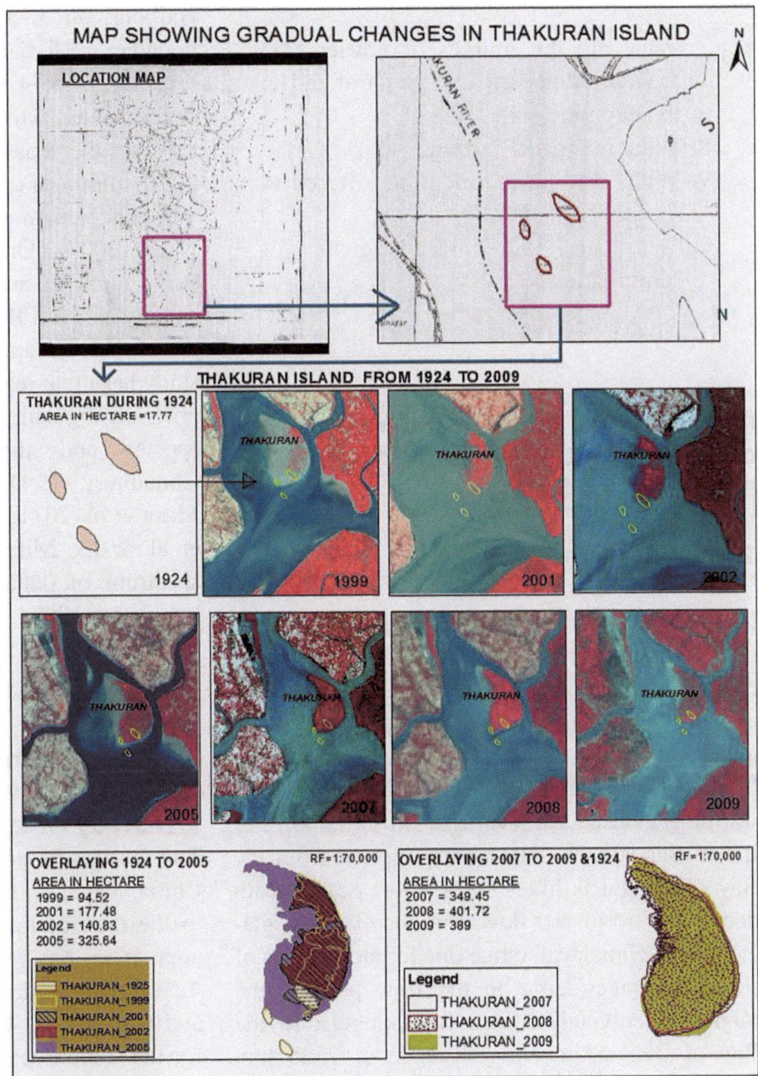

Fig. 4.8 Accretion of Thakuran Island due to sediment deposition

Brain Churners

1. How is the principle of constant proportions between major ions of the seawater violated in the coastal zone and estuaries?
2. What factors control the salinity of nearshore region?
3. How does sea level rise alter the salinity in the nearshore region and estuaries?
4. Compare between the solubility of oxygen and carbon dioxide in seawater with depth.
5. What is the main cause behind the presence of oxygen minimum layer at a depth around 800 m?
6. How is the pH of seawater affected due to the rise of carbon dioxide in the atmosphere and seawater? What are the possible implications of the gradual acidification of seawater?

(continued)

7. What are the sources of nutrients in seawater and how are they utilized in the euphotic zone?
8. What is organic rafting?
9. When can the sediment be referred to as ooze?
10. Who are the main agents of bioturbation?

Annexure 4A: Interannual Variation of Salinity in the Inshore Region of the Bay of Bengal

1. Introduction

India is a large country with 329 million hectares of geographical area and is situated in the tropics between 8°4′ and 37°6′ N latitude and 68°7′ and 97°25′ E longitude. The country has been identified as one of the 27 countries which are the most vulnerable to the impacts of global warming-induced sea level rise (Mitra 2000). For Indian Sundarbans, this salinity ingress from the Bay of Bengal is likely to be more severe with decreasing freshwater flow from the Gangotri glacier of the Himalayan range due to intervention of several barrages/dams on the flow path of the mighty river Ganges from the Gangotri to the Bay of Bengal, covering distance of 2525 km. The siltation of the Bidyadhari river since the late fifteenth century has decreased the freshwater flow in the central sector of Indian Sundarbans to a considerable level (Mitra et al. 2009, 2011; Sengupta et al. 2013; Roy Chowdhury et al. 2014). In addition to this, there are reports of rising sea level at 3.14 mm/year (Hazra et al. 2002) in Indian Sundarbans, which is also a primary factor influencing the spatio-temporal variation of salinity. UNESCO (2007) reported that 45 cm rise in sea level (likely by the end of the twenty-first century, according to the IPCC), combined with other forms of anthropogenic stress on the Sundarbans, could lead to the destruction of 75 % of the Sundarban mangroves. Diamond Harbour, an area just adjacent to the northern boundary of Indian Sundarbans, exhibits a net sea level rise of 5.74 mm/year (considering the subsidence value), which is much higher compared to several other coastal cities of India like Mumbai (1.20 mm/year), Kochi (1.75 mm/year) and Vishakhapatnam (1.09 mm/year) (Jagtap and Nagle 2007). On this background the present study aims to analyse the decadal variation of salinity since 1984 in Indian Sundarban region located at the apex of the Bay of Bengal. The study has great relevance as salinity is the primary criterion regulating the distribution of mangrove species and their growth (Chaudhuri and Choudhury 1994; Mitra and Banerjee 2005; Mitra et al. 2011; Banerjee et al. 2013; Sengupta et al. 2013; Mitra 2013). The entire biological spectrum of deltaic Sundarbans along with the livelihood of the local people is also influenced by salinity of the ambient media.

2. Materials and Methods

2.1. Study Area

The Ganges is an international river shared by China, Nepal, India and Bangladesh. With regard to the distribution of the 109.5×10^6 ha basin area, India has 79 %, Nepal 14 %, Bangladesh 4 % (this is equivalent to 37 % of Bangladesh) and China 3 %. The river has great importance for the socio-economy of the co-basin countries. It is estimated that about 410×10^6 people are directly or indirectly dependent on the Ganges river.

The river Ganges is thus a trans-boundary river of Asia which flows through India and Bangladesh Sundarbans, a mangrove-dominated delta complex in the inshore region of the Bay of Bengal. The 2525 km (1569 mile) river rises in the western Himalayas in the Indian state of Uttarakhand and flows south and east through the Gangetic plain of North India into Bangladesh, where it empties into the Bay of Bengal. It is the third largest river by discharge. At the apex of the Bay of Bengal (in the Indian side), a delta has been formed which is

Table 4A.1 Sampling stations in the western, central and eastern sectors of Indian Sundarbans in the lower Gangetic delta region

Sectors		Sampling stations	Latitude	Longitude
Western sector	**Stn. 1**	Chemaguri (W_1)	21°38′25.86″N	88°08′53.55″E
	Stn. 2	Saptamukhi (W_2)	21°40′02.33″N	88°23′27.18″E
	Stn. 3	Jambu Island (W_3)	21°35′42.03″N	88°10′22.76″E
	Stn. 4	Lothian (W_4)	21°38′21.20″N	88°20′29.32″E
	Stn. 5	Harinbari (W_5)	21°44′22.55″N	88°04′32.97″E
	Stn. 6	Prentice Island (W_6)	21°42′47.88″N	88°17′55.05″E
Central sector	**Stn. 7**	Thakuran Char (C_1)	21°49′53.17″N	88°31′25.57″E
	Stn. 8	Dhulibasani (C_2)	21°47′06.62″N	88°33′48.20″E
	Stn. 9	Chulkathi (C_3)	21°41′53.62″N	88°34′10.31″E
	Stn. 10	Goashaba (C_4)	21°43′50.64″N	88°46′41.44″E
	Stn. 11	Matla (C_5)	21°53′15.30″N	88°44′08.74″E
	Stn. 12	Pirkhali (C_6)	22°06′00.97″N	88°51′06.04″E
Eastern sector	**Stn. 13**	Arbesi (E_1)	22°11′43.14″N	89°01′09.04″E
	Stn. 14	Jhilla (E_2)	22°09′51.53″N	88°57′57.07″E
	Stn. 15	Harinbhanga (E_3)	21°57′17.85″N	88°59′33.24″E
	Stn. 16	Khatuajhuri (E_4)	22°03′06.55″N	89°01′05.33″E
	Stn. 17	Chamta (E_5)	21°53′18.56″N	88°57′11.40″E
	Stn. 18	Chandkhali (E_6)	21°51′13.59″N	89°00′44.68″E

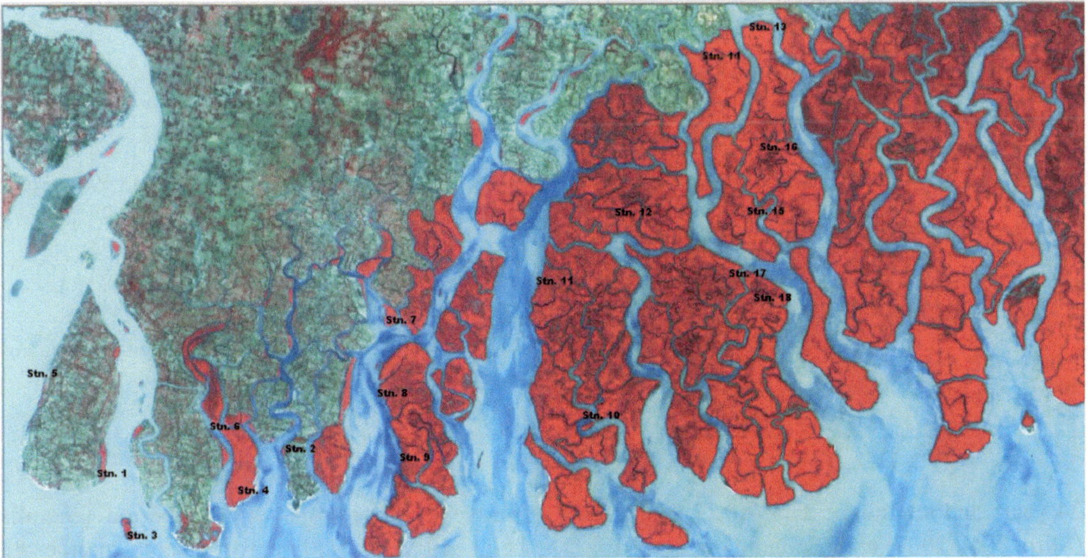

Fig. 4A.1 Location of sector-wise sampling stations in Indian Sundarbans; the *red colour* indicates the mangrove vegetation

recognized as one of the most diversified and productive ecosystems of the tropics and is referred to as the Indian Sundarbans. The deltaic complex has an area of 9630 km^2 and houses about 102 islands (Mitra 2000). 18 sampling sites were selected, 6 each in the western, central and eastern sectors of Indian Sundarbans (Table 4A.1, Fig. 4A.1). We demarcated the three sectors of Indian Sundarbans on the basis of our primary surface water salinity data of

Annexure 4A.1 Annual variation of surface water salinity in western Indian Sundarbans during 30 years' (1984–2013) observation period

	Station					
Year	Chemaguri	Saptamukhi	Jambu Island	Lothian	Harinbari	Prentice Island
	88°09′46.64′	88°22′59.30″	88°10′29.25″	88°21′23.93″	88°04′10.83″	88°17′39.77″
	21°38′54.86″	21°40′27.78″	21°35′22.55″	21°38′05.04″	21°44′22.16″	21°42′44.89″
1984	21.76	20.89	23.47	22.89	16.18	22.02
1985	21.01	20.94	22.89	22.64	16.02	22.56
1986	20.02	21.06	23.86	22.45	15.95	22.39
1987	20.39	21.33	23.02	23	15.28	22.89
1988	21	20.49	23.11	23.05	15.84	23
1989	21.23	20.88	23.95	23	15.77	22.48
1990	20.16	21.08	22.79	22.44	14.98	22.31
1991	20.14	21.26	22.48	22.13	15.00	22.08
1992	20.01	22.49	22.16	22.00	14.69	21.97
1993	19.34	22.13	22.44	21.95	14.75	21.82
1994	19.31	20.77	22.39	21.89	14.83	21.64
1995	19.42	21.45	22.18	21.96	15.12	21.48
1996	19.00	22.02	22.07	21.00	14.20	20.75
1997	18.65	22.43	21.95	21.41	14.05	20.82
1998	18.43	22.61	21.68	20.59	13.98	20.44
1999	18.21	22.90	21.79	20.33	13.16	20.68
2000	18.05	22.43	20.63	20.31	13.00	20.26
2001	17.79	22.56	21.92	20.37	12.49	20.13
2002	17.14	22.79	21.00	20.10	12.82	20.02
2003	17.03	22.81	20.49	20.02	12.73	19.75
2004	17.56	22.34	20.32	19.85	11.65	19.65
2005	16.83	21.95	20.17	19.54	11.49	19.00
2006	16.98	22.46	20.68	19.91	11.33	18.59
2007	17.02	22.99	19.95	19.08	10.49	18.92
2008	16.75	23.01	19.89	19.12	10.66	18.98
2009	16.45	23.42	20.01	19.00	10.42	18.68
2010	16.29	23.83	19.95	18.95	9.85	18.48
2011	16.04	23.98	19.83	18.88	9.79	17.95
2012	15.96	24.02	19.76	18.79	9.23	17.49
2013	15.75	24.11	19.41	18.60	9.10	17.53

24 years and secondary data (of 27 years) from (Mitra et al. 2009) and (Sengupta et al. 2013).

2.2. Data Sources and Quality

We considered a data set of 30 years (Vide Annexures 4A.1, 4A.2, 4A.3) in this first-order analysis as per the minimum standard norm of climate-related researches. The World Meteorological Organization and the Intergovernmental Panel on Climate Change (IPCC 2007; Smith 2001) define 'climate' as the average state of the weather over time with the period generally being 30 years (although for some marine climate parameters such as storminess, longer averages are required) (Zhang et al. 2000).

More than two decades of data (1984–2013) were compiled from the archives of the Department of Marine Science, University of Calcutta, for this study. A number of studies on different aspects of the Sundarban complex have been published over the years, which include description of the data (and methods) at different times

Annexure 4A.2 Annual variation of surface water salinity in central Indian Sundarbans during 30 years' (1984–2013) observation period

Year	Station					
	Thakuran	Dhulibasani	Chulkathi	Goashaba	Matla	Pirkhali
	88°31′53.16″	88°34′55.85″	88°34′33.58″	88°46′43.52‴	88°44′01.30″	88°51′01.43″
	21°49′36.45″	21°47′12.88″	21°41′22.09″	21°43′37.53″	21°54′07.00″	22°01′12.92″
1984	21.67	21.93	22.01	21.95	21.61	21.59
1985	21.94	22.06	22.43	22.13	21.9	21.88
1986	21.86	22.19	22.5	22.23	21.8	21.76
1987	22.49	22.89	22.99	22.92	22.36	22.3
1988	22.2	22.46	22.78	22.58	22.17	22.1
1989	22.86	23.01	23.5	23.41	22.8	22.76
1990	23.15	23.89	23.97	23.91	23.10	23.06
1991	23.17	23.91	24.02	24.00	23.12	23.09
1992	23.80	24.01	24.20	24.15	23.76	23.69
1993	23.49	23.98	24.09	24.00	23.45	23.40
1994	23.96	24.16	24.65	24.55	23.91	23.88
1995	24.70	25.11	25.43	25.33	24.65	24.60
1996	24.73	25.12	25.62	25.41	24.70	24.68
1997	24.88	24.92	25.01	24.97	24.82	24.80
1998	25.06	25.78	25.98	25.65	25.00	24.98
1999	25.13	25.67	25.78	25.71	25.06	25.00
2000	25.01	25.98	26.01	25.99	24.98	24.95
2001	25.73	25.99	26.43	26.09	25.70	25.68
2002	26.07	26.71	26.92	26.83	26.00	25.86
2003	26.16	26.79	26.85	26.80	26.12	26.10
2004	26.30	26.43	26.79	26.54	26.27	26.24
2005	27.09	28.03	28.13	28.10	27.00	26.90
2006	27.19	27.42	27.85	27.52	27.10	27.05
2007	27.40	27.95	28.01	27.97	27.37	27.34
2008	27.50	27.92	28.05	28.01	27.38	27.31
2009	28.02	28.19	28.83	28.76	28.00	27.95
2010	28.13	28.91	28.97	28.93	28.10	28.06
2011	28.88	28.90	28.92	28.89	28.81	28.79
2012	28.69	28.76	28.81	28.70	28.62	28.60
2013	28.90	28.93	28.94	28.90	28.88	28.85

for more than two decades (Chakraborty and Choudhury 1985; Mitra et al. 1987, 1992, 2009; Mitra and Choudhury 1994; Saha et al. 1999; Banerjee et al. 2002, 2003, 2013; Mondal et al. 2006; Sengupta et al. 2013). Real-time data (through field sampling by the authors) were also collected simultaneously since 1998 from 18 sampling stations in the lower Gangetic region during high-tide condition to assure quality and continuity to the data bank. For each observational station, at least 30 samples were collected within 500 m of each other and the mean value of 30 observations was considered for statistical interpretations.

2.3. Measurement of Surface Water Salinity

In situ surface water salinity was estimated from the selected stations during high-tide condition with the help of a refractometer (VEE GEE STX-3). For cross-checking, water samples from the selected stations were brought to the

Annexure 4A.3 Annual variation of surface water salinity in eastern Indian Sundarbans during 30 years' (1984–2013) observation period

Year	Station					
	Arbesi	Jhilla	Harinbhanga	Khatuajhuri	Chamta	Chandkhali
	89°00′03.15″	88°56′04.13″	88°59′05.32″	89°01′03.12″	88°57′26.52″	89°00′50.92″
	22°12′07.44″	22°10′20.55″	21°59′17.79″	22°03′09.52″	21°53′05.91″	21°51′31.73″
1984	15.39	15.82	16.46	16.49	17.08	17.19
1985	15.52	15.39	16.19	16.93	17.15	17.38
1986	15.06	15.16	16.21	16.46	17.09	16.85
1987	15.08	15.05	15.53	16.35	16.23	16.98
1988	14.95	14.91	15.7	15.79	16.11	16.43
1989	14.1	14.84	15.34	15.86	14.99	15.66
1990	14.22	14.66	15.83	15.79	14.63	15.31
1991	14.15	14.65	15.11	15.77	14.48	14.98
1992	14.08	13.13	14.05	14.94	14.29	15.06
1993	13.44	13.79	14.21	14.12	15.02	15.15
1994	13.89	13.46	13.96	14	14.37	14.88
1995	13.53	13.31	13.44	13.85	14.49	14.93
1996	13.3	13.39	13.98	14.01	14.56	14.99
1997	13.56	13.21	13.77	14.63	14.81	15.16
1998	13.18	13.84	13.68	13.98	14.47	14.88
1999	12.95	13.68	13.43	14.51	14.39	14.53
2000	12.63	13.1	13.41	14	14.22	14.01
2001	12.49	13.32	12.55	12.86	13.99	14.16
2002	12.33	12.49	12.2	12.43	13.45	13.98
2003	12.46	12.3	12.67	12.93	13.33	13.85
2004	12.73	12.15	12.54	12.75	13.67	13.96
2005	11.89	12.69	12.29	12.48	13.85	13.53
2006	11.93	11.54	12.88	13.08	13.43	13.31
2007	11.97	11.86	12.49	12.97	13.39	13.45
2008	12	11.93	12.66	12.85	13.17	13.28
2009	11.41	12	12.41	12.63	13.1	13.86
2010	11.31	11.99	11.59	12.44	13.04	13.09
2011	11.26	11.41	11.93	12.29	12.85	13.03
2012	11.19	11.32	11.41	12.08	12.49	12.85
2013	11.21	11.17	11.65	11.95	14.33	12.97

laboratory in ice-freezed condition and analysed for chlorinity by argentometric method and converted into salinity through standard equation.

2.4. Statistical Analysis

Time series analysis was performed to forecast the trend of surface water salinity on the basis of the past 30 year's real-time data. Exponential smoothing method produces maximum likelihood estimates and can reflect the future trend of the selected variable. This approach was adopted to forecast the values for surface water salinity in the ambient media of the sampling station till 2043.

3. Result

It is interesting to note the significant spatio-temporal variation of surface water salinity in the study region. In the western sector, the salinity decrease ranged from 0.58 psu/year (at Jambu

Island) to 1.46 psu/year (at Harinbari). Although station 2 (Saptamukhi) is situated in the western sector, but the salinity has increased by 0.51 psu/year (Fig. 4A.2). Considering all the six stations in the western sector, the decrease of salinity is 0.63 psu/year, which represents a decrease of 7.50 psu/decade. The salinity has decreased from 17.30 % (in Jambu Island) to 43.76 % (in Harinbari) over a period of 30 years (Fig. 4A.2).

The exponential smoothing method that produces maximum likelihood estimate of the variable predicts a salinity value of 13.05 psu in 2043 (Fig. 4A.3), which is a decrease of 38.4 % since 1984 (over a span of 60 years).

The central sector presents a completely reverse picture in terms of aquatic salinity. Irrespective of stations, salinity has increased (Fig. 4A.4) between the range 1.05 psu/year (in Chulkathi) and 1.12 psu/year (in Matla and Pirkhali). Considering the salinity values of selected six stations, the increase is 1.09 psu/year, which is equivalent to 13.05 psu/decade. The percentage of salinity increase in this sector ranges from 31.49 psu (in Chulkathi) to 33.64 psu (in Matla) with an average of increase of 32.62 % over a period of 30 years (Fig. 4A.4).

Considering the observed data set of 30 years (1984–2013), we predict that salinity will be around 36 psu after a period of 30 years in the central sector of Indian Sundarbans (Fig. 4A.5), which is an indication of alarming hypersaline condition (a rise by 67.1 %) in 2043 in this sector.

In the eastern sector, salinity has decreased (Fig. 4A.6), which ranges from 0.54 psu/year (in Chamta) to 0.98 psu/year (in Jhilla). Considering all the six stations in eastern Indian Sundarbans, the average decrease of salinity is 0.86 psu/year, equivalent to a decadal decrease of 10.30 psu. Over a period of 30 years, the average percentage decrease of salinity is 25.66 psu (Fig. 4A.6).

On the basis of observed data, the prediction of salinity in 2043 is around 7.54 psu (Fig. 4A.7), which is decrease of 52.4 % considering a time span of 60 years.

4. Discussion

The results of the long-term observed data on surface water salinity clearly confirm significant spatio-temporal variations of the salinity in the study region ($p < 0.01$). Basically a bell-shaped salinity profile (Fig. 4A.8) can be a representation for the region with a hypersaline environment in the central sector (mean salinity = 25.43 ± 2.24 psu) between two hyposaline sectors, viz. western (mean salinity = 19.46 ± 3.46 psu) and eastern (mean salinity = 13.85 ± 1.48 psu).

The bell-shaped salinity profile in the present study region is not merely a representation of salinity pattern, but it can be a test bed for future climate-related research due to following reasons:

1. Presence of unique mangrove-centric gene pool in the deltaic complex [from microbes to Royal Bengal tiger (*Panthera tigris tigris*)] primarily influenced by salinity.
2. Ecosystem services of the system to about 4.2 million people dwelling in the delta region.
3. No trans-boundary-related research has yet been taken up considering Sundarbans as an integrated system, although Farakka discharge (a trans-boundary anthropogenic component) has great influence on the mangrove health and livelihood of the integrated Sundarbans.

Similar profile is also observed in the Bangladesh part of Sundarbans, where three salinity zones have been identified, viz. less saline zone (5–15 ppt), moderately saline zone (15–25 ppt) and strong saline zone (25–30 ppt) based on degree of salinity (CEGIS 2006).

The hyposaline environment of western Indian Sundarbans may be attributed to Farakka barrage discharge situated in the upstream region of Ganga–Bhagirathi–Hooghly river system. 10-year surveys (1999-2008) on water discharge from Farakka dam revealed an average discharge of $(3.7 \pm 1.15) \times 10^3$ m^3/s. Higher discharge values were observed during the monsoon with

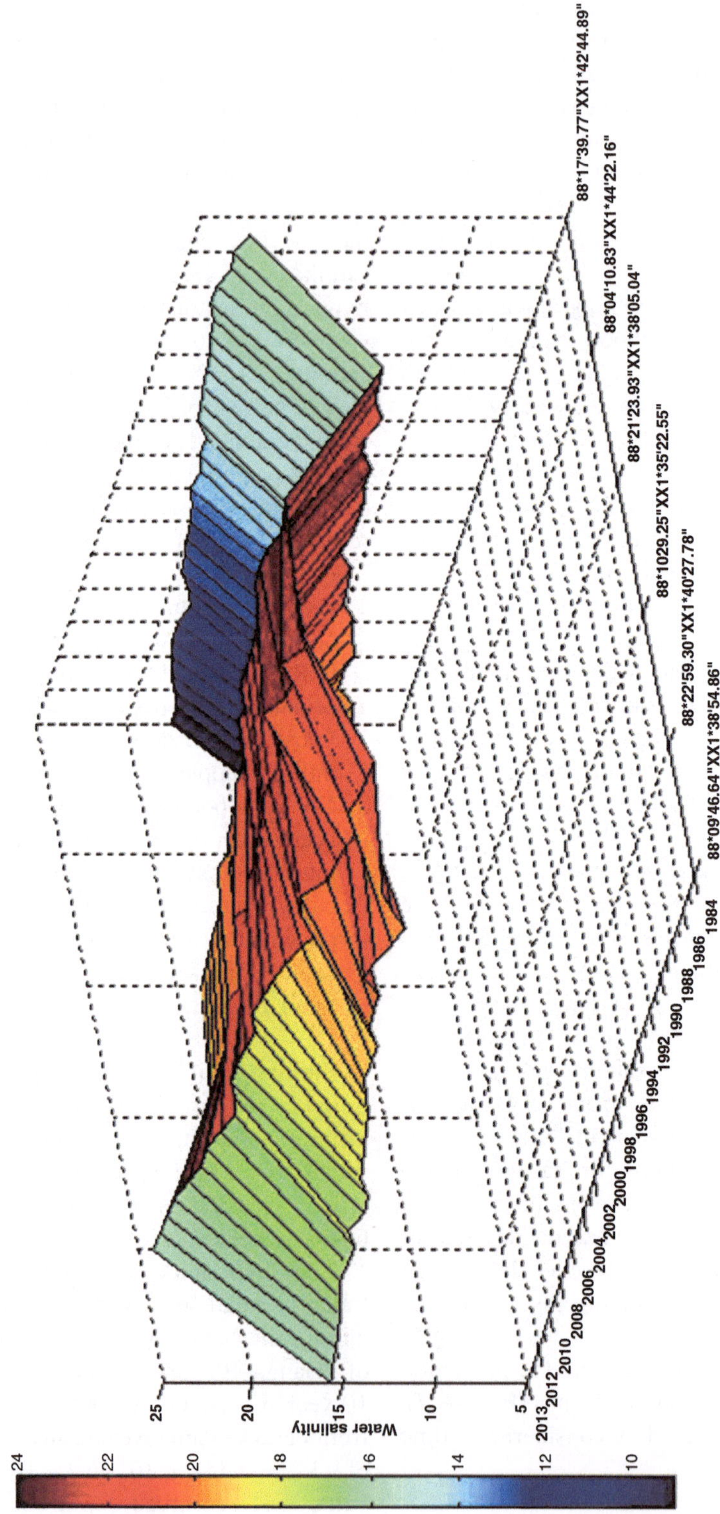

Fig. 4A.2 Spatio-temporal variation of salinity in western Indian Sundarbans

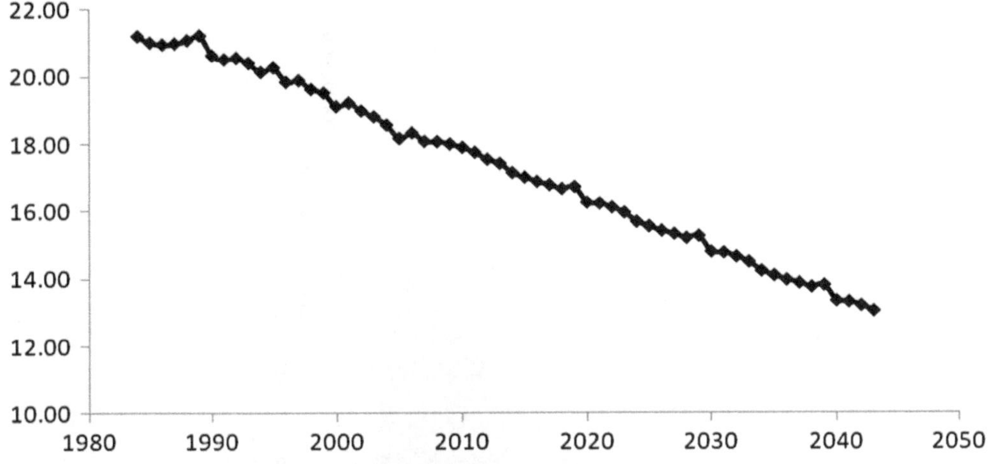

Fig. 4A.3 Future trend of surface water salinity in western Indian Sundarbans

an average of $(3.81 \pm 1.23) \times 10^3$ m^3/s and the maximum of the order 4524 m^3/s during freshet (September). Considerably lower discharge values were recorded during premonsoon with an average of $(1.18 \pm 0.08) \times 10^3$ m^3/s and the minimum of the order 846 m^3/s during May. During postmonsoon discharge, values were moderate with an average of $(1.98 \pm 0.97) \times 10^3$ m^3/s (Mitra 2013).

The central sector represents a hypersaline environment due to complete obstruction of the freshwater flow from the upstream region owing to Bidyadhari siltation since the late fifteenth century (Chaudhuri and Choudhury 1994; Banerjee et al. 2013; Mitra 2013; Sengupta et al. 2013). The Matla estuary in the central Indian Sundarbans cannot be referred to as an ideal estuary as there is no head on discharge or dilution of the system with freshwater. Thus, Matla can be designated as a tidal channel, whose survival depends on the tidal flow from the Bay of Bengal.

The eastern sector of the Indian sector exhibits a low-saline profile possibly due to interconnection with several creeks and channels of the Harinbhanga estuary (the aquatic border of India and Bangladesh Sundarbans) with the tributaries of Bangladesh Sundarbans that arise from Padma–Meghna (Fig. 4A.9) river system.

On the basis of significant spatio-temporal variations of salinity and its future trend, we recommend a trans-boundary coordinated programme of long-term research linking monitoring, process studies and numerical modelling on the foundation of a diverse, interdisciplinary, multi-institution approach and establishment of a strong institutional network between researchers and decision makers of India and Bangladesh.

Annexure 4B: Litmus Test of Climate Change in Indian Sundarbans

1. Introduction

Climate change is one of the most critical global challenges in the present *era*. Recent events like flood, hurricanes, tornadoes and forest fires have emphatically demonstrated our growing vulnerability to climate change. Climate change impacts encompass several sectors like agriculture—further endangering food security, sea level rise and the accelerated erosion of coastal zones, increasing intensity of natural hazards, species extinction and the spread of vector-borne diseases. The impact of climate change on the aquatic ecosystem is an interlinked event between the melting of polar ice or glaciers feeding the rivers and the alteration of salinity in the riverine and estuarine waters of the tropics and temperate and subtemperate zones. Thus, climate change is not one story, but many parallel stories of several

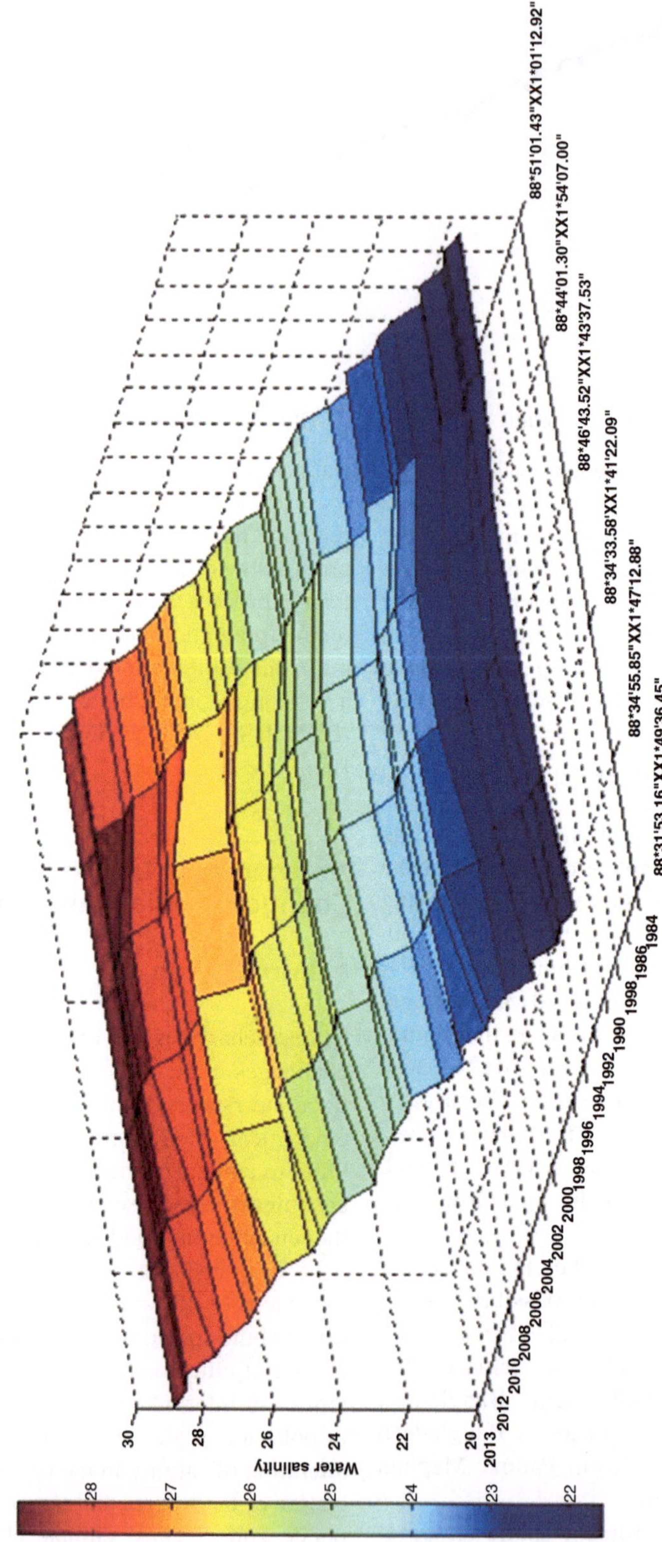

Fig. 4A.4 Spatio-temporal variation of salinity in central Indian Sundarbans

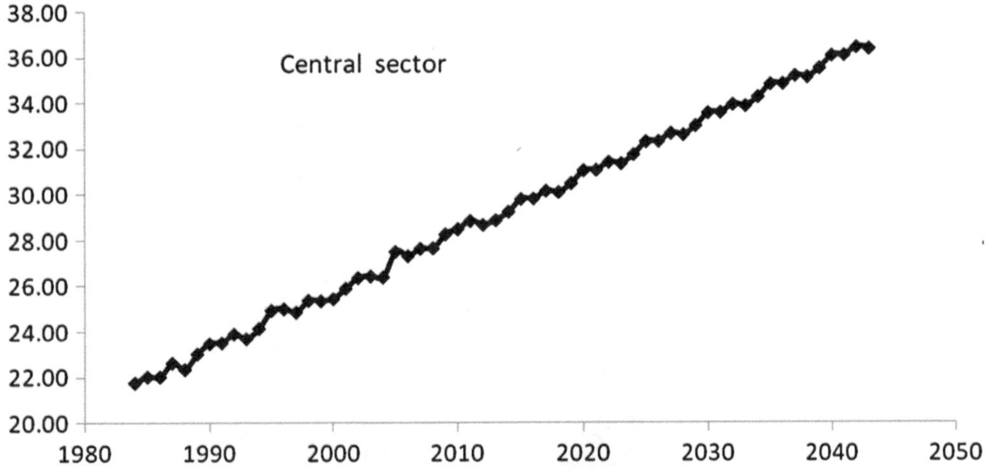

Fig. 4A.5 Future trend of surface water salinity in central Indian Sundarbans

ecosystems—only sporadically connected so far without concrete database. We have attempted to interlink the source (Himalayan glacier) and the sink ecosystems in this article to detect the abiotic indicators of climate change at the sink (aquatic system of Indian Sundarbans).

2. Problem Statement

Sea levels on the Indian subcontinent are increasing at the rate of about 2.5 mm/year; the rate of increment is greater in the eastern coast, with an estimated sea level rise of about 3.14 mm/year. This suggests that mean annual sea levels in the Indian subcontinent will be some 3 cm higher in 2012 and 15 cm higher in 2060 than what it was during 2000. The Indian Sundarbans in the northeast coast of the country, at the apex of the Bay of Bengal, is an extremely dynamic deltaic lobe sustaining a wide spectrum of mangrove flora and fauna. It is estimated that sea level in this deltaic lobe has increased by about 15 cm since the 1950s, and this has been correlated with changes in the pattern and rates of erosion and accretion in the islands of the Indian Sundarbans. Such geophysical phenomena may not only pose serious impact on the adjacent aquatic system by way of altering turbidity, nutrient budget, salinity, pH, etc., but the phenomenon has every possibility to shift the biodiversity spectrum.

The geographical and ecological profiles of the western and central parts of Indian Sundarbans are contrasting as the rivers in the western part (Hugli and Murlganga) are connected to the Himalayan glaciers through Ganga–Bhagirathi system, whereas rivers in the central sector (around Matla) are all tide-fed. Hence, to achieve a realistic picture of interannual variation of hydrological parameters, we collected data after segmentation of the deltaic lobe on the basis of dilution of the system by freshwater (Fig. 4B.1). The article also attempts to link up the salinity profile of the mangrove-dominated deltaic Indian Sundarbans with the retreating glaciers in the Himalayan range as the main estuary in this zone (Hugli) is a direct continuation of the Ganga–Bhagirathi system, the origin of which is the Himalayan glacier.

3. Observations: West Versus Central

The surface water temperatures in both the sectors have shown a rising trend (Fig. 4B.2), a signal of global warming in this part of the Indian subcontinent at the apex of the Bay of Bengal.

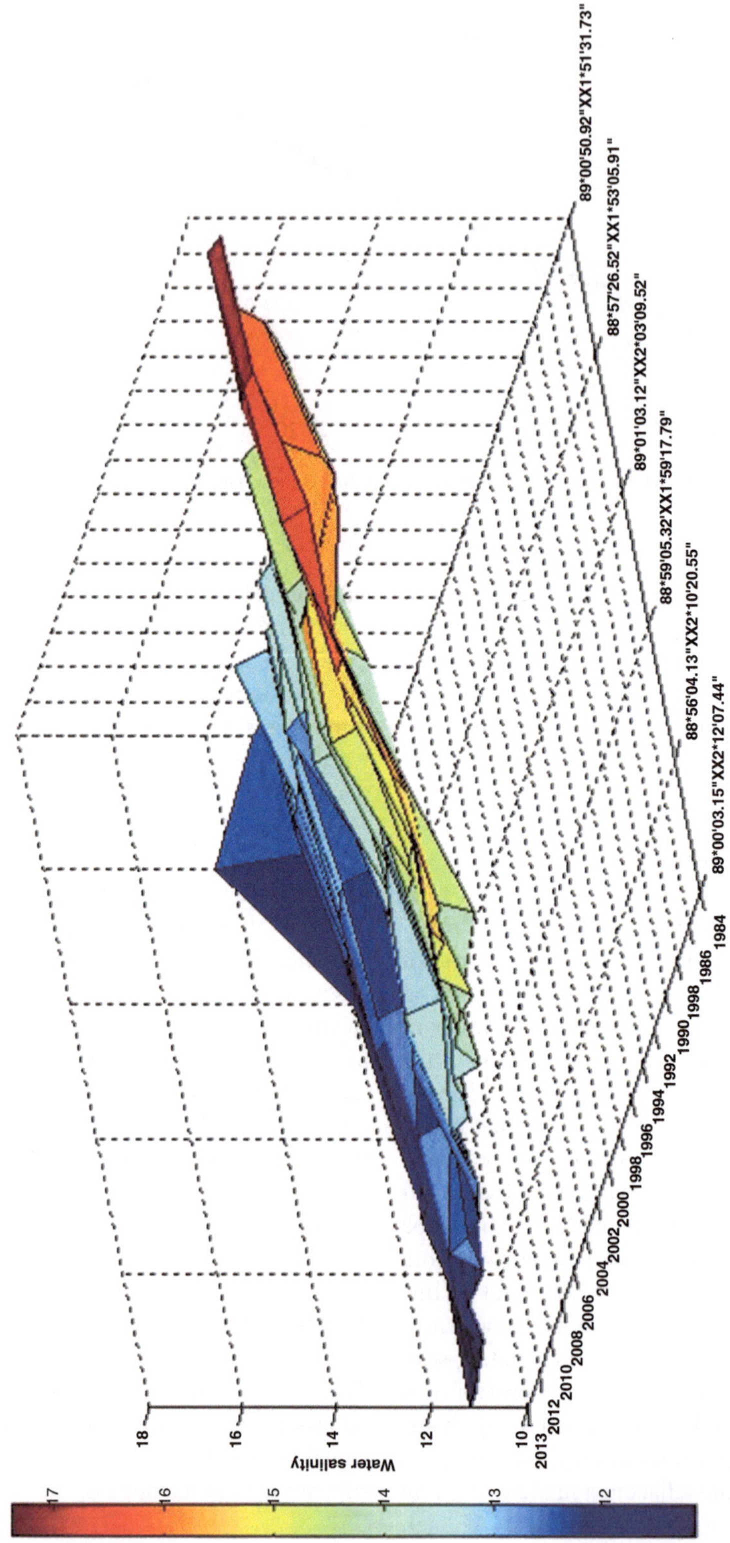

Fig. 4A.6 Spatio-temporal variation of salinity in central Indian Sundarbans

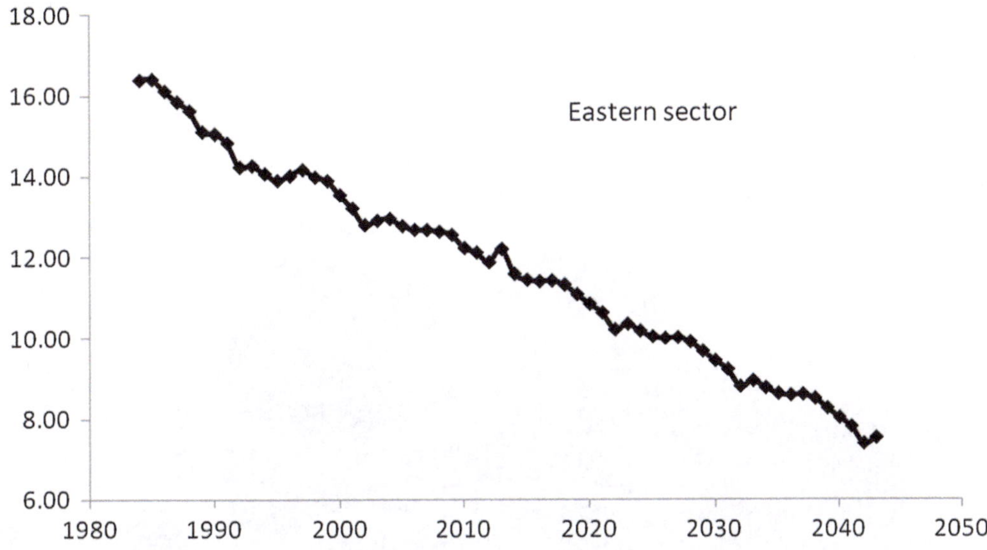

Fig. 4A.7 Spatio-temporal variation of salinity in eastern Indian Sundarbans

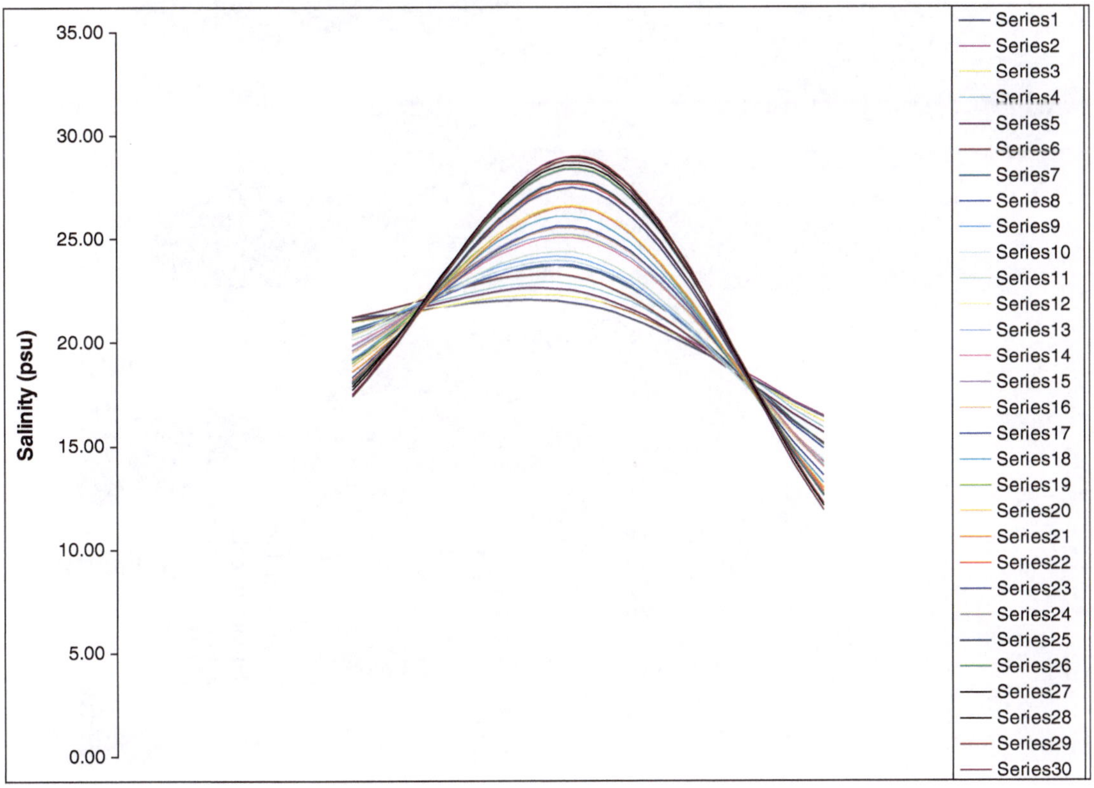

Fig. 4A.8 Bell-shaped nature of salinity profile of Indian Sundarbans based on 9200 readings; 30 series represent 30 consecutive years (1984–2013)

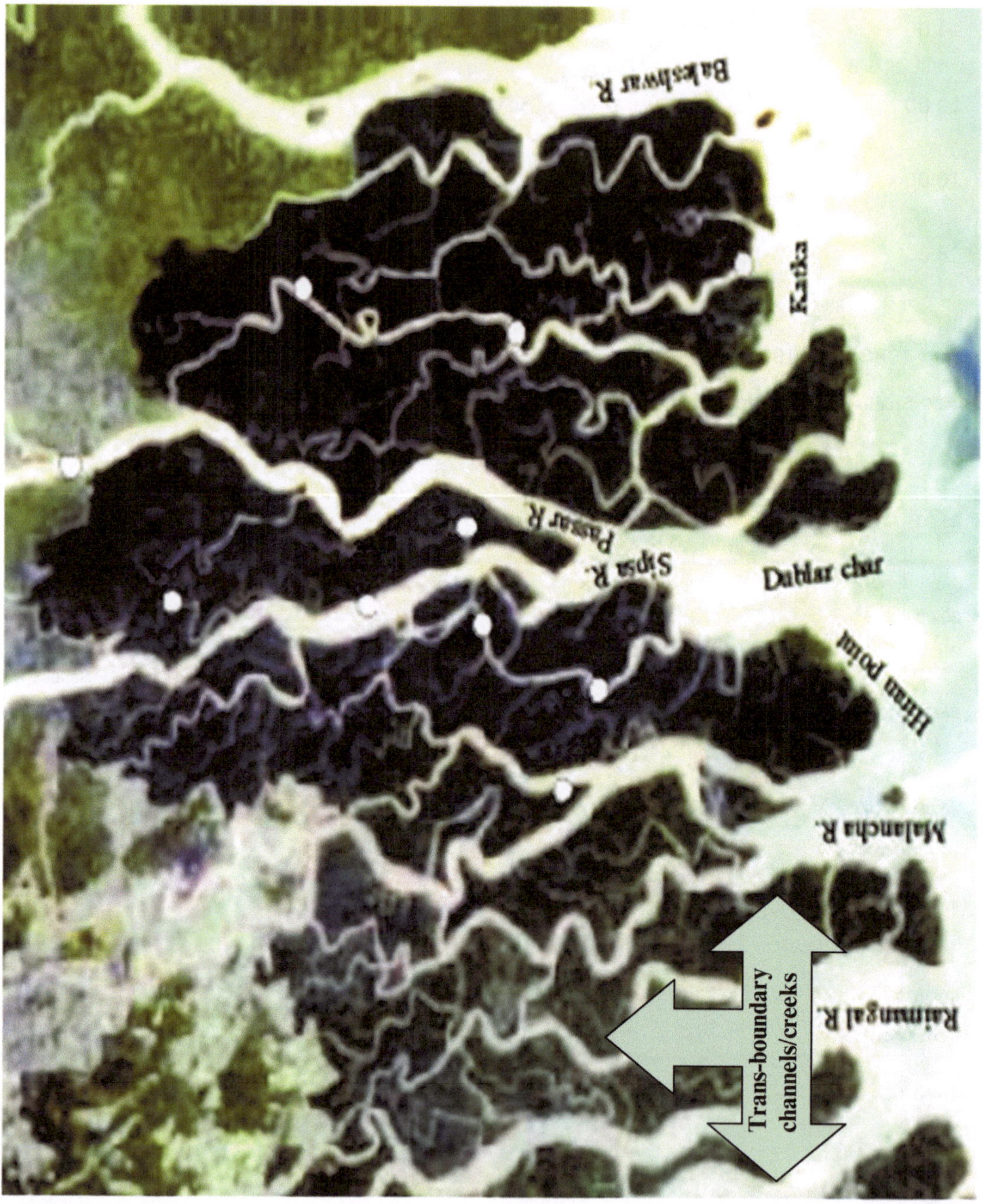

Fig. 4A.9 Trans-boundary channels feeding freshwater to eastern sector of Indian Sundarbans

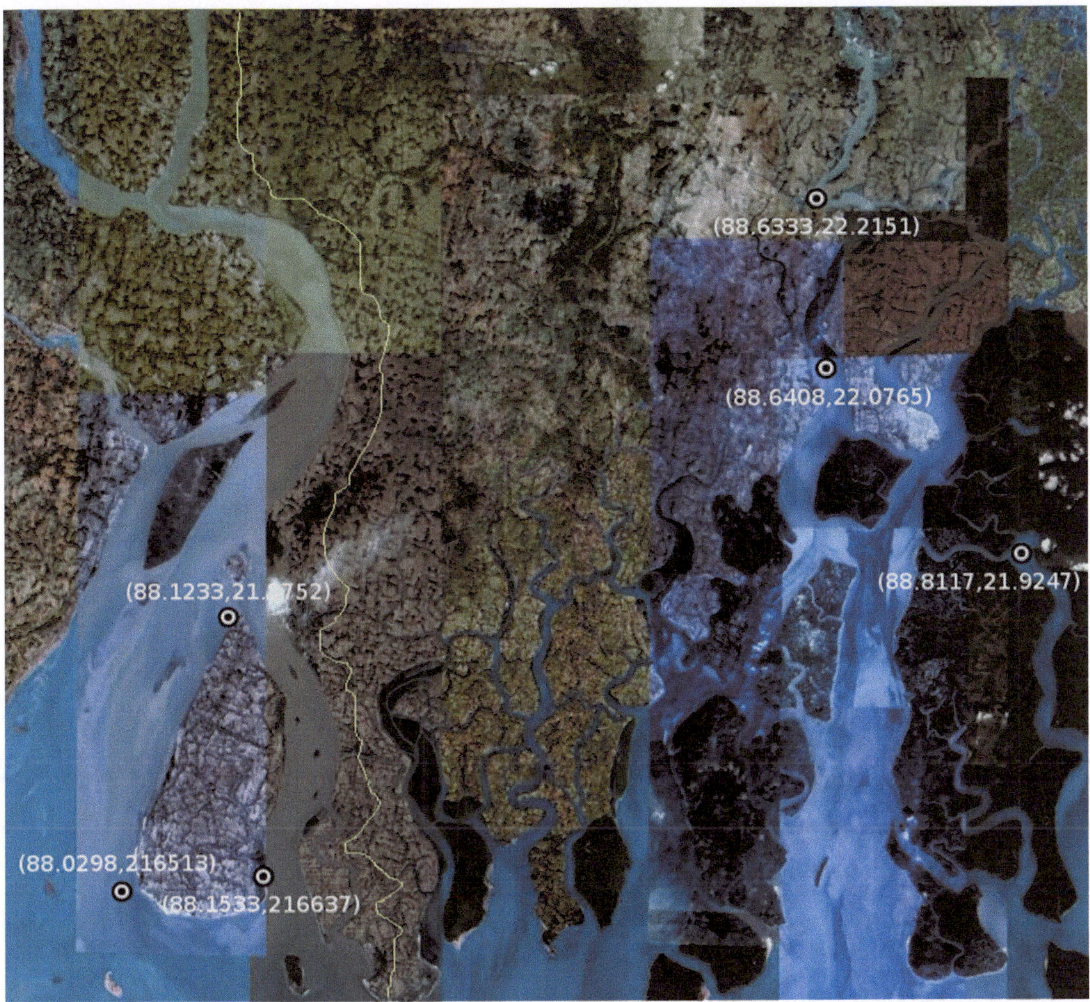

Fig. 4B.1 Location of sampling stations (three each in the western and central sector) in the mangrove-dominated Indian Sundarbans

The pulse of the phenomenon of temperature rise is visualized in two contradictory ways in both the sectors due to variation in geographical features. The rivers in the western sector of Indian Sundarbans (Hugli and Muriganga), being the continuation of Ganga–Bhagirathi system, receive the snowmelt water of Himalaya. The impact of temperature rise is perhaps already upon the Himalayas. The 30.2 km-long Gangotri glacier is receding rapidly: the rate of retreat in the last three decades has been found to be more than three times the rate during the earlier 200 years or so. The average rate of recession has been evaluated by comparing the snout position on 1985 topo-sheet map and the 2001 satellite panchromatic imagery, and the results show that the average recession for this period is about 23 m/year (Hasnain 1999, 2000, 2002). The pictorial representation of historical evidences and recent data on Gangotri glacier retreat in a research by Jeff Kargel, geologist of USGS, also supports the increased rate of retreat of the Gangotri (Fig. 4B.2).

The ex situ effect of Himalayan deglaciation is confirmed not only by lowering trend of salinity (Fig. 4B.3) but also through gradual increase of dissolved oxygen (Fig. 4B.4), in spite of rapid industrialization and urbanization in the western part of Indian Sundarbans receiving freshwater supply through Ganga–Bhagirathi conveyer belt

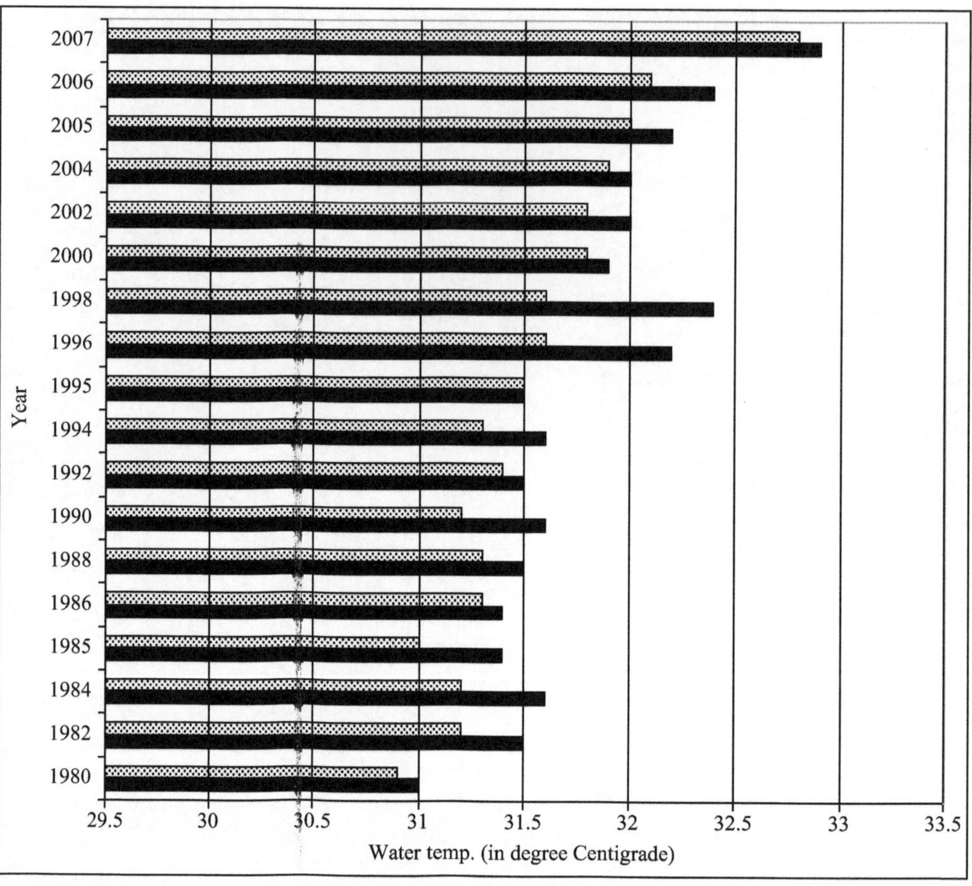

Fig. 4B.2 Increasing trend of surface water temperature (in °C) in Indian Sundarbans. The temperature in the western sector is represented by *dotted graphs* and that of the central sector by *black graphs*

system. The change of hydrological parameters in the western sector is also regulated by Farakka barrage, which was constructed on the Ganga river in April 1975 to augment water supply to the Calcutta port. The project has brought about a significant increase in freshwater discharge in its distributary, the Hugli estuary (Sinha et al. 1996). Such trends of lowering salinity and increasing dissolved oxygen level are absent in and around the Matla river in the central sector, and here the footprint of global warming is felt through rising tidal amplitude, increased salinity (Fig. 4B.4) and lower dissolved oxygen in the river mouths (Fig. 4B.5) due to intrusion of saline water from the adjacent Bay of Bengal region.

The results of different footprints of climate change may pose serious problems in livelihood and biodiversity spectrum of the area. Increased flooding in the adjoining cities and towns of western Sundarbans was already observed since last few decades. The livelihood of the people in the eastern sector is also under threat due to salinization of agricultural land, loss of commercially important fishes in the catch basket, vanishing of freshwater-loving Sundari (*Heritiera fomes*) trees (Fig. 4B.6), etc. While sea level rise and subsequent change in hydrology have every possibility to hinder economic profile of the deltaic complex, the impacts are likely to be hardest felt by the people living below poverty line. Pressure on livelihoods may force poor landless island dwellers of the eastern Indian Sundarbans to migrate in the western part or adjacent cities like Kolkata, Howrah and the

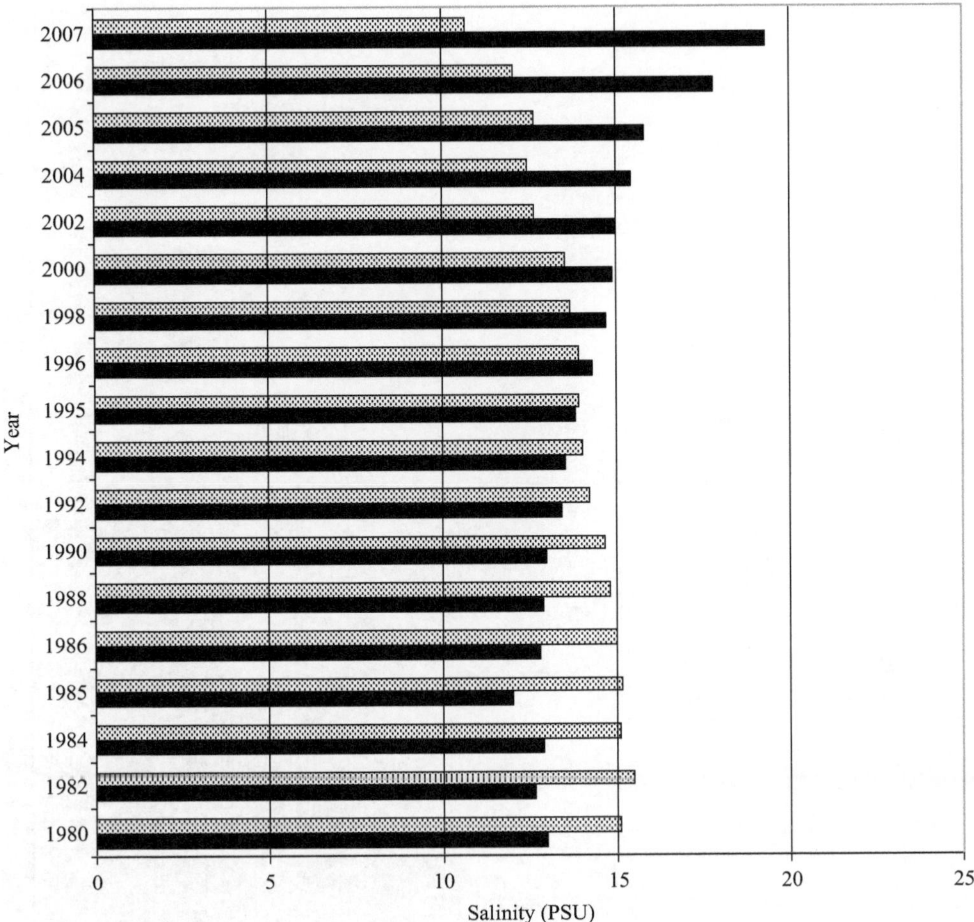

Fig. 4B.3 Salinity (in PSU) fluctuation due to global warming in central (rising trend, marked in *black*) and western (decreasing trend, marked in *dots*) Indian Sundarbans

newly developing Haldia industrial complex. This will definitely magnify the magnitude of vulnerability as some 25–40 % of the urban population in developing countries already lives in impoverished slums, with little or no access to water and sanitation (World Bank 1997; IPCC 2001b).

Annexure 4C: Impact of Climate Change on Fish Community, A Case Study from Indian Sundarbans

Global, regional and local effects of climate change on biotic community are ultimately the sum of processes which act on individual organisms. Perceptible changes are evident in the past few decades in the climate of the planet Earth as manifested by increase in air and water temperature. This has triggered the alteration of salinity profile in marine and estuarine compartment and more specifically at the river mouth and estuarine systems that are connected to glaciers. The fish communities thriving in these dynamic systems shift or orient (adapt) themselves in response to ecological conditions. The mangrove ecosystem of Indian Sundarbans is an ideal zone for such study as the western and eastern sectors of the deltaic lobe are drastically different from each other with respect to salinity owing to the connections of the rivers in the western zone (Hugli and Muriganga) with the Himalayan glaciers. Construction of the

Fig. 4B.4 Skeletal remains of a wasted glacier near Everest

Farakka barrage on the Ganga river in April 1975 to augment water supply to the Calcutta port has brought about a significant increase in freshwater discharge in its distributary, the Hugli estuary. The rivers in the eastern sector on the other hand have lost their connections with Ganga–Bhagirathi system in the course of time and are now tide-fed in nature. This variation probably caused a compositional variation in fish community. The western sector showed the presence of more economically important fish species in comparison to trash fishes that may be attributed to decreasing trend of salinity in this zone (Table 4C.1) in recent times.

In the eastern sector, the ingression of seawater and resultant salinity increase has completely reversed the picture with more quanta of trash fishes in comparison to economically important species (Table 4C.2). We computed the Shannon–Wiener species diversity index (H) with the collected samples (collection time was 18.03.2007–21.02.2008) for both commercial and trash varieties with a sample size of

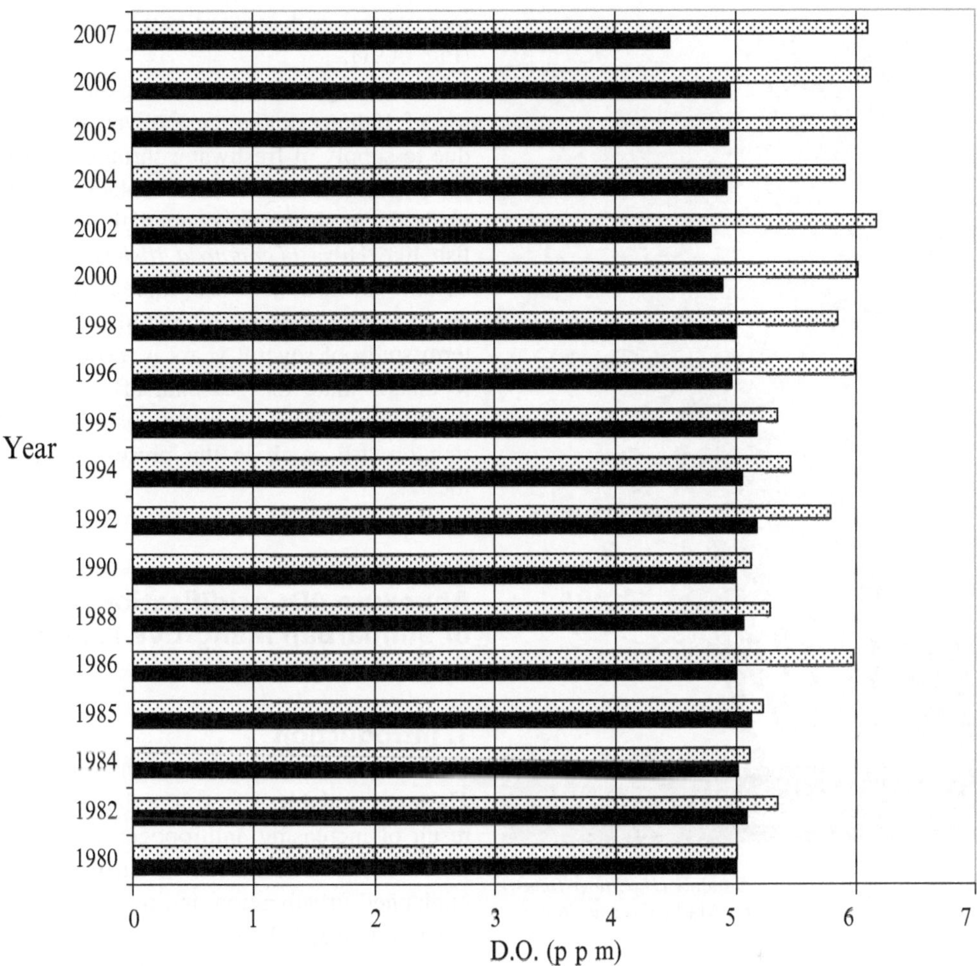

Fig. 4B.5 Dissolved oxygen (in ppm) fluctuation due to global warming in central (decreasing trend, marked in *black*) and western (increasing trend, marked in *dots*) Indian Sundarbans

100 and 50 kg for commercially important and trash fishes, respectively, and documented more diversity of trash fishes in the eastern sector, which may be attributed to intrusion of saline water from the adjacent Bay of Bengal region in the south.

The rate of climate change may thus be a major determinant of the abundance and distribution of new populations. Rapid change from physical forcing usually will favour production of smaller, low-priced, opportunistic species that discharge large numbers of eggs over long periods. Reports of decline of species numbers in fish due to increase of salinity have been published by several workers (Carpelan 1967; Copeland 1967; Hammer 1986). The main causes behind the alteration of fish community structure due to increase in salinity (a consequence of seawater ingression because of warming effect) are:

- Reproductive failure
- Interaction of other environmental parameters with salinity to cause excessive mortality (synergistic effect)
- Loss of primary food supply due to exceedance of salinity tolerance for that organism
- Direct mortality due to exceedance of salinity tolerance

Fig. 4B.6 Freshwater-loving Sundari (*Heritiera fomes*) is gradually vanishing from the Matla riverine zone: a threat posed by rising salinity?

The warning bell of climate change is also perceived in the Hilsa population. The Indian shad *Tenualosa ilisha* is one of the important food fish of India, Pakistan, Bangladesh and Arabian Gulf area. It constitutes a fishery of considerable economic weightage and importance in almost all the river systems and estuaries of Gangetic delta. The species is anadromous in nature with feeding grounds in the sea and spawning grounds in the upstream region of estuaries. The recent trend of increasing salinity in the central sector of Gangetic delta has dramatically modified their migratory path towards the west—where the salinity is relatively low. The landing stations in the central sector of Indian Sundarbans witness severe crisis of the species, whereas the western sector exhibits an increasing trend in shad catch in recent times (Fig. 4C.1).

According to the researchers, the western sector of Indian Sundarbans, lowering of salinity due to supply of freshwater through Hugli (after the commissioning of the Farakka barrage) has made the environment congenial particularly for fish like Hilsa (*Tenualosa ilisha*). The landing volume of fish has also increased during the post-Farakka period (Sinha et al. 1996). A long-term study of several years is, however, needed to discriminate the seasonal effect (feature) of fish diversity from the impact of salinity fluctuation on fish catch in the present geographical locale.

Annexure 4D: Acidification of Sundarban Mangrove Estuarine System

1. Introduction

There are reliable scientific evidences that, as a result of increasing anthropogenic carbon dioxide emissions due to intense industrialization and unplanned urbanization and tourism, absorption of carbon dioxide by the oceans and estuaries has considerably increased the average oceanic acidity from pre-industrial levels. However, uncertainties exist in the data structure because of several natural and man-made factors (Mitra 2013). According to the Intergovernmental Panel on Climate Change (IPCC), continuing carbon dioxide emissions in line with current trends could make the oceans up to 150 % more acidic by 2100 than they were at the beginning of the Anthropocene. Acidification decreases the ability of the ocean to absorb additional atmospheric carbon dioxide, which implies that future carbon dioxide emissions are likely to lead to more rapid global warming.

Ocean acidification is also problematic because of its negative effects on marine ecosystems, especially marine calcifying organisms, and marine resources and services

Table 4C.1 Salinity (in ‰) variation in the western Indian Sundarbans after the commissioning of the Farakka barrage

Place	Pre-Farakka (1960–1961)	Post-Farakka (1985)	Post-Farakka (1995)	Post-Farakka (2005)[a]
Kakdwip	32.80	15.10	13.93	8.56
Canning	28.00	28.90	26.70	11.34

Source: Sanyal et al. (2007), In: Sundarban Wetlands (edited by Dr. Madhumita Mukherjee); published by Department of Fisheries, Aquaculture, Aquatic resources and Fishing Harbour, Govt. of West Bengal
[a]Survey conducted by the authors

Table 4C.2 Mean value of Shannon–Wiener species diversity index (H) computed from 1-year survey (18.03.2007–21.02.2008) on fish catch by local fishermen in Indian Sundarbans

Station	Commercial variety	H (sample size = 100 kg)	Trash variety	H (sample size = 50 kg)
Namkhana (in western Indian Sundarbans)	*Pama pama*, *Polynemus paradiseus*, *Arius jella*, *Tenualosa ilisha*, *Sillaginopsis panijus*, *Osteogeneious militaris* and *Polydactylus indicus*	3.187	*Thryssa* sp., *Stolephorus* sp., *Harpodon nehereus* and *Cynoglossus* sp.	1.895
Bali Island (in eastern Indian Sundarbans)	*Pama pama*, *Polynemus paradiseus*, *Arius jella* and *Sillaginopsis panijus*	2.014	*Thryssa sp.*, *Stolephorus* sp., *Harpodon nehereus* and *Cynoglossus* sp.	3.961

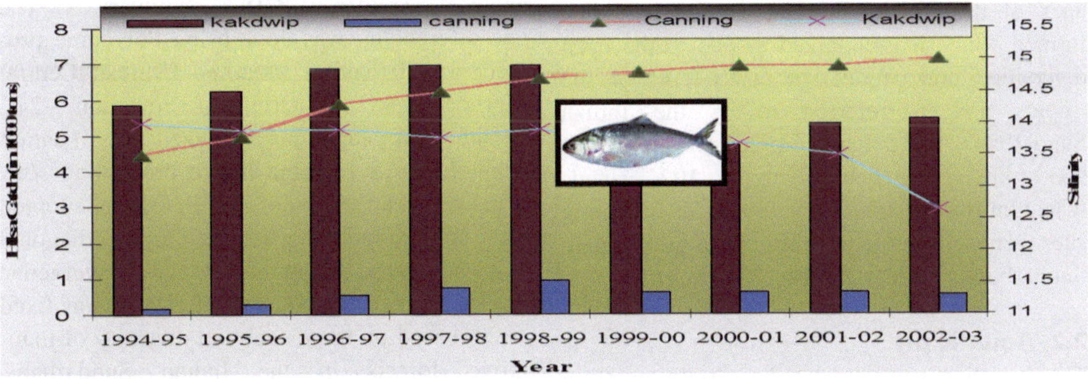

Fig. 4C.1 Hilsa (*Indian shad*) catch (*bars*) showing inverse relationship with aquatic salinity (*line*)

upon which human societies largely depend such as energy, water and fisheries. The effect of ocean acidification on the carbon system, its consequences for the ability of the ocean to take up carbon, its impact on marine ecosystems and the time involved in recovery of the ocean ecosystems have been investigated on a variety of timescales. The coastal vegetation (commonly referred to as blue carbon bank) also has a regulatory impact on aquatic pH.

It is now well established that the average pH of the world ocean surface waters has already fallen by about 0.1 units from an average value of about 8.21–8.10 since the beginning of the industrial revolution (http://mcbi.marineconservation.org/publications/pub_pdfs/feely_etal_2008_pices.pdf).

The estuaries of Indian Sundarbans is a perfect test bed to carry on such experiments because of its proximity to highly urbanized city of Kolkata (the direct cause of carbon dioxide emission) (Mitra et al. 1992; Mitra et al. 2010) and presence of about 2110 km^2 of mangroves (Sengupta et al. 2013; Raha et al. 2013). On this background the present study was conducted to evaluate the alteration of pH in six stations of eastern Indian Sundarbans situated within the dense mangrove forest for a period of 30 years (1984–2013). Another important aim of the study is to analyse the role of mangrove vegetation in regulating the trend of acidification in the present geographical locale.

2. Materials and Methods

2.1. Study Site Description

The mighty river Ganga emerges from a glacier at Gangotri, about 7010 m above mean sea level in the Himalayas, and flows down to the Bay of Bengal covering a distance of 2525 km. At the apex of the Bay of Bengal, a delta has been formed which is recognized as one of the most diversified and productive ecosystems of the tropics and is referred to as the Indian Sundarbans. The deltaic complex has an area of 9630 km^2 and houses about 102 islands (Chaudhuri and Choudhury 1994). Six sampling sites were selected in the eastern sector of Indian Sundarbans (Table 4D.1, Fig. 4D.1).

2.2. Aquatic pH

The sampling strategy aimed to study temporal and spatial variability of surface water pH in the estuarine waters of eastern part of Indian Sundarbans. Accordingly, systematic sampling was carried out in June (premonsoon season) for 30 consecutive years (1984–2013) during high-tide period in six selected stations. The pH of the surface water was measured with a portable pH meter (Model BST-BT65; sensitivity = ±0.01). Twenty-five readings were recorded from each site and the mean value was considered for statistical analyses.

Table 4D.1 Sampling stations with coordinates in western Indian Sundarbans

Sampling stations	Latitude	Longitude
Arbesi (Stn. 1)	22°12′07.44″N	89°00′03.15″E
Jhilla (Stn. 2)	22°10′20.55″N	88°56′04.13″E
Harinbhanga (Stn. 3)	21°59′17.79″N	88°59′05.32″E
Khatuajhuri (Stn. 4)	22°03′09.52″N	89°01′03.12″E
Chamta (Stn. 5)	21°53′05.91″N	88°57′26.52″E
Chankhali (Stn. 6)	21°51′31.73″N	89°00′50.92″E

2.3. Mangrove Vegetation Mapping Through Remote Sensing

AWiFS data for June 2010 from IRS-P6 satellite was requisitioned from NRSC Hyderabad for the scene (row 55, path 107) with the primary objective of carrying out rapid assessment of mangrove pool. The digital data were registered with geo-referenced vector coverage data, encompassing the Sundarban Reserved Forest, using ArcInfo GIS software. The geo-referencing was done using Polyconic projection with Datum Modified Everest, Central Meridian −88° and Origin of Projection −24°. In the next stage, unsupervized classification using ERDAS Imagine image processing software was carried out. The algorithm adopted was 'initializing the means' along diagonal axis, standard deviation = 2 and convergence threshold of 0.980. Number of classes was fixed at 20, based on the extent and diversity of mangrove forests in the Indian Sundarbans. Attributes of the classified images were then grouped into different classes like dense mangrove, open mangrove, mudflat/tidal creeks and water with variable sediment load (Fig. 4D.1). The classifications were subsequently confirmed through ground truth verification and earlier reference maps. The classification was followed by the extraction of the forest area enclosing the sampling stations. Area of mangroves around each station was calculated to evaluate the mangrove pool (comprising of dominant mangrove

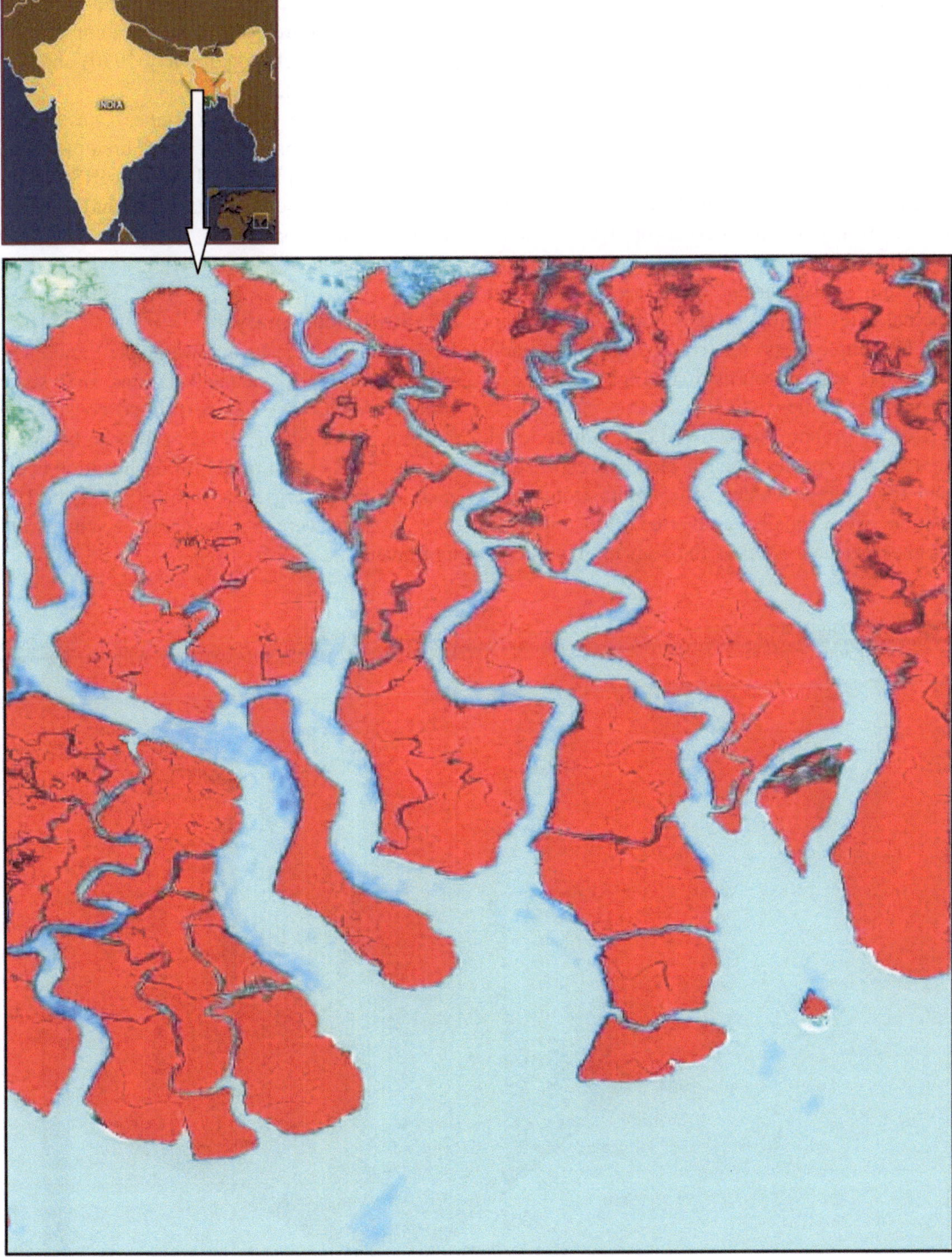

Fig. 4D.1 Location of sampling stations in the eastern Indian Sundarbans; the *red colour* indicates the mangrove vegetation

species) in tonnes. Five dominant species were selected as per the order of relative abundance of each mangrove species in the sampling plots.

2.4. Mangrove Vegetation Pool

The estimation was carried out by the researcher from 02.06.2010 to 18.06.2010 during the low-tide period. In each station, selected forest patches were ~12 years old. Fifteen sample plots (10 × 10 m) at an interval of 100 m were established (in the river bank) through random sampling in the various qualitatively classified biomass levels for each station. The mean (of 15 sample plots) relative abundance of each mangrove species was evaluated for the order of dominance in the study area.

Above-ground biomass (AGB) of individual trees of the species in each plot was estimated, and the average values of 15 plots from each station were finally converted into biomass (tonnes per ha). AGB is the sum total of stem, branch and leaves of the tree. The biomass of above-ground structures was estimated as per the standard procedures outlined for stem (Husch et al. 1982), branch (Chidumaya 1990) and leaf (Mitra et al. 2011).

The mangrove pool (in tonnes) was finally estimated by the product of forest area (assessed from AWiFS data retrieved through IRS-P6 satellite) around the selected station (in ha) and total biomass of dominant mangrove species (in tonnes per ha).

2.5. Statistical Analysis

To assess whether surface water pH varied significantly between years and stations, analysis of variance (ANOVA) was performed considering the data collected for 30 years. It is to be noted that every data point of pH for each station is the mean of 25 readings. Possibilities less than 0.05 ($p < 0.05$) were considered statistically significant.

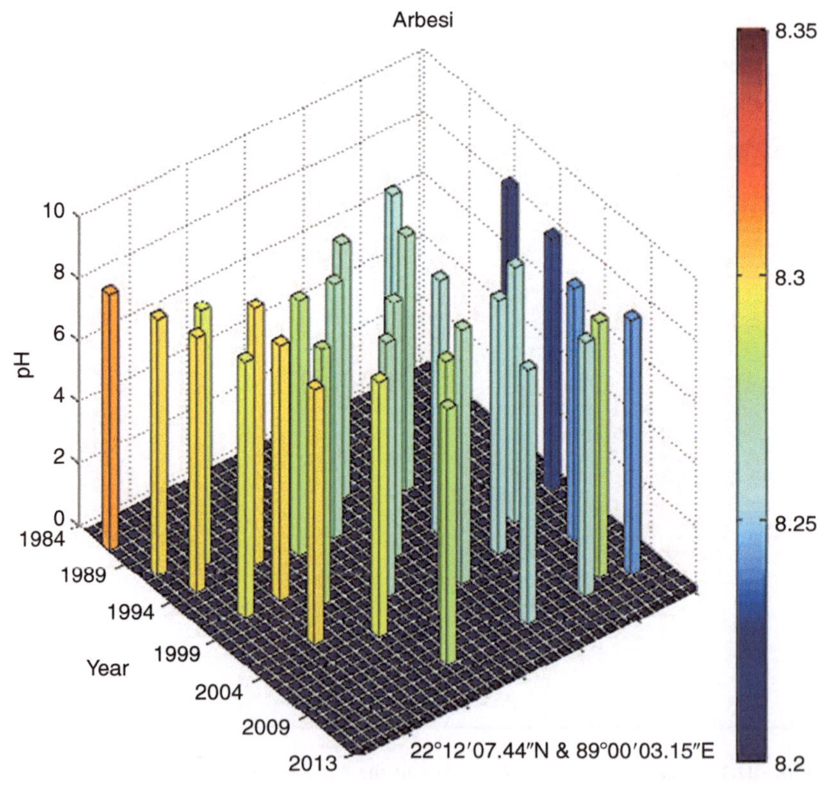

Fig. 4D.2 Temporal variation of aquatic pH at station 1

The regression equation between mangrove forest area and aquatic pH was assessed by considering the mangrove forest area for the year 2010 (y) and average aquatic pH of six selected stations for the same year. All statistical calculations were performed with SPSS 9.0 for Windows.

3. Results

The temporal variations of aquatic pH in all the six selected stations are presented in Figs. 4D.2, 4D.3, 4D.4, 4D.5, 4D.6 and 4D.7.

Table 4D.2 presents the percentage of pH reduction in the surface water of selected stations. The table shows that in all the selected stations the pH has decreased within the range 0.36 % (at Stn. 6) to 1.08 % (at Stn. 1) over a period of 30 years. The rate of decrease per decade ranges from 0.012 (at Stn. 6) to 0.036 (at Stn. 1). The spatial order of aquatic pH in the present study area is Arbesi (Stn. 1) > Jhilla (Stn. 2) > Harinbhanga (Stn. 3) = Khatuajhuri (Stn. 4) = Chamta (Stn. 5) > Chandkhali (Stn. 6). This spatio-temporal variation is also confirmed through ANOVA (Table 4D.3).

4. Discussion

The average decrease of aquatic pH in eastern sector of Indian Sundarbans is 0.026, which is quiet close to the observations recorded in several oceans and estuaries (IPCC 2007; Doney et al. 2012; Kim et al. 2013; http://www.espl.noaa.gov/gmd/ccgg/trends; http://hahara.soest.hawaii.edu) (Table 4D.4).

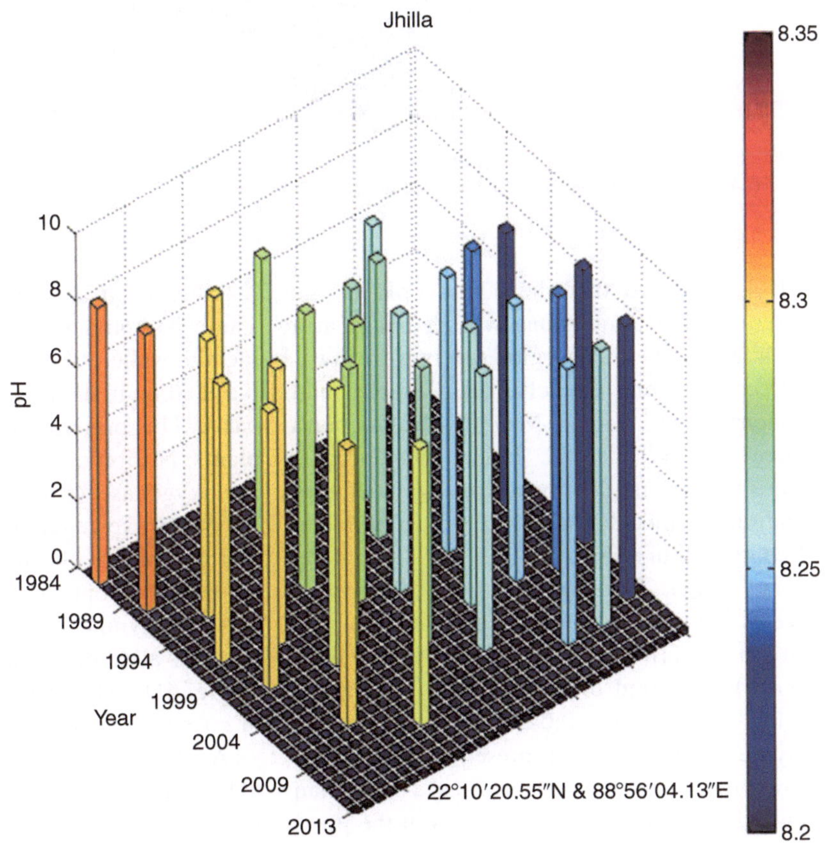

Fig. 4D.3 Temporal variation of aquatic pH at station 2

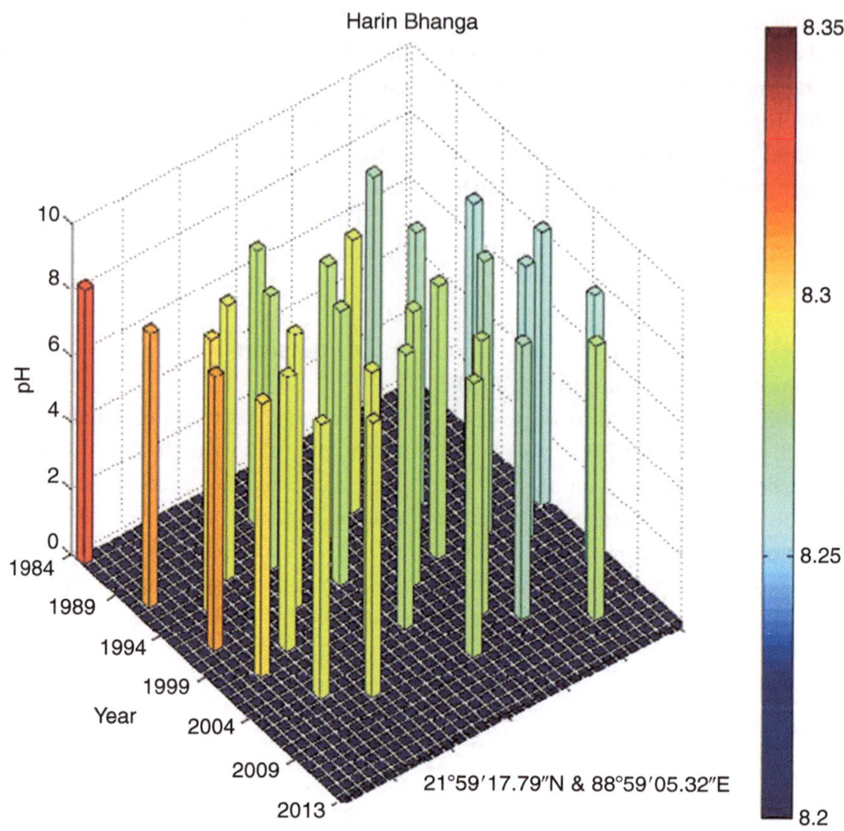

Fig. 4D.4 Temporal variation of aquatic pH at station 3

It is expected that the aquatic pH in the eastern Indian Sundarbans will decrease at a much faster rate as the region receives freshwater supply from the channels, creeks and tributaries of Padma–Meghna–Brahmaputra system, unlike central Indian Sundarbans, where there is no supply of freshwater due to siltation of the Bidyadhari river in the late fifteenth century (Mitra et al. 2009; Sengupta et al. 2013). The present study, however, reveals a slow pace of acidification of estuarine water of eastern Indian Sundarbans (0.026 unit/decade) compared to the estuaries of western Indian Sundarbans (0.04 unit/decade). This may be primarily due to presence dense mangrove forest in the eastern part of Indian Sundarbans. The dominant species in the present study area are *Sonneratia apetala*, *Avicennia alba*, *Avicennia officinalis*, *Avicennia marina* and *Excoecaria agallocha*. These halophytes absorb carbon dioxide from the atmosphere and the water column and store it as carbon in their vegetative and reproductive parts. The principal carbon-fixing enzyme in plants is ribulose-1,5-bisphosphate carboxylase/oxygenase (RuBisCo). The magnitude of the mangrove vegetation pool thus exerts a regulatory role on the pH of the estuarine water as these halophytes absorb carbon dioxide for photosynthesis and shifts the equilibrium towards alkalinity. The scatter plot presented in Fig. 4D.8 also depicts the positive influence of coastal vegetation on aquatic pH.

A close observation on Figs. 4D.2, 4D.3, 4D.4, 4D.5, 4D.6 and 4D.7 exhibits sudden rise of aquatic pH irrespective of stations during 2009. This was the effect of AILA, a supercyclone that passed across the Gangetic

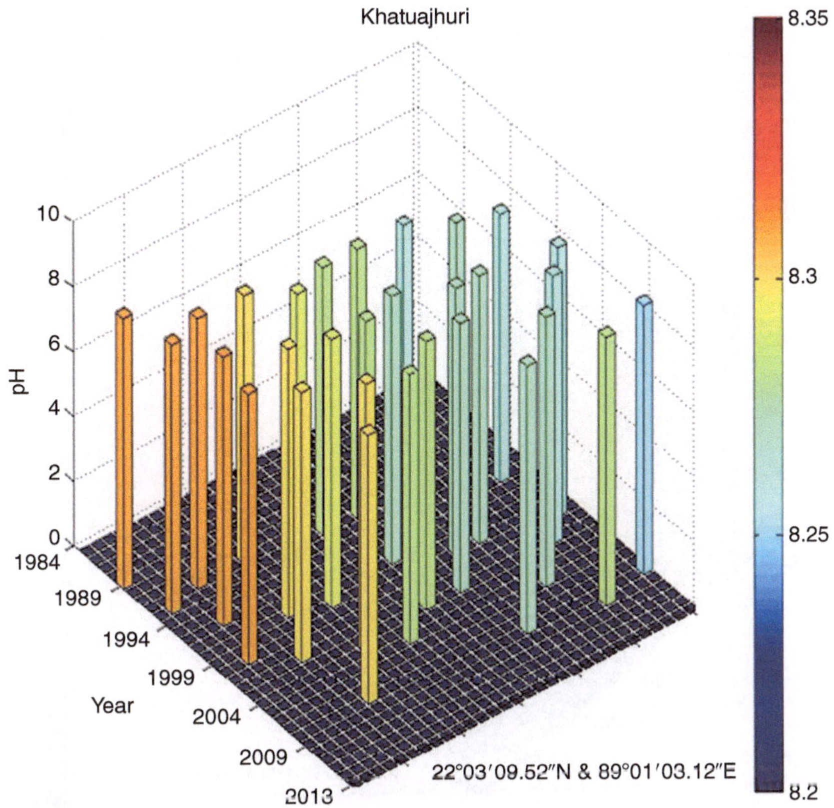

Fig. 4D.5 Temporal variation of aquatic pH at station 4

delta with a speed of 110 km/hr during 25 May 2009. This supercyclone caused significant rise of surface water salinity and pH in the estuaries of the Indian Sundarbans (Mitra et al. 2011).

The overall result of the study suggests that the pH of the estuarine water in eastern Indian Sundarbans is regulated by the synergistic effect of the halophytes and magnitude of mixing of seawater from the Bay of Bengal. Thus, conservation of coastal vegetation and mangrove afforestation must be given priority to prevent the process of acidification.

5. Conclusion

This study has demonstrated that the change of aquatic pH in the eastern sector of Indian Sundarbans is controlled chiefly by the estuarine mixing process and process of halophyte interaction with the soil chemistry in coastal sectors of the Bay of Bengal. Thus, in the future, additional research on this sensitive process is needed, and more mangrove afforestation is needed to be introduced in order to safeguard the nature from the danger of further acidification.

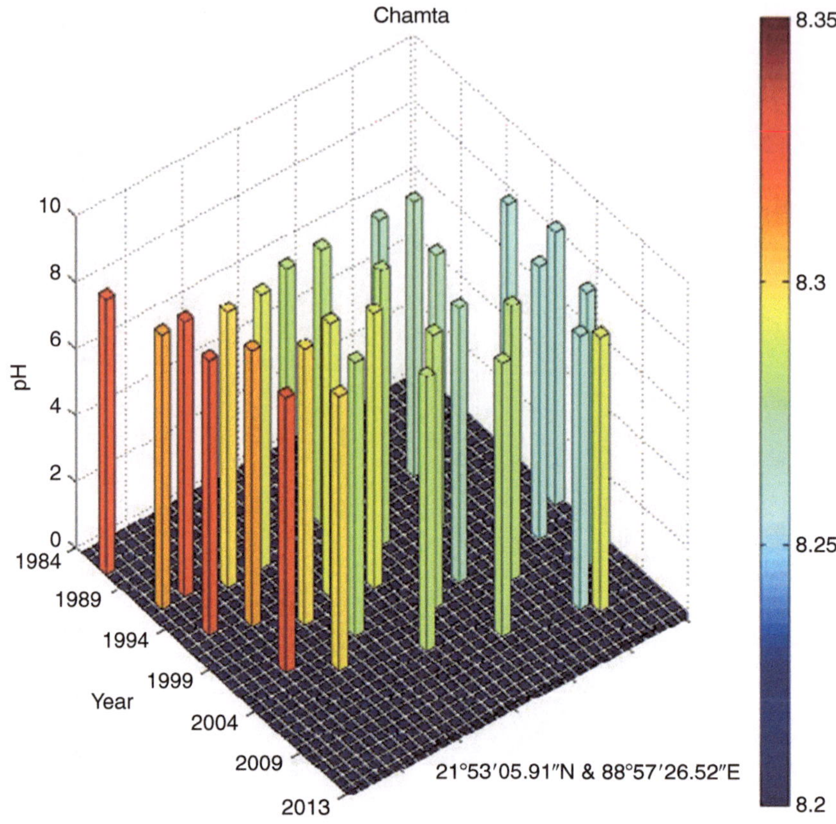

Fig. 4D.6 Temporal variation of aquatic pH at station 5

6. Future Issues

Key recommendations for future action in order to fully understand the trend of estuarine acidification are as follows:

1. Quantify further the biological and biogeochemical response to estuarine acidification from the organismal to the mangrove ecosystem level.
2. Integrate natural and social sciences to help mitigate estuarine acidification and develop adaptation strategies, taking into account the socio-economic impacts on natural resources and human communities. The full cost of abating carbon dioxide emissions (from the adjacent cities of Kolkata, Howrah and Haldia industrial zone), several brick kilns located in the area and carbon capture and storage should be also considered.
3. Monitor the environmental services of endemic coastal vegetation like *Porteresia coarctata* (salt marsh grass), mangroves (all 34 species), seaweeds (*Enteromorpha intestinalis*, *Ulva lactuca*, *Catenella repens*, etc.) and other mangrove associates in mitigating the acidification process.
4. Facilitate dissemination and capacity building to help deliver scientific knowledge-based advice to researchers and policymakers, to share best practices among researchers and success stories with island dwellers of Sundarbans highlighting the role of mangroves in maintaining the estuarine water quality.

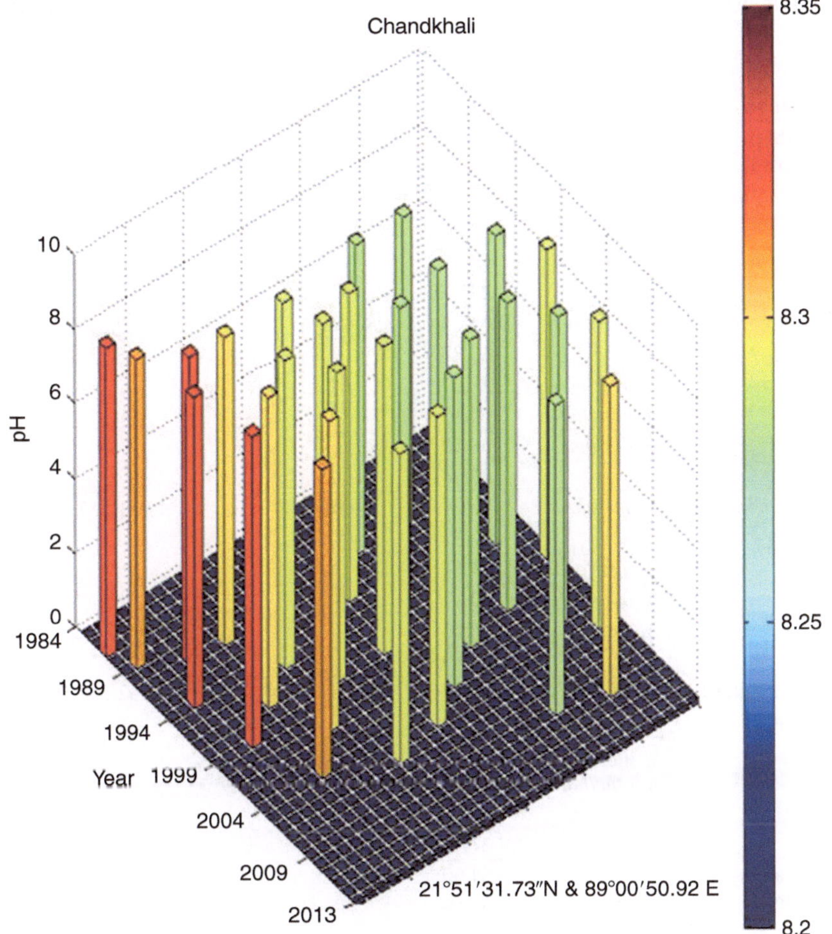

Fig. 4D.7 Temporal variation of aquatic pH at station 6

Table 4D.2 Mangrove pool (tonnes), % pH reduction in the eastern Indian Sundarbans

Station	Total biomass of dominant species (tonnes/ha)		Forest area (ha)	Mangrove pool (tonnes)	% pH reduction
	Date type: primary		Date type: primary	Date type: primary	Date type: primary
Arbesi (Stn. 1)	*S. apetala*	67.12	7200	1,971,000	1.08
	A. alba	62.20			
	E. agallocha	21.79			
	A. officinalis	63.56			
	A. marina	59.08			
		273.75			
Jhilla (Stn. 2)	*S. apetala*	65.33	8387	2,309,863.67	0.96
	A. alba	63.88			
	E. agallocha	21.97			
	A. officinalis	64.13			
	A. marina	60.10			
		275.41			
Harinbhanga (Stn. 3)	*S. apetala*	66.21	7465	1,971,357.20	0.72
	A. alba	59.31			
	E. agallocha	21.84			
	A. officinalis	62.17			
	A. marina	54.55			
		264.08			
Khatuajhuri (Stn. 4)	*S. apetala*	66.18	8666	2,402,648.50	0.72
	A. alba	64.77			
	E. agallocha	21.58			
	A. officinalis	64.97			
	A. marina	59.75			
		277.25			
Chamta (Stn. 5)	*S. apetala*	64.99	12,955	3,641,002.75	0.72
	A. alba	66.09			
	E. agallocha	22.02			
	A. officinalis	67.94			
	A. marina	60.01			
		281.05			
Chandkhali (Stn. 6)	*S. apetala*	62.14	9635	2,536,895.50	0.36
	A. alba	60.54			
	E. agallocha	22.87			
	A. officinalis	62.58			
	A. marina	55.17			
		263.30			

Table 4D.3 ANOVA results showing spatio-temporal variation of aquatic pH

Variable	F_{cal}	F_{crit} ($p < 0.05$)
Aquatic pH		
Between stations	35.2747	2.2766
Between years	34.4408	1.5458

Table 4D.4 Reduction of surface water pH in different stations around the world

Location	Rate of decrease in pH
Ulleung Basin East/Japan Sea	0.04 pH units/decade since 1958
ALOHA	0.02 pH units/decade since 1989
HOT, BATS, ESTOC	0.02 pH units since 1980
3200 km north near Hawaii	0.026 pH units since 15 years
Western sector of Indian Sundarbans	0.024 pH units since 1984

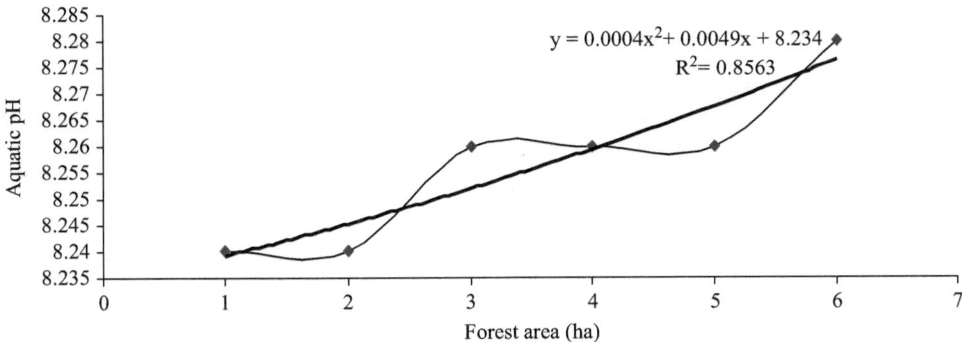

Fig. 4D.8 Interrelationships between forest area and aquatic pH in eastern Indian Sundarbans

References

Bramlette, M. N., & Bradley, W. H. (1940). *Geology and biology of north Atlantic deep-sea cores between Newfoundland and Ireland. Part I. Lithology and geological interpretations* (United States Geological Survey, professional paper, 196-A, pp. 1–34). Reston: United States Geological Survey.

Chowdhury, N. T. (2010). Water management in Bangladesh: An analytical review. *Water Policy, 12*(1), 32–51. IWA Publishing, uncorrected Proof.

Cleveland, C. G., Townsend, A. R., Schimel, D. S., Fisher, H., Howarth, R. W., Hedin, L. O., et al. (1999). Global patterns of terrestrial biological nitrogen (N_2) fixation on natural ecosystems. *Global Biogeochemical Cycles, 13*, 623–645.

Galloway, J. N., Schlesinger, W. H., Levy, C., Michaels, A., & Schnoor, J. L. (1995). Nitrogen fixation: Atmospheric enhancement and environmental response. *Global Biogeochemical Cycles, 9*, 235–252.

Gripenberg, S. (1934). *A study of the sediments of the north Baltic and adjoining seas* (Thalass. Inst. Mere. Fennia, Vol. 60, No. 3). Helsingfors: Tekijä. 231 pp.

Hedley, M. J., & Sharpley, A. N. (1998). Strategies for global nutrient cycling. In L. Currie (Ed.), *Long-term nutrient needs for New Zealand's primary industries: Global supply, production requirements and environmental constraints* (pp. 70–95). Palmerston North: The Fertilizer and Lime Research Centre, Massey University.

Howarth, R. W., Jensen, H. S., Marino, R., & Postma, H. (1995). Transport to and processing of phosphorus in near shore and oceanic waters. In H. Tiessen (Ed.), *Phosphorus in the global environment* (pp. 323–345). New York: Wiley.

Islam, M. R. (2004). *Where land meets the sea: A profile of coastal zone of Bangladesh* (pp. 1–317). Dhaka: The University Press Ltd.

Islam, S. N., & Gnauck, A. (2008). Mangrove wetland ecosystems in Ganges-Brahmaputra delta in Bangladesh. *International Journal of Frontier Earth Science China, 2*(4), 439–448. Springer.

Joseph, P. S. (2006). The environmental management the better supply of fresh water in transboundary river: The Ganges could run dry. In J. G. Perez (Ed.), *Proceedings of III international symposium on transboundary waters management-overcoming water management boundaries, 30 May–2 June.* Ciudad Real: Universidad de Castilla-La Mancha.

Matson, P. A., Parton, W. J., Power, A. G., & Swift, M. J. (1997). Agricultural intensification and ecosystem properties. *Science, 277*, 504–509.

Murray, J., & Hjort, J. (1912). *The depths of the ocean* (p. 821). London: Macmillan.

National Research Council. (2000). *Clean coastal waters: Understanding and reducing the effects of nutrient pollution.* Washington, DC: National Academy Press.

Radezewski, O. E. (1939). Eolian deposits in marine sediments. In P. D. Trask (Ed.), *Recent marine sediments* (pp. 496–502). Tulsa: The American Association of Petroleum Geologists.

Riley, J. P., & Skirrow, G. (1975). *Chemical oceanography* (2nd ed., Vol. 1, p. 366). London: Academic Press Ltd.

Schmitt, R. W., Bogden, P. S., & Dorman, C. E. (1989). Evaporation minus precipitation and density fluxes for the North Atlantic. *Journal of Physical Oceanography, 19*, 1208–1221.

Smil, V. (2001). *Enriching the earth.* Cambridge, MA: MIT Press.

Twenhofel, W. H. (1932). *Treatise on sedimentation* (2nd ed., p. 926). Baltimore: Williams and Wilkins.

Vitousek, P. M., Aber, J., Bayley, S. E., Howarth, R. W., Likens, G. E., Matson, P. A., et al. (1997). Human alteration of the global nitrogen cycle: Causes and consequence. *Issues in Ecology, 1*, 1–15.

Annexure 4A References

Banerjee, K., Mitra, A., Bhattacharyya, D. P., & Choudhury, A. (2002). Role of nutrients on phytoplankton diversity in the north–east coast of the Bay of Bengal. In A. Kumar (Ed.), *Ecology and ethology of aquatic biota* (pp. 102–109). New Delhi: Daya Publishing House.

Banerjee, K., Mitra, A., & Bhattacharyya, D. P. (2003). Phytopigment level of the aquatic subsystem of Indian Sundarbans at the apex of Bay of Bengal. *Sea Explorers, 6*, 39–46.

Banerjee, K., Sengupta, K., Raha, A., & Mitra, A. (2013). Salinity based allometric equations for biomass estimation of Sundarban mangroves. *Biomass and Bioenergy, 56*, 382–391.

CEGIS. (2006). *Impacts of sea level rise on landuse suitability and adaptation options*. Draft final report. Submitted to the Ministry of Environment and Forest, Government of Bangladesh and United Nations Development Programme (UNDP) by Centre for Environmental Geographic Information Services (CEGIS), Dhaka.

Chakraborty, S. K., & Choudhury, A. (1985). *Distribution of fiddler crabs in Sundarbans mangrove estuarine complex, India*. Proceedings of national symposium on biology, utilization and conservation of mangroves, Department of Ocean Development, India, pp. 467–472.

Chaudhuri, A. B., & Choudhury, A. (1994). *Mangroves of the Sundarbans, India*. Bangkok: IUCN.

Hazra, S., Ghosh, T., Dasgupta, R., & Sen, G. (2002). Sea level and associated changes in Sundarbans. *Science and Culture, 68*, 309–321.

IPCC (Intergovernmental Panel on Climate Change). (2007) *Climate change 2007: Impacts, adaptation and vulnerability*. Contribution of Working Group II to the fourth assessment report of the Intergovernmental Panel on Climate Change, Geneva, 976 pp.

Jagtap, T. G., & Nagle, V. L. (2007). Response and adaptability of mangrove habitats from Indian subcontinent to changing climate. *Ambio, 36*(4), 328–334.

Mitra, A. (2000). The Northeast coast of the Bay of Bengal and deltaic Sundarbans. In C. Sheppard (Ed.), *Seas at the millennium—An environmental evaluation* (Chap. 62, pp. 143–157). Coventry: University of Warwick/Elsevier Science.

Mitra, A. (2013). *Sensitivity of mangrove ecosystem to changing climate* (p. 323). New Delhi: Springer. doi:10.1007/978-81-322-1509-7: 323

Mitra, A., & Banerjee, K. (2005). In C. S. R. Banerjee (Ed.), *Living resources of the seas: Focus Sundarbans*. WWF-India, Canning Field Office, 24 Parganas(S), West Bengal, 96 pp.

Mitra, A., & Choudhury, A. (1994). Heavy metal concentrations in oyster *Crassostrea cucullata* of Henry's Island, India. *Journal of Ecobiology, 6*(2), 157–159.

Mitra, A., Ghosh, P. B., & Choudhury, A. (1987). *A marine bivalve Crassostrea cucullata can be used as an indicator species of marine pollution*. Proceedings of national seminar on estuarine management, Department of Ocean Development, India, pp. 177–180.

Mitra, A., Choudhury, A., & Zamaddar, Y. A. (1992). Effects of heavy metals on benthic molluscan communities in Hooghly estuary. *Proceedings of the Zoological Society, 45*, 481–496.

Mitra, A., Banerjee, K., Sengupta, K., & Gangopadhyay, A. (2009). Pulse of climate change in Indian Sundarbans: A myth or reality. *National Academy of Science Letters, 32*, 1–7.

Mitra, A., Sengupta, K., & Banerjee, K. (2011). Standing biomass and carbon storage of above-ground structures in dominant mangrove trees in the Sundarbans. *Forest Ecology and Management, 261* (7), 1325–1335. Elsevier doi:10.1016/j.foreco.2011.01.012.

Mondal, K., Mukhopadhyay, S. K., Biswas, H., De, T. K., & Jana, T. K. (2006). Fluxes of nutrients from the tropical River Hooghly at the land–ocean boundary of Sundarbans, NE Coast of Bay of Bengal, India. *Journal of Marine Systems, 62*, 9–21.

Roy Chowdhury, M., Zaman, S., Jha, C. S., Sengupta, K., & Mitra, A. (2014). Mangrove biomass and stored carbon in relation to soil properties: A case study from Indian Sundarbans. *International Journal for Pharmaceutical Research Scholars, 3*(3), 58–69.

Saha, S. B., Mitra, A., Bhattacharyya, S. B., & Choudhury, A. (1999). Heavy metal pollution in Jagannath canal, an important tidal water body of the north Sundarbans aquatic ecosystem of West Bengal. *Indian Journal of Environmental Protection, 19*(11), 801–804.

Sengupta, K., Roy Chowdhury, M., Bhattacharya, S. B., Raha, A., Zaman, S., & Mitra, A. (2013). Spatial variation of stored carbon in Avicennia alba of Indian Sundarbans. *Discovery Nature, 3*(8), 19–24.

UNESCO. (2007). *Case studies on climate change and world heritage*. Paris: UNESCO World Heritage Centre.

Zhang, K. Q., Douglas, B. C., & Leatherman, S. P. (2000). Twentieth-century storm activity along the US east coast. *Journal of Climate, 13*, 1748–1761.

Annexure 4B References

Hasnain, S. I. (1999). *Himalayan glaciers: Hydrology and hydrochemistry*. New Delhi: Allied Publisher.

Hasnain, S. I. (2000). *Status of the glacier research in the HKH region 2000*. Kathmandu: ICIMOD.

Hasnain, S. I. (2002). *Himalayan glaciers meltdown: Impact on South Asian Rivers* (IAHS publication 274, pp. 1–7). Wallingford: IAHS.

IPCC. (2001). In J. T. Houghton et al. (Eds.), *Climate change, 2001: The scientific basis. Contribution of Working Group I to the third assessment report of the Intergovernmental Panel on Climate Change* (881 pp.). Cambridge: University of Cambridge.

Sinha, M., Mukhopadhyay, M. K., Mitra, P. M., Bagchi, M. M., & Karmakar, H. C. (1996). Impact of Farakka

Barrage on the hydrology and fishery of Hooghly Estuary. *Estuaries, 19*(3), 710–722.

World Bank. (1997). *World development indicators.* Washington, DC: World Bank. 231 pp.

Annexure 4C References

Carpelan, L. H. (1967). Invertebrates in relation to hypersaline habitats. Invertebrates in supersaline waters. *University Texas Contribution Marine Science, 12* (21), 219–229.

Copeland, B. J. (1967). Environmental characteristics of hypersaline lagoons. *University of Texas Contribution Marine Science, 12*, 207–218.

Hammer, U. T. (1986). *Saline lake ecosystems of the world.* Dordrecht: Dr. W Junk Publishers.

Sanyal, P., Mukherjee, M., & Das, K. L. (2007). Status of sharks, rays and dolphins of Sunderbans with special views on marine reserve. In M. Mukherjee (Ed.), *Sunderban wetlands.* Kolkata: Department of Fisheries, Aquaculture, Aquatic Resources and Fishing Harbours, Govt. of West Bengal. 256 pp.

Sinha, M., Mukhopadhyay, M. K., Mitra, P. M., Bagchi, M. M., & Karmakar, H. C. (1996). Impact of Farakka Barrage on the hydrology and fishery of Hooghly Estuary. *Estuaries, 19*(3), 710–722.

Annexure 4D References

Chaudhuri, A. B., & Choudhury, A. (1994). *Mangroves of the Sundarbans India.* Bangkok: IUCN.

Chidumaya, E. N. (1990). Above ground woody biomass structure and productivity in a Zambezian woodland. *Forest Ecology Management, 36*, 33–46.

Doney, S., Ruckelshaus, M., Duffy, E., Barry, J., Chan, F., English, C., Galindo, H., Grebmeier, J., Hollowed, A., Knowlton, N., Polovina, J., Rabalais, N., Sydeman, W., & Talley, L. (2012). Climate 10 change impacts on marine ecosystems. *Annual Review of Marine Science, 4*, 1–27.

http://hahara.soest.hawaii.edu.

http://mcbi.marineconservation.org/publications/pub_pdfs/feely_etal_2008_pices.pdf.

http://www.espl.noaa.gov/gmd/ccgg/trends.

Husch, B., Miller, C. J., & Beers, T. W. (1982). *Forest mensuration.* New York: Ronald Press.

Intergovernmental Panel on Climate Change (IPCC). (2007). *Climate change 2007: The physical science basis.* Cambridge: Cambridge University Press. 996 pp.

Kim, J. Y., Kang, D. J., Lee, T., & Kim, K. R. (2013). Rapid increasing trend of CO_2 and ocean acidification in the surface water of the Ulleung Basin. East/Japan Sea inferred from the observations from 1995 to 2004. *BGD, 10*, 9573–9602.

Mitra, A. (2013). *Sensitivity of mangrove ecosystem to changing climate.* New Delhi: Springer. ISBN: 978-81-322-1508-0 (Print) 978-81-322-1509-7 (Online).

Mitra, A., Choudhury, A., & Zamaddar, Y. A. (1992). Seasonal variations in metal content in the gastropod *Cerithedia (Cerithideopsis) cingulata. Proceedings of the Zoological Society, 45*, 497–500.

Mitra, A., Banerjee, K., Sengupta, K., & Gangopadhyay, A. (2009). Pulse of climate change in Indian Sundarbans: A myth or reality? *National Academy of Science Letters, 32*, 1–2.

Mitra, A., Mondal, K., & Banerjee, K. (2010). Concentration of heavy metals in fish juveniles of Gangetic delta of West Bengal, India. *Research Journal Fisheries and Hydrobiology, 5*(1), 21–26.

Mitra, A., Sengupta, K., & Banerjee, K. (2011). Standing biomass and carbon storage of above ground structures in dominant mangrove trees in the Sundarbans. *Forest Ecology Management, 261*, 1325–1335.

Raha, A. K., Bhattacharyya, S. B., Zaman, S., Banerjee, K., Sengupta, K., Sinha, S., Sett, S., Chakraborty, S., Datta, S., Dasgupta, S., Chowdhury, M. R., Ghosh, R., Mondal, K., Pramanick, P., & Mitra, A. (2013). Carbon census of mangrove vegetation in Indian Sundarbans. *Journal of Energy and Environmental Science, 127*, 345–354.

Sengupta, K., Roy Chowdhury, M., Bhattacharya, S. B., Raha, A., Zaman, S., & Mitra, A. (2013). Spatial variation of stored carbon in *Avicennia alba* of Indian Sundarbans. *Discovery Nature, 3*(8), 19–24.

Producers of the Marine and Estuarine Ecosystems

5

Contents

In the marine and estuarine ecosystems, the producer communities act as the converter of solar energy into other utilizable forms of energy. However, in terms of productivity (preferably net primary productivity), the order is estuaries, swamps, and marshes > coastal zone > open ocean (Fig. 5.1).

It is interesting to note that highest primary production in the marine and estuarine systems mostly occur along continental margins and shallow seas (Fig. 5.2). This is attributed to input of nutrients through continental run-off coupled with penetration of sunlight due to shallow depth, both of which favour the growth and production of phytoplankton.

The producer community consists of organisms that are capable of synthesizing their own food through photo- or chemosynthesis. Most of the primary production in the marine and estuarine ecosystems is carried out by phytoplankton. Apart from phytoplankton that manufacture their organic food through photosynthesis (in the photic zone), there are several strains of bacteria that can prepare their food through the process of chemosynthesis. Chemosynthetic bacteria (also known as chemoautotrophs) oxidize inorganic compounds such as ammonia, nitrites and sulphides and trap the small amount of energy released from these oxidations to use in the reactions that synthesize carbohydrates. It has been discovered that chemosynthetic bacteria support a biological community near deep

A. Mitra, S. Zaman, *Basics of Marine and Estuarine Ecology*,
DOI 10.1007/978-81-322-2707-6_5

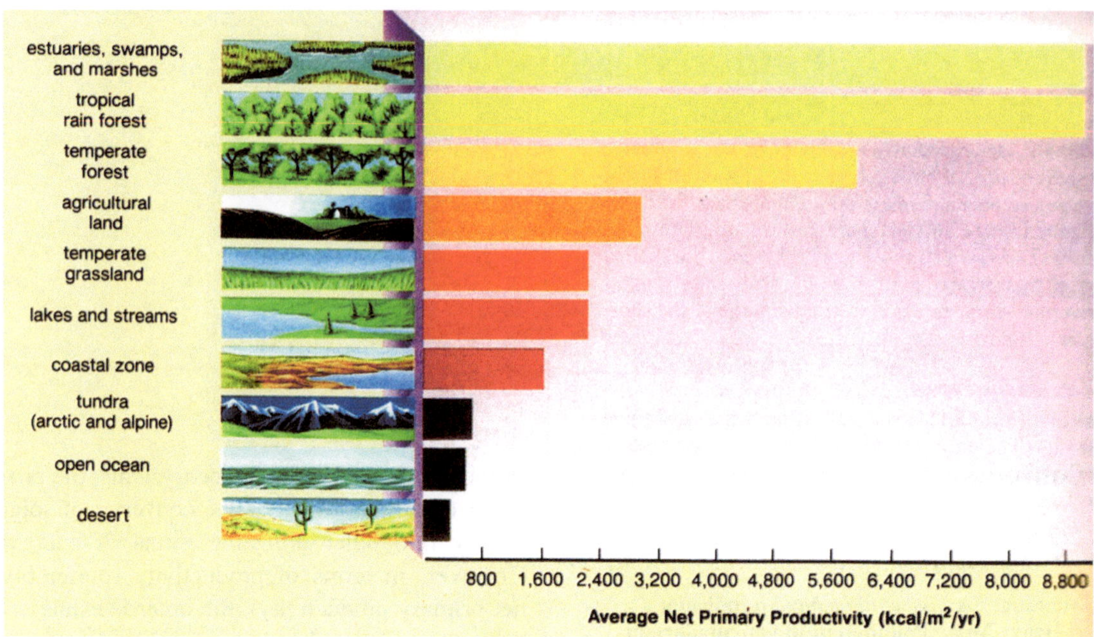

Fig. 5.1 Average net primary productivity in different ecosystems

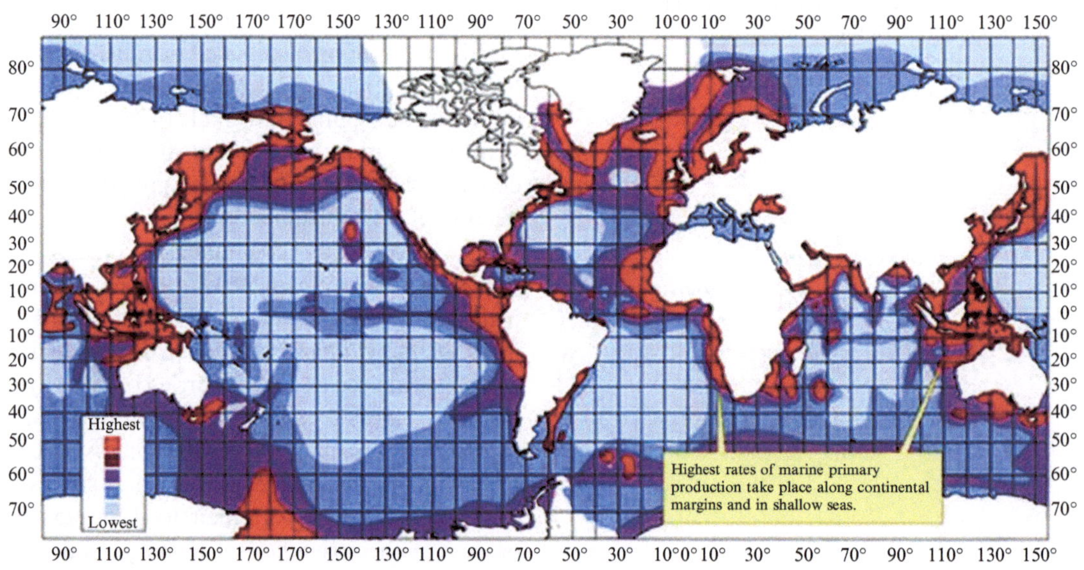

Fig. 5.2 Highest primary productivity (marked with *red colour*) is observed along continental margins

ocean ridges and vents. At the mid-ocean ridge system, hot minerals spew out of the inner Earth, providing hydrogen sulphide to chemosynthetic bacteria. The organic molecules produced by the bacteria support the growth of giant tube worms, clams and crabs (Fig. 5.3).

Several groups of bacteria, including the purple bacteria, the green bacteria and cyanobacteria, are able to photosynthesize. Purple bacteria and green bacteria are usually strict anaerobes (organisms that live in environments that lack oxygen) and are found in areas of the

Fig. 5.3 Chemosynthetic bacteria survive in the deep sea vent zones and are capable of synthesizing food from carbon dioxide and hydrogen sulphide using the energy derived from chemical reactions

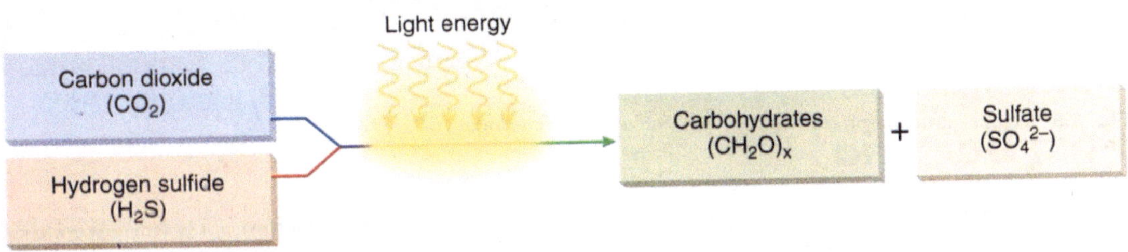

Fig. 5.4 Free oxygen is not generated during food preparation by *purple* and *green* bacteria

marine environment such as mudflats that provide light, anaerobic conditions and sulphur-containing compounds. These organisms cannot use water in their photosynthetic process, and therefore they do not produce any oxygen (Fig. 5.4). Instead, they often use hydrogen sulphide (H_2S) and produce elemental sulphur or sulphate. Purple bacteria and green bacteria are an important source of food for some zooplankton and filter feeders.

The seaweeds, seagrasses, salt marsh grasses, sand binders and mangroves are also among the producer list of the marine and estuarine ecosystems (Figs. 5.5, 5.6, 5.7 and 5.8). Not all these categories of producers are found in the beaches, coastal zones or estuaries uniformly throughout the world. The pattern and distribution of these producers is largely controlled by edaphic and climatic factors. *Ipomoea pes-caprae*, commonly known as sand binder, is abundant in Sundarban delta preferably in regions where the sediment is dominated by sand. Their elongated root system helps to bind the sediment particles and is very effective in controlling erosion (Fig. 5.8). However in coastal areas of Florida, the species is completely absent.

5.1 Phytoplankton Community

Phytoplanktons are taxonomically diverse, single-celled organisms that populate the upper sunlit layer of nearly all aquatic ecosystems on Earth. They are free floating and cannot move against the current. Net annual photosynthesis by ocean phytoplankton alone is of similar magnitude as that by all terrestrial plants (Field et al. 1998; Behrenfeld et al. 2001) and it plays a *vital* role in biospheric carbon cycling. Ocean productivity is strongly impacted by climate variations (Gregg et al. 2005; Behrenfeld et al. 2006a), but quantifying this impact and predicting future

Fig. 5.5 Thallus structure of seaweed (*Ulva lactuca*) in the intertidal zone

Fig. 5.6 Salt marsh grasses contribute to primary production in the coastal and estuarine system

Fig. 5.7 Mangrove vegetation contribute to primary production in the coastal and estuarine system

Fig. 5.8 *Ipomoea pes-caprae* commonly known as sand binder is an important producer in deltaic Sundarbans

change requires an understanding of environmental factors regulating phytoplankton light-use efficiencies and growth (Behrenfeld et al. 2008). However it is extremely difficult to quantify the physiological variability of phytoplankton with respect to time and space. In this context chlorophyll level of aquatic phase is often used as proxy that is detected through remote sensing.

Fig. 5.9 Common marine phytoplankton with unique shape and configuration

As with terrestrial plants, oxygenic photosynthesis in phytoplankton involves light harvesting and electron transport between two pigmented reaction centres: photosystem II (PSII), which is solely responsible for oxygen evolution, and photosystem I (PSI), where light energy is captured in chemical form. Both photosystems contribute to cellular chlorophyll concentration and light absorption, but in vivo fluorescence emanates almost exclusively from PSII (Falkowski and Kiefer 1985). Artificially stimulated fluorescence has accordingly been used as a non-invasive probe for studying phytoplankton and PSII functioning (Cullen 1982; Cleveland and Perry 1987; Krause and Weis 1991; Falkowski and Kolber 1995; Behrenfeld et al. 2006b). Fluorescence emission under natural sunlight is also detectable in subsurface and above-surface upwelling irradiance spectra. Incident irradiance and pigment absorption strongly influence this "natural fluorescence" signal (Babin et al. 1996; Cullen et al. 1997; Huot et al. 2007), which was proposed originally as a tool for deriving phytoplankton photosynthetic rates and can register physiological variability. Such field observations were instrumental to the development of satellite fluorescence detection capabilities.

Phytoplankton sometimes may cause adverse impact on the marine and estuarine environment. During excessive bloom of phytoplankton, the light energy is intercepted, which could otherwise reach fixed plants like eel grass (*Zostera* sp.) and kelp. Furthermore, when the phytoplankton eventually die back and break down, an excessive amount of oxygen is required to fuel this process, and hence areas may become deprived of oxygen. Excessive nutrients, and/or changes in their relative concentrations, may be one factor in a chain of events leading to changes in the species composition of the phytoplankton communities. Increased occurrence of toxic algal blooms may accelerate toxin production. Toxic phytoplankton, when consumed by shellfish or other species, can affect the marine food chain, including poisoning of seabirds, mammals and even humans.

Nearly all marine plants, whether unicellular or multicellular, even those attached to substrata (sessile) or free floating, pass some part of their life cycle in floating condition as phytoplankton. However, those organisms which always remain planktonic throughout the life cycle are (1) diatoms, (2) dinoflagellates, (3) coccolithophores, (4) selective species of blue-green algae and (5) some species of green algae. The members of phytoplankton community have unique size, shape and morphological features (Fig. 5.9).

5.1.1 Diatoms

Diatoms are characterized by the presence of shell or frustules made up of translucent silica. The cell wall of diatom has two parts resembling a pillbox bottom and lid. The lid is called the *epitheca* and the bottom is known as *hypotheca*. These shells have great importance from the geological point of view and constitute the diatomaceous crust. The diatoms exhibit remarkable varieties and forms and many species possess beautifully sculptured shells. Few diatoms commonly found in the marine and estuarine waters are presented in Table 5.1.

Depending on the nature of valves and pattern of ornamentation in the valve surface, the diatoms are grouped into *centric* and *pennate* diatoms. The major differences between these two groups are given in Table 5.2.

5.1.2 Dinoflagellates

These are important producers of the marine environment and rank second in importance in the economy of the sea. Typically, these are unicellular; some are naked while others are armoured with plates of cellulose. The dinoflagellates possess two flagella for locomotion. Several of them are luminescent and produce light. Few species of marine and estuarine dinoflagellates are presented in Table 5.3.

5.1.3 Coccolithophores

These are among the smallest category of phytoplankton having a size range between 5 and 20 μ. Some coccolithophores have flagella, while others are devoid of them. Their soft bodies are shielded by tiny, calcified circular plates or shields of various designs. These are normally found in the open sea, but their profuse occurrence has been recorded in coastal waters. They form important diet components of filter feeding animals. *Coccolithus* sp. and *Isochrysis galbana* are common coccolithophores of oceans, seas and estuaries.

5.1.4 Blue-Green Algae

These include both unicellular and multicellular organisms. The blue colour in them is due to the presence of a pigment known as phycocyanin. Of the various organisms belonging to this category, the most important is *Trichodesmium erythraeum* because in certain seasons of the year its biomass increases greatly resulting in the formation of clumps.

5.1.5 Green Algae

Microscopic green algae present in the planktonic community largely occur in coastal waters. The green colour in them is due to the presence of chloroplasts. They are widely distributed in the warmer (tropical) seas and only few species are found in the Arctic and Antarctic oceans. Some common species of microscopic green algae that are planktonic in nature are *Chlorella marina*, *Chlorella salina*, etc.

5.1.6 Classification of Phytoplankton

The phytoplankton community consists of a variety of organisms, namely, diatoms, dinoflagellates, blue-green algae, silicoflagellates, coccolithophores, etc., which ranges in terms of size from 0.001 to 0.2 mm.

Phytoplankton may be classified variously from different angles which are discussed here:

(a) On the basis of size, the phytoplankton may be grouped under five categories (Table 5.4).
(b) Phytoplankton can also be classified on the basis of the cell characteristics (Table 5.5).

Table 5.1 Common diatoms in marine and estuarine ecosystem

S. no.	Species	Microscopic view
1.	*Coscinodiscus eccentricus*	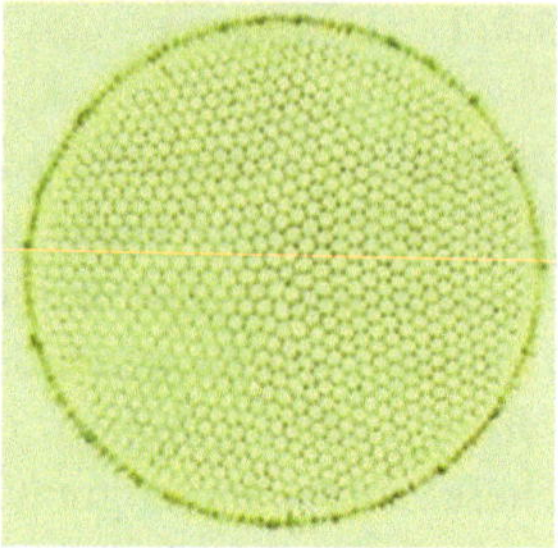
2.	*Coscinodiscus jonesianus*	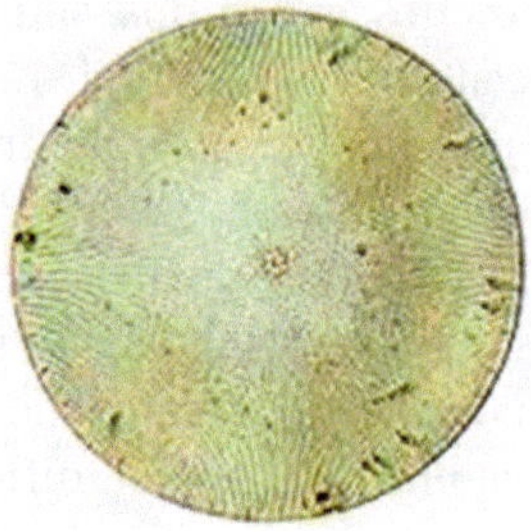
3.	*Coscinodiscus lineatus*	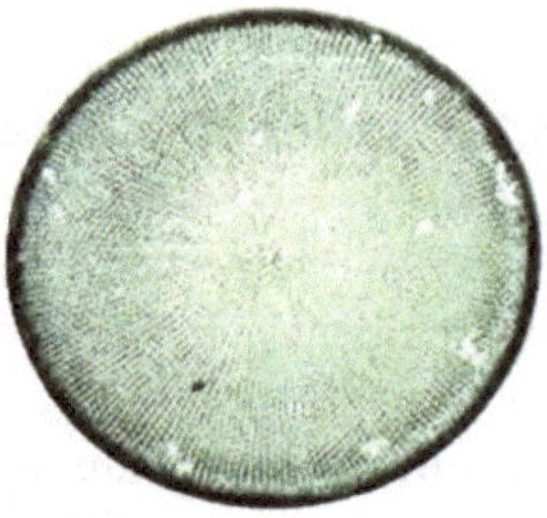
4.	*Coscinodiscus radiates*	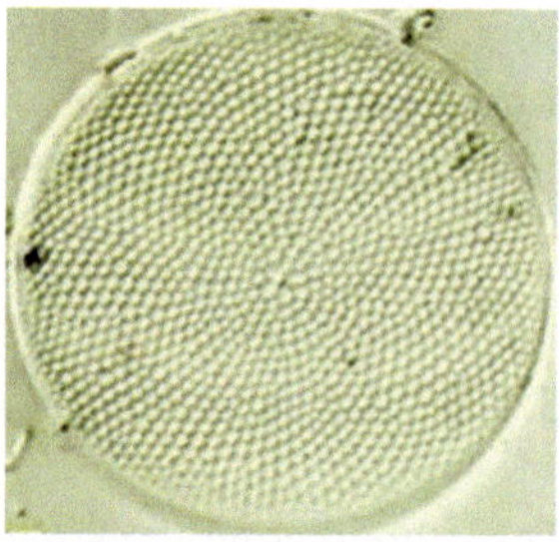
5.	*Coscinodiscus gigas*	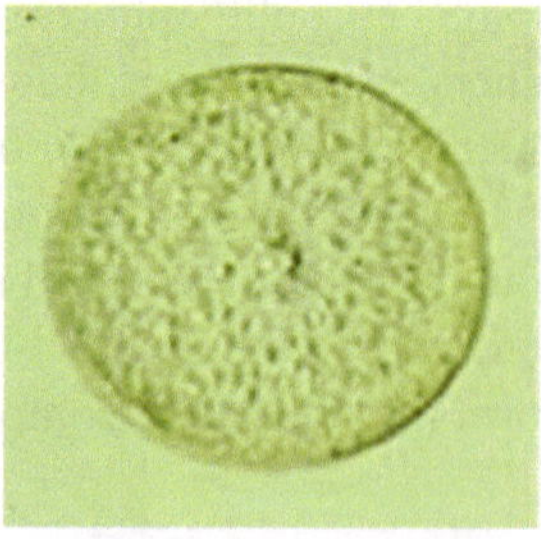

Table 5.2 Differences between centric and pennate diatoms

Point	Centric diatom	Pennate diatom
Cell shape	Discoid, solenoid or cylindrical	Elongated and fusiform, oval, sigmoid or roughly circular
Ornamentation	Radial in nature, i.e. the pattern of markings is radiating from the centre	Bilateral in nature, i.e. the markings are on either side of the apical (main) axis

5.2 Macrophytes

5.2.1 Seaweed Community

Seaweeds or macroalgae grow attached to rocks, shells or any solid objects (may be mangrove twig) (Fig. 5.10). Sometimes these are found in floating condition at the water's edge or scattered on the shore, but these are practically dislodged from their respective substrates.

Seaweeds are attached by *holdfast* that anchors the plant to a solid base or substrate. The holdfast is not a root in real sense; it does not absorb water or nutrients. Above the holdfast is a stem-like portion known as *stipe*. It acts as the flexible connection between the holdfast and the blade and ranges from very short length to about 35 m. The *blades* of the seaweeds are the photosynthetic portions, just like the leaves of the trees.

Based on the colour of their pigmentation, the seaweeds are broadly classified into various categories like Chlorophyceae (green), Phaeophyceae (brown), Rhodophyceae (red), etc.

Chlorophyceae These types of seaweeds are green in colour, owing to the presence of chlorophyll. They are abundantly distributed in the littoral zone of the marine ecosystem. Some examples of green seaweeds are *Codium* sp., *Enteromorpha intestinalis* (Fig. 5.11), *Enteromorpha compressa*, *Ulva lactuca*, *Chaetomorpha* sp., *Cladophora* sp., etc.

(a) *Phaeophyceae:* The brown seaweeds are distinguished by their colour, which varies from olive green through light golden to a deep shade of brown. Motile reproductive cells are commonly found in the brown algae, e.g. kelps like *Postelsia* sp., *Nereocystis* sp., *Macrocystis* sp., and *Laminaria* sp. and *Fucus* sp., *Padina gymnospora* and *Rosenvingea intricata*. Since brown algae live primarily in shallow waters or on shoreline rocks, they exhibit adaptations that protect them from the constant pounding of the waves. The body of brown algae is very flexible, which allow them to bend or orient with the wave action. Unlike red and green algae, brown algae usually do not reproduce by fragmentation because the tissues are too highly specialized to regenerate new parts. A major exception to this rule is *Sargassum* sp. (Fig. 5.12).

(b) *Rhodophyceae:* These seaweeds posses bright pink colour caused by the biloprotein pigments, r-phycoerythrin and r-phycocyanin. The freshwater forms however are of bluish-green in colour. Red algae produce large amount of polysaccharides around their cells. Several of these polysaccharides are commercially important. The majority of the marine forms occur from low-tide marks to greater depths up to 100 m beneath the surface of the sea, e.g. *Porphyra* sp., *Chondrus crispus*, *Catenella repens* (Fig. 5.13), etc.

Seaweeds have been used for centuries to treat a variety of illness. Asiatic cultures used seaweeds as long as 300 B.C. to treat glandular disorders, such as goitre. A goitre is an enlargement of the thyroid, a gland located in the neck. The thyroid gland requires iodine to produce its secretions, and when an individual's diet is deficient in iodine, the gland frequently enlarges. Since many seaweeds concentrate iodine from seawater in their tissues, consumption of seaweed treats the condition.

Table 5.3 Common dinoflagellates in marine and estuarine waters

Sl. no.	Species	Microscopic view
1.	*Prorocentrum gracile*	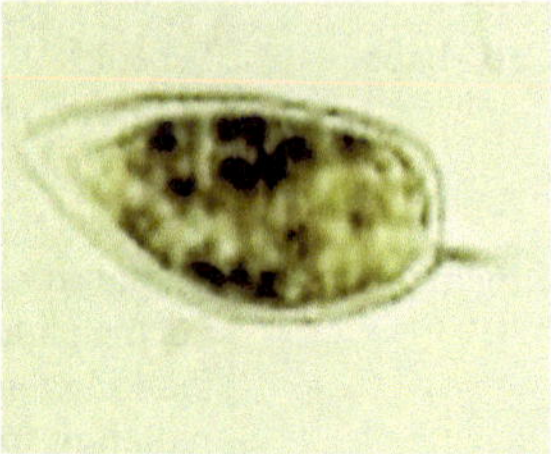
2.	*Prorocentrum micans*	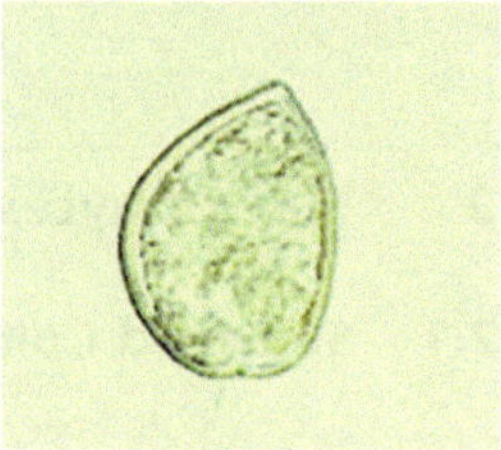
3.	*Prorocentrum concavum*	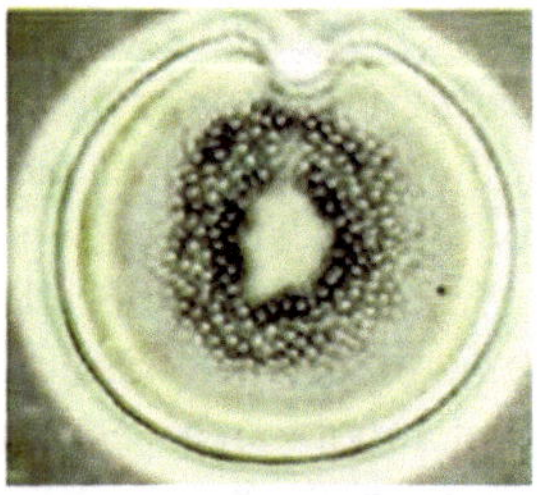
4.	*Ceratium furca*	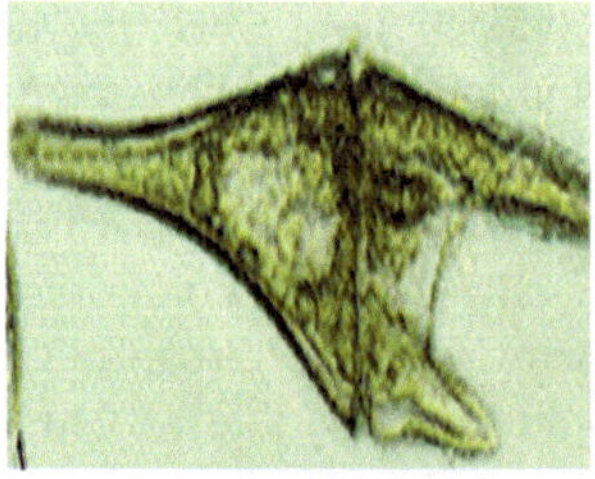
5.	*Ceratium fusus*	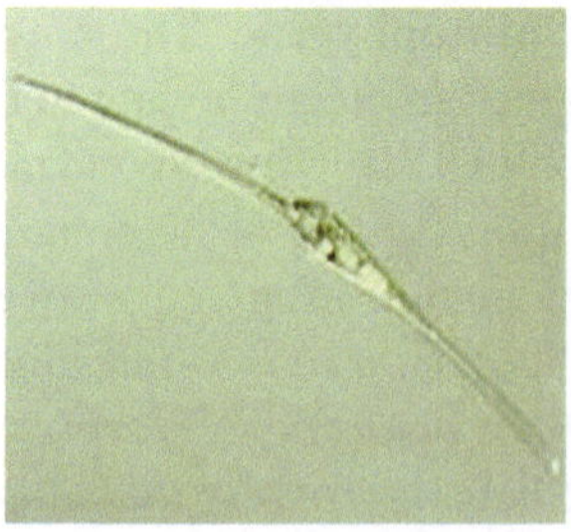

The ancient Romans used seaweeds for treating burn and rashes and healing wounds. The slimy mucilage that coats the blades of many seaweeds effectively blocks air and microorganisms from reaching an affected area. This relieves some of the discomfort, helps prevent infection and promotes healing. The red alga *Porphyra* contains vitamin C and was used by English sailors to prevent scurvy, a disease caused by vitamin C deficiency. *Porphyra* was more readily available than citrus fruits during ocean voyages and did not spoil as quickly. Several species of red algae were used to eliminate parasitic worms from the intestines of affected individuals. Kainic acid, which is extracted from the red alga *Digenia*, is still used for this purpose.

In the past, polysaccharides called phycocolloids (fy-koh-KAHL-loydz) that were isolated from red algae were used to treat a variety of intestinal ailments. Phycocolloids dissolve slowly and are not digested. These characteristics allow them to coat the lining of the stomach and intestines so that material in the digestive tract, such as acids, will not irritate it, thus alleviating some of the discomfort related to ulcers and stomach aches. Phycocolloids were used to treat constipation because they promote retention of fluid in the large intestine, which helps make faecal material easier to move. Today products from red algae, such as agar and carrageenan, are used in the treatment of ulcers. The pharmaceutical industry uses a variety of polysaccharides from red algae in the coating of pills and the production of time-release capsules.

Probably the most important algal product used in medicine and research today is agar, which is used for culturing microorganisms.

Table 5.4 Classification of phytoplankton on the basis of size

Plankton category	Maximum dimension (μm)
Ultraplankton	<2
Nanoplankton	2–20
Microplankton	20–200
Macroplankton	200–2000
Megaplankton	>2000

Table 5.5 Classification of phytoplankton on the basis of cell characteristics

Class	Common name	Area(s) of predominance	Common genera
Cyanophyceae (cyanobacteria)	Blue-green algae	Tropical	*Oscillatoria, Synechococcus*
Rhodophyceae	Red algae	Cold temperate	*Rhodella*
Cryptophyceae	Cyptomonads	Coastal	*Cryptomonas*
Chrysophyceae	Chrysomonads Silicoflagellates	Coastal Cold waters	*Aureococcus* *Dictyocha*
Bacillariophyceae (Diatomophyceae)	Diatoms	All waters, *esp*. coastal	*Coscinodiscus, Chaetoceros, Rhizosolenia*
Raphidophyceae	Chloromonads	Brackish	*Heterosigma*
Xanthophyceae	Yellow-green algae	Brackish	Very rare
Eustigmatophyceae	Yellow-green algae	Estuarine	Very rare
Prymnesiophyceae	Coccolithophorids Prymnesiomonads	Oceanic Coastal	*Emiliania* *Isochrysis* *Prymnesium*
Euglenophyceae	Euglenoids	Coastal	*Eutreptiella*
Prasinophyceae	Prasinomonads	All waters	*Tetraselmis* *Micromonas*
Chlorophyceae	Green algae	Coastal	Rare
Pyrrophyceae (Dinophyceae)	Dinoflagellates	All waters, esp. warm	*Ceratium* *Gonyaulax* *Protoperidinium*

Fig. 5.10 Seaweed on a mangrove twig

Fig. 5.11 *Enteromorpha intestinalis*: a *green* seaweed

Fig. 5.12 *Sargassum* sp.: a common *brown* seaweed

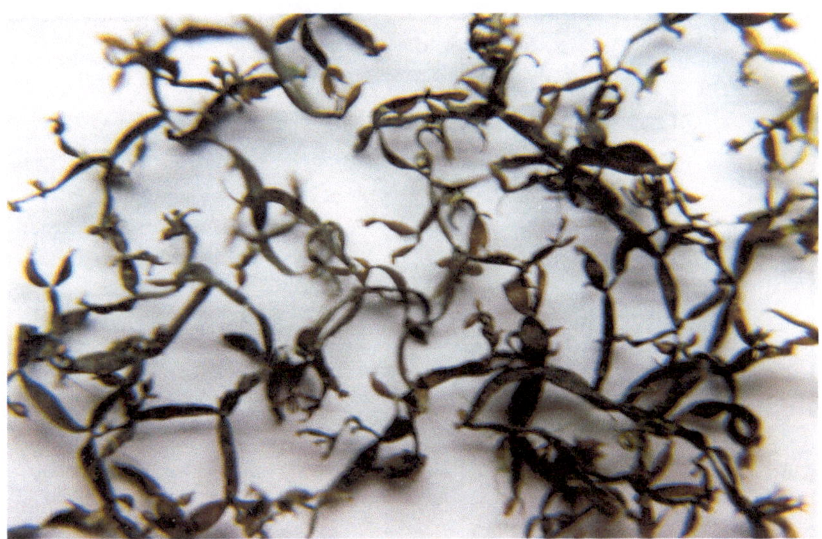

Fig. 5.13 *Catenella repens*: a *red* seaweed

Agar can withstand high temperatures so it can be sterilized. It is porous so that it allows the movement of nutrients. It is solid at room temperature and resists decomposition by most microorganisms. These characteristics make agar an ideal medium for growing bacteria and fungi for study and research.

Seaweed beds also act as significant sink of carbon dioxide. A study conducted by Chung et al. (2013) shows that seaweed beds can remove substantial amount of carbon dioxide. The researchers observed that a pilot coastal carbon dioxide removal belt (*CCRB*) firm can draw approximately 10 tons of CO_2/ha/year. Increment in biomass accumulation served as success indicator of the process. There was also a simultaneous decrease in the amount of dissolved inorganic carbon in the ambient water column. In order to translate this unique ecosystem service of seaweed into reality, the researchers also designed a mid-water rope culture system for pilot CCRB firm along the southern coast of Korea (Fig. 5.14).

Taxonomic position and salient features of some common seaweed found in the mangrove ecosystem of Indian Sundarbans are listed here.

Type I

Systematic Position
Division—Chlorophyta
Class—Chlorophyceae
Order—Ulvales
Family—Ulvaceae
Genus—*Enteromorpha*
Species—*intestinalis*

Salient Features

1. Plant body is tubular, more or less compressed, constricted and coiled in the form of intestine.
2. Thallus dark green in colour and found attached to the substratum with the help of primary attaching cell.
3. Presence of numerous multinucleated rhizoids growing from lower cell of the thallus.
4. Cells of the thallus are small and elongated.

Type II

Systematic Position
Division—Chlorophyta
Class—Chlorophyceae
Order—Ulvales
Family—Ulvaceae
Genus—*Enteromorpha*
Species—*compressa*

Salient Features

1. Plant body is tubular, more or less compressed and constricted.

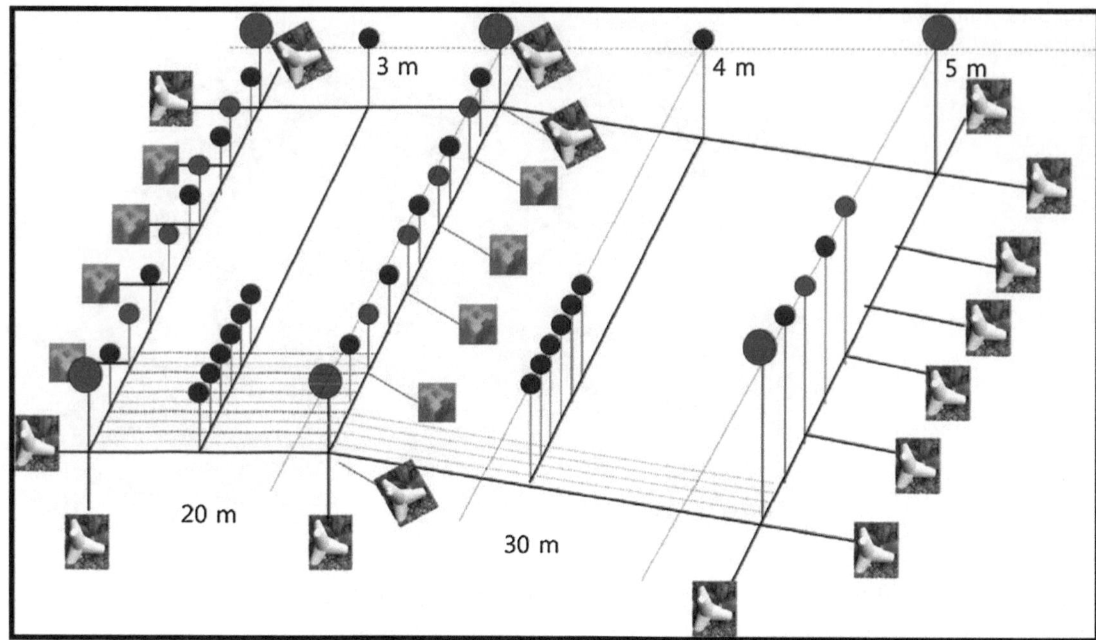

Fig. 5.14 Schematic structural diagram of mid-water rope-culture system for pilot seaweed CCRB farm installed along southern coast of Korea

2. Thallus light green in colour and found attached to the substratum with the help of primary attaching cell.
3. Presence of numerous multinucleated rhizoids growing from lower cell of the thallus.
4. Cells of the thallus are small and round.

Type III

Systematic Position
Division—Chlorophyta
Class—Chlorophyceae
Order—Ulvales
Family—Ulvaceae
Genus—*Ulva*
Species—*lactuca*

Salient Features
1. Plant body is tubular, more or less compressed, flattened leaf-like.
2. Thallus dark green in colour and found attached to the substratum with the help of primary attaching cell.
3. Presence of numerous multinucleated rhizoids growing from lower cell of the thallus.
4. Cells of the thallus are small and ovoid.

Type IV

Systematic Position
Division—Chlorophyta
Class—Chlorophyceae
Order—Cladophorales
Family—Cladophoraceae
Genus—*Rhizoclonium*
Species—*hookeri*

Salient Features
1. Filaments rigid, dark green in colour.
2. Thallus simple, intertwined to form a fleecy layer with numerous rhizoids.
3. Branches are formed at right angles to the main axis.
4. Cells measure about 170–82 μ diameter and 72–105 μ long.

Type V

Systematic Position
Division—Chlorophyta
Class——Chlorophyceae
Order—Cladophorales
Family—Cladophoraceae

Genus—*Rhizoclonium*
Species—*riparium*

Salient Features

1. Filaments pale green, expanded on the substrate.
2. Filaments flexuous, intertwined into a fleece with thin cell wall.
3. Frequent rhizoid branches.
4. Vegetative cells measure about 23–26 μ; diameter usually 1–2 times as long as broad.

Type VI

Systematic Position
Division—Chlorophyta
Class—Chlorophyceae
Order—Cladophorales
Family—Cladophoraceae
Genus—*Chaetomorpha*
Species—*aerea*

Salient Features

1. Filaments are unbranched.
2. Filaments 50–55 μ in diameter and 1–1.5 times long.
3. Rhizoids present at the base are unbranched and form disc-like in appearance.
4. Cell wall thick, hyaline and lamellate.

Type VII

Systematic Position
Division—Chlorophyta
Class—Chrysophyceae
Order—Vaucheriaceae
Family—Cladophoraceae
Type—*Vaucheria* sp.

Salient Features

1. Thallus filamentous, aseptate and laterally and irregularly branched.
2. Filaments usually attached by branched rhizoids.
3. Filaments cylindrical with numerous small chloroplasts towards the exterior boundering the central vacuole.
4. Chloroplasts are without pyrenoids.

Type VIII

Systematic Position
Division—Chlorophyta
Class—Rhodophyceae
Order—Gigartinales
Family—Rhabdoniaceae
Genus—*Catenella*
Species—*repens*

Salient Features

1. Plants with repent and assurgent branches.
2. Branching is ditrichotomous below but clearly pinnate above.
3. The axis and branches divided into dorsiventrally compressed ellipsoid to ovate segment three to five times longer than broad.
4. The haptera terminating into uncorticated flagellar outgrowth chiefly formed at points of forking and not in the plane of branching of the thallus.

Type IX

Systematic Position
Division—Chlorophyta
Class—Rhodophyceae
Order—Ceramiales
Family—Rhodomelaceae
Type—*Bostrychia* sp.

Salient Features

1. Filiform black or dull purplish.
2. Thallus with erect branches often distinguishable.
3. Rhizoids polysiphonous, ordinarily regular and bilaterally branched.
4. Branches with several cells of equal length being disposed about the central axis or these pericentral cells regularly transversely divided.

Type X

Systematic Position
Division—Chlorophyta
Class—Rhodophyceae
Order—Ceramiales
Family—Delesseriaceae

Genus—*Caloglossa*
Species—*leprieurie*

Salient Features

1. Plants dorsiventral, spreading or somewhat erect up to 4 cm across deep reddish-violet in colour.
2. Blades 1.5 mm broad, constricted at the forking and elsewhere.
3. Individual segments are lanceolate, 4–6 mm long, sometimes linear-attenuate, rarely ovate.
4. Rhizoids and secondary segments or blades formed at the constrictions; blades formed here are irregularly branched.

The macroalgae are noted for their primary production. Efficiencies of solar energy trapped showed a maximum in *Enteromorpha intestinalis* (0.64 %) and *Ulva lactuca* (0.43 %) with an average of 0.35 % by this group. A research conducted on this aspect indicates that in the deltaic complex of Indian Sundarbans, *Enteromorpha intestinalis* and *Ulva lactuca* are the most productive species, followed by *Enteromorpha prolifera* and *Rhizoclonium grande* (Chaudhuri and Choudhury 1994). The gross and net primary productions and energetics of benthic macroalgae in this mangrove-dominated ecosystem are highlighted in Tables 5.6 and 5.7.

5.2.2 Mangroves

Mangroves are the salt-tolerant forest ecosystems found mainly in the tropical and subtropical intertidal regions of the world. They consist of swamps, forestland within and its surrounding waterbodies. It is a matter of great surprise that mangrove floral species thrive luxuriantly in saline habitat, which is basically physiologically dry in nature. The mangroves not only stabilize the shoreline and act as a bulwark against the encroachment by the sea, but they also act as the abode of several species of finfish, nursery of a wide range of finfish and shellfish juveniles and biopurifying matrix of wastes generated as a result of industrialization and urbanization. In mangrove ecosystem, different kinds of unrelated fauna and flora get themselves adapted to thrive under the influence of tidal inundation and brackish water. Thus this ecosystem acts as the zone of adaptive convergence, which is a vital issue in the sphere evolutionary biology.

Mangroves are circumtropical in distribution, and this forest community occupies approximately 75 % of the total tropical coastline. Northern extension of this coastline occurs in Japan (31°22′N) and Bermuda (32°20′N), whereas southern extensions are in New Zealand (38°03′S), Australia (38°45′S) and on the east coast of South Africa (32°59′S). Globally, mangroves are distributed in 112 countries and territories. It is interesting to note that mangrove

Table 5.6 Gross primary production (GPP) and energetics of benthic macroalgae

Species	GPP gC/m^2/day	Glucose g/m^2/day	Energy Kcal/m^2/day	Efficiencies (%)
Chlorophyceae				
Enteromorpha intestinalis	3.84	9.60	35.91	0.64
E. prolifera	2.10	2.75	10.29	0.19
Ulva lactuca	2.54	6.35	23.75	0.43
Rhizoclonium grande	0.84	2.10	7.86	0.14
Rhodophyceae				
Bostrychea radicans	2.26	3.15	12.78	0.21
Bostrychea sp.	0.61	2.53	5.72	0.10
Catenella nipae	2.14	2.85	10.66	0.19
C. adnata	0.24	0.60	2.25	0.04
C. leprieurii	0.27	0.68	2.55	0.05
Gracilaria verrucosa	0.96	2.40	8.98	0.16
Total	**12.80**	**32.01**	**119.72**	**2.14**
Mean	**2.28**	**3.20**	**12.97**	**0.22**

Table 5.7 Net primary production (NPP) and energetics of benthic macroalgae

Species	NPP g/m^2/day	Glucose g/m^2/day	Energy Kcal/m^2/day	(%) of GPP	Net efficiencies
Chlorophyceae					
Enteromorpha intestinalis	3.26	8.150	30.48	85.00	0.55
E. prolifera	0.98	2.450	9.14	89.09	0.16
Ulva lactuca	2.25	5.630	22.04	88.58	0.38
Rhizoclonium grande	0.32	0.800	2.99	80.01	0.05
Rhodophyceae					
Bostrychea radicans	0.86	2.154	8.04	68.25	0.15
Bostrychea sp.	0.38	0.950	3.56	62.29	0.06
Catenella nipae	0.86	2.150	8.04	75.43	0.15
C. adnata	0.12	0.300	2.12	50.00	0.02
C. leprieurii	0.21	0.530	2.97	77.77	0.03
Gracilaria verrucosa	0.62	2.550	5.80	64.98	0.10
Total	**9.86**	**24.660**	**92.18**	**76.99**	**2.65**
Mean	**0.97**	**2.470**	**9.22**	**77.18**	**0.17**

plants are not native to the Hawaiian Islands—six species have been introduced there since the year 1900. The mangrove diversity is more in Southeast Asian countries (Fig. 5.15). The region holds nearly 75 % of the world's mangrove species with the highest species diversity found in Indonesia with 45 species, followed by Malaysia (36 species) and Thailand (35 species). India is no less in terms of the number of mangrove species (34 species of true mangroves) and hence is considered as one of the mega biodiversity countries in the world.

The total global coverage of mangroves has been variously estimated as 14–15 million hectares (Schwamborn and Saint-Paul 1996), 10 million hectares (Bunt 1992) and 24 million hectares (Twilley et al. 1992). Spalding et al. (1997) gave a recent estimation of global mangrove coverage around 18 million hectares with 42.4 % in South and Southeast Asia and additional 23.5 % in Indonesia.

The 15 most mangrove-rich countries and their cumulative percentage are highlighted in Table 5.8.

In the Indian Ocean region, the mangroves are found in a variety of coastal settings, ranging from arid areas through estuaries, lagoons and deltas to coastal fringes. The functional types of mangroves in the Indian Ocean region are:

1. Over-wash mangrove forests: small mangrove islands, frequently over-washed by the tides.
2. Fringing mangrove forests: found along the waterways, influenced by daily tides.
3. Basin mangrove forests: stunted mangroves, located in the interior of swamps.
4. Hammock mangrove forests: similar to basin type, but existing in more elevated sites.
5. Scrub mangrove forests: dwarf stands of mangroves, existing on flat coastal fringes.

The sheltered coasts support a luxuriant growth of mangroves and a higher biodiversity and this is because of the favourable conditions such as muddy sediment, frequent water exchange, high rainfall and high humidity, prevailing in the areas. The best examples are mangroves of Sundarbans (India and Bangladesh), Malaysia and Indonesia. In contrast, the arid regions of Arabian Gulf countries, Pakistan and Gujarat (India), where the sediment is sandy, highly saline and poor in nutrients, have only dwarf mangrove stands.

The mangrove forests are highly productive ecosystems with productivity about 20 times more than the average oceanic production (Gouda and Panigrahy 1996). Moreover, it is a 'detritus-based' ecosystem unlike other coastal ecosystems, which are usually 'plankton based'.

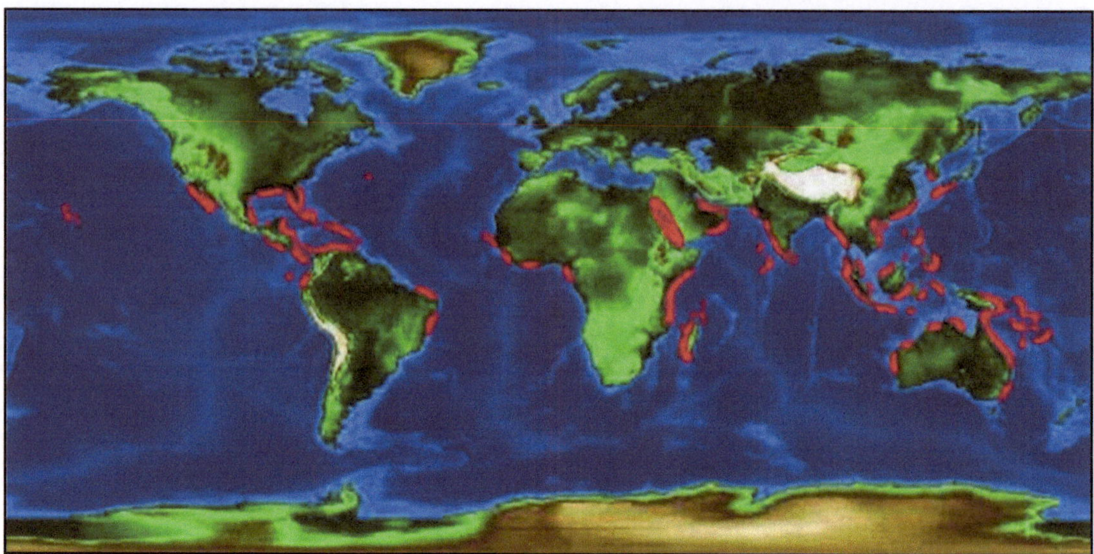

Fig. 5.15 Global distribution of mangrove diversity

Table 5.8 Fifteen most mangrove-rich countries of the world

Sl. No.	Country	Area (ha)	% of global total
1.	Indonesia	3,112,989	22.6
2.	Australia	977,975	7.1
3.	Brazil	962,683	7.0
4.	Mexico	741,917	5.4
5.	Nigeria	653,669	4.7
6.	Malaysia	505,386	3.7
7.	Myanmar	494,584	3.6
8.	Papua New Guinea	480,121	3.5
9.	Bangladesh	436,570	3.2
10.	Cuba	421,538	3.1
11.	India	368,276	2.7
12.	Guinea Bissau	338,652	2.5
13.	Mozambique	318,851	2.3
14.	Madagascar	278,078	2.0
15.	Philippines	263,137	1.9

The detritus supplied by the ecosystem saturates the ambient water with nutrients, which triggers the growth and development of planktonic community in the waterbodies on which the fishery resource is also dependent. The greatest concentration of mangrove species is observed usually at the mouth of tidal creeks and rivers where salt- and freshwater mix in ideal proportion and floodwaters deposit plenty of material to build up the banks. This unique coastal ecosystem of the world sustains a rich spectrum of floral and faunal community in and around its vicinity. The mangroves enrich the coastal waters with nutrients, yield commercial forest products, protect coastlines and support coastal fisheries (Kathiresan and Bingham 2001). Generally, the mangrove vegetations are well adapted to extreme conditions of salinity, tides, winds and temperature, although they show a preference for freshwater. There are no floral groups in the plant kingdom, which possess such well-organized and highly developed morphological, biological and physiological as well as ecological adaptations to extreme environmental conditions.

Mangrove plants tolerate salinity of the soil and water through three basic processes as listed here:

(i) *Salt excretion:* Mangrove plants take saline water as such through roots. However, in the tissues of some species of mangroves, only water molecules and essential salts are retained. Excess salts are excreted through *salt glands* that are present in the leaves. The salt-excreting species of mangrove community like *Avicennia alba*, *Aegiceras corniculatum*, *Acanthus ilicifolius*, *Aegialitis rotundifolia*, etc., regulate their internal salt levels through foliar glands. In salt-secreting (excreting) mangroves, the NaCl concentration of xylem sap is relatively high, about one tenth of the concentration of salt in seawater. So, the salt-excreting species allow more salt into the xylem than do the non-excretors but still exclude about 90 % of the salts (Scholander et al. 1962; Azocar et al. 1992). Salt is only partially excluded at the roots. The absorbed salt is primarily excreted metabolically via specialized salt glands in the leaves. The salt in solution can crystallize by evaporation and either can be blown away or washed off. Since, in salt-excreting mangroves, superfluous salts are excreted by guttation through special salt glands, all these salt-excreting halophytes are often referred to as *crinohalophytes*.

   It is interesting to note that salt excretion is an active process, as evidenced by ATPase activity in the plasmalemma of the excretory cells (Drennan and Pammenter 1982). The process is probably regulated by leaf hypodermal cells, which may store salt as well as water (Balsamo and Thomson 1995).

(ii) *Salt exclusion:* In some of the mangrove plants, the roots possess an *ultrafiltration* mechanism called *reverse osmosis* by which water and salts in the seawater are separated in the root zone itself and only water is taken inside and the salts are rejected. Many mangrove species can exclude 90 % of salt in the ambient seawater or estuarine system (http://www.epa.qld.gov.au/nature_conservation/habitats/mangroves_andwetlands/mangroves). *Rhizophora mucronata*, *Ceriops decandra*, *Bruguiera gymnorrhiza*, *Kandelia candel*, etc., are few salt excluders of mangrove community. Scholander (1968) demonstrated experimentally that the salt separation process in mangroves occurs at or near the root surface. This is mediated by physical processes alone, since it is not inhibited by poisons or high temperature, which may cause an inhibitory effect on metabolic process. In the root area the physical mechanism for salt separation involves ultrafiltration which occurs either at the root surface (epiblema) or at the root endodermis. However, the latter region might be the most preferable site (Tomlinson 1986) because the ultimate absorbing roots in most of the mangroves lack root hairs (e.g. capillary rootlets of *Rhizophora* sp.). This indicates that the absorbing area of mangroves is reduced in comparison to non-mangrove plants.

(iii) *Salt accumulation:* In this type of mangrove plants, the species possess neither salt glands nor ultrafiltration system, but they have the capacity to accumulate a large amount of salts in their leaves. This imparts succulence to their leaves. *Sonneratia apetala*, *Lumnitzera racemosa*, *Excoecaria agallocha*, *Sesuvium portulacastrum*, *Suaeda maritima* and *Suaeda nudiflora* are included in this category. Leaf succulence in mangroves has a simple explanation in terms of salt balance. The osmotic potential of the leaf cells of mangroves is high (Scholander et al. 1964) which is essential if mangroves are to draw water from the sea with its high negative water potential. However, Scholander (1968) noted that the salt concentration of mangrove leaves remain constant and independent of age. Measurement of salt content in xylem sap demonstrates incomplete salt exclusion at the roots. But mangroves accumulate salt, and so this accumulation is partly compensated by salt glands, mainly in the less efficient salt excluders. Since salt concentration is constant and independent of leaf age, salt must accumulate by an increase in the volume of the leaf cells inducing succulence. The leaf succulence in mangroves may therefore be accepted as a part of their adaptation in an environment that provides ample water at the expense of some compensation for high aquatic salinity.

Studies on salt tolerance in *Aegiceras corniculatum* and *Sesuvium portulacastrum* generated a few interesting findings: (i) NaCl salinity has considerable effect on the degree of succulence. With the increase of NaCl salinity in the ambient media, the mass and volume of the leaves increases due to increment in the water content. (ii) Effect of NaCl and Na_2SO_4 is more pronounced in *Sesuvium* sp. than *Aegiceras* sp. (iii) In *Sesuvium* sp., effect of chloride salinity is more prominent than that of sulphate salinity. (iv) NaCl is the most effective salt in promoting succulence (Van Eijk 1939). Succulence is due to expansion of the cell wall leading to increase in the size of cells. (v) Accumulation of NaCl is more in *Sesuvium* than *Aegiceras* due to their difference in the mode of salt regulation. (vi) Chlorophyll content decreases sharply at high concentration of NaCl in both the plant species. (vii) High concentration of Na_2SO_4 stimulates the synthesis of chlorophyll in *Aegiceras corniculatum* but inhibits the same in *Sesuvium* sp.

Some common mangrove species (true mangroves) abundantly found in Indian Sundarbans along with their identifying features and ecosystem services are highlighted in Table 5.9.

The growth, distribution and survival of mangroves are primarily regulated by salinity, although edaphic factors play important roles in the process. A detailed study conducted on five dominant mangrove species in Indian Sundarbans (Vide Annexure 5B) clearly confirms the influence of aquatic salinity on the biomass, carbon content and litter biomass of mangrove species. The study also reflects difference in salinity tolerance among the true mangrove species, which can be used as a road map in the scenario of sea level rise and subsequent alteration of salinity.

5.2.3 Seagrass and Salt Marsh Grass

Seagrasses are specialized marine flowering plants that are found abundantly in the nearshore region (Fig. 5.16). They are marine plants that generally inhabit the protected shallow waters of temperate and topical coastal areas. These plants are not true grasses, and they present several genera that appear to be more closely related to members of the lily family. The most extensive areas of seagrasses are found in the tropics. They are concentrated in two major regions of the Indo-Pacific, as well as both coasts of the Americas. Seagrasses do not thrive in areas of low light intensity. If the water in a seagrass bed becomes too turbid (clouded with sediments), it can destroy the seagrasses and their dependent organisms. A massive die-off of seagrasses took place in Maryland's Chesapeake Bay estuary during the 1960s, largely due to diminished light associated with excessive sediment run-off from the land surrounding the bay.

Seagrasses are hydrophytes, which means they generally live submerged beneath the water. To survive in their subtidal habitat, these plants must be adapted to a saline environment with wave action and tidal currents. They must also be able to carry out pollination and seed dispersal underwater.

The major stems of seagrasses, called rhizomes or long shoots, grow horizontally,

Table 5.9 Common mangroves of Indian Sundarbans

Mangrove species	Identifying features	Ecosystem services
 *Aegiceros corniculatum* **Common name:** Khalsi	1. Shrub tolerant to high saline areas 2. Bark reddish-brown with leaves elliptical, leaf-tip notched, cuneate at base 3. Fruit green to reddish in maturation, sharply curved 4. Fragrant white flowers, curved yellow	1. Fruit is poisonous to fish due to presence of saponin and tannin and hence used in aquacultural farms to kill crabs and unwanted predatory fishes 2. Bark is a good source of tannin 3. The flowers of these trees are good sources of high-quality honey
 *Avicennia alba* **Common name:** Kalo baen	1. Trees are tolerant to high salinity, pneumatophores spongy, narrowly pointed with slender stilt roots 2. Bark dark brown or even black 3. Leaves lanceolate to elliptical, leaf-tip acute, lower surface silver grey to white 4. Curved fruit with relatively long beak	1. Used for fodder and fuel 2. Paste prepared from seeds helps to relieve small pox ulceration 3. Wood ash after burning wood is given in dysentery 4. Acts as unique agent of bioremediation by accumulating Zn, Cu and Pb in the body tissues (Interested readers/ researchers may Vide Annexure 5A)
 *Avicennia marina* **Common name:** Piara baen	1. Trees are tolerant to high salinity 2. Pneumatophores pencil like 3. Bark yellowish brown 4. Leaves elliptical, leaf-tip rolling, lower surface white to light grey 5. Inflorescence terminal or axillary, orange yellow in colour	1. Bark decoction applied on "Haza" (a kind of sore due to wet sand) for curing purpose 2. Paste prepared from seeds helps to relieve small pox ulceration 3. Resinous exude is used for birth control purposes 4. Efficient in bioaccumulation of Zn in the body tissues

(continued)

Table 5.9 (continued)

Mangrove species	Identifying features	Ecosystem services
 *Avicennia officinalis* **Common name:** Sada baen	1. Trees are tolerant to high salinity, pneumatophores pencil like 2. Bark yellowish brown. Leaves elliptical, leaf-tip rolling, lower surface white to light grey 3. Inflorescence terminal or axillary, orange yellow	1. Used for firewood 2. Ash obtained after burning the dried bark is used as antacid 3. Paste of green (raw) fruits is applied on boils and skin diseases (like scabies) for the purpose of curing 4. Efficient bioaccumulator of heavy metals like Cu and Pb
 *Sonneratia apetala* **Common name:** Keora	1. Trees are with long, corky, forked pneumatophores 2. Stem light brown in colour 3. Leaves thick, coriaceous, narrowly elliptic oblong tapering towards apex 4. Flowers are cream coloured in axillary cymes with globose berry seated in flattened calyx tube	1. Wood used in house building and packing cases 2. Yields excellent fuel 3. Fruit contains 28/100 g of vitamin C and hence helps to fight cough and cold 4. Efficient bioaccumulator of heavy metals (Zn, Cu, Pb and Mn)
 *Bruguiera gymnorrhiza* **Common name:** Kankra	1. Trees generally found on elevated interior parts of mangrove forest with prominent buttress roots 2. Bark dark grey. Leaves simple, elliptical-oblong, leathery and leaf-tip acuminate 3. Flowers axillary, single with red calyx, red in colour and almost 16 lobed; fruits are cigar shaped, stout and dark green	1. Bark decoction with coconut oil is applied to cure dermatological problems 2. Unique source of tannin (36 %) 3. Wood used for house post 4. Yields excellent fuel

(continued)

Table 5.9 (continued)

Mangrove species	Identifying features	Ecosystem services
 *Ceriops decandra* **Common name:** Goran	1. Stilt roots arising from pyramidal stem base 2. Light grey barked stem 3. Leaves elliptic-oblong, emarginate at apex, cuneate at base 4. Flowers axillary in condensed cymes 5. Fruit is berry, dark red when mature, warty towards tip, ridged, not hanging down	1. Source of tannin (29 %) 2. Provides fuel charcoal 3. Used for boat building 4. Obstetric and haemorrhage cases are treated with an infusion of *Ceriops* bark
 *Excoecaria agallocha* **Common name:** Genwa	1. Prominent main root absent, many laterally spreading snake like roots producing elbow-shaped pegs 2. Bark greyish. Poisonous milky latex highly irritating to eyes 3. Leaves light green with wavy margin 4. Catkin inflorescence terminal or axillary, orange yellow	1. Latex applied on leprotic wounds 2. Leaf decoction is claimed as medicine in epilepsy 3. Heart wood and pneumatophores give scent, but not commercially used 4. Efficient bioaccumulator of heavy metals from the ambient media
 *Heritiera fomes* (Sterculiaceae) **Common name:** Sundari	1. Trees with numerous peg-like pneumatophores and blind root suckers 2. Young branches covered with shining golden-brown scales 3. Leaves elliptic with lower surface shining with silvery scales 4. Flowers golden yellow with reddish tinge inside and fruits sub-globose, woody, indehiscent with longitudinal and transverse ridges	1. Wood fetches a high price in the international market, as it is of demand for making furniture 2. Indicator of salinity profile of an area 3. Source of tannin (11 %). Rich in astaxanthin

(continued)

Table 5.9 (continued)

Mangrove species	Identifying features	Ecosystem services
Nypa fruticans **Common name:** Golpata	1. Palm tree like appearance with no aerial roots 2. Leaves lanceolate, palm leaves arising from root stock, leaf tip acute 3. Flowers female in globose head, males in catkin-like, brick red to yellow; fruit dark brown or brick red, globose, pericarp fleshy, fibrous	1. Ash prepared after burning tender leaves is used in stomach ulcers 2. Dried leaves are often used for making roofs in thatched houses 3. Source of alcohol 4. Rich in astaxanthin

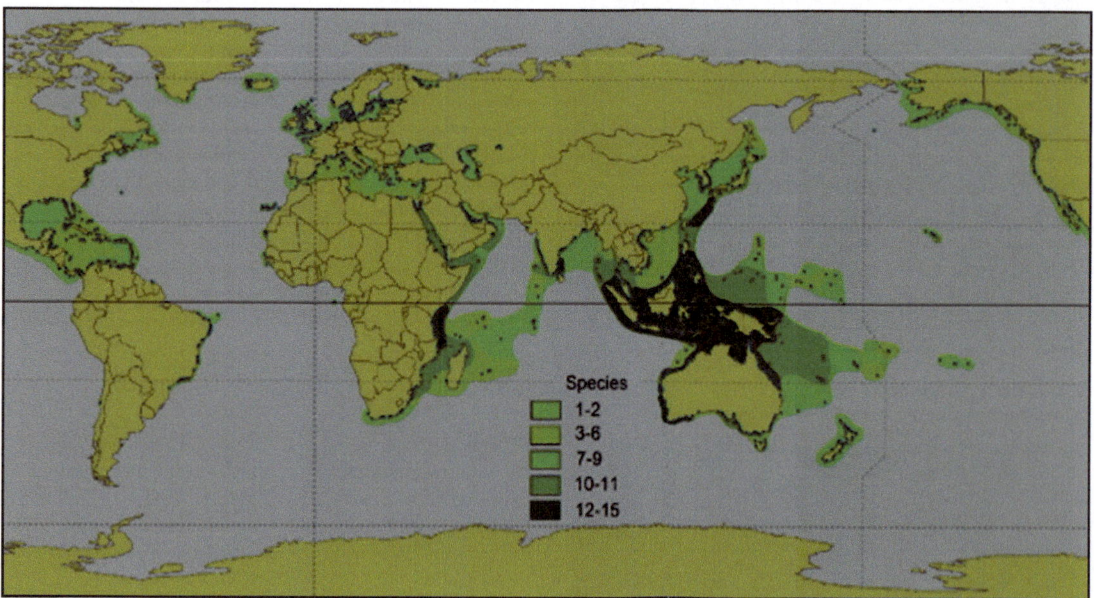

Fig. 5.16 Global distribution of seagrass

usually just below the surface of the bottom sediments. The rhizomes and roots of seagrasses help to stabilize the bottom and, along with the leaves, help trap large amounts of sediment. The leaves are either flat or oval and ribbon shaped or cylindrical and are flexible to better withstand water movement while remaining erect. Leaves either develop directly from the rhizomes or from small vertical side stem called short shoots. Roots develop from both the rhizomes and the base of short shoots. Unlike the fibrous roots of terrestrial grasses, the roots of seagrasses are usually thicker and more fleshy. The flowers, normally inconspicuous, small and pale white, are usually located at the bases of the leaves. The stamens (male flower parts) and pistils (female flower parts) generally extend above the petals of the flower. As a rule, the pollen is released from the stamens in long, gelatinous strands that are carried by water currents to the pistils.

Seagrasses belong to the families Hydrocharitaceae and Potamogetonaceae. There are about 13 genera and 58 species of seagrass available all over the world. Of these, six genera (*Amphibolis, Heterozostera, Phyllospadix, Posidonia, Pseudalthenia* and *Zostera*) are mostly restricted to temperate seas and the remaining seven genera (*Cymodocea, Enhalus, Halodule, Halophila, Syringodium, Thalassia* and *Thalassodendron*) are distributed in tropical seas. Seagrass species have worldwide distribution. They are found in tropical (hot), temperate (cool) and the edge of the Arctic (freezing) regions. They thrive luxuriantly in coastal water region. Seagrasses are mostly found in patches, but these patches can expand to form huge seagrass beds or meadows. The beds may sustain one species of seagrass or multiple species.

Seagrass meadows are essential coastal ecosystems that provide many ecosystem services such as improved water quality and light availability, increases in biodiversity and habitat, sediment stabilization and carbon and nutrient accumulation (Hemminga and Duarte 2000; Orth et al. 2006; McGlathery et al. 2012). Seagrasses serve as breeding grounds and nurseries for important marine organisms. Recently, seagrass meadows have been acknowledged for their carbon storage potential, and it has been estimated that globally as much as 19.9 Pg of organic carbon are stored in seagrass meadows (Fourqurean et al. 2012). Seagrass meadows cover only 0.1 % area of the world's ocean floor yet account for 10–18 % of the total oceanic carbon burial, accumulating carbon at rates of 48–112 Tg C year^{-1} (Duarte et al. 2005; Kennedy et al. 2010; Mcleod et al. 2011).

We studied the stored carbon in three seagrass species (namely, *Cymodocea serrulata, Thalassia hemprichii* and *Halophila ovalis*) in the Gulf of Mannar region (Fig. 5.17) of India.

The Gulf of Mannar extends from Rameswaram Island in the north to Kanyakumari in the south in the southeast art of Indian subcontinent. It consists of a chain of 21 islands stretching from Mandapam to Tuticorin. The islands are located approximately 2–10 km from the mainland and are classified based on their proximity to mainland into four major groups: Tuticorin, Vembar, Keezhakarai and Mandapam groups. The region was declared a Marine Biosphere Reserve in 1989 under the Man and Biosphere Programme of UNESCO in lieu of its invaluable biodiversity and multiple-use status. The 21 islands are protected under the Gulf of Mannar Marine National Park.

It is observed that in the seagrass species collected from three stations in the Gulf of Mannar region, the ratio of AGB and BGB ranges between 1:1.2 and 1:1.30. Because of relatively more biomass in the below-ground structures, the stored carbon in the below-ground structures (BGC) is relatively more than above-ground carbon (AGC) (Tables 5.10 and 5.11).

The soil associated with the seagrass meadow is also a unique reservoir of carbon. In addition to the decomposition of the seagrass species, associated flora (blue carbon) and fauna also contribute organic matter to the intertidal mudflats. We also monitored organic carbon level in the soil associated with the seagrass species and observed considerable carbon percentage in the associated soil (Table 5.12).

The seagrass bed is often affected by physical and biological factors. Migration of sand dune often affects the coverage and growth pattern of some species of seagrass, whereas consumption by herbivores can greatly reduce the standing stock of the community. Bioturbation (e.g. feeding by stingray) also contributes greatly to the destruction of seagrass meadows, and in some areas, decapods burrowing creates favourable environment for storm events that can tear the rhizome mats of seagrass.

Globally, seagrass ecosystems are declining in area by about 5 % per year due to anthropogenic stresses, including decreased water quality and increased water temperatures (Orth et al. 2006; Mcleod et al. 2011; Waycott et al. 2009), and this decline could result in the release of large amounts of stored carbon Duarte et al. (2005). In order to partially mitigate seagrass decline, restoration in areas with

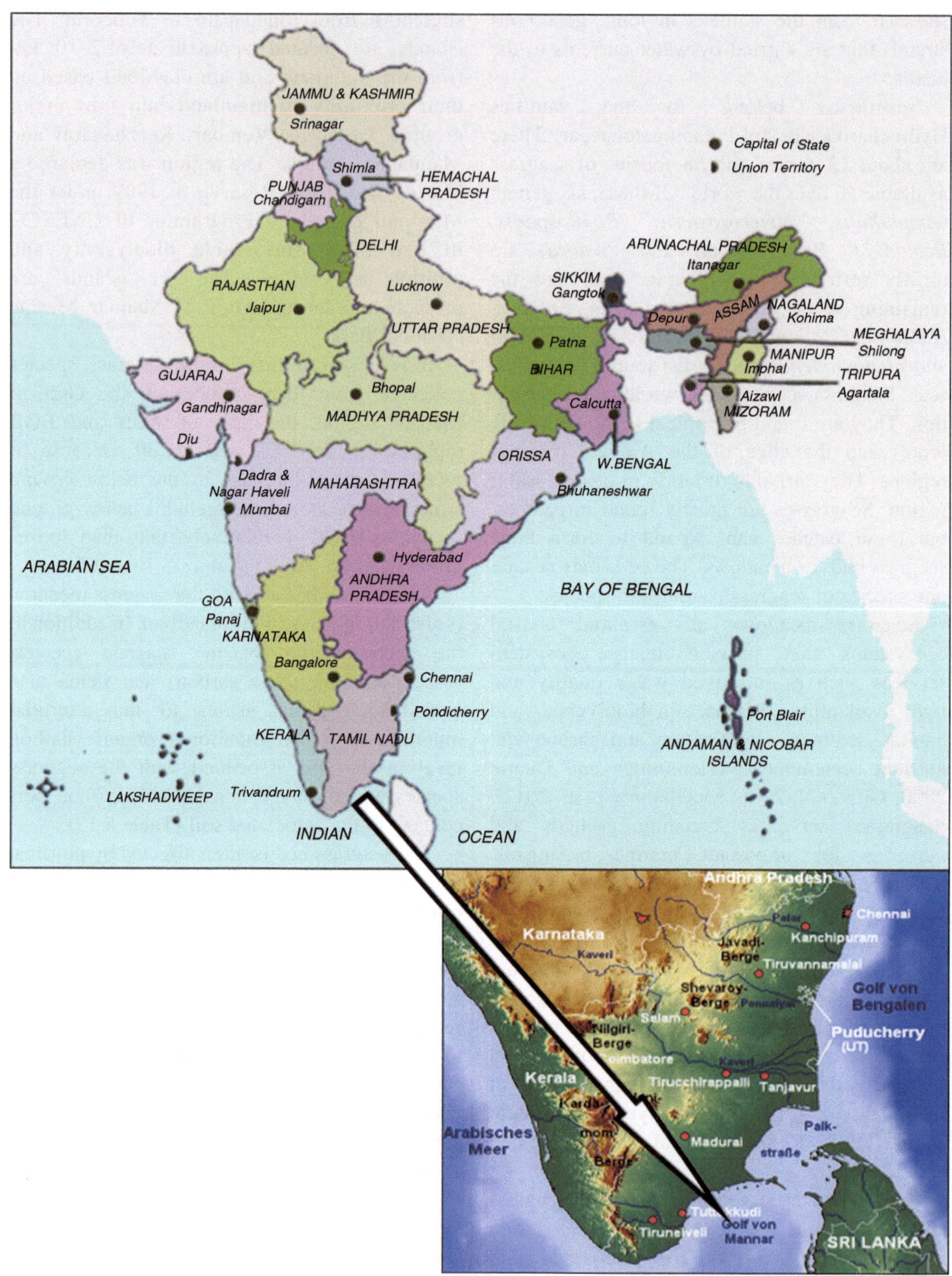

Fig. 5.17 Gulf of Mannar region

Table 5.10 Species-wise above-ground biomass (AGB) and above-ground carbon (AGC) per unit area in Gulf of Mannar

Species	AGB (gm dry wt. m^{-2})			AGC (gm dry wt. m^{-2})		
	Stn 1	Stn 2	Stn 3	Stn 1	Stn 2	Stn 3
 *Cymodocea serrulata*	115.23 (49.6 %)	109.60 (49.7 %)	101.85 (49.9 %)	57.15	54.47	50.82
 *Thalassia hemprichii*	56.98 (48.5 %)	49.79 (48.9 %)	38.64 (47.9 %)	27.63	24.35	18.51
 *Halophila ovalis*	28.87 (41.23 %)	25.19 (43.05 %)	19.49 (45.16 %)	11.90	10.84	8.80

suitable habitat is an effective option that has the potential to re-establish lost carbon stores and sinks, as well as other important ecosystem services seagrass meadows provide.

Plants such as cordgrass (*Spartina*), rushes (*Juncus)* and pickle weed (*Salicornia*) are adapted to life in salt marshes. These plants generally grow in the middle to upper intertidal zones, where they are protected from the actions of waves and tidal currents and where they are not completely submerged. Like seagrasses and mangroves, they are considered facultative halophytes, meaning that they can tolerate salty conditions as well as freshwater. Since they live in sediments with a high salt content, salt marsh plants tend to lose water to their environment by osmosis and display adaptations similar to those of desert plants to help them retain water. Both cordgrass and rushes have leaves covered by a thick cuticle to help retard water loss, and the stems and leaves contain well-developed vascular tissues for the efficient transport of water within the plant's tissues. Many of the plants that inhibit salt marshes, such as pickle weed, have thick, water-retaining, succulent leaves similar to those of plants that live in hot, arid terrestrial habitats.

Salt marsh grasses are special type of halophytes that are adapted to continual and periodic flooding. They are found primarily throughout the tropical, temperate and subarctic regions. The tide is the dominating characteristic of a salt marsh. The salinity of the aquatic phase defines the plants and animals species that can survive in the marsh area. The vertical range of the tide determines flooding depths and thus the height of the vegetation, and the tidal cycle controls

Table 5.11 Species-wise below-ground biomass (BGB) and below-ground carbon (BGC) per unit area in Gulf of Mannar

Species	BGB (gm dry wt. m^{-2})			BGC (gm dry wt. m^{-2})		
	Stn 1	Stn 2	Stn 3	Stn 1	Stn 2	Stn 3
Cymodocea serrulata	144.04 (48.4 %)	140.29 (48.6 %)	132.41 (49.1 %)	69.72	68.18	65.01
Thalassia hemprichii	71.79 (47.9 %)	63.23 (48.2 %)	49.84 (48.5 %)	34.39	30.48	24.17
Halophila ovalis	36.31 (39.69 %)	33.14 (38.43 %)	24.75 (38.11 %)	14.41	12.74	9.43

Table 5.12 Soil organic carbon (SOC) level associated with the seagrass species

Species	SOC (%)		
	Stn 1	Stn 2	Stn 3
Cymodocea serrulata	1.98	1.36	1.09
Thalassia hemprichii	1.07	0.98	0.89
Halophila ovalis	1.13	0.97	1.01

how often and how long vegetation is submerged. Two areas are delineated by the tide: the low marsh and the high marsh. The low marsh generally floods and drains twice daily with the rise and fall of the tide; the high marsh, which is at a slightly higher elevation, floods less frequently.

Plants of the salt marsh have two types of roots: one that act to anchor the plant and another, short-lived type that functions in nutrient absorption. Stems develop from rhizomes and produce the leaves. Like seagrasses, salt marsh plants usually have shallow roots and rhizomes. This arrangement helps to stabilize coastal sediments and prevent shoreline erosion.

Salt marshes usually are developed on a sinking coastline, originating as mudflats in the shallow water of sheltered bays, lagoons and estuaries or behind sandbars. They are formed where salinity is high, ranging from 20 to 30 psu. Proceeding up the estuary, there is a

Fig. 5.18 Salt marsh grass *Porteresia coarctata*: a mangrove associate species

transitional zone where salinity ranges from 20 to less than 5 psu. In the upper estuary, where river input dominates, the water has only a trace of salt. This varying salinity produces changes in the marsh—in the kinds of species and also in their number. Typically, the fewest species are found in the hypersaline zone and the greatest numbers of species are found in the freshwater tidal marsh. In Indian Sundarban region, salt marsh grass *Porteresia coarctata* grows in Nayachar Island (21°45′24″N and 88°15′24″E), where the salinity of water touches almost 2 psu during monsoon. Also the species is abundant in the intertidal mudflats of islands located in the high-saline zone like Sagar Island (21°39′04″N and 88°01′47″E), Gosaba (22°15′45″N and 88°39′46″E) and Satjelia Island (22°11′52″N and 88°50′43″E), etc., where the salinity varies between 15 and 28 psu. In this mangrove-dominated ecosystem, the species is the neighbour of mangroves (Fig. 5.18) and plays an important role to maintain stability of the islands (Mitra and Banerjee 2005). It is the pioneer species in the process of ecological succession. Studies carried out by researchers showed that salinity has a regulatory influence on the biomass and carbon content of *P. coarctata*.

The salt marsh is one of the most productive ecosystems in nature and serves as a sediment sink, a nursery habitat for finfish and shellfish species, a feeding and nesting site for waterfowl and shorebirds, a habitat for numerous unique plants and animals, a nutrient source, a reservoir for storm water, an important erosion control component and a site for aesthetic pleasures.

Brain Churners

1. Why are continental margins more productive than open oceans?
2. Do you consider chemosynthetic bacteria as producers of marine ecosystem?
3. Phytoplankton may cause harm to marine environment during bloom phase. Explain the statement.
4. Why do seaweeds under Rhodophyceae exhibit pinkish colour?
5. Why is the holdfast of seaweeds not a root in real sense?

(continued)

6. How does salinity regulate the growth and biomass of mangroves?
7. How do mangroves adapt to saline environment?
8. How does seagrass differ from salt marsh grass?
9. What are the causes of rapid decline of seagrass ecosystem?
10. Why is the mangrove ecosystem regarded as the region of adaptive convergence?

Annexure 5A: *Avicennia alba*: An Indicator of Heavy Metal Pollution in Indian Sundarban Estuaries

1. Introduction

Avicennia alba is a common mangrove floral species in Indian Sundarbans where the soil is inundated (twice a day) with saline water of 2.0–28 psu. The species can also withstand submergence with saline water for a long period of time due to its inherent capacity to tolerate high levels of salinity and submergence. The species is endowed with a unique system of ion influx–efflux regulation (Mitra et al. 2004).

The coastal zones and estuaries, which are the primary habitat of this mangrove species, are exposed to effluents from chemical industries everyday (Manahan 1994; Krishnamurti and Viswanathan 1991). As a consequence, nutrients and contaminants like heavy metals, pesticides and halogenated hydrocarbons are transported into the intertidal mudflats of mangrove forest (Delaune et al. 1990) since they are partly bound to suspended particles. The mangrove forests adjacent to the highly urbanized and industrialized cities and towns often receive large pollutant loads, including discharges of heavy metals from industry and transportation activities. In the intertidal mudflats, water-soluble metals and exchangeable metals are the most available and precipitated inorganic compounds; metal complexes with large molecular weight, humus materials and metals adsorbed to hydrous oxides are also possibly available (Gambrell 1994; Williams et al. 1994). *A. alba*, being an endemic and dominant species of the Sundarban region, is exposed to heavy metal pollution, as the region is contaminated with conservative pollutants (Mitra 1998; Mitra et al. 2011; Banerjee et al. 2012; Mitra and Ghosh 2014). On this background the present study was carried out to detect the concentrations of heavy metals in *A. alba* in three stations located in Indian Sundarbans.

2. Materials and Methods

2.1. Sampling of *A. alba*

Three stations (Fig. 5A.1) were selected in the central part of Indian Sundarbans: Canning, Stn.1 (22°18′37″N; 88°40′36″E); Chhotomollakhali, Stn.2 (22°10′21.74″N; 88°53′55.18″E); and Bali, Stn.3 (22°04′35.17″N; 88°44′55.70″E). The species collected from the selected stations during low tide were brought to laboratory, washed and dried with tissue paper and stored at −20 °C for further analysis.

2.2. Analysis of Dissolved Zn, Cu and Pb

10 l Teflon-lined Glo-Flo bottles, filled with Teflon taps and deployed on a rosette or on Kevlar line, were used for collecting surface water samples. Nucleopore filters (0.4 μm pore diameter) was used to filter the collected water samples and the aliquots of the filters were acidified with sub-boiling distilled nitric acid to a pH of about 2.0 and stored in clean polyethylene bottles of low density. Using dithiocarbamate complexation and subsequent extraction into Freon TF, followed by back extraction as per the procedure of Danielsson et al. (1978), dissolved heavy metals were separated and concentrated from the ambient water of the selected stations. Zn, Cu and Pb in the extracts were detected using atomic absorption spectrophotometer (Perkin Elmer Model 3030). The quality aspect of dissolved heavy metal determination is confirmed by good agreement

between our values and reported for certified reference seawater materials (CASS 2) (Table 5A.1).

2.3. Analysis of Biologically Available Zn, Cu and Pb

Collection of sediment samples from surface (cm depth) was done by scrapping using a precleaned and acid-washed plastic scale and stored immediately in clean polyethylene bags, which were sealed. Double-distilled water free of metal was used to wash the collected samples and oven-dried at 05 °C for 5–6 h. The samples were freed from visible shells or shell fragments, ground to powder in a mortar and stored in acid-washed polyethylene acids. As per the standard procedure of Malo (1977), the analyses of biologically available metals were done after redrying the samples, from which 1 g was taken and digested with 0.5 (N) HCl. Polyethylene containers were used to store the resulting solution. Atomic absorption spectrophotometer was used for determining the metal concentrations. In the reagent blank, no detectable trace metals were found. To assure the quality of the data, analysis of the NIES Sargasso sample was carried out that revealed least standard deviation between the observed values and certified values (Table 5A.2).

2.4. Analysis of Tissue Zn, Cu and Pb

20 g leaf and stem samples were oven-dried at 105 °C overnight to a constant weight. 1 g of dried sample of each of the vegetative part was digested with a mixture of 10 ml nitric acid and perchloric acid (3:1) following the method as outlined by Lithor (1975) till a clear solution was obtained. The resulting solution was made up to a constant volume with 0.05 N nitric acid. Each of the vegetative samples was analysed for Zn, Cu and Pb against standard concentration of each metal on a Perkin–Elmer Atomic Absorption Spectrophotometer (Model 3030) equipped with a HGA 500 graphite furnace atomizer and a deuterium background corrector. Blank correction was done to bring accuracy to the results.

2.5. Data Analysis

Statistical software SPSS 16.0 was applied to determine the interrelationships of heavy metal concentrations in *A. alba* and ambient environment by Pearson correlation coefficient analysis. Statistical significance was tested at 95 % confidence level. The computations of correlation coefficients were done considering the mean values of selected variables for 2 consecutive years ($n = 24$).

3. Results

3.1. Dissolved Heavy Metal

The dissolved heavy metals followed the order Zn > Cu > Pb irrespective of all stations. Dissolved Zn ranged from 213.64 ± 57.67 ppb (at Bali) to 551.01 ± 60.39 ppb (at Canning) during 2011 and from 232.77 ± 57.76 ppb (at Bali) to 570.14 ± 60.39 ppb (at Canning) during 2012. Dissolved Cu ranged from 79.83 ± 54.19 ppb (at Bali) to 159.47 ± 11.09 ppb (at Canning) during 2011, and in 2012 it ranged from 94.95 ± 54.19 ppb (at Bali) to 173.09 ± 12.47 ppb (at Canning). Dissolved Pb ranged from 8.41 ± 3.76 ppb (at Bali) to 35.41 ± 4.29 ppb (at Canning) during 2011 and from 16.69 ± 4.95 ppb (at Bali) to 45.18 ± 4.35 ppb (at Canning) during 2012 (Figs. 5A.2a, 5A.2b and 5A.2c).

3.2. Biologically Available Heavy Metals

Biologically available heavy metals in sediment followed the same order as that of water, i.e. Zn > Cu > Pb irrespective of all stations. Sediment Zn ranged from 40.84 ± 10.09 ppm (at Bali) to 78.01 ± 9.43 ppm (at Canning) during 2011 and from 49.87 ± 10.09 ppm (at Bali) to 87.03 ± 9.43 ppm (at Canning) in 2012. In case of sediment, Cu ranged from 10.68 ± 2.84 ppm (at Bali) to 26.86 ± 2.76 ppm (at Canning) during 2011, and in 2012 it ranged from 15.64 ± 2.84 ppm (at Bali) to 32.41 ± 3.63 ppm (at Canning). Pb in sediment ranged from 6.26 ± 1.83 ppm (at Bali) to 12.65 ± 2.33 ppm (at Canning) during 2011 and from 8.70 ± 2.01 ppm (at Bali) to

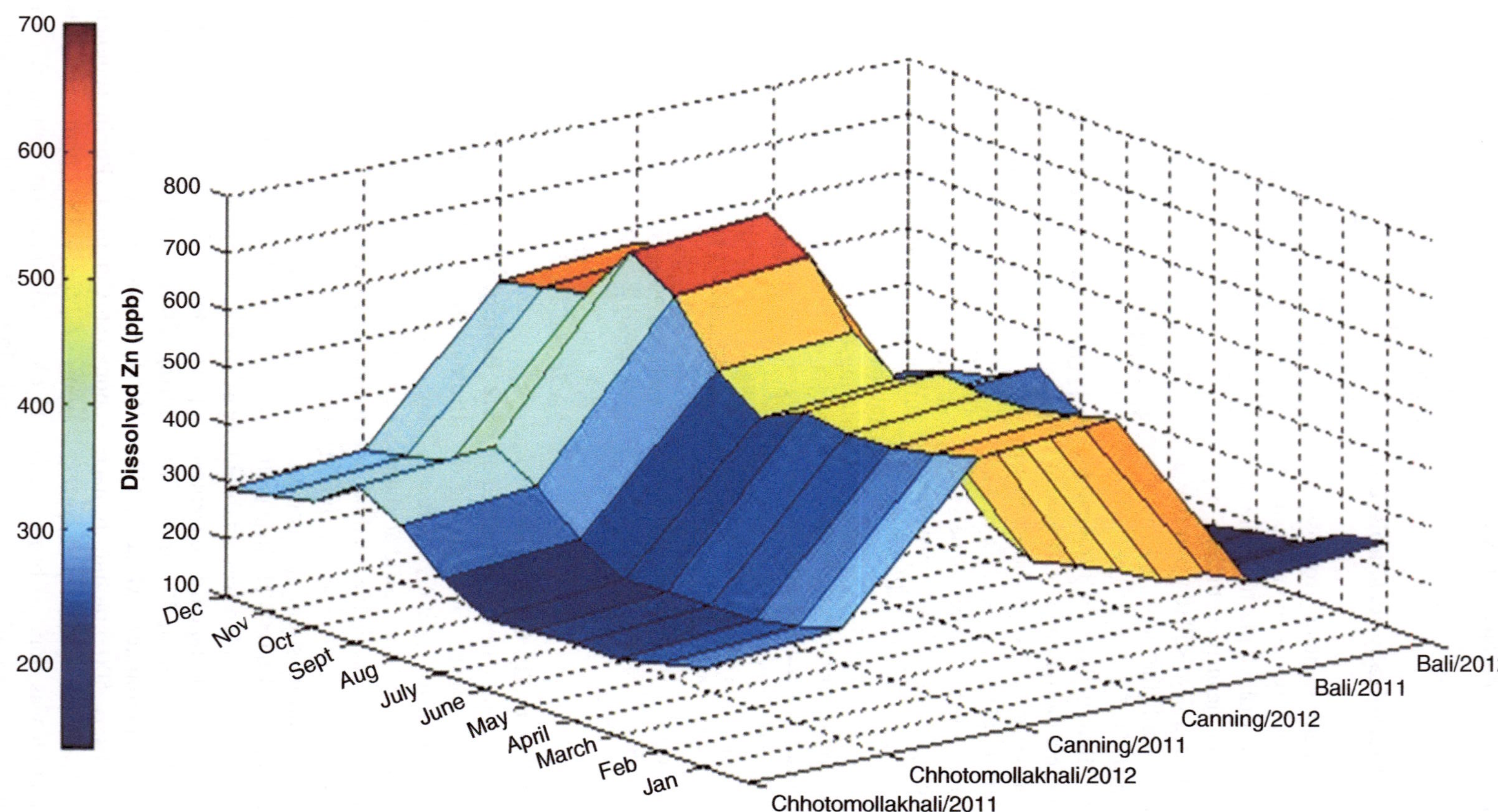

Fig. 5A.1 Map of Indian Sundarbans showing the sampling stations

Table 5A.1 Analysis of reference material for nearshore seawater (CASS 2)

Element	Certified value ($\mu g\ l^{-1}$)	Laboratory results ($\mu g\ l^{-1}$)
Zn	1.97 ± 0.12	2.01 ± 0.14
Cu	0.675 ± 0.039	0.786 ± 0.058
Pb	0.019 ± 0.006	0.029 ± 0.009

Table 5A.2 Analysis of reference material (NIES Sargasso sample) for sediments obtained from the National Institute of Environmental Studies, Japan

Element	Certified value ($\mu g\ g^{-1}$)	Laboratory results ($\mu g\ g^{-1}$)
Zn	28.6	26.2
Cu	14.9	13.7
Pb	2.4	2.9

13.96 ± 2.37 ppm (at Canning) during 2012 (Figs. 5A.3a, 5A.3b and 5A.3c).

3.3. Bioaccumulation Pattern

In *A. alba* samples, the heavy metals varied as per the order Zn > Cu > Pb. This sequence is uniform in all the three selected stations. In the present study, the mean concentration of Zn in leaf ranged from 6.60 ppm (at Bali) to 11.66 ppm (at Canning) during 2011, and in 2012 the mean concentration ranged from 7.13 ppm (at Bali) to 13.84 ppm (at Canning). Mean concentration of Cu ranged from 5.75 ppm (at Bali) to 8.21 ppm (at Canning) in 2011 and 6.12 ppm (at Bali) to 10.05 ppm (at Canning) in 2012. Mean value of Pb in *A. alba* leaf ranged from 1.79 ppm (at Bali) to 2.44 ppm (at Canning) in 2011 and from 1.85 ppm (at Bali) to 2.98 ppm (at Canning) in 2012 (Fig. 5A.4a).

In the same species, the mean concentration of Zn in stem ranged from 9.89 ppm (at Bali) to 17.84 ppm (at Canning) during 2011, and in 2012 the mean concentration ranged from 11.23 ppm (at Bali) to 21.33 ppm (at Canning). Mean Cu concentration ranged from 7.89 ppm (at Bali) to 12.54 ppm (at Canning) in 2011 and 8.33 ppm (at Bali) to 15.75 ppm (at Canning) in 2012. Pb ranged from mean value of 1.95 ppm (at Bali) to 2.80 ppm (at Canning) in 2011 and from 2.02 ppm (at Bali) to 3.59 ppm (at Canning) in 2012 (Fig. 5A.4b).

4. Discussion

Metal pollution in the estuarine, harbour and coastal environment is usually caused by land run-off, mining activities, shipping and dredging activities and anthropogenic inputs (Panigrahy et al. 1997). Sediments in much affected domains not only record its history but also indicate the degree of pollution (Sahu and Bhosale 1991). The mechanism of accumulation of pollutants in the sediments is strongly controlled by the nature of substrate as well as the physico-chemical conditions controlling dissolution and precipitation (Panigrahy et al. 1997). Even though many of the organic compounds occurring in the marine sediments are known to form complexes with metal ions in solution, it has been observed that humic substances are responsible for these strong interactions (Manskaya and Drozdova 1968). The physico-chemical variables also play a major role in the compartmentalization and speciation of metal in the coastal environment. It has been reported that apart from anthropogenic influences, variables like aquatic salinity and pH regulate the process of precipitation of metallic compounds on the seabed and sediment.

Heavy metals have contaminated the aquatic environment in the present century due to intense industrialization and urbanization. The Gangetic delta is no exception to this usual trend. The rapid industrialization and urbanization of the city of Kolkata (formerly known as Calcutta), Howrah and the newly emerging Haldia complex in the maritime state of West Bengal has caused considerable ecological imbalance in the adjacent coastal zone (Mitra and Choudhury 1992; Mitra 1998). The Hooghly estuary, situated on the western sector of the Gangetic delta, receives drainage from these adjacent cities, which have sewage outlets into the estuarine system. The chain of factories and industries situated on the western bank of the Hooghly estuary is a major cause behind the gradual transformation of this beautiful ecotone into stinking cesspools of the megapolis (Mitra and Choudhury 1992). The lower part of the estuary has multifarious industries such as paper, textiles, chemicals,

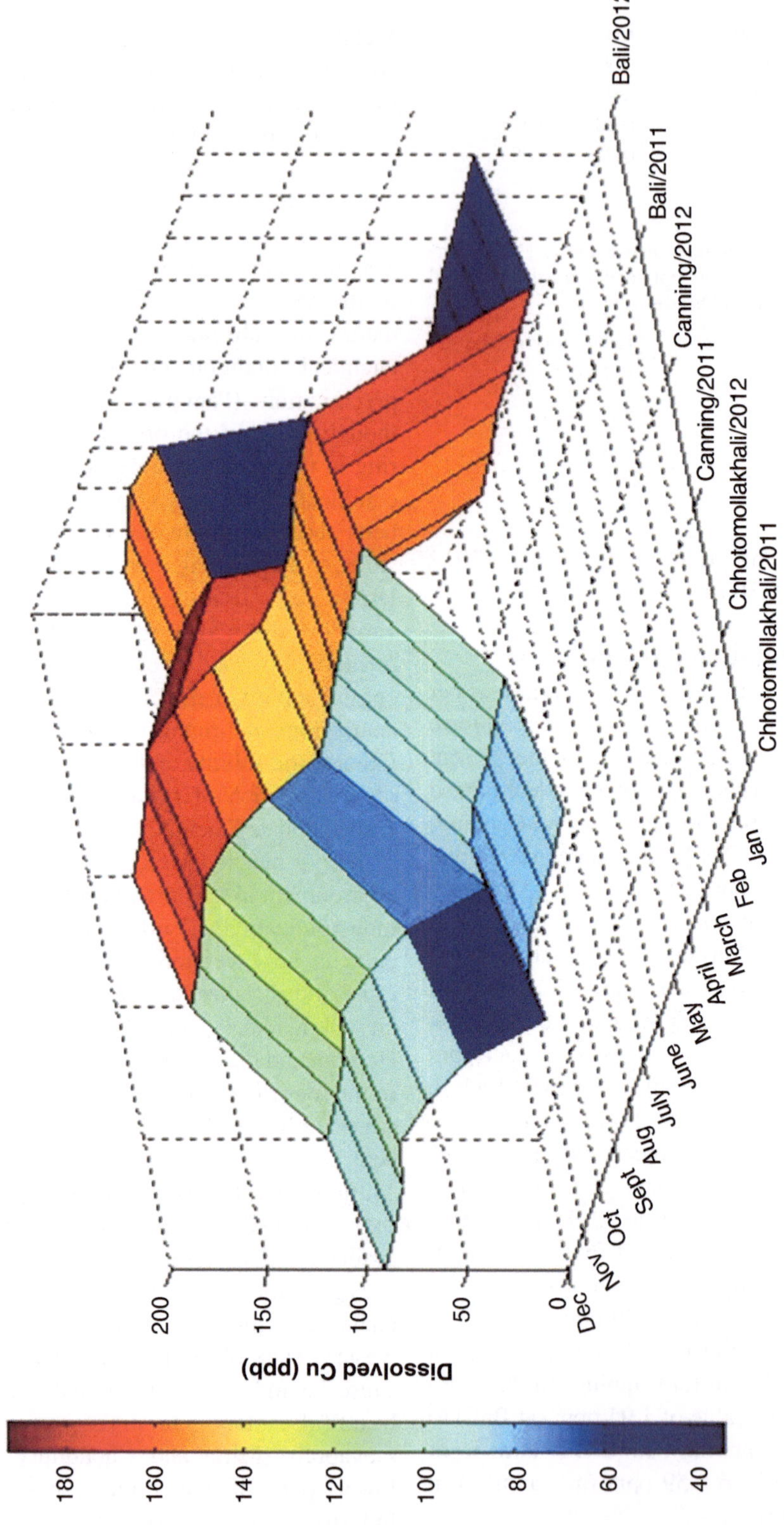

Fig. 5A.2a Dissolved Zn in three stations of Indian Sundarbans during 2011–2012

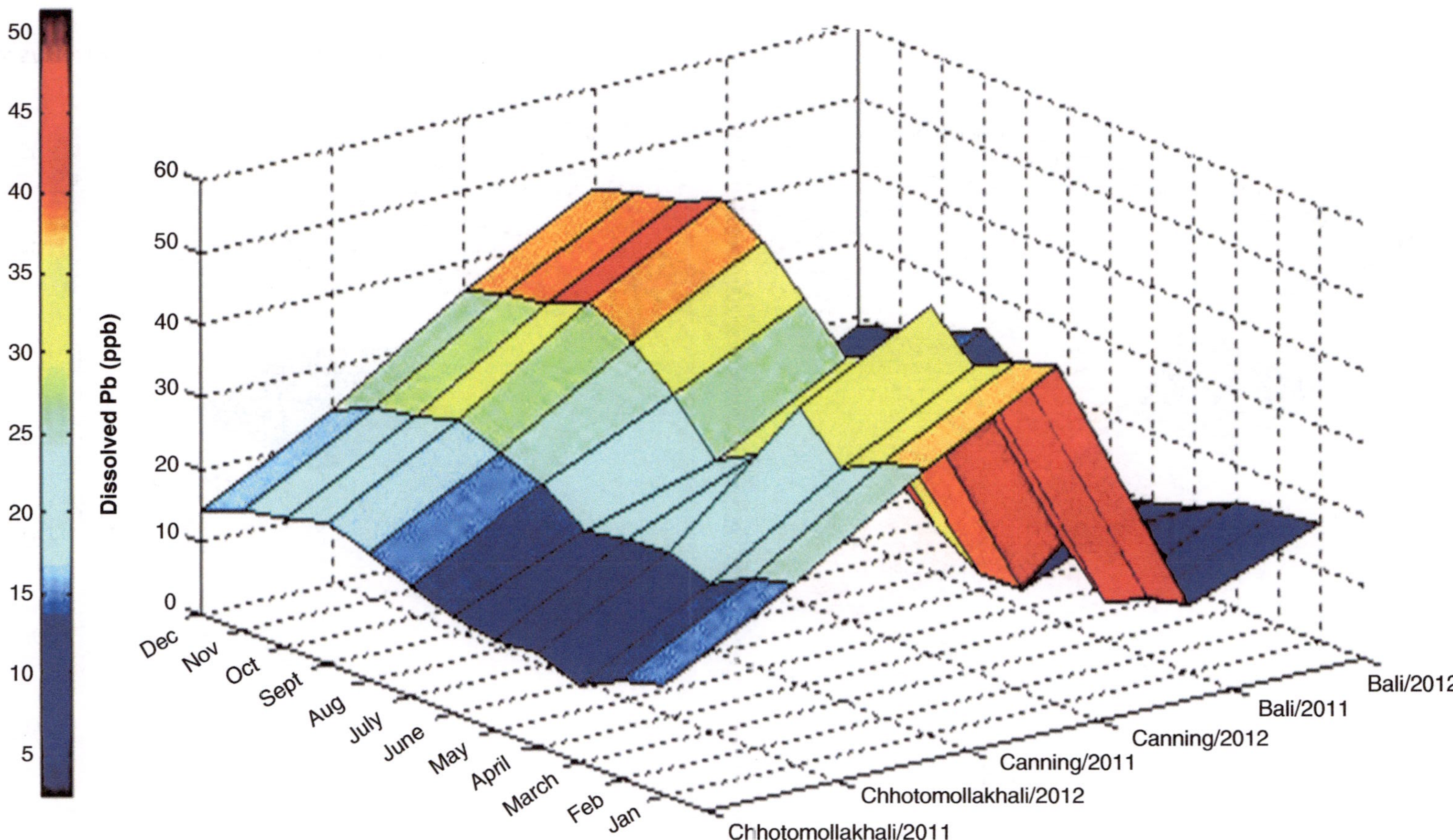

Fig. 5A.2b Dissolved Cu in three stations of Indian Sundarbans during 2011–2012

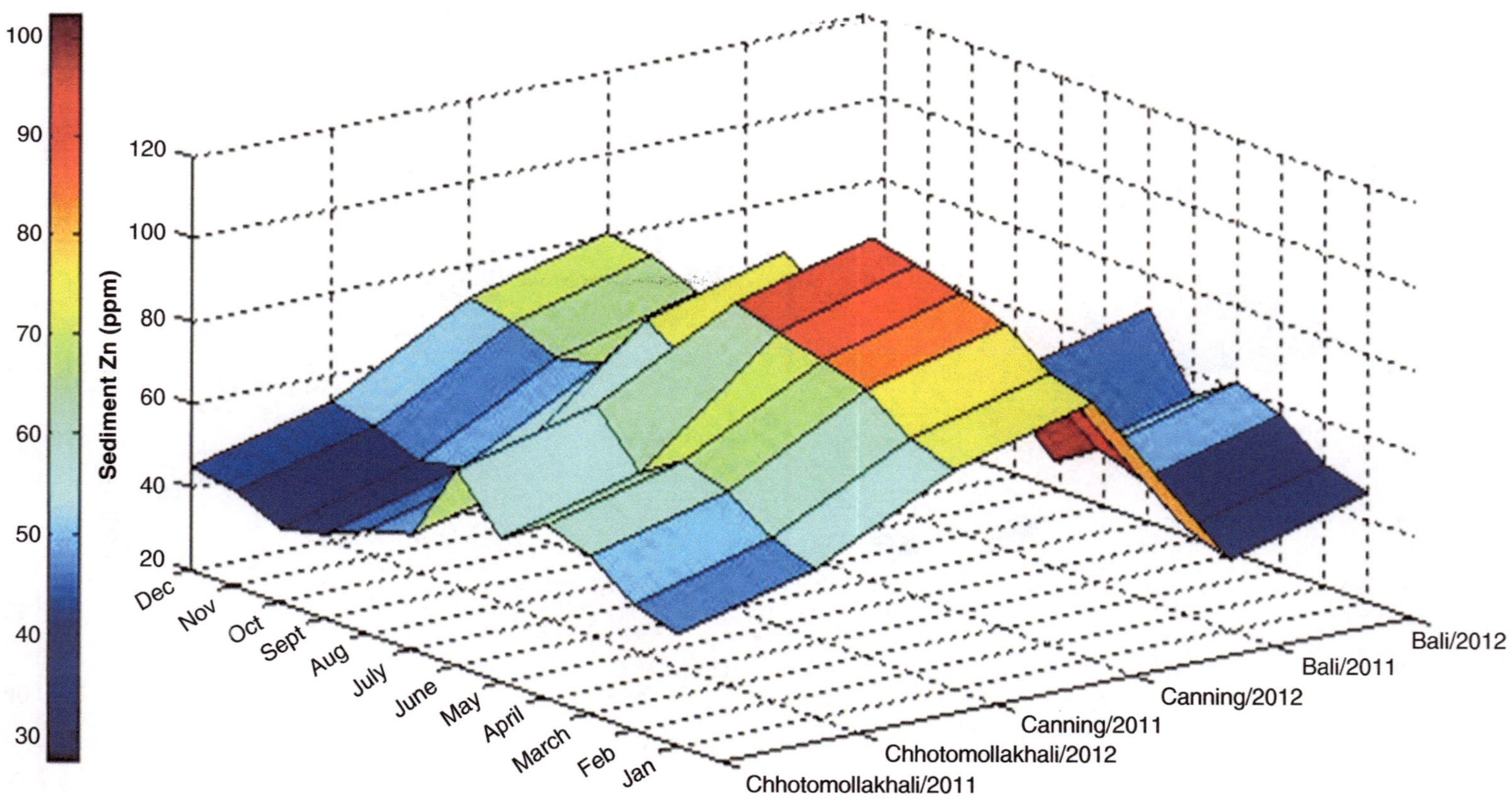

Fig. 5A.2c Dissolved Pb in three stations of Indian Sundarbans during 2011–2012

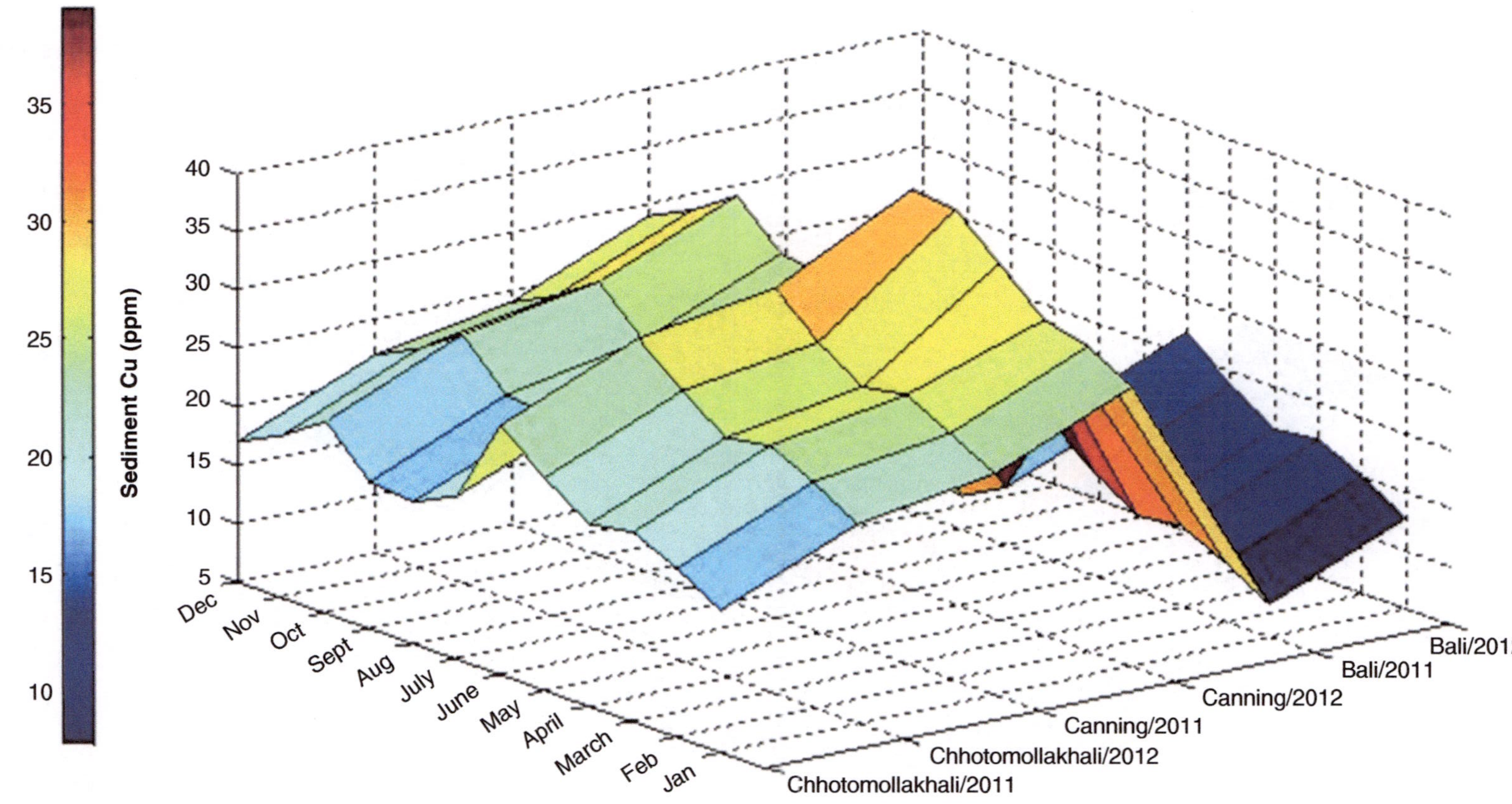

Fig. 5A.3a Biologically available Zn in surface sediments of three stations of Indian Sundarbans during 2011–2012

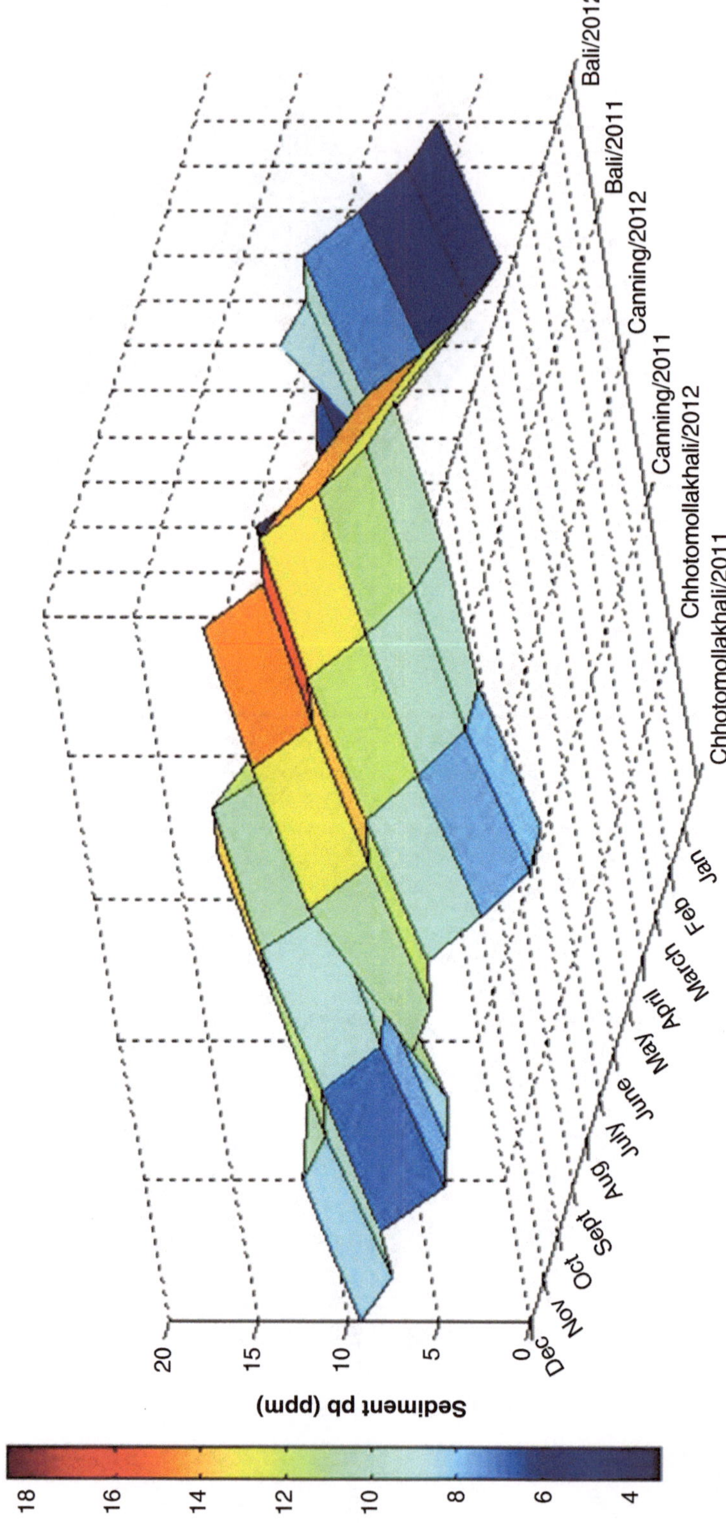

Fig. 5A.3b Biologically available Cu in surface sediments of three stations of Indian Sundarbans during 2011–2012

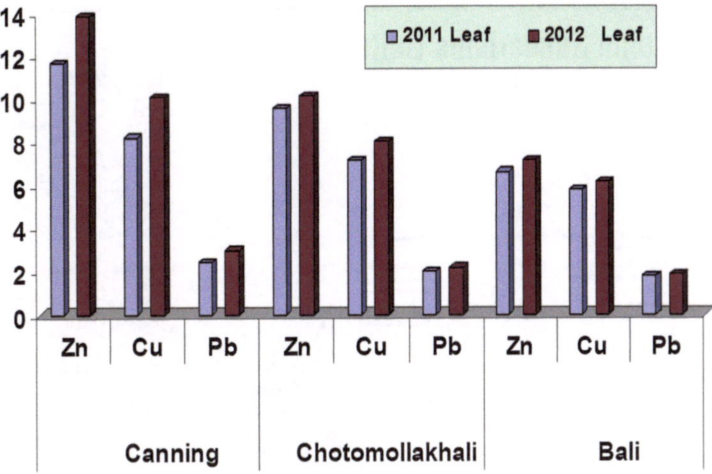

Fig. 5A.3c Biologically available Pb in surface sediments of three stations of Indian Sundarbans during 2011–2012

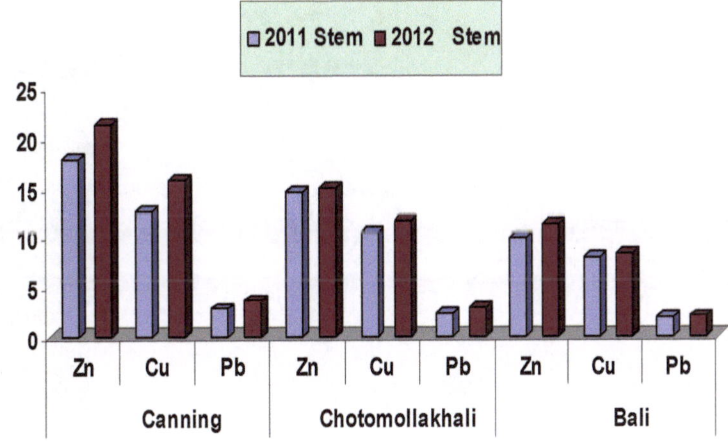

Fig. 5A.4a Mean concentrations ($n = 12$ for each year) of heavy metals (in ppm dry wt.) in *A. alba* leaf during 2011 and 2012

pharmaceuticals, plastic, shellac, food, leather, jute, tyres and cycle rims (UNEP 1982). These units are point sources of heavy metals in the estuarine water.

Of the three metals studied in the present work, Zn and Cu are essential elements while Pb is a non-essential element for most of the living organisms (Trieff 1980).

The main sources of zinc in the present geographical locale are the galvanization units, paint-manufacturing units and pharmaceutical processes. The main sources of Cu in the coastal waters are antifouling paints (Goldberg 1975), particular type of algaecides used in different aquaculture farms, paint-manufacturing units, pipeline corrosion and oil sludges (32–120 ppm). Ship bottom paint has been found to produce very high concentration of Cu in seawater and sediment in harbours of Great Britain and southern California (Bellinger and Benhem 1978; Young et al. 1979). The most toxic of these heavy metals is Pb, which finds its way in coastal waters through the discharge of industrial wastewaters, such as from painting, dyeing, battery-manufacturing units, oil refineries, etc. Antifouling paints used to prevent growth of marine organisms at the bottom of the boats and trawlers also contain lead as an important component. These paints are designed to constantly leach toxic metals into the water to kill organisms that may attach to bottom of the boats, which ultimately is transported to the sediment and aquatic compartments. Lead also

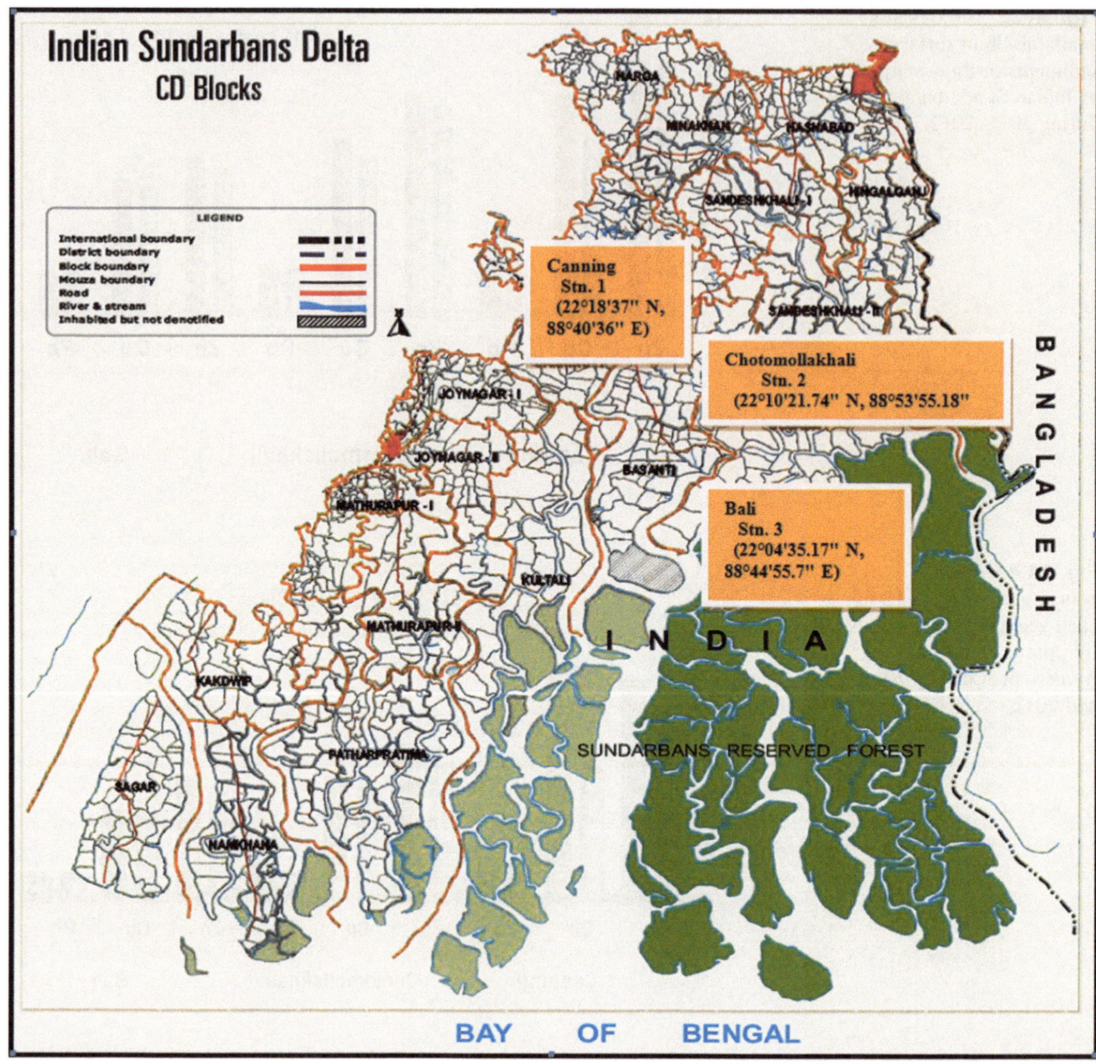

Fig. 5A.4b Mean concentrations ($n = 12$ for each year) of heavy metals (in ppm dry wt.) in *A. alba* stem during 2011 and 2012

enters the oceans and coastal waters both from terrestrial sources and atmosphere and the atmospheric input of lead aerosols can be substantial. The sampling area is exposed to all these activities being proximal to the highly urbanized city of Kolkata, Howrah and the newly emerging Haldia port-cum-industrial complex.

Wetland vegetations are known to absorb and accumulate metals from contaminated sediment (Giblin et al. 1980; Kraus et al. 1986; Kraus 1988; Sanders and Osman 1985). The absorption of contaminants is one reason that wetlands are being used for wastewater treatment. Metals taken up by plants are also capable of re-entering wetland systems through excretion from leaf salt glands (Kraus et al. 1986; Kraus 1988). Metals present within the water column may be carried through open stoma of the plant. This could result in the adhesion of the metals to the outer surface of the leaf. In the present study, the significant positive correlations ($p < 0.05$) observed between dissolved heavy metals and plant tissue metal confirm the adhesive characteristics of the metals (Table 5A.3). Also, significant negative

Table 5A.3 Interrelationships between heavy metal concentrations in *A. alba* and ambient environment

Year	Station				
		r-value			
	Combination	Canning	Chotomollakhali	Bali	*p*-value (for 3 stations)
2011	Tissue Zn × Dissolved Zn	0.8123	0.8864	0.7985	<0.01
	Tissue Cu × Dissolved Cu	0.7654	0.8032	0.7911	<0.01
	Tissue Pb × Dissolved Pb	0.6854	0.7039	0.8105	<0.01
	Tissue Zn × Sediment Zn	−0.5999	−0.6343	−0.7107	<0.01
	Tissue Cu × Sediment Cu	−0.7185	−0.8004	−0.7397	<0.01
	Tissue Pb × Sediment Pb	−0.6115	−0.7559	−0.8023	<0.01
2012	Tissue Zn × Dissolved Zn	0.7644	0.8123	0.6994	<0.01
	Tissue Cu × Dissolved Cu	0.7145	0.8323	0.6558	<0.01
	Tissue Pb × Dissolved Pb	0.6948	0.7732	0.9144	<0.01
	Tissue Zn × Sediment Zn	−0.7184	−0.8176	−0.7739	<0.01
	Tissue Cu × Sediment Cu	−0.7149	−0.8028	−0.6997	<0.01
	Tissue Pb × Sediment Pb	−0.7314	−0.6954	−0.6249	<0.01

relationships between tissue metals and biologically available metals in sediments signify that the primary source of selected heavy metals in *A. alba* is the aquatic phase and not the sediment.

5. Looking Forward

In the coastal environment, mangroves play a major role in regulating the nutrient balance of the zone. It can absorb excess nutrients, reduce suspended solids and sequester other pollutants from aquatic phase. Therefore, in many parts of the globe, this unique ecosystem is presently used as the sink of waste generated from urban, agricultural and industrial sectors located nearby.

During the past few decades, both natural and constructed wetlands were considered as low-cost, easy maintained, simple and effective alternatives for the treatment of municipal, industrial and agricultural effluents (Tam and Yao 1998). The mangrove wetlands are no exceptions to this bio-purification activity. The mechanisms of mangroves to improve the health of the ambient aquatic phase involve interactive processes of plants, soil/sediment and microorganisms (Kanokporn et al. 2002). Plants' uptake of nutrients and transfer of oxygen to the rhizosphere by leakage from roots serve as matrix for growth of microorganisms and stimulate more microbial activities (Conley et al. 1991). Microorganisms regulate the decomposition of complex organic matter in the mangrove ecosystem and result in nitrogen turnover. The soil/sediment of mangroves not only provides a habitat for micro- and macro-flora and fauna that are involved in chemical transformation but plays a significant role in its ability to retain certain chemicals.

It has been recorded that mangroves possess high capacity to retain pollutants, which may be attributed to their presence in anaerobic and reduced conditions, periodically flooded by tides and high clay and organic matter content (Tam and Yao 1998). However the impact of wastewater to the mangrove ecosystem is a matter of great concern than the efficiency of wetlands in improving water quality. The productivity of mangroves may increase due to discharge anthropogenic wastes and this process is beneficial particularly in those areas where nutrient status is low (Wong et al. 1995). This research article thus reflects another face of mangroves, which has been paid minimum attention but can successfully be used to control anthropogenic wastes.

Annexure 5B: Regulatory Role of Salinity on Biomass and Carbon Content in Mangroves of Lower Gangetic Delta

1. Introduction

The general consensus among climate researchers and environmentalists is that increased levels of greenhouse gases (GHGs) from human activities and luxurious lifestyles, burning fossil fuels and massive deforestation in many regions of the world are changing the climate of the planet Earth. CO_2 plays the major role in absorbing outgoing terrestrial radiation and contributes about half of the total greenhouse effect. Between 1850 and 1900, around 100 gigatons of carbon was released into the air just for land-use changes (Pandey 2002). Most of the increase has been since 1940 (Hair and Sampson 1992). The atmospheric CO_2 concentration is currently rising by 4 % per decade. Worldwide concern about climate change has created increasing interest in trees to help reduce the level of atmospheric CO_2 (Dwyer et al. 1992). Forests are most critical components for taking carbon out of circulation for long periods of time. Of the total amount of carbon tied up in earthbound forms, an estimated 90 % is contained in the world's forests, which includes trees, forest floor (litter) and forest soil. For each cubic foot of merchantable wood produced in a tree, about 33 lb. (14.9 kg) of carbon is stored in total tree biomass (Sampson et al. 1992). Tropical forests in general are a disproportionately important component in the global carbon cycle and are thought to represent 30–40 % of the terrestrial net primary production (Clark et al. 2001). Although the area covered by mangrove ecosystems represent only a small fraction of tropical forests, their position at the terrestrial–ocean interface and potential exchange with coastal water suggests these forests make a unique contribution to carbon biogeochemistry in coastal ocean (Twilley et al. 1992). Mangrove ecosystems thrive along coastlines throughout most of the tropics and subtropics. About 75 % of tropical and subtropical countries of the world comprise of mangrove forests (William 2005). These intertidal forests play important ecological and socioeconomic roles by acting as a nutrient filter between land and sea (Robertson and Phillips 1995), contributing to coastline protection (Vermatt and Thampanya 2006), providing commercial fishery resources (Constanza et al. 1997) and nursery grounds for coastal fishes and crustaceans. The coastal zone ($<$200 m depth), covering ~7 % of the ocean surface (Gattuso et al. 1998), has an important role in the oceanic carbon cycle, and various estimates indicate that the majority of mineralization and burial of organic carbon, as well as carbonate production and accumulation, take place in the coastal ocean (Gattuso et al. 1998; Mackenzie et al. 2004). The potential impact of mangrove on coastal zone carbon dynamics has been a topic of intense debate during the past decades. The 'outwelling' hypothesis, first proposed for mangroves by Odum (1968) and Odum and Heald (1972), suggested that a large fraction of the organic matter produced by mangrove trees is exported to the coastal ocean, where it forms the basis of a detritus food chain and thereby supports coastal fisheries. A number of recent studies have indicated that a direct trophic link between mangrove forest production and offshore secondary production is unlikely for many mangrove systems. Despite the large number of case studies dealing with various aspects of organic matter cycling in mangrove systems (Kristensen et al. 2008), there is very limited consensus on the carbon-sequestering potential of mangroves.

The present study is an attempt to establish a baseline data set of the carbon content in the mangrove ecosystem of Indian Sundarbans that has received the crowns of World Heritage Site and Biosphere Reserve in 1987 and 1989 respectively by UNESCO, owing to its unique biological productivity, taxonomic diversity and aesthetic beauty. To preserve the ecosystem in its pristine form, mangrove plantation is carried out on regular basis in the entire Gangetic delta complex. An accurate estimate of carbon storage and sequestration is essential for any project related to plantation particularly in the sector of social forestry. In context to mangrove-dominated Gangetic delta region, this is extremely important as several government, nongovernment organizations and even foreign donors are participating in the mangrove afforestation programme, owing to extreme vulnerability of the system to sea level rise, erosion and tidal surges (Hazra et al 2002; Mitra and Banerjee 2004). The ability of these plantations to sequester carbon has generated a lot of interest, since carbon sequestration projects in developing nations could receive investments from companies and governments wishing to offset their emissions of greenhouse gases through the Kyoto Protocol's Clean Development Mechanism (Fearnside 1999). Carbon registries typically segregate a number of carbon pools within a mangrove forest that can be identified and quantified. These carbon pools are categorized in a variety of ways, but typically include four major compartments. The total carbon in a mangrove system is the summation of above-ground biomass, below-ground biomass, litter and soil. The mangrove ecosystem is unique in terms of carbon dynamics as the litters and detritus contributed by the floral species are exported to adjacent waterbodies in every tidal cycle.

In this study, the above-ground stem, branch and leaf biomass and litter and soil were analysed for carbon content in two different physiographic settings in and around Indian Sundarbans. The difference is caused by freshwater supply from Himalayan glaciers (largest glacial coverage ~34,660 km^2) through Farakka barrage in the western part of Gangetic delta. The barrage was constructed in 1975 to ensure availability of water to the riverine ports. The Ganga–Bhagirathi–Hugli river system in the western part of Indian Sundarbans is therefore appropriately diluted in relation to mangrove growth. In contrast, the Matla river in the central sector is disconnected to the Himalayan glaciers' freshwater due to heavy siltation of the Bidyadhari river since late fifteenth century and is now primarily tide-fed. This difference created a contrasting natural laboratory for identifying climatic signals in salinity profile and mangrove growth leading to variation in carbon pool under different environmental conditions.

2. Methods

2.1. Study Site Description

Two sampling sites were selected each in the western and central sectors and around Indian Sundarbans, a Gangetic delta at the apex of the Bay of Bengal (Fig. 5B.1). This deltaic complex has an area of 9630 Km^2 and houses 102 islands. The western sector of the deltaic lobe receives the snowmelt water of mighty Himalayan glaciers after being regulated through several barrages on the way. The central sector, on the other hand, is fully deprived from such supply due to heavy siltation and clogging of the Bidyadhari channel in the late fifteenth century (Chaudhuri and Choudhury 1994). The station in the western part lies at the confluence of the river Hugli (a continuation of Ganga–Bhagirathi system) and Bay of Bengal. The site is locally known as Sagar South (88°01′47.28″ N latitude and 21°31′4.68″ E longitude). In the central sector, the sampling station was selected at Canning (88°40′36.84″ N latitude and 22°18′37.44″ E longitude), adjacent to tide-fed Matla river. Samplings in both these sectors were carried out in low-tide period during January 2012.

In each sector, plot size of 10 × 10 m was selected for the study and the average readings were documented from 15 such plots. The mean relative abundance of each species was evaluated for the order of dominance of mangrove species at the study sites.

The above-ground biomass (AGB) of individual trees of five dominant species, namely, *Sonneratia apetala, Avicennia alba, Avicennia marina, Avicennia officinalis* and *Excoecaria agallocha*, in each plot was estimated as per the standard procedure stated here, and the average values of 15 plots were finally converted into biomass (in tonnes) per hectare in the study area. Litter production studies were carried out in both the sectors through net collection method and organic carbon in the soil substratum was

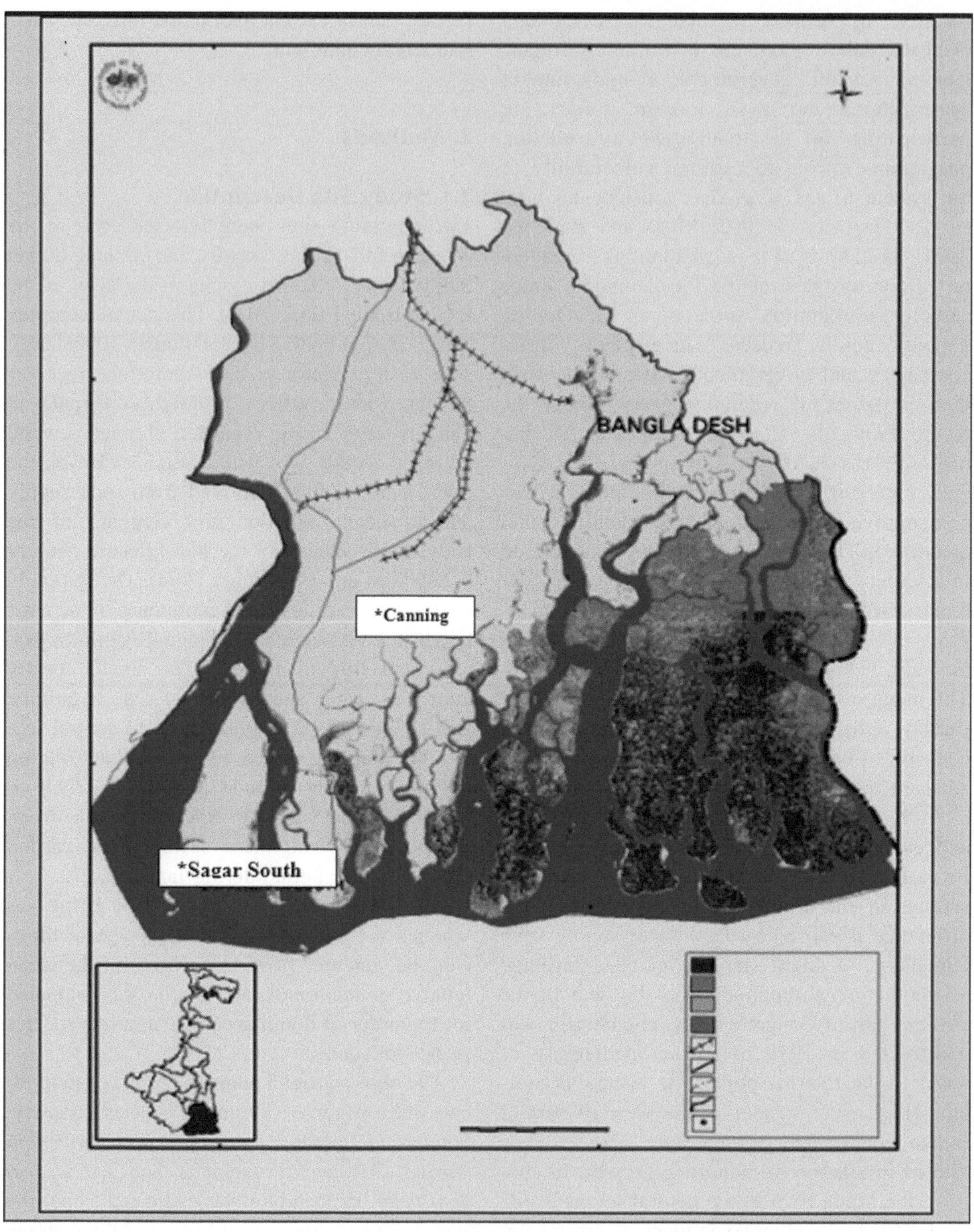

Fig. 5B.1 Location of sampling stations in the western and central sectors of Indian Sundarbans

analysed following the modified method of Walkley and Black (1934).

2.2. Above-Ground Stem Biomass Estimation

The above-ground (stem) biomass of individual trees of each species in every plot was estimated using non-destructive method in which the diameter at the breast height (DBH) was measured with a calliper and height with Ravi's multimeter. Form factor was determined as per the expression outlined by Koul and Panwar (2008) with Spiegel relascope to find out the tree volume (V) using the standard formula given by Pressler (1895) and Bitterlich (1984). Specific gravity (G) was estimated taking the stem cores, which was further converted into stem biomass (B_S) as per the expression $B_S = GV$. The expression for V is FHΠR^2, where F is the form factor, R is the radius of the tree derived from its DBH and H is the height of the target tree.

2.3. Above-Ground Branch Biomass Estimation

The total number of branches irrespective of size was counted on each of the sample trees. These branches were categorized on the basis of basal diameter into three groups, viz. <5, 5–10 and >10 cm. Fresh weight of two branches from each size group was recorded separately. Dry weight of branches was estimated using the equation of Chidumaya (1990).

Total branch biomass (dry weight) per sample tree was determined as per the expression

$$B_{\text{db}} = n_1 b_{\text{w1}} + n_2 b_{\text{w2}} + n_3 b_{\text{w3}} = \sum n_i b_{\text{w}i}$$

where B_{db} is the dry branch biomass per tree, n_i the number of branches in the ith branch group, b_{wi} the average weight of branches in the ith group and $i = 1, 2, 3, \ldots$the branch groups ($i = 3$ in the present study). This procedure was followed for all the dominant mangrove species separately in both the sectors of the study area.

2.4. Above-Ground Leaf Biomass Estimation

Leaves from ten branches (of all the three size groups) of individual trees of each species were removed. One tree of each species per plot was considered for estimation. The leaves were weighed and oven-dried separately to a constant weight at 80 ± 5 °C. The species-wise leaf biomass was then estimated by multiplying the average biomass of the leaves per branch with the number of branches in a single tree and the average number of trees per plot as per the expression

$$L_{\text{db}} = n_1 Lw_1 N_1 + n_2 Lw_2 N_2 + \cdots + n_5 Lw_5 N_5$$

where L_{db} is the dry leaf biomass of dominant mangrove species per plot, $n_i \ldots n_5$ are the number of branches of each tree of five dominant species, $Lw_1 \ldots Lw_5$ are the average dry weight of leaves removed from ten branches of each of the five species and N_1 to N_5 are the number of trees per species in the plot.

2.5. Litterfall Estimation

Litterfall was determined by setting 15 rectangular traps (3 × 3 m) in all the 15 plots in each sector. The traps were made of 1 mm mesh size nylon screen, through which rainwater can pass (Brown and Lugo 1984). The traps were positioned above the high tide level (Jeffrie and Tokuyama 1998) and contents of all the 15 traps per sector were collected and brought to the laboratory after duration of 1 month. The collected materials were segregated into leaves and miscellaneous fraction that comprised of fruits, twigs, stipules, flowers, etc. The materials were dried separately to a constant weight 80 ± 5 °C. Finally the mean weight per plot was estimated for both the western and central sectors in the study area and transformed into gm^{-2} day^{-1} unit.

2.6. Carbon Estimation in Trees and Litter

Direct estimation of percent carbon was done by a CHN analyser. For this a portion of fresh sample of stem, branch and leaf from 30 trees (two trees/species/plot) of individual species (covering all the 15 plots) was oven-dried at 70 °C, randomly mixed and ground to pass through a 0.5 mm screen (1.0 mm screen for leaves). The carbon content (in %) was finally analysed on a LECO® CHN-600 analyser. For litter, the same procedure was followed after oven-drying the net collection at 70 °C.

2.7. Organic Carbon Analysis in Soil

Soil samples from the upper 5 cm were collected from all the 15 plots and dried at 60 °C for 48 h. For analysis, visible plant particles and other organisms (like molluscs, crabs, decaying bodies of fishes, etc.) were handpicked and removed from the soil. After sieving the soil through a 2 mm sieve, we ground the samples of the bulk soil (50 g from each plot) finely in a ball mill. The fine dried sample was randomly mixed to get a sector-wise representative picture of the study site. Modified version of Walkley and Black method (1934) was then followed to determine the organic carbon of the soil in %.

3. Results and Discussion

The biomass and productivity of mangrove forests have been studied mainly in terms of wood production, forest conservation and ecosystem management (Putz and Chan 1986; Tamai et al. 1986; Komiyama et al. 1987; Clough and Scott 1989; McKee 1995; Ong et al. 1995). The contemporary understanding of the global warming phenomenon, however, has generated interest in the carbon-stocking ability of mangroves. The carbon sequestration in this unique producer community is a function of biomass production capacity, which in turn depends upon interaction between edaphic, climate and topographic factors of an area. Hence, results obtained at one place may not be applicable to another. Therefore region-based potential of different land types needs to be worked out. In the present study, the results obtained have been compared with other regions of the world to evaluate the potential of Indian Sundarban mangrove as carbon sink on the background of changing scenario of the climate. The present sectorial case study has also been undertaken with the aim to visualize the impact of salinity on the biomass and carbon budget of mangrove system.

3.1. Relative Abundance

Nine species of true mangroves were documented in the selected plots in the western sector, but in the central sector only six species were recorded. The mean order of abundance of these species was *Sonneratia apetala* (27.08) > *Excoecaria agallocha* (18.75) > *Avicennia alba* (14.58 %) > *Avicennia marina* (12.5 %) = *Avicennia officinalis* (12.5 %) > *Acanthus ilicifolius* (6.25 %) > *Aegiceras corniculatum* (4.17 %) > *Bruguiera gymnorrhiza* (2.08 %) = *Xylocarpus moluccensis* (2.08 %) in the western sector, but order in central sector was *Excoecaria agallocha* (23.68 %) > *Avicennia alba* (21.05 %) > *Avicennia marina* (15.79 %) = *Avicennia officinalis* (15.79 %) > *Sonneratia apetala* (13.16 %) > *Acanthus ilicifolius* (10.53 %) (Table 5B.1). Few mangrove associate floral species (like *Porteresia coarctata*, *Suaeda* sp., etc.) were also documented in the plots. On the basis of relative abundance of the true mangrove species, only five dominant species, namely, *Avicennia alba*, *Avicennia marina*, *Excoecaria agallocha*, *Sonneratia apetala* and *Avicennia officinalis*, were considered for carbon stock estimation in their respective above-ground biomass. In both these sectors, the forests were 12 years old, but high salinity in the central sector probably created a stress to the growth of the floral species.

3.2. Above-Ground Stem Biomass

In the western sector, the above-ground stem biomass of the dominant mangrove trees were 104.09, 14.09, 27.20, 21.37 and 21.46 t ha^{-1} for *Sonneratia apetala*, *Excoecaria agallocha*, *Avicennia alba*, *Avicennia marina* and *Avicennia officinalis*, respectively, but in the central sector, these values were much lower, exhibiting 21.68, 9.27, 15.56, 11.93 and 6.18 t ha^{-1} for *Sonneratia apetala*, *Excoecaria agallocha*, *Avicennia alba*, *Avicennia marina* and *Avicennia officinalis*, respectively (Table 5B.2). The values in the

Table 5B.1 Relative abundance of mangrove species (mean of 15 plots) in the study area

Species	No./100 m^2		Relative abundance (%)	
	Western sector	Central sector	Western sector	Central sector
Sonneratia apetala	13	5	27.08	13.16
Excoecaria agallocha	9	9	18.75	23.68
Avicennia alba	7	8	14.58	21.05
Avicennia marina	6	6	12.5	15.79
Avicennia officinalis	6	6	12.5	15.79
Acanthus ilicifolius	3	4	6.25	10.53
Aegiceras corniculatum	2	ab	4.17	–
Bruguiera gymnorrhiza	1	ab	2.08	–
Xylocarpus moluccensis	1	ab	2.08	–

'*ab*' means absence of the selected species in the selected plots of the study site

western sector are similar to the data of Komiyama et al. (2008) in a secondary mangrove (*Ceriops tagal*) forest at Southern Thailand.

The relatively higher stem biomass of similar aged trees in the western sector may be attributed to optimum hydrological and soil characteristics contributed by the river Ganges. Mangroves, in general, prefer brackish water environment, and in extreme saline condition stunted growth is observed (Mitra et al. 2004). The western sector of Indian Sundarbans provides a congenial environment for mangrove sustenance due to freshwater discharge from Farakka barrage in the Hugli estuarine system. Five-year surveys (1999–2003) on water discharge from Farakka barrage revealed an average discharge of $(3.4 \pm 1.2) \times 10^3$ m^3s^{-1}. Higher discharge values were observed during the monsoon with an average of $(3.2 \pm 1.2) \times 10^3$ m^3s^{-1} and the maximum of the order 4200 m^3s^{-1} during freshet (September). Considerably lower discharge values were recorded during premonsoon with an average of $(1.2 \pm 0.09) \times 10^3$ m^3s^{-1} and the minimum of the order 860 m^3s^{-1} during May. During postmonsoon, discharge values were moderate with an average of $(2.1 \pm 0.98) \times 10^3$ m^3s^{-1}. The lower Gangetic deltaic lobe also experiences considerable rainfall (1400 mm average rainfall). This causes a considerable volume of surface run-off from the 60,000 km^2 catchment areas of Ganga–Bhagirathi–Hugli system and their tributaries. All these factors (dam discharge + precipitation + runoff) increase the dilution factor of the Hugli estuary in the western part of Indian Sundarbans—a condition for better growth and increase of mangrove biomass. The central sector, on contrary, does not receive the freshwater discharge on account of siltation of Bidyadhari river which may be accounted for low above-ground stem biomass of the selected mangrove species inhabiting the zone.

3.3. Above-Ground Branch Biomass

The branch biomass of mangroves showed marked differences between the trees of western and central sectors. In western sector, the values were 42.64, 6.30, 12.42, 10.08 and 9.23 t ha^{-1}, and in central sectors the values were 9.03, 3.81, 6.30, 5.25 and 2.59 t ha^{-1} for *Sonneratia apetala*, *Excoecaria agallocha*, *Avicennia alba*, *Avicennia marina* and *Avicennia officinalis* respectively (Table 5B.2). The branch biomass in the western sector is almost similar to the values in a secondary mangrove (*Ceriops tagal*) forest at Southern Thailand as documented by Komiyama et al. (2000). Stunted branches of mangroves of central sector may again be related to high salinity in this sector (Mitra et al. 2009).

3.4. Above-Ground Leaf Biomass

The leaf biomass of the trees in the western and central sectors were 22.88 and 4.33 t ha^{-1} respectively for *Sonneratia apetala*, 3.22 and 1.85 t ha^{-1} respectively for *Excoecaria agallocha*, 7.07 and 2.96 t ha^{-1} respectively for *Avicennia alba*, 4.83 and 2.20 t ha^{-1} respectively for *Avicennia marina* and 5.46 and 1.24 t ha^{-1} respectively for *Avicennia officinalis*

Table 5B.2 Above-ground biomass (*t*/ha) of five dominant mangrove species in the intertidal

Mangrove vegetative part	*Sonneratia apetala*		*Excoecaria agallocha*		*Avicennia alba*		*Avicennia marina*		*Avicennia officinalis*	
	Western sector	Central sector	Western sector	Central sector	Western sector	Central sector	Western sector	Central sector	Western sector	Central sector
Stem	104.09	21.68	14.09	9.27	27.20	15.56	21.37	11.93	21.46	6.18
Branch	42.64	9.03	6.30	3.81	12.42	6.30	10.08	5.25	9.23	2.59
Leaf	22.88	4.33	3.22	1.85	7.07	2.96	4.83	2.20	5.46	1.24
Total (AGB)	169.61	35.04	23.61	14.93	46.69	24.82	36.28	19.38	36.15	10.01

(Table 5B.2). The values in the western sector are comparatively similar to the records of other workers like 12.1–15.0 t ha^{-1} in *Avicennia* forests (Briggs 1977), 6.2–20.2 t ha^{-1} in *Rhizophora apiculata* young plantations (Aksomkoae 1975), 13.3 t ha^{-1} in *Rhizophora* patch (de la Cruz and Banaag 1967) and 8.1 t ha^{-1} in a matured *Rhizophora* forest (Tamai et al. 1986).

3.5. Litter Production

Average values of total litter, leaf litter and miscellaneous litterfall (comprised of twigs, stipules, flowers and fruits) are shown in Fig. 5B.2. The biomass of total litter is more in the western sector in comparison to central part of Indian Sundarbans. The leaf litter accounted for nearly 70 % and 64 % of the total litter in the western and central sectors respectively.

Although we have not studied the litterfall throughout the year, a significant difference was observed between western and central sectors of the study area with respect to quantum and rate of litter production. The value in the western sector is comparable to the data of several workers. Twilley et al. (1986) reported that the total annual litterfall of mixed mangrove forest of *Avicennia germinans*, *Rhizophora mangle* and *Laguncularia racemosa* in South Florida was 8.68 t ha^{-1} $year^{-1}$ (in Fort Myers) and 7.51 t ha^{-1} $year^{-1}$ (at Rookery Bay). Steinke and Charles (1984) reported the total annual litterfall of mangrove forest in the Mgeni estuary was 8.61 t ha^{-1} $year^{-1}$. Kishimoto et al. (1987) reported that the litterfall of mangrove stands on Iriomote Island (Japan) was 7.5 and 8.8 t ha^{-1} $year^{-1}$ in *Rhizophora stylosa* and *Bruguiera gymnorrhiza* community, respectively. The annual litterfall across broad geographic boundaries are reported as 7–12 t dry weight ha^{-1} $year^{-1}$ (Duke et al. 1981; Twilley et al. 1986; Hardiwinoto et al. 1989; Lee 1990; Gong and Ong 1990; Mall et al. 1991 and Mmochi 1993). In context to Indian mangrove system, the mangrove litter production was recorded as 7.50 tonnes/ha/year in Pichavaram at Tamil Nadu (Krishnamurthy 1985), in which leaf biomass amounts to about 80–90 % (Yadav and Choudhury 1985). Assuming hypothetical situation of uniformity in litterfall through seasons, our data may be interpolated to yield an annual litter production of 3.19 t ha^{-1} in the western sector and 1.33 t ha^{-1} in the central sector respectively. The lower value of litter production in the central Indian Sundarbans may be attributed to the trend of rising salinity due to siltation of Bidyadhari river in the present geographical locale (Chaudhuri and Choudhury 1994). The growth, survival and biomass of mangroves depend on appropriate dilution of the brackish water system with freshwater. The central sector of Indian Sundarbans hardly witness such dilution as the freshwater discharge of the Ganga–Bhagirathi system cannot reach the area due to clogging of the Bidyadhari river by silt and solid wastes (Mitra et al. 2009). The

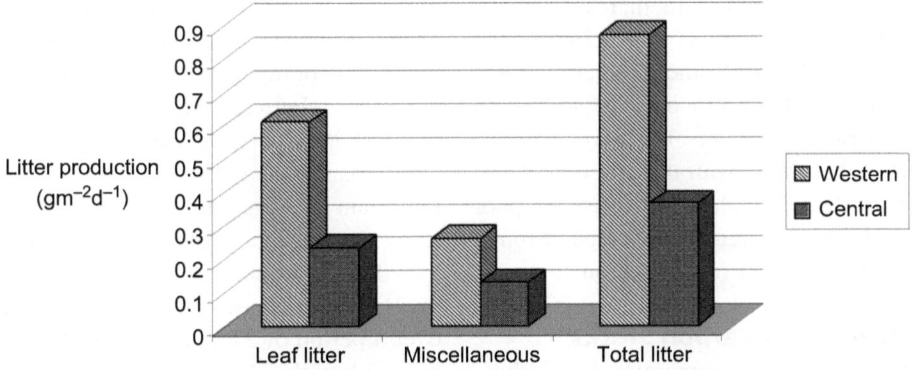

Fig. 5B.2 Variation in leaf litter, miscellaneous litter and total litter in the western and central sectors of Indian Sundarbans

rivers in the study area are noted for their silt-carrying potential. It has been reported that each year Ganga and Brahmaputra bring around 166.70 crore tonnes of silt that has created the present Gangetic delta and the building process is still ongoing.

3.6. Soil Organic Carbon

The values of organic carbon were 2.78 % in the western sector and 0.58 % in the central sector. These values are indicators of mangrove growth, biomass, decay and litterfall for a particular site. Carbon fixed within plant biomass ultimately enters within the soil, where it may reside for hundreds of years. The ability of soil to store this additional carbon, however, is highly controversial, because there are two contrasting ways in which the increased input of carbon may be processed in the soil. First, the extra-fixed carbon may become soil organic carbon. Second, this readily available source of carbon may stimulate soil microbial processes by providing substrates that enhance decomposition of the organic matter through the so-called priming effect (Peterson et al. 1997). Strong evidence for a long-term sink for increased atmospheric CO_2 in soils is still lacking (Schlesinger 1990; Schimel 1995; Canadell et al. 1995). Our study indicates that high saline soil is relatively poor sink of CO_2, which may be attributed to either poor growth of mangroves (Mitra et al. 2004) or low fertility of the soil in terms of nitrogen that acts as retarding factor for plant growth. Canadell et al. (1995) opined that soil quality may influence sequestration of carbon in response to increased atmospheric CO_2. Soil fertility may control the carbon inputs into the soil, since CO_2 enrichment can stimulate plant growth only in soils with adequate nutrients (Egli et al. 1998). Absence of nutrient in the soil of central sector may therefore be considered as plausible cause of poor plant growth in the area as reflected through comparatively low soil organic carbon content.

3.7. Comparison of Carbon Stocks

Mangroves are unique storehouse for carbon. The global storage of carbon in mangrove biomass is estimated to be 4.03 pg, 70 % of which occurs in coastal margins from 0° to 10° latitude (Twilley et al. 1992). For the present study, the results of carbon stock in the above-ground biomass of the selected species are shown in Table 5B.3. Species-wise carbon content are in the order *Sonneratia apetala* > *Avicennia alba* > *Avicennia marina* > *Avicennia officinalis* > *Excoecaria agallocha* in the western sector and *Sonneratia apetala* > *Avicennia alba* > *Avicennia marina* > *Excoecaria agallocha* > *Avicennia officinalis* in the central sector. The % of carbon in the mangrove litter was 31.8 and 29.3 in the western and central sectors respectively. On the basis of the % carbon and average daily production values, the carbon stock of the litter were 1.01 and 0.39 t ha^{-1} year $^{-1}$ in the western and central sectors respectively (Table 5.7). The soil organic carbon also exhibited similar trend with higher value in the western sector (2.78 %) than that of the central region (0.58 %). Considering the carbon pool in the above-ground biomass of the dominant mangrove species and total litter and assuming seasonal uniformity in carbon stock, the corresponding CO_2 equivalents ha^{-1} year $^{-1}$ in western and central sectors of Indian Sundarbans were 477.98 t and 225.13 tonnes respectively (Table 5B.4), which are effective figures when the present trend of atmospheric CO_2 rise is 4 % per decade (Hyun-Kil and Gregory McPherson 2001). These figures can be manipulated through effective soil management, tidal interactions (through artificial canalization) and proper dilution of the system with freshwater, which are important requisites for accelerating the biomass of mangrove species. The data generated in the present geographical locale show significant variations between the two sectors. The hypiersalinity of the central part of Indian Sundarbans may be considered as one of the important reason for such shortfall. Records show that surface water salinity has increased by 40.46 % in central sector and decreased by 46.21 % in western sector of Indian Sundarbans over a period of 27 years (1980–2007), which is the result of the blockage of freshwater flow from western side of Indian Sundarbans to central sector (Mitra et al. 2009). Higher salinity has

Table 5B.3 Above-ground carbon stock (t/ha) of five dominant mangrove species in the intertidal

Mangrove vegetative part	*Sonneratia apetala*		*Excoecaria agallocha*		*Avicennia alba*		*Avicennia marina*		*Avicennia officinalis*	
	Western sector	Central sector	Western sector	Central sector	Western sector	Central sector	Western sector	Central sector	Western sector	Central sector
Stem	43.51	8.63	5.78	3.81	11.07	6.32	8.48	4.89	9.14	2.61
Branch	18.34	3.61	2.63	1.55	5.25	2.52	4.19	2.09	3.98	1.03
Leaf	9.38	1.69	1.29	0.73	2.80	1.16	2.02	0.88	2.38	0.50
Total (AG carbon stock)	71.23	13.93	9.70	6.09	19.12	10.00	14.69	22.55	15.50	4.14

Table 5B.4 Carbon stock and CO_2 equivalent in AGB of dominant mangrove species and litter in t ha^{-1} $year^{-1}$ in the western and central sectors of Indian Sundarbans

Sampling station	Component	AGB	Litter	Interpretation
Sagar south (88°01′47.28″N latitude and 21°31′4.68″E longitude) in the western sector	C	130.24	1.01	Dilution of the Hugli river by ice melt water from Himalaya through barrage regulation
	CO_2 equivalent	477.98	11.70	
Canning (88°40′36.84″N latitude and 22°18′37.44″E longitude) in the central sector	C	56.71	0.39	Disconnection of freshwater supply due to massive siltation in the Bidyadhari river leading to higher aquatic salinity
	CO_2 equivalent	225.13	4.91	

therefore reduced the floral growth and subsequent litter production and organic carbon in soil of central sector of Indian Sundarbans. Considering the ecological significance of mangroves, policy must be implemented (both at regional and global level) to preserve and restore the system, which have been destroyed and damaged in many parts of the globe by activities like dredging, urbanization, draining, construction of shrimp farms and tourism units, sea level rise, etc.

In the present framework, interlinking of the tide-fed rivers of the central portion with the Ganga–Bhagirathi–Hugli river system in the western part might serve as an effective management strategy for accelerating the mangrove plant biomass and subsequent rate of carbon sequestration by the mangrove system in the central sector around the Matla river.

References

Azocar, A., Rada, F., & Orozco, A. (1992). Water relations and gas exchange in two mangrove species with contrasting mechanisms of salt regulation. *Ecotropicos, 5*, 11–19.

Babin, M., Morel, A., & Gentili, B. (1996). Remote sensing of sea surface sun-induced chlorophyll fluorescence: Consequences of natural variations in the optical characteristics of phytoplankton and the quantum yield of chlorophyll a fluorescence. *International Journal of Remote Sensing, 17*, 2417–2448.

Balsamo, R. A., & Thomson, W. W. (1995). Salt effects on membranes of the hypodermis and mesophyll cells of *Avicennia germinans* (Avicenniaceae): A freeze-fracture study. *American Journal of Botany, 82*, 435–440.

Behrenfeld, M. J., Randerson, J. T., McClain, C. R., Feldman, G. C., Los, S., Tucker, C., Falkowski, P. G., Field, C. B., Frouin, R., Esaias, W., Kolber, D., & Pollack, N. (2001). Biospheric primary production during an ENSO transition. *Science, 291*, 2594–2597.

Behrenfeld, M. J., O'Malley, R., Siegel, D. A., McClain, C. R., Sarmiento, J. L., Feldman, G. C., Milligan, A. J., Falkowski, P. G., Letelier, R., & Boss, E. S. (2006a). Climate-driven trends in contemporary ocean productivity. *Nature, 444*, 752–755.

Behrenfeld, M. J., Worthington, K., Sherrell, R. M., Chavez, F. P., Strutton, P., McPhaden, M., & Shea, D. M. (2006b). Controls on tropical Pacific Ocean productivity revealed through nutrient stress diagnostics. *Nature, 442*, 1025–1028.

Behrenfeld, M. J., Halsey, K., & Milligan, A. J. (2008). Evolved physiological responses of phytoplankton to their integrated growth environment. *Philosophical Transactions of the Royal Society of London, 363*, 2687–2703. doi:10.1098/rstb.2008.0019.

Briggs, S. V. (1977). Estimates of biomass in a temperate mangrove community. *Australian Journal of Ecology, 2*, 369–373.

Bunt, J. S. (1992). Introduction. In A. I. Robertson & D. M. Alongi (Eds.), *Tropical mangrove ecosystem* (pp. 1–6). Washington, DC: American Geophysical Union.

Chaudhuri, A. B., & Choudhury, A. (1994). *Mangroves of the Sundarbans* (Vol. 1, p. 165). Bangkok: IUCN—The World Conservation Union.

Chung, I. K., Oak, J. H., Lee, J. A., Shin, J. A., Kim, J. G., & Park, K.-S. (2013). Installing kelp forests/seaweed beds for mitigation and adaptation against global warming: Korean project overview. *ICES Journal of Marine Science Advance Access*. doi:10.1093/icesjms/fss206. pp. 1–7.

Cleveland, J. S., & Perry, M. J. (1987). Quantum yield, relative specific absorption and fluorescence in nitrogen limited, *Chaetoceros gracilis*. *Marine Biology, 94*, 489–497.

Cullen, J. J. (1982). The deep chlorophyll maximum: Comparing vertical profiles of chlorophyll a. *Canadian Journal of Fishery and Aquatic Science, 39*, 791–803.

Cullen, J. J., Ciotti, A. M., Davis, R. F., & Neale, P. J. (1997). Relationship between near-surface chlorophyll and solar-stimulated fluorescence: Biological effects, Ocean Optics XIII. *Proceedings SPIE, 2963*, 272–277.

Drennan, P., & Pammenter, N. W. (1982). Physiology of salt secretion in the mangrove *Avicennia marina* (Forsk.) Vierh. *New Phytology, 91*, 1000–1005.

Duarte, C. M., Middelburg, J. J., & Caraco, N. (2005). Major role of marine vegetation on the oceanic carbon cycle. *Biogeosciences, 2*, 1–8. doi:10.5194/bgd-2-1-2005.

Falkowski, P. G., & Kiefer, D. A. (1985). Chlorophyll a fluorescence in phytoplankton: Relationship to photosynthesis and biomass. *Journal of Plankton Research, 7*, 715–731.

Falkowski, P. G., & Kolber, Z. (1995). Variations in chlorophyll fluorescence yields in phytoplankton in the world oceans. *Australian Journal of Plant Physiology, 22*, 341–355.

Field, C. B., Behrenfeld, M. J., Randerson, J. T., & Falkowski, P. G. (1998). Primary production of the biosphere: Integrating terrestrial and oceanic components. *Science, 281*, 237–240.

Fourqurean, J. W., Duarte, C. M., Kennedy, H., Marba, N., & Holmer, M. (2012). Seagrass ecosystems as a globally significant carbon stock. *National Geoscience, 5*, 505–509. doi:10.1038/NGEO1477.

Gouda, R., & Panigrahy, R. C. (1996). Ecology of phytoplankton in coastal water of Gopalpur, East coast of India. *Indian Journal of Marine Science, 2*, 13–18.

Gregg, W. W., Casey, N. W., & McClain, C. R. (2005). Recent trends in global ocean chlorophyll. *Geophysics Research Letter, 32*, L03606. doi:10.1029/2004GL021808.

Hemminga, M. A., & Duarte, C. M. (2000). *Seagrass ecology*. Cambridge: Cambridge University Press.

Huot, Y., Babin, M., Bruyant, F., Grob, C., Twardowski, M. S., & Claustre, H. (2007). Relationship between photosynthetic parameters and different proxies of phytoplankton biomass in the subtropical ocean. *Biogeosciences, 4*, 853–868.

Kathiresan, K., & Bingham, B. L. (2001). Biology of mangrove ecosystems. *Advances in Marine Biology, 40*, 81–251.

Kennedy, H., Beggins, J., Duarte, C. M., Fourqurean, J. W., & Holmer, M. (2010). Seagrass sediments as a global carbon sink: Isotopic constraints. *Global Biogeochemical Cycles, 24*, 1–8.

Krause, G. H., & Weis, E. (1991). Chlorophyll fluorescence and photosynthesis: The basics. *Annual Review Plant Physiology Plant Molecular Biology, 42*, 313–349.

McGlathery, J. K., Reynolds, L. K., Cole, L. W., Orth, R. J., & Marion, S. R. (2012). Recovery trajectories during state change from bare sediment to eelgrass dominance. *Marine Ecology Programme Series, 448*, 209–221. doi:10.3354/meps09574.

Mcleod, E., Chmura, G. L., Bouillon, S., Salm, R., & Bjork, M. (2011). A blueprint for blue carbon: Toward an improvement understanding of the role of vegetated coastal habitats in sequestering CO_2. *Frontal Ecology Environment, 9*, 552–560. doi:10.1890/110004.

Mitra, A., & Banerjee, K. (2005). In Published by WWF India (Eds.), *Living resources of the sea: Focus Indian Sundarbans* (p. 96). West Bengal: Col S.R. Banerjee, carry field office, 24 Parganas (S).

Orth, R. J., Carruthers, T. J. B., Dennison, W. C., Duarte, C. M., & Fourqurean, J. W. (2006). A global crisis for seagrass ecosystems. *Bioscience, 56*, 987–996. Pub Med: 16775979.

Scholander, P. F. (1968). How mangroves desalinate water. *Physiologia Plantarum, 21*, 251–261.

Scholander, P. F., Hammel, H. T., Hemmingsen, E., & Garey, W. (1962). Salt balance in mangroves. *Plant Physiology, 37*, 722–729.

Scholander, P. F., Hammel, H. T., Hemmingsen, E. A., & Bradstreet, E. D. (1964). Hydrostatic pressure and osmotic potential in leaves of mangroves and some other plants. *Proceedings of the National Academy of Sciences, USA, 52*, 119–125.

Schwamborn, R., & Saint-Paul, U. (1996). Mangroves—forgotton forests? *Natural Resources and Development, 43/44*, 13–36.

Spalding, M., Blasco, F., & Field, C. (1997). *World mangrove atlas* (p. 178). Okinawa: The International Society for Mangrove Ecosystems.

Tomlinson, P. B. (1986). *The botany of mangroves*. London: Cambridge University Press.

Twilley, R. R., Chen, R. H., & Hargis, T. (1992). Carbon sinks in mangrove forests and their implications to the carbon budget of tropical coastal ecosystems. *Water, Air, and Soil Pollution, 64*, 265–288.

Van Eijk, M. (1939). Analyse der Wirkung des NaCl auf die Entwicklung Sukhulenz and Transpiration bei *Salicornia herbacea*, Sowie untersuchungen uber den Einfluss der Salzauf Nahme auf die Wurzelatmung bei *Aster tripolium*. Recueil des Travaux. *Botaniques Neerlandais, 36*, 559–657.

Waycott, M., Duarte, C. M., Carruthers, T. J. B., Orth, R. J., & Dennison, W. C. (2009). Accelerating loss of seagrass across the globe threatens coastal ecosystems. *Proceedings of National Academy of Science, USA, 106*, 12377–12381.

Internet References

http://www.epa.qld.gov.au/nature_conservation/habitats/mangroves_andwetlands/man groves

Annexure 5A: References

Banerjee, K., Roy Chowdhury, M., Sengupta, K., Sett, S., & Mitra, A. (2012). A. Influence of anthropogenic and natural factors on the mangrove soil of Indian Sundarban Wetland. *Archives of Environmental Science, 6*, 80–91.

Bellinger, E., & Benhem, B. (1978). The levels of metals in Dockyard sediments with particular reference to the contributions from ship bottom paints. *Environmental Pollution Assessment, 15*(1), 71–81.

Conley, L. M., Dick, R. I., & Lion, L. W. (1991). An assessment of the root zone method of waste-water treatment. *Research Journal of Water Pollution Control Federation, 63*(3), 239–247.

Danielsson, L. G., Magnusson, B., & Westerlund, S. (1978). An improved metal extraction procedure for the determination of trace metals in seawater by atomic absorption spectrometry with electrothermal atomization. *Analytical Chemical Acta, 98*, 45–57.

Delaune, R. D., Pezeshki, S. R., Pardue, J. H., Whitcomb, J. H., & Patrick, W. H., Jr. (1990). Some influences of sediment addition to a deteriorating salt marsh in the Mississippi River Deltaic plain: A pilot study. *Journal of Coastal Research, 6*, 181–188.

Gambrell, R. P. (1994). Trace and toxic metals in wet-lands: A review. *Journal of Environmental Quality, 23*, 883–891.

Giblin, A. E., Bourg, A., Valiela, I., & Teal, J. M. (1980). Uptake and losses of heavy metals in sewage sludge by a New England saltmarsh. *American Journal of Botany, 67*, 1059–1068.

Goldberg, E. D. (1975). The mussel watch—A first step in global marine monitoring. *Marine Pollution Bulletin, 6*, 111.

Kanokporn, B., Somkiat, P., & Pipat, P. (2002). The use of a mangrove plantation as a constructed wetland for municipal wastewater treatment. *Journal of Science Research, Chula. University, 27*, 1.

Kraus, M. L. (1988). Accumulation and excretion of five heavy metals by the saltmarsh cordgrass *Spartina alterniflora. Bulletin of the New Jersey Academy of Sciences, 33*, 39–43.

Kraus, M. L., Weis, P., & Crow, J. H. (1986). The excretion of heavy metals by the salt marsh cord grass, *Spartina alterniflora*, and Spartina's role in mercury cycling. *Marine Environmental Research, 20*, 307–316.

Krishnamurti, C. R., & Viswanathan, P. (Eds.). (1991). *Toxic metals in the environment* (p. 246). New Delhi: Tata McGraw-Hill Publishing Company Ltd.

Lithor, G. (1975). *Methods for detection measurement and monitoring of water pollution* (p. 41). Rome: FAO.

Malo, B. A. (1977). Partial extraction of metals from aquatic sediments. *Environmental Science and Technology, 11*, 277–288.

Manahan, S. E. (1994). *Environmental chemistry* (4th ed., p. 597). Boston: Willard Grant Press.

Manskaya, S. M., & Drozdova, T. V. (1968). Geochemistry of organic substances. In L. Shapiro & I. A. Breger (Eds.), *Geochemistry of organic substances* (International series of monographs in earth sciences, Vol. 28). NewYork: Pergamon. 345p.

Mitra, A. (1998). Status of coastal pollution in West Bengal with special reference to heavy metals. *Journal Indian Ocean Studies, 5*(2), 135–138.

Mitra, A., & Choudhury, A. (1992). Trace metals in macrobenthic molluscs of the Hooghly estuary. *Indian Marine Pollution Bulletin UK, 26*(9), 521–522.

Mitra, A., & Ghosh, R. (2014). Bioaccumulation pattern of heavy metals in commercially important fishes in and around Indian Sundarbans. *Global Journal of Animal Scientific Research, 2*(1), 33–44.

Mitra, A., Banerjee, K., & Bhattacharya, D. P. (2004). *The other face of mangroves*. West Bengal: Department of Environment, Government of West Bengal.

Mitra, A., Chowdhury, R., & Banerjee, K. (2011). Concentrations of some heavy metals in commercially important finfish and shellfish of the River Ganga. *Environmental Monitoring and Assessment*. doi:10.1007/s10661-011-2111-x.

Panigrahy, P. K., Nayak, B. B., Acharya, B. C., Das, S. N., Basu, S. C., & Sahoo, R. K. (1997). Evaluation of heavy metal accumulation in coastal sediments of northern Bay of Bengal. In C. S. P. Iyer (Ed.), *Advances in environmental science* (pp. 139–146). New Delhi: Educational Publishers and Distributors.

Sahu, K. C., & Bhosale, U. (1991). Heavy metal pollution around the island city of Bombay, India, part-I: Quantification of heavy metal pollution of aquatic sediments and recognition of environment discriminates. *Chemical Geology, 91*, 263–268.

Sanders, J. G., & Osman, R. W. (1985). Arsenic incorporation in a saltmarsh ecosystem. *Estuarine and Coastal Shelf Science, 20*, 387–392.

Tam, N. F. Y., & Yao, M. W. Y. (1998). Normalization and heavy metal contamination in mangrove sediments. *Science of the Total Environment, 216*(1–2), 33–39.

Trieff, R. A. (1980). *Environment and health*. Ann Arbor Science Publishers Inc., The Butterworth Group.

UNEP. (1982). *Pollution and the marine environment in the Indian Ocean*. Geneva: UNEP Regional Seas Programme Activity Centre.

Williams, T. P., Bubb, J. M., & Lester, J. N. (1994). Metal accumulation within salt marsh environments: A review. *Marine Pollution Bulletin, 38*, 277–290.

Wong, Y. S., Lam, C. Y., Che, S. H., Li, X. R., & Tam, N. F. Y. (1995). Effect of wastewater discharge on nutrient contamination of mangrove soil and plants. *Hydrobiologia, 295*, 243–254.

Young, D. R., Alexander, G. V., & McDermott-Ehrlich, D. (1979). Vessel related contamination of southern California harbours by copper and other metals. *Marine Pollution Bulletin, 10*, 50–56.

Annexure 5B: References

Aksomkoae, S. (1975). *Structure regeneration and productivity of mangroves in Thailand*. PhD dissertation, Michigan State University, pp. 1–109.

Bitterlich, W. (1984). The relaskop idea slough: Commonwealth agricultural Bureause, Farnham Royal, England. *Indian Forester, 127*(2), 144–153.

Brown, S., & Lugo, A. E. (1984). Biomass of tropical forests: A new estimate based on forest volumes. *Science, 223*, 1290–1293.

Canadell, J. G., Pitelka, L. F., & Ingram, J. S. I. (1995). The effects of elevated [CO_2] on plant-soil carbon below-ground: A summary and synthesis. *Plant and Soil, 187*, 391–400.

Chaudhuri, A. B., & Choudhury, A. (1994). *Mangroves of the Sundarbans* (Vol. 1, p. 165). Bangkok: IUCN—The World Conservation Union.

Chidumaya, E. N. (1990). Aboveground woody biomass structure and productivity in a Zambezian woodland. *Forest Ecology and Management, 36*, 33–46.

Clark, D. A., Brown, S., Kiicklighter, D. W., Chambers, J. Q., Thomlinson, J. R., Ni, J., & Holland, E. A. (2001). Measuring net primary production in forest: An evaluation and synthesis of existing field data. *Ecological Applications, 11*, 371–384.

Clough, B. F., & Scott, K. (1989). Allometric relationship for estimating above ground biomass in six mangrove species. *Forest Ecology and Management, 27*, 117–127.

Constanza, R., et al. (1997). The value of the world's ecosystem service and natural capital. *Ecological Economics, 25*, 3–15.

de la Cruz, A. A., & Banaag, B. F. (1967). The ecology of a small mangrove patch in Matabungkay Beach Batangas Province. *Natural and Applied Science Bulletin, 20*, 486–494.

Duke, N. C., Bunt, J. S., & Williams, W. T. (1981). Mangrove litterfall in north-eastern Australia, annual totals by component in selected species. *Australian Journal of Botany, 29*, 547–553.

Dwyer, J. F., Mcpherson, E., Gregory, S., Herbert, W., & Rowan, A. (1992). Assessing the benefits and costs of the urban forest. *Journal of Arboriculture, 18*, 227–234.

Egli, P., Maurer, S., Gunthardt-Georg, M., & Korner, C. (1998). Effects of elevated CO_2 and soil quality on leaf gas exchange and above ground growth in beech-spruce model ecosystems. *New Phytologist, 140*, 185–196.

Fearnside, P. M. (1999). Forests and global warming mitigation in Brazil opportunities in the Brazilian forest sector for responses to global warming under the "clean development mechanism". *Biomass and Bioenergy, 16*, 171–189.

Gattuso, J. P., Frankignoulle, M., & Wollast, R. (1998). Carbon and carbonate metabolism in coastal ecosystems. *Annual Review of Ecological Systems, 29*, 405–434.

Gong, W. K., & Ong, J. E. (1990). Plant biomass and nutrient flux in a managed mangrove forest in Malaysia. *Estuarine, Coastal and Shelf Science, 31*, 519–530.

Hair, D., & Sampson, R. N. (1992). Climate change-history, prospects, and possible impacts. In D. Hair & R. N. Sampson (Eds.), *Forests and global change volume one: Opportunities for increasing forest cover* (pp. 1–10). Washington, DC: American Forests.

Hardiwinoto, S., Nakasuga, T., & Igarashi, T. (1989). Litter production and decomposition of mangrove forest at Ohura Bay, Okinawa. *Research Bulletins of the College Experiment Forest Hokkaido University, 46* (3), 577–594.

Hazra, S. I., Ghosh, T., Dasgupta, R., & Sen, G. (2002). Sea level and associated changes in the Sundarbans. *Science and Culture, 68*, 309–321.

Hyun-Kil, J., & Gregory McPherson, E. (2001). Carbon storage and flux in urban residential greenspace. *Journal of Environmental Management, 61*, 165–177.

Jeffrie, F. M., & Tokuyama, A. (1998). Litter production of mangrove forests at the Gesashi River. *Bulletin of the College of Science, University of the Ryukyus, 65*, 73–79.

Kishimoto, T., Miyajima, H., Nakasuga, T., & Baba, S. (1987). Zonation of mangrove forest (III) litterfall and its seasonal change. *Transactions of the Japanese Forest Society, 98*, 301–302 (in Japanese).

Komiyama, A., Ogino, K., Aksomkoae, S., & Sabhasri, S. (1987). Root biomass of a mangrove forest in southern Thailand 1. Estimation by the trench method and the zonal structure of root biomass. *Journal of Tropical Ecology, 3*, 97–108.

Komiyama, A., Havanond, S., Srisawatt, W., Mochida, Y., Fujimoto, K., & Ohnishi, T. (2000). Top root biomass ratio of secondary mangrove (*Ceriops tagal* (Perr.) C.B Rob.) forest. *Forest Ecology and Management, 139*, 127–134.

Komiyama, A., Ong, J. E., & Poungparn, S. (2008). Allometry, biomass, and productivity of mangrove forests: A review. *Aquatic Botany, 89*, 128–137.

Koul, D. N., & Panwar, P. (2008). Prioritizing land management option for carbon sequestration potential. *Current Science, 95*(5), 658–663.

Krishnamurthy, K. (1985). *The changing landscape of the Indian mangroves*. Proceedings natural symposium biology, utilization and conservation of mangroves, Tamil Nadu, pp. 119–126.

Kristensen, E., Bouillon, S., Dittmar, T., & Marchand, C. (2008). Organic matter dynamics in mangrove ecosystems. *Aquatic Botany, 89*(2), 201–219.

Lee, Y. S. (1990). Primary productivity and particulate organic matter flow in an estuarine a mangrove wetland in Hongkong. *Marine Ecology, 106*, 453–463.

Mackenzie, F. T., Lerman, A., & Andersson, A. J. (2004). Past and present of sediment of carbon biogeochemical cycling models. *Biogeosciences, 1*, 11–32.

Mall, L., Singh, V. P., & Garge, A. (1991). Study of biomass, litterfall, litter decomposition and soil respiration monogeneric and mixed mangrove forest of Andaman Island. *Tropical Ecology, 32*, 235–249.

McKee, K. L. (1995). Interspecific variation in growth biomass partitioning and defensive characteristics of neotropical mangrove seedlings response to light and nutrient availability. *American Journal of Botony, 82*, 299–307.

Mitra, A., & Banerjee, K. (2004). In Col S. R. Banerjee (Ed.), *Living resources of the sea: Focus Indian Sundarbans* (p. 96). West Bengal: Canning Field Office, 24 Parganas (S).

Mitra, A., Banerjee, K., & Bhattacharya, D. P. (2004). *The other face of mangroves*. West Bengal: Department of Environment, Government of West Bengal.

Mitra, A., Banerjee, K., Sengupta, K., & Gangopadhyay, A. (2009). Pulse of climate change in Indian Sundarbans: A myth or reality? *National Academy Science Letters, 32*, 1–2.

Mmochi, A. J. (1993). *Ecology of mangrove ecosystems: Role of mangrove in dissolved inorganic nutrient fluxes, sediment budgets and litter supplies to Gesashi Bay, Higashi Village, Okinawa, Japan*. Thesis Master of Science, College of Science University of the Ryuykyus, p. 82.

Odum, E. P. (1968). *A research challenge: Evaluating the productivity of coastal and estuarine waters*. Proceedings of the 2nd sea grant conference, University of Rhode Island Kingston, pp. 63–64.

Odum, W. E., & Heald, E. J. (1972). Trophic analysis of an estuarine mangrove community. *Bulletin of Marine Science, 22*, 671–738.

Ong, J. E., Gong, W. K., & Clough, B. F. (1995). Structure and productivity of a 20-year old stand of *Rhizophora apiculata* BL mangrove forest. *Journal of Biogeography, 55*, 417–424.

Pandey, D. N. (2002). Global climate change and carbon management in multifunctional forests. *Current Science, 83*, 593–602.

Peterson, E., Hall, J. M., Rattray, E. A. S., Griffiths, B. S., Ritz, K., & Killham, K. (1997). Effect of elevated CO^2 on rhizosphere carbon flow and soil microbial processes. *Global Change Biology, 3*, 363–377.

Pressler, M. (1895). Das Gesetz der stambildung Leipzig, p. 153.

Putz, F. E., & Chan, H. T. (1986). Tree growth dynamics and productivity in a mature mangrove forest in Malaysia. *Forest Ecology and Management, 17*, 211–230.

Robertson, A. I., & Phillips, M. J. (1995). Mangroves as filters of shrimp pond effluent: Prediction and biogeochemical research needs. *Hydrobiology, 295*, 311–321.

Sampson, R. N., Moll Gary, A., & Kielbaso, J. J. (1992). Opportunities to increase urban forests and the potential impacts on carbon storage and conservation. In D. Hair & R. Neil Sampson (Eds.), *Global change volume one: Opportunities for increasing forest cover* (pp. 51–72). Washington, DC: American Forests.

Schimel, D. S. (1995). Terrestrial ecosystems and the carbon cycle. *Global Change Biology, 1*, 77–91.

Schlesinger, W. H. (1990). Evidence from chronosequence studies for a low carbon storage potential of soils. *Nature, 348*, 232–234.

Steinke, T. D., & Charles, L. M. (1984). Productivity and phenology of *Avicennia marina* and *Bruguiera gymnorrhiza* in Mgeni estuary, South Africa. In H. J. Teas (Ed.), *Physiology and management of mangroves* (pp. 25–36). The Hague: Dr. W. Junk Publisher.

Tamai, S., Nakasuga, T., Tabuchi, R., & Ogino, K. (1986). Standing biomass of mangrove forests in Southern Thailand. *Journal of Japanese Forest Society, 68*, 384–388.

Twilley, R., Lugo, A. E., & Zucca, C. P. (1986). Litter production and turnover in basin mangrove forest in South- West Florida, USA. *Ecology, 67*(3), 670–683.

Twilley, R. R., Chen, R. H., & Hargis, T. (1992). Carbon sinks in mangrove forests and their implications to the carbon budget of tropical coastal ecosystems. *Water, Air, and Soil Pollution, 64*, 265–288.

Vermatt, J. E., & Thampanya, U. (2006). Mangroves mitigate tsunami damage: A further response. *Estuarine, Coastal and Shelf Science, 69*, 1–3.

Walkley, A., & Black, I. A. (1934). An examination of the Degtjareff method for determining soil organic matter, and a proposed modification of the chromic acid titration method. *Soil Science, 37*, 29–38.

William, N. (2005). Tsunami insight to mangrove value. *Current Biology, 15*(3), 73.

Yadav, V. K., & Choudhury, A. (1985). *Litter production in mangrove forest of Lothian island in Sunderbans*. Proceedings natural symposium biology, utilization and conservation of mangroves, West Bengal, pp. 227–229.

6 Consumers of the Marine and Estuarine Ecosystems

Contents

The spectrum of life in the blue soup of the planet Earth starts with the primary producers, which encompass phytoplankton, seaweeds, salt marsh grass, seagrass and mangroves. The energy (through nutrition) and survival of consumers are direct function of the standing stock of primary producers. The consumers of oceans, seas, bays and estuaries feed on primary producers and acquire energy for performing various life processes. Depending on the environmental conditions, the food chains may be short or long. In extreme types of environments like Arctic or Antarctic, very short food chains are observed. Food chains basically represent complex interrelationships among organisms, in which case it is more appropriate to designate the pattern as food web (Fig. 6.1). Species may, however, change levels in the food chain or web at difference stages of their life cycle or consumers may feed more than one level.

The transference of food energy and nutrients from the members of the lowest trophic levels to highest trophic levels is represented through food chains and food webs. The members of lowest trophic levels in marine and estuarine waters are phytoplankton and other primary producers. The zooplankton occupy the second tier in food chains/food webs and the highest tiers are occupied by sharks, whales, etc. Sea birds are also within the domain of highest level consumer in marine and estuarine food webs.

A. Mitra, S. Zaman, *Basics of Marine and Estuarine Ecology*,
DOI 10.1007/978-81-322-2707-6_6

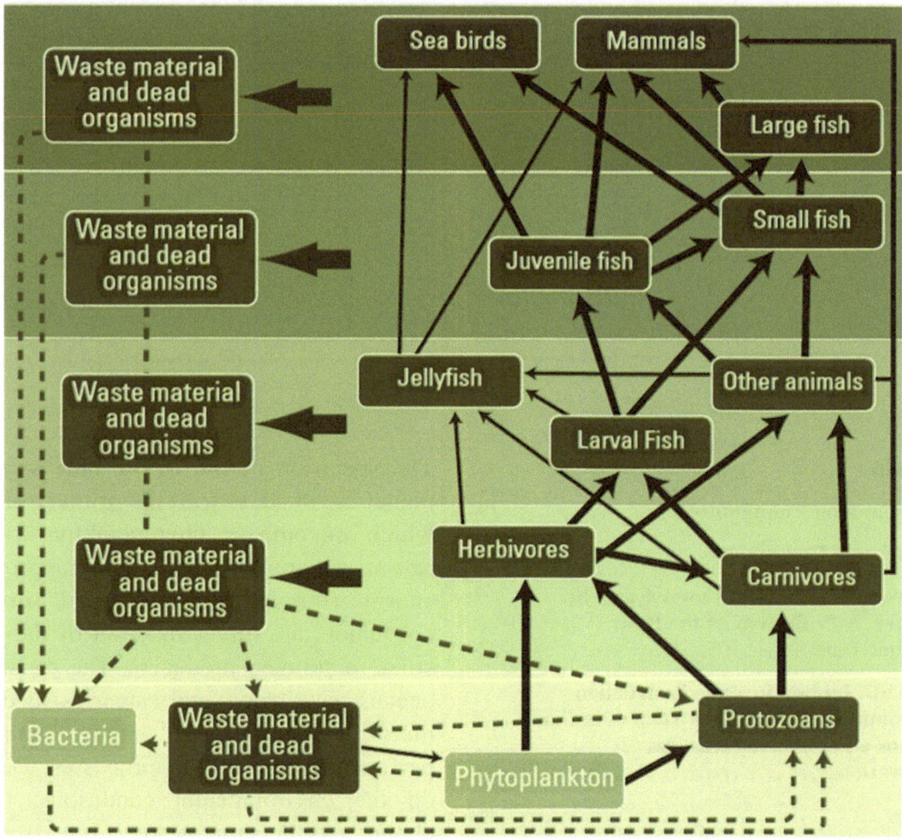

Fig. 6.1 A common food web in marine ecosystem

It is to be mentioned that the numbers of biomass of organisms decrease while approaching from the lowest to the highest trophic levels.

In marine and estuarine ecosystems, the upper trophic levels are occupied by sharks, skates and rays (Fig. 6.2). Marine mammals like dolphins and whales are also included in the higher trophic level of marine food web.

A brief account of the commonly observed secondary/tertiary consumers in the marine and estuarine ecosystems (that are primarily vertebrates) is highlighted in Table 6.1.

6.1 Zooplankton Community

Zooplankton community encompasses free-floating tiny animals that cannot move against the currents of marine and estuarine ecosystems. They include tiny animals in general, but bigger animals like jellyfishes are also included within zooplankton. Unlike phytoplankton, which can manufacture their own food, zooplankton are heterotrophic nature. Many species of zooplankton are herbivorous (that consume phytoplankton), while many others are carnivorous (that consume smaller zooplankton) and omnivorous (that feed on phytoplankton and zooplankton) (Fig. 6.3).

Zooplankton, which are the consumers in the second tier of marine and estuarine food chains, can also be subdivided into holoplankton and meroplankton depending on the time they spend in free-floating form in their life cycles. The holoplankton spend their entire life cycle in the aquatic phase as seen in case of copepods. The meroplankton, on the other hand, spend a part their life cycle as plankton. After a certain period of their life, they switch over to benthic or nektonic forms. The veligar larvae of oysters are

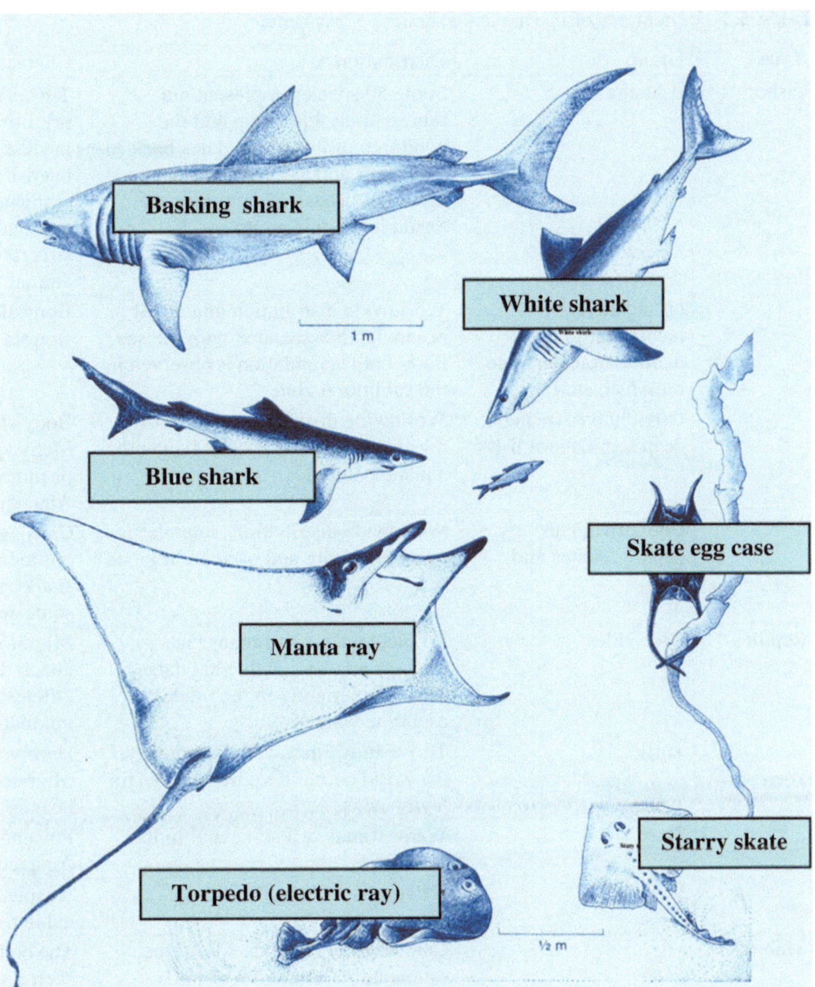

Fig. 6.2 Sharks and skates

unique example in this context. After spending a part of their life as larvae, they settle on hard substrate as spats, which finally grow into adult (Fig. 6.4).

Most of the ichthyoplankton also become nektonic during their adult phase, when they can swim against the current. The planktonic and nektonic phases of bony fishes are highlighted in Figs. 6.5 and 6.6, respectively.

Zooplankton are found in almost all the layers of the photic zone of the ocean. They are potentially limited by two factors in the coastal and estuarine zones: firstly by turbidity which can limit phytoplankton production and thus restrict the ration supply for the zooplankton community and secondly by currents which, particularly in small estuaries are dominated by high river flow that usually carries the zooplankton out to the sea. We conducted a monitoring programme on the zooplankton diversity in the East Coast of India during 4–20 March 2014 and observed 88 species of zooplankton along with various fish larvae (ichthyoplankton). In this investigation sampling was carried out along the mouth of five estuaries, namely, Hooghly (in West Bengal), Mahanadi (in Odisha), Rushikulya (in Odisha), Godavari (in Andhra Pradesh) and Krishna (in Andhra Pradesh). These rivers drain huge amount of fresh water and sediment to the bay, particularly in the monsoonal season, and thus regulate the primary and secondary productivity in the near-shore aquatic ecosystem of the

Table 6.1 Vertebrates of marine and estuarine ecosystems

Type	Group	Distribution	Characteristics
Fishes	Agnatha	Some 50 species at present are representing this group and the evolution of this group dates back to Cambrian about 550 million years ago. Hag fishes and lampreys are examples of this group	This class encompasses the most primitive varieties of fish, mainly the jawless lampreys and hagfish. Both hagfish and lampreys possess elongated and eel-like body without any scales. A sucking disc-like structure surrounds the mouth of this animal
	Osteichthyes (substratum-based: halibut, flounder, sole, rock fish, etc.)	Worldwide distribution and dwell in ocean depth associated with the sea floor. Large population is observed in the sublittoral zone	Bony skeleton with scales and fins. Bottom and near-bottom feeders
	Osteichthyes (pelagic, deep sea: lantern fish)	Worldwide distribution and mostly dwell between 200 and 3000 m in the aquatic column	Bony skeleton with scales and fins. Many species exhibit the phenomenon of bioluminescence. Abundant and small in size
	Chondrichthyes (sharks, skates and rays)	Worldwide distribution; subpolar to tropics; inshore and offshore regions	Cartilaginous skeleton. Sensory abilities are exceptional. Large sharks and rays are plankton feeders; some species are predators
Reptiles	Crocodiles	Tropical waters and mangrove swamps are the suitable habitats. Thrive luxuriantly in the muddy estuarine waters	Migratory in nature. Ferociously attack their prey in the territory. Often seen basking on the intertidal mudflats
	Turtles	Thrive in warmer waters throughout the world ocean. Require beaches for laying eggs	Herbivores and carnivores. Migratory in nature. Long-lived, but heavily impacted by humans
	Snakes	Warm waters of Pacific and Indian Oceans	Poisonous in most cases. Fish eaters. Adapted for swimming and diving
	Lizards	Galápagos Islands, Pacific	Adapted for swimming in marine habitat. Herbivorous in nature
Seabirds		Comprises only about 3 % of the estimated 9700 species of birds. Seabirds are distributed from pole to pole, and their impact on marine life is significant. Most of the species feed on fish, squid and bottom invertebrates, but some feed on plankton	The bodies of sea birds are adapted in such a way that the heat energy is conserved within their body. For this purpose, the entire body is covered with feathers, which are waterproof in nature. An oily substance, secreted from the gland situated at the basal region of the tail, serves the purpose of waterproofing. Birds preen by rubbing the oil into their feathers with their beaks. Flight is made easier by their light, hollow bones. Furthermore, their eggs have hard shells that are more resistant to water loss than those of reptiles
Mammals			
Whales	Toothed	Worldwide distribution; polar to tropic ocean and coastal waters. Some species exhibit migratory behaviour	Carnivores and prey on fish, squid and seals. Included in this group are sperm and killer whales, dolphins and porpoises
	Baleen	Worldwide distribution from polar to tropic open ocean and coastal waters. Some species exhibit migratory behaviour	Includes the great whales. Many populations are getting depleted by whaling industry. Krill feeders

(continued)

Table 6.1 (continued)

Type	Group	Distribution	Characteristics
Pinnipeds	True seals	Distributed mainly between polar to midlatitudes; require shore or sea ice areas	No external ears. Torpedo shape. Non-rotating flippers. Consume a variety of vertebrates. Harbour seal, elephant seal are typical examples
	Eared seals	Mainly distributed between polar to midlatitudes; require shore or sea ice areas	External ears, longer necks and rotating flippers are present. Consume a variety of invertebrates and vertebrates. Examples are sea lion and northern fur seal
Walrus		Distributed in the Arctic regions only; need shore and sea ice areas	Tusks in males and females. Up to 2 tonnes in size with rotating flippers. Absence of external ears. Feeds on shellfish
Sea cows	Dugongs	Distributed in tropical regions, Australia, Southeast Asia and Africa coastal areas and common to Pacific	Docile, slow-moving herbivores and prefer to eat seagrass. They have fluke-like tails and tusks and are smaller than manatees. Dugongs are confined to saltwater only. Heavily impacted by humans
	Manatees	Manatees are found in the Caribbean towards South America, Africa and Gulf of Mexico coastal areas	The manatees are herbivores and depend on marine vegetation for their nutrition. They are extremely slow moving in nature. They are different from dugongs with respect to length of snout. Manatees possess snout of shorter length compared to the snout of dugongs. Paddle-like tails are also the characteristic features of manatees. Manatees have incisors. Heavily impacted by humans
Sea otters		Distributed in the sub-Arctic and midlatitude, cold coastal waters	No blubber layer. Fur is the only insulation. Eat and sleep in the aquatic sub-system. Feed on shellfish, crabs and sea urchins

bay. Zooplankton samples were collected at each station from surface water by horizontal hauling of zooplankton net (mouth area 0.25 m^2, mesh size 800 μm) for 10 min and preserve in 4 % neutral formaldehyde. In the laboratory, zooplankton samples were subsampled with a help of Folson plankto spitter for quantitative and qualitative analysis. An aliquot of the sample was taken from the subsample and observed under an inverted microscope (Cippon; Model No. 21033) for identification and counting. The numerical abundance values were represented in nos./100 m^3. Relative abundance was computed from total density and the density of each group. Different groups/species of zooplankton were identified referring standard literatures (Kasturirangan 1963; Wimpenny 1966; Newell and Newell 1977; Conway et al. 2003). Finally the Shannon–Wiener species diversity index was computed on the basis of relative abundance of each species and total number of individuals of all the species (Table 6.2). The main purpose of this study was to evaluate the congenial environment for zooplankton. On this background the order of best to worst environment were Godavari (4.149) > Mahanadi (4.039) > Krishna (3.897) > Rushikulya (3.774) > Hooghly (3.606). The high pollution load in the Hooghly estuary (Mitra and Choudhury 1994a; Mitra 1998; Ray Chaudhuri et al. 2014; Mitra

Fig. 6.3 Zooplankton community

Fig. 6.4 Adult oysters (sedentary in nature) on hard substrata (pillars of jetties)

and Zaman 2015) may be attributed to lowest species diversity index of zooplankton. Thus zooplankton also act as proxy to health of the aquatic ecosystem.

The zooplankton biomass can increase the fishery productivity because they chiefly consume the primary producers (phytoplankton) and form the major food source for members of higher trophic levels in which several species of osteichthyes and chondrichthyes exist. Despite several uses of zooplankton in terms of energy flow through aquatic food chain, they are often

Fig. 6.5 Planktonic stage of osteichthyes

Fig. 6.6 Nektonic stage of osteichthyes

destroyed in huge quantum in some regions of India and Bangladesh during the process of screening the coastal/estuarine waters for tiger prawn seeds. India researchers documented the destruction of several species of ichthyoplankton (a major component of the zooplankton community) during tiger prawn seed collection, and this operation is carried out due to lack of hatcheries in some maritime states of Indian subcontinent like West Bengal, Odisha, etc., because of low salinity. A detailed report of ichthyoplankton destruction along with remedial measures is highlighted as Annexure 6B.

Zooplankton are classified according to their habitat, depth distribution, size and duration of planktonic life (Tables 6.3, 6.4, 6.5 and 6.6), Fig. 6.7.

6.1.1 Zooplankton: Do They Maintain Uniformity in Their Life Timeline?

Zooplankton do not maintain uniformity in their life history. Some species (as stated in Table 6.6) remain as free-floating organisms throughout

Table 6.2 Population density of zooplankton (No./100 m^2) and Shannon–Wiener species diversity index of the zooplankton community in East Coast of Indian subcontinent

S.no.	Species	Hooghly	Mahanadi	Rushikulya	Godavari	Krishna
Protozoa						
Acantharia						
1.	*Acanthometron* sp.	0	0	0	34	112
Ciliata						
2.	*Tintinnopsis tocantinensis*	145	0	0	67	0
3.	*Tintinnopsis nordqvisti*	129	0	0	83	0
Cnidaria						
Hydroidomedusae						
4.	*Cladonema* sp.	0	0	0	0	3
5.	*Podocoryne* sp.	0	0	0	91	0
6.	*Phialella quadrata*	0	0	12	4	31
Siphonophorae						
7.	*Agalma elegans*	44	67	56	29	60
8.	*Sulculeolaria turgida*	0	0	87	11	13
9.	*Diphyes bojani*	37	11	10	3	41
10.	*Muggiaea* sp.	7	6	0	76	81
11.	*Bassia bassensis*	0	48	0	56	45
Ctenophora						
12.	*Pleurobrachia pileus*	12	60	70	87	39
13.	*Beroe* sp.	29	48	23	61	32
Annelida						
14.	Polychaete larva	30	78	155	105	45
Arthropoda						
Copepoda (Calanoida)						
15.	*Nannocalanus minor*	77	19	56	65	58
16.	*Mesocalanus tenuicornis*	20	40	40	20	55
17.	*Eucalanus* sp.	0	0	30	0	0
18.	*Eucalanus attenuatus*	11	20	23	0	38
19.	*Acrocalanus gracilis*	6	34	50	76	65
20.	*Paracalanus aculeatus*	157	85	109	60	93
21.	*P. parvus*	47	55	16	67	69
22.	*Paracalanus* sp.	28	30	80	56	99
23.	*Euchaeta marina*	81	78	78	75	14
24.	*Euchaeta indica*	45	78	10	56	31
25.	*Scolecithrix danae*	0	0	0	0	0
26.	*Centropages orsini*	60	40	76	71	93
27.	*C. furcatus*	67	50	99	88	50
28.	*Centropages tenuiremis*	10	40	24	19	75
29.	*Centropages dorsispinatus*	0	0	80	0	0
30.	*Centropages* sp.	98	119	33	87	97
31.	*Pseudodiaptomus aurivilli*	0	0	0	0	0
32.	*P. serricaudatus*	0	0	0	0	0
33.	*Temora discaudata*	23	10	51	51	79
34.	*T. turbinata*	46	86	82	20	90
35.	*Metacalanus aurivilli*	20	40	33	19	78
36.	*Candacia discaudata*	18	45	34	133	66
37.	*Paracandacia simplex*	12	13	9	103	90
38.	*Labidocera acuta*	85	54	0	44	92
39.	*Labidocera minuta*	0	48	0	0	0

(continued)

Table 6.2 (continued)

S.no.	Species	Hooghly	Mahanadi	Rushikulya	Godavari	Krishna
40.	*Calanopia minor*	39	33	0	51	63
41.	*Pontellina plumata*	31	26	53	42	50
42.	*Acartia centrura*	53	29	81	46	90
43.	*Acartia spinicauda*	48	29	11	51	62
44.	*Acartia danae*	46	48	23	37	88
45.	*Tortanus barbatus*	39	42	51	79	93
Copepoda (Cyclopoida)						
46.	*Corycaeus catus*	61	21	13	12	87
47.	*Corycaeus longistylis*	16	55	32	20	60
48.	*Farranula gibbula*	39	21	0	53	74
49.	*Copilia quadrata*	25	81	24	19	75
50.	*Farranula curta*	53	19	0	61	89
51.	*Corycaeus typicus*	70	55	3	22	80
52.	*Oithona brevicornis*	23	11	45	49	101
53.	*O. spinirostris*	63	83	29	75	84
54.	*Oithona* sp.	69	52	45	12	91
55.	*Oncaea venusta*	53	36	25	87	25
56.	*Oncaea conifera*	30	40	48	23	31
57.	*Oncaea* sp.	0	0	60	45	19
Copepoda (Harpacticoida)						
58.	*Microsetella norvegica*	77	93	23	41	88
59.	*M. rosea*	0	0	3	15	95
60.	*Macrosetella gracilis*	0	60	0	80	99
61.	*Clytemnestra scutellata*	0	8	0	19	53
62.	*Euterpina acutifrons*	0	0	17	0	0
Cladocera						
63.	*Evadne nordmanni*	0	43	0	56	95
64.	*Evadne tergestina*	77	68	83	17	56
65.	*Penilia avirostris*	0	55	72	40	83
Ostracoda						
66.	*Macrocypridina castanea*	132	34	50	67	69
67.	*Conchoecia elegans*	68	45	33	51	91
Amphipoda						
68.	*Hyperia* sp.	29	43	0	14	84
Isopoda						
69.	Isopod larvae	0	0	0	48	79
Decapoda						
70.	*Lucifer penicillifer*	0	0	0	63	53
71.	*Sergestes* sp.	0	0	0	91	88
72.	Brachyuran megalopa larva	203	116	79	114	109
73.	Protozoea of *Lucifer*	0	0	0	133	1397
74.	Protozoea of *Penaeus indicus*	41	33	29	21	97
75.	Zoea larva of *Elamen* sp.	46	84	13	30	90
76.	Zoea larva of porcellanid crab	41	33	20	56	82
Other crustaceans						
77.	Larvae of euphausiid	41	33	25	50	76
78.	Alima larva of *Squilla*	19	46	53	26	88
79.	Cyprid larvae	50	40	51	66	84

(continued)

Table 6.2 (continued)

S.no.	Species	Hooghly	Mahanadi	Rushikulya	Godavari	Krishna
Mollusca						
80.	Veliger larva	85	111	148	66	130
Pteropoda						
81.	*Creseis acicula*	33	29	17	44	51
82.	*Creseis* sp.	0	95	0	80	105
Echinodermata						
83.	Ophiopluteus larva	84	66	43	93	104
Chaetognatha						
84.	*Sagitta bipunctata*	302	221	197	116	413
85.	*Sagitta enflata*	29	102	56	54	61
Phoronida						
86.	Actinotroch larvae	21	33	7	84	116
Brachiopoda						
87.	Brachiopod larvae	0	5	0	0	23
Bryozoa						
88.	*Cyphonautes* larva	0	0	0	0	54
Chordata						
89.	Ichthyoplankton	127	106	383	85	99
N		3607	3585	3371	4321	7308
H		3.606	4.039	3.774	4.149	3.897

Table 6.3 Classification of zooplankton on the basis of habitat

Type	Description
Oceanic plankton	These are marine zooplankton that inhabit beyond the continental shelf
Neritic plankton	These zooplankton inhabit waters overlying continental shelves. These waters are often very productive as they receive the run-off from the adjacent landmasses that triggers the phytoplankton growth in these regions
Brackish water plankton	These zooplankton inhabit estuarine regions, where there is a continuous mixing of freshwater and sea water. The zooplankton species of this category have wide range of tolerance to different dilution factors. Such zooplankton are very common in the shrimp culture farms and form important diet of the prawns

Table 6.4 Classification of zooplankton on the basis of depth distribution

Type	Description
Neuston	The zooplanktons of this category are restricted at the top few millimetres (usually 10 mm) of the surface micro layer
Pleuston	These are widely distributed at the surface of the sea (with parts of the body sometime projecting above the water)
Epipelagic	These are distributed between 0 and 300 m water column, e.g., siphonophores, arrow worms, etc.
Mesopelagic	The zooplankton of this category are restricted within the depth 300–1000 m, e.g., euphausiids, chaetognath, etc.
Bathypelagic	These are restricted within the depth 1000 and 3000 m, e.g., foraminifera, euphausiids etc.
Abyssopelagic	The waters overlying the vast abyssal plains of the ocean are inhabited by a variety of zooplankton species, which is often referred to as abyssopelagic zooplankton. These zooplankton are thus restricted between 3000 and 4000 m

Source: Santhanam and Srinivasan (1998)

Table 6.5 Classification of zooplankton on the basis of size

Type	Size range
Nannozooplankton	<20 μm
Microzooplankton	20–200 μm
Mesozooplankton	200 μm–2 mm
Macrozooplankton	2–20 mm
Megazooplankton	>20 mm

Source: Santhanam and Srinivasan (1998)

the body of phytoplankton. These areas exhibit dense population of fishes as revealed from the Potential Fishing Zone (PFZ) maps.

The body of fishes is spindle shaped or streamlined due to which they can penetrate and move swiftly through the water column. The fins with rays provide force to the fishes to move against the currents. The fins also help the fishes to orient their direction and maintain the body

Table 6.6 Classification of zooplankton on the basis of duration of planktonic life

Type	Description
Holoplankton	This group includes organisms which are planktonic throughout their life cycle, e.g., tintinnids, cladocerans, copepods, chaetognaths, etc. Figure 6.8 shows a very common holoplankton found in the marine and estuarine waters of tropics
Meroplankton	This group encompasses those organisms which remain free floating only during certain phase of their life cycle, e.g., larvae of benthic invertebrates and fish larvae (ichthyoplankton). In the estuaries of Indian Sundarbans, larvae of prawn (*Macrobrachium rosenbergii*) are found in plenty (Fig. 6.9), which finally become a semi-benthic species in the adult stage (Fig. 6.10)

Source: Santhanam and Srinivasan (1998)

their entire life cycle, while others switch off to different modes like nekton or benthos. The former are designated as holoplankton, while the latter meroplankton.

balance.

Fish constitute the largest and most diverse group of marine vertebrates.

They are taxonomically separated into three classes: (i) Agnatha, (ii) Osteichthyes and (iii) Chondrichthyes.

6.2 Vertebrate Community

Animals with bony vertebral column that surround a nerve cord are referred to as vertebrates. Vertebrates are also characterized with the presence of a brain enclosed within the skull. Present-day vertebrates include fishes, amphibians, reptiles, birds and mammals.

6.2.1 Fishes

Fishes dominate the nekton of the oceans, seas and estuaries. They are widely distributed at all depths in the oceans, seas, bays and estuaries. The distribution and abundance of fishes are largely controlled by the primary producers, which provide nutrition to the fish directly or indirectly (through zooplankton). The areas of upwelling in the oceans are rich in nutrients like nitrate, phosphate, silicate, etc., which constitute

6.2.1.1 Agnatha

The Agnatha are the most primitive types of fishes which encompasses lampreys and hag fishes. These fishes lack jaws (Fig. 6.11).

The evolution of Agnatha occurred during Cambrian period about 550 million years ago. There are about 50 species of Agnatha in the present time. The bodies of the representatives belonging to Agnatha are eel-like without any bony scales. There is contrasting variation in the mode of nutrition between the hagfishes and lampreys. The hagfishes depend on dead or dying organism for their food by burrowing into the internal part of their prey. The lampreys on the other hand are parasitic in nature and get attached to the body of the host by their suckers. They cut the flesh of the host and feed on their soft parts and body fluids. The lampreys spend a considerable portion of lives in freshwater system.

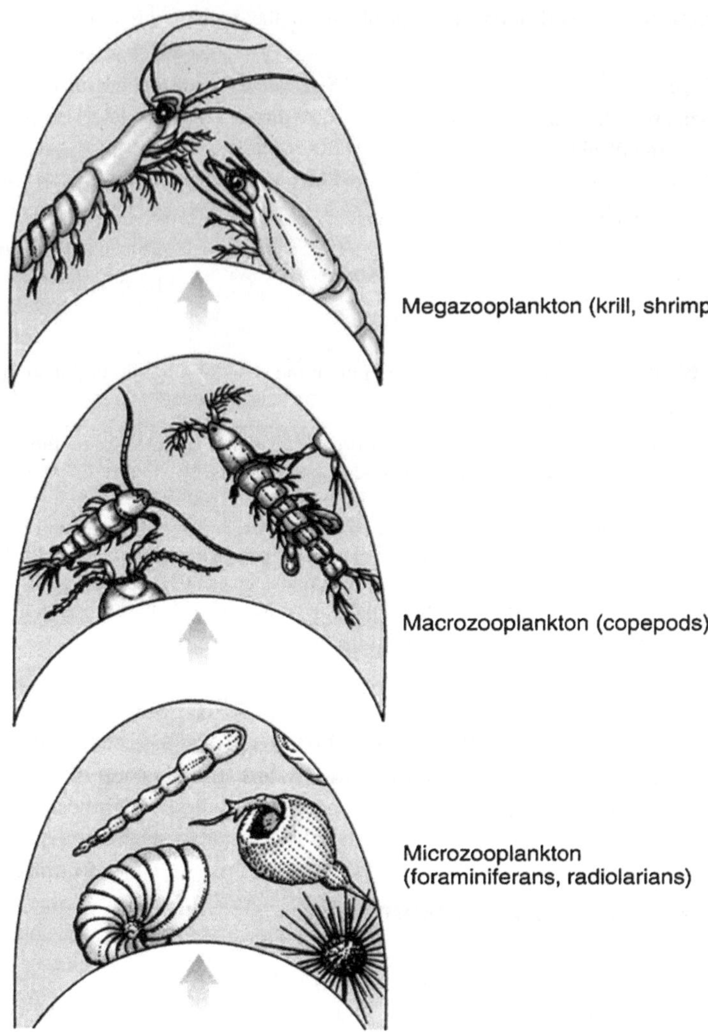

Fig. 6.7 Variation of zooplankton size

6.2.1.2 Osteichthyes

Osteichthyes are bony fishes that are widely distributed in the surface and column of the waters in oceans and estuaries. They have a streamlined body, which enable them to travel over long distance against the current. Most of the commercially important fishes available in the market belong to the class Osteichthyes.

Estuaries are rich in fish resources. Most of the commercial fishes with high economic value are available in the estuarine waters. Representative bony fishes that are fished commercially from the waters of mangrove-dominated Sundarbans in the lower Gangetic delta are shown in Fig. 6.12a, b. The aquatic subsystem of deltaic Sundarbans is the dwelling spot, nursery and breeding ground of a wide variety of finfish and shellfish. However, with respect to ecological tolerance, a large fraction of the fish species in the mangrove-dominated estuarine complex are euryhaline in nature and move freely from the upper stretch of minimum salinity to lower stretch of maximum salinity.

Several workers have depicted the taxonomic diversity of fish species in the aquatic subsystem of mangrove-dominated Indian Sundarbans deltaic complex. Pillay (1967) estimated the species number to be more than 120. Jhingran (1982) documented a total of 172 species and stated that the diversity is comparatively more in the

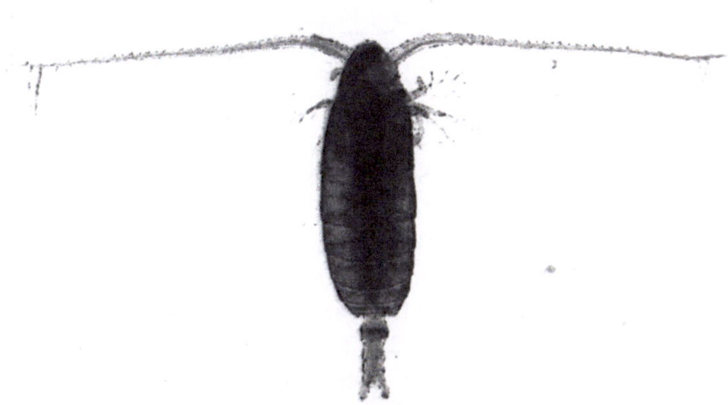

Fig. 6.8 Copepod: a common holoplankton

Fig. 6.9 Larval stage (plankton) of *Macrobrachium rosenbergii*

Fig. 6.10 Adult stage (prefers substrate) of *Macrobrachium rosenbergii*

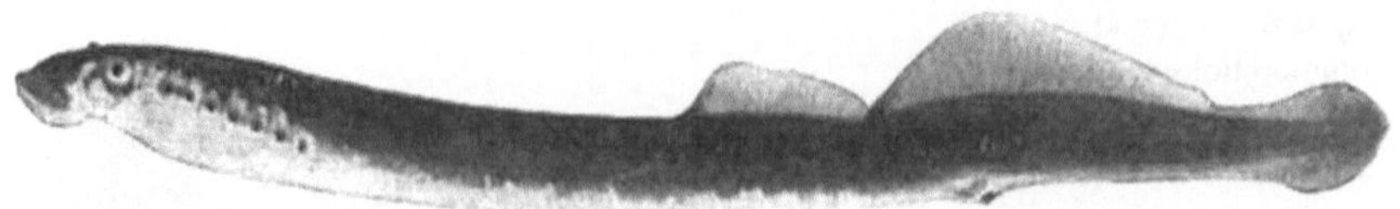

Fig. 6.11 Agnatha

high saline zone of Indian Sundarbans. His estimate reveals 73 species of freshwater origin and 99 species of marine/higher salinity origin. Mandal and Nandi (1989) documented 141 species under 100 genera, while Chaudhuri and Choudhury (1994) recorded 250 species under 96 genera in the aquatic sub-system of Indian Sundarbans. Khan (2003) recorded 107 species from Sundarban Biosphere Reserve region, but this figure does not include the species restricted in the low saline upper zone of the Hooghly–Matla estuarine complex. The fish fauna of the estuarine waters around Sundarbans has been classified into residents and transients (migrants). The species whose individuals of different sizes are present during all the months of the year in any zone of the estuary are referred to as resident species. The important resident species of fish are *Mugil parsia*, *Mugil tade*, *Polynemus paradiseus*, *Polydactylus indicus*, *Otolithoides biauritus*, *Lates calcarifer*, *Hilsa toli*, *Arius jella*, *Harpodon nehereus*, *Setipinna taty*, *Ilisha elongata*, *Setipinna phasa*, *Coilia ramcarati*, *Pama pama* and *Sillaginopsis panijus*. The transient or migratory fishes enter and stay in the Bay of Bengal associated estuaries for a short period. Depending on their migratory pattern and direction, the migrants may be divided into three categories (Jhingran 1982):

1. Marine forms that migrate upstream and spawn in freshwater areas of the estuary like *Tenualosa ilisha*, *Polynemus paradiseus*, *Sillaginopsis panijus and Pama pama*.
2. Freshwater species, which spawn in saline area of the estuary like *Pangasius pangasius*.
3. Marine species, that spawns in less saline water of the estuary like *Arius jella*, *Osteogeneious militaris* and *Polydactylus indicus*.

The representatives of osteichthyes that live in the deeper waters greater than 300 m are low in terms of taxonomic diversity compared to those that inhabit the surface waters. In fishes dwelling in the deeper part of the ocean, bioluminescence is often witnessed. Lantern fish is a prominent example in this context. Bioluminescence is the result of the oxidation of luciferin in presence of the enzyme luciferase.

In the dim transitional layer between 200 and 1000 m (600–3000 ft), the waters support vast schools of small luminous fish. The many species of *Cyclotbone* are believed to be the most common fish in the sea. Deeper-living species are black, while the shallower-living species are silvery, to blend with the dim light. The lantern fish has a worldwide distribution; some 200 species are distinguished by the pattern of light organs along their sides. They are a major item in the diet of tuna, squid and porpoises. *Stomias*, a fish with a huge mouth, long, pointed teeth and light organs along its sides, is also found here, as is the large-eyed hatchel fish. These fish prey on the great clouds of euphausiids and copepods found at this depth.

In perpetually dark zones are predators with highly specialized equipment for catching their prey. These fish are equipped with light-producing organs used as lures, large teeth that in some species fold backwards toward the gullet so their prey cannot escape and gaping mouths with jaws that unhinge to accommodate large fish. Among the most famous of these predators is *Saccopharynx* with its funnel-like throat and tapering body. When the stomach is empty, the fish appears slender, but it expands to accept anything the great mouth can swallow. The female angler fish has a dorsal fin modified into a fishing rod with an illuminated lure dangling just above her jaws.

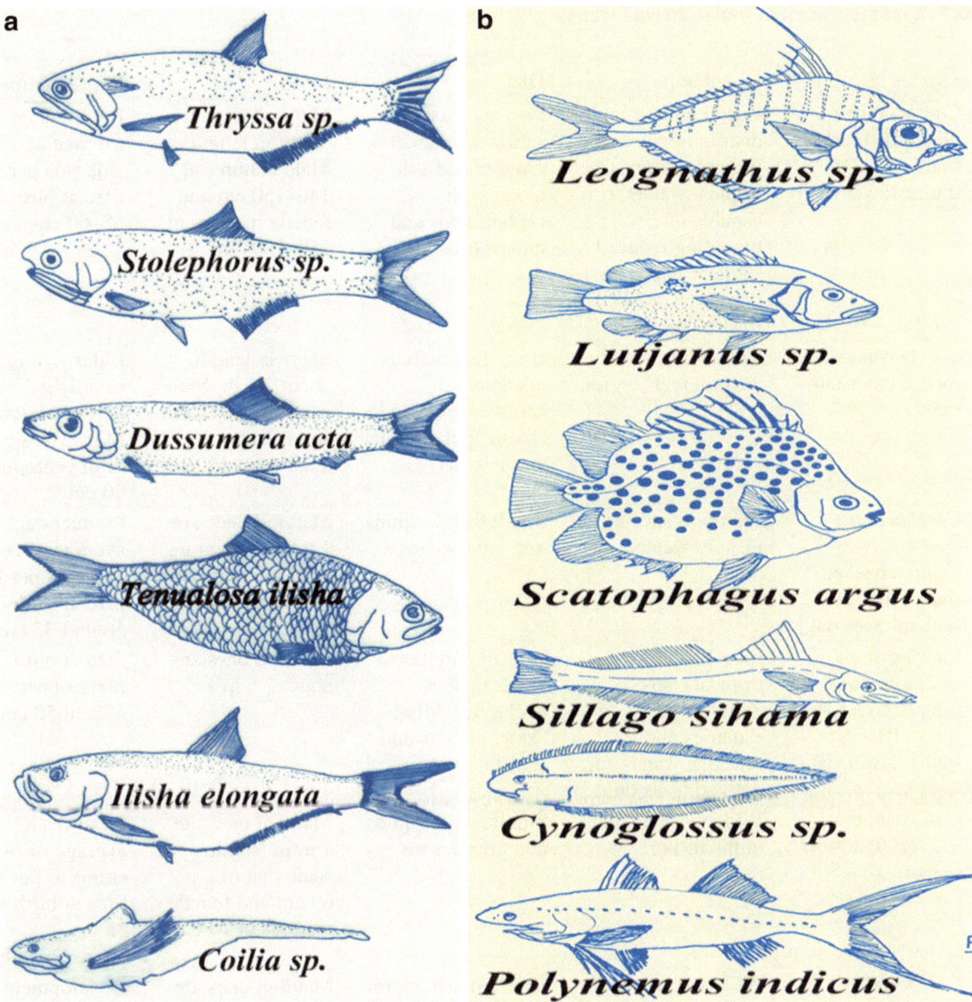

Fig. 6.12 (**a**) Commercially important fin fish species of deltaic Sundarbans. (**b**) Commercially important fin fish species of deltaic Sundarbans

6.2.1.3 Chondrichthyes

The chondrichthyes comprised of sharks, skates and rays have cartilaginous body. Sharks have spindle shaped body with powerful muscles to swim against the current. Skates and rays, on the other hand, have flattened bodies and lead a bottom-dwelling life.

Sharks and rays typically have internal fertilization and low fecundity, producing only small numbers of relatively large eggs. Most sharks and all rays give birth to live young. Skates lay their eggs in protective cases that are attached to a substrate and the young hatch from these within a few weeks or months. There is an increasing demand for shark meat and shark fins, the latter being considered a delicacy in Asia. Shark fin soup is a delicious dish in many hotels. Numbers of sharks have sharply and rapidly declined in regions where fishing has intensified, and many sharks are also captured incidentally in commercial fishing operations for other species.

In the Indian coastal waters, the common shark species are highlighted in Table 6.7.

As stated earlier the chondrichthyes (preferably the sharks) are presently under threat due to several reasons Sharks have wide use starting from food to pharmaceutical industries. The

Table 6.7 Common shark in Indian coastal waters

Sl. no.	Species	Distribution	Diet	Length	Reproduction
1.	*Carcharhinus limbatus* (blacktip shark) [IUCN status: lower risk]	Cosmopolitan in distribution near the inshore regions of tropical waters. It is capable of tolerating reduced salinity, but never penetrates into freshwater	Fish such as sardines, mackerel, croaker and sole along with cephalopods and crustaceans	Matured ones are 2.5 m in length. Male matures at 140–150 cm and female matures at 150–160 cm	Produces an average of six embryos per litter. Size at birth is 55–60 cm
2.	*Carcharhinus sorrah* (spot-tail shark)	Commonly found in coral reefs region	Bony fish such as mackerels, sardines, etc., and shellfish like squids and prawns	Short in length, about 2.5 m. Male matures at 115 cm and female matures at 120 cm	Litter size is 2–6 usually delivered during March to May. The size at birth is around 40 cm
3.	*Carcharhinus dussumieri* (white-cheeked shark) [IUCN status: lower risk]	Common species in inshore waters	Small fishes, squids and crustaceans	Matured ones are 1 m. Male matures at 65 cm and female matures at 75 cm	Produces an average of two embryos per litter. Size at birth is around 35 cm
4.	*Carcharhinus melanopterus* (blacktip reef shark) [IUCN status: lower risk]	Indo-Pacific tropical shark migrates into estuaries and brackish waters for delivering its pups	Fish-like mullets, silver bellies, anchovies, hilsa, skate, prawns and squilla	Matured ones are about 2.5 m	Size at birth ranges between 45 and 50 cm
5.	*Carcharhinus macloti* (hardnose shark)	Widely distributed in the inshore region	Small fishes, squids and crustaceans	Matured ones are 1 m in length. Male matures at 60 cm and female matures at 70 cm	Produces an average of two embryos per litter. Size at birth is 35 cm approximately
6.	*Galeocerdo cuvier* (tiger shark) [IUCN status: lower risk]	Widely distributed tropical shark. It is capable of cruising in mid-ocean and shows nocturnal movement into bays and estuaries	Eels, cat fish, parrot fish, flat fish, flying fish, skates, rays, marine turtle, sea snake, sea birds, sea lion, dolphin, etc.	Matured ones are about 7.4 m in length	Development is oviparous and the litter size is between 10 and 82. The size at birth is 50–75 cm
7.	*Scoliodon laticaudus* (Indian dog shark) [IUCN status: lower risk]	West and south coasts of India, dominant in East Coast and in Indian Sundarbans	Small fishes, crustaceans and squids	The majority of males are 50–55 cm and females are about 65 cm. Males and females mature at 30 and 35 cm, respectively	Produces up to 20 embryos per litter. The size at birth is 14.5 cm
8.	*Rhizoprionodon acutus* (Brazilian sharpnose shark)	Abundant in the shore waters, particularly in the West Coast of India during September to February and East Coast during summer months	Small fishes, squids, cuttle fish, crabs and shrimps	Matured ones are about 1 m in length	Produces an average of 2–6 embryos per litter, which are 26–27 cm long

(continued)

Table 6.7 (continued)

Sl. no.	Species	Distribution	Diet	Length	Reproduction
9.	*Sphyrna lewini* (hammerhead shark) [IUCN status: vulnerable]	Abundantly found in the Indian seas, notable for its unique migratory behaviour	Sardines, anchovies, mackerel, eel, milkfish and sole fish. It also feeds on sharks and rays	Matured ones are up to 4.2 m in length	Development is viviparous and produces an average of 15–30 embryos per litter. The size at birth is 45–55 cm
10.	*Rhincodon typus* (whale shark) [IUCN status: vulnerable]	Appears occasionally at Indian coastal waters	Filter feeder and is believed to sieve plankton as small as 1 mm diameter through the fine mesh of their gill rakers	Largest length up to 1200 cm but mean length is around 700 cm	Ovoviviparous; embryo length varies from 48 to 58 cm
11.	*Atelomycterus marmoratus* (coral cat shark)	Appears occasionally at Indian coastal waters	Squids, crabs, etc.		
12.	*Glyphis gangeticus* (river shark) [IUCN status: extremely endangered; Wildlife Protection Act: Schedule I]	Hooghly–Ganges river system of West Bengal coast	Fish	Maximum length is around 200 cm. Matured males are about 178 cm	Viviparous. Newly born individuals are 56–61 cm long
13.	*Glyphis glyphis* (sharptooth shark or speartooth shark)	Confined to turbid waters of rivers, estuaries and inshore waters of coastal West Bengal	Fish	Minimum length is around 100 cm	Live bearing probably with yolk-sac placenta
14.	*Glyphis siamensis* (Irrawaddy river shark)	Confined to turbid waters of rivers, estuaries and inshore waters of coastal West Bengal	Fish	Minimum length is around 63 cm	Live bearing probably with yolk-sac placenta

Source: Mitra and Banerjee (2005)

value of shark (in Indian currency) is shown in Table 6.8.

Besides the fish meat, the shark meat has become popular as a delicious food. Nowadays shark fins are exported to foreign countries. Shark fins are usually obtained from species like *Sphyrna zygaena*, *Rhizoprionodon acutus*, *Scoliodon laticaudus*, *Carcharhinus melanopterus*, etc. The shark community, in the present *era*, is under threat due to various indirect and direct benefits obtained from shark (Table 6.9).

In coastal West Bengal (a maritime state in the northeast coast of India), sharks are trapped mainly for the purpose of meat, fins and liver. The catch of shark has progressively increased since 1994 in the coastal and estuarine sectors of the state (Fig. 6.13a–h).

Incidence of poaching whale shark in recent year (2003) has also been recorded from the Digha coast of West Bengal (Fig. 6.14).

Table 6.8 Valuation of shark parts

Product	Minimum (Rs/kg)	Maximum (Rs/kg)
Shark meat	25	30
Shark bones	70	75
Shark fins	280	340

Source: http://www.tao.org/DOCREP/005/X3690E/x3690elq.htm

Deterioration of water quality due to severe anthropogenic pressure is also a major threat to shark community. In the last few decades, the aquatic phase in and around Indian Sundarbans has undergone considerable change due to overcrowding of several small- and medium-scale industrial units, tourism units, brick fields and several fishing vessels and trawlers (that release considerable concentrations of Zn, Cu and Pb during the process of conditioning with antifouling paints). Many species of shark (particularly *Glyphis* sp.) cannot resist such alteration of water quality due to poor adaptive power.

At the regional level, the following steps are useful for shark conservation:

- A national fishery management policy has to be formulated and tuned to meet the guidelines of species conservation. The interaction and coordination between national survey, academic institutes, research organizations, forest department and fisherman co-operative societies should be improved. Implementation of modern technologies has to be encouraged for judicious exploitation of natural resources. Awareness has to be created *among fishermen* for their active participation in planning, management and implementation of shark conservation programme. Mechanism needs to be evolved to monitor the shark catch by vessel exclusively meant for the purpose.
- Strict regulations are required to control fishing of cartilaginous species at regional level (particularly those enlisted as endangered, vulnerable, rare, threatened and commercially threatened). This could be achieved by regulating the number and size of the mechanized crafts and strict monitoring at the landing stations. New gear has to be certified by government agencies before introduction and code and mesh size are to be strictly regulated. The best method for checking overfishing is to declare off season especially during spawning and breeding months.
- Training to the forest personnel, fisherman and fishery personnel is needed so that they are in a stage to identify the species and know their status (vulnerable/endangered/critically endangered/threatened) in the conservation matrix.
- Decentralization of shipping activity and adopting a leak proof oil transmitting and carrier system would be advisable. Heavy penalty should be imposed on polluting vessels and fitness certificate should be granted after checking of oil and fuel. The coastal industries should be strictly advised to maintain the effluent discharge norms and a technical committee should be formed comprising representatives of forest department, fishery department, state pollution control board, academic institutes and research organizations to undertake regular water quality monitoring.
- Specialized fishing gears and crafts are used to trap the sharks in coastal West Bengal. Cod nets and trawl nets are commonly used by the fishermen to catch the shark species, although hook and line gear are also used. Strict checking should be done on these gear types to monitor the shark catch both qualitatively and quantitatively.

6.2.2 Reptiles

There are around 7000 living species of reptiles, which includes lizards, snakes, turtles and crocodiles. Their dry skin is covered with scales to prevent water loss. Their eggs have a leathery shell that prevents them from drying out, and therefore reptiles can lay their eggs on land. Like most fishes, reptiles are poikilotherms and

Table 6.9 Uses of different shark parts

Shark parts	Uses
Shark meat	It is a delicious food, rich in protein, mineral and iodine
Shark fin and fin rays	Fin and fin rays are used for preparing soup, which is a delicious and healthy food
Shark cartilage	The cartilage of shark contains collagens, chondromucoids and chondroalbumin. The proximate analysis shows that shark cartilage contains 41 % ash, large amounts of Ca and P, 39 % protein, 12 % carbohydrate, 7 % water, 1 % fibre and about 0.3 % fat. This unique food value is extremely good for human health
Shark liver	The Indian sharks usually contain 2–180 kg of liver depending upon their size and the season. The livers weigh 10–25 % of the sharks' body weight and contains 60–70 % oil. Shark liver oil is used in the tanning and textile industries, as a lubricant and also as a rich source of Vitamin A. Components of shark oil are said to cure certain diseases, relieve pain and improve health. Traditional uses in folk medicines have been cited from Japan, China and Spain. **Squalene**, extracted from the shark liver oil, is a biosynthetic precursor of cholesterol, but still it is claimed that it helps to normalize the blood cholesterol level in people who consume large amounts of food rich in fat (http://www.tao.org/DOCREP/005/X3690E/x3690elq.htm)
Shark hide	Surface skin of shark is known as *shagroon*, which is a rough leather with dermal denticles embedded in the skin. It is used for rasping and polishing. A rare and expensive product known as the *boroso* leather can be obtained by polishing the denticles to a high gloss. The hide can be converted to a fancy leather by removing dermal denticles and used in manufacturing of shoes, wallets, hand bags, purses, dress belts, etc.

ectotherms, commonly called 'cold blooded'. Like other poikilotherms, their metabolic rate—and therefore activity level—varies with temperature; they usually get sluggish in the cold. This tends to keep them out of cold regions, especially on land because the air temperature fluctuates more widely than does the ocean temperature.

Reptiles are abundantly found in the oceans and estuaries. Some common reptiles of marine and estuarine ecosystems are gavial, turtle, marine iguana, saltwater crocodiles, etc. The population of marine reptiles is today under threat due to factors like pollution, urbanization, habitat destruction/degradation and destructive fishing methods. In addition to these factors, the demand for turtle products by human beings has also posed an adverse impact on the turtle population. Turtle eggs and turtle meat have great demand throughout the Pacific, and nests are regularly raided by poachers. Hawksbill turtles are killed by poachers in order to get their shell, which is used to make combs, boxes and jewelry. The skins of turtles are used for making leather belts. The fats and oils of turtles are often used for making cosmetics.

6.2.2.1 Marine Crocodiles

The *saltwater crocodile* (*Crocodylus porosus)* is the largest of all living reptiles, which inhabits mangrove swamps and estuaries in the eastern Indian Ocean, Australia and some of the western Pacific islands. They have the broadest distribution of any modern crocodile ranging from the eastern coast of India throughout most of Southeast Asia stretching south and northern Australia and even beyond the East Coast of Africa and as far as the coastal water off Japan. Saltwater crocodiles live mostly on the coast but also are found in rivers and are known to venture into the open sea. They are among the most aggressive of all marine animals and are known to attack and eat people. Where they occur, they are more feared than sharks.

Their average life span is 70 years. The average length of male saltwater crocodile varies between 430 and 520 cm, while females are relatively smaller with length between 290 and 310 cm. The average weight of males varies between 400 and 1500 kg. The saltwater crocodiles are found in the mangrove swamps, estuaries, deltas, lagoons and river mouths (Fig. 6.15).

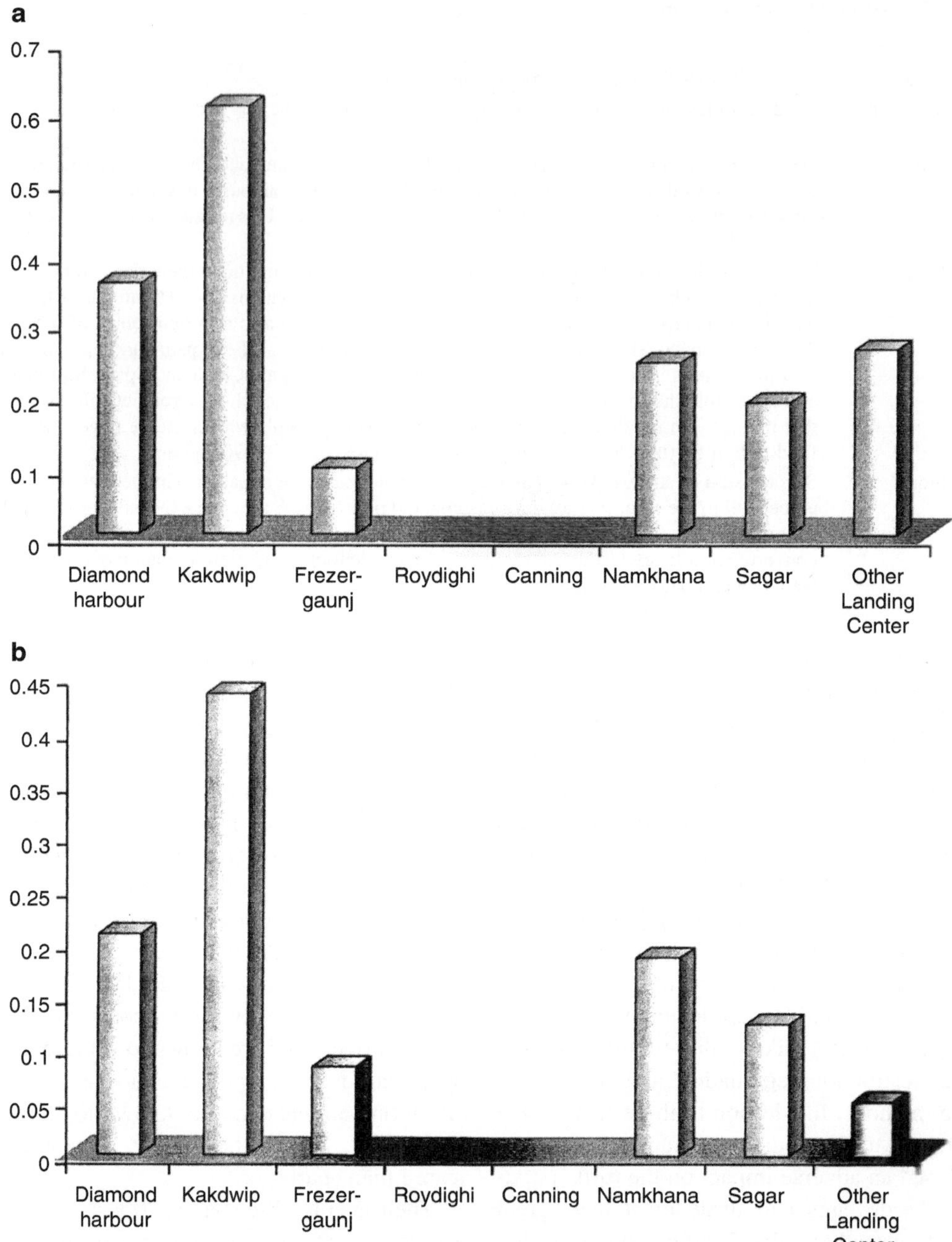

Fig. 6.13 (**a**) Catch of shark (in 000' tonnes) in major fish landing stations of West Bengal during 1994–1995. (**b**) Catch of shark (in 000' tonnes) in major fish landing stations of West Bengal during 1995–1996. (**c**) Catch of shark (in 000' tonnes) in major fish landing stations of West Bengal during 1996–1997. (**d**) Catch of shark (in 000' tonnes) in major fish landing stations of West Bengal during 1997–1998. (**e**) Catch of shark (in 000' tonnes) in major fish landing stations of West Bengal during 1998–1999. (**f**) Catch of shark (in 000' tonnes) in major fish landing stations of West Bengal during 1999–2000. (**g**) Catch of shark (in 000' tonnes) in major fish landing stations of West Bengal during 2000–2001. (**h**) Catch of shark (in 000' tonnes) in major fish landing stations of West Bengal during 2001–2002

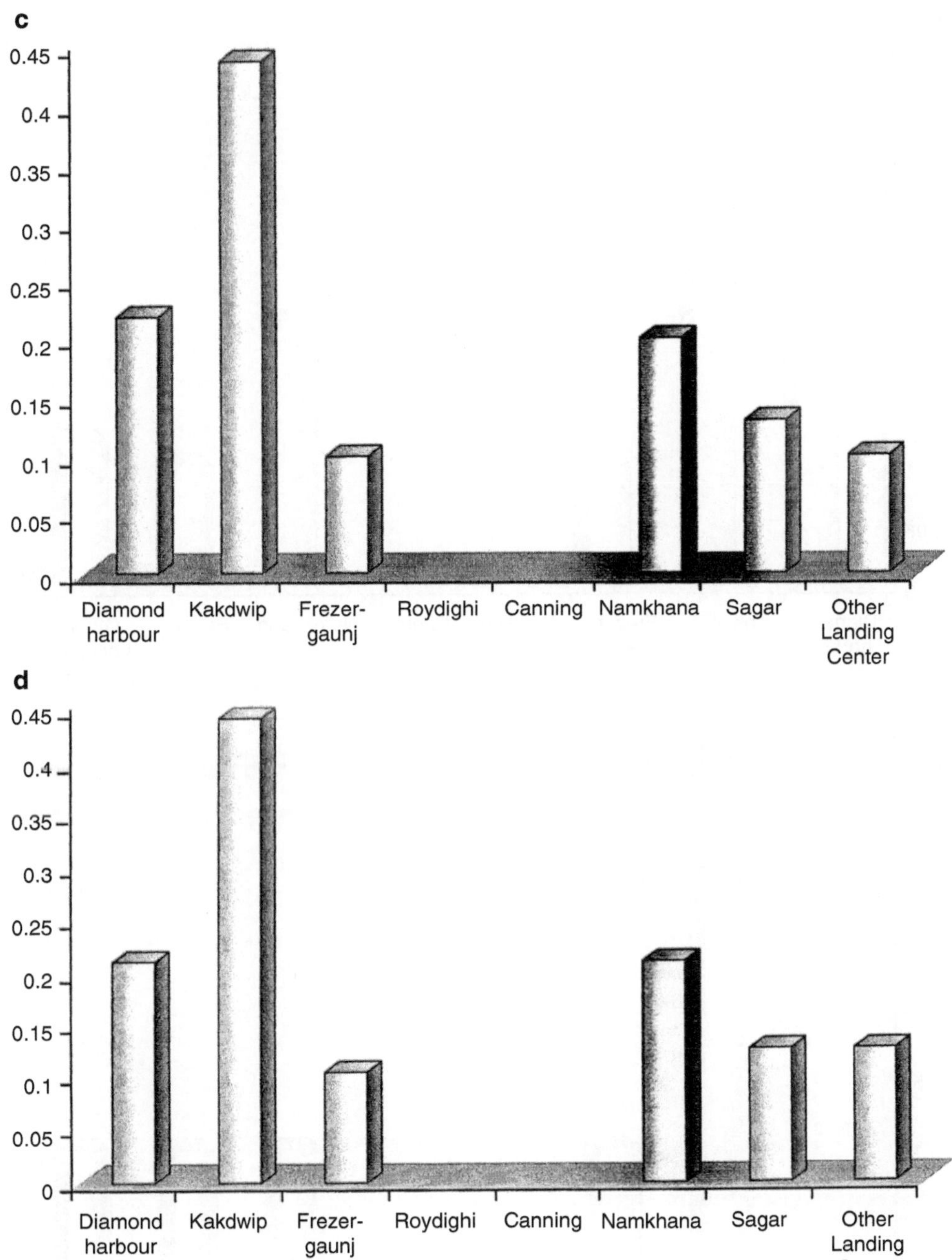

Fig. 6.13 (continued)

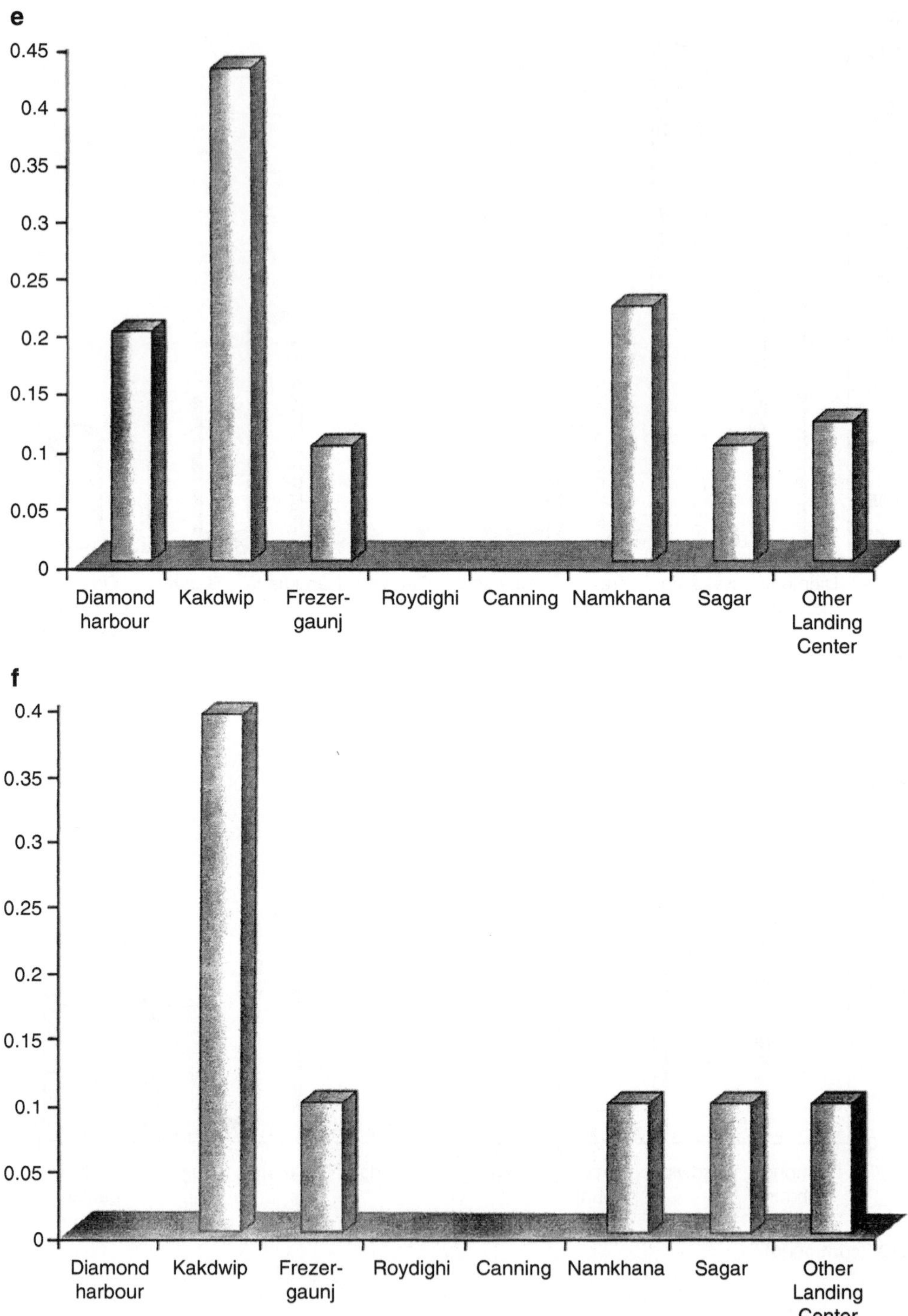

Fig. 6.13 (continued)

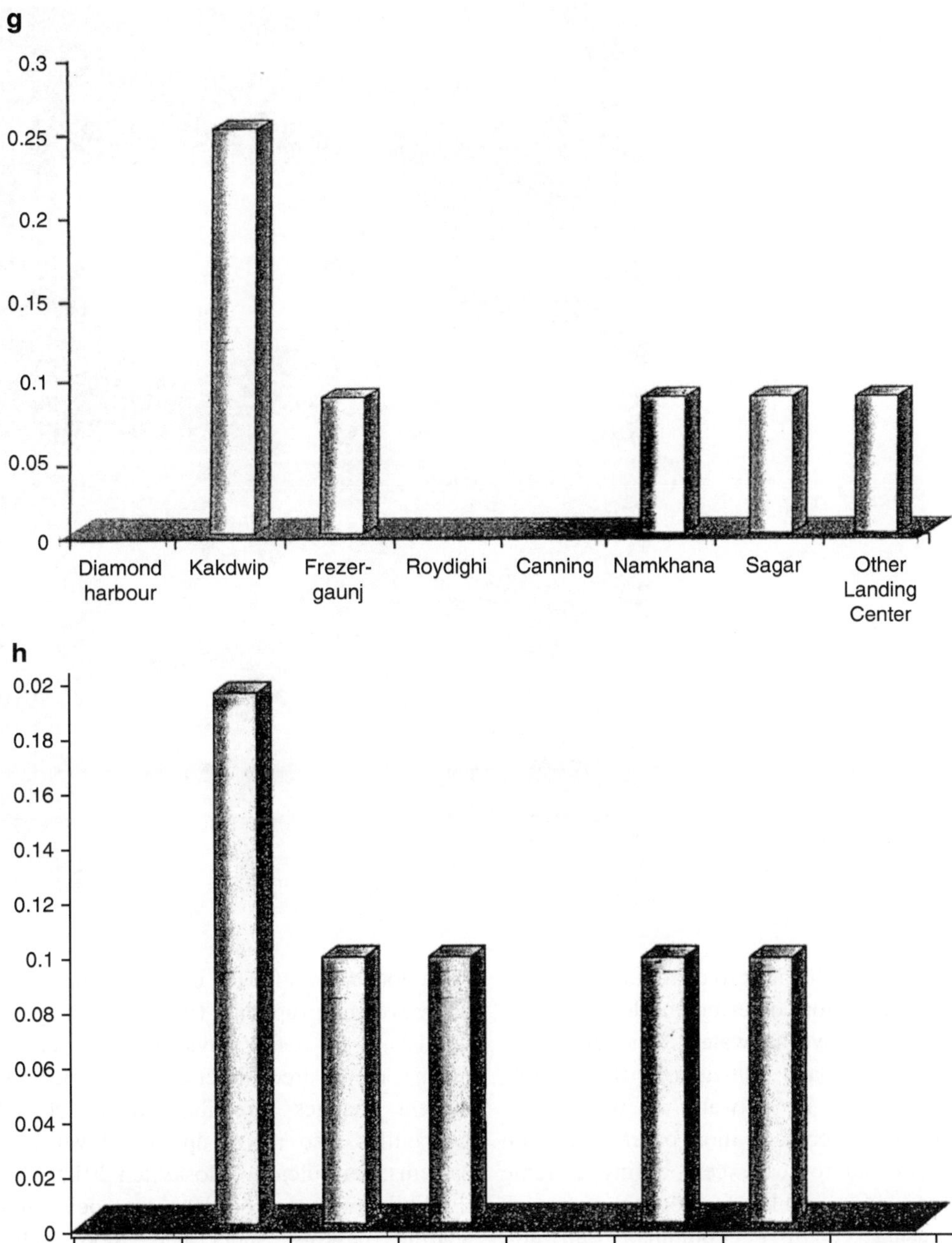

Fig. 6.13 (continued)

Fig. 6.14 Sharks caught illegally for fins

The primary behaviour to distinguish the saltwater crocodile from other crocodiles is its tendency to occupy saltwater. Though other crocodiles also have salt glands that enable them to survive in saltwater, a trait which alligators do not possess, most other species do not venture out to sea except during extreme conditions. In a similar fashion to migratory birds using thermal columns, saltwater crocodiles use ocean currents to travel long distances (Moskvitch 2010). In a study, 20 crocodiles were tagged with satellite transmitters; 8 of these crocodiles ventured out into open ocean, in which one of them travelled 590 km (370 mi) in 25 days. Another specimen, a 4.84-m-long male, travelled 411 km (255 mi) in 20 days. Without having to move around much, sometimes simply by floating, the current-riding behaviour allows for the conservation of energy. They even interrupt their travels, residing in sheltered bays for a few days, when the current is against the desired direction of travel, until the current changes direction (Moskvitch 2010). Crocodiles also travel up and down in river systems, periodically (Moskvitch 2010).

While most crocodilians are social animals sharing basking spots and food, saltwater crocodiles are more territorial and are less tolerant of their own kind; adult males will share territory with females, but drive off rival males. Saltwater crocodiles mate in the months of monsoon, the period characterized with lowest aquatic salinity. They lay eggs in a nest consisting of a mound of mud and vegetation. The female guards the nest and hatchlings from predators.

Fig. 6.15 Crocodiles on the mudflat of Indian Sundarbans

A study of seasonal saltwater crocodile behaviour in Australia indicated that they are more active and more likely to spend time in the water during the Australian summer; conversely, they are less active and spend relatively more time basking in the sun during the winter (Grigg et al. 1998). Saltwater crocodiles, however, are among the most active of all crocodilians, spending more time cruising and active, especially in water. They are much less terrestrial than most species of crocodiles, spending little time on land other than basking. At times, they can spend weeks at sea in search of land, and in some cases, barnacles have been observed growing on crocodile scales, indicative of the long periods they spend at sea (Linnaeus 1758). They drink saltwater eliminating the excess salt through salt glands on their tongues. These reptiles live in burrows along the shore, where they make their nest and lay eggs.

Several species of crocodiles like American crocodile (*Crocodylus acutus*), Nile crocodile (*Crocodylus niloticus*) and Asian saltwater crocodile (*Crocodylus porosus*) are best suited to the marine environment.

The researchers of Forest Department of the State of West Bengal (India) cited considerable number of crocodiles during 2009–2011 in the Reserve Forest area of Indian Sundarbans (Fig. 6.16). Occasionally the estuarine crocodiles of Sundarbans stray to village pond and are rescued. People believe that the straying is mainly to eat the fishes of the village ponds. A total of 69.23 % crocodile straying in villages takes place in the month of September and October. This is the time when crocodiles are reported to start making nest to lay egg and are on search of safe nesting sites.

Saltwater crocodiles have a strong tendency to consider human beings in their territory as their lucrative prey. They have a long history of attacking human beings who stray into their territory. It has been reported by marine biologists that prawn seed collectors of Sundarbans estuaries, who are mostly women, are attacked by crocodiles during the time of dragging nets in the muddy estuarine water for tiger prawn seed collection (Fig. 6.17). It is surprising that although saltwater crocodiles are relatively lethargic in nature, but as predators they display extreme reflex in terms of speed and attack. They are capable of explosive burst of speed when launching an attack from water. They can also swim at a speed of 24–30 km/h if they target for their prey in their surroundings.

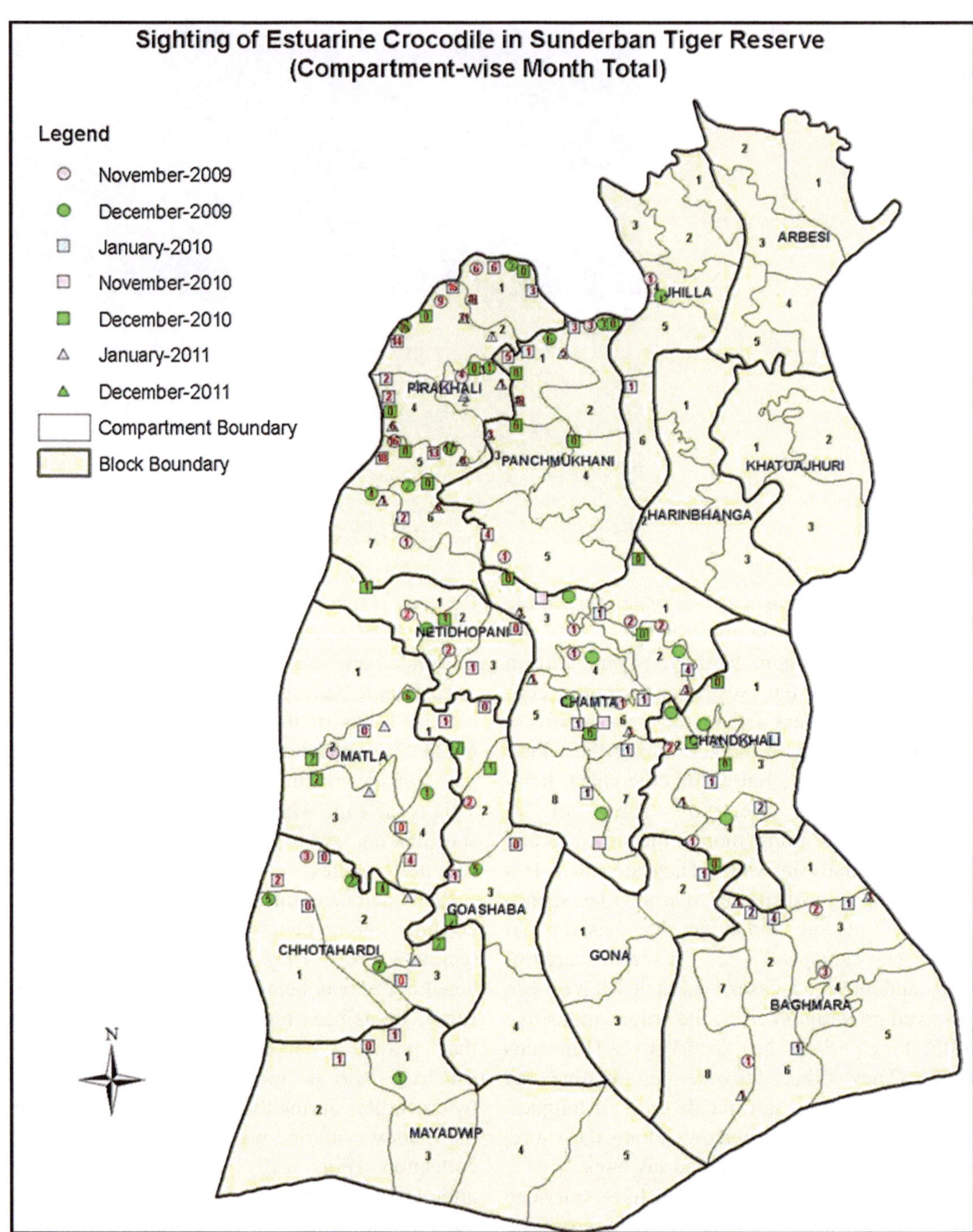

Fig. 6.16 Areas of citation for crocodiles in the Reserve Forest area of Indian Sundarbans

They feed mainly on fishes and are sometimes known to attack and kill sharks close to their own size. They are very aggressive and often attack and kill human beings within their range.

The species was indiscriminately killed for the purpose of making luxury goods from its skin in the late 1960s and early 1970s. The level of poaching became so severe that the

Fig. 6.17 Prawn collectors of Indian Sundarbans are prone to crocodile attack

population subsequently declined, making the species endangered.

6.2.2.2 Sea Turtles

Sea turtles belong to an ancient group of reptiles who primarily live in the ocean, seas, bays and estuaries. Their bodies are enclosed by an armour-like shell, or carapace, that is fused to the backbone. They have a carapace or upper shell and a lower shell known as plastron; unlike land turtles, sea turtles cannot retreat into their shell. They have paddle-like flippers, which help them for propelling through the water. They also breathe air and hence sea turtles come to the water surface when they need to breathe.

There are only seven species of sea turtles (though some biologists recognize eight), which live primarily in warm waters. Six of them (the hawk bill turtle, green turtle, flatback turtle, the loggerhead turtle, kemp's ridley turtle and olive ridley turtle) have shells made up of hard scutes while leatherback turtle has a leathery carapace made up of connective tissue. Sea turtles range in size from 2 to 6 ft depending on the species and age. Kemp's ridley turtle is the smallest while the olive ridley turtle is the largest.

The remaining species of sea turtles have diverse diets, feeding on soft, bottom invertebrates like sponges, soft corals and jellyfishes as well as hardier invertebrates like crabs and molluscs. The jaws of most species are adapted for crushing. The leatherback jaws, however, are scissor-like and are used to capture jellyfishes.

All sea turtles must return to land to reproduce. They migrate long distances to lay their eggs on remote sandy beaches, and they were doing so millions of years before humans appeared on the scene. Green turtles still gather to nest on beaches on the East Coast of Central America, Northern Australia, Southeast Asia, Ascension Island (in the middle of the South Atlantic) and a few other locations. Marine biologists have tagged adult sea turtles at Ascension and have found that the turtles regularly cross 2200 km (1360 mi) of open water to their feeding grounds along the coast of Brazil, a journey that takes a little more than 2 months. Though we are still not sure how they find their way, there is evidence that they do it by sensing the Earth's magnetic field.

The breeding, nesting and hatching and migration of sea turtles are interesting areas of research. All sea turtles are oviparous. The sexual maturity may take 5–35 years depending on the species. During breeding period males and females migrate to the breeding grounds which are close to the nesting areas. Males and females mate offshore and females travel to nesting areas to lay their eggs. It is interesting to note that female sea turtles return to the same beaches where they were born. Even it may be 30 years later and the beach may have greatly changed. The female sea turtles crawl up on the beach, dig pits of about 1 ft deep with their flippers and finally dig nests for the eggs with their hind flippers. After the eggs are laid, the female sea turtles cover the nests with the hind flippers and pack the sand down. The eggs require 45–70-day incubation period. This period is much influenced by the ambient temperature of the sediment in which the eggs are laid. Temperature determines the gender of the hatchlings. Cooler temperature favours the development of more males while warmer temperature favours the development of more females. It may be an interesting research project if the rising trend of temperature in tropical estuaries like estuaries of Indian Sundarbans has any potential implications on the sexual pattern of sea turtle hatchlings. The pattern of temperature rise in the estuaries of the Indian Sundarbans is presented as Annexure 6A. The researchers have found that eggs placed in the centre of the nest are more likely to hatch females, while eggs on periphery of the nest are more likely to hatch males. The migratory behaviour of sea turtles is also unique. Sea turtles migrate long distance between feeding and nesting grounds and prefer warmer waters when the seasons change.

The salient features of seven major turtle types are discussed separately.

Leatherback (*Dermochelys coriacea*)

The largest sea turtle is the leatherback (*Dermochelys coriacea)* (Fig. 6.18). This turtle's shell consists of a single piece with five ridges and is distinctive from other turtles that have plated shells. Instead of a solid shell, they have a series of small bones buried in the dark skin, forming distinct longitudinal ridges. Its length and weight reach over 6 ft and 2000 pounds, respectively. T leatherbacks are an open-water species, the widest ranging of all marine reptiles, and are rarely seen except on nesting beaches. They are deep divers—one individual was

Fig. 6.18 Leatherback turtle (*Dermochelys coriacea*)

Fig. 6.19 Green turtle (*Chelonia mydas*)

tracked diving to 640 m (2100 ft). Their diet consists largely of jellyfishes. Like other species of sea turtles, leatherbacks are in danger of disappearing forever.

Green Turtle (*Chelonia mydas*)

Green turtles (*Chelonia mydas)* were once found in coastal waters throughout the tropics. Their shells may grow to 1 m (40 in.) in length. They feed mostly on seagrasses and seaweeds. Like all turtles, green turtles lack teeth, but they have strong biting jaws. The green turtle (Fig. 6.19) is large in size about 3 ft long. Green turtles weigh up to 350 pounds and their carapace can be many colours, including shades of black, grey, green, brown or yellow.

Loggerhead (*Caretta caretta*)

The loggerhead turtles (Fig. 6.20) are reddish brown in colour possessing a very large head. Their length can extend up to 3.5 ft. The body weight of the loggerhead turtles range around 400 pounds. Crabs, jellyfish and molluscs constitute the diet of loggerhead turtles.

Fig. 6.20 Loggerhead turtle (*Caretta caretta*)

Fig. 6.21 Hawksbill turtle (*Eretmochelys imbricate*)

Hawksbill (*Eretmochelys imbricate*)

The hawksbill turtle (*Eretmochelys imbricate*) is smaller, and the shell is reddish brown with yellow streaks. It uses its beak-like mouth to feed on encrusting animals (sponges, sea squirts, barnacles) and seaweeds. The length of the hawksbill turtle (Fig. 6.21) is about 3.5 ft. The body weight is around 180 pounds. The name hawksbill turtle may have originated from the shape of their beak which is similar to the beak of a raptor. The beautiful patterns of the carapace of the hawksbill turtle have attracted the poachers who often kill these turtles. The hawksbill turtles are abundantly found in tropical waters and their preferred diet are the sponges.

Kemp's Ridley (*Lepidochelys kempii*)

Kemp's ridley turtle (Fig. 6.22) has a length of about 2.5 ft. and weight ranges around 90 pounds. In fact these turtles are very small in size. Their preferred food materials are crabs and they are found in coastal waters of temperate to subtropical regions.

Olive Ridley (*Lepidochelys olivacea*)

Olive ridley turtles (Fig. 6.23) possess olive coloured shells. They are not very big in size and weighs approximately 60–90 pounds. Their diet includes crabs, shrimps, lobsters, jellyfish and even algae. During the phase of nesting, the females come in thousands and have mass nesting aggregations (*arribadas*) on the coasts of Central America and East India.

Fig. 6.22 Kemp's Ridley turtle (*Lepidochelys kempii*)

Fig. 6.23 Olive Ridley turtle (*Lepidochelys olivacea*)

Flatback (*Natator depressus*)

The flattened carapace is perhaps the origin of the name flatback turtles (Fig. 6.24). They are widely distributed in the coastal waters of Australia and feed on sea cucumbers, soft corals, bivalves and gastropods.

6.2.2.3 Marine Lizard

The *marine iguana* (*Amblyrhynchus subcristatus*), the only marine lizard (Fig. 6.25), is native to the Galápagos Islands. It spends most of its time basking in large groups on rocks along the coast, warming up after swimming in the cold water. It eats seaweeds and can dive as deep as 10 m (33 ft) to graze.

These large lizards are descendents of green, vegetarian iguanas that still inhabit the tropical forests of the mainland. It is believed that in the distant past, chunks of riverbank from Central America may have broken loose and carried into the sea along with different other flora and

Fig. 6.24 Flatback turtle (*Natator depressus*)

fauna into the Galápagos Islands, which might be the probable cause of this animal to have settled here. The iguanas travelled and wandered in the islands in search of food and vegetation that they used to feed in the forest. The Galápagos Islands was dramatically different from the forests, but it had adequate resources for its survival. The conditions of the island thus favoured the species to survive and propagate. The unusual lifestyle of the marine iguana was due to its adaptability in precarious condition, when it had to feed on the marine algae, which got exposed only during low tide. Although it was a foreign food, its subsequent generations resisted and persisted this condition. This caused the present-day iguana to be completely different from its relatives on the mainland.

The marine iguana is 3 ft long lizard and is entirely black; some are mottled red and black showing some hint of green during the breeding season. This dark colouration allows them to absorb more heat energy to raise their body temperature so that they can swim and feed in the cold Pacific waters. They are very good swimmers, using lateral undulations of their body and tail to propel them through waters. They avoid heavy surf and rarely venture more than 10 m from shore. When leaving the water, they tend to ride in with the swell and then swiftly crawl up the rocks. If they do not find their territory immediately, they touch the rocks with their tongues and carry scent to a receptor in the roof of the mouth. When they locate their own scent, they follow it to their territory, where they rest on the rocks, lying almost motionless above the high tide.

Fig. 6.25 Marine iguana (*Amblyrhynchus subcristatus*)

6.2.2.4 Sea Snakes

Approximately 55 species of sea snakes are found in the tropical Indian and Pacific oceans. Their bodies are laterally flattened and the tail paddle-shaped for swimming (Fig. 6.26). Most are 1–1.3 m (3–4 ft) long. Practically all sea snakes lead a totally marine existence. They mate in the ocean and are ovoviviparous, giving birth to live young. A few species, however, still come ashore to lay their eggs. Like all snakes, sea snakes are carnivores. Most feed on bottom fish, a few specializing in fish eggs. They are closely related to cobras and their allies, the most venomous of all snakes. Sea snakes are among the most common of all venomous snakes, and their bites can be fatal to humans. Fortunately, they are rarely aggressive, and the mouth is too small to get a good bite. Most casualties, swimmers accidentally stepping on them and fishers removing them from nets have been reported in Southeast Asia. Sea snakes are also victims of overexploitation. They are hunted for their skins, and some species have become rare.

The coral reef snakes or sea snakes inhabit marine environments in most of the period of their life cycle. The origin of coral reef snakes occurred in the terrestrial environment. Then they migrate to the aquatic system in course of time. Today except the genus *Laticauda*, other

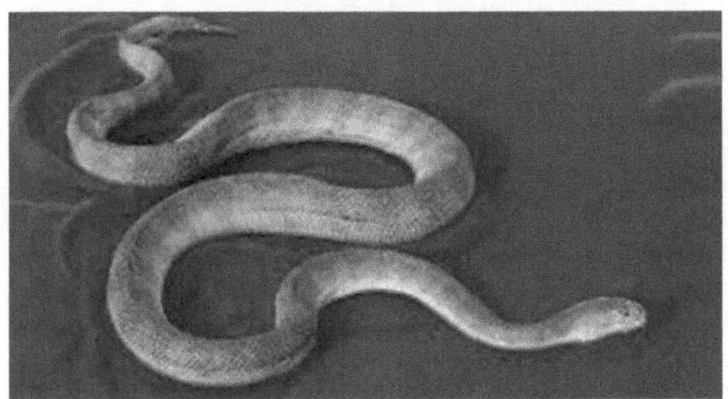

Fig. 6.26 Sea snakes

members of sea snakes or coral reef snakes are able to move on land. They are abundantly distributed in the warm coastal waters from the Indian Ocean to the Pacific. The coral reef snakes have paddle-like tails with laterally compressed body very similar to the structure of eel. Representatives of *Laticauda* possess venom. They are among the most completely aquatic of all air-breathing vertebrates (Parker and Grandison 1977; Mehrtens 1987). Currently, 17 genera are described as sea snakes, comprising 62 species (Elapidae 2007a, b).

Sea snakes are generally reluctant to bite, (Stidworthy 1974; Fichter 1982) and are usually considered to be mild tempered, although there is variation among species and individuals (Hydrophiidae 2007). Some species, such as *Pelamis platurus*, which feed by simply gulping down their prey, are more likely to bite when provoked because they seem to use their venom more for defence. Others, such as *Laticauda*, use their venom for prey immobilization; these snakes are often handled without concern by local fishermen, who unravel and toss them back into the water barehanded when the snakes become entangled in fishing nets (Stidworthy 1974; Sea snakes at FAO 2007). Species reported as much more aggressive include *Aipysurus laevis*, *Astrotia stokesii*, *Enhydrina schistosa* and *Hydrophis ornatus* (US Navy 1991).

Sea snakes (Fig. 6.26) appear to be active both day and night. In the morning, and sometimes late in the afternoon, they can be seen at the surface basking in the sunlight, and they will dive when disturbed (Stidworthy 1974). They have been reported swimming at depths of over 90 m (300 ft) and can remain submerged for as long as a few hours, possibly depending on temperature and degree of activity (Fichter 1982; Hydrophiidae 2007).

Some common sea snakes along with their geographical distribution are highlighted in Table 6.10.

6.2.3 Sea Birds

Birds are homeotherms, commonly referred to as 'warm blooded'. They are also endotherms. This feature has allowed the sea birds to thrive in a wide variety of environments. Their bodies are covered with waterproof feathers that help conserve body heat. Waterproofing is provided by oil from a gland above the base of the tail. Birds preen by rubbing the oil into their feathers with their beaks. Flight is made easier by their light, hollow bones. Furthermore, their eggs have hard shells that are more resistant to water loss than those of reptiles.

Sea birds are birds that spend a significant part of their lives at sea and feed on marine organisms. Most breed in large colonies on land, mate as lifelong pairs and take care of their young. True sea birds have webbed feet for swimming.

Sea birds have evolved from several different groups of land birds. As a result, they differ widely in their flying skills, feeding mechanisms and ability to live away from land. Though comprising only about 3 % of the estimated 9700 species of birds, seabirds are distributed from pole to pole, and their impact on marine life is significant. Most are predators of fish, squid and bottom invertebrates, but some feed on plankton. Seabirds have amazing appetites. They need a lot of food to supply the energy required to maintain their body temperature.

The sea birds constitute about 3 % of the world's bird. About 260–285 sea birds have been documented. The birds which are efficiently adapted to marine environment are albatrosses, auks, petrels, penguins, etc. However, birds like sandpipers and plovers are incapable of swimming and are seen on seashore and intertidal mudflats.

The aquacultural farms along the coastline/mangrove intertidal mudflats are the home ground of several species of shorebirds (Fig. 6.27) that feed on shrimps and crabs.

The variety of seabirds may be grouped into categories like true shorebirds, diving shorebirds, diving pelagic birds, etc.

6.2.3.1 True Shorebirds

The true shore birds include sandpiper, oystercatcher, snowy egret, some types of ducks, etc. They are adapted to survive and grab their food materials from the sandy and muddy beaches. The beak of sandpiper (*Calidris* sp.) is narrow and pointed for penetrating into the sand in search of small invertebrates preferably the infauna. The sandpipers usually remain in small flocks.

The oystercatcher (*Haematopus* sp.) has a large body size with a knife-like beak meant for consuming various types of molluscs. The oystercatchers are mainly seen in marshy land and sandy beaches.

The snowy egret (*Egretta thula* sp.) are seen usually in the intertidal mudflats and salt marshes. They possess long legs, long neck and a pointed bill. They prefer small fishes like herons. In tropical regions flamingoes are seen in the shallow aquatic zones. They have pointed and bent beaks which enable them to capture small invertebrates from the water.

6.2.3.2 Diving Shore Birds

These birds are capable of diving to grab their prey. The diving shore birds include cormorant, common tern, brown pelican, osprey, etc. The diving shore birds are equipped with excellent vision, which enable them to locate their prey while flying.

6.2.3.3 Diving Pelagic Birds

Many seabirds that nest on islands and along coastlines actually spend most of their lives at sea. The pelagic, or open ocean, diving seabirds include a variety of types, from sparrow-sized storm petrels to auks, gannets, puffins, petrels, shearwaters, guillemots and eagle-sized albatrosses. Some of these birds, such as the shearwaters and storm petrels, migrate thousands of kilometres each year as they follow schools of fish or drifting plankton. Probably the most oceanic of all the seabirds is the *wandering albatross* (*Diomedea exulans*) of the South Pacific. The albatross, among the largest of all seabirds with a wingspan of about 3 m, is adept at gliding effortlessly on air currents over the ocean. The albatross may spend 3 or 4 years at sea before returning to its birth island to breed. During this time, the bird rarely stops flying or gliding and may actually circle the entire globe.

6.2.3.4 Penguins

Penguins are the most aquatic of all seabirds (Fig. 6.28). There are about 18 species of penguins; all but one species live in the southern hemisphere. They vary in height from about one third of a metre to more than 1 m tall. Penguins have no flight feathers and are completely flightless. However, they are excellent swimmers and divers. The smaller wings of a penguin function as flippers that can propel the bird through the water at speeds of up to 24 km per hour. Penguins have dense bones with considerable specific gravity, which give them the weight necessary for deep dives.

Seabirds are indicators of ocean productivity and rank second in order of consumption of large planktonic crustaceans, just after marine mammals. The marshy area of mangrove-dominated Sundarbans is the homeland of a variety of seabirds like herons, cormorants, egrets, kingfishers, storks and darters. About 250 species of birds have been documented in this deltaic lobe at the apex of Bay of Bengal, of which a sizable proportion is migratory. A number of species of eagles, including the white-bellied sea eagle, crested serpent eagle and Pallas's fishing eagle, and owls such as the barn owl, scops owl, spotted owl and brown fish owl are also found in the mangrove trees. On the intertidal

Table 6.10 Some common sea snakes and their geographical distribution

Common name	Geographical distribution
Spiny-headed sea snake or horned sea snake	Gulf of Thailand, South China Sea, the Strait of Taiwan, and the coasts of Guangdong, Indonesia, the Philippines, New Guinea, New Caledonia, Australia (Northern Territory, Queensland, Western Australia)
Olive sea snake	South China Sea, Gulf of Thailand, and coasts of Australia (Northern Territory, Queensland, West Australia), New Caledonia, Loyalty Islands, southern New Guinea, Indonesia, western Malaysia and Vietnam
Stokes' sea snake	Coastal areas from west India and Sri Lanka through Gulf of Thailand to South China Sea, west Malaysia, Indonesia, east to New Guinea, north and east coasts of Australia, the Philippines
Turtlehead sea snake	The coasts of Timor (Indonesian Sea), New Caledonia and Australia (Northern Territory, Queensland, West Australia) and in the Southeast Asian sea along the coasts of China, Taiwan, Japan and the Ryukyu Islands; in the Persian Gulf (Oman, United Arab Emirates, etc.), south to the Seychelles and Madagascar
Beaked sea snake	Southeast Asia (Pakistan, India, Bangladesh, Myanmar, Thailand, Vietnam), Australia (Northern Territory, Queensland), New Guinea and Papua New Guinea
Grey's mudsnake	Northwestern Australia
Port Darwin mud snake	Northern Australia, southern New Guinea
Sea snake	Indo-Australian and Southeast Asian waters (Hydrophiidae 2007)
Jerdon's sea snake	Southeast Asian waters (Hydrophiidae 2007)
Bighead sea snake	Indian Ocean
Spine-bellied sea snake, Shaw's sea snake	Persian Gulf to Indian Ocean, South China Sea, Indo-Australian Archipelago and the western Pacific (Hydrophiidae 2007)
Sea kraits	Southeast Asian and Indo-Australian waters
Northern mangrove sea snake	Northern Australia, southern New Guinea
Hediger's snake	Bougainville Island, Solomons
Yellow-bellied sea snake	From the Persian Gulf to the Indian Ocean, the South China Sea and northeast to the coastal region of Fujian and Strait of Taiwan
Anomalous sea snake	South China Sea (Malaysia, Gulf of Thailand), Indian Ocean (Sumatra, Java, Borneo)

Fig. 6.27 Shorebirds are attracted in shrimp farms for abundant food supply

Fig. 6.28 Penguin with no flight feathers

mudflats, different species of plovers and sandpipers are common. Marsh birds like herons, egrets, bitterns, storks and rails are resident birds and are found throughout the year. Migratory flocks of sandpipers, redshanks and whimlocks can be seen during winter. Table 6.11 lists few frequently sighted avifauna in Indian Sundarbans.

Seabirds are often known for their excellent migratory behaviour. American subspecies *(Calidris canutus rufa)* spends the austral summer at the southern tip of Argentina, feeding primarily on young mussels in intertidal areas. It migrates northwards in March, stopping to feed on clams, mussels and worms along the South American coast. By late May, over 100,000 birds flock in Delaware Bay, along the eastern seaboard of the United States. Their arrival coincides with the breeding cycle of horseshoe crabs (Fig. 6.29) that come ashore in thousands to lay millions of eggs. Their eggs provide a high energy food source for the birds, and this fuels their remaining flight to islands in the Canadian Arctic.

Nowadays several advanced technologies are adapted by fishermen to catch fishes. Nets of various types are used to capture commercially important fishes. In North Atlantic and Pacific, the drift nets cause mortality and injury to a large number of birds during diving. Decrease in food is also one of the reasons behind the reduction in bird population. Several sea birds and shorebirds are adversely affected by oil pollution (Hiscock 1997). The floating oil sticks to the feathers of the birds making them unable to fly. The study on impact of climate change on sea bird population is still under progress. However, anthropogenic factors like construction of hotels, refineries, power plants and aquacultural farms in the coastal region have greatly affected the shore bird population.

6.2.4 Marine Mammals

Marine mammals spend all their lives in the ocean, or they may return to land for mating and giving birth to their young ones. The breeding grounds of several marine mammals like humpback whale and grey whale are restricted in the coastal region, while majority of the whales have their breeding grounds in the oceanic water (Table 6.12).

There are *three* orders of mammals that have evolved from different terrestrial ancestors and independently adapted to life in the sea. These three orders include, respectively, the whales, dolphins and porpoises; the seals, sea lions and

Table 6.11 Common avifauna of Indian Sundarbans

Vernacular name	Scientific name
Resident	
Little cormorant	*Phalacrocorax niger*
Great cormorant	*P. carbo*
Grey heron	*Ardea cinerea*
Purple heron	*A. purpurea*
Indian pond heron	*Ardeola grayii*
Cattle egret	*Bubulcus ibis*
Open-bill stork	*Anastomus oscitans*
Great egret	*Egretta alba*
Intermediate egret	*E. intermedia*
Little egret	*E. garzetta*
Black-crowned night heron	*Nycticorax nycticorax*
Greater adjutant	*Leptoptilos dubius*
Black-necked stork	*Xenorhynchus asiaticus*
Black-headed ibis	*Threskiornis melanocephala*
White-breasted water hen	*Amaurornis phoenicurus*
Bronze-winged jacana	*Metopidius indicus*
Pheasant-tailed jacana	*Hydrophasianus chirurgus*
Black-capped kingfisher	*Halcyon pileata*
Collared kingfisher	*H. chloris*
Brown-winged kingfisher	*Pelargopsis amauroptera*
Migratory species	
Spot-billed pelican	*Pelecanus philippensis*
Northern pintail	*Anas acuta*
Common teal	*A. crecca*
Tufted duck	*Aythya fuligula*
Common pochard	*A. ferina*
Ruddy shelduck	*Tadorna ferruginea*
Whimbrel	*Numenius phaeopus*
Black-tailed godwit	*Limosa limosa*
Little stint	*Calidris minuta*
Great knot	*C. tenuirostris*
In shallow water	
Eastern golden plover	*Pluvialis dominica*
Kentish plover	*Charadrius alexandrinus*
Eurasian curlew	*Numenius arquata*
Wood sandpiper	*Tringa glareola*
Terek sandpiper	*Xenus cinereus*
Reclaimed area	
Indian darter	*Anhinga melanogaster*
Chestnut bittern	*Ixobrychus cinnanoneus*
Black bittern	*Butoridea stellaris*
Spoonbill	*Platalea leucorodia*
Red-wattled lapwing	*Vanellus indicus*
Grey-headed lapwing	*V. cinereus*
Herring gull	*Larus argentatus* (rare visitor)
Great crested tern	*Sterna bergii* (rare visitor)
Lesser crested tern	*S. bengalensis* (rare visitor)
Sooty tern	*S. fuscata* (rare visitor)

(continued)

Table 6.11 (continued)

Vernacular name	Scientific name
Red jungle fowl	*Gallus gallus (found in forest areas only)*
Swamp partridge	*Francolinus gularis*
Greater coucal	*Centropus sinensis*
Occasional visitors to the estuary	
Osprey	*Pandion haliaetus*
Black kite	*Milvus migrans*
Brahminy kite	*Haliastur indus*
White-bellied sea eagle	*Haliaeetus leucogaster*
Crested serpent eagle	*Spilornis cheela*
Spotted dove	*Streptopelia chinensis*
Yellow-footed pigeon	*Treron phoenicoptera*
Large Indian parakeet	*Psittacula eupatria*
Rose-ringed parakeet	*P. krameri*
Barn owl	*Tyto alba*
Brown fish owl	*Bubo zeylonensis*
Spotted owlet	*Athene brama*
Magpie robin	*Copsychus saularis*
Asian paradise flycatcher	*Terpsiphone paradise*
Black drongo	*Dicrurus adsimilis*
Treepie	*Crypsirina vagabonda*
Grey shrike	*Lanius excubitor*
Common swallow	*Hirundo rustica*
Indian roller	*Coracias bengalensis*

Fig. 6.29 Horseshoe crabs: eggs of this living fossil serve as the fuel for migratory seabirds

walruses; and the dugongs, manatees and sea cows. All share the mammalian characteristics of being warm blooded (homoiothermic) and nursing their young, and they all rely on breathing air.

6.2.4.1 Whales, Dolphins and Porpoises

The largest group of marine mammals is the cetaceans (order Cetacea) consisting of whales, dolphins and porpoises. The bodies of cetaceans are streamlined and look remarkably fish-like.

Table 6.12 Distribution and food habits of whales

Whale	Distribution	Food
Toothed whales		
Sperm	Worldwide; breeding herds in tropic and temperate regions	Squid, fish
Ballen whales		
Blue	Worldwide; large north–south migrations	Krill
Finback	Worldwide; large north–south migrations	Krill and other plankton, fish
Humpback	Worldwide; large north–south migrations along coasts	Krill, fish
Right	Worldwide; cool temperate	Copepods and other plankton
Sei	Worldwide; large north–south migrations	Copepods and other plankton, fish
Grey	North Pacific; large north–south migrations along coasts	Benthic invertebrates
Bowhead	Arctic; close to edge of ice	Krill
Bryde's	Worldwide; tropic and warm temperate regions	Krill
Minke	Worldwide; north–south migrations	Krill

This is a dramatic example of convergent evolution, in which different species develop similar structures because they have similar lifestyles. Though they superficially resemble fishes, cetaceans breathe air and will drown if trapped below the surface. They are 'warm blooded', have hair (though scanty) and produce milk for their young. Cetaceans have a pair of front flippers, but the rear pair of limbs has disappeared. Actually, the rear limbs are present in the embryo, but fail to develop. In adults they remain only as small, useless bones. Like fishes, many cetaceans have a dorsal fin, another example of convergent evolution. The muscular tail ends in a pair of fin-like, horizontal flukes. Blubber provides insulation and buoyancy; body hair is practically absent. Cetacean nostrils differ from those of other mammals. Rather than being on the front of the head, they are on top, forming a single or double opening called the blowhole. There are approximately 90 species of cetaceans. They are all marine except for five species of freshwater dolphins. Cetaceans are divided into two groups: *the toothless, filter-feeding whales* and the *toothed, carnivorous whales*, a group that includes the dolphins and porpoises.

There are about 76 species under the order Cetacea and include mammals like whales, porpoises, dolphins, etc. They originated in the terrestrial environment and migrated to the marine compartment about 55 million years ago. The largest representative under this order is the blue whale, which has a body length of about 35 m.

6.2.4.2 Baleen Whales

Baleen whales form a separate suborder (*Mysticeti*) of about ten species. Like the largest of the sharks, most of these immense whales feed primarily on zooplankton that they strain through specialized horny plates called baleen or whalebone.

The blue whale (*Balaenoptera musculus*), which is actually blue grey, is the largest of all. Males average 25 m (80 ft), and there is a record of a female 33.5 m (110 ft) long (90–140 tonnes), but the record is an estimated 178,000 kg (200 tonnes). The blue whale, the fin whale (*B. physalus*), and the minke whale (*B. acutorostrata*—together with five other related species) are known as the rorquals. Two of these whales were recognized as separate species in 2003. The rorquals and the humpback whale (*Megaptera novaeangliae*, which is often included among the rorquals) feed by gulping up schools of fish and swarms of krill. The lower part of the throat expands when feeding, hence the distinctive, accordion-like grooves on the underside of these whales. Krill is the most important part of the rorqual diet, especially in the Southern Hemisphere. Humpback whales often herd fishes like herring and mackerel by blowing curtains of bubbles around them. The right whales (*Eubalaena*, *Caperaea*) and the bowhead whale (*Balaena mysticetus*) feed by swimming along the surface with their huge mouths open. They have the largest baleen plates of the whales but the finest bristles. This allows them to filter small plankton like copepods and some krill. Grey whales (*Eschrichtius robustus*) are primarily bottom feeders. When examined, their stomachs contain mostly amphipods that inhabit soft bottoms. Greys stir up the bottom with their pointed snouts and then filter the sediment, leaving characteristic pits on the bottom. Most appear to feed on their right side because the baleen on this side is more worn. Some, however, are 'left-handed' and feed on the left side. A 10-week-old female kept in captivity in San Diego, California, ate over 815 kg (1800 lb) of squid every day, gaining weight at the rate of 1 kg (2.2 lb) an hour.

6.2.4.3 Toothed Whales

There are 66 species of cetaceans under the sub-order Odontoceti. The members of this group have teeth and possess a single blowhole unlike two blowholes that are found in baleen whales. They are fond of squid and several categories of fishes. They possess the power of locating their prey through eco-mechanism.

The largest toothed whale is the sperm whale (*Physeter catodon*). The other toothed whales are much smaller than the great whales. One is the killer whale, or orca (*Orcinus orca),* a magnificent black and white predator with a taste for seals, sea lions, penguins, fishes, sea otters and even other whales. They use their white bellies to flash and frighten herring schools and their flukes to stun the fish. Killer whales are most common in cold water but are found around the world. Killer whales have a nasty reputation, but there are very few confirmed cases of their attacking humans in the wild. One of the most enigmatic of whales is the narwhal (*Monodon monoceros*), a small Arctic whale having a long, spiral tusk that can be as long as 2.7 m (9 ft). Narwhal tusks washed ashore apparently gave rise to the legend of the unicorn. The tusk, found only in males, appears to be related to the establishment of a hierarchy of dominance. It has been recently shown, however, that the tusks have a network of small, fluid-filled tubes, which are connected to the narwhal's nervous system, and it has been suggested that they may sense the water for the detection of prey or other chemical cues.

The small-toothed whales are called *dolphins* or *porpoises*. Dolphin and porpoise are common names of two subtly different groups of small odontocete whales.

The term dolphin is generally used for small cetaceans having a slender body and beak-like snout and porpoise for animals having a stocky body and blunt snout. *Dolphins* are strongly social animals. They exhibit problem-solving skills, have long periods to mature with many learning experiences and are capable of intraspecies and inter-species cooperation. Dolphins are usually larger and have an extended bottlelike jaw filled with sharp round teeth. The small jumping whales in most oceanarium are dolphins and killer whales. However, this terminology has not been strictly followed in naming the species.

Some species of dolphins typically possess a distinctive snout, or beak and a perpetual 'smile'. They often travel in large groups called pods, herd, or schools. They like to catch rides along the bows of boats or around great whales. The bottlenose dolphin (*Tursiops truncatus*) is the dolphin seen in marine theme parks and oceanaria around the world. The spinner dolphin (*Stenella longirostris*) is so named because of its spectacular twisting jumps in air. It is one of the species of dolphins that get caught in the nets of tuna fishers. This happens because the tuna and dolphins eat the same fish and often occur together.

Habitat wise, the cetaceans of the Indo-Malayan region fall under three categories: (i) the river dolphin, (ii) coastal and estuarine forms and (iii) marine species which live in deep waters. The latter may be resident of tropical waters or those, which mainly live in cold waters but migrate seasonally towards the tropics like *Balaenoptera musculus*, *Balaenoptera physalus*, *etc*.

The common species of dolphin found in the Gangetic stretch adjacent to coastal Bay of Bengal are Gangetic dolphin (*Platanista gangetica*) and Irrawaddy dolphin (*Orcaella brevirostris*). Gangetic dolphin is restricted in freshwater zone, whereas Irrawaddy dolphin is widely visible in the brackish water in and around the deltaic Sundarbans.

Technically, porpoises comprise only a small group of blunt-nosed whales, but in some places the name *porpoise* is given to some of the dolphins. Porpoise, as a term, refers to smaller members of the group, which have spade-shaped teeth, a triangular dorsal fin and a smooth front end tapering to a point. Black finless porpoise (*Neomeris phocaenoides*) is often sighted in the brackish waters of Indian Sundarbans.

6.2.4.4 Seals, Sea Lions and the Walrus

These marine mammals constitute the order Pinnipedia and are characterized with the presence of feather-like foot, which help in swimming. Included in this order are the mammals like seals, walruses and sea lions. Compared to whales these animals spend a considerable portion of their time on land or on ice floes. They utilize this space for breeding and nesting. Their preferred food materials include fish and squid, but the tusks of walruses are used to dig molluscs and other benthic organisms from the seashore.

Seals and related forms are marine mammals that have paddle-shaped flippers for swimming but still need to rest and breed on land. Their streamlined bodies are adapted for swimming. Most pinnipeds live in cold water. To keep warm, they have a thick layer of fat, called blubber, under their skin. Besides acting as insulation, it serves as a food reserve and helps provide buoyancy. Some pinnipeds also have bristly hair for added protection against the cold. Many of them are quite large, which also helps conserve body heat because large animals have less surface area for their size than small animals and therefore lose less body heat.

The largest group of pinnipeds, including some 19 species, is the seals. Seals are distinguished by having rear flippers that cannot be moved forward. On land they must move by pulling themselves along with their front flippers. They swim with powerful strokes of the rear flippers. Harbour seals (*Phoca vitulina*) are common in both the North Atlantic and North Pacific. Elephant seals (*Mirounga*) are the largest pinnipeds. Males, or bulls, reach 6 m (20 ft) in length and can weigh as much as 3600 kg (4 tonnes). One unusual seal is the crab-eater seal (*Lobodon carcinophagus*), which feeds on Antarctic krill. These seals strain krill from the water with their intricately cusped, sieve-like teeth. Unlike most seals, monk seals (*Monachus*) live in warm regions. The Mediterranean (*M. monachus*) and Hawaiian (*M. schauinslandi*) monk seals are now endangered. A third species, the Caribbean monk seal (*M. tropicalis*), was last seen in 1952. Seals have been hunted for their skin and meat and for the oil extracted from their blubber. The Marine Mammal Protection Act of 1972 extends protection to all marine mammals and restricts the sale of their products in the United States. For some seals, however, this protection has not been enough.

Sea lions, or *eared seals*, and the related *fur seals* are similar to seals except that they have external ears. They can also move their rear flippers forwards, so they can use all four limbs to walk or run on land. The front flippers can be rotated backwards to support the body, permitting the animal to sit on land with its neck and head raised. Sea lions are graceful and agile swimmers, relying mostly on their broad front flippers. Adult males are much bigger than females, or cows, and have a massive head with a hairy mane. There are six species of sea lions and nine of fur seals. The most familiar of all is the California sea lion (*Zalophus californianus*) of the Pacific coast of North America. These sea lions are the trained barking circus "seals" that do tricks for a fish or two. Fur seals, like the northern fur seal (*Callorhinus ursinus*), were once almost exterminated for their thick fur. Now most are protected around the world, though some species are still hunted. Population of the Steller sea lion (*Eumetropias jubatus*) in Alaska has decreased by 80 % since the 1970s, most probably because of a decline in the fish they eat as a result of an increase in commercial fishing.

The *walrus* (*Odobenus rosmarus*) is a large pinniped with a pair of distinctive tusks protruding down from the mouth. It feeds mostly on bottom invertebrates, particularly clams. It was once thought that the walrus used its tusks to dig up food, but there is no evidence for this. Instead, these pinnipeds apparently suck up their food as they move along the bottom. The stiff whiskers of the snout probably act as feelers. The tusks are used for defence and to hold or anchor to ice.

6.2.4.5 Manatees and Dugong

Manatees and dugongs belong to the mammalian order Sirenia and are also known as *sea cows*, or sirenians. They have a pair of front flippers but no rear limbs. They swim with up-and-down strokes of their paddle-shaped, horizontal tail. The round, tapered body is well padded with blubber. They have wrinkled skin with a few scattered hairs. Sirenians are gentle creatures. They usually live in groups. Manatees and dugongs are thought to have been highly social animals, as old records report huge congregations of these animals before hunting decimated their numbers. The few remaining individuals have tended to become solitary or form only small family groups. Like elephants and other large mammals, they reproduce slowly, typically one calf every 3 years. They are the only herbivorous aquatic mammals, and they rely on larger plants, not algae, for nourishment. Their food requirements restrict them to living in shallow coastal waters, estuaries and rivers. Their large lips are used to feed on seagrasses and other aquatic vegetation. Seagrass beds, their source of food, are being destroyed at an alarming rate by anchor and boat damage and by excess silt and nutrients from land, which often results from deforestation and intensive farming. Only four species remain, and all are in danger of extinction. All four species of this order reside in warm waters and do not come on to land.

All sirenians are large. Dugongs may reach 3 m (10 ft) and 420 kg (930 lb), whereas manatees are 4.5 m (almost 15 ft) in length and weigh about 600 kg (1320 lb). A fifth species of sirenian, Steller's sea cow *(Hydrodamalis gigas)*, became extinct within historical times which grew up to 7.5 m (25 ft) long. The sirenians have been particularly vulnerable to hunting pressure because of their inshore habitats and their slow and placid behaviour, and they are prized for their meat, oil and hides in many cultures. At one time, dugongs had a widespread distribution which included Atlantic waters; today, they are restricted to the Indian and Pacific oceans. All three species (*Trichechus*) of manatees are found only in tropical Atlantic waters; one is restricted to the Amazon and the others inhabit shallow coastal waters and rivers from Florida to West Africa. The dugong (*Dugong dugon*) is strictly marine and survives from East Africa to some of the western Pacific islands. Its numbers are critically low throughout most of its range.

6.2.4.6 Sea Otter and Polar Bear

Sea otters are members of the weasel or mustelid family. Though there is doubt about the pinnipeds, the sea otter (*Enhydra lutris*) is definitely a member of the order Carnivora. It also differs from other marine mammals in lacking a layer of blubber. Therefore, they rely on their fur to keep warm. If their fur is oiled, it loses its insulating qualities and the sea otters soon chill.

Insulation from the cold is provided by air trapped in its dense fur. Their fur actually consists of two layers, an undercoat and longer guard hairs. This system traps a layer of air next to their skin so their skin does not get wet. In fact, at 850,000 to one million hairs per square inch, they have the thickest fur of any mammal. Sea otters are usually dark brown, often with lighter guard hairs. This splendid, dark brown fur, unfortunately, attracted hunters. Alaskan sea otters tend to have lighter fur on their heads.

Sea otters are the smallest marine mammal. The female weight ranges between 35 and 60 pounds and males weigh between 70 and 90 pounds. In the Alaskan Sea, the sea otters are relatively bigger with males having weight around 100 pounds. The preferred habitats of sea otter are the kelp beds. Their body is covered with soft beautiful fur.

The population of sea otter is presently under threat due to over hunting and killing by poachers. The beautiful fur of sea otters is greed to the hunters and poachers. Oil spill has considerable adverse impact on the sea otter population. Often after massive oil spill, sea otters are found covered with dark oil.

Sea otters were slaughtered to near extinction until they became protected by an international agreement in 1911. The sea otter was then able to slowly expand from the few individuals that had managed to survive in some remote locations. Their numbers, however, have been declining in recent years, particularly as they have slightly declined since peaking in 2004, mostly as a result of disease.

Sea otters are playful and intelligent animals. They spend most or all of their time in the water, including breeding and giving birth. Sea otters need to eat 25–30 % of their body weight per day, so they spend a lot of time looking for food. They satisfy their ravenous appetites with sea urchins, abalone, mussels, crabs, other invertebrates and even fishes. The kelp acts as an anchor that the sea otters use to wrap themselves in when they are resting. Sea otters are social animals, with females and pups spending time together in one group and males in another. The furry pup is constantly groomed and nursed by its mother.

The polar bear (*Ursus maritimus*) is the second member of the order Carnivora that inhabits the marine environment. Polar bears are semi-aquatic animals that spend a good part of their lives on drifting ice in the Arctic. They feed primarily on seals, which they stalk and capture as the seals surface to breathe or rest.

Brain Churners

1. Why regions of high primary production coincide with large populations of animals in that sea?
2. What are the basic factors controlling the abundance of zooplankton in the estuarine region?
3. How are holoplankton different from meroplankton?
4. How myctophids produce light in the deeper part of the ocean?
5. What are the major threats on sharks?
6. How is the excess salt eliminated by saltwater crocodiles?
7. How ambient temperature influences the sex of the turtle hatchlings in the beach?
8. What is the scientific basis of the coincidence of arrival of *Calidris canutus rufa* in Delaware Bay with the post-mating period of horseshoe crab?
9. How are the crustaceans adopted to marine environment?
10. How is the kelp community associated with the sea otters?

Annexure 6A: Warming of Lower Gangetic Delta Water, A Projection on the Basis of Real-Time Data

1. Introduction

The carbon dioxide concentration in the atmosphere of India is gradually increasing over a period of time. The mid-tropospheric carbon dioxide during 2002–2008 retrieved from Atmospheric Infrared Sounder (AIRS) on board Advanced Microwave Sounding Unit (AMSU-A) was analyzed over the Indian subcontinent and surrounding oceans. Important features exhibited by the observations are the strong seasonal and latitudinal gradient modulated by strong monsoonal activity over the study region (Fig. 6A.1). Further analysis suggests that atmospheric carbon dioxide concentration has increased linearly from 372 ppm in 2002 to 386 ppm in 2008 at the rate of 2.05 ppm/year during past 6 years with strong seasonal variation over the land and relatively weaker seasonal variability over the ocean (Figs. 6A.1 and 6A.2).

Carbon dioxide, being a heavier gas, forms a blanket on the Earth's surface that retains solar radiation. The global air temperature is, therefore, increasing due to increase in GHGs mainly carbon dioxide (Solomon and IPCC 2007).

There have been many studies related to long-term trends in surface air temperature in India. During 1901–2003 average long-term trend of 0.05 °C/decade was found in the all-India annual mean surface air temperature (Kothawale and Kumar 2005). The recent period of 1971–2003 has seen a relatively accelerated warming of 0.22 °C/decade. Recently, surface air temperatures in different parts of India were analysed for the period of 1901–2003 (Ganguly and Iyer 2008) and an increasing trend

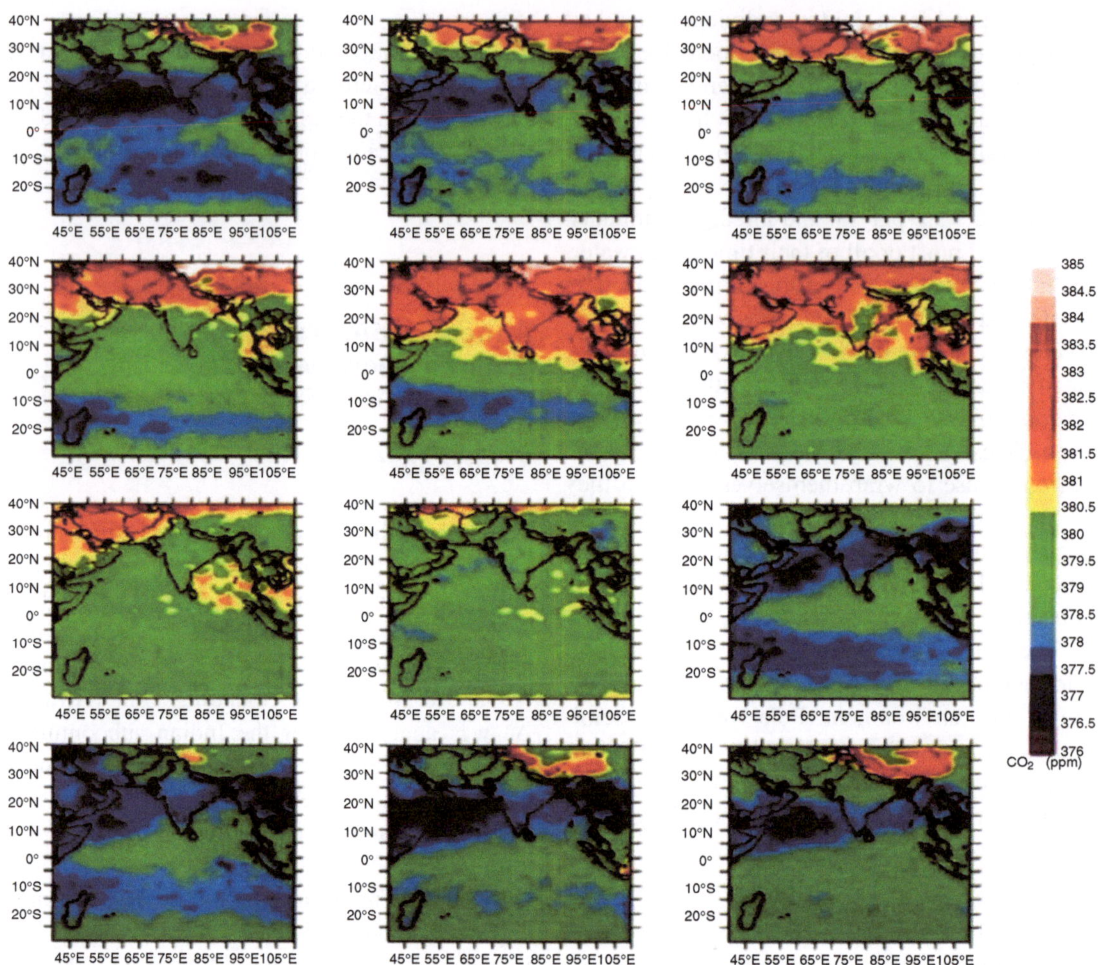

Fig. 6A.1 Monthly climatology of mid-troposphere carbon dioxide over India and surrounding oceans based on AIRS observation during 2003–2008 (*Source*: National Carbon Project Status Report 2010)

(0.03–0.09 °C/decade) was found over India, except in the northwestern part (with decreasing trend of −0.01 °C/decade). However, there have been limited studies related to long-term changes in surface water temperature in the lower Gangetic delta region although a significant transfer of heat and water vapour occurs at the air–water interface.

West Bengal, a maritime state in the northeast part of India, is the domain area of this first-order analysis. The percentage of increase of carbon dioxide emission in West Bengal is 50.79 % from 1980 to 2000 (Ghoshal and Bhattacharyya 2010). The increase in carbon dioxide in the air is reflected through gradual increase of air and surface water temperature in the lower Gangetic delta region. The implications of such rise in surface water temperature are serious on different tiers of biodiversity, as the region sustains some 34 true mangrove species (Mitra 2000) and some 62 mangrove associate species (Mitra and Pal 2002).

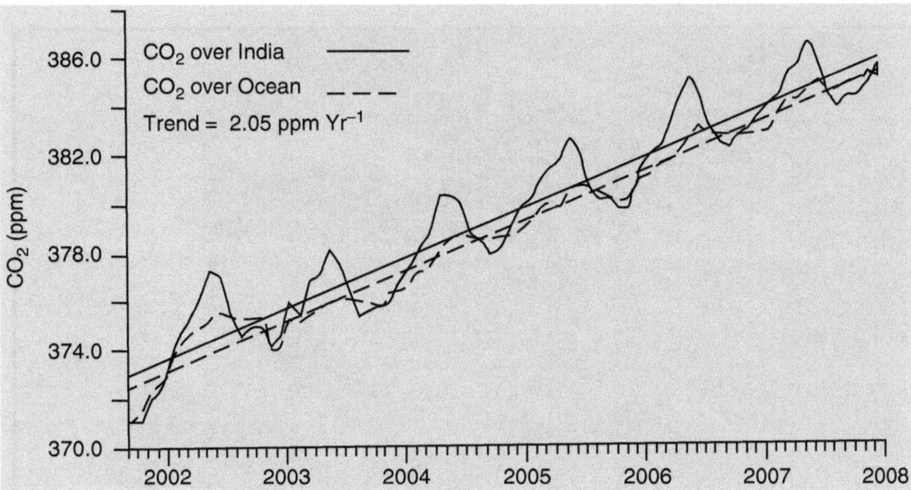

Fig. 6A.2 Trend of mean atmospheric carbon dioxide concentration over India and surrounding oceans (*Source*: National Carbon Project Status Report 2010)

Table 6A.1 Sampling stations in the western, central and eastern sectors of Indian Sundarbans in the lower Gangetic delta region

Sectors		Sampling stations	Latitude	Longitude
Western sector	**Stn. 1**	Chemaguri (W_1)	21°38′25.86″N	88°08′53.55″E
	Stn. 2	Saptamukhi (W_2)	21°40′02.33″N	88°23′27.18″E
	Stn. 3	Jambu Island (W_3)	21°35′42.03″N	88°10′22.76″E
	Stn. 4	Lothian (W_4)	21°38′21.20″N	88°20′29.32″E
	Stn. 5	Harinbari (W_5)	21°44′22.55″N	88°04′32.97″E
	Stn. 6	Prentice Island (W_6)	21°42′47.88″N	88°17′55.05″E
Central sector	**Stn. 7**	Thakuran Char (C_1)	21°49′53.17″N	88°31′25.57″E
	Stn. 8	Dhulibasani (C_2)	21°47′06.62″N	88°33′48.20″E
	Stn. 9	Chulkathi (C_3)	21°41′53.62″N	88°34′10.31″E
	Stn. 10	Goashaba (C_4)	21°43′50.64″N	88°46′41.44″E
	Stn. 11	Matla (C_5)	21°53′15.30″N	88°44′08.74″E
	Stn. 12	Pirkhali (C_6)	22°06′00.97″N	88°51′06.04″E
Eastern sector	**Stn. 13**	Arbesi (E_1)	22°11′43.14″N	89°01′09.04″E
	Stn. 14	Jhilla (E_2)	22°09′51.53″N	88°57′57.07″E
	Stn. 15	Harinbhanga (E_3)	21°57′17.85″N	88°59′33.24″E
	Stn. 16	Khatuajhuri (E_4)	22°03′06.55″N	89°01′05.33″E
	Stn. 17	Chamta (E_5)	21°53′18.56″N	88°57′11.40″E
	Stn. 18	Chandkhali (E_6)	21°51′13.59″N	89°00′44.68″E

2. Materials and Methods

2.1. Study Area

The mighty river Ganga emerges from a glacier at Gangotri, about 7010 m above mean sea level in the Himalayas, and flows down to the Bay of Bengal covering a distance of 2525 km. At the apex of Bay of Bengal, a delta has been formed which is recognized as one of the most diversified and productive ecosystems of the tropics and is referred to as the Indian Sundarbans. The deltaic complex has an area of 9630 km^2 and houses about 102 islands (Mitra 2000). Eighteen sampling sites were selected, six each in the western, central and eastern sectors of Indian Sundarbans (Table 6A.1, Fig. 6A.3).

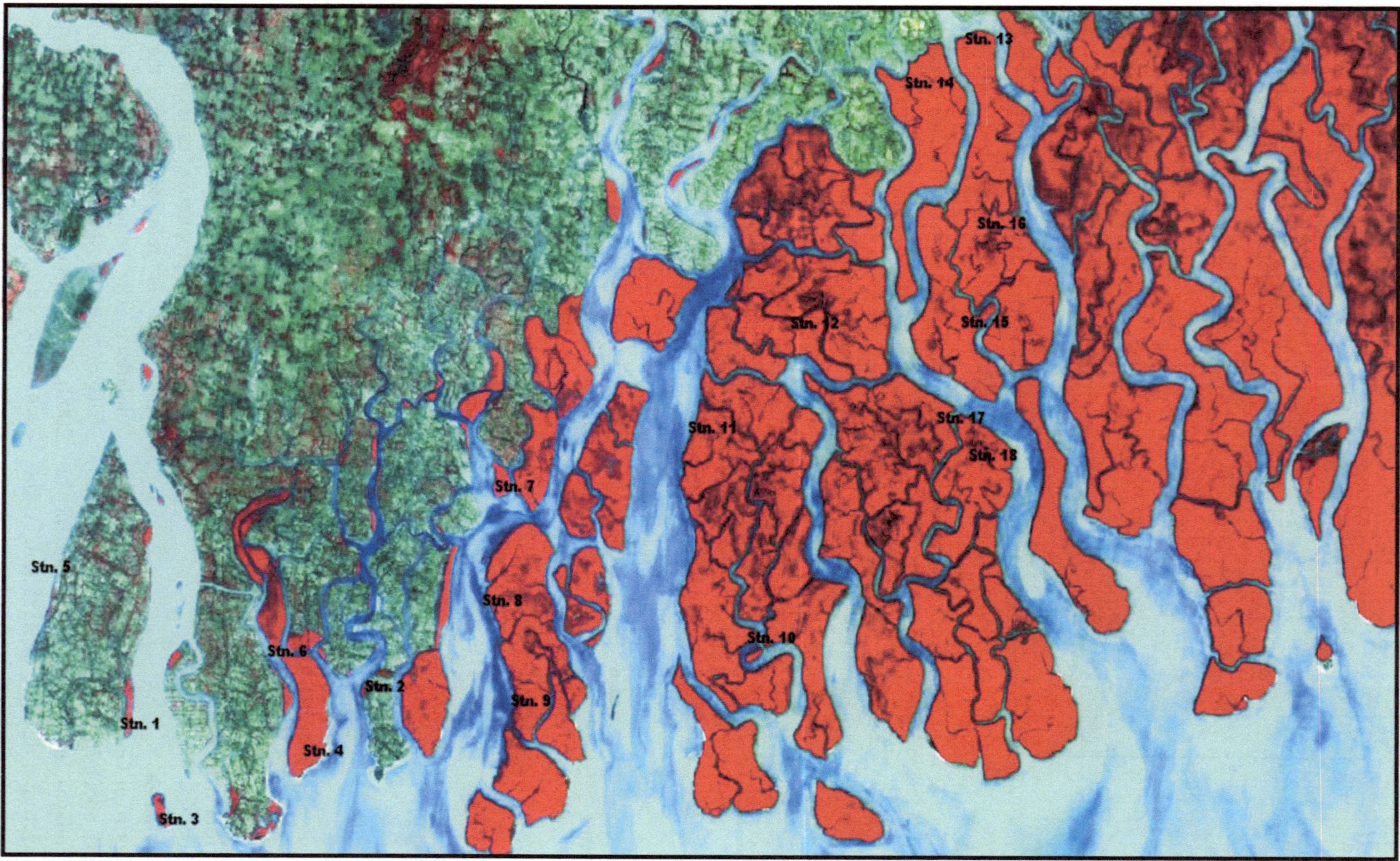

Fig. 6A.3 Location of sector-wise sampling stations in Indian Sundarbans; the *red colour* indicates the mangrove vegetation

We demarcated the three sectors of Indian Sundarbans on the basis of our primary surface water salinity data of 24 years and secondary data (of 27 years) from Mitra et al. (2009) and Sengupta et al. (2013).

2.2. Data Sources and Quality

We considered a data set of 30 years in this first-order analysis per the minimum standard norm of climate-related researches. The World Meteorological Organization and the Intergovernmental Panel on Climate Change (IPCC) define "climate" as the average state of the weather over time with the period generally being 30 years (although for some marine climate parameters such as storminess, longer averages are required) (Zhang et al. 2000).

More than two decades of data (1984–2013) were compiled from the archives of the Department of Marine Science, University of Calcutta, for this study. A number of studies on different aspects of the Sundarban complex have been published over the years, which include description of the data (and methods) at different times over the past three decades (Mitra et al. 1987, 1992, 2009; Chakraborty and Choudhury 1985; Mitra and Choudhury 1994a; Saha et al. 1999; Banerjee et al. 2002a, 2003, 2013; Mondal et al. 2006; Sengupta et al. 2013). Real-time data (through field sampling by the authors) were also collected simultaneously since 1998 from 18 sampling stations in the lower Gangetic region during high-tide condition to assure quality and continuity to the data bank. For each observational station, at least 30 samples were collected within 500 m of each other and the mean value of 30 observations was considered for statistical interpretations.

2.3. Measurement of Surface Water Temperature

The surface water temperature in the selected sampling stations was measured with a portable thermometer (sensitivity $= \pm 0.02$).

2.4. Statistical Analysis

Time series analysis was performed to forecast the trend of surface water temperature on the basis of the past 30 year's real-time data. Exponential smoothing method produces maximum likelihood estimates and can reflect the future trend of the selected variable. This approach was used to forecast the values for surface water temperature in the ambient media of the sampling station till 2043.

3. Results

3.1. Surface Water Temperature: Real-Time Data

The surface water temperature exhibited an increasing trend in all the three sectors of lower Gangetic region. In the western sector, the rate of increase ranges between 0.04 °C/year (at Saptamukhi) to 0.06 °C/year (at Jambu Island) (Fig. 6A.4). In the central sector, the rate of increase ranges from 0.04 °C/year (at Dhulibasani) to 0.05 °C/year (at Thakuran, Chulkathi, Goasaba and Pirkhali) (Fig. 6A.5). The eastern sector, adjacent to Bangladesh Sundarbans, exhibited similar increase in trend with values ranging from 0.05 °C/year (at Harinbhanga, Khatuajhuri, Chamta and Chandkhali) to 0.06 °C/year (at Arbesi) (Fig. 6A.6). The overall increase in surface water temperature during the study period ranges from 3.73 % (at Saptamukhi) to 5.59 % (at Jambu Island) in the study region. The long-term data of 30 years reveals an increasing trend in all the three sectors, but the average (mean of 6 sampling stations) rate of increase is relatively low in the western sector (4.47 %) compared to the central (4.55 %) and eastern (4.92 %) sectors.

3.2. Surface Water Temperature: Forecast Data

The forecast method predicts that in the western, central and eastern sectors of the lower Gangetic region, the surface water temperature can reach up to 34.9, 35.0 and 35.2 °C, respectively, during 2043 (Figs. 6A.7, 6A.8 and 6A.9) compared to the present average value of 32.8 °C, 32.6 °C and 32.7 °C, respectively. Thus after a period of 30 years from April 2013, there will be a rise of 6.40 %, 7.36 % and 7.65 % in the western, central and eastern sectors, respectively, in deltaic complex.

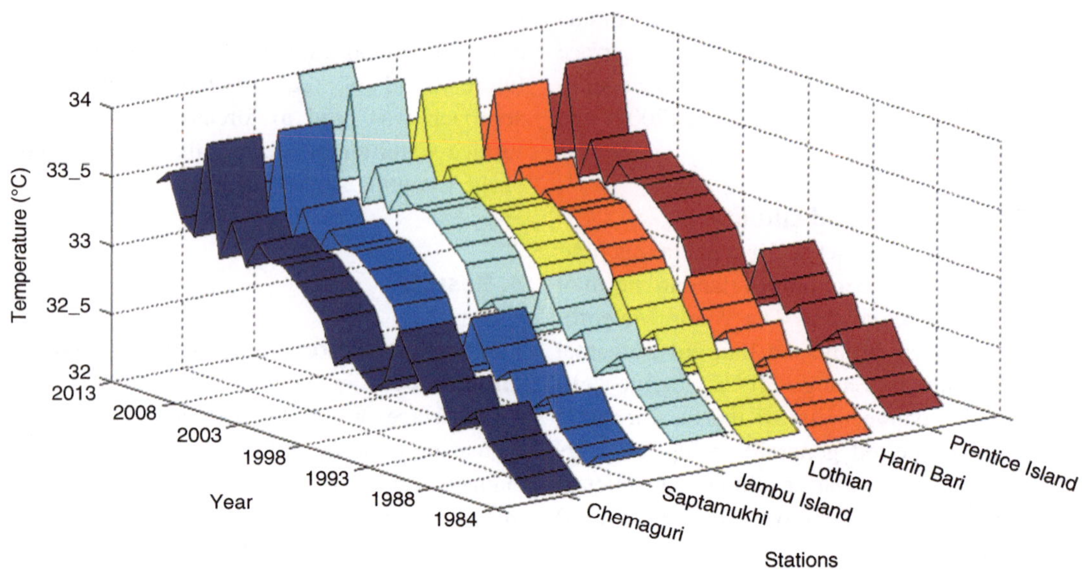

Fig. 6A.4 Increasing trend of surface water temperature in six stations of western Indian Sundarbans

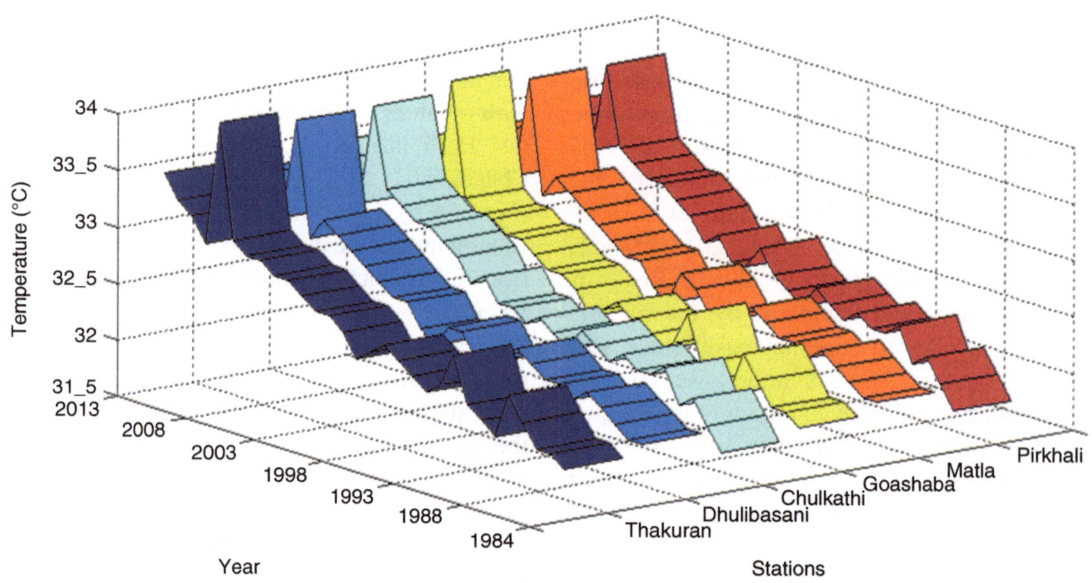

Fig. 6A.5 Increasing trend of surface water temperature in six stations of central Indian Sundarbans

4. Discussion

Climate variability has become an important topic of scientific pursuit during the past few decades, intimately linking the economy of a nation with its climate resources. In addition to land and atmosphere, waters of ocean, seas, coasts, bays and estuaries are important components of the Earth's climate system. The correlations between surface water temperature and surface air temperature are generally high for all averaging periods and seasons indicating the high level of thermal communication at the interface.

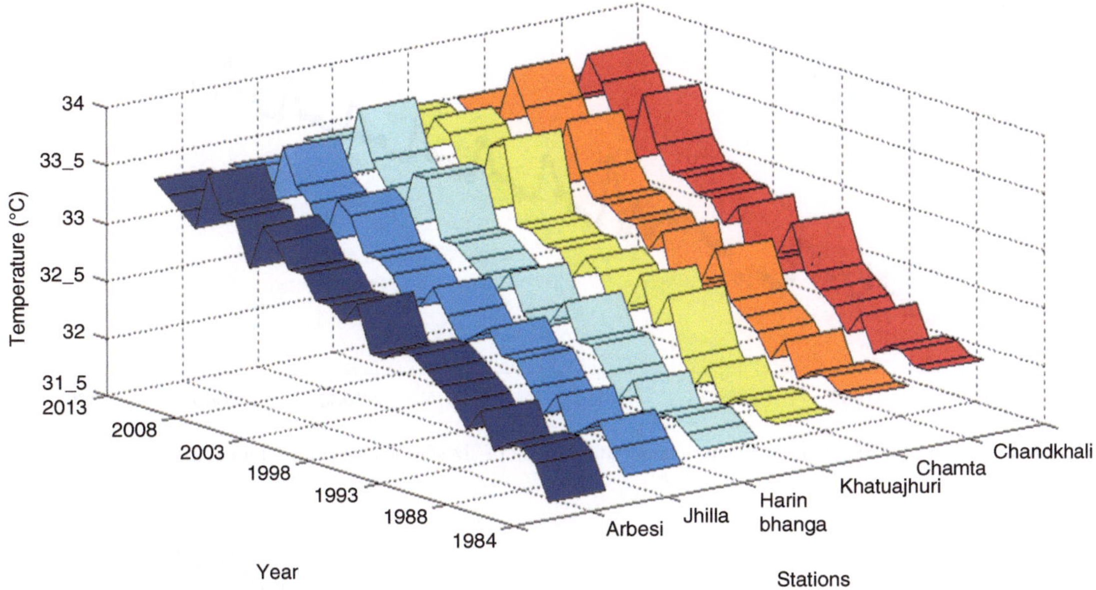

Fig. 6A.6 Increasing trend of surface water temperature in six stations of eastern Indian Sundarbans

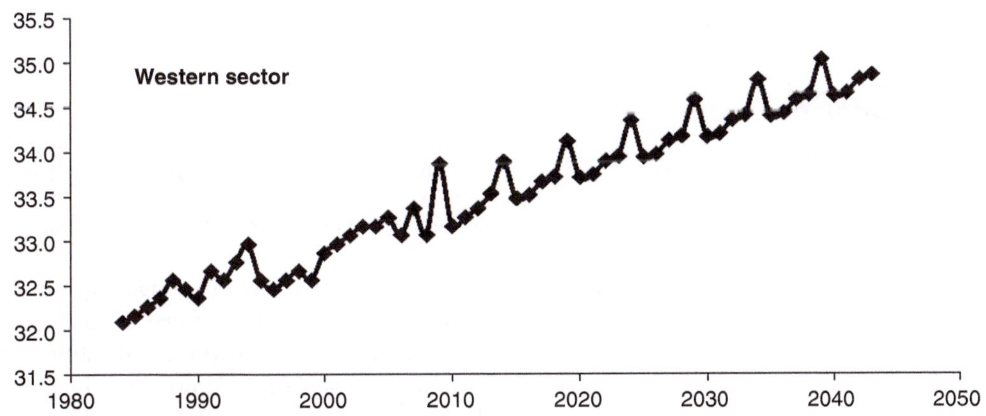

Fig. 6A.7 Forecast values of surface water temperature (mean of six stations) in western Indian Sundarbans

The ocean processes are slower and large-scale changes in the ocean that occur on monthly to multi-decadal timescales as compared to changes occurring in the atmosphere.

India is surrounded by oceans and convection over these oceans plays a major role in the monsoon rainfall over India. A recent study by Barnett et al. (2005) has shown that oceans are not only becoming warmer at the surface, but there is also a penetration of human-induced warming into the deeper parts of the oceans. An understanding of the sea-surface temperature (SST) is therefore essential to know the pulse of climate change at the local and regional levels.

The trend of increase of water temperature has several implications like alteration of circulation pattern as well as the rate of biogeochemical and ecological processes that determine the water quality with the increase of water temperature marine and estuarine species ranges will likely expand towards environments that are presently cooler (IPCC 2007). If dispersal capabilities are

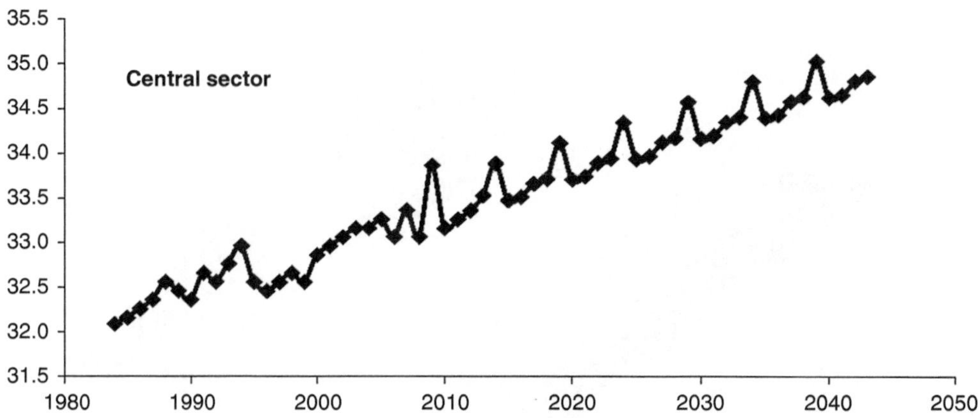

Fig. 6A.8 Forecast values of surface water temperature (mean of six stations) in central Indian Sundarbans

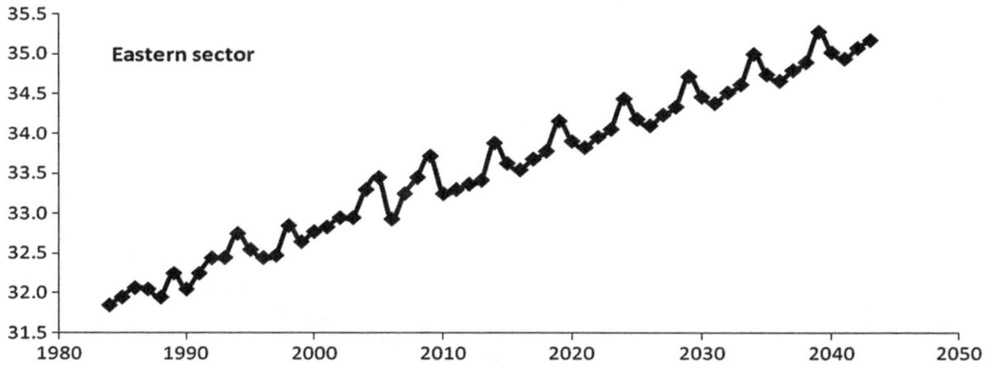

Fig. 6A.9 Forecast values of surface water temperature (mean of six stations) in eastern Indian Sundarbans

limited due to lack of suitable habitat (in terms of thermal statistics), local extirpations and extinctions are likely to occur (Thomas et al. 2004). Temperature rise may also have profound impact on commercial fisheries through continued shifts in distribution and alteration in community interactions (Perry et al. 2005). In many regions like Gulf of Mexico and Mid- and South Atlantic shorelines, extensive algal blooms are reported due to rise of water temperature, which have adverse impact on living resources, local economies and public health (Cambers et al. 2007). In the present geographical local, the increase of surface water temperature is more in the central (7.36 %) and eastern (7.65 %) sectors compared to the western (6.40 %). This variation may result in the shifting of nektonic species from the central and eastern sectors of Indian Sundarbans to the western sector, which is relatively congenial in terms of thermal condition (as revealed from the present data) and salinity (Mitra et al. 2009, 2011; Sengupta et al. 2013). However, still there is a lack of information on how tropical fish will respond to temperature increases (Cambers et al. 2007), although few studies in the present geographical level points towards more fish diversity in the western sector compared to the central sector of the lower Gangetic region (Mitra 2013).

5. Conclusion

Based on an analysis of roughly 30,000 datasets, the IPCC (2007) concluded that 85 % of the physical and biological changes in natural systems observed globally since 1970 were consistent with the responses that would be expected to accompany atmospheric warming. However, our understanding on the impact of warming of ocean and estuarine water on biodiversity are in its infancy, and quantification of the impacts through proper scaling is even in an embryo stage. The present findings could cover the decadal variation of surface water temperature in three sectors of lower Gangetic delta and concludes that these local level slight variation of water temperature may be attributed largely to anthropogenic factors, like more dilution in the western sector (on account of Farakka discharge) compared to central and eastern sectors, where the freshwater availability is practically negligible due to siltation of the Bidyadhari river since the late fifteenth century. We also hypothesize the migration of nektonic species from the high thermal (central and eastern sectors) to the low thermal region (western sector), but could not confirm the hypothesis due to absence of baseline data on nektonic domain.

Annexure 6B: Finfish Juvenile Destruction in and Around Mangrove Ecosystem of Indian Sundarbans and Possible Measures of Conservation

1. Introduction

Ichthyoplankton and fish juveniles are the major biotic components of the world ocean (Verlecar and Parulekar 2001) and play a vital role in regulating the marine food chain. They are widely distributed in the brackish water system and mangrove-dominated coastal zone owing to the presence of mangrove vegetation-derived nutrients and congenial environmental conditions for their growth and survival. It is reported that the mangrove ecosystem serves as the nursery of a large spectrum of fish juveniles (Banerjee et al. 2002; Mitra 2000; Mitra and Pal 2002; Mitra et al. 1992). This unique ecosystem forms an ideal ecological reservoir for the pelagic community because the production of leaf litter and detrital matter from mangrove plants fulfils the nutritional requirements of prawn juveniles, crab megalops and juveniles of several species of osteichthyes. Due to this fact, the mangrove ecosystem is recognized as the world's most potential nursery and breeding ground (Mitra and Pal 2002).

The Indian Sundarbans lie at the apex of the Bay of Bengal and sustain the famous mangrove ecosystem of the tropics. The islands and the associated waters, creeks and estuaries of the deltaic lobe are the nurseries for many coastal and oceanic species of fish that support important subsistence and commercial fisheries in the northern Bay of Bengal. The diverse vegetation of the Sundarbans forests includes about 36 true mangrove species that provide a unique mix of habitats for wildlife, which includes tiger, deer, wild boar, monkey, turtles, terrapins, dolphins, estuarine crocodiles and several species of birds, many of which are migratory. This region is considered as one of the most backward regions in the maritime state of West Bengal and is home to about 4.2 million people, of whom 56 % are landless. Literacy rates are below 35 %; most communities do not have electricity or safe drinking water and agricultural productivity is lower than the state average. Inadequate infrastructure, poor communication facilities, lack of access to clean drinking water, health and education services and a fragile and limited natural resource base have contributed to a low level of development and high poverty incidence in the region.

A diminishing natural resource base is threatening the ecological integrity of Indian Sundarbans and the livelihood of the inhabitants. At present, the major environmental-cum-socio-economic issue in this mangrove ecosystem is the proliferation of the shrimp farms. In the absence of any hatchery in the entire state, the supply of tiger prawn seeds to these farms is done through the wild collection from the estuarine and coastal waters. Thousands of women and children collect wild tiger prawn seeds by employing nets of a particular mesh size to haul in the drifting community, irrespective of the tides, and throw away the major portion of the haul (containing the juveniles of finfish and shellfish) after sorting out the post larvae (or seeds) of tiger prawn. This practice results in a great loss of pelagic biodiversity (in terms of stock) which might cause an adverse effect on the delicate ecological balance of the system (Trivedi et al. 1994; Nageswara 2004). A negative impact also becomes obvious in the prospect of demersal fisheries of the state of West Bengal due to this indiscriminate destruction of the various species of fishes at their juvenile stages. With this background, the present paper aims to highlight the magnitude of finfish juvenile diversity loss in different seasons of the year during 2004–2013. This occupation has spread its root to such an extent that about 95 % of the coastal population living below poverty line is engaged with this destructive activity. It is very difficult to implement a total ban on this activity, as these huge masses will face extreme economic marginalization. So, our present paper suggests few alternative livelihood schemes, which may establish a harmony between the biological diversity and socio-economic profile of the area.

2. Description of the Study Area

West Bengal is a maritime state in the northeast part of the Indian subcontinent adjacent to Bangladesh. The coastal zone of this state spreads over an area of 10,158.22 km^2 and is restricted within the latitude 21°30′N–22°30′N and longitude 87°25′E–89°10′E. The river Harinbari or Heronbhanga of the Indian Sundarbans (India– Bangladesh border) is the easternmost border, while the New Digha Coast of Orissa-West Bengal border constitutes the western boundary of coastal West Bengal. With considerable degree of marine characteristics in the major portion of the ecosystem, the important morphotypes of coastal West Bengal are sandflats, coastal dunes, beaches, mudflats, estuaries, creeks, inlets and mangrove flats. There is a drastic variation of salinity and dilution factor between different horizontal transects of the ecosystem. The dilution factor, salinity and pH reveal significant variations at the same time in different locations of the coastal zone (Mitra 2000). The recent emergence of Haldia industrial complex and various fish landing stations in the coastal zone has also opened the gateway of input of various categories of wastes in this ecosystem.

Geographically, the study area encompasses three major districts of the state of West Bengal, namely, 24 Parganas (North), 24 Parganas (South) and Medinipur (East). The Indian Sundarbans which has been declared as the World Heritage Site by UNESCO in the year of 1989 fall within the North and South 24 Parganas districts. Three sampling stations were selected in and around this deltaic ecosystem, namely, Diamond Harbour (station 1), Sagar Lighthouse (station 2) and Junput (station 3). Each of these sampling stations is markedly different from the other with respect to aquatic salinity due to their location and proximity to the Bay of Bengal.

Diamond Harbour (station 1) is situated in the low saline upper stretch of Hooghly estuary, just outside the northern boundary of Indian Sundarbans. The station is very near to the Haldia port-cum-industrial complex. Salinity of surface water is minimum around the station owing to its location far away from the Bay of Bengal in the extreme upstream region and also due to huge freshwater discharge from the Hooghly river, which is perennial in nature. The station has three mangrove species along with few mangrove associates like *Porteratia coarctata* (salt marsh grass) and seaweeds (*Enteromorpha* sp.).

Sagar Lighthouse (station 2) is situated in the southwestern tip of the Sagar Island and falls in the western sector of Indian Sundarbans. The station has rich mangrove vegetation and extensive mudflats. Although there are no industrial activities in this station, but the presence of sizable number of shrimp farms (presently carrying on shrimp culture by traditional method with a very low stocking density of prawn seeds) has enriched the surrounding water with nutrients and organic load.

Junput (Station 3) is situated in the Medinipur (East) district of coastal West Bengal, which is noted for its high aquatic salinity owing to its proximity to the Bay of Bengal. The extremely high salinity has posed an inhibitory effect on the growth and survival of mangroves in this station (Mitra et al. 2004a). Existence of saltpans has made the soil of the area hyper saline in nature. Although the station has no industry around its vicinity, but the presence of Digha tourist centre and Shankarpur fishing harbour close to the station has multiplied the anthropogenic pressure around the zone.

In all these stations, tiger prawn seed collection is a very common scene in which the coastal and estuarine waters are screened in search of target species (seeds of *Penaeus monodon*) by using nets of very fine mesh size. This activity traps a varied spectrum of several finfish juveniles (nontarget species), which are subsequently thrown away and destroyed.

3. Materials and Methods

Finfish juvenile samples were collected monthly during January 2004 to December 2013 from tiger prawn seed collectors. Collection of ten nets was randomly mixed and a constant weight of 10 g was preserved in 4 % formaldehyde. The samples were identified from the Zoological Survey of India and its count was done using hand lens. Triplicate counts were done to reach the maximum accuracy level and the mean value of the species count was finally used to enumerate the community structure of finfish juvenile in the study area through computation of Shannon–Wiener species diversity index (9) per the AAMSTAT software developed by BOOLEAN LOGIC PRIVATE LIMITED in 2005.

4. Results and Discussion

A. Root Cause of Tiger Prawn Seed Collection

1. *Mushrooming of shrimp culture activities:* The aquatic sub-subsystem of coastal West Bengal is highly congenial for the culture of tiger prawn not only because of the aquatic salinity, tidal flushing (which favours the process of water exchange), but also because of the rich phytoplankton diversity (102 species documented so far) in the optimum level of aquatic nutrients (Mitra et al. 2004b). About 33,000 ha in North 24 Parganas and 12,000 ha in South 24 Parganas districts have been devoted for shrimp culture activities, which have high demand for tiger prawn seeds for economic viability. This demand is mitigated through wild harvest of PL_{20} (larval stage 20 days from the hatching date) from estuaries, coastal wetland and brackish water creeks (Banerjee et al. 2005a, b; Mitra et al. 2005a).
2. *Absence of hatchery:* Currently, there are few operational shrimp nurseries in West Bengal, but no ideal hatchery in real sense (i.e. spawning of adult to produce nauplii). Low salinity is the biggest problem for shrimp hatchery in West Bengal coast, particularly in Sundarbans. In the absence of any hatchery, the demand for tiger prawn seeds has become very acute and the entrepreneurs have no option, but to depend on wild harvest of tiger prawn seeds.
3. *Socio-economic profile:* The socio-economic profile of coastal West Bengal is not very encouraging; particularly, in Indian Sundarbans, it is highly distressed. Recently, the population has touched the figure of 4.2 million, which is predominated by farming and fishing community. More than 90 % of the farming community is either small or marginal farmers and about 50 % survives below the poverty level. This sizable chunk of population engages themselves in screening the estuarine water from dawn to dusk in search of the target species which brings immediate cash to them. The price of tiger prawn seeds ranges from Rs. 200 to Rs. 800 per 1000 depending on the season and demand (Mitra and Bhattacharyya 2003).

B. Community Structure of Finfish Juvenile

The coastal zone of West Bengal is one of the most biologically productive, taxonomically diverse and aesthetically celebrated ecotone of the country, which is sustaining the famous mangrove ecosystem of Indian Sundarbans (Bhattacharya et al. 1999). The pelagic system around the coastal zone is highly productive due to considerable concentrations of nutrients derived from anthropogenic origin, run-off from the adjacent landmasses, litters and detritus of mangrove vegetation, etc.

Table 6B.1 exhibits the monthly average of standing stock (N) and Shannon–Wiener species diversity index (H) of finfish juvenile species in the three sampling stations. Members of genus *Coilia* sp., *Thryssa* sp., *Mugil* sp., *Tenualosa* sp., *Liza* sp., *Scatophagus* sp., *Stolephorus* sp., *Cynoglossus* sp. and *Sillago* sp. are dominant (in terms of standing stock or biomass) in all the stations (Table 6B.2) and in all the months of the study period, which suggests the wide range of tolerance of this genus in the present geographical locale. The study clearly depicts the regulation of finfish juveniles by aquatic salinity. Similar observations were also recorded by earlier workers in this study area (Bhattacharya et al. 1999; Panja 2004; Mitra et al. 1998, 2000; Niyogi et al. 1998). The regulatory role of salinity on finfish juvenile species composition and abundance was also confirmed in a South African case study (Whitfield 1999). The diversity of finfish juvenile is highest at station 3 (Junput) which is followed by station 2 (Sagar Lighthouse) and station 1 (Diamond Harbour). The highest values of H and N at Junput confirm the positive role of aquatic salinity in maintaining the fish germplasm in the study area. Diamond Harbour is an extremely low saline zone in the upstream area with an average aquatic salinity of 1.7 ‰ during premonsoon, 0°‰ during monsoon and 1.3 ‰ during postmonsoon (Mitra et al. 2005b), and hence chance of getting stenohaline fish juvenile species is rare. Irrespective of significant variation of species number and diversity (H) among the selected stations, the present geographical locale is an abode of several species of finfish and shellfish and nursery of fish juveniles (Bhattacharyya et al. 2003). Destruction of this juvenile stock may therefore create an adverse impact on the pelagic and demersal fish production status of the state in near future (Trivedi et al. 1994).

C. Conservation Measures

The wasted juvenile species of osteichthyes have considerable total economic value and hence the destructive operation rate needs an immediate retardation through few suggestive measures as pinpointed below:

- Development of alternative livelihood (like piggery, Campbell culture, broiler rearing, sewing, apiculture, fish feed manufacturing, handicrafts, spice grinding, etc.) to defray the prawn seed collectors from the present destructive operation. Such livelihood schemes have already been implemented by WWF-India at Bali and Chotomollakhali islands of Indian Sundarbans since 2003 and considerable success has been obtained.
- Setting up of tiger prawn seed bank (nursery) in the high saline zone of Sundarbans and linking the same with the hatcheries present in the maritime states of South India like Orissa, Andhra Pradesh and Tamil Nadu.
- Increase of awareness programme among the prawn seed collectors. Several academic institutes and reputed NGOs of the state have initiated this task through several projects financed by Ministry of Environment and Forests, Govt. of India, Dept. of Science and Technology, Govt. of India, Dept. of Environment, Govt. of West Bengal, etc.
- Encouraging the people for canal excavation to harvest the rain water which can help to develop the second cropping during summer season. Four canals have already been excavated in the Chotomollakhali islands of Indian Sundarbans with the funding and assistance of WWF-India and mono-cropping pattern in the zone has been replaced by multiple cropping systems.
- Promotion and subsequent scaling up of non-conventional pilot projects like seaweed culture, oyster culture, ornamental fishery, biofertilizer preparation, etc., to broaden the avenue of alternative livelihood for future.

Table 6B.1 Monthly average standing stock (N) and Shannon–Wiener species diversity index of finfish juvenile species (H) of three selected sampling stations during January 2004 to December 2013

Year	Monthly average standing stock (N) of finfish species									Monthly average of Shannon–Wiener species diversity index of finfish species (H)								
	stn1			stn2			stn3			stn1			stn2			stn3		
	Prm	Mon	Pom	Prm	Mon	Pom	Prm	Mon	Pom	Prm	Mon	Pom	Prm	Mon	Pom	Prm	Mon	Pom
1994	392	161	245	1598	134	522	1593	260	767	2.9151	2.2994	2.7801	3.5291	1.7486	3.0118	3.4839	2.2867	3.2634
1995	321	133	291	1315	122	422	1303	213	625	2.9178	2.3014	2.7791	3.5297	1.5557	3.0087	3.4825	2.2792	3.2665
1996	200	82	183	1273	99	416	1290	213	631	2.8946	2.2715	2.7675	3.5291	1.6186	3.0187	3.4842	2.2894	3.2781
1997	195	79	175	1279	99	413	1229	201	596	2.8816	2.2440	2.7362	3.5097	1.5998	3.0374	3.4795	2.2942	3.2786
1998	215	89	197	1090	\84	361	1055	172	651	2.9286	2.3276	2.8140	3.5277	1.4978	3.0275	3.4813	2.2973	3.2809
1999	217	89	198	716	56	233	679	123	333	2.9295	2.3268	2.8054	3.5234	1.6254	3.0432	3.4616	2.2846	3.2878
2000	224	93	204	639	51	209	601	103	291	2.9197	2.3078	2.7919	3.5353	1.6378	3.0421	3.4647	2.2933	3.2508
2001	221	92	201	545	42	180	521	89	255	2.9253	2.3098	2.7977	3.5415	1.6021	3.0330	3.4847	2.3029	3.2921
2002	225	93	204	910	178	442	934	153	440	2.9310	2.3078	2.7919	3.4058	2.5097	3.1565	3.4986	2.2077	3.3045
2003	389	38	67	936	73	304	1020	79	367	2.3682	1.4809	1.9693	3.5312	1.6087	3.0308	3.4273	1.8401	3.0912

stn station, *Prm* premonsoon, *Mon* monsoon, *Pom* postmonsoon

Table 6B.2 Distribution of finfish juvenile species in 10 g composite wasted sample collected from three selected sampling stations during January 2004 to December 2013

Sl. no.	Species	Station 1	Station 2	Station 3
1.	*Coilia* sp.	+	+	+
2.	*Thryssa hamiltonii (Grey)*	+	+	+
3.	*Thryssa baelama*	+	+	+
4.	*Torquigener oblongus*	+	+	+
5.	*Rhinomugil corsula*	+	+	+
6.	*Mugil cephalus*	+	+	+
7.	*Sillaginopsis panijus*	+	+	+
8.	*Sillago sihama*	+	+	+
9.	*Sillago soringa*	+	+	+
10.	*Zenarchopterus dispar*	+	+	+
11.	*Glossogobius quiris*	+	+	+
12.	*Pseudapocryptes lanceolatus*	+	+	+
13.	*Eupleurogrammus glossodon*	+	+	+
14.	*Pseudorhombus* sp.	+	+	+
15.	*Pisodonophis boro*	+	+	+
16.	*Tenualosa ilisha*	+	+	+
17.	*Cynoglossus arel*	+	+	+
18.	*Cynoglossus* sp.	+	+	+
19.	*Leiognathus blochii*	–	+	+
20.	*Leiognathus equulus*	+	+	+
21.	*Hilsa* sp.	+	+	+
22.	*Scatophagus argus*	+	+	+
23.	*Liza parsia*	+	+	+
24.	*Liza tade*	+	+	+
25.	*Stolephorus commersonii*	+	+	+
26.	*Stolephorus baganensis*	+	+	+
27.	*Stolephorus kammalensis*	+	+	+
28.	*Lutjanus johnii*	+	+	+
29.	*Setipinna taty*	+	+	+
30.	*Lagocephalus lunaris*	–	+	+
31.	*Escualosa thoracata*	–	+	+
32.	*Epinephelus tauvina*	–	+	+
33.	*Epinephelus coioides*	–	+	+
34.	*Sphyraena* sp.	–	+	+
35.	*Pomadasys* sp.	–	+	+
36.	*Sardinella longiceps*	–	+	+
37.	*Periophthalmus* sp.	–	+	+
38.	*Macrognathus* sp.	–	+	+
39.	*Tetradon cutcutia*	–	+	+
40.	*Bregmaceros meclellandii*	–	+	+
41.	*Ichthyocampus carce*	–	+	+
42.	*Stigmatogobius* sp.	–	+	+
43.	*Channa* sp.	–	+	–
44.	*Kurtus indicus*	–	+	–
45.	*Harpodon nehereus*	–	+	–
46.	*Moringua raitaborua*	–	+	–
47.	*Suggrundus rodricensis*	–	+	–

'+' refers to presence of finfish species
'–' refers to absence of finfish species

Again, a new concept of conservation is soon going to crystallize in the Indian Sundarbans regions in the form of 'Rural Technology Park'. In this unit, apart from awareness generation programme through adult education, some experimental ponds will be designed to monitor the survival rate of fish juveniles which are thrown away and wasted during wild tiger prawn seed collection. It is expected that such approach may reduce the mortality percentage of fish species in their juvenile stage (Mitra and Banerjee 2004).

Acknowledgement

The authors gratefully acknowledge the scientists of the Zoological Survey of India for identifying the fish juveniles and Mr. Uttam Kumar Panja, Deputy Director, Fisheries Department, Govt. of West Bengal, for overall domain knowledge on the problem.

References

Chaudhuri, A. B., & Choudhury, A. (1994). *Mangroves of the Sundarbans, India.1*. Bangkok: IUCN.

Conway, D., White, R. G., Hugues-dit-Ciles, J., Gallienne, C. P., & Robins, D. B. (2003). Guide to the coastal and surface zooplankton of the South-Western Indian Ocean. UK-Defra Darwin initiative project 162/09/004 zooplankton of the Mascarene Plateau. *Marine Biological Association of the United Kingdom Occasional Publication, 1*, 294–318.

Elapidae. (2007a). *Integrated taxonomic information system*. Retrieved August 7, 2007.

Elapidae. (2007b). *Elapidae at the new reptile database*. Accessed 12 August 2007.

Fichter, G. S. (1982). *Poisonous snakes. A first book*. Franklin Watts. ISBN: 0-531-04349-5, 66pp.

Grigg, G. C., Seebacherd, F., Beard, L. A., & Morris, D. (1998). Thermal relations of large crocodiles, Crocodylus porosus, free-ranging in a naturalistic situation. *Proceedings of the Royal Society B: Biological Sciences, 265*(1407), 1793–1799. doi:10.1098/rspb.1998.0504. PMC 1689355.

Hiscock, K. (1997). Conserving biodiversity in North-East Atlantic marine ecosystems. In R. F. G. Ormond, J. D. Gage, & M. V. Angel (Eds.), *Marine biodiversity. Patterns and processes* (pp. 415–427). Cambridge: Cambridge University Press.

Jhingran, V. G. (1982). *Fish and fisheries of India* (p. 666). New Delhi: Hindustan Publication Corporation.

Kasturirangan, L. R. (1963). *A key to the identification of the more common planktonic Copepoda of Indian coastal waters* (p. 87). New Delhi: Indian National Committee on Oceanic Research/Council of Scientific and Industrial Research.

Khan, R. A. (2003). *Fish faunal resources of Sundarban estuarine system with special reference to the Biology of some commercially important species* (Records of the Zoological Survey of India. Occasional paper, Vol. 209, pp. 1–150).

"*Lepas anatifera* Linnaeus., 1758". WallaWalla. Retrieved from December 2, 2011.

Mandal, A. K., & Nandi, N. C. (1989). *Fauna of Sundarban mangrove ecosystem, West Bengal, India* (Fauna of conservation areas, Vol. 3). Kolkata: Zoological Survey of India. 116 pp.

Mehrtens, J. M. (1987). *Living snakes of the world in color*. New York: Sterling Publishers. 480 pp. ISBN 0-8069-6460-X.

Mitra, A. (1998). Status of coastal pollution in West Bengal with special reference to heavy metals. *Journal of Indian Ocean Studies, 5*(2), 135–138.

Mitra, A. (2013). *Sensitivity of mangrove ecosystem to changing climate*. New Delhi: Springer. ISBN 10: 8132215087.

Mitra, A., & Banerjee, K. (2005). *Living resources of the sea: Focus Indian Sundarbans*. New Delhi: WWF-India Secretariat, Sunderbans Landscape Project. 120.

Mitra, A., & Choudhury, A. (1994). Dissolved trace metals in surface waters around Sagar Island, India. *Journal of Ecobiology, 6*(2), 135–139

Mitra, A., & Zaman, S. (2015). *Blue carbon reservoir of the blue planet* (pp. 1–299). New Delhi: Springer, Book ID 322214_1_En Book, Book doi: 10.1007/978-81-322-2107-4; ISBN: 978-81-322-2106-7; Copyright Holder Name Springer India Copyright Year 2015.

Moskvitch, K. (2010, June 7). *BBC news: Crocodiles 'surf' long distance on ocean currents*. Bbc.co.uk. Retrieved June 12, 2013.

Newell, G. E., & Newell, R. C. (1977). *Marine plankton—A practical guide* (p. 244). London: Hutchinson Educational Ltd.

Parker, H. W., & Grandison, A. G. C. (1977). *Snakes: A natural history* (2nd ed., p. 108). Ithaca: British Museum (Natural History) and Cornell University Press. 16 plates. LCCCN 76-54625. ISBN: 0-8014-1095-9 (cloth), ISBN: 0-8014-9164-9 (paper).

Pillay, T. V. R. (Ed.). (1967). *FAO fishery report*, (44), 1 (2): 55 p. Proceedings of the FAO world symposium on warm-water Pond Fish Culture, Rome, 18–25 May 1966.

Ray Chaudhuri, T., Fazli, P., Zaman, S., Pramanick, P., Bose, R., & Mitra, A. (2014). Impact of acidification on heavy metals in Hooghly Estuary. *Journal of*

Harmonized Research in Applied Sciences, 2(2), 91–97.

Santhanam, R., & Srinivasan, A. (1998). *Distribution and biomass of copepods in the coral reef areas of Tuticorin, South India*. Paper presented in the conference on coastal zone Canada'98, Victoria, during August.

Sea snakes at Food and Agriculture Organization of the United Nations. Accessed 7 August 2007.

Stidworthy, J. (1974). *Snakes of the world*. New York: Grosset & Dunlap Inc.. 160 pp.

The Hydrophiidae at Cyberlizard's home pages. Accessed 12 August 2007.

U.S. Navy. (1991). *Poisonous snakes of the world. US Government*. New York: Dover Publications Inc. 203 pp. ISBN 0-486-26629-X.

Wimpenny, R. S. (1966). *The plankton of the sea* (p. 426). London: Faber and Faber Ltd.

Internet References

http://www.tao.org/DOCREP/005/X3690E/x3690elq.htm

Annex 6A: References

Banerjee, K., Mitra, A., Bhattacharyya, D. P., & Choudhury, A. (2002). Role of nutrients on phytoplankton diversity in the north–east coast of the Bay of Bengal. In Arvind Kumar (Ed.), *Ecology and ethology of aquatic biota* (pp. 102–109). New Delhi: Daya Publishing House

Banerjee, K., Mitra, A., & Bhattacharyya, D. P. (2003). Phytopigment level of the aquatic subsystem of Indian Sundarbans at the apex of Bay of Bengal. *Sea Explorers, 6*, 39–46.

Banerjee, K., Sengupta, K., Raha, A., & Mitra, A. (2013). Salinity based allometric equations for biomass estimation of Sundarban mangroves. *Biomass and Bioenergy, 56*, 382–391.

Barnett, T. P., Pierce, D. W., Achuta Rao, K. M., Gleckler, P. J., Santer, B. D., Gregory, J. M., & Washington, W. M. (2005). Penetration of human-induced warming into the world's oceans. *Science, 309*, 284–287.

Cambers, G., Claro, R., Juman, R., & Scott, S. (2007). *Working group report, climate change impacts on coastal and marine biodiversity*. Caribbean Natural Resources Institute, West Indies, 87 pp.

Chakraborty, S. K., & Choudhury, A. (1985). *Distribution of fiddler crabs in Sundarbans mangrove estuarine complex, India*. Proceedings of national symposium on biology, utilization and conservation of mangroves, pp. 467–472.

Ganguly, N. D., & Iyer, K. N. (2008). Long term variations of surface air temperatures during summer in India. *International Journal of Climatology, 29*(5), 735–746. doi:10.1002/joc.1748.

Ghoshal & Bhattacharyya. (2010). *Carbon dioxide emissions, India, 1980–2000*. ISRO Geosphere Biosphere Program (IGBP), National Carbon Project (NCP), status report- soil and vegetation-atmosphere fluxes, 20 March, 2010, Indian Institute of Remote Sensing (NRSC), Indian Space Research Organisation, Department of Space, Government of India, Dehradun.

IPCC (Intergovernmental Panel on Climate Change). (2007). *Climate cange 2007: Impacts, adaptation and vulnerability, contribution of working group II to the fourth assessment report of the intergovernmental panel on climate change* (976 pp.). Cambridge: University of Cambridge.

Kothawale, D. R., & Kumar, R. (2005). On the recent changes in surface temperature trends over India. *Geophysical Research Letters, 32*, L18714. doi:10.1029/2005GL023528.

Mitra, A. (2000). Chapter 62: The Northeast coast of the Bay of Bengal and deltaic Sundarbans. In C. Sheppard (Ed.), *Seas at the Millennium—An environmental evaluation* (pp. 143–157). Amsterdam: Elsevier Science.

Mitra, A., & Choudhury, A. (1994). Heavy metal concentrations in oyster *Crassostrea cucullata* of Henry's Island, India. *Journal of Ecobiology, 6*(2), 157–159.

Mitra, A., & Pal, S. (2002). *The oscillating mangrove ecosystem in the Indian Sundarbans*. Kolkata: WWF, WBSO.

Mitra, A., Ghosh, P. B., & Choudhury, A. (1987). *A marine bivalve Crassostrea cucullata can be used as an indicator species of marine pollution*. Proceedings of national seminar on Estuarine Management, Trivandrum, pp. 177–180.

Mitra, A., Choudhury, A., & Zamaddar, Y. A. (1992). Effects of heavy metals on benthic molluscan communities in Hooghly estuary. *Proceedings of Zoological Society, 45*, 481–496.

Mitra, A., Banerjee, K., Sengupta, K., & Gangopadhyay, A. (2009). Pulse of climate change in Indian Sundarbans: A myth or reality. *National Academy of Science Letters, 32*, 1–7.

Mitra, A., Sengupta, K., & Banerjee, K. (2011). Standing biomass and carbon storage of above-ground structures in dominant mangrove trees in the Sundarbans. *Forest Ecology and Management, 261* (7), 1325–1335 (Elsevier doi:10.1016/j.foreco.2011.01.012).

Mondal, K., Mukhopadhyay, S. K., Biswas, H., De, T. K., & Jana, T. K. (2006). Fluxes of nutrients from the tropical river Hooghly at the land–ocean boundary of

Sundarbans, NE Coast of Bay of Bengal, India. *Journal of Marine Systematics, 62*, 9–21.

National Carbon Project Status Report on Soil and Vegetation—Atmosphere Fluxes. (2010). pp. 13–14.

Perry, A. L., Low, P. J., Ellis, J. R., & Reynolds, J. D. (2005). Climate change and distribution shifts in marine fishes. *Science, 308*, 1912–1915.

Saha, S. B., Mitra, A., Bhattacharyya, S. B., & Choudhury, A. (1999). Heavy metal pollution in Jagannath canal, an important tidal water body of the north Sundarbans aquatic ecosystem of West Bengal. *Indian Journal of Environmental Protection, 19*(11), 801–804.

Sengupta, K., Roy Chowdhury, M., Bhattacharya, S. B., Raha, A., Zaman, S., & Mitra, A. (2013). Spatial variation of stored carbon in *Avicennia alba* of Indian Sundarbans. *Discovery Nature, 3*(8), 19–24.

Solomon, S. D., & IPCC (2007). *Summary for policymakers*. Climate Change 2007: The physical science basis, contribution of working group I to the fourth assessment report of the Intergovernmental Panel on Climate Change, Cambridge University Press, Cambridge/New York.

Thomas, C. D., Cameron, A., Green, R. E., Bakkenes, M., Beaumont, L. J., Collingham, Y. C., Erasmus, B. F. N., De Siqueira, M. F., Grainger, A., Hannah, L., Hughes, L., Huntley, B., Van Jaarsveld, A. S., Midgley, G. F., Miles, L., Ortega-Huerta, M. A., Peterson, A. T., Phillips, O. L., & Williams, S. E. (2004). Extinction risk from climate change. *Nature, 427*(6970), 145.

Zhang, K. Q., Douglas, B. C., & Leatherman, S. P. (2000). Twentieth-century storm activity along the US East Coast. *Journal of Climatology, 13*, 1748–1761.

Annex 6B: References

Banerjee, K., Mitra, A., Bhattacharyya, D. P., & Choudhury, A. (2002). Role of nutrients on phytoplankton diversity in the north–east coast of the Bay of Bengal. In Arvind Kumar (Ed.), *Ecology and ethology of aquatic biota* (pp. 102–109). New Delhi: Daya Publishing House.

Banerjee, K., Das, S., & Mitra, A. (2005a). *Temporal loss of ichthyoplankton due to wild harvest of prawn seeds in Indian Sundarbans*. Seshiyana, India (In press).

Banerjee, K., Bhattacharyya, D. P., & Mitra, A. (2005b). *Tiger prawn seed catch in coastal waters: Estimation of ichthyoplankton community damage*. SIOS. (In press).

Bhattacharya, M., Banerjee, K., Roy, A., Mitra, A., & Bhattacharyya, D. P. (1999). Preliminary study on the impact of oil and grease on finfish juveniles around Haldia Port-cum-industrial complex. *Journal of the Inland Fisheries Society of India, 31*(2), 36–38.

Bhattacharyya, D. P., Mahapatra, B., Banerjee, K., & Mitra, A. (2003). Application of statistics in marine biology—A right choice is required. *Sea Explorers, 6*, 13–20.

Mitra, A. (2000). Chapter 62: The Northeast coast of the Bay of Bengal and deltaic Sundarbans. In C. Sheppard (Ed.), *Seas at the Millennium—An environmental evaluation* (pp. 143–157). Amsterdam: Elsevier Science.

Mitra, A., & Banerjee, K. (2004). Physiology of mangroves. In C. S. R. Banerjee (Ed.), *Living resources of the seas: Focus Sundarbans*. Canning Town: WWF-India, Canning Field Office, 24 Parganas(S), W.B. 96pp.

Mitra, A., & Bhattacharyya, D. P. (2003). Environmental issues of shrimp farming in mangrove ecosystems. *Journal of Indian Ocean Studies, 11*(1), 120–129.

Mitra, A., & Pal, S. (2002). *The oscillating mangrove ecosystem in the Indian Sundarbans*. Kolkata: WWF, WBSO.

Mitra, A., Choudhury, A., & Zamaddar, Y. A. (1992b). Effects of heavy metals on benthic molluscan communities in Hooghly estuary. *Proceedings of Zoological Society, 45*, 481–496.

Mitra, A., Niyogi, S., & Choudhury, A. (1998). Impact of wild harvest of prawn seeds on the community structure of juvenile fish in the coastal part of West Bengal. In S. Z. Qasim & G. S. Roonwal (Eds.), *Living resources of India's exclusive economic zone* (pp. 86–96). New York: Omega Scientific Publisher.

Mitra, A., Sasmal, S. K., Choudhury, A., & Bhattacharyya, D. P. (2000). Seasonal variations of ichthyoplankton in mangrove ecosystem of Indian Sundarbans. *Indian Journal of Environment and Ecoplanning, 3*(3), 535–538.

Mitra, A., Banerjee, K., & Bhattacharyya, D. P. (2004a). *The other face of mangroves* (1st ed.). West Bengal: Department of Environment, Government of West Bengal, India.

Mitra, A., Banerjee, K., & Gangopadhyay, A. (2004b). *Introduction to marine plankton*. New Delhi: Daya Publishing House.

Mitra, A., Banerjee, K., Chakraborty, R., Banerjee, A., Mehta, N., & Berg, H. (2005a). *Study on the water quality of the shrimp culture ponds in Indian Sundarbans*. Indian Science Cruiser. (In press).

Mitra, A., Banerjee, K., Raha, A. K., Ghosh, S., & Bhattacharyya, D. P. (2005b). Status of shark community in coastal West Bengal. *Indian Science Cruiser, 19*(2), 1–9.

Nagaeswara Rao, C. A. (2004). Faunal diversity: Estuarine ecosystem. *EPTRI-ENVIS Newsletter, 10*(3), 8–11.

Niyogi, S., Mitra, A., Saha, S., & Choudhury, A. (1998). Inter-relationship between diversity of prawn juveniles and mangroves in the coastal zone of West Bengal. In S. Z. Qasim & G. S. Roonwal (Eds.), *Living resources of India's exclusive economic zone* (pp. 97–102). New Delhi: Omega Scientific Publisher.

Panja, U. (2004). *Shrimp seed collection and ecological crop loss in coastal West Bengal, World Environment day special*. West Bengal: WWF-India.

Shannon, C. E., & Weiner, V. (1949). *The mathematical theory of communication*. Urbana: University of Illionis Press. 117pp.

Trivedi, S., Mitra, A., Choudhury, A., Gupta, A., Singh, B., & Choudhuri, A. (1994). *A case study of the loss of bio-diversity during prawn seed collection from the Hooghly Estuary, India*. Proceedings of the national convention of environment of India- Challenges for the 21st century; Institution of Public Health Engineers, Kolkata, India, pp. T-V/16-T-V/24.

Verlecar, X. N., & Parulekar, A. H. (2001). Chapter 10: Primary productivity. In R. Sengupta & E. Desa (Eds.), *The Indian Ocean—A perspective* (Vol. 2, pp. 397–415). Oxford: IBH Publishers Co. Pvt. Ltd.

Whitfield, A. K. (1999). Ichthyofaunal assemblages in estuaries: A South African case study. *Reviews in Fish Biology and Fisheries, 9*, 151–186.

Decomposers of the Marine and Estuarine Ecosystems 7

Contents

Decomposers are widely distributed in the salty blue soup of the planet Earth and occupy a key position in an ecological food chain/web. They are considered as 'cleaners' of the ecosystem as they are capable of degrading complex organic matter in to simpler forms. The vast volume of saltwater may be the reason behind the presence of wide variety and large number of decomposers in the marine and estuarine ecosystems.

Unlike terrestrial organisms, marine flora and fauna are exposed to different degrees of pressure, light, salinity, oxygen level (dissolved oxygen), etc. Hence, they are adapted to these pronounced variations of environmental variables. Each level in the food chain needs to be balanced for the survival and existence of life. So just as phytoplankton (the foundation of ocean food web) are important to provide energy for higher level consumers, equally important is the role of decomposers to provide energy by decomposition of dead organic matter and releasing nutrients and making it available to the phytoplankton and clean up the ocean biome.

7.1 Overview of Marine and Estuarine Microbes

The marine and estuarine environment is the dwelling place of diverse groups of microorganisms like bacteria, filamentous fungi, yeasts, microalgae and protozoa, which inhabit

A. Mitra, S. Zaman, *Basics of Marine and Estuarine Ecology*,
DOI 10.1007/978-81-322-2707-6_7

various types of niches. They are distributed at the surface of the sea as neuston (also known as pleuston), at the photic zone of the pelagic region as plankton or at the epibiotic habitats (as attached communities). Epibiotic habitats may be inanimate or animate. They are also present inside the tissues of other marine organisms (endobiotic habitats). The endobiotic usually denotes the environment within the tissues of other larger organisms. Here the relationship with the host may be beneficial (mutualism), may be detrimental to the host (parasitism) or may cause diseases (pathogenesis). A research programme undertaken by Jana et al. (2013) revealed the presence of microbes in the edible oyster *Saccostrea cucullata* (Fig. 7.1); (Vide Annexure 7A).

The intertidal zone of the marine environment is rich in mangroves, seagrass, salt marsh grass, seaweeds, etc., which offer unique dwelling sites for marine/estuarine microbes. Researchers observed that grasses possess a narrower range of microorganisms than seaweeds. It has been observed by marine microbiologists that cordgrass (*Spartina alterniflora*) is colonized initially by fungi (*Sphaerulina pedicellata*), whereas eelgrass (*Zostera marina*) mostly possesses the pinnate diatom (*Cocconeis scutellum*). On seaweeds (Fig. 7.2), microorganisms like diatom, yeast and bacteria thrive luxuriantly.

Fig. 7.1 Oyster tissue: site for microbes

The mangrove swamps also sustain a wide range of microbes. A large number of bacteria have been isolated from a wide variety of substrates from Indian Sundarbans mangrove ecosystem (Bhowmik et al. 1985). These include 48 bacterial strains which have been isolated from the litter and detritus of mangrove origin. Species of bacteria identified include *Brevibacterium lypolyticum, B. marinopiscosum, B. minutiferula, B. sociovivum, B. stationis, Clostridium carnis, C. pectinivorum, Kurthia bessonii, K. zopfii, Lactobacillus brevis, L. delbrueckii, L. fermentum, L. plantarum* and *Listiria monocytogenes*. Micrococcus was found to be predominant. Among the fungi, the genera *Aspergillus, Collectotrichium, Fusarium* and *Helminthosporium* were notably common. Bacteria identified in decomposed litter have also been found within the gut tissues of detritivorous benthic fauna that rely on the substrate for their food source. Bacteria have been identified from the haemocoelic fluid of *Boleophthalmus* spp. and the gut of *Mystus gulio* and *Uca* spp. It is contended that this unique association of different bacterial strains both in litter and in the detritivore benthic fauna may have ecological implications within the food web through microbial metabolic processes.

Researches have been undertaken on the mangrove litter, soil bacteria and fungi (microbes) from different locations within the littoral zone of mangroves (Chaudhuri and Choudhury 1994). Site selected for investigation included western, central and eastern sectors of Indian Sundarbans mangroves. The study revealed that 12 species of bacterial strain were dominant. Most of the strains were gram-positive forms, while gram-negative forms were rare (Chaudhuri and Choudhury 1994). Samples obtained from the ridges, slopes and mudflats at each site indicated that the bacterial populations were comparatively high within the ridges and lowest in the mudflats. There were also indications that samples collected from the eastern sector and from the south (facing the Bay of Bengal) contained lower numbers of microbial forms than those

Fig. 7.2 Seaweed beds are the hot spots of several microbial strains

from other areas. Fungal populations of *Penicillium* and *Aspergillus* were very common in these areas.

Higher microbial counts, particularly of bacteria, were recorded in the upper littoral zone. These observations may have resulted from an accumulation of rich detritus content, caused by the mangrove litter and reduced flooding by tidal water in this region. By comparison, successive low counts of bacteria in the mid-littoral and lower zones suggest less litter and increased tidal interplay in these zones. It was also observed that salinity in the upper littoral zone was always high. From these results, it may be inferred that the bacterial populations in the coastal zone and estuary are directly influenced by the salinity and the amount of litter content in the substratum.

Studies have been undertaken to elucidate the relationship between the different types of soil with their associated benthic organisms and leaf litter decompositions, to assess the particular roles played by various microbes, fungi and bacteria. Analysis of litter and soil samples from selected mangrove areas revealed that there were marked differences in microbial composition in undisturbed mangrove soils, such as the reserve forest area, compared to the species composition found in reclaimed mangrove areas, such as on Sagar Island, a site which has been completely altered to agricultural farmland and shrimp culture activities.

Populations of fungi, such as *Actinomycetes*, were common and substantially more abundant in the reclaimed areas than in the reserve forest. Differences in the microbial populations were also noted in the *Avicennia*- and *Rhizophora*-dominated sectors of the forest, the count being 7.7×10^4 per gram of soil and 5×10^4 per gram of soil, respectively. All the bacterial forms isolated from these samples were highly tolerant to the prevailing oscillations in salinity. Similarly, the bacterial population of 7.7×10^3 per gram sampled in the *Avicennia*-dominated soil was also higher than the sample population of 4×10^3 per gram from the soil occupied by *Rhizophora*. These results helped in documenting the existing results of functional groups of microbes such as cellulose decomposers, denitrifiers and nitrogen fixers in different forest soils.

Fig. 7.3 Estuary adjacent to mangrove-dominated intertidal mudflats provides different salinity-based niches for survival and growth of microbes

The estuarine environment associated with mangroves is highly favourable for the survival of diverse strains of microbes (Fig. 7.3). This is because the constantly changing environmental parameters (due to tidal actions or land run-off) can create a wide diversity of ecological niches in this brackish water ecosystem (Atlas 1998). Estuaries have different types of salinity ranging from 0 psu to 32 psu. The freshwater, brackish water and saline water of estuaries offer suitable habitats for different categories of decomposers. This form of ecological partitioning reduces exploitative competition and enhances growth of different types of microbial communities (Campbell 1993).

In estuarine system autochthonous biological activity modifies the mixing of freshwater and seawater and generates pronounced biological and chemical gradients in estuarine water. The decomposition of microbes in such a dynamic system is mostly regulated by salinity, nutrient concentration, organic matter composition and bacterivore community composition (Barcina et al. 1997; Giovannoni and Rappe 2000). Such changes in environmental conditions, when recreated in mesocosm and microcosm experiments, caused shifts in the phylogenetic composition of bacterioplankton communities (Gasol et al. 2002; Lebaron et al. 2001; Schafer et al. 2001; Van Hannen et al. 1999).

A large number of scientific literatures are available on estuarine microbial diversity where the microbes have been observed to survive and adapt under different salinity conditions (Bidle and Fletcher 1995; Bouvier and Del Giorgio 2002; De Bie et al. 2001; Hollibaugh et al. 2000; Murray et al. 1996; Troussellier et al. 2002), but few reports have provided evidence of a unique estuarinc bacterioplankton community. This is partly due to the dynamic nature of estuaries and the difficulty in distinguishing estuarine populations from those that washes in from adjacent environments. Crump et al. (1999) identified putative estuarine bacteria associated with particles in the Columbia River estuarine turbidity maximum (ETM) by

Fig. 7.4 Fish landing stations often contaminate estuarine system with microbial load

comparing environmental clone libraries of PCR-amplified 16S ribosomal DNA (rDNA) from the river, the estuary and the coastal ocean. Similarly, Hollibaugh et al. (2000) demonstrated the mixing of bacterial communities in the ETM of the San Joaquin River and San Francisco Bay system by characterizing communities at three sampling stations using denaturing gradient gel electrophoresis (DGGE) of PCR-amplified 16S rDNA. Selje and Simon (2003) used this same technique, but with greater spatial resolution (six sampling stations), in the Weser River estuary and concluded that a distinct microbial community resides in the brackish section of the system. These three studies demonstrated the presence of river and coastal ocean bacteria in estuaries and suggested that the development of unique estuarine bacterial communities may be related to the relatively long residence time of particles and particle-attached bacteria in some ETMs.

The residence time of water and free-living bacteria, however, can be too short in some estuaries, relative to bacterial growth rate, for such a shift to occur. In the Rhone River plume, where water residence time is less than 6 h, bacterioplankton appeared to be a mixture of Rhone River and Mediterranean Sea bacterioplankton (Troussellier et al. 2002). Similarly, in the Columbia River estuary, where water residence time averages 1–2 days and bacterial production is low, a mixture of freshwater and marine populations dominated the free-living bacterioplankton community (Crump et al. 1999). Thus, bacterial growth and time are the two vital parameters that govern the alteration of community composition of microbes.

The microbial population in the coastal zone is associated with anthropogenic activities like rapid rate of urbanization, unplanned tourism, mushrooming of shrimp farms (that often release wastes), industrial effluent discharges or decomposition of fishes in fish landing stations (that are mostly constructed in the river mouths and estuaries; see Fig. 7.4). A detailed study of monthly variation of bacterial content in the Hooghly estuary, the westernmost estuaries of Indian Sundarbans, is highlighted as Annexure 7B.

It has been observed by researchers that the more we approach towards offshore regions from the continental margins, the microbial load gradually decreases. The magnitude of human activities and its effects may be a cause for this. In the inshore region of Bay of Bengal, a case study conducted by the present authors during 1990–2012 in three stations, viz., Harinbari (inshore region; marked as 1 in Fig. 7.5), Chemaguri (midshore region; marked as 2 in Fig. 7.5) and Jambu island (offshore region;

Fig. 7.5 Map showing three selected stations in the inshore, midshore and offshore region of Indian Sundarbans

marked as 3 in Fig. 7.5), exhibited unique spatial variation of nitrate and phosphate level. The gradual decrease of nitrate and phosphate levels (which are primarily liberated from sewage) while approaching from the inshore to the offshore waters (Figs. 7.6, 7.7, 7.8, 7.9, 7.10 and 7.11) speaks in favour of severe anthropogenic pressure along the nearshore waters.

The gradual increase of nitrate and phosphate in the lower Gangetic delta region since the last three decades (Figs. 7.12 and 7.13) is another prominent cause behind the increase of microbial load in the estuaries of Indian Sundarbans.

Microorganisms are also distributed in the deep-sea sediments. Deming and Colwell (1985) used epifluorescence microscopy to determine the vertical distribution of bacteria in deep-sea sediments. Thus using cores collected at depths exceeding 4000 m, it was recorded that bacterial populations at the surface layer of sediment amounted to 4.65×10^8 bacteria/g dry weight. However, there was a doubling in numbers to 8.29×10^8 bacteria/g dry weight at a sediment depth of 3 cm followed by a progressive decline to 1.7×10^7 bacteria/g dry weight in a core sample at 15 cm from the surface of the sediment. Parallel results were obtained in a second core collected from a similar depth. Higher counts of approximately 3.07×10^{10} bacteria/g dry weight were recorded from faecal pellets. These counts were 9–72-fold higher than in the underlying surface sediment (Deming 1985).

The deep-sea environment is thus an important habitat of marine microorganisms. Turner (1979) observed bacteria on the surface of faecal pellets and concluded that many of the deep-sea microorganisms have their origin at the surface layer of the ocean. This view has been supported to some extent by the results of experiments, which demonstrated an enhanced rate in metabolic activity of marine microorganisms with a reduction of pressure (Jannasch and Wirsen

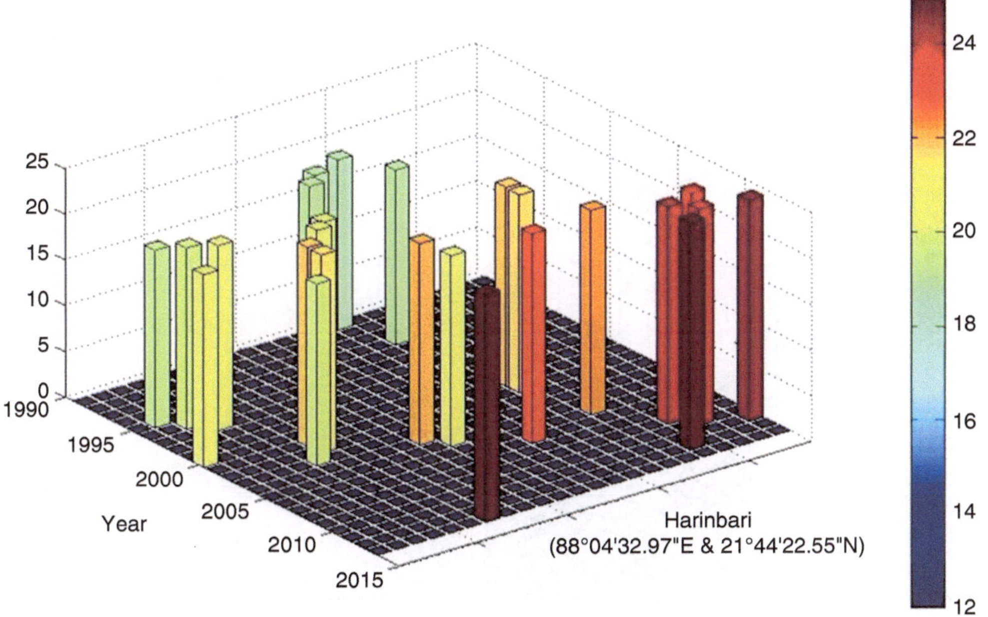

Fig. 7.6 Nitrate level (in μgatl^{-1}) in the inshore region of Indian Sundarbans (Graph designed by Dr. Nibedita Mukhopadhyay, Environmentalist)

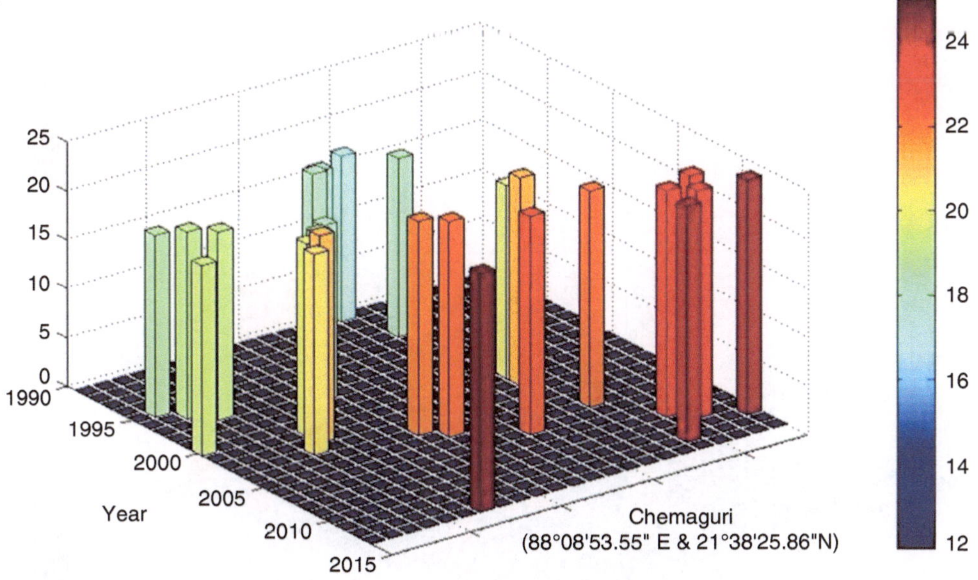

Fig. 7.7 Nitrate level (in μgatl^{-1}) in the midshore region of Indian Sundarbans (Graph designed by Dr. Nibedita Mukhopadhyay, Environmentalist)

1982). This helps to draw an inference that the activity of marine microorganisms is more in the shallow water and decreases with the increase of depth and pressure. Some interesting facts have been documented with respect to distribution of microbes in the deep-sea environment, particularly around the vent region. The deep-sea vents occur in the ocean floor where the ocean crustal

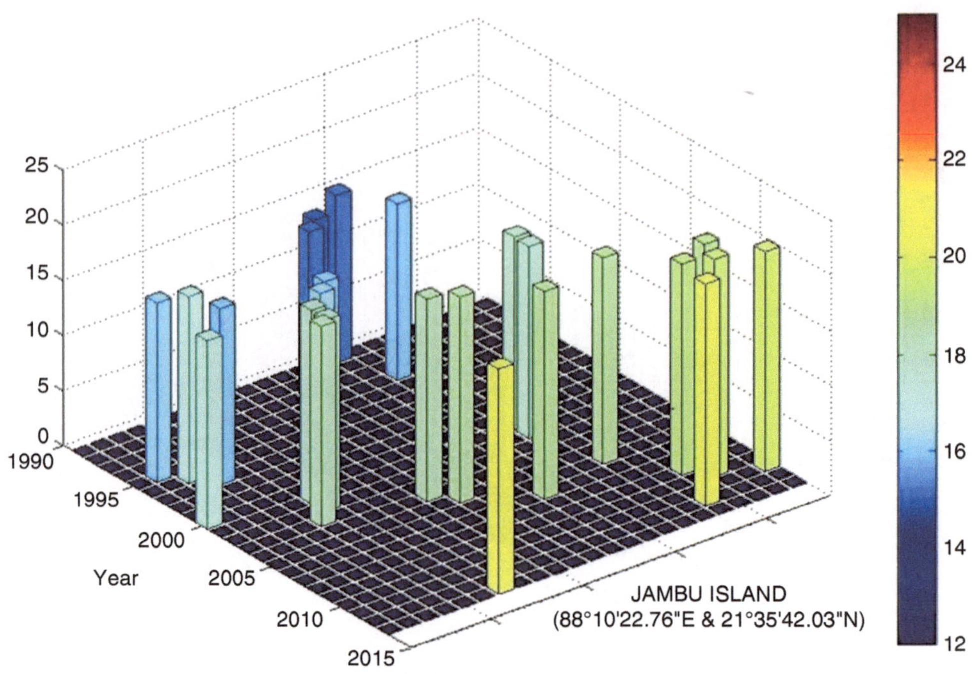

Fig. 7.8 Nitrate level (in µgatl^{-1}) in the offshore region of Indian Sundarbans (Graph designed by Dr. Nibedita Mukhopadhyay, Environmentalist)

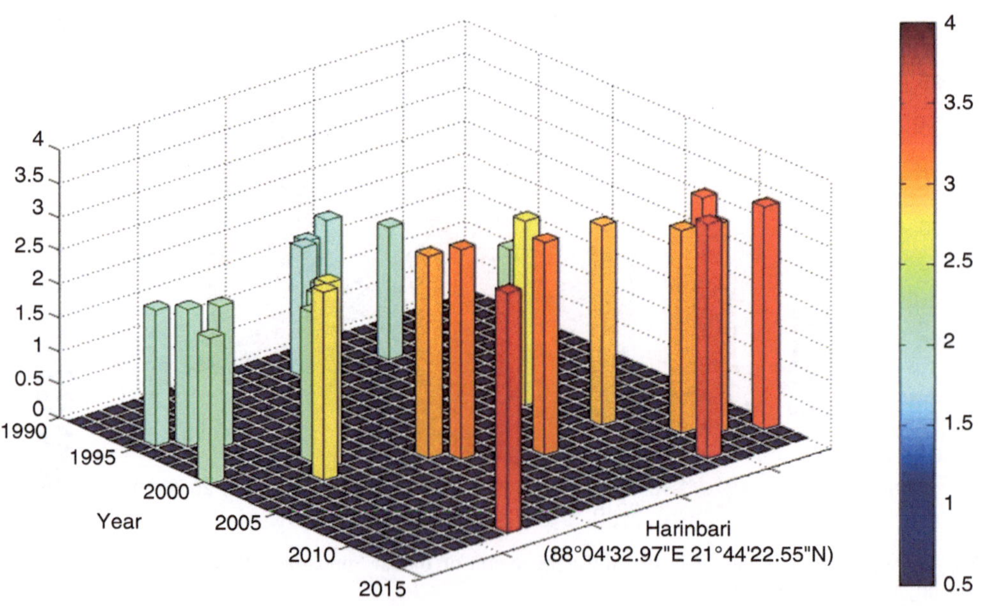

Fig. 7.9 Phosphate level (in µgatl^{-1}) in the inshore region of Indian Sundarbans (Graph designed by Dr. Nibedita Mukhopadhyay, Environmentalist)

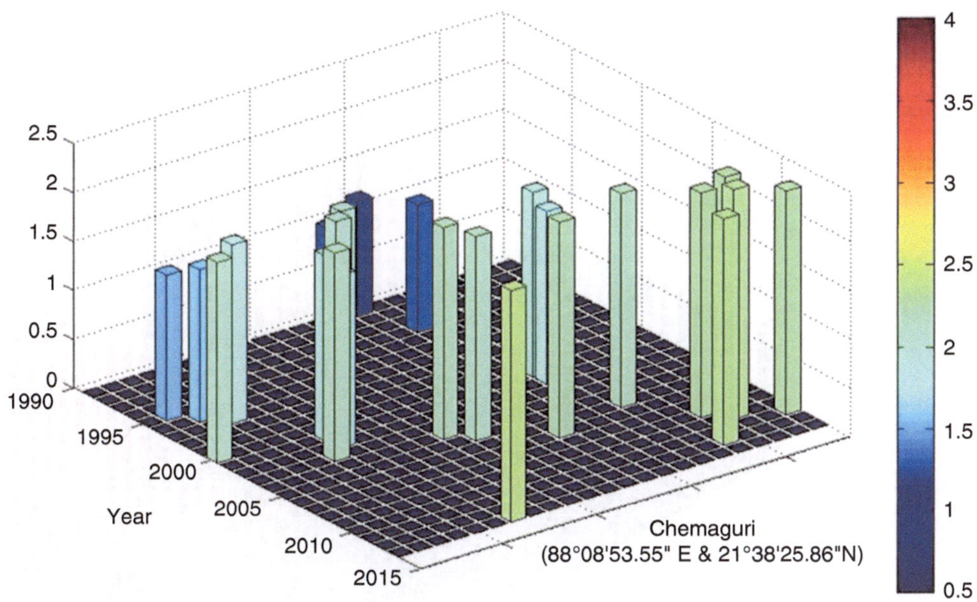

Fig. 7.10 Phosphate level (in μgatl^{-1}) in the midshore region of Indian Sundarbans (Graph designed by Dr. Nibedita Mukhopadhyay, Environmentalist)

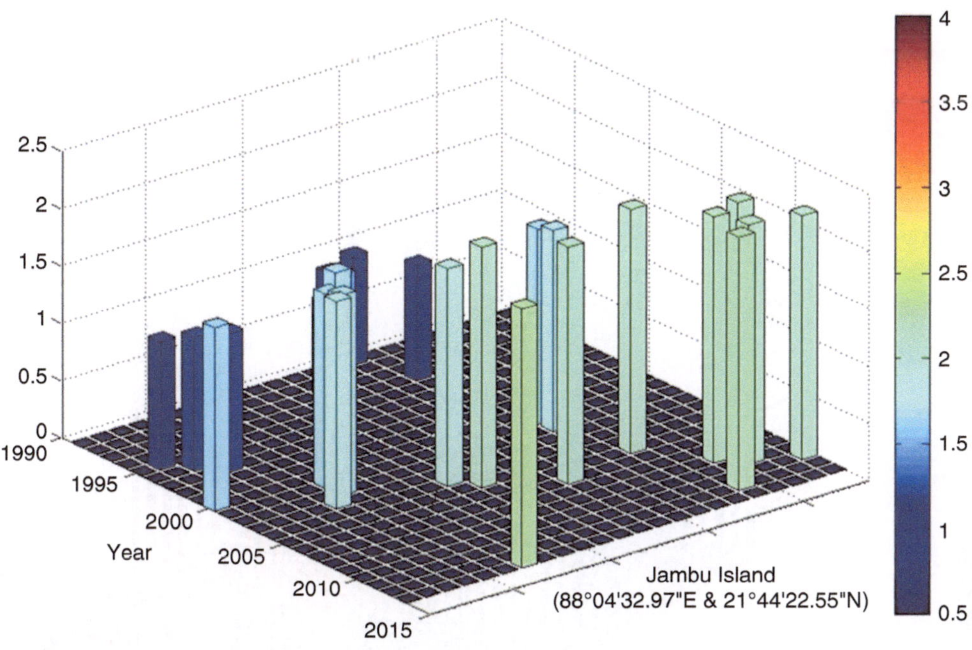

Fig. 7.11 Phosphate level (in μgatl^{-1}) in the offshore region of Indian Sundarbans (Graph designed by Dr. Nibedita Mukhopadhyay, Environmentalist)

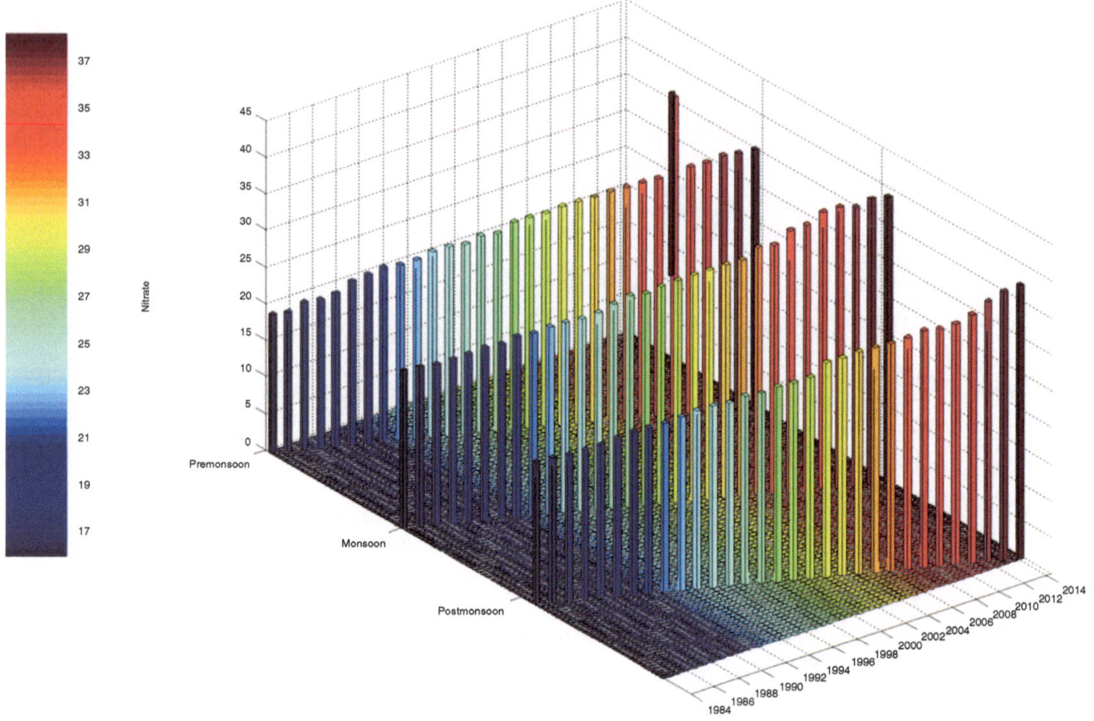

Fig. 7.12 Decadal variation of nitrate (in μgat l^{-1}) in the Hooghly estuary of lower Gangetic delta

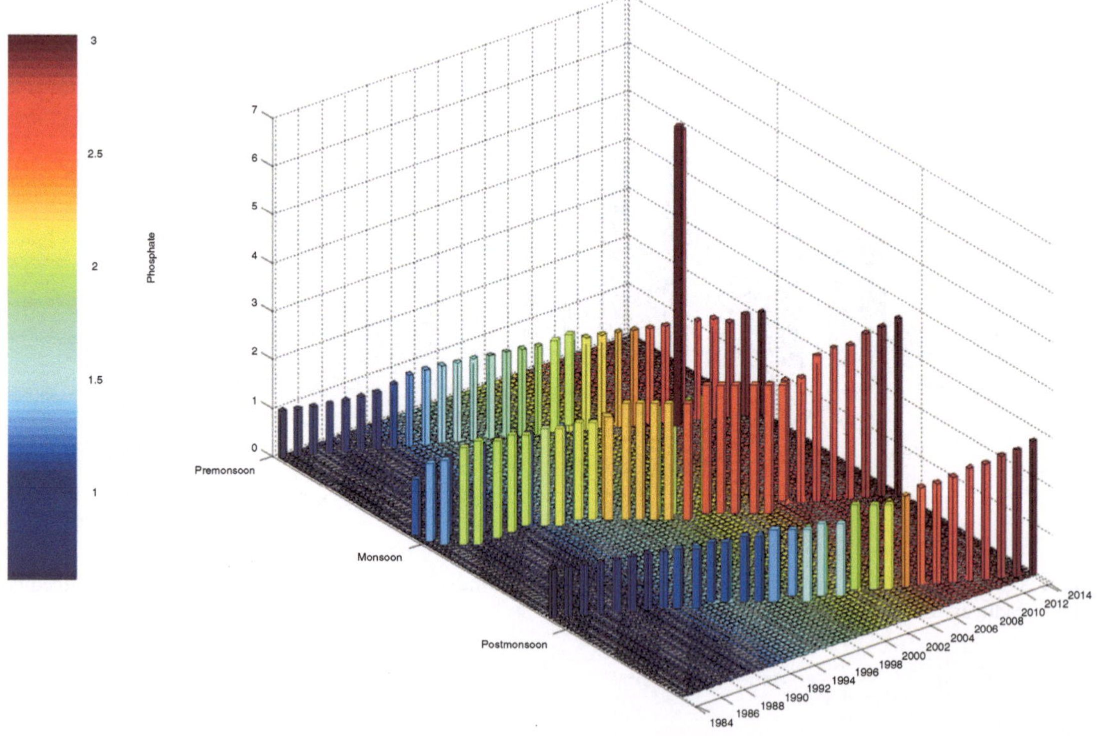

Fig. 7.13 Decadal variation of phosphate (in μgat l^{-1}) in the Hooghly estuary of lower Gangetic delta

plates spread apart and cause plumes of hot lava to erupt into the ocean. The high concentrations of electron-rich elemental compounds are very congenial for the growth and survival of eubacteria such as the chemoautotrophs (Campbell 1993). Each type of bacteria has special adaptations that enable the organisms to obtain metabolic energy and to withstand the extreme environmental conditions of high pressure and temperature (over 100 °C). The concentration of microbial population drops dramatically as one move away from the deep-sea vent region (Atlas 1998). There are heterotrophic bacteria in the seafloor sediments, which feed on photoautotrophic cyanobacteria that drift down attached to sediment particles.

Hydrothermal vents mostly found in the trench regions also offer unique survival tents for microbes. The gradients of temperature and nutrients that exist at hydrothermal systems provide a great diversity of habitats for microbes suspended in the surrounding heated waters and in sediments and attached to surfaces of the chimneys. Many of these are hyperthermophilic bacteria and archaea, which can grow at temperatures up to 121 °C, while others grow at lower temperatures further from the fluid emissions. Molecular studies are revealing an astonishing diversity of such organisms, many of which have biotechnological applications. The microbiology of the deep subsurface rocks beneath vents is also now under investigation, and many novel microbes and metabolic processes are being discovered.

7.1.1 Marine Organisms as Habitats of Microorganisms

Microbial biofilms are formed on all kinds of animals, seaweeds and coastal vegetation. The biofilms provide a congenial environment for the growth and survival of marine and estuarine microbes through promotion of processes like colonization, selective colonization, etc. Many microbes are extremely choosy in nature, allowing only certain microbes to colonize and inhibit the colonization of others through secretion of chemicals. Many microalgae such as diatoms and dinoflagellates harbour bacteria on their surfaces or as endosymbionts within their cells. Intimate associations between bacterial and archaeal cells are also being revealed by new imaging techniques (Table 7.1). Blue carbon community also sustains a dense population of bacteria (up to 10^6 per cm^2) on their surface. In this context it is to be noted that the term blue carbon refers to coastal vegetation that encompasses mangroves, seagrasses, salt marsh grass, seaweeds, etc.

7.1.2 Identification of Marine and Estuarine Microorganisms: A Molecular Genetic Approach

The twenty-first century is the *era* of innovation of scientific methodologies and technologies. Today, the classification, identification and characterization of marine and estuarine microbes are done through genotypic and proteomics technologies. These technologies are superior to the traditional methods as they can generate rapid results and produce the best level of discrimination among strains of microbes.

The genotypic technique of identifying bacterial strains is based on profiling and organisms' genetic material mainly through its DNA. Genotypic techniques are superior to phenotypic methods because they are independent of the physiological state of an organism and are not influenced by constituents of the medium. The genotypic techniques are not at all influenced by the growth phase of the microbes.

Genotypic microbial identification can be carried out by (I) fingerprinting-based methodologies and (II) sequence-based techniques. Each of these techniques is discussed here separately in Tables 7.2 and 7.3.

7.2 Importance of Marine and Estuarine Microbes

Marine and estuarine microorganisms comprise a comparatively untapped reservoir of

Table 7.1 Size range of some representative marine microbes

Organism	Characteristics size	(μm)[a]	Volume (μm^3)[b]
Parvovirus	Icosahedral DNA virus infecting shrimp	0.02	0.000004
Coccolithovirus	Icosahedral DNA virus infecting *Emiliania huxleyi*	0.17	0.003
Thermodiscus	Disc-shaped hyperthermophilic *archaea*	0.2 × 0.08	0.003
Pelagibacter ubique[c]	Crescent-shaped *bacteria* ubiquitous in ocean plankton (cultured example of SAR11 clade)	0.1 × 0.9	0.01
Prochlorococcus	Cocci. Dominant photosynthetic ocean *bacteria*	0.6	0.1
Ostreococcus	Cocci. Prasinophyte alga. Smallest known eukaryote	0.8	0.3
Vibrio	Curved rods. Bacteria common in coastal environments and associated with animals	1 × 2	2
Pelagomonas calceolata	Photosynthetic flagellate adapted to low light	2	24
Pseudo-nitzschia	Pennate diatom which produces toxic domoic acid	5 × 80	1600
Staphylothermus marinus	Cocci. Hyperthermophilic *archaea*	15	1800
Thiploca auracae	Filamentous. Sulphur *bacteria*	30 × 53	30,000
Lingulodinium polyedrum	Bioluminescent bloom-forming dinoflagellate	40	65,000
Beggiatoa	Filamentous. Sulphur *bacteria*	50 × 160	314,000
Epulopiscium fishelsoni	Rods. Bacteria symbiotic in fish gut	80 × 600	3,000,000
Thiomargarita namibiensis	Cocci. Sulphur *bacteria*	300[d]	14,137,100

[a]Approximate diameter × length; where one value is given, this is the diameter of spherical virus particles or cells
[b]Approximate values, calculated assuming spherical or cylindrical shapes
[c]*Candidatus*; provisional taxonomic name—see Glossary
[d]Cells up to 750 μm have been recorded

commercially valuable compounds. Many bacteria are able to produce and secrete polymers and enzymes. Some marine bacteria are potent producers of DNase, lipase, alginases and proteases. Hence, the economic backbone of any country may be strengthened if special thrust is given to certain applied sector of microbiology like production of biochemical compounds, enzymes, single-cell protein and pharmaceutical compounds.

The various advantages that can be derived from marine microbes are discussed here in brief.

7.2.1 Production of Antibiotics

Rosenfeld and Zobell (1947) described the production of antibiotics by marine bacteria. Interestingly, in their study the majority of the antibiotic producers were equated with *Bacillus* and *Micrococcus*, which are usually regarded as terrestrial organisms rather than representative of the true marine microfauna. Further work by Krassil'nikova (1961) and Buck et al. (1962) confirmed antibiosis among marine bacteria, with the latter study pointing to inhibitory effects against yeasts. Then a proliferation of research resulted in numerous publications, starting in 1966. Of these, the work of Buckholder et al. (1966) is relevant insofar as characterization of the inhibitory compound revealed a novel chemical structure. This was confirmed by Lovell (1966) as 2,3,4-tribromo-5(1′-hydroxy-2′,4′-dibromophenyl)-pyrrole. The compound was recovered from bacteria, which had been isolated from *Thalassia* sp. (turtle grass) collected in the vicinity of Puerto Rico.

Within a decade of this work, another inhibitory bromopyrrole was recovered from marine bacteria. Using nonselective isolation techniques, Anderson et al. (1974) recovered an antibiotic producing purple-pigmented

Table 7.2 Common fingerprinting methods used in identifying bacteria

Method	Technology	Application	Database
Repetitive element polymerase chain reaction (rep-PCR)	Polymerase chain reaction (PCR) primers target specific repetitive elements randomly distributed in the chromosomes of bacteria and archaea	Identification at the species and strain levels	User created; commercial individual database available
Amplified fragment length polymorphism	Restriction digestion of chromosomal DNA is followed by PCR using adapters coupled to the restriction sites	Identification at the species and strain levels; can be used for both bacteria and archaea	User created
Riboprinting	Restriction digest of chromosomal DNA is followed by probing genes	Identification at the species and strain levels; often used in quality control and assurance	User created; commercial universal database available
Random amplification of polymorphic DNA	A set of arbitrary short primers are used to randomly amplify short stretches of chromosomal DNA	Comparison of strains of known species	User created
Pulsed-field gel electrophoresis	Chromosomal DNA is cut into large fragments with rare-cutting restriction enzymes, and fragment is determined	Typing of pathogenic bacteria	Public universal database administered by the Centers for Disease Control and Prevention (www.cdc.gov/pulsenet)
Multiplex PCR	Multiple PCR primers are used for diagnostic genes	Identification of multiple species in mixed samples, such as food and clinical specimens	User created

Table 7.3 Common DNA sequencing methods used in identifying bacteria

Method	Technology	Application	Database
Small-subunit ribosomal gene sequencing (SSU rDNA)	Conserved primers are used to amplify and then sequence the SSU rDNA gene; sequences are then compared with database	The current gold standard in bacterial identification and determination of evolutionary relationships; may not distinguish strains or species within a genus	Public universal databases available, including the Ribosomal Database Project (http://rdp.cme.msu.edu) and greengenes (http://greengenes.ilb.gov)
Multilocus sequence typing (MLST)	DNA sequencing of a specific subset of conserved and semiconserved genes for a given species is followed by a comparison of concatenated sequences	Typing of pathogenic bacteria; epidemiology	Public universal database available (administered by www.mist.net)
Multilocus sequence analysis	DNA sequencing or a specific subset of conserved genes is followed by a comparison of concatenated sequences; this method typically uses fewer genes than MLST does (only two or three)	Provides more robust identification at the species level than traditional SSU rRNA gene sequencing	User created; based on BLAST (Basic Local Alignment Search Tool) searches against GenBank

Chromobacterium, designated strain 1-L-133, from seawater in the North Pacific. The organism was described as motile, gram-negative rods, which produce oxidase but not catalase or indole, hydrolyzed starch and gave negative responses to the methyl red test and Voges–Proskauer reaction. From these traits, the organism was equated with *Chromobacterium* and, in particular, considered to resemble *C. marinum.*

Interest in the production of antibiotics by marine gram-negative bacteria did not stop, and the researches continued. Lemos et al. (1985) examined the microflora from seaweeds (*Enteromorpha intestinalis, E. compressa, Fucus ceranoides, Pelvetia canaliculata* and *Ulva lactuca)*, which were collected in Spain from the intertidal zone at low tide. The seaweed was rinsed in sterile seawater to remove superfluous organisms, and a 1 cm^2 area was swabbed. This material was spread over the surface of marine 2216E agar plates, with incubation at 20 °C for 7 days. Antibacterial activity was demonstrated in 38/224 (17 %) isolates derived from 62 samples of seaweed, with the majority originating from *E. intestinalis.*

7.2.2 Production of Antitumour Compounds

Flavobacterium uliginosum is a marine strain from which a water-soluble compound named Marinactan was produced. Okami (1986) discussed the isolation of a polysaccharide from marine *Flavobacterium* with marked activity in mice against sarcoma-180 solid tumour virus (s-180). From the seaweed disturbed along the Sagami Bay, Japan, bacterial isolations were carried out on a nonselective nutrient medium with incubation at 27 °C for 1–3 days. Isolated colonies were examined for the ability to produce polysaccharide on a medium containing sugar after incubation at 27 °C for 2 days.

Novel bioactive compound has also been isolated from a very important group of microalgae, the diatoms. This algal class has traditionally been regarded as providing the bulk of the food that sustains the marine food chains to top consumers, which includes variety of fishes of commercial importance. However, this beneficial role has been extended on the basis of recent laboratory findings showing that diatoms produce antibiotic compounds that block embryogenesis in copepod and sea urchin eggs and arrest proliferation of human carcinoma cells (Miralto et al. 1999).

7.2.3 Production of Enzyme

Marine and estuarine microorganisms are unique sources of enzymes. Already it has been established that some marine bacteria produce copious quantities of alginate, lyases and chitinases, which may warrant commercial exploitation. For that matter, deep-sea thermophilic bacteria may provide useful sources of heat-stable enzymes (Deming 1986). Okami (1986) recovered a useful isolate of *Bacillus circulans* (No MT-GT2) from marine mud in Tokyo Bay, which produced an enzyme capable of hydrolyzing glucan. Enzyme production, which was induced by the presence of glucan in the medium, was maximal after 6 days. Fractionation was performed by chromatography on DEAF-Sephadex eluted with potassium chloride followed by further chromatography on DE-32 cellulose, which was eluted with acetate buffer. Finally gel filtration was achieved with Sephadex G-150. From this procedure a novel enzyme was recovered which degraded glucans consisting α1, 3 and α1, 6 linkages. The optimum temperature and pH for activity were 35 °C and 6.2–6.7, respectively. It is speculative whether or not this enzyme could be used for prevention/reduction of dental problems but its incorporation into toothpastes for this purpose is a possibility.

7.2.4 Bioremediation of Petroleum Hydrocarbon

Bioremediation is the utilization of a microorganism to remove pollutants from the environment. It is in fact an acceleration of the natural fate of biodegradable pollutants and hence can be regarded as green solution to oil pollution. Bioremediation is a necessary and cost-effective method of removing certain environmental pollutants that adversely affect human health or environmental quality. The enormous natural quality of diverse microorganisms to degrade numerous organic compounds (ranging from petroleum hydrocarbon to chlorinated solvents) and to transform various inorganic substances to metals forms the basis of bioremediation. The metabolic activities of microorganisms are used to change an undesirable chemical into one that has less objectionable properties, for example, changing a toxic pesticide or a carcinogenic petroleum hydrocarbon into nontoxic carbon dioxide and water. To date, most bioremediation projects have relied on the use of naturally occurring microorganisms (often the indigenous microorganisms) at contaminated sites. Presently some new projects are attempting to use genetically engineered species to degrade the pollutants at a much faster rate or to grow under more adverse conditions. This research is very important in the present *era*, as many pollutants are nonaqueous and often occur in the environment that does not favour microbial growth and biodegradative properties. Petroleum hydrocarbon is a widespread environmental pollutant that is amendable to removal by bioremediation.

Bioremediation of oil pollution usually relies on modifying the environment so that the growth of indigenous hydrocarbon degrading microorganism is stimulated. Since microorganism requires nitrogen, phosphorus and other mineral nutrient for incorporation into biomass, the availability of these nutrients within the area of hydrocarbon degradation is critical. In this situation the addition of nitrogen and phosphorus containing fertilizer overcomes the nutritional limitation for microbial growth, because petroleum contains concentration of these substances well below those needed for microbial growth. The typical ratio of carbon to nitrogen in a microbial cell is 10:1 and has carbon to phosphorus 30:1. Besides nitrogen and phosphate, rapid hydrocarbon degradation requires molecular oxygen because the initial steps in biodegradation of hydrocarbon by most microorganisms such as *Pseudomonas* sp. involve the incorporation of hydrocarbon of oxygen by oxygenase.

7.2.5 Degradation of Mangrove Litter

The genus *Aspergillus* plays a major role in the degradation process of mangrove litter. Till date, four species of *Aspergillus* have been identified from lower Gangetic delta complex which are the main drivers of litter degradation. An experiment in which the degradation process was allowed to take place for 30 days found that the genus *Aspergillus* comprised about 75 % numerically of the total fungal population, while the remaining 25 % comprised species belonging to the genus *Helminthosporium* (Chaudhuri and Choudhury 1994). A long-term field-based experiment conducted by the present authors in Indian Sundarbans after litter collection and subsequent analyses confirms that the role of bacteria in the microbial degradation of mangrove litter is also significant (Figs. 7.12 and 7.13). Cellulose, hemicelluloses, cellobiose, pectin, lignin, chitin, lipid and protein are some of the essential litter substances which are exposed to microbial degradation (Biswas et al. 1986). These molecules are broken down by the enzymatic activity of specific bacteria (Figs. 7.14 and 7.15).

Most of the bacteria isolated in this study belong to ammonifying groups and some belong to nitrifying groups. A few strains that were isolated were chitinolytic bacteria, suggesting that these bacteria assist in the breakdown of chitin found in a variety of marine and estuarine arthropod sources. The above findings indicate that bacteria are involved in the mineralization process of the soil and the adjacent coastal waters as well as play an important role in the breakdown of litter in the mangrove substrate.

Fluctuations in fungal populations in response to changes in some leaf constituents during degradation in the natural environment have been studied by Misra et al. (1985). The study found that the *Aspergillus* genus was predominant

Fig. 7.14 Net for mangrove litter collection

Fig. 7.15 Mangrove litter collection in a net

throughout the decaying process. The carbon, hydrocarbon and wax ester contents decreased during the process of degradation. This suggests that the invading microbes utilized these constituents as a carbon source. A decrease in cellulose indicated that the microbes use cellulolytic enzymes. The decrease in the constituents was proportional to the increase in nitrogen and protein content produced during the decay process. This may be due to an increase in numbers of microorganisms within the leaves, or alternatively, the organisms may be nitrogen-fixing types. The decomposed leaves are rich in protein and carbohydrates and serve as food for the detritus-feeding benthic community.

7.2.6 Drugs from Marine Microbes

Marine microbes are the sources of several diverse categories of bioactive substances from which unique pharmaceutical products can be synthesized. The contribution of probiotic bacteria, such as lactobacilli and bifidobacteria, is mainly in the control of pathogenic microbes, through production of antibacterial protein, namely, bacteriocin (DeVugst and Vandamme 1994; Kathiresan and Thiruneelakandan 2008) and anticancer substances (Wollowski et al. 2001). The dietary supplements of lactobacilli are reportedly decreasing the induction of experimental colon cancer (Goddin and Gorbach 1992). They stimulate and modulate the mucosal immune system by reducing the production of pro-inflammatory cytokines through actions on NFκB pathways, increasing production of anti-inflammatory cytokines such as IL-10 and host defence peptides such as β-defensin 2, enhancing IgA defences and influencing dendritic cell maturation as well as modulation of cell proliferation and apoptosis through cell responses to short-chain fatty acids (Devine and Marsh 2009). Thiocoraline is a novel bioactive depsipeptide isolated from *Micromonospora marina*, a marine microorganism located in the Mozambique Strait that inhibits RNA synthesis. The bioactive compound is also selectively cytotoxic against lung and colon cancer cell lines as well as melanoma. Interestingly, the compound exerts preferential antiproliferative effects in colon cancer cell lines with defective p53 systems (Erba et al. 1999). Thiocoraline represents a model of an anticancer agent acquired from marine microorganisms and illustrates how the problems of drug supply can be overcome by artificial culture (Harvey 1975).

A rich profile of biologically active metabolites is described from filamentous fungi of terrestrial origin, especially from just three genera: *Penicillium, Aspergillus* and *Fusarium* (Lene 1996). However, the marine fungi are least studied than terrestrial counterparts and other ecological groups. Obligate marine fungi are still an unexplored resource, although, marine facultative fungi have been studied due to their production of new metabolites which are not found in terrestrial fungi. Recently, more interest has been generated on studying biologically active metabolites from higher fungi (Basidiomycetes), endophytic fungi and filamentous fungi from marine habitats, the symbiotic lichens. In one study, the lignicolous fungus *Leptosphaeria oraemaris* (Pleosporaceae) yielded leptosphaerin (Schiehser et al. 1986; Pallenberg and White 1986). A further study of the same fungal species yielded none of the previously found metabolites, but the polyketides, leptosphaerolide, its *o*-dihydroquinone derivative and leptosphaerodione (Guerriero et al. 1991). This leads to a conclusion that the production of secondary metabolites might be highly dependent on the culture conditions and the origin of the strains. To produce these metabolites and to maximize the potential chemical diversity, they need to be grown in various nutrient-limited media. For example, media for *Penicillium* spp. that are deficient in carbon can produce penicillins, those that are phosphorus limited can produce cephalosporins and vancomycin and those that are nitrogen limited can produce carbapenems (Lawrence 1999). Marine-derived fungi are known to be a source of antioxidative natural products: (i) acremonin

A from *Acremonium* sp. (Abdel-Lateff et al. 2002) and (ii) xanthone derivative from *Wardomyces anomalus* (Abdel-Lateff et al. 2003).

Marine blue-green algae (*Cyanobacteria*) are considered to be one of the potential organisms which can be the richest sources of known and novel bioactive compounds including toxins with potential for pharmaceutical applications (Thajuddin and Subramanian 2005; Jha and Zi-Rong 2004). Some of the marine cyanobacteria appear to be potential sources for large-scale production of vitamins (B complex, E) of commercial interest. Scytonemin is a protein serine/threonine kinase inhibitor (Stevenson et al. 2002a), isolated from the cyanobacterium *Stigonema* sp., and this compound is a yellow-green ultraviolet sunscreen pigment, known to be present in the extracellular sheaths of different genera of aquatic and terrestrial blue-green algae. Scytonemin regulates mitotic spindle formation as well as enzyme kinases involved in cell cycle control, and the compound also inhibits proliferation of human fibroblasts and endothelial cells. Thus, scytonemin may provide an excellent drug as protein kinase inhibitors to have antiproliferative and anti-inflammatory activities (Stevenson et al. 2002b).

More than 50 % of the marine cyanobacteria are potentially exploitable for extracting bioactive substances which are effective in either killing the cancer cells by inducing apoptotic death or effecting the cell signalling through activation of the members of protein kinase-c family of signalling enzymes. The cell extracts of *Calothrix* isolates inhibit the growth in vitro of a chloroquine-resistant strain of the malarial parasite, *Plasmodium falciparum,* and of human HeLa cancer cells in a dose-dependent manner (Rickards et al. 1999). Bioassay-directed fractions of the extracts have led to their isolation and structural characterization of calothrixin A (I) and B (II), pentacyclic metabolites with an indole (3, 2-j) phenanthridine alkaloids which exert their growth inhibitory effects at nanomolar concentrations (Rickards et al. 1999). Another compound, curacin A, isolated from the organic extracts of Curacao collections of *Lyngbya majuscula* is an exceptionally potent antiproliferative agent as it inhibits the polymerization of the tubulin, and it also displays the inhibitory activity selectively on colon, renal and breast cancer-derived cell lines (Carte 1996).

Brain Churners

1. What are the different categories of microbes found in the marine and estuarine ecosystems?
2. Define neuston.
3. Name the fungus associated with cordgrass (*Spartina alterniflora*).
4. Why do the mangroves offer diverse category of microbes?
5. What is microbial biofilm?
6. State the technology for repetitive element polymerase chain reaction for identifying bacteria.
7. What are the common DNA sequencing methods for identifying bacteria?
8. Name the microbial strain for the production of antitumour compounds (Marinactan).
9. How microbes are used for bioremediation?
10. State the importance of curacin A.

Annexure 7A: Study on the Quality of Edible Oyster in Indian Sundarbans with Respect to Coliform Load

1. Introduction

The Indian Sundarbans is one of the most biologically productive, taxonomically diverse, mangrove-dominated ecosystems of the tropics (Mitra et al. 1992), which has been declared as a World Heritage Site in 1987 by UNESCO. The deltaic lobe is a unique genetic reservoir sustaining a wide spectrum of commercially important finfish and shellfish. In recent times, thrust has been given on the culture of edible oyster, *Saccostrea cucullata*, as alternative livelihood scheme for the local people (Mitra and Banerjee

2005) It is therefore extremely important to evaluate the quality of meat of these commercially important bivalve species in terms of coliform load. Necessity of such work arises mainly due to discharge of anthropogenic wastes in coastal areas due to which the zone becomes vulnerable in terms of microbial load (Glasoe and Aimee 2004). The microbes from the ambient media often get accumulated in shellfish because of their filter-feeding activity. Under favourable condition, a single large oyster may filter up to 5 l of water per hour. Such a species is thus an effective filter device of particles that enter in the estuarine environment through urban, industrial and municipal wastes (Pommepuy 1996). Considering this property of oyster, some directives have been given in connection to marketing of live bivalves. An EC Directive (91/492/EEC) has defined the health conditions for production and marketing of live bivalves. All Member States are required to classify their harvesting areas into one of three categories according to the level of faecal indicators present in shellfish samples. Shellfish from Category A can be placed directly on the market. They must meet a standard of no more than 230 *Escherichia coli* per 100 g shellfish flesh (or 300 FC/100 g) as well as other standard for specific pathogens (such as *Salmonella*), chemicals and algal biotoxins. Shellfish from category B must be purified before marketing, and shellfish from category C must be placed again in clear water for 2 months prior to marketing (Pommepuy 1996) Although such types of classification do not exist in Indian coastal and estuarine shellfish culture area, but a survey was conducted in three sampling stations of Indian Sundarbans during April, 2007 to assess the health of aquatic subsystem and edible oyster in terms of microbial load.

2. Materials and Methods

The present investigation was carried out during the month of April, 2007 at three different stations, namely, Namkhana, Frasergaunje and Sajnekhali. The sampling stations were selected considering the magnitude of anthropogenic pressure. Station 1 (Namkhana) is situated in the western sector of the Sundarbans, which is not only an important fish landing station but also receives the wastewater from Kolkata and nearby Haldia port–cum–industrial complex. Station 2 (Frasergaunje) is also an official fish landing station of the state of West Bengal, but in addition to this activity, the water of this station receives the discharge of several hotels and tourism units located at Bakkhali station. Station 3 (Sajnekhali) is situated in the eastern sector of Indian Sundarbans, which is noted for its wilderness. Anthropogenic stress is minimum in this sampling station owing to presence of mangrove forest. Water samples were collected using water sampler and the sediment samples were collected with the help of the Petersen grab. The water and sediment samples for microbial analysis were immediately transferred into the sterile bottles, and central portions of the sediment samples were aseptically taken and put into sterile polythene bags and transported to the laboratory under ice for bacteriological examinations. Oyster samples were collected from the intertidal zone of the selected sampling stations for carrying out microbial load analysis in terms of total coliform and faecal coliform.

For bacterial analysis, the oyster samples were accurately weighed and blended with 0.1 % peptone buffer and 3 % NaCl diluent for 1 min and finally inoculated taking different dilutions. The incubation was done at 37 °C for 24 h, and the result was expressed from MPN index per gram basis. For bacteriological analysis of water and sediment samples, the standard method as stated APHA 20th Edition, 2001 was followed.

3. Results and Discussion

The station-wise order of microbial contamination in the study area is Namkhana > Frasergaunje > Sajnekhali. This spatial variation may be attributed to the degree of anthropogenic stress. Namkhana and Frasergaunje, being the fish landing sites, are constantly exposed to wastes of complex nature. In addition to decomposed fish products, these sampling

Fig. 7A.1 Availability of different aquacultural products in the local market for internal consumption

Table 7A.1 Total coliform load (five test tube method) in water samples collected from the sampling stations during April 2007

Sl. no.	Stations	MPN/100 ml in LST	MPN/100 ml in BGL
1	Namkhana	>1600	>1600
2	Frasergaunje	1600	1600
3	Sajnekhali	550	350

Table 7A.2 Total coliform load (three test tube method) in sediment samples collected from the sampling stations during April 2007

Sl. no.	Stations	MPN/g in LST	MPN/g in BGL
1	Namkhana	30	30
2	Frasergaunje	24	24
3	Sajnekhali	9.5	9.5

stations are also contaminated with zinc, copper and lead released from antifouling paints required for conditioning fishing vessels and trawlers. Sajnekhali, on the other hand, is a wildlife sanctuary with minimum environmental stress. The area sustains unique mangrove vegetation, which acts as agents of bioremediation. The mode of activities and degree of anthropogenic stress have been reflected through microbial load

The result indicates an alarming situation with respect to coliform load in the shellfish tissues sampled from Namkhana and Frasergaunje. Hence, not only depuration but also a proper feasibility report is needed to initiate oyster culture in these areas. Oyster being an edible product needs continuous monitoring with respect to coliform load to overcome the barrier of consumer acceptability, which may otherwise pose an adverse effect on the human health (consumers).

Today, shellfish industry has gained considerable momentum, and different types of molluscs are widely available in the markets (Fig. 7A.1).

Marketing of oyster or any aquacultural products (both for internal consumption and export) is a function of purity of the cultured species. Under such circumstances, results of the present work may serve as baseline information for initiating oyster industry in the maritime state of West Bengal (Tables 7A.1, 7A.2, 7A.3, 7A.4, 7A.5 and 7A.6).

Table 7A.3 Total coliform load (three test tube method) in edible oyster samples collected from the sampling stations during April 2007

Sl. no.	Stations	MPN/g in LST	MPN/g in BGL
1	Namkhana	>110	>110
2	Frasergaunje	110	110
3	Sajnekhali	46	45

Table 7A.4 Total faecal coliform load (five test tube method) in water samples collected from the sampling stations during April 2007

Sl. no.	Stations	MPN/100 ml in EC
1	Namkhana	40
2	Frasergaunje	50
3	Sajnekhali	4

Table 7A.5 Total faecal coliform load (three test tube method) in sediment samples collected from the sampling stations during April 2007

Sl. no.	Stations	MPN/g in EC
1	Namkhana	24
2	Frasergaunje	15
3	Sajnekhali	2.3

Table 7A.6 Total faecal coliform load (three test tube method) in edible oyster samples collected from the sampling stations during April 2007

Sl. no.	Stations	MPN/g in EC
1	Namkhana	>110
2	Frasergaunje	110
3	Sajnekhali	46

Annexure 7B: Study of the Microbial Health in and Around the Lower Stretch of Hooghly Estuary

1. Introduction

Microorganisms such as bacteria, fungi, actinomycetes, etc., are widely distributed in the water and sediment of marine and brackish water environments. They have far reaching effects on the biological as well as the geochemical systems. They also play an important role in the decompositions of organic matter, dissolution of inorganic insoluble salts and regeneration of nutrients. The activities of the total heterotrophic bacteria and the relative abundance reflect the hydrographic structure or the nature and nutrient concentrations in the aquatic environment (Oppenheimer and Wood 1962).

The overloading of nutrients and organic load also provides a favourable environment for the growth and survival of a wide spectrum of microbial strains. The high population densities and activities often common in the coastal areas result in pollution and release of contaminated wastewater. Pathogenic microorganisms such as bacteria and viruses, abundant in human wastes, are often discharged into natural waters with little or no treatment. Survival of microbes in waters depends on many parameters such as biological (interaction with other bacteria) and physical factors (temperature). Numerous studies have been carried out in coastal areas over long periods of time, demonstrating the various abiotic environmental conditions (fluxes, currents, presence of mud and silt, etc.) due to which the distribution of microbes is affected. The undertreated effluents from the coastal population and discharges from industrial belt regions often pose an adverse impact on marine and estuarine species. The members like *Salmonella* sp., *E. coli*, faecal coliform, etc., can multiply and survive in the estuarine environment for weeks. The *Enterobacteriaceae* (*Salmonella*, *E. coli*, etc.) occur in the water as a result of contamination from the animal or human origin. This contamination has been normally associated with faecal contamination or pollution of natural waters or water environments, where these organisms survive for a long time (months) or through direct contamination of products during processing. In the entire Gangetic plain, it is the river Hugli that is subjected to heavy pollution load from the industrialized and highly urbanized cities of the Kolkata and Howrah. The discharges from the port–cum–industrial complex of Haldia have aggravated the magnitude of pollution. The marine ecosystem nearest to the city of Kolkata is the Indian Sundarbans, which is the most biologically productive, taxonomically diverse and aesthetically celebrated ecotone in the Indian subcontinent. The untreated and the undertreated sewage of the city of Kolkata and Howrah is responsible for the microbial load. Among microbial flora, the presence of pathogens such as *Salmonella*, *Streptococcus* sp., *Vibrio* sp. *and E. coli* has been determined. It is thus clear that the present zone of investigation is under severe stress due to municipal discharge containing appreciable amount of sewage generated from municipal and several categories of anthropogenic microbial activities.

2. Aims and Objectives

The present study aims to evaluate the microbial load (total coliform and faecal coliform) in water sample in and around lower stretch of Hooghly river estuary. The area is stressed due to industrial and anthropogenic activities. Now-a-days bacterial indicators are measured instead of pathogenic organisms because the indicators are safer and can be measured with faster, less expensive methods than the pathogen of concern (McGee et al. 1997).

The main objectives of the present study are highlighted:

- To monitor monthly variation of physico-chemical variables during the study periods
- To observe the spatial variation of the selected physico-chemical variables in the study area
- To scan the microbiological parameters during 3 months of the study period

- To monitor the spatial variation of the microbiological parameter in the study area

3. Physiography

3.1. The Ecological Profile

The Indian Sundarbans Delta (ISD) is part of the delta of the Ganga–Brahmaputra–Meghna (GBM) basin in Asia. The Sundarbans shared between Indian and Bangladesh is home to one of the largest mangrove forests in the world. The ISD spread over about 9630 km^2 between 21°40′04″ N and 22°09′21″N latitude and 88°01′56″ E and 89°06′01″ E longitude is the smaller and western part of the complete Sundarbans Delta.

The Indian Sundarbans Delta is bounded by the Ichamati–Raimangal River in the east, by the Hooghly River in the west, by the Bay of Bengal in the South and by the Dampier-Hodges line drawn in 1829–1830 in the north. A little over half of this area has human settlements on 54 deltaic Islands; the remaining portion is under mangrove vegetation. Soils of ISD are principally alfisols (older and alluvial soil) and aridisols (coastal saline soil).

The landscape is characterized by a web of tidal water systems. The average tidal amplitude is between 3.5 m and 5 m, with the highest amplitudes in July–August and the lowest in December–January. Of the eight rivers that dominate the landscape, only the Hugli and Ichamati–Raimangal carry freshwater flow of some significance. Being the moribund part of the lower delta plain of the GBM system, the ISD is experiencing both declining freshwater supplies and net erosion, as has been recorded since 1969 (Hazra et al. 2002; Hazra et al. 2010).

The Indian Sundarbans at the apex of the Bay of Bengal (between 21°13′N and 22°40′N latitude and 88°03′E to 89°07′E longitude) is located on the southern fringe of the state of West Bengal (a maritime state in the north-east coast of India). The area of the Indian Sundarbans is 9630 Km^2. of which the forest area is about 4200 Km^2. The region is bordered by Bangladesh in the East, the Hooghly river (a continuation of the Ganges river) in the west, Dampier and Hodges Line in the north and the Bay of Bengal in the south. With a considerable degree of maritime characteristics in major portion of the ecosystem, the important morphotypes of deltaic Sundarbans are beaches, mud flats, coastal dunes, sand flats, estuaries, creeks, inlets and mangrove swamps.

The rivers are the live matrix of deltaic complex, on which the unique spectrum of biological diversity is embedded. In Indian Sundarbans, approximately 2069 Km^2 area is occupied by tidal river system or estuaries which finally end up in the Bay of Bengal. The deltaic complex of Indian Sundarbans is also noted for its seasonality in terms of climatic condition and wind action as highlighted here in brief. Frequent nor'westers are also common in the premonsoon season.

3.2. Climate of Indian Sundarbans

The deltaic lobe of Indian Sundarbans experiences a moderate type of climate because of its location adjacent to the Bay of Bengal as well as due to regular tidal flushing in the estuaries. Wave actions, micro- and macrotidal cycles and long-shore currents are recorded in most of the islands of the ecosystems. Coastal processes are very dynamic and are accelerated by tropical cyclones, which is locally called 'Kal Baisakhi' (nor'wester). The seasonal climate in study area may be conveniently categorized into premonsoon (March–June), monsoon (July–October) and postmonsoon (November–February). Each season has a characteristics feature of its own, which is very distinct and unique. The oscillations of various physico-chemical variables in different seasons of the year are discussed here in brief.

3.3. Wind

The direction and velocity of wind system in the coastal West Bengal are mainly controlled by the north-east and southwest monsoons. The wind from the north and north-east commences at the beginning of October and continues till the end of March. The months of January and February are relatively calm with an average wind speed around 3.5 km/h.

Violent wind speed recommences from the southwest around the middle of March and continues till September. During this period, several low-pressure systems occur in this region, a number of which take the form of depressions and cyclonic storms of varying intensity. The air temperature of Sundarbans area varies from 19.0 °C to 34.0 °C and velocity of wind from 0.85 to 4.54 m/s (Saha et al. 1998).

3.4. Waves and Tides

The wind is the basic driving force for generating surface waves in the coastal zone of West Bengal. Sea waves in this region rarely become destructive except during cyclonic storms. During nor'westers the wind speed rises above 100 km/h. and is usually accompanied by huge tidal waves. When the cyclonic incidences coincide with the spring tides, wave height can rise over 5 m above the mean sea level. Ripple waves appear in the months of October, November and December when wind-generated wave height varies approximately between 0.20 and 0.35 m. In the months of April to August, large wavelets are formed in the shelf region, and they start breaking when they approach towards the coastal margin. Wave height rises up to 2 m during this period, which causes maximum scoring of land masses. The average tidal amplitude in the estuaries of the Sundarbans ranges from 3.5 to 4.0 m. Wave actions, micro- and macrotidal cycles and long-shore currents are recorded in most of the islands in this ecosystem.

3.5. Surface Water Temperature

In coastal West Bengal, the seasonal variation of surface water temperature is not so drastic between premonsoon and monsoon seasons. The premonsoon period (March to June) is characterized by a mean surface water temperature around 34 °C. The monsoon period (July to October) shows a surface water temperature around 32 °C (mean), and the postmonsoon period (November to February) is characterized by cold weather with a mean surface water temperature around 23 °C.

3.6. Rainfall

The average annual rainfall in deltaic Sundarbans region is 1920 mm. Rainfall is usually maximum during the month of August/ September, and the monsoon period lasts from July to October. The southwest wind triggers the precipitation in the monsoon period with an average rainfall of about 165 mm (Human Development Report, South 24 Parganas, 2009). The postmonsoon (November to February) is characterized by negligible rainfall, and the premonsoon period (March to June) is basically dry but occasionally accompanied by rains and thunderstorms.

4. Materials and Methods

The present programme encompasses the evaluation of *microbial load (water)* and some related *physico-chemical variables* such as:

(a) Surface water temperature
(b) Surface water salinity
(c) Surface water pH
(d) Surface water dissolved oxygen (DO)
(e) Surface water nitrate (NO_3^-)
(f) Surface water phosphate (PO_4^{2-})
(g) Surface water silicate (SiO_2)

The work was carried out on a monthly basis from February 2013 to April 2013. Samplings have been carried out at eight different stations in Indian Sundarbans, namely:

1. Haldia
2. Daimond Harbour
3. Lot 8
4. Kachuberia
5. Chemaguri
6. Sagar Island
7. Namkhana
8. Frasergaunje

The entire work procedure has been divided into four procedural phases as mentioned below:

Table 7B.1 Sampling stations with coordinates

Station no.	Name of the station	Coordinates
Stn. 1	Haldia	22°1′31″N/88°3′29″E
Stn. 2	Diamond Harbour	22°11′04.2″N/ 88°11′22.2″E
Stn. 3	Lot 8	21°53′25.6″N/ 88°09′58.4″E
Stn. 4	Kachuberia	2152′26.50″N/ 8808′04.43″E
Stn. 5	Chemaguri	2139′58.15″N/ 8810′07.03″E
Stn. 6	Sagar Island	21°38′54.37″N/ 88°03′06.17″E
Stn. 7	Namkhana	21°76′N/88°23′E
Stn. 8	Frasergaunje	21°36′55.72″N/ 88°12′33.15″E

- *Phase A:* Site selection.
- *Phase B:* Analysis of physico-chemical variables of water.
- *Phase C*: Analysis of microbial load (total bacterial count, total coliform, faecal coliform, *E. coli*, *Vibrio* sp., *Streptococcus* sp., *Salmonella* sp.) in ambient water media.
- *Phase D:* Statistical analysis.

Phase A: Site Selection

The first phase of the work involves selection of eight sampling stations (Table 7B.1) in the deltaic region of Sundarbans (Figs. 7B.1A and 7B.1B).

Phase B: Analysis of Physico-chemical Variables of Water

(a) *Surface water temperature:* The surface water temperature was measured using 0–100°C mercury thermometer.

(b) *Surface water salinity:* The surface water salinity was recorded by means of an optical refractometer (Atago, Japan) and cross-checked in the laboratory by employing more Knudsen method (Strickland and Parsons 1972). The correction factor was found out by titration of the silver nitrate ($AgNO_3$) solution against standard seawater (IAPO Standard Seawater Service Charlottenlund, Slot Denmark, Chlorinity 19.376 ppt).

(c) *Surface water pH:* The surface water pH was measured by using a portable pH-meter sensitivity $= \pm 0.02$.

(d) *Surface water dissolved oxygen (DO):* The surface water DO was measured by DO meter in the field and subsequently cross-checked in the laboratory by Winkler's method.

(e) *Surface water nutrient analysis:* Surface water was collected for nutrient analysis in cleaned Tarsons bottles and transported to the laboratory in iced freeze condition. Triplicate samples were collected from same collection sites to maintain the quality of the data. The standard spectrophotometric method of Strickland and Parsons (1972) was adopted to determine the nutrient concentration in surface water.

(i) *Nitrate analysis:* Nitrate was analyzed by oxidizing it to nitrite by means of passing the sample with ammonium chloride buffer through a glass column packed with amalgamated cadmium filings and finally treating the samples with sulphonyl amide. The resultant diazonium ion was coupled with N-(1-naphthyl)-ethylene diamine to give an intensely pink azo dye.

(ii) *Phosphate analysis:* Determination of the phosphate was carried out by treatment of an aliquot of the sample with an acidic molybdate reagent containing ascorbic acid and a small proportion of potassium antimony tartrate.

(iii) *Silicate analysis:* Dissolved silicate was determined by treating the sample with acidic molybdate reagent. The resultant silico-molybdic acid was reduced to molybdenum blue complex by ascorbic acid and incorporating the oxalic acid to prevent the formation of similar blue complex phosphate.

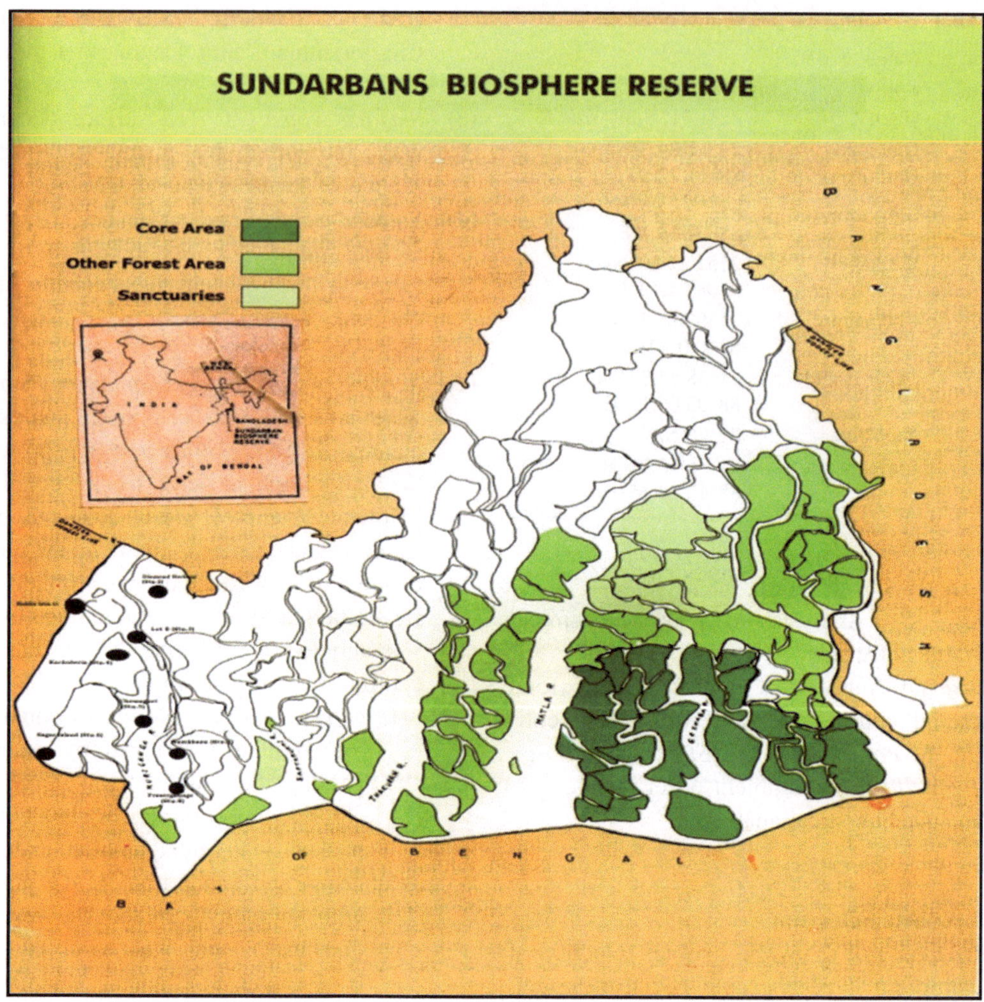

Fig. 7B.1A Map of Indian Sundarbans

Phase C: Analysis of Microbial Load in Ambient Water

1. *Sampling:*

(i) *Sampling of the water:* Water samples were collected fortnightly aseptically in sterilized glass container (sterilized in autoclave) with utmost care from February 2013 to April 2013. The collected samples were immediately transferred in icebox and brought to the laboratory for further analysis.

(ii) *Preparation of culture media for the microbial analysis :*

(a) *Preparation of the Lauryl Tryptose Broth (LTB) for presumptive test:* In order to prepare the LTB, at first a dehydrated amount of ingredients for single strength (SS) and double strength (DS) was required to dissolve separately in each 1 l of sterilized distilled water, and it was thoroughly mixed and slightly heated by proper swirling. The pH was adjusted up to 6.8 ± 0,2 by

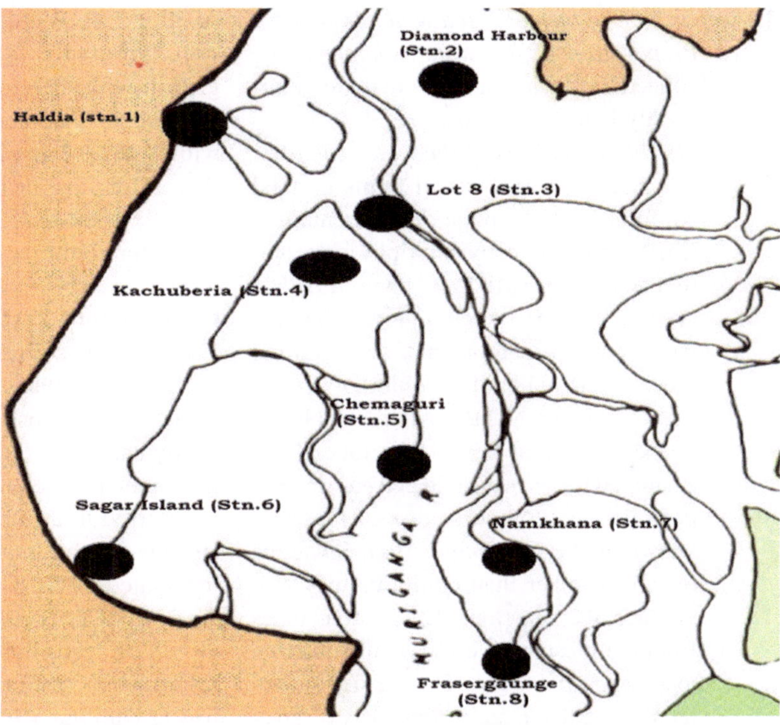

Fig. 7B.1B Map showing sampling stations

Table 7B.2 Ingredients of LTB

Composition	Dose
Tryptose	20 g
Lactose	5 g
Dipotassium hydrogen phosphate	2.75 g
Potassium dihydrogen phosphate	2.75 g
Sodium chloride	5 g
Sodium lauryl sulphate	0.10 g
Sterilized distilled water	1000 ml

Table 7B.3 Ingredients of BGLB

Composition	Dose
Peptone	10 g
Lactose	10 g
Ox gall	20 g
Brilliant green	0.133 g
Sterilized distilled water	1000 ml

either 0.1(N) sodium hydroxide (NaOH) or 0.1 (N) hydrochloric acid (HCl). After that it was distributed as required (10 ml SS and 10 ml DS) in test tube containing inverted Durham's tube and then placed in the autoclave for sterilization at 121°C and 15 lbs for 15 min. The general ingredients of the LTB are given in Table 7B.2.

(b) *Preparation of Brilliant Green Lactose Bile Broth (BGLB) for confirmed test:* At first the required amount of the dehydrated ingredients was dissolved in 1 l of sterilized distilled water which was thoroughly mixed and slightly heated by proper swirling, and then pH was adjusted up to 7.2 ± 0.2 by either 0.1(N) sodium hydroxide (NaOH) or 0.1 (N) hydrochloric acid (HCl). After that it was distributed in test tubes (10 ml each) containing inverted Durham's tube and then placed in autoclave for sterilization at 121 °C and 15 lbs for 15 min. The general ingredients of BGLB are as follows (Table 7B.3):

2. *Preparation of the collected water samples:*
   (i) *Preparation of the water samples:* The collected water samples were mixed thoroughly before analysis.
   (ii) *Microbial analysis of the water samples:* For microbial analysis in terms of total coliform load, the most probable number (MPN) procedure by multiple fermentation techniques (MFT) is stated in APHA (1998). The techniques involve inoculating the sample and/ or its several dilutions in a liquid medium of Lauryl Tryptose Broth (LTB). After completion of the incubation period, the tubes were examined for growth, acid and gas production by the coliform organisms. This test is known as the *presumptive test.* Since the organisms other than coliforms may also produce the reaction, the positive tubes from the presumptive test were subjected to a *confirmatory test.* The density of bacteria was calculated on the basis of positive and negative combination of the tubes. For water samples, the results were expressed in MPN/100 ml (APHA 1998).

Presumptive Test for Total Coliform

Presumptive test for total coliform and Lauryl Tryptose Broth was used as culture media. For analysis of water five test tubes each of 10 ml, 1 ml, 0.1 ml sample portions were used as the presumptive test.

First set of contained five numbers of 10 ml (DS) broth tubes. Second and third sets contained ten numbers of 10 ml (SS) broth tubes for analysis of water. Each tube in a set of five 10 ml, 1 ml and 0.1 ml of water sample was inoculated in the first, second and third sets of media tube, respectively, and mixed thoroughly. In each case a controlled set was run parallelly. The inoculated test tubes were incubated at 36 ± 1 °C after 24 ± 2 h, and the inoculated tubes were examined for growth of gas and acidic reaction. If there was no gas and acid reaction, the tubes were re-examined and re-incubated at the end of 48 ± 2 h. Within each tube, Durham's tubes were invertedly placed to show the bacterial growth with the emission of gas. Production of gas bubbles and acids with growth in the tubes within 48 ± 2 h contributes presumptive reaction. After the incubation period of 48 h, the number of positive tubes were counted and preceded for confirmatory test.

Confirmatory Test for Total Coliform

For the total coliform test, the culture medium used was Brilliant Green Lactose Bile Broth (BGLB). The positive presumptive tubes were gently shaken with a sterile loop (3.5 mm-5 mm in diameter); one or two loopfuls of culture were transferred to a test tube containing BGLB with an invertedly placed Durham's tube. The inoculated BGLB tubes were incubated at 36 ± 1 °C. Formation of any gas within 48 ± 2 h constituted the confirmed test. The results were obtained in MPN/100 ml by comparing with the MPN table.

Phase D: Statistical Analysis

In order to find the differences between months and stations, *ANOVA* was done using Excel under Windows 2007.

5. Results

Marine and estuarine ecosystems are being threatened by the discharge of untreated sewage wastes and industrial effluents which ultimately affects the sustainability of living resources and public health.

Some microbial pathogens in the coastal environment are indigenous to the oceans, including *Vibrios*, whereas others like *Escherichia coli, Salmonella* sp. *and Shigella* sp. are allochthonous which are introduced through agricultural, urban surface run-off, wastewater discharges and from domestic and wild animals. Most of the *Vibrios* and *Salmonella* sp. are pathogenic to humans and some have fatal infections (Blake et al. 1980; Grimes 1975; Carlson et al. 1968; Gerba and Schaiberger 1975).

Infections with *Vibrios* are known to be associated with either consumption of seafood or exposure to marine environment (Raveendran et al. 1990). The presence of faecal coliforms forms representative for the assessment of coastal recreational water quality. The present investigation highlights the occurrence distribution pattern of enteric pathogens in marine water. It also evaluates the influence of anthropogenic inputs and raw sewage on the incidence of these bacteria in and around Indian Sundarbans.

The microbial load is also influenced by physico-chemical variables like temperature, salinity, pH, etc. The level of DO also fluctuates depending on the microbial load and action. The present dissertation was therefore undertaken to focus the spatio-temporal variations of physico-chemical and microbiological parameter as highlighted here:

5.1 Physico-chemical Parameters

5.1.1 Surface Water Temperature

In February 2013, the surface water temperature ranged from 26.8 ± 0.1 °C to 27.1 ± 0.1 °C during the study period. The station-wise order of surface water temperature is Frasergaunje (Stn. 8) > Namkhana (Stn. 7) = Sagar Island (Stn. 6) = Lot 8 (Stn. 3) > Chemaguri (Stn. 5) = Haldia (Stn. 1) > Diamond Harbour (Stn. 2) = Kachuberia (Stn. 4) (Fig. 7B.2).

In March 2013, the surface water temperature ranged from 28.8 ± 0.1 °C to 30.7 ± 0.1 °C during the study period. The station-wise order of surface water temperature is Frasergaunje (Stn. 8) > Namkhana (Stn. 7) > Sagar Island (Stn. 6) > Kachuberia (Stn. 4) = Haldia (Stn. 1) > Lot 8 (Stn. 3) = Diamond Harbour (Stn. 2) > Chemaguri (Stn. 5) (Fig. 7B.3).

In April 2013, the surface water temperature ranged from 32.2 ± 0.2 °C to 33.1 ± 0.2 °C during the study period. The station-wise order of surface water temperature is Frasergaunje (Stn. 8) > Namkhana (Stn. 7) = Sagar Island (Stn. 6) > Kachuberia (Stn. 4) = Chemaguri (Stn. 5) > Lot 8 (Stn. 3) > Haldia (Stn. 1) > Diamond Harbour (Stn. 2) (Fig. 7B.4).

5.1.2 Surface Water Salinity

In February 2013, the surface water salinity ranged from 1.05 ± 0.05 psu to 20.89 ± 0.34 psu during the study period. The station-wise order of surface water salinity is Frasergaunje (Stn. 8) > Namkhana (Stn. 7) > Sagar Island (Stn. 6) > Chemaguri (Stn. 5) > Kachuberia (Stn. 4) > Lot 8 (Stn. 3) > Diamond Harbour (Stn. 2) > Haldia (Stn. 1) (Fig. 7B.5).

In March 2013, the surface water salinity ranged from 2.89 ± 0.05 psu to 24.89 ± 0.20 psu during the study period. The station-wise order of surface water salinity is Frasergaunje (Stn. 8) > Namkhana (Stn. 7) > Sagar Island (Stn. 6) > Chemaguri (Stn. 5) > Kachuberia (Stn. 4) > Lot 8 (Stn. 3) > Diamond Harbour (Stn. 2) > Haldia (Stn. 1) (Fig. 7B.6).

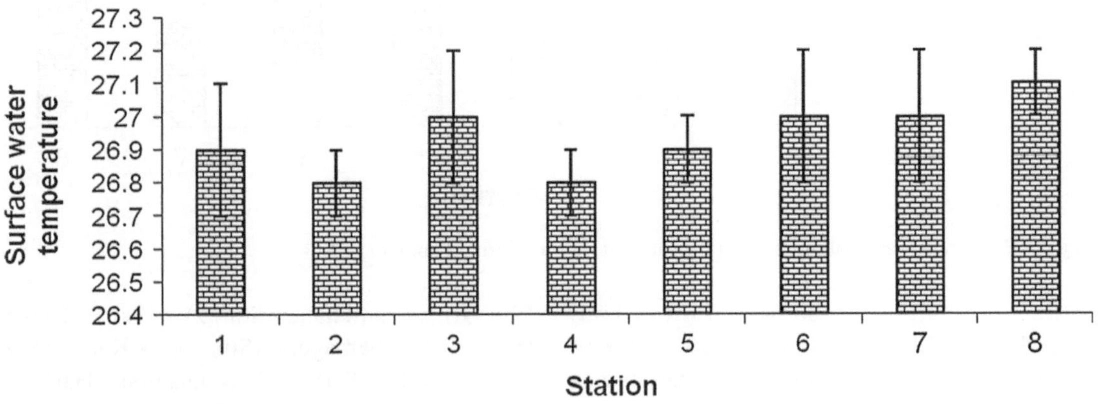

Fig. 7B.2 Surface water temperature (°C) in the selected stations during February 2013

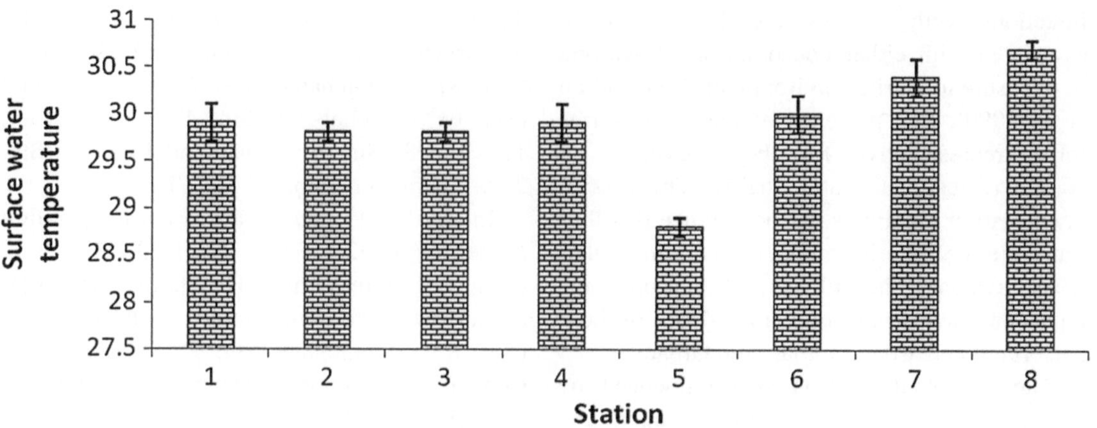

Fig. 7B.3 Surface water temperature (°C) in the selected stations during March 2013

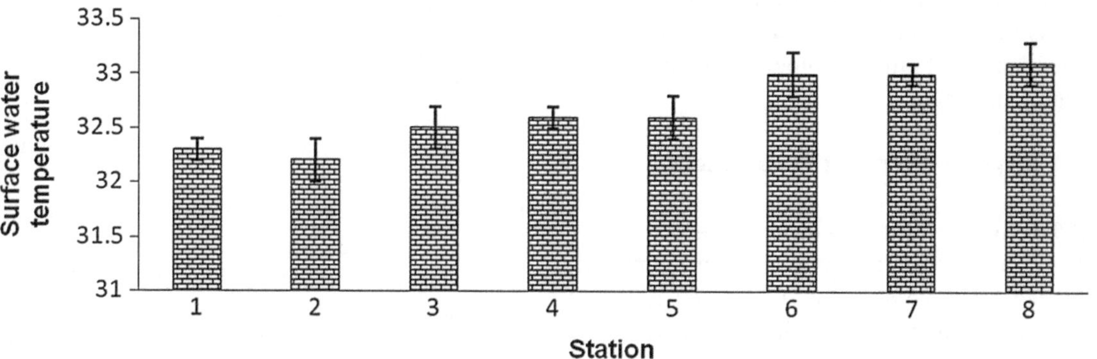

Fig. 7B.4 Surface water temperature (°C) in the selected stations during April 2013

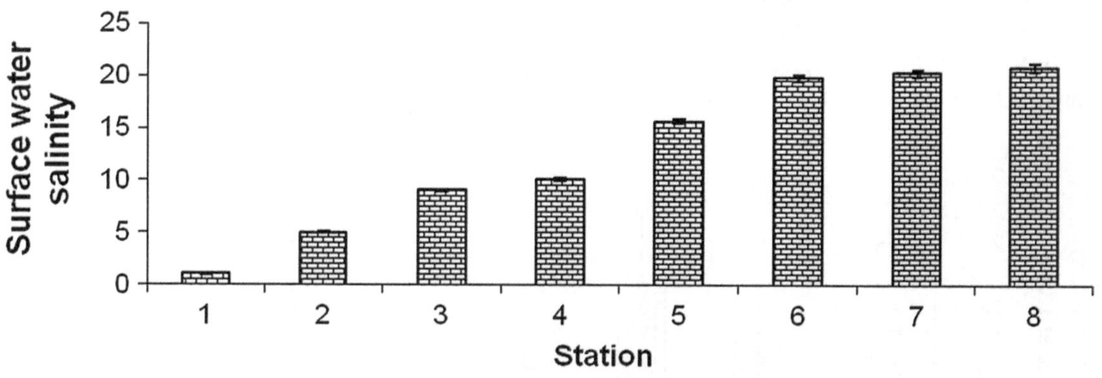

Fig. 7B.5 Surface water salinity (‰) in the selected stations during February 2013

In April 2013, the surface water salinity ranged from 7.12 ± 0.08 psu to 29.55 ± 0.05 psu during the study period. The station-wise order of surface water salinity is Frasergaunje (Stn. 8) > Namkhana (Stn. 7) > Sagar Island (Stn. 6) > Chemaguri (Stn. 5) > Kachuberia (Stn. 4) > Lot 8 (Stn. 3) > Diamond Harbour (Stn. 2) > Haldia (Stn. 1) (Fig. 7B.7).

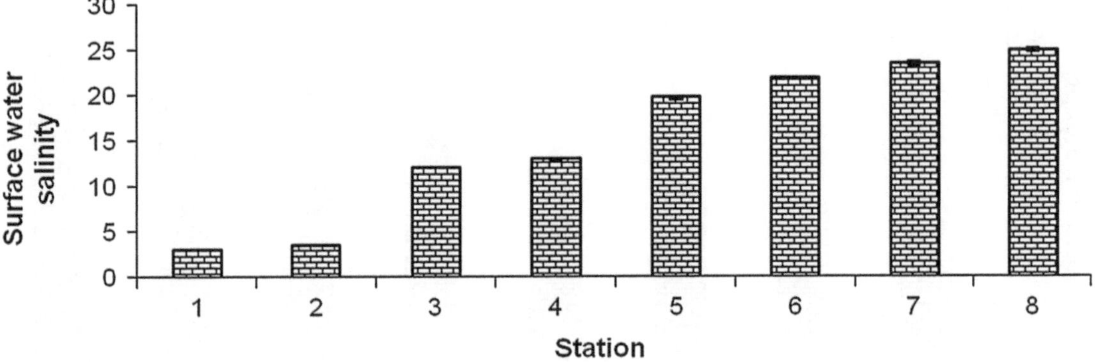

Fig. 7B.6 Surface water salinity (‰) in the selected stations during March 2013

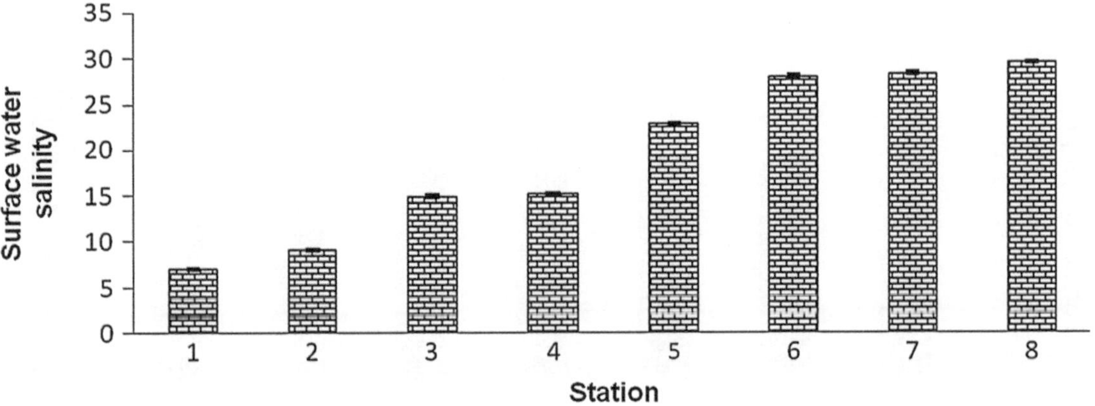

Fig. 7B.7 Surface water salinity (‰) in the selected stations during April 2013

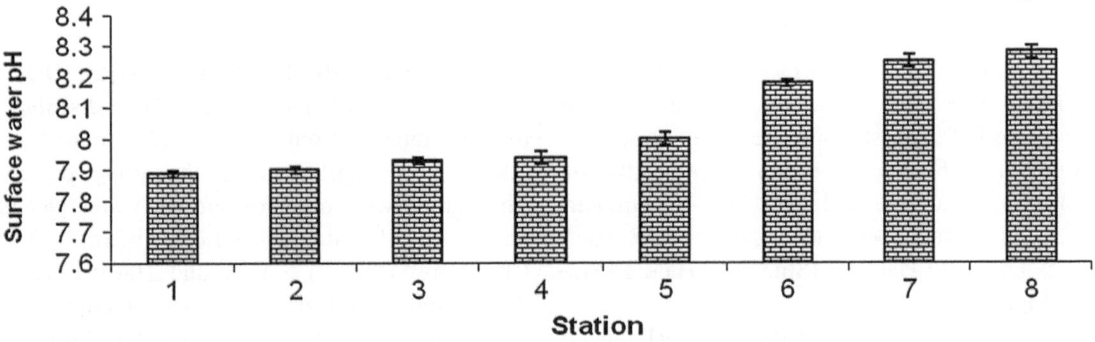

Fig. 7B.8 Surface water pH in the selected stations during February 2013

5.1.3 Surface Water pH

In February 2013, the surface water pH ranged from 7.89 ± 0.01 to 8.28 ± 0.02 during the study period. The station-wise order of surface water pH is Frasergaunje (Stn. 8) > Namkhana (Stn. 7) > Sagar Island (Stn. 6) > Chemaguri (Stn. 5) > Kachuberia (Stn. 4) > Lot 8 (Stn. 3) > Diamond Harbour (Stn. 2) > Haldia (Stn. 1) (Fig. 7B.8).

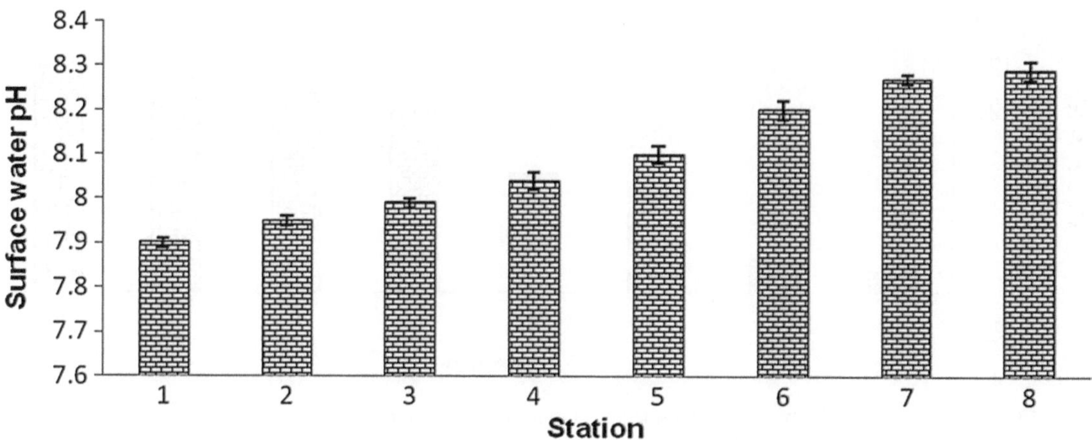

Fig. 7B.9 Surface water pH in the selected stations during March 2013

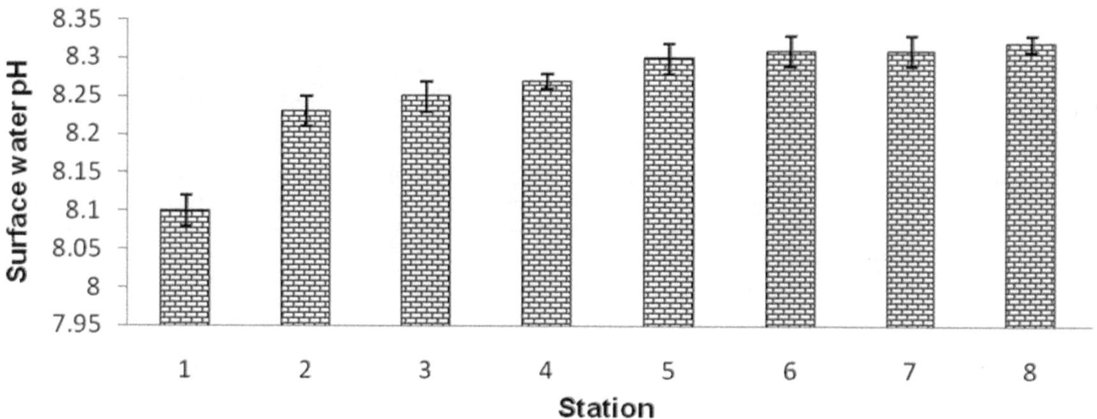

Fig. 7B.10 Surface water pH in the selected stations during April 2013

In March 2013, the surface water pH ranged from 7.90 ± 0.01 to 8.29 ± 0.02 during the study period. The station-wise order of surface water pH is Frasergaunje (Stn. 8) > Namkhana (Stn. 7) > Sagar Island (Stn. 6) > Chemaguri (Stn. 5) > Kachuberia (Stn. 4) > Lot 8 (Stn. 3) > Diamond Harbour (Stn. 2) > Haldia (Stn. 1) (Fig. 7B.9).

In April 2013, the surface water pH ranged from 8.10 ± 0.02 to 8.32 ± 0.01 during the study period. The station-wise order of surface water pH is Frasergaunje (Stn. 8) > Namkhana (Stn. 7) = Sagar Island (Stn. 6) > Chemaguri (Stn. 5) > Kachuberia (Stn. 4) > Lot 8 (Stn. 3) > Diamond Harbour (Stn. 2) > Haldia (Stn. 1) (Fig. 7B.10).

5.1.4 Surface Water Dissolved Oxygen (DO)

In February 2013, the surface water dissolved oxygen ranged from 4.99 ± 1.23 (mg/L) to 5.98 ± 1.35 (mg/L) during the study period. The station-wise order of surface water DO is Sagar Island (Stn. 6) > Lot 8 (Stn. 3) > Chemaguri (Stn. 5) > Diamond Harbour (Stn. 2) > Namkhana (Stn. 7) > Frasergaunje (Stn. 8) > Kachuberia (Stn. 4) > Haldia (Stn. 1) (Fig. 7B.11).

In March 2013, the surface water dissolved oxygen ranged from 4.93 ± 1.23 (mg/L) to 6.02 ± 1.03 (mg/L) during the study period. The station-wise order of surface water DO is Namkhana (Stn. 7) > Sagar Island (Stn. 6) > Chemaguri (Stn. 5) > Kachuberia (Stn. 4) >

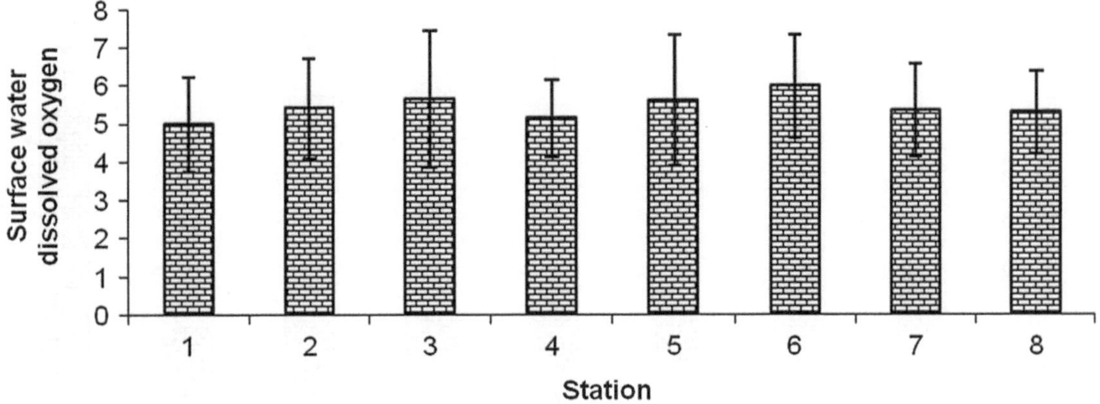

Fig. 7B.11 Surface water DO (mg/L) in the selected stations during February 2013

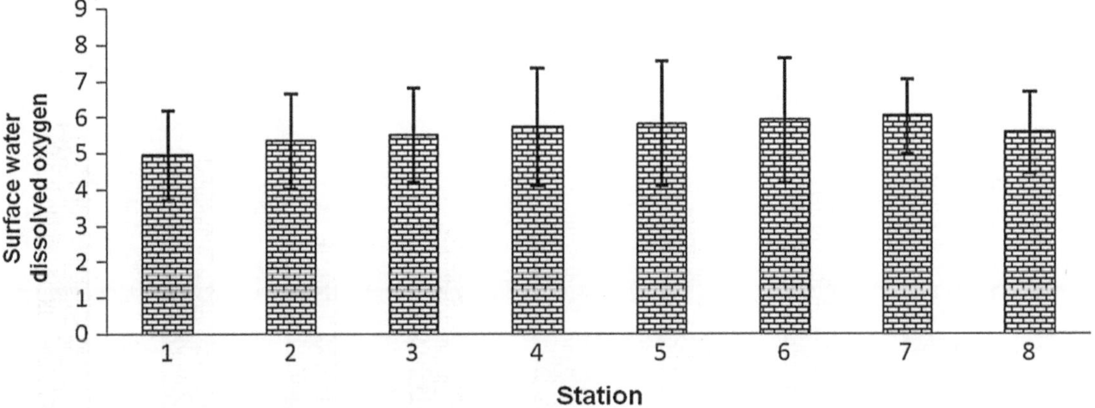

Fig. 7B.12 Surface water DO (mg/L) in the selected stations during March 2013

Frasergaunje (Stn. 8) > Lot 8 (Stn. 3) > Diamond Harbour (Stn. 2) > Haldia (Stn. 1) (Fig. 7B.12).

In April 2013, the surface water dissolved oxygen ranged from 4.88 ± 1.33 (mg/L) to 5.45 ± 1.13 (mg/L) during the study period. The station-wise order of surface water DO is Chemaguri (Stn. 5) > Sagar Island (Stn. 6) > Kachuberia (Stn. 4) > Frasergaunje (Stn. 8) > Diamond Harbour (Stn. 2) > Lot 8 (Stn. 3) > Haldia (Stn. 1) > Namkhana (Stn. 7) (Fig. 7B.13)

5.1.5 Surface Water Nitrate

In February 2013, the surface water nitrate ranged from 18.32 ± 1.04 (mg/L) to 29.88 ± 1.31 (mg/L) during the study period. The station-wise order of surface water nitrate is Frasergaunje (Stn. 8) > Haldia (Stn. 1) > Namkhana (Stn. 7) > Lot 8 (Stn. 3) > Diamond Harbour (Stn. 2) > Kachuberia (Stn. 4) > Chemaguri (Stn. 5) > Sagar Island (Stn. 6) (Fig. 7B.14).

In March 2013, the surface water nitrate ranged from 15.30 ± 0.76 (mg/L) to 26.33 ± 1.04 (mg/L) during the study period. The station-wise order of surface water nitrate is Haldia (Stn. 1) > Frasergaunje (Stn. 8) > Namkhana (Stn. 7) > Lot 8 (Stn. 3) > Diamond Harbour (Stn. 2) > Kachuberia (Stn. 4) > Chemaguri (Stn. 5) > Sagar Island (Stn. 6) (Fig. 7B.15).

In April 2013, the surface water nitrate ranged from 14.21 ± 0.76 (mg/L) to 24.33 ± 1.01

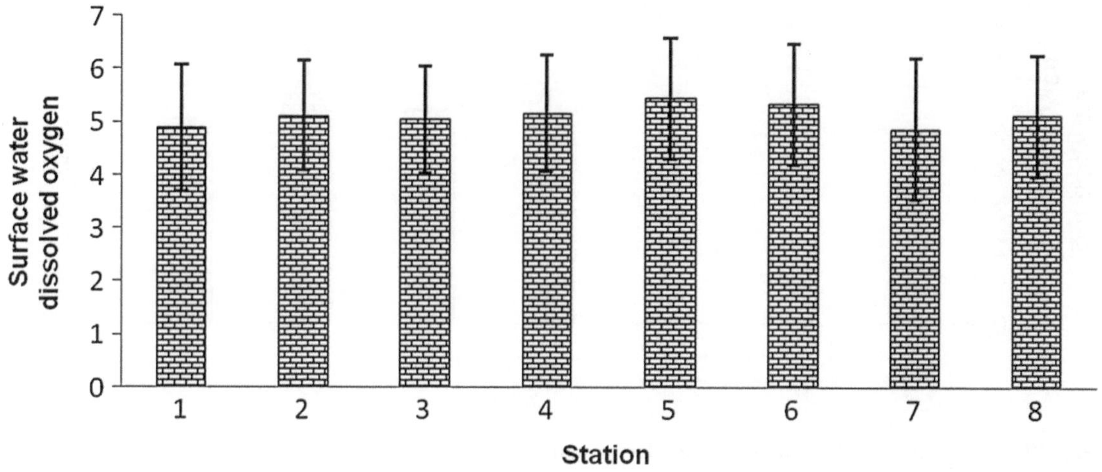

Fig. 7B.13 Surface water DO (mg/L) in the selected stations during April 2013

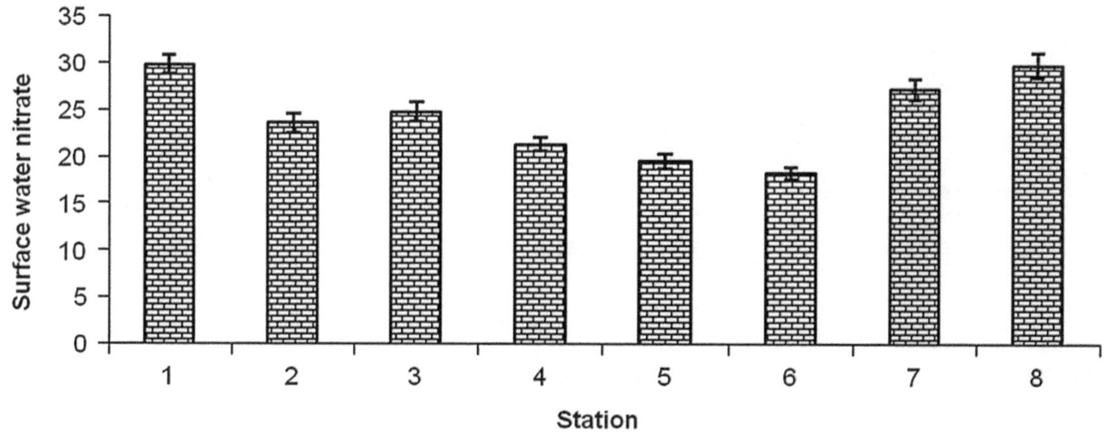

Fig. 7B.14 Surface water nitrate (mg/L) in the selected stations during February 2013

(mg/L) during the study period. The station-wise order of surface water nitrate is Haldia (Stn. 1) > Frasergaunje (Stn. 8) > Namkhana (Stn. 7) > Diamond Harbour (Stn. 2) > Lot 8 (Stn. 3) > Chemaguri (Stn. 5) > Kachuberia (Stn. 4) > Sagar Island (Stn. 6) (Fig. 7B.16).

5.1.6 Surface Water Phosphate

In February 2013, the surface water phosphate ranged from 1.32 ± 0.49 (mg/L) to 3.14 ± 0.91 (mg/L) during the study period. The station-wise order of surface water phosphate is Haldia (Stn. 1) > Diamond Harbour (Stn. 2) > Lot 8 (Stn. 3) > Kachuberia (Stn. 4) > Frasergaunje (Stn. 8) > Chemaguri (Stn. 5) > Namkhana (Stn. 7) > Sagar Island (Stn. 6) (Fig. 7B.17).

In March 2013, the surface water phosphate ranged from 1.29 ± 0.10 (mg/L) to 2.83 ± 0.81 (mg/L) during the study period. The station-wise order of surface water phosphate is Haldia (Stn. 1) > Diamond Harbour (Stn. 2) > Lot 8 (Stn. 3) > Kachuberia (Stn. 4) > Frasergaunje (Stn. 8) > Chemaguri (Stn. 5) > Namkhana (Stn. 7) > Sagar Island (Stn. 6) (Fig. 7B.18).

In April 2013, the surface water phosphate ranged from 1.09 ± 0.11 (mg/L) to 2.11 ± 0.81 (mg/L) during the study period. The station-wise order of surface water phosphate is Haldia (Stn.

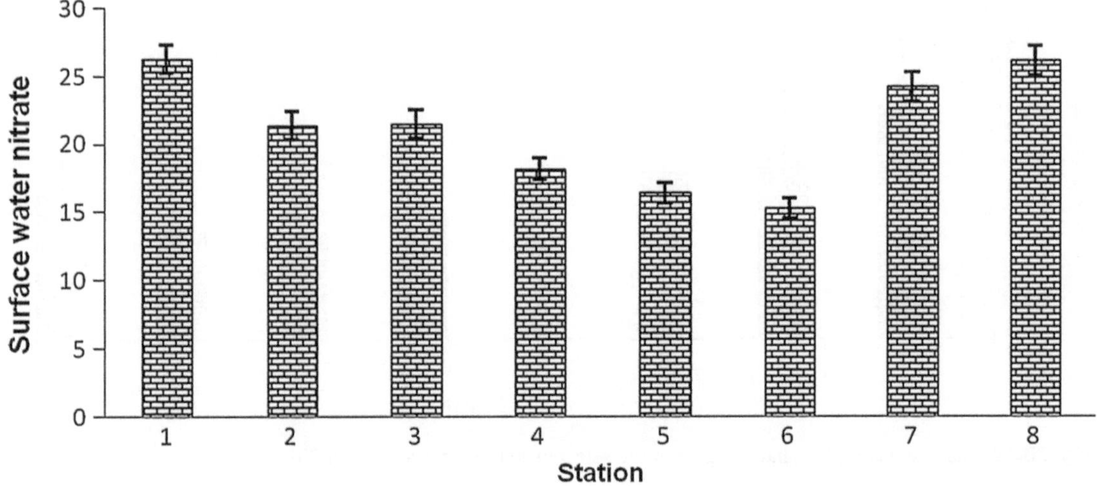

Fig. 7B.15 Surface water nitrate (mg/L) in the selected stations during March 2013

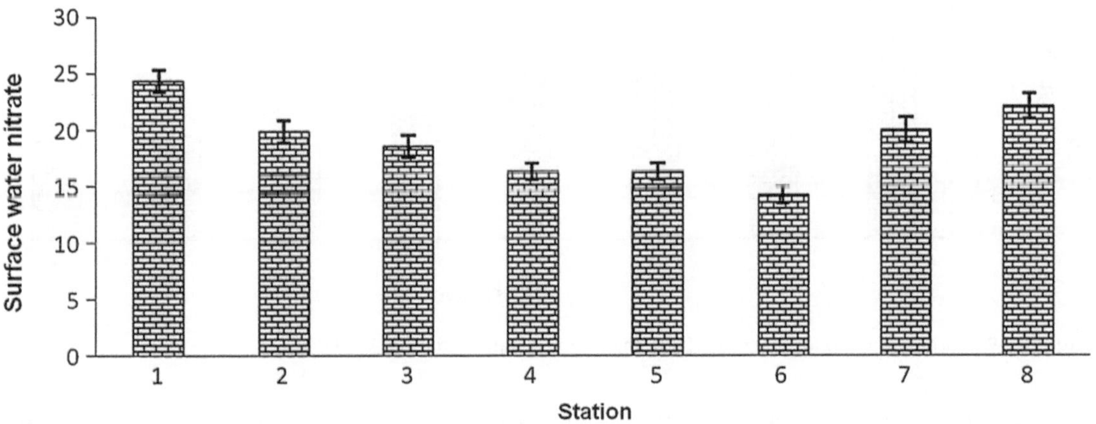

Fig. 7B.16 Surface water nitrate (mg/L) in the selected stations during March 2013

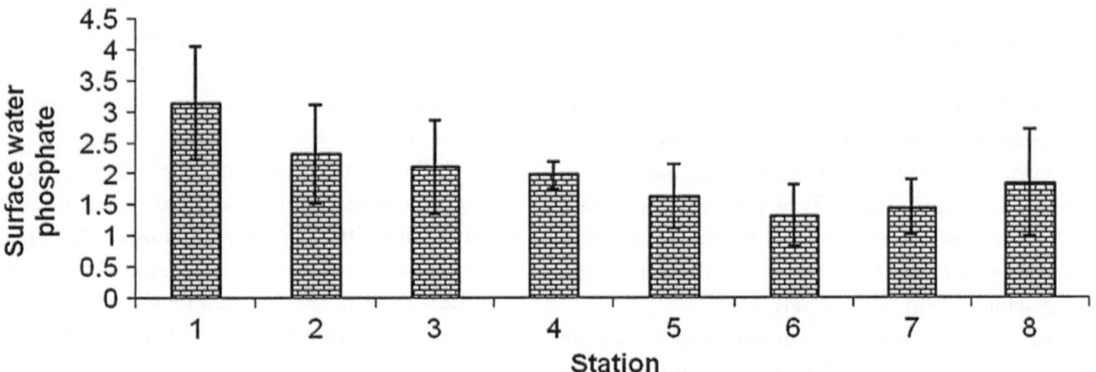

Fig. 7B.17 Surface water phosphate (mg/L) in the selected stations during February 2013

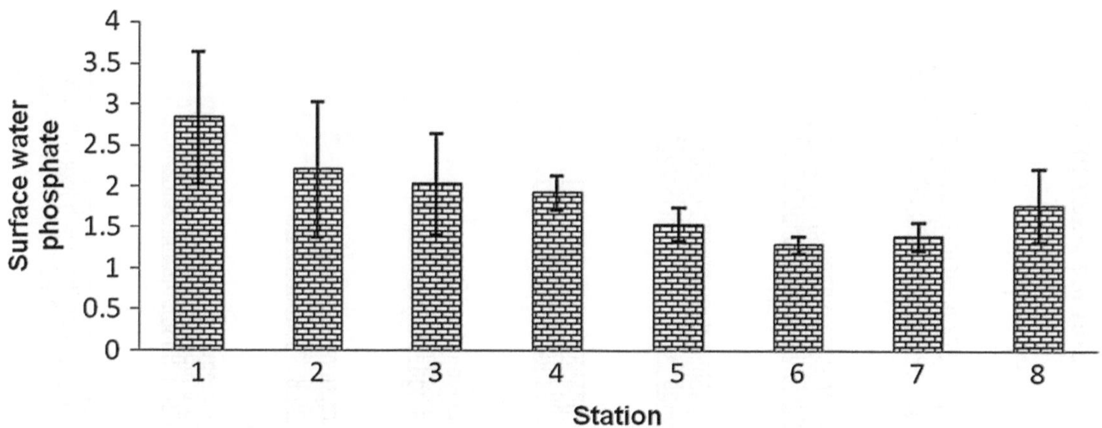

Fig. 7B.18 Surface water phosphate (mg/L) in the selected stations during March 2013

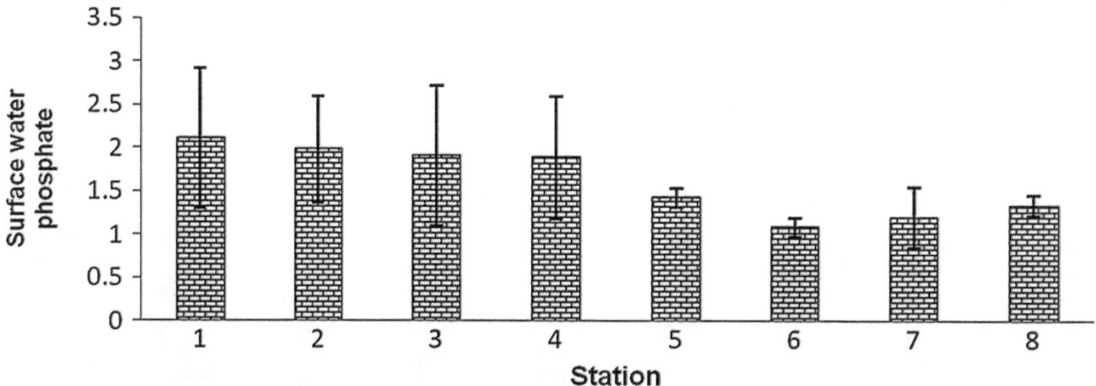

Fig. 7B.19 Surface water phosphate (mg/L) in the selected stations during April 2013

1) > Diamond Harbour (Stn. 2) > Lot 8 (Stn. 3) > Kachuberia (Stn. 4) > Chemaguri (Stn. 5) > Frasergaunje (Stn. 8) > Namkhana (Stn. 7) > Sagar Island (Stn. 6) (Fig. 7B.19).

5.1.7 Surface Water Silicate

In February 2013, the surface water silicate ranged from 49.84 ± 3.42 (mg/L) to 81.22 ± 3.93 (mg/L) during the study period. The station-wise order of surface water silicate is Sagar Island (Stn. 6) > Namkhana (Stn. 7) > Frasergaunje (Stn. 8) > Chemaguri (Stn. 5) > Kachuberia (Stn. 4) > Lot 8 (Stn. 3) > Diamond Harbour (Stn. 2) > Haldia (Stn. 1) (Fig. 7B.20).

In March 2013, the surface water silicate ranged from 44.32 ± 3.42 (mg/L) to 76.57 ± 1.99 (mg/L) during the study period. The station-wise order of surface water silicate is Sagar Island (Stn. 6) > Namkhana (Stn. 7) > Chemaguri (Stn. 5) > Frasergaunje (Stn. 8) > Lot 8 (Stn. 3) > Diamond Harbour (Stn. 2) > Kachuberia (Stn. 4) > Haldia (Stn. 1) (Fig. 7B.21).

In April 2013, the surface water silicate ranged from 41.90 ± 3.42 (mg/L) to 66.44 ± 2.05 (mg/L) during the study period. The station-wise order of surface water silicate is Namkhana (Stn. 7) > Frasergaunje (Stn. 8) > Sagar Island (Stn. 6) > Chemaguri (Stn. 5) > Lot 8 (Stn. 3) > Kachuberia (Stn. 4) > Diamond Harbour (Stn. 2) > Haldia (Stn. 1) (Fig. 7B.22).

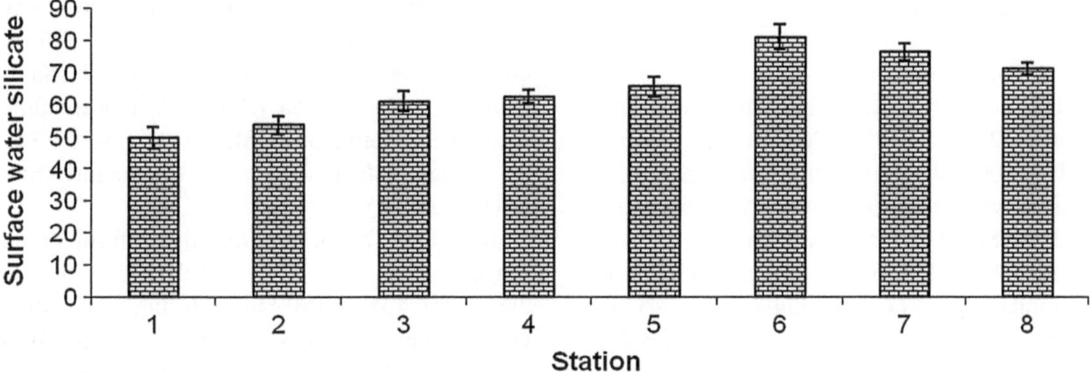

Fig. 7B.20 Surface water silicate (mg/L) in the selected stations during February 2013

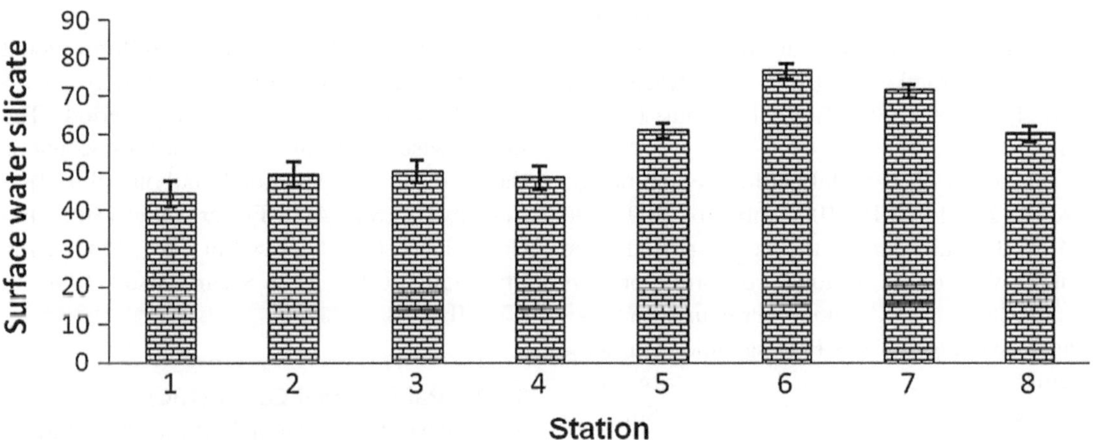

Fig. 7B.21 Surface water silicate (mg/L) in the selected stations during March 2013

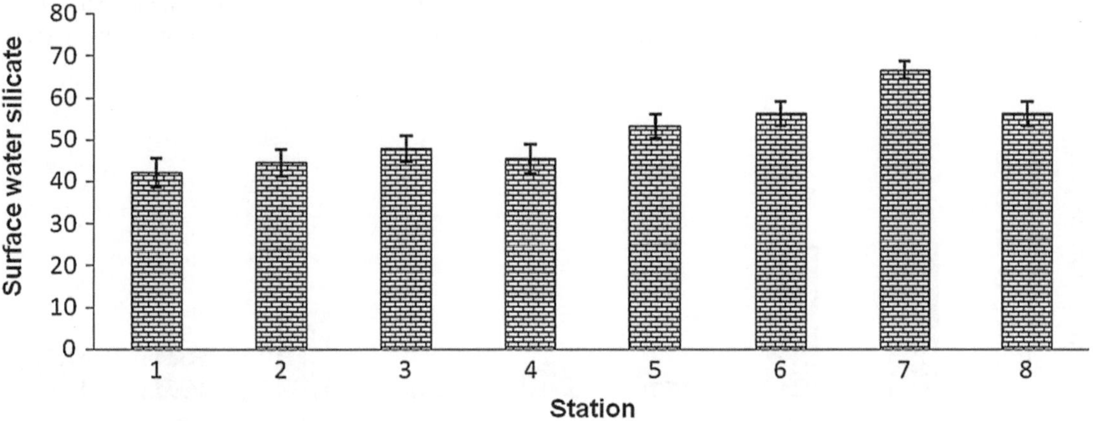

Fig. 7B.22 Surface water silicate (mg/L) in the selected stations during April 2013

5.2 Microbial Load

5.2.1 Total Coliform

In February 2013, the total coliform count ranged from 22 ± 4 (MPN/100 ml) to 391 ± 80 (MPN/100 ml) during the study period. The station-wise order of total coliform count is Haldia (Stn. 1) > Diamond Harbour (Stn. 2) > Sagar Island (Stn. 6) > Frasergaunje (Stn. 8) > Namkhana (Stn. 7) > Kachuberia (Stn. 4) > Lot 8 (Stn. 3) > Chemaguri (Stn. 5) (Fig. 7B.23)

In March 2013, the total coliform count ranged from 20 ± 5 (MPN/100 ml) to 356 ± 80 (MPN/100 ml) during the study period. The station-wise order of total coliform count is Haldia (Stn. 1) > Diamond Harbour (Stn. 2) > Sagar Island (Stn. 6) > Frasergaunje (Stn. 8) > Namkhana (Stn. 7) > Kachuberia (Stn. 4) > Lot 8 (Stn. 3) > Chemaguri (Stn. 5) (Fig. 7B.24).

In April 2013, the total coliform count ranged from 12 ± 1 (MPN/100 ml) to 309 ± 80 (MPN/100 ml) during the study period. The station-wise order of total coliform count is Haldia (Stn. 1) > Diamond Harbour (Stn. 2) > Sagar Island (Stn. 6) > Frasergaunje (Stn. 8) > Namkhana (Stn. 7) > Kachuberia (Stn. 4) > Chemaguri (Stn. 5) > Lot 8 (Stn. 3) (Fig. 7B.25).

5.2.2 Faecal Coliform (FC)

In February 2013, the faecal coliform count ranged from 2 ± 1 (MPN/100 ml) to 354 ± 190 (MPN/100 ml) during the study period. The station-wise order of faecal coliform count is Haldia (Stn. 1) > Diamond Harbour (Stn. 2) > Kachuberia (Stn. 4) > Frasergaunje (Stn. 8) > Namkhana (Stn. 7) > Lot 8 (Stn. 3) > Sagar Island (Stn. 6) > Chemaguri (Stn. 5) (Fig. 7B.26).

In March 2013, the faecal coliform count ranged from 10 ± 1 (MPN/100 ml) to 300 ± 80 (MPN/100 ml) during the study period. The station-wise order of faecal coliform count is Haldia (Stn. 1) > Diamond Harbour (Stn. 2) > Kachuberia (Stn. 4) > Frasergaunje (Stn. 8) > Namkhana (Stn. 7) > Lot 8 (Stn. 3) > Chemaguri (Stn. 5) = Sagar Island (Stn. 6) (Fig. 7B.27).

In April 2013, the faecal coliform count ranged from 5 ± 1 (MPN/100 ml) to 267 ± 75 (MPN/100 ml) during the study period. The station-wise order of faecal coliform count is Haldia (Stn. 1) > Diamond Harbour (Stn. 2) > Kachuberia (Stn. 4) > Frasergaunje (Stn. 8) > Namkhana (Stn. 7) > Lot 8 (Stn. 3) > Chemaguri (Stn. 5) > Sagar Island (Stn. 6) (Fig. 7B.28).

5.2.3 Total Bacterial Count (TBC)

In February 2013, the total bacterial count ranged from 5.76 ± 0.01 (CFU × 10^6/ml) to 17.11 ± 0.12 (CFU × 10^6/ml) during the study period. The station-wise order of TBC is Frasergaunje (Stn. 8) > Namkhana (Stn. 7) > Kachuberia (Stn. 4) > Lot 8 (Stn.

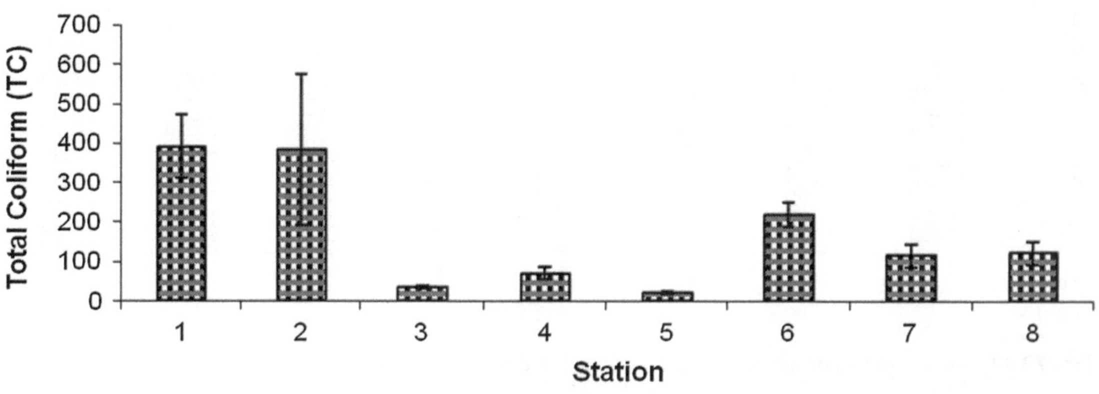

Fig. 7B.23 Total coliform count (MPN/100 ml) in the selected stations during February 2013

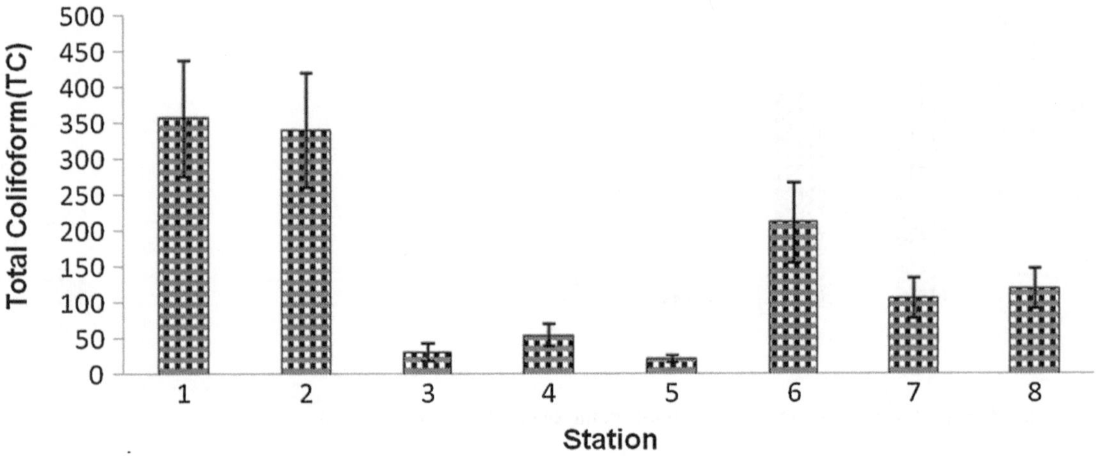

Fig. 7B.24 Total coliform count (MPN/100 ml) in the selected stations during March 2013

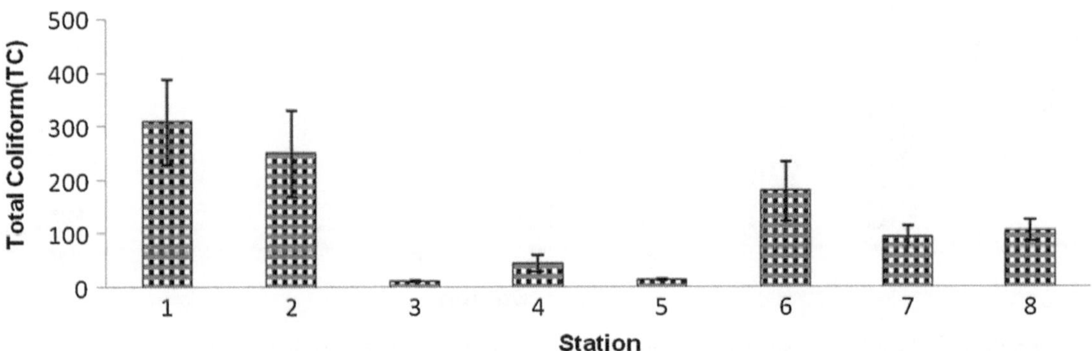

Fig. 7B.25 Total coliform count (MPN/100 ml) in the selected stations during April 2013

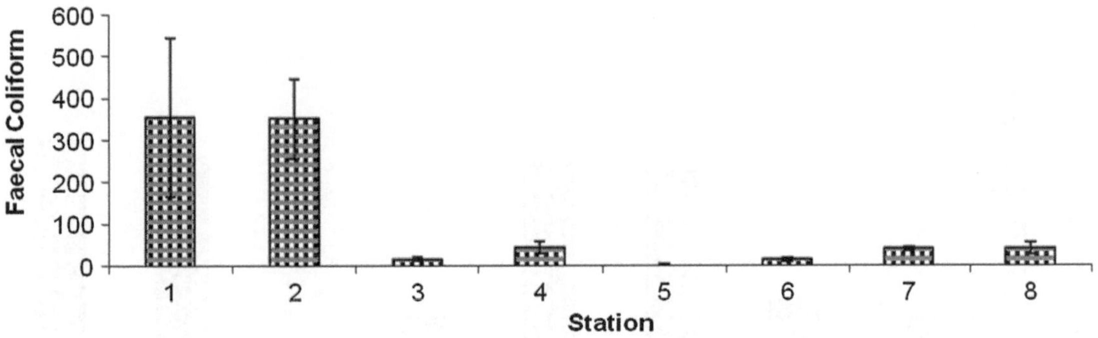

Fig. 7B.26 Faecal coliform count (MPN/100 ml) in the selected stations during February 2013

3) > Diamond Harbour (Stn. 2) > Haldia (Stn. 1) > Chemaguri (Stn. 5) > Sagar Island (Stn. 6) (Fig. 7B.29).

In March 2013, the total bacterial count ranged from 0.68 ± 0.01 (CFU $\times$ 10^6/ml) to 16.94 ± 0.02 (CFU $\times$ 10^6/ml) during the study period. The station-wise order of TBC is Frasergaunje (Stn. 8) > Namkhana (Stn. 7) > Lot 8 (Stn. 3) > Kachuberia (Stn. 4) > Diamond Harbour (Stn. 2) > Haldia (Stn. 1) >

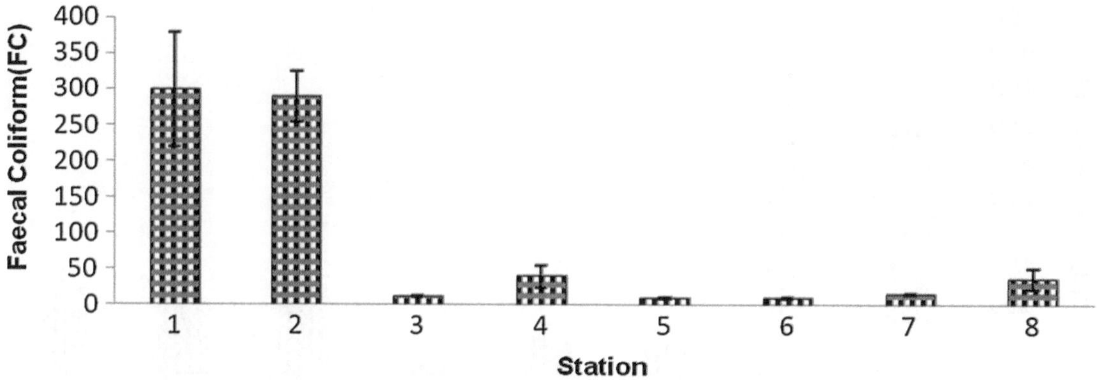

Fig. 7B.27 Faecal coliform count (MPN/100 ml) in the selected stations during March 2013

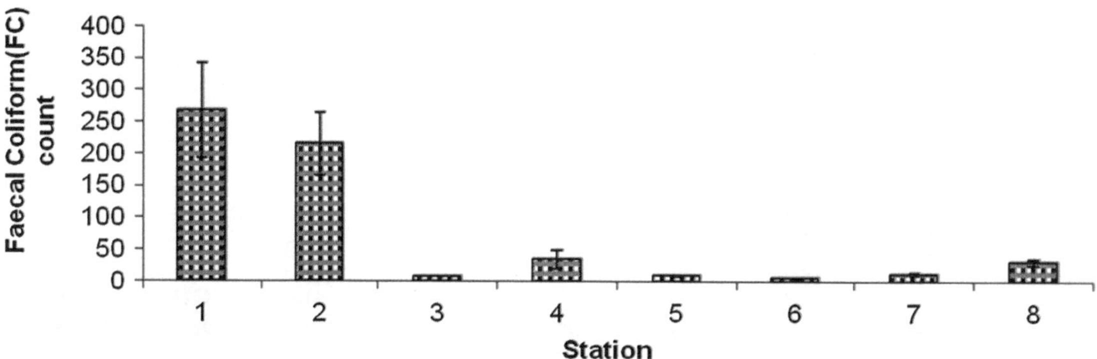

Fig. 7B.28 Faecal coliform count (MPN/100 ml) in the selected stations during April 2013

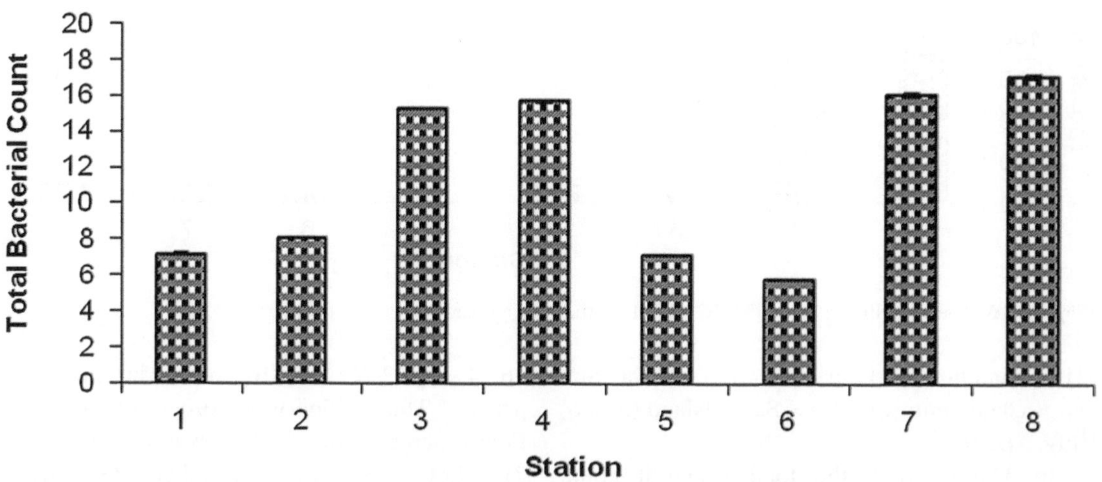

Fig. 7B.29 TBC (CFU $\times$ 10^6/ml) in the selected stations during February 2013

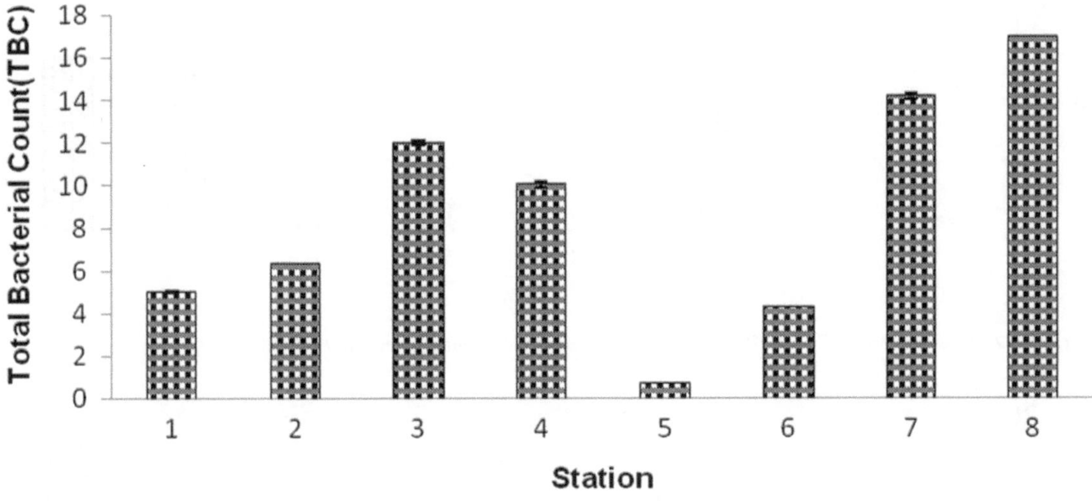

Fig. 7B.30 TBC (CFU × 10^6/ml) in the selected stations during March 2013

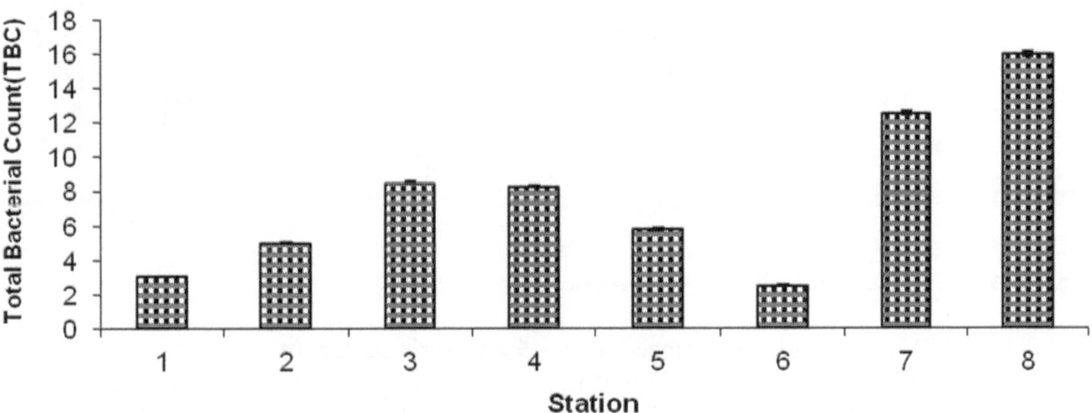

Fig. 7B.31 TBC (CFU × 10^6/ml) in the selected stations during April 2013

Sagar Island (Stn. 6) > Chemaguri (Stn. 5) (Fig. 7B.30).

In April 2013, the total bacterial count ranged from 2.50 ± 0.01 (CFU × 10^6/ml) to 16.00 ± 0.15 (CFU × 10^6/ml) during the study period. The station-wise order of TBC is Frasergaunje (Stn. 8) > Namkhana (Stn. 7) > Lot 8 (Stn. 3) > Kachuberia (Stn. 4) > Chemaguri (Stn. 5) > Diamond Harbour (Stn. 2) > Haldia (Stn. 1) > Sagar Island (Stn. 6) (Fig. 7B.31).

5.2.4 *E. coli* Count

In February 2013, the *E. coli* count ranged from 0.01 ± 0.001 (CFU × 10^6/ml) to 19.89 ± 0.12 (CFU × 10^6/ml) during the study period. The station-wise order of *E. coli* count is Frasergaunje (Stn. 8) > Namkhana (Stn. 7) > Lot 8 (Stn. 3) > Chemaguri (Stn. 5) > Haldia (Stn. 1) > Sagar Island (Stn. 6) > Diamond Harbour (Stn. 2) > Kachuberia (Stn. 4) (Fig. 7B.32).

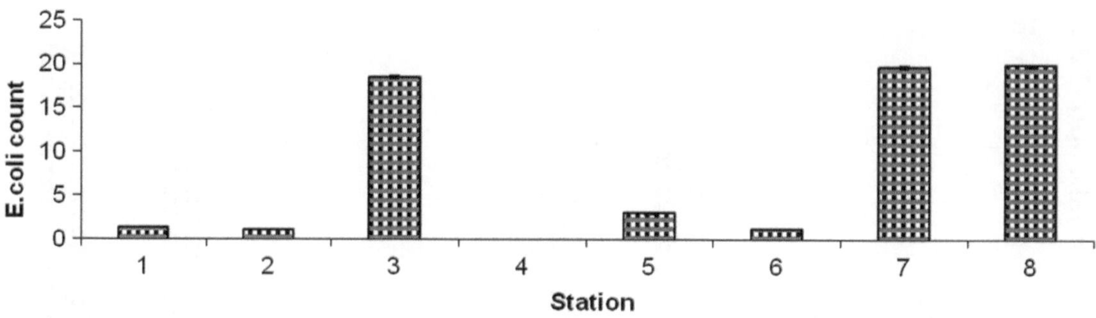

Fig. 7B.32 *E. coli* count (CFU × 10^6/ml) in the selected stations during February 2013

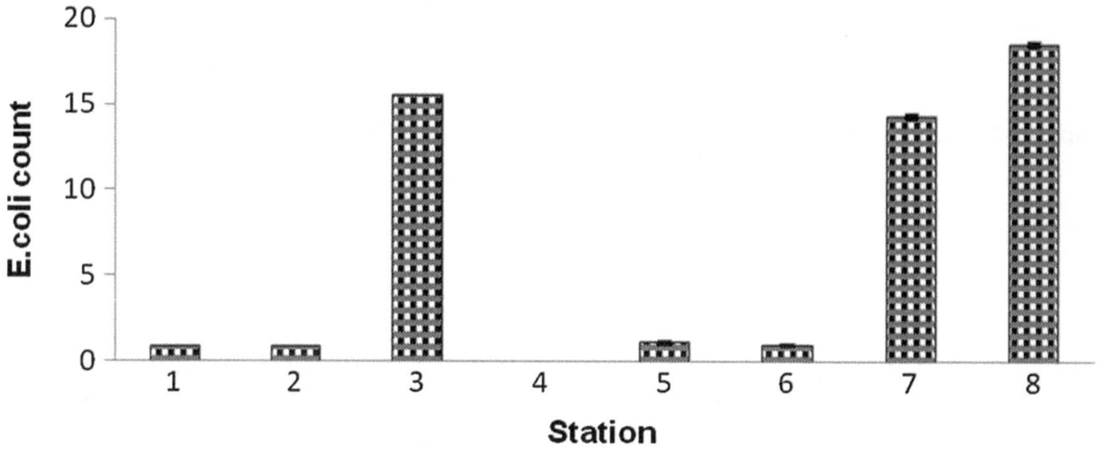

Fig. 7B.33 *E. coli* count (CFU × 10^6/ml) in the selected stations during March 2013

In March 2013, the *E. coli* count ranged from 0.002 ± 0.001 (CFU × 10^6/ml) to 18.55 ± 0.12 (CFU × 10^6/ml) during the study period. The station-wise order of *E. coli* count is Frasergaunje (Stn. 8) > Lot 8 (Stn. 3) > Namkhana (Stn. 7) > Chemaguri (Stn. 5) > Sagar Island (Stn. 6) > Haldia (Stn. 1) > Diamond Harbour (Stn. 2) > Kachuberia (Stn. 4) (Fig. 7B.33).

In April 2013, the *E. coli* count ranged from 0.01 ± 0.001 (CFU × 10^6/ml) to 17.89 ± 0.05 (CFU × 10^6/ml) during the study period. The station-wise order of *E. coli* count is Frasergaunje (Stn. 8) > Namkhana (Stn. 7) > Lot 8 (Stn. 3) > Chemaguri (Stn. 5) > Diamond Harbour (Stn. 2) > Sagar Island (Stn. 6) > Haldia (Stn. 1) > Kachuberia (Stn. 4) (Fig. 7B.34).

5.2.5 *Vibrio* sp. Count

In February 2013, the *Vibrio* sp. count ranged from 0.10 ± 0.01 (CFU × 10^6/ml) to 2.20 ± 0.09 (CFU × 10^6/ml) during the study period. The station-wise order of *Vibrio* sp. count is Chemaguri (Stn. 5) > Sagar Island (Stn. 6) > Lot 8 (Stn. 3) > Namkhana (Stn. 7) > Frasergaunje (Stn. 8) > Haldia (Stn. 1) > Diamond Harbour (Stn. 2) > Kachuberia (Stn. 4) (Fig. 7B.35).

In March 2013, the *Vibrio* sp. count ranged from 0.00 (CFU × 10^6/ml) to 16.0 ± 0.11 (CFU × 10^6/ml) during the study period. The station-wise order of *Vibrio* sp. count is Chemaguri (Stn. 5) > Namkhana (Stn. 7) > Frasergaunje (Stn. 8) > Lot 8 (Stn. 3) > Haldia (stn1) > Diamond Harbour (Stn. 2) > Kachuberia (Stn. 4) > Sagar Island (Stn. 6) (Fig. 7B.36).

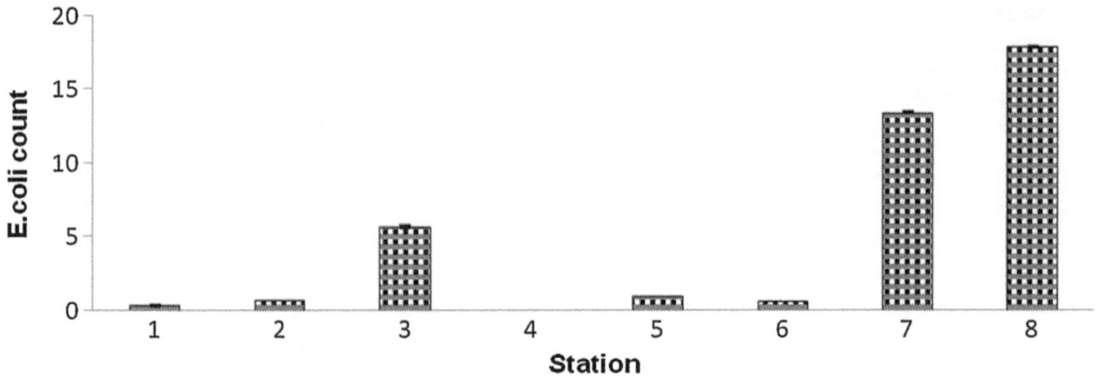

Fig. 7B.34 *E. coli* count (CFU × 10^6/ml) in the selected stations during April 2013

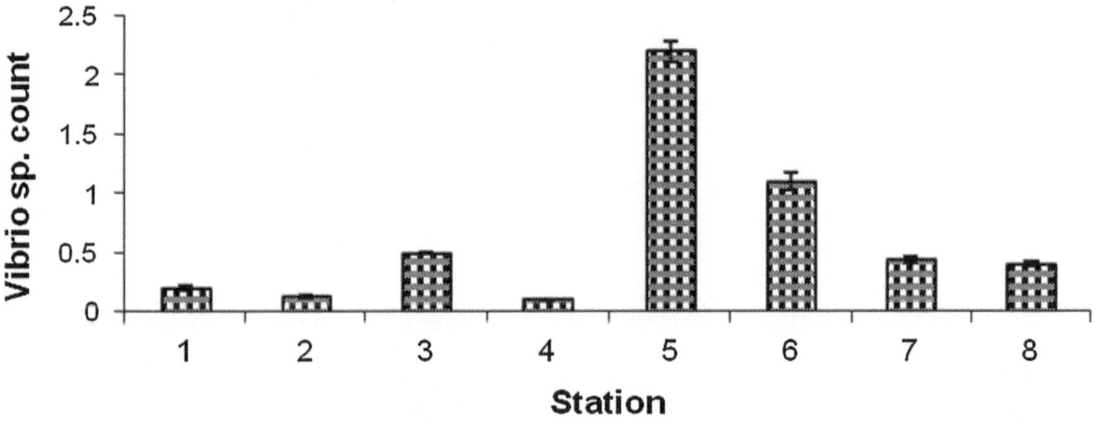

Fig. 7B.35 *Vibrio* sp. count (CFU × 10^6/ml) in the selected stations during February 2013

In April 2013, the *Vibrio* sp. count ranged from 0.00 (CFU × 10^6/ml) to 0.90 ± 0.01 (CFU × 10^6/ml) during the study period. The station-wise order of *Vibrio* sp. count is Chemaguri (Stn. 5) > Namkhana (Stn. 7) > Frasergaunje (Stn. 8) > Haldia (stn1) > Lot 8 (Stn. 3) > Diamond Harbour (Stn. 2) = Kachuberia (Stn. 4) > Sagar Island (Stn. 6) (Fig. 7B.37).

5.2.6 *Streptococcus* sp. Count

In February 2013, the *Streptococcus* sp. count ranged from 0.00 (CFU × 10^6/ml) to 0.33 ± 0.02 (CFU × 10^6/ml) during the study period. The station-wise order of *Streptococcus* sp. count is Frasergaunje (Stn. 8) > Namkhana (Stn. 7) > Haldia (Stn. 1) > Diamond Harbour (Stn. 2) = Lot 8 (Stn. 3) > Kachuberia (Stn. 4) = Chemaguri (Stn. 5) = Sagar Island (Stn. 6) (Fig. 7B.38).

In March 2013, the *Streptococcus* sp. count ranged from 0.00 (CFU × 10^6/ml) to 0.30 ± 0.01 (CFU × 10^6/ml) during the study period. The station-wise order of *Streptococcus* sp. count is Frasergaunje (Stn. 8) > Haldia (stn1) > Namkhana (Stn. 7) > Diamond Harbour (Stn. 2) > Lot 8 (Stn. 3) > Kachuberia (Stn. 4) = Chemaguri (Stn. 5) = Sagar Island (Stn. 6) (Fig. 7B.39).

In April 2013, the *Streptococcus* sp. count ranged from 0.00 (CFU × 10^6/ml) to 0.30 ± 0.01 (CFU × 10^6/ml) during the study period. The station-wise order of *Streptococcus* sp. count is Frasergaunje (Stn. 8) > Namkhana (Stn. 7) = Haldia (Stn. 1) > Diamond Harbour (Stn. 2) > Lot 8 (Stn. 3) > Kachuberia (Stn.

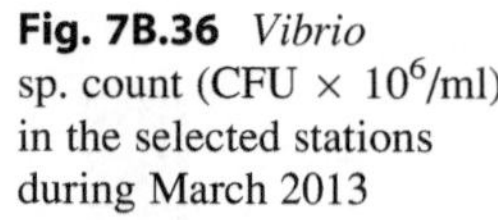

Fig. 7B.36 *Vibrio* sp. count (CFU × 10^6/ml) in the selected stations during March 2013

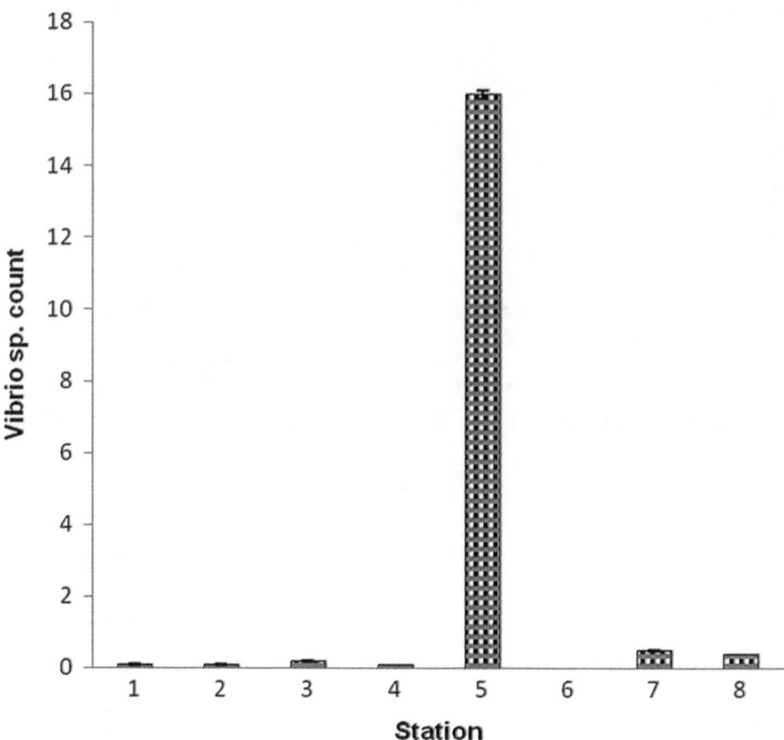

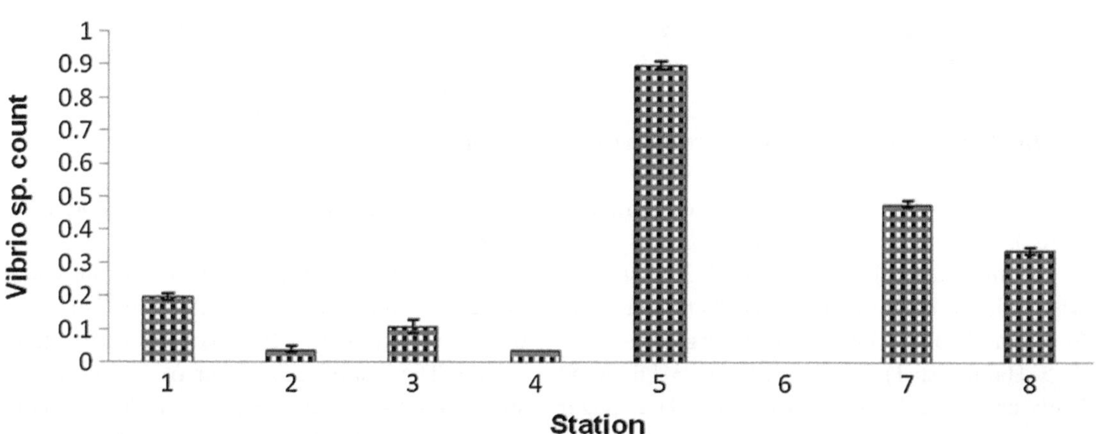

Fig. 7B.37 *Vibrio* sp. count (CFU × 10^6/ml) in the selected stations during April 2013

4) = Chemaguri (Stn. 5) = Sagar Island (Stn. 6) (Fig. 7B.40).

5.2.7 *Salmonella* sp. Count

In February 2013, the *Salmonella* sp. count ranged from 0.02 ± 0.001 (CFU × 10^6/ml) to 0.82 ± 0.04 (CFU × 10^6/ml) during the study period. The station-wise order of *Salmonella* sp. count is Frasergaunje (Stn. 8) > Namkhana (Stn. 7) > Haldia (Stn. 1) > Diamond Harbour (Stn. 2) > Lot 8 (Stn. 3) > Chemaguri (Stn. 5) > Sagar Island (Stn. 6) > Kachuberia (Stn. 4) (Fig. 7B.41).

In March 2013, the *Salmonella* sp. count ranged from 1.65 ± 0.01 (CFU × 10^6/ml) to 90.8 ± 0.03 (CFU × 10^6/ml) during the study

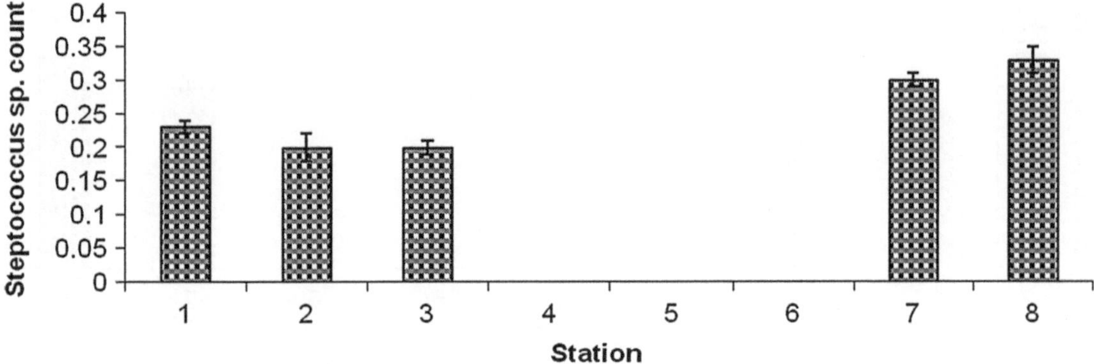

Fig. 7B.38 *Streptococcus* sp. count (CFU × 10^6/ml) in the selected stations during February 2013

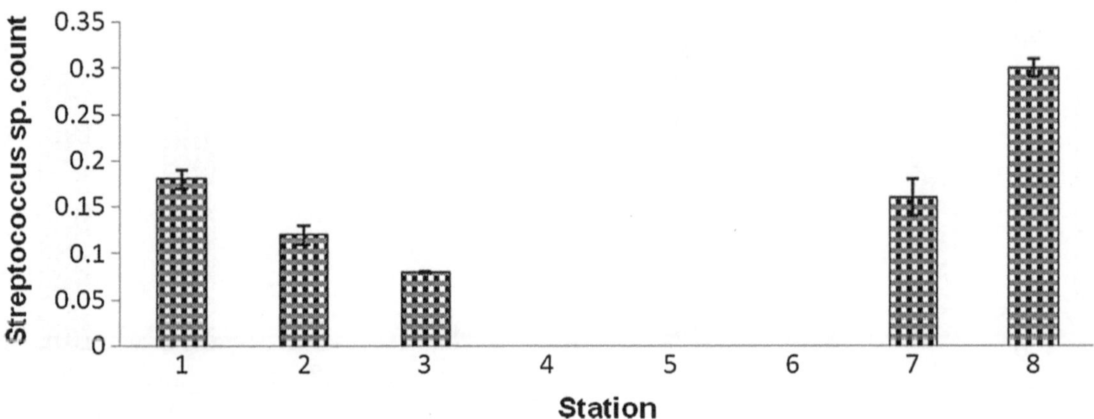

Fig. 7B.39 *Streptococcus* sp. count (CFU × 10^6/ml) in the selected stations during March 2013

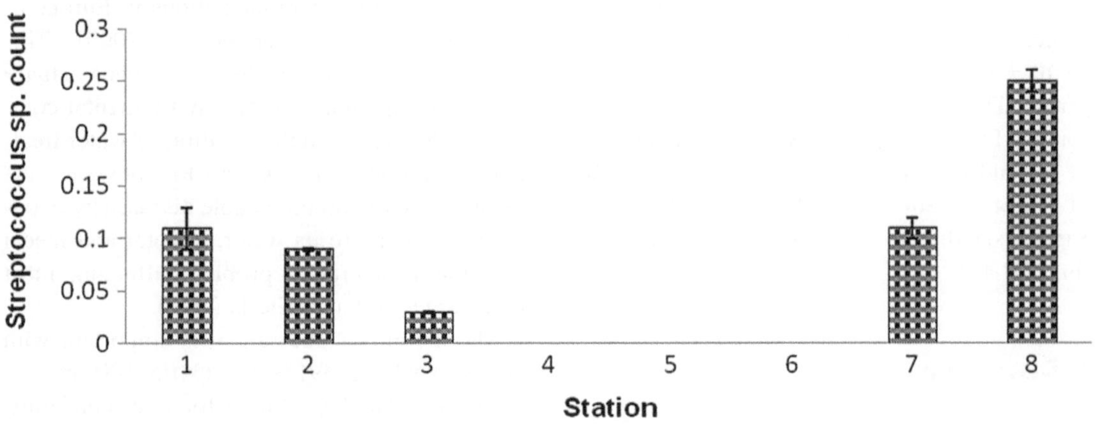

Fig. 7B.40 *Streptococcus* sp. count (CFU × 10^6/ml) in the selected stations during April 2013

period. The station-wise order of *Salmonella* sp. count is Lot 8 (Stn. 3) > Frasergaunje (Stn. 8) > Namkhana (Stn. 7) > Haldia (Stn. 1) > Diamond Harbour (Stn. 2) > Chemaguri (Stn. 5) > Sagar Island (Stn. 6) > Kachuberia (Stn. 4) (Fig. 7B.42).

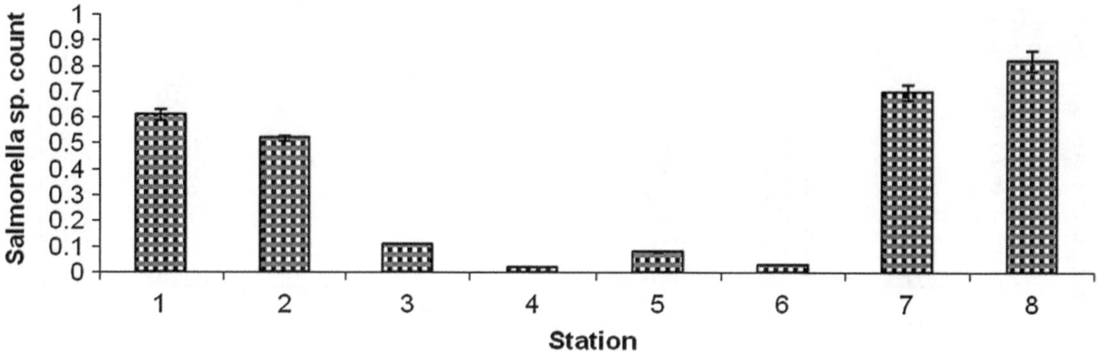

Fig. 7B.41 *Salmonella* sp. count (CFU × 10^6/ml) in the selected stations during February 2013

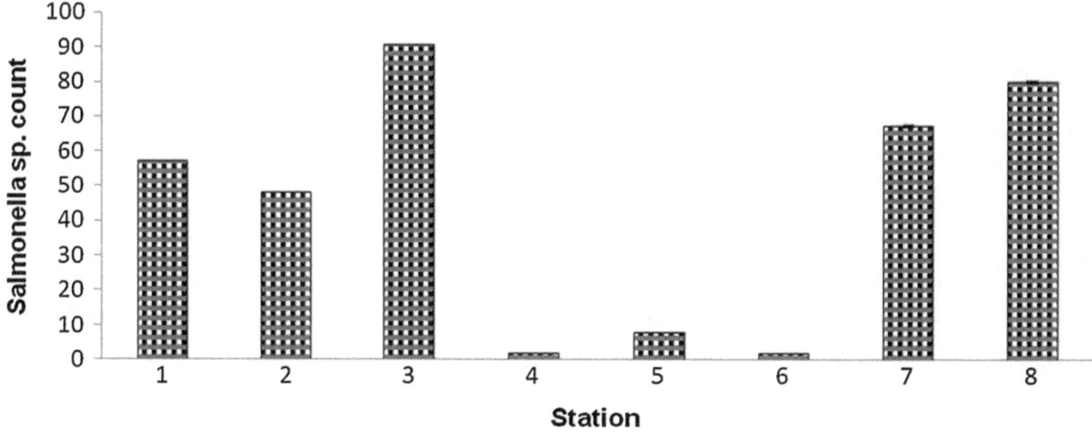

Fig. 7B.42 *Salmonella* sp. count (CFU × 10^6/ml) in the selected stations during March 2013

In April 2013, the *Salmonella* sp. count ranged from 1.25 ± 0.01 (CFU × 10^6/ml) to 75.40 ± 0.13 (CFU × 10^6/ml) during the study period. The station-wise order of *Salmonella* sp. count is Frasergaunje (Stn. 8) > Namkhana (Stn. 7) > Haldia (stn1) > Diamond Harbour (Stn. 2) > Lot 8 (Stn. 3) > Chemaguri (Stn. 5) > Sagar Island (Stn. 6) > Kachuberia (Stn. 4) (Fig. 7B.43).

6. Discussion

Microbiological water quality investigations of lotic and lentic ecosystems are very rare, despite their importance in accompanying the role of large water bodies for cases of recreation, tourism and aquaculture. An attempt to adequately monitor scientific data in large-scale river bodies is a priority to some organizations in Europe as stipulated by Kirschner et al. (2009). This microbiological data in the Sundarbans estuary gives a strong signal to the environmental community to embark on the creation of water treatment facilities in order to prevent the transmission of communicable diseases by population that explore its water, as total and faecal pollution is a crucial problem affecting most urban water systems (Eleria 2002).

The monthly values are very important with values reaching 391 ± 80 (MPN/100 ml) in February at Haldia (Stn. 1) for total coliforms. These pathogens will keep on accumulating in the open system (Bell et al. 1994). The values obtained could be spatio-temporarily linked to the number of visitors in this ecosystem and also the role played by point and non-point sources in the biocontamination of aquatic

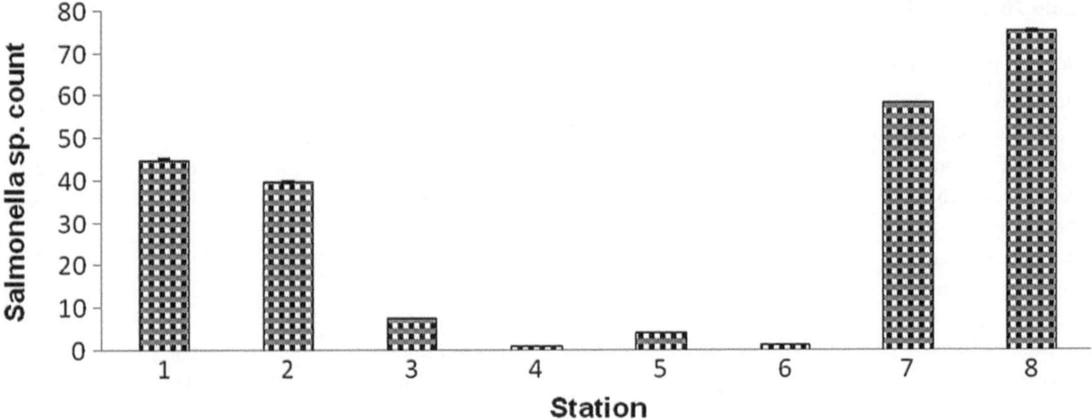

Fig. 7B.43 *Salmonella* sp. count (CFU × 10^6/ml) in the selected stations during April 2013

ecosystem (Cieslak et al. 1993; Kelsey et al. 2004). These pathogens could be free living, particle associated or in an intermediary state, depending on the organic and inorganic condition of the medium as stipulated in the findings of Basemer et al. (2005).

Rainfall–storm water run-off is a significant source of pollutants to the river, which can include bacteria, viruses and sediment, to which the substrate pollutants attached as indicated by Mallin et al. (2000). Storm rainfall characteristics and conditions prior to the storms are significant factors in the transport and concentrations of pollutants in the river. Stream flow–river flow is the primary transport media of faecal coliform bacteria (Christensen et al. 2000).

Among the diseases associated with poor microbial water quality, those causing dehydrating diarrhoea are of critical importance as they could lead to death within 48 h after the initial symptoms as analyzed in the findings of Manja et al. (1982). These extreme cases are more predominant in countries where overcrowding and poor sanitary conditions are the norm (Francy et al. 2002). The presence of faecal coliforms indicates the contamination of water with faecal waste that may contain other harmful or disease-causing microorganisms, including bacteria, viruses, protozoa or other infectious agents (Brewster et al. 1994). Drinking water contaminated with these organisms can cause stomach and intestinal illness including diarrhoea and nausea.

The Hooghly estuary is the lifeline of the highly urbanized city of Kolkata and supports industry of crucial economic importance. The surrounding area is a complex mixture of commercial, industrial, agricultural and residential development. The watershed provides important services for drinking water, wildlife habitat, recreation (swimming, fishing, boating), pilgrimage and transportation. The mixed use of the watershed results in a complex pattern of waste and pollutant input that alters ecosystem health. Microbes play an important role in determining water quality (nutrient concentration, clarity, oxygen levels, pathogen load) by controlling the internal transformations, but their activity is modulated by the system's variable environmental conditions.

The present dissertation focuses the following points:

- There is an increasing trend in surface water temperature, salinity, pH and silicates while approaching from upstream to downstream region. This may be because of the effects of the tidal action from Bay of Bengal, which is in the south of selected stations. The surface water dissolved oxygen concentration, nitrate and phosphate did not show any general spatial trend. The anthropogenic activities basically control the concentration of dissolved

Table 7B.4 ANOVA: showing variation of hydrological parameters between stations and months (for physico-chemical parameters)

Variables	F_{cal}	F_{crit}
Surface water temperature		
Between stations	2.71	2.76
Between months	719.21	3.74
Surface water salinity		
Between stations	169.15	2.76
Between months	72.29	3.74
Surface water pH		
Between stations	10.39	2.76
Between months	23.61	3.74
Surface water dissolved oxygen		
Between stations	3.75	2.76
Between months	9.68	3.74
Surface water nitrate		
Between stations	56.75	2.76
Between months	70.48	3.74
Surface water phosphate		
Between stations	28.34	2.76
Between months	11.09	3.74
Surface water silicate		
Between stations	21.57	2.76
Between months	28.33	3.74

Table 7B.5 ANOVA: showing variation of microbial load between stations and months (for microbial parameters)

Variables	F_{cal}	F_{crit}
Total coliform (TC) count		
Between stations	105.62	2.76
Between months	8.06	3.74
Faecal coliform (FC) count		
Between stations	77.94	2.76
Between months	3.83	3.74
Total bacterial count (TBC)		
Between stations	25.69	2.76
Between months	11.41	3.74
E. coli count		
Between stations	33.28	2.76
Between months	3.47	3.74
Vibrio sp. count		
Between stations	1.58	2.76
Between months	0.91	3.74
Streptococcus sp. count		
Between stations	21.13	2.76
Between months	8.72	3.74
Salmonella sp. count		
Between stations	2.47	2.76
Between months	8.28	3.74

oxygen, nitrate and phosphate through sewage and other waste disposals. Haldia (Stn. 1), Namkhana (Stn. 7) and Frasergaunje (Stn. 8) sustain port, industries, hotels and tourism units and fish landing stations. These point sources generate wastes of complex characters due to which the nitrates and phosphates exhibited comparatively higher values in these stations.

- ANOVA revealed significant monthly variations ($p < 0.01$) in surface water temperature, salinity, pH, dissolved oxygen, nitrate, phosphate and silicate. However, the statistical difference is not significant between stations in case of surface water temperature as F_{cal} (2.71) is less than F_{crit} (2.76). In case of other physico-chemical variables like surface water salinity, pH, DO, nitrate, phosphate and silicate, significant differences between stations were observed (Table 7B.4). All these are related to nature and magnitude of human/anthropogenic activities.
- ANOVA revealed significant monthly variations ($p < 0.01$) in total coliform count, faecal coliform count, total bacterial count, *E. coli* count, *Vibrio* sp. count, *Streptococcus* sp. count and *Salmonella* sp. count (exceptions are observed in case *of E. coli* count and *Vibrio* sp. count). However, the statistical difference is not significant between stations in case of *Vibrio* sp. count as F_{cal} (1.58) is less than F_{crit} (2.76). In case of *Salmonella* sp. count, the statistical difference between stations is also not significant as F_{cal} (2.47) is lower than F_{crit} (2.76) (Table 7B.5).

7. Conclusion

It is very difficult to come to a solid conclusion with a meagre data of 3 months. A long-term study is required (at least 2 years) to monitor the seasonal effects of physico-chemical parameters.

With these snapshots of 3 months, it is very clear that the unplanned urbanization, tourism units and agricultural activities are mainly responsible for deterioration of water quality along the Hooghly estuary stretches. It is interesting to note that the salinity, pH and silicate level increases from upstream to downstream, that is, from Haldia to Frasergaunje. This is exclusively the marine effect of the Bay of Bengal. However, the microbial parameters in some stations like Haldia (Stn. 1), Diamond Harbour (Stn. 2), Namkhana (Stn. 7) and Frasergaunje (Stn. 8) increase abruptly because of release of untreated waste from hotels, fish landing stations and market places. These activities have not only increased the coliform load in the water bodies, but also pathogenic strains, such as *Salmonella* sp. and *Vibrio* sp., have also been observed in stations like Chemaguri (Stn. 5), Namkhana (Stn. 7) and Frasergaunje (Stn. 8) where shrimp culture (*Penaeus monodon)* activities are a major issue. Overstocking of tiger prawn feed and periodic release of the wastewater in the surrounding estuarine is also one of the important reasons for enhanced microbial load in these stations.

Continuous monitoring of the system and strict regulation by concerned government departments and agencies are essential to restore the ecological health of the system.

References

Abdel-Lateff, A., Konig, G. M., Fisch, K. M., Holler, U., Jones, P. G., & Wright, A. D. (2002). New antioxidant hydroquinone derivatives from the algicolous marine fungus *Acremonium* sp. *Journal of Natural Products, 65*(11), 1605–1611.

Abdel-Lateff, A., Klemke, C., Konig, G. M., & Wright, A. D. (2003). Two new xanthone derivatives from the algicolous marine fungus Wardomyces anomalus. *Journal of Natural Products, 66*(5), 706–708.

Anderson, R. J., Wolfe, M. S., & Faulkner, D. J. (1974). Autotoxic antibiotic production by a marine *Chromobacterium. Marine Biology, 27*, 281–285.

Atlas, R. (1998). Chapter 2: Microbial evolution and biodiversity. In *Microbial ecology: Fundamentals and applications* (4th ed., p. 694). Menlo Park: Benjamin Cummings.

Barcina, I., Lebaron, P., & Vives-Rego, J. (1997). Survival of allochthonous bacteria in aquatic systems: A biological approach. *FEMS Microbiology Ecology, 23*, 1–9.

Bhowmik, S. K., Roy, G. N., & Barman, S. (1985). *A list 24 strains of bacteria from the decomposed litter of mangrove areas of Sundarbans complex*. National Symposium on Biology, Utilization and Conservation of Mangroves. Abstract No. 72, p. 38.

Bidle, K. D., & Fletcher, M. (1995). Comparison of - free-living and particle-associated bacterial communities in the Chesapeake Bay by stable low-molecular-weight RNA analysis. *Applied Environmental Microbiology, 61*, 944–952. [PMC free article] [PubMed].

Biswas, A., Bhowmik, S. K., & Roy, G. N. (1986). Chitinolytic bacteria of mangrove swamps of the Sundarbans. In L. J. Bhosale (Ed.), *The Mangroves* (pp. 382–384). Kolhapur: Shivaji University Press.

Bouvier, T. C., & Del Giorgio, P. A. (2002). Compositional changes in free-living bacterial communities along a salinity gradient in two temperate estuaries. *Limnological Oceanography, 47*, 453–470.

Buck, J. D., Meyers, S. P., & Kamp, K. M. (1962). Marine bacteria with antiyeast activity. *Science (New York), 138*, 1339–1340.

Buckholder, P. R., Pfister, R. M., & Leitz, F. H. (1966). Production of a pyrrole antibiotic by a marine bacterium. *Applied Microbiology, 14*, 649–653.

Campbell, N. (1993). Microbes: Concepts and applications. In *Biology* (3rd ed., p. 1190). Redwood City: Benjamin Cummings.

Carte, B. K. (1996). Biomedical potential of marine natural products. *BioScience, 46*(4), 271–286.

Chaudhuri, A. B., & Choudhury, A. (1994). *Mangroves of the Sundarbans*. Bangkok: IUCN.

Crump, B. C., Armbrust, E. V., & Baross, J. A. (1999). Phylogenetic analysis of particle-attached and free-living bacterial communities in the Columbia River, its estuary, and the adjacent coastal ocean. *Applied Environmental Microbiology, 65*, 3192–3204. [PMC free article] [PubMed].

De Bie, M. J. M., Speksnijder, A., Kowalchuk, G. A., Schuurman, T., Zwart, G., Stephen, J. R., Diekmann, O. E., & Laanbroek, H. J. (2001). Shifts in the dominant populations of ammonia-oxidizing beta-subclass Proteobacteria along the eutrophic Schelde estuary. *Aquatic Microbial Ecology, 23*, 225–236.

Deming, J. W. (1985). Bacterial growth in deep-sea sediment trap and box core samples. *Marine Ecology-Progress Series, 25*, 305–312.

Deming, J. W. (1986). The biotechnological future for newly described, extremely thermophilic bacteria. *Microbial Ecology, 12*, 111–119.

Deming, J. W., & Colwell, R. R. (1985). Observations of barophilic microbial activity in samples of sediment and intercepted particulates from the Demerara Abyssal Plain. *Applied and Environmental Microbiology, 50*, 1002–1006.

Devine, D. A., & Marsh, P. (2009). Prospects for the development of probiotics and prebiotics for oral applications. *Journal of Oral Microbiology, 1*, 1–11.

DeVugst, L., & Vandamme, E. J. (1994). *Bacteriocins of lactic acid bacteria. Microbiology genetic application* (Vol. 75, pp. 140174–140179). London: Blackie Acadamic & Profession.

Erba, E., Bergamaschi, D., & Ronzoni, S. (1999). Mode of action at thiocoraline, a natural marine compound with anti-tumour activity. *British Journal of Cancer, 80*(7), 971–980.

Gasol, J. M., Comerma, M., Garcia, J. C., Armengol, J., Casamayor, E. O., Kojecka, E., & Simek, K. (2002). A transplant experiment to identify the factors controlling bacterial abundance, activity, production, and community composition in a eutrophic canyon-shaped reservoir. *Limnological Oceanography, 47*, 62–77.

Giovannoni, S. J., & Rappe, M. S. (2000). Evolution, diversity, and molecular ecology of marine prokaryotes. In D. L. Kirchman (Ed.), *Microbial ecology of the oceans* (pp. 47–84). New York: Wiley.

Goddin, B. R., & Gorbach, S. L. (1992). Probiotics for humans. In R. Fuller (Ed.), *Probiotics* (pp. 355–376). London: Chapman and Hall.

Guerriero, A., Amrosio, M. D., Cuomo, V., & Pietra, F. (1991). A novel, degraded polyketidic lactone, leptosphaerolide, and its likely diketone precursor, leptosphaerodione, isolation from cultures of teh marine ascomycete Leptosphaeria oeaemaris (Linder). *Helvetica Chimica Acta, 74*, 1445.

Hazra, S., Ghosh, T., Das Gupta, R., & Sen, G. (2002). Sea level and associated changes in the Sundarbans. *Science and Culture, 68*(9–12), 309–321.

Hazra, S., Samanta, K., Mukhopadhyay, A., & Akhand, A. (2010). *Temporal change detection (2001–2008) of the Sundarban*. New Delhi: WWF-India.

Hollibaugh, J. T., Wong, P. S., & Murrell, M. C. (2000). Similarity of particle-associated and free-living bacterial communities in northern San Francisco Bay, California. *Aquatic Microbial Ecology, 21*, 103–114.

Jana, H., Maity, C., Das, A., Pati, B. R., Mitra, A., & Mondal, K. C. (2013). Investigating the effect of storage temperature and hot-water treatment on the microbial dynamics in edible oyster (*Saccostrea cucullata*). *International Journal of Postharvest Technology and Innovation, 3*(4), 382–391.

Jannasch, H. W., & Wirsen, C. O. (1982). Microbial activities in undercompressed and decompressed deep-seawater samples. *Applied and Environmental Microbiology, 43*, 1116–1124.

Jha, R. K., & Zi-Rong, X. (2004). Biomedical compounds from marine organisms. *Marine Drugs, 2*, 123–146.

Kathiresan, K., & Thiruneelakandan, G. (2008). Prospects of lactic acid bacteria of marine origin. *Indian Journal of Biotechnology, 7*(2), 170–177.

Krassil'nikova, E. N. (1961). Antibiotic properties of micro-organisms isolated from various depth of world oceans. *Microbiologiya, 30*, 545–550.

Lawrence, R. N. (1999). Rediscovering natural product biodiversity. *Drug Discovery Today, 4*(10), 449–451.

Lebaron, P., Servais, P., Troussellier, M., Courties, C., Muyzer, G., Bernard, L., Schafer, H., Pukall, R., Stackebrandt, E., Guindulain, T., & Vives-Rego, J. (2001). Microbial community dynamics in Mediterranean nutrient-enriched seawater mesocosms: Changes in abundances, activity and composition. *FEMS Microbiology Ecology, 34*, 255–266. [PubMed].

Lemos, M. L., Teranzo, A. E., & Barja, J. L. (1985). Antibiotic activity of epiphytic bacteria isolated from intertidal seaweeds. *Microbial Ecology, 11*, 149–163.

Lene, L. (1996). Microbial metabolites-an infinite source of novel Chemistry. *Pure and Applied Chermistry, 68*, 745–748.

Lovell, F. M. (1966). The structure of a bromine rich antibiotic. *Journal of the American Chemical Society, 88*, 4510–4511.

McGee, C. D., Leecaster, M. K., Vainik, P. M., Noble, R. T., Walker, K. O., & Weisberg, S. B. (1997). Comparison of bacterial indicator measurements among southern California Marine Monitoring. Southern California Water Research Program Annual Report 1997/1998, California, pp. 187–198.

Miralto, A., Barone, G. R., Poulet, S. A., Lanora, A., Russo, G. L., Buttino, I., Mazzarella, G., Laabir, M. C., & Giacobbe, M. G. (1999). The insidious effect of diatoms on copepod reproduction. *Nature, 402*, 173–176.

Misra, S., Dutta, J., Ghosh, A., & Choudhury, A. (1985). Role of hydrophobic components of leaves in the adaption of plants to periodic submersion by tidal water in mangrove ecosystem. *Journal of Chemical Ecology, 11*(3), 339–343.

Murray, A. E., Hollibaugh, J. T., & Orrego, C. (1996). Phylogenetic compositions of bacterioplankton from two California estuaries compared by denaturing gradient gel electrophoresis of 16S rDNA fragments. *Application of Environmental Microbiology, 62*, 2676–2680. [PMC free article] [PubMed].

Okami, Y. (1986). Marine micro-organisms as a source of bioactive agents. *Microbial Ecology, 12*, 65–78.

Pallenberg, A. J., & White, J. D. (1986). The synthesis and absolute configuration of (+)-leptosphaerin. *Tetrahedron Letters, 27*(46), 5591–5594.

Rickards, R. W., Rothschild, J. M., & Willis, A. C. (1999). Calothrixins A and B, novel pentacyclic metabolites from Calothrix cyanobacteria with potent activity against malaria parasites and human cancer cells. *Tetrahedron, 55*(47), 13513–13520.

Rosenfeld, W. D., & Zobell, C. E. (1947). Antibiotic production by marine microorganisms. *Journal of Bacteriology, 54*, 393–398.

Saha, S. B., Bhattacharyya, S. B., Basu, S., Mitra, A., Zamadar, Y. A., & Choudhury, A. (1998). Primary production and ecological efficiency of brackishwater shrimp culture in the vicinity of Sundarbans mangrove ecosystem. *Journal of Aquaculture in the Tropics, 13*(2), 151–158.

Schafer, H., Bernard, L., Courties, C., Lebaron, P., Servais, P., Pukall, R., Stackebrandt, E., Troussellier, M., Guindulain, T., Vives-Rego, J., & Muyzer, G. (2001). Microbial community dynamics in Mediterranean nutrient-enriched seawater mesocosms: Changes in the genetic diversity of bacterial populations. *FEMS Microbiology Ecology, 34*, 243–253. [PubMed].

Schiehser, G. A., White, J. D., Matsumoto, G., Pezzanite, J. O., & Clardy, J. (1986). The structure of leptosphaerin. *Tetrahedron Letters, 27*(46), 5587–5590.

Selje, N., & Simon, M. (2003). Composition and dynamics of particle-associated and free-living bacterial communities in the Weser estuary, Germany. *Aquatic Microbial Ecology, 30*, 221–237.

Stevenson, C. S., Capper, E. A., & Roshak, A. K. (2002a). Scytonemin—A marine natural product inhibitor of kinases key in hyperproliferative inflammatory diseases. *Inflammation Research, 51*(2), 112–114.

Stevenson, C. S., Capper, E. A., & Roshak, A. K. (2002b). The iden-tification and characterization of the marine natural product scytonemin as a novel antiproliferative pharmacophore. *Journal of Pharmacology and Experimental Therapeutics, 303*(2), 858–866.

Thajuddin, N., & Subramanian, G. (2005). Cyanobacterial biodi-versity and potential applications in biotechnology. *Current Science, 89*(1), 47–57.

Troussellier, M., Schafer, H., Batailler, N., Bernard, L., Courties, C., Lebaron, P., Muyzer, G., Servais, P., & Vives-Rego, P. (2002). Bacterial activity and genetic richness along an estuarine gradient (Rhone River plume, France). *Aquatic Microbial Ecology, 28*, 13–24.

Turner, J. T. (1979). Microbial attachment to copepod faecal pellets and its possible ecological significance. *Transactions of the American Microscopic Society, 98*, 131–135.

Van, H. E. J., Mooij, W., Van, M. P., Gons, A. H. J., & Laanbroek, H. J. (1999). Detritus-dependent development of the microbial community in an experimental system: Qualitative analysis by denaturing gradient gel electrophoresis. *Application of Environmental Microbiology, 65*, 2478–2484. [PMC free article] [PubMed].

Wollowski, I., Rechkemmer, G., & Pool-Zobel, B. L. (2001). Protective role of probiotics and prebiotics in colon cancer. *American Journal of Clinical Nutrition, 73*(2), 451–455.

Internet References

http://greengenes.ilb.gov
http://rdp.cme.msu.edu and greengenes.
www.mist.net

Annexure 7A: References

APHA (American Public Health Association). (2001). *Standard methods for the examination of water and wastewater* (20th ed.). Washington, DC: APHA.

Glasoe, S., & Aimee, C. (2004). *Coastal urbanization and microbial contamination of shellfish growing areas* (pp. 1–28). Olympia, WA: Puget Sound Action Team.

Mitra, A., & Banerjee, K. (2005). *Living resources of the sea: Focus Indian Sundarbans.* Published by WWF-India Secretariat, Sunderbans Landscape Project, New Delhi, p. 120.

Mitra, A., Choudhury, A., & Zamaddar, Y. A. (1992). Seasonal variations in metal content in the gastropod *Cerithedia (Cerithideopsis) cingulata. Proceedings of the Zoological Society, 45*, 497–500.

Pommepuy, M. (1996). Behaviour of pathogenic microorganisms in coastal areas. *Applied Enviromental Microbiology, 62*, 4621–4626.

Annexure 7B: References

APHA (American Public Health Association). (1998). *Standard methods for the examination of water and wastewater* (20th ed., pp. 1–200). Washington, DC: American Public Health Association, American Water Works Association, and the Water Environment Federation.

Basemer, K., Markus, M. M., Jesus, M. A., Gerhard, J. H., & Peter, P. (2005). Complexity of Bacterial communities in a River-floodplain System (Danube, Austria). *Applied Environmental Microbiolgy, 71*, 609–620.

Bell, B. P., Goldoft, M., & Griffin, P. M. (1994). A multistate outbreak of *Escherichia coli* O157:H7-associated bloody diarrhea and hemolytic uremic syndrome from hamburgers: The Washington experience. *JAMA, 272*, 1349–1353.

Blake, P. A., Weaver, R. E., & Hollis, D. G. (1980). Diseases of humans (other than cholera) caused by vibrios. *Annual Review Microbiology, 34*, 341–367.

Brewster, D. H., Brown, M. I., Robertson, D., Houghton, G. L., Bimson, J., & Sharp, J. C. M. (1994). An outbreak of *Escherichia coli* O157 associated with a children's paddling pool. *Epidemiology Infection, 112*, 441–447.

Carlson, G. F., Jr., Woodard, F. E., Wentworth, D. F., & Sproul, O. J. (1968). Virus inactivation on clay particles in natural waters. *Journal of Water Pollution, Control Federation, 40*, R89–R106.

Christensen, V., Jian, X., & Ziegler, A. (2000). *Regression analysis and real-time water quality monitoring to estimate constituent concentrations, loads and yields in the Little Arkansas River, South Central Kansas, 1995–1999.* USGS Water Resources Investigations report 00–4126, Lawrence.

Cieslak, P. R., Barrett, T. J., & Griffin, P. M. (1993). *Escherichia coli* O157:H7 infection from a manured garden. *Lancet, 342*, 367.

Eleria, A. L. (2002). *Forecasting fecal coliform bacteria in the Charles River Basin*. Master's thesis, Tufts University, Medford, pp. 5–38.

Francy OS, Gifford AM, Darner RA, 2002. *Escherichia coli at Ohio bathing beaches: distribution, sources, wastewater indicators, and predictive modelling*. In cooperation with the Ohio Water Department Authority, U.S. Dept. of the Interior, U.S. Geological Survey, Columbus/Denver, p. 121.

Gerba, C. P., & Schaiberger, G. E. (1975). Effect of particulates on virus survival in seawater. *Journal of Water Pollution Control Federation, 47*, 93–103.

Grimes, D. J. (1975). Release of sediment-bound fecal coli-forms by dredging. *Applied Microbiology, 29*, 109–111.

Kelsey, H., Porter, D. E., Scott, G., Neet, M., & White, D. (2004). Using geographic information systems and regression analysis to evaluate relationships between land use and fecal coliform bacterial pollution. *Journal of Expedition Marine Biology Ecology, 298*, 197–209.

Kirschner, A. K. T., Gerhard, G. K., Branko, V., Robert, L. M., Regina, S., & Andreas, H. F. (2009). Microbiological water quality along the Danube River: Integrating data from two whole-river surveys and a transnational monitoring network. *Water Research, 43*(15), 3673–3684. doi:10.1016/j.watres. 2009.05.034.

Mallin, M. A., Williams, K. E., Esham, E. G., & Low, R. P. (2000). Effect of human development on bacteriological water quality in coastal watersheds. *Ecology Applied, 10*(4), 1047–1056.

Manja, K. S., Maurya, M. S., & Rao, K. M. (1982). A simple test for the detection of faecal pollution in drinking water. *Bulletin World Health Organisation, 60*(5), 797–801.

Oppenheimer, C., & Wood, E. J. F. (1962). Note on the effect of contamination on a marine slough and the vertical distribution of unicellular plants in the sediment. *Z. allg. Microbiology, 2*, 45–47.

Raveendran, O., Gore, P. S., Iyer, T. S. G., Varma, P. R. G., & Sankaranarayanan, V. N. (1990). Occurrence of enteric bacteria in seawater and mussels along the south west coast of India. *Indian Jounal of Marine Science, 19*, 282–284.

Strickland, J. D. H., & Parsons, T. R. (1972). A practical handbook of seawater analysis. 2 (Ed.). *Journal of Fish Research Board Canada, 167*, 1–310.

8 Adaptation of Marine and Estuarine Organisms

Contents

8.1 Adaptation of Organisms in the Intertidal Zone

The intertidal zone can be considered as the zone of intersection between land and the sea, which remains inundated during high tide and becomes naked (exposed to air only) during low tide. The region may be rocky (Fig. 8.1), sandy (Fig. 8.2) or muddy (Fig. 8.3).

The intertidal zone may be conveniently divided into upper intertidal, mid-intertidal, low intertidal and subtidal zones (Fig. 8.4). The upper intertidal zone receives very little water and subtidal zone remains under water in most of the time. The spray or splash zone is followed by the upper intertidal zone. For convenience of the readers, each of these zones is discussed here separately.

Spray or Splash Zone This zone extends from the highest reach of spray and storm waves to the average height of the high tides. It is usually dry with poor biodiversity. Species found in the splash zone might include small barnacles (Fig. 8.5) and gastropods. Mangrove species with xerophytic adaptation like *Acanthus ilicifolius* is abundantly found in this zone.

Upper or High Intertidal Zone This zone is covered with water during highest tide and is the survival ground of acorn barnacles, sea

A. Mitra, S. Zaman, *Basics of Marine and Estuarine Ecology*,
DOI 10.1007/978-81-322-2707-6_8

Fig. 8.1 Rocky intertidal zone

anemones (Fig. 8.6), shore crabs, etc. Several species of seaweeds are also found in this zone.

In the tropical belt, several mangrove associate species like *Suaeda* sp. (Fig. 8.7) and *Porteresia coarctata* (Fig. 8.8) are abundantly distributed in the upper or high intertidal zone.

Mid-intertidal Zone This zone extends from just below average sea level to the upper limit of the average lowest tides (i.e. it is exposed at low tides—usually twice a day). This zone has rich taxonomic diversity and sustains organisms like mussels, oysters, anemones, chitons (Fig. 8.9), sponges and several species of macroalgae.

Low Intertidal Zone This zone remains under water during maximum period and is exposed only during lowest tide. Chitons, crabs and sea urchins (Fig. 8.10) are the common fauna in this zone. Among the flora, kelp is predominant.

Subtidal The zone remains under saline water and sustains floral species like kelp (Fig. 8.11), bottom-dwelling invertebrates and several species of fishes.

The presence or absence of water, the oscillation and range of temperature, wave action, variation in salinity (saltiness), exposure to light and other factors determine the type of organisms that are able to survive comfortably in each zone. In general, physical factors, especially exposure to drying, limit how far up on shore an organism can live. An organism's lower limit is often determined by competition or predators living in the lower zone.

Organisms of the marine and estuarine organisms orient themselves in different ways in different categories of intertidal habitats, e.g. the mode of adaptation of organisms inhabiting hard substratum (like rocky shore) is totally different from those inhabiting soft bottom (sandy or muddy substratum). Thus, the

Fig. 8.2 Sandy intertidal zone

Fig. 8.3 Muddy intertidal zone

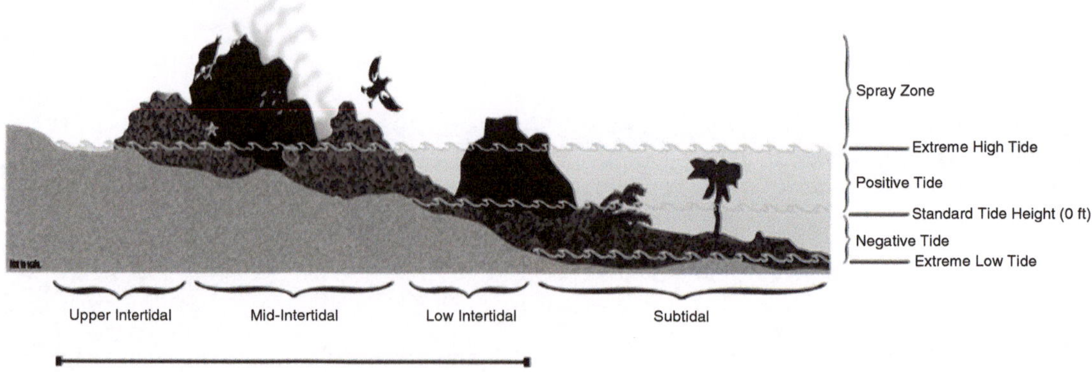

Fig. 8.4 Zonation of the intertidal zone

Fig. 8.5 Spray or splash zone with small barnacles

nature of the substratum is a determining factor in the process of adaptation.

8.1.1 Adaptation of Biota Inhabiting Rocky Shore

Rocky shore provides a relatively stable surface for attachment of organisms. As the tide retreats from a rocky shore, the higher regions of the coast become completely exposed to air. On close examination of this region, horizontal bands or zones inhabited by specific set of community are observed. The segregation of organisms into such definite bands or zones on the rocky substratum is referred to as **zonation**. The uppermost area of the rocky shore, which is covered only by the highest (spring) tides and is exposed to spray from the splashing of waves, is the **supratidal** or **splash zone**. The **subtidal zone** is the region of the shore that is covered by water, even during low tide. The **intertidal zone** lies between the supratidal zone and subtidal zone and is exposed and covered during low and high tides respectively on regular basis. The intertidal rocky shore is characterized by the

Fig. 8.6 Upper or high intertidal zone with anemones

Fig. 8.7 Abundance of *Suaeda maritima* in the upper or high intertidal zone

abundance of light, water, nutrients, dissolved oxygen, etc. There is luxuriant floral diversity in the intertidal rocky shore, which offers shelter to a wide variety of animal species. The animals are usually sessile and sedentary in nature. Hence, development of attachment organs has become important criteria of the organisms inhabiting the zone. In addition the rocky shore organisms have developed unique adaptations to get rid of desiccation. Few important adaptive modifications are discussed here:

Fig. 8.8 *Porteresia coarctata* in the upper or high intertidal zone

Fig. 8.9 Upper or high intertidal zone with chiton

1. To get rid from the strong wave actions, most of the rocky shore organisms have developed a very powerful holdfast or any other attaching organs. Due to this sedentary mode of life, several morphological changes have taken place in the organisms of this zone such as the following:
   (a) *Mytilus* sp. and *Ostrea* sp. have lost their locomotory organs.
   (b) Ascidians and polyzoans have developed efficient sense organs.
   (c) *Balanus* and bivalves have developed thick test.
   (d) *Mytilus* sp. have developed special organs for attachment in the form of byssus threads.
   (e) Rock urchins hollow out cavities in the rocks.
   (f) Sea stars use suction cups on the ends of their tube feet to cling to the surface of rocks.
2. To conserve water and to avoid desiccation, animals have developed dwelling habits in crevices. These shelters not only help in preventing desiccation but also from escaping the pounding action of the waves.
3. During unfavourable environmental conditions, many animals (mainly gastropods) are able to remain for long periods in a condition of suspended activity.
4. Animals like limpets remain in a fixed position relative to water level for a given tidal height by the help of scar marks. Scars due to

Fig. 8.10 Low intertidal zone with sea urchin

Fig. 8.11 Subtidal zone with floral species

adhesion to the rocky surface mark their home territory or home base. Several interesting mechanisms of attachment are observed in animals inhabiting rocky shore. Sessile animals, which permanently attach themselves to one spot for life, use various cementing substances. Spirorbid polychaetes attach temporarily by mucus secreted from ventral glands. They then spread the contents of the shell gland over the body by rolling from side to side (Nott 1973). A further secretion from the ventral gland (now on the upper surface) is used to create a primary mucus tube. After metamorphosis, a calcareous tube secreted by the collar of the animal replaces this. This tube is firmly cemented to the substratum.

5. To get rid from the pounding action of the waves, many animals (particularly gastropods) take shelter in the seaweed community of the rocky shore, which acts as cushion. It has been found that the number of rocky individuals per 1000 g of seaweeds varies from 44 to about 13,000. In the *Enteromorpha* bed of Sundarbans, *Littorina* spp. are found in plenty.
6. To reduce friction many rocky shore animals have developed flattened body such as *Oscarella* sp., leaf-like turbellarians,

tunicates and certain crabs like *Sesarma* sp. In molluscs, the shells of species in areas with considerable water movement tend to be flatter, with large apertures providing a greater attachment area to the substratum.

7. Many rocky shore animals have developed a peculiar mechanism for shifting their orientation in response to tidal cycle. The intertidal periwinkle *Littorina* sp. orients their body in response to high tide and low tide.
8. The animals living on exposed rocks have developed various defence mechanisms such as spicules in sponges and stinging cells in coelenterates.
9. Many animals of the rocky shore exhibit the phenomenon of concealing mimicry in which the body colour matches with the background substratum. Crabs like *Sesarma* sp. are notable example in this context.
10. Many organisms like the fiddler crabs have developed certain adaptations to cope with the tidal cycle that enable the animals to exploit the advantages of living in the stressful condition. They have internal biological clocks that help to perform their feeding activity during the most favourable period of tidal cycle.
11. Many plants develop strong holdfast and float to escape from the pounding action of the waves. The large brown algae (*Laminaria* sp.) have this type of adaptation.
12. Many plants develop a complex tissue differentiation as anatomical adaptation for the rocky intertidal zone. In kelp, the tissue differentiation is very complex with an outer epidermis, outer and inner cortex, and a central medulla. Trumpet hyphae and sieve like elements are common in the medulla. It has been found that the medullary cells are the sites of translocation in Kelps (Schmitz and Srivastava 1980).
13. Different seaweeds may cope with the same wave conditions in different ways, as shown by *Lessonia nigrescens* and *Durvillaea antarctica*, both of which occur in the exposed sublittoral zone of central and southern Chile. *Lessonia* has a strong, stiff stipe that bends with the flow. In contrast, the elastic stipe and stretchy blade of *Durvillaea* sp. allow it to align with the flow completely. Flexible stipes that can be bent parallel with water movements reduce the stress on the thallus and result in the plant being closer to the rock where movement is less. These structures respond best to the chaotic multidirectional water movements typical of exposed shores. In the North Atlantic, flexible elastic stipes are found in species from wave-exposed sites, such as *Alaria esculenta*.
14. Avoiding dislodgement by wave action is a major priority for rocky shore species, but still all seaweeds require some degree of water movement to break down the boundary layer around the thallus. Materials have to diffuse in and out of the plant through this layer of slowly moving fluid, so that the thicker the boundary layers the slower the uptake of materials. In still water, the boundary layer can be several millimetres thick. Many species have a surface of spiny outgrowths (*Macrocystis pyrifera*), wavy margins (*Laminaria saccharina*) or even holes or undulations in the blade (*Agarum cribrosum*, *Laminaria saccharina*) which are thought to enhance the turbulence in water flowing over the lamina and hence allow greater uptake of raw materials for photosynthesis. There is also evidence that fucoids grow hair-like protrusions during periods of low nutrient concentration and that these are involved in nutrient uptake (Hurd et al. 1993).

8.1.2 Adaptation of Biota Inhabiting Soft Substratum

The term soft substratum in marine ecology refers to intertidal zone with dominancy of sand, silt, clay, etc. Both sandy and muddy beaches are thus encompassed within the domain

Fig. 8.12 Muddy substratum with mangrove trees

of soft substratum. The muddy substratum sustains a wide spectrum of vegetation, mostly mangroves (Fig. 8.12).

Sandy shores occupy maximum percentage of the world shoreline and are made up of sand, gravel or shells. The environment is extremely dynamic in nature where the sand particles are always in a state of motion due to action of waves, tides and air. Fine-grained sand beaches tend to be quite flat (Fig. 8.13).

Sandy beaches are soft shores that are formed by deposition of particles that have been carried by water currents from other areas. The transported material is in part derived from the erosion of shores, but the major part is derived from the land and transported by rivers to the sea. The two main types of beach material are quartz (silica) sands of terrestrial origin and carbonate sands of marine origin. The carbonate sand is weathered from mollusc shells and skeletons of other animals. Other materials include heavy minerals, basalt (volcanic origin) and feldspar.

Mudflats also known as tidal flats are coastal wetlands that form when mud is deposited by tides or rivers. They are found in sheltered areas such as bays, bayous, lagoons and estuaries. Mudflats may be viewed geologically as exposed layers of bay mud, resulting from deposition of estuarine silts, clays and marine animal detritus. Most of the sediment within a mudflat is within the intertidal zone, and thus the flat is submerged and exposed approximately twice daily. In the past tidal flats were considered unhealthy, economically unimportant areas and were often dredged and developed into agricultural land. Tidal flats, along with intertidal salt marshes and mangrove forests, are important ecosystems. They usually support a large population of wildlife and are a key habitat that allows tens of millions of migratory shorebirds to migrate from breeding sites in the northern hemisphere to nonbreeding areas in the southern hemisphere. They are often of vital importance to migratory birds, as well as certain species of crabs, molluscs and fish. The mudflats in Sundarbans,

Fig. 8.13 Flat sandy beach

at the apex of Bay of Bengal, are the home ground of Royal Bengal tiger (*Panthera tigris tigris*) (Fig. 8.14) and their prey base like deer, monkeys, etc. (Fig. 8.15).

The organisms thriving on/in the soft substratum adapt themselves on the basis of the nature and composition of the soft substratum. For convenience of the readers, the adaptations of organisms inhabiting sandy or muddy substratum are discussed separately.

8.1.2.1 Adaptation of Biota Inhabiting Sandy Shore

A sandy shore would seem at first to be a rigorous habitat, with poor biodiversity, but it is better populated than first appearances indicate. Most of the larger animals are specialized burrowers, and the diatoms, amphipods and other infauna, which live among the sand grains, would not be noticed at first sight. Some common sand burrowers are *Callinectes* sp., *Ogyris* sp., *Lepidopa* sp., *Calappa* sp., *Ovalipes* sp., *Emerita* sp., *Arenaeus* sp., *Lysiosquilla* sp., *Callianassa* sp., etc. The sandy shore is highly unstable due to movement of the sediments by current and tides.

The most important factor in determining the distribution of life on sandy beaches is wave action, because the waves regulate the availability of the amount of moisture and oxygen to the organisms. In addition the nutritional requirements of the sandy shore organisms is also fulfilled by waves as they carry plankton and remove large amounts of organic material from the beach, leaving little for scavengers to consume. For an organism to survive in this hostile environment, the wave action must be able to meet the organism's specific needs for food, oxygen and waste removal. Based on the magnitude and type of wave action, sandy beaches can be classified into four broad divisions (Table 8.1).

The unfavourable situation in the sandy shore is usually avoided by biota through the process of burrowing. Most intertidal macrofauna of sandy

Fig. 8.14 Royal Bengal tiger on Sundarban mudflats

Fig. 8.15 Deer population on the intertidal mudflats of Sundarbans

Table 8.1 Classification of sandy beaches by exposure

Criterion	Exposed	Semi-exposed	Protected	Highly protected
Wave action	High	Considerable	Low	None
Wave types	Plunging, surging	Spilling	–	–
Slope	Steep	Gradual	Gradual	Flat
Width	Narrow	Wide	Various	Extensive
Oxygen	Highest	–	–	Lowest
Moisture	Lowest	–	–	Highest
Particle size	Coarser	Finer	Finer	Varies
Organic matter	Lowest	–	–	Highest
Permanent burrows	No	No	Few	Very frequently
Plankton diversity	Maximum	Considerable	Medium	Low
Bioturbation	Very less frequent	Less frequent	Few	Very frequently

shore exploit the physical properties of the sediment in particular dilatancy and thixotropy, to burrow. When the water content becomes less than 22 % by weight, any force applied upset the close packing of the grains, and the interstitial water is no longer able to fill all spaces between the particles; the sediment becomes hard and resistant to shear forces. This effect is referred to as **dilatancy**. In contrast, when the water content is above 25 %, there is a reduction in resistance with increasing rates of shear, and the sediment becomes liquefied slurry of grains and water. This condition is called **thixotropy**. Most burrowing animals use repeated small penetrations to displace sediments, making use of their thixotropic property. Repeated small agitations of the sediment result in the reduction of the ratio of resistance to penetration by a factor of 10. Subsequently anchorage is achieved by dilatancy and the process is then repeated. The molluscan species thriving in the sand–mud mixed substratum exhibits habitat selection with respect to texture of the underlying matrix. The distribution of infauna is regulated by the grain size or texture of the bottom. Determining the sand–silt–clay ratio has considerable predictive value as to the kinds of organisms to be expected (Odum 1971). The method of feeding by the benthos undergoes an interesting change along a sand–mud gradient. The filter-feeding organisms dominate the sandy matrix, while deposit feeders are common on the silty or muddy substrates.

Some adaptations to cope with the unstable condition of the sandy shore are listed here:

1. During unfavourable environmental conditions, many animals are able to remain for a long period in a condition of suspended activity.
2. Due to development of burrowing habit, the organisms of the sandy shore have developed certain adaptive features like development of digging organs as found in *Natica* sp. and *Nassa* sp., development of ciliary mode of feeding as seen in *Macoma birmanica*, development of highly efficient respiratory devices, etc.
3. The process by which the burrowing animals solve the problem of respiration is highly interesting. The decapod, *Albunea* sp., holds both its long antennae together to form a long tube through which water is sucked for breathing purpose. *Arenicola* has a series of gills on the side of the middle portion of the body, which are very efficient for water exchange with the ambient aquatic phase.
4. To carry out the process of burrowing very efficiently, *Solen* and *Donax* have flattened shape. *Natica* and *Nassa* have expanded foot for the purpose of burrowing and creeping. In some polychaetes, the blisters are very poorly developed to avoid friction during the time of their entry inside the sandy substratum.

5. Burrowing molluscs have unique food-taking capacity by means of filtering mechanism. The inhalant siphon helps to draw food from the upper aquatic phase as seen in *Scrobicularia* sp.
6. Burrowing animals like *Arenicola* sp. can load oxygen in their body tissues even at very low pressure.
7. Many burrowing animals swallow the sand containing organic detritus as a source of their energy, e.g. *Arenicola*, *Synapta*, *Balanoglossus*, etc.
8. Many animals resist the water loss by reducing their body surface area. Sea anemone is a notable example in this context, which can contract their body volume 8.5 times of the initial volume.
9. In some sandy intertidal zone, the impact of current on the substratum is significant. Because of the wave and current action, sediments constantly shift providing an unstable environment for sandy beach organisms. However, an adaptation is noticed in some animals like *Dotilla* sp. against the constant shifting of the bottom sediments. These animals can rotate itself within a fraction of second as soon the waves and currents expose them.
10. Certain animals like *Donax* sp. exhibit a peculiar shifting behaviour in response to wave action. These animals need moist sand for survival and cannot stand thrashing of the waves. Hence, as the wave approaches, these animals emerge from the sand bed, in response to instability of the sand–water mixture, and are carried shorewards by the wave. As the tidal amplitude falls, the animals again emerge by pushing their foot downwards and are carried down by the beach to the moist zone. It has been found that only 0.87 cal of energy is expended per day for this migratory behaviour (Ansell and Trueman 1973).
11. The infauna of the sandy beaches may not suffer the acute desiccation as experienced by the rocky shore species, but they too suffer the problem of low level of oxygen, which becomes more severe with depth and with finer particles. Most species have appropriate physiological mechanism or behaviours for coping with the problem of low levels of oxygen at low tide. Many animals have respiratory pigments such as haemoglobin of polychaetes and haemocyanin of decapod crustaceans. Other species are able to regulate their oxygen consumption and reduce their activity to minimize demand at low tide. In more stable sediments, many species construct semi-permanent burrows, which provide access to better-aerated surface water at low tide. Finally as is the case of many rocky shore species, several species have to resort to anaerobic metabolism for some part of the low tide period.
12. The oxygen concentration is extremely low below the chemocline of the redox potential discontinuity (RPD), but sulphide concentration is extremely high, which poses an adverse impact on the animal community. Many of the more familiar infauna, such as crustaceans, molluscs and annelids, do not occur within this black layer, unless they possess some kind of burrow or tube which gives them access to the upper sediment and so to oxygenated water. However, many meiofaunal species regularly occur in the black layer, and some groups, such as the Gnathostomulida and the turbellarian groups' Solenomorphidae and Catenulida, seem restricted to it. The inhabitants of this black zone and the chemocline may represent a self-contained sulphide ecosystem or **thiobios** (Fenchel and Redl 1970; Boaden and Platt 1971; Boaden 1989). The organisms found in the overlying oxic sediment, many of which are absent from the thiobios, for example, Proseriata (Turbellaria), Nemertini, Tardigrada and Archiannelida, have been termed the **oxybios**. The thiobios are assumed to possess physiological adaptations to a lack of oxygen and to the presence of toxic compounds such as hydrogen sulphide and

ammonia. However, Reise and Ax (1979) have argued convincingly that the division of sediments into an upper oxic zone and a lower sulphidic region is a gross oversimplification. Many macrofaunal species excavate burrows, which cross the chemocline and create oxic microhabitats within the black zone. While all classic thiobiota seem to live where the oxygen concentration is effectively zero, Meyers et al. (1987) found that species were zoned in a continuous fashion along the oxygen and sulphide concentration gradients, so that it is probably not sensible to make a clear distinction between an oxybios and a thiobios. It is also possible that the thiobiotic meiofauna can utilize extremely low concentrations of oxygen that occur in reduced sediments, but which are not possible to measure at present.

13. The mole crab *Emerita* sp. exhibits a unique sinking behaviour. The species is capable of sinking backwards into the sand within a few seconds. These crabs feed by extending their feathery antennae above the sand and collecting plankton from the flowing water during the ingression of tide.
14. The activity of intertidal organisms is keyed to the movement of the tides. As the tide moves in, the pace of life begins to quicken. During high tide, bivalves, such as cockles (*Cardium*), tellins and surf clams (*Spisula*), project their siphons from their burrows and begin to filter the water for food and bathe their gills with oxygen.
15. Amphipods and isopods inhabiting sandy substratum show highly rhythmic behaviour with a strong endogenous component. The rhythms are such that animals emerge from the sand, feed, and then return to burrow or hide in the strand-line material at the appropriate tidal level. The midshore isopod *Eurydice pulchra* has tidal, daily and semilunar rhythms. It emerges from the sand where the tide comes in, feeds while swimming in the water column, ceases activity before the retreating tide and burrows into the sand for the duration of the tide-out period. Activity is greater during the night than in daytime and is also greater on spring, than neap, tides (review: Naylor 1985).

The semiterrestrial amphipod *Talitrus* has a circadian rhythm, foraging at night (Williams 1983). On these excursions, it goes down the beach into the intertidal zone before returning to the top of the shore. It also exhibits a semilunar rhythm, being more active on neap tides, thus avoiding being swept away on spring tides. In addition to the temporal pattern of activity, the animals orientate themselves to return to the correct position. Transplant experiments to either side of narrow isthmuses have shown that *Talitrus* will hop off in the wrong direction if taken to a beach where strand line is in the opposite direction (Ugolini et al. 1986). Visual cues, such as the light/dark boundary on the horizon, have been implicated in this behaviour. Interestingly, the orientation behaviour also has a rhythmic component, with an exploratory and a homing phase. Various cues such as changes in pressure, salinity, agitation, immersion/emersion or light/dark associated with environmental cycles have been shown to entrain rhythms. The accepted general view is of an approximate endogenous pacemaker, which modulates activity patterns and is continually reset by environmental cues to local time.

8.1.2.2 Adaptation of Biota Inhabiting Muddy Shore

The muddy environment in the marine and estuarine zone is composed of several categories of sediments and is usually dominated by silty-clay type. The substratum is loose unlike rock. Hence, an interesting adaptation is noticed in the muddy shore biota in respect to their reduction of the body pressure. The muddy substratum is also the zone of maximum amount of detritus, and the animals derive their energy mainly from the decomposed parts of the halophytic species. Since the detritus is dominated by a large group of microbes, therefore most of the animals having their habitat in the muddy substratum secrete a type of mucous having antibacterial property.

Some important adaptations of the biota inhabiting this zone are highlighted here:

1. The body of most animals is very soft; even if they possess shell, they are very thin. Nudibranchs are very common in this type of substratum.
2. The presence of a long siphon is an important characteristic of the animal (particularly infauna) present in this habitat. Some molluscs (like *Mya* sp., *Macoma birmanica*, etc.) have very long inhalant and exhalant siphons.
3. Mud dwellers are characterized by very ill-developed nervous system, e.g. some sense organs like eye are either degenerated or absent.
4. Mud dwellers reduce their pressure on the loose substratum through the presence of a broad foot.
5. The body colour of many animals also matches with that of the substratum. *Boleophthalmus* sp., *Nerita articulata* and *Onchidium* sp. are some of the animals having their body colour greyish brown. This helps them to escape from their predators.
6. Animals inhabiting muddy substratum are usually detritivores, i.e. the energy for sustaining life is obtained from the decomposed parts of mangroves and its associates, seaweeds, salt marsh grass, etc.
7. Many animals develop a tube dwelling habit and these tubes are analogous to horizontal tubes.
8. Many animals are also burrowing in nature. *Peachia* sp. remains burrowed up to 30 cm below the surface and the long tentacles reach out the burrow for capturing prey.
9. Mangroves are the most important biotic component inhabiting the muddy substratum in coastal zones. Salt in high concentrations in plant tissues seems to pose an adverse impact (although the physiological reason is not known clearly) and must be largely excluded. All mangroves exclude most of the salt in seawater. Thus, mangroves are endowed with a unique system of ion influx–efflux regulation, and accordingly three categories of mangroves have been recognized (Walter 1961), namely, salt-excluding, salt-excreting and salt-accumulating types.

In salt-excluding species, the root system possesses an ultrafiltration mechanism, which is just like an insurance of this particular group to dominate in the mangrove community. Many mangrove species can exclude 90 % of salt in seawater. *Rhizophora mucronata*, *Ceriops candolleana*, *Bruguiera gymnorrhiza*, *Kandelia candel*, etc., are few salt excluders of mangrove community.

The salt-excreting species of mangrove community like *Avicennia alba*, *Avicennia marina*, *Avicennia officinalis*, *Aegiceras corniculatum*, *Acanthus ilicifolius*, etc., regulate their internal salt levels through foliar glands. However, salt-accumulating species has the ability to accumulate high concentration of salts in their cells and tissues, which imparts succulence. *Sonneratia apetala*, *Lumnitzera racemosa*, *Excoecaria agallocha*, *Sesuvium portulacastrum*, *Suaeda maritima* and *Suaeda nudiflora* are included in this particular group of mangrove vegetation.

In salt-secreting (excreting) mangroves such as *Avicennia* sp., *Acanthus* sp., *Aegiceras* sp. and *Aegialitis* sp., the NaCI concentration of xylem sap is relatively high, about one tenth of the concentration of salt in seawater. So, the salt-excreting species allow more salt into the xylem than do the non-excretors, but still exclude about 90 % of the salts (Scholander et al. 1962; Azocar et al. 1992). Salt is only partially excluded at the roots. The absorbed salt is primarily excreted metabolically via specialized salt glands in the leaves. The salt in solution can crystallize by evaporation and either can be blown away or washed off. Since, in salt-excreting mangroves, superfluous salts are excreted by guttation through special salt glands, all these salt-excreting halophytes are often referred as crinohalophytes.

It is interesting to note that salt excretion is an active process, as evidenced by ATPase activity in the plasmalemma of the excretory cells (Drennan et al. 1992). The process is probably regulated by leaf hypodermal cells, which may store salt as well as water (Balsamo and Thomson 1995).

In non-secretors, e.g. *Rhizophora* sp. and *Sonneratia* sp., xylem sap has a salt concentration less than 1/100th of that of seawater, but still about 10 times more concentrated than that of non-mangrove plants such as *Hibiscus* sp., *Eugenia* sp., etc. Although non-secretory mangroves have no specialized mechanism for active secretion of salts, they lose some salt through leaf surfaces especially by cuticular transpiration. The easiest mechanism for the elimination of salt in all mangroves is simply by loss of plant parts, notably leaves. Many species of mangroves cope with the salt by concentrating it in bark or older leaves, which carry it with them when they drop. *Lumnitzera* sp, *Avicennia* sp, *Ceriops* sp. and *Sonneratia* sp. exhibit this interesting feature.

Several researchers demonstrated experimentally that the salt separation process in mangroves must occur at or near the root surface and is mediated by physical processes alone, since it is not inhibited by poisons or high temperature, which may cause an inhibitory effect on metabolic process. This physical process in root area for salt separation involves ultrafiltration mechanism which occurs either at the root surface (epiblema) or at the root endodermis, but the latter region might be the most preferable site (Tomlinson 1986) because the ultimate absorbing roots in most of the mangroves lack root hairs (e.g. capillary rootlets of *Rhizophora* sp.). This indicates that the absorbing area of mangroves is reduced in comparison with non-mangrove plants.

The most distinctive trichome (appendages which are epidermal in origin) that develops in certain mangrove leaves is the structure for secreting certain ions like Na^+ and Cl^-. These form a general class of secretory structures referred to as 'salt glands' by Fahn (1979). Salt glands are abundant on leaves of this particular group of mangroves (salt secretors), but are not necessarily equally frequent on upper and lower surfaces.

The intertidal zone is today experiencing sea level rise and subsequent rise in salinity. It is an interesting area of research that how the organisms are coping with the event of rising salinity. We conducted some important experiments on mangroves to observe which species is better suited to hypersaline condition and observed that the endemic mangrove species of Indian Sundarbans named *Heritiera fomes* (locally known as Sundari and where from the name Sundarbans probably has originated) is on the verge of extinction (Fig. 8.16).

Steps have been taken to conserve the species through mass plantation in place like Bonnie camp in the central Indian Sundarbans. Before planting in the field, the seedlings are acclimatized in earthen pots (Fig. 8.17).

We also observed that the adaptive capacity of *Nypa fruticans* (a freshwater-loving mangrove flora) is much less compared to *Avicennia* spp. or *Excoecaria agallocha*, and on increasing the salinity stress, the chlorophyll level decreased significantly in *N. fruticans* (***Vide*** Annexure 8A), thus indicating towards the poor coping potential of the species in climate change-induced sea level rise.

8.2 Adaptation of Organisms Inhabiting Deep Sea

The mystery of the deep ocean has attracted man since time immemorial and after a long preparation backed up with scientific technology, two scientists and a pilot ventured in the spring of 1977 through a small spherical submarine, and the Alvin into the green Pacific swells 274 km north-east of Galapagos islands and due west of Columbia, South America. Entering into this strange world, John Edmond and biologist John Corliss discovered an entirely new ecosystem based on the heat energy from hydrothermal vents. The living community in this surprising ecosystem encompassed yellow vent mussels, crabs, large vent clams and tube worms with red plumes. The environment in the deep ocean is totally different from the surface. There is no sunlight to support photosynthesis, and the cold temperatures and extreme pressures at great depths are thought to limit the variations in the biotic community. Although the conditions of the deep sea are adverse, they are stable. This has allowed several thousand species to adapt to this habitat with a wide range of morphological, physiological and biochemical adaptations. Resource limitation is an important feature of the deep oceanic environment because of extremely low

Fig. 8.16 *Heritiera fomes* (Sundari) is gradually vanishing from the central sector of Gangetic delta due to rising salinity

temperature; some building materials like calcium carbonate are metabolically expensive in terms of secretion and maintenance. Under this restricted energy budget, along with a combination of cold temperature and short nutrients supply, dwarfism is noticed in some infauna that secretes calcium carbonate. Gigantism, on the other hand, occurs mainly in the epifaunal sector, and it is achieved through some unique low-energy constructional scheme. The apparently gigantic fauna of the deep sea are thus in true sense caloric dwarf. Associated with the morphological peculiarities are other adaptations like low metabolic rate, bioluminescence, low reproductive potential and use of skeletal materials other than calcium carbonate. All these marvellous specializations achieved by the members of the deep-sea biotic communities are discussed here.

8.2.1 Morphological Adaptations

Gigantism and bizarreness are the two striking features of deep-sea fauna. Gigantism implies larger than the normal size, and to achieve this enhanced body volume, the organisms often deviate from the standard body plans or morphologies of familiar or typical living relatives giving rise to bizarreness in the animal kingdom. In fact the large body volume is an adaptive measure to retain heat to withstand the chilling cold of the deep oceanic environment. The large size is attained through several ways like the elongation of appendages and sense organs, extreme flatness, mineralized lattice construction, stalk elongation, agglutination and big bag construction (Table 8.2).

Physiological Adaptations

The main physiological adaptations of the deep-sea animals include:

- Low basal metabolic rate
- Low metabolic activity levels, lethargy and vertical orientation
- Low reproductive potential with eggs few in number
- Slow gametogenesis
- Late reproductive maturity

Fig. 8.17 Acclimatization of *Heritiera fomes* (Sundari) marked by 'E' (means experimental) before mass plantation as part of species conservation programme

- Slow embryonic development
- Low growth rates
- Reduced biomass/volume ratio

The low basal metabolic rates in deep-sea animals were confirmed from the in situ experiment of oxygen uptake by Smith and Teal (1973) who demonstrated reduced rates of respiration by abyssal meiofauna.

The rates of metabolic activities related to growth (particularly when calcium carbonate is concerned), feeding and reproduction appear to be low, but they are difficult to assess in terms of their relative importance. Some organisms are consistently lethargic, while others alternate between periods of very low and normal metabolic activity. *Peniagone diaphana* is a deep-sea elasipod holothurians, which are free-swimming forms that orient vertically with oral tentacles hanging downwards in a lethargic state near the bottom. This orientation involves negligible expenditure of energy, and thus low metabolic activity level is achieved.

The deep-sea animals also possess low reproductive potential and exhibit slow gametogeneses which are the various ways to restrict or minimize the expenditure of energy. Sanders (1977) and Allen (1979) reported a strange incidence (adaptation) in relation to gonadial development of a deep-sea bivalve *Tindaria callistiformis*. They found that this minute bivalve exhibits no gonadial development during the first 30–40 years of its life and their sexual maturity is attained at 100+ years. The process of embryonic development is very slow in certain deep oceanic organisms, which requires a very low expenditure of energy.

The growth rate in most deep-sea animals is extremely slow owing to low metabolic rates. The deposit-feeding bivalve of the deep-sea *Tindaria callistiformis* (recovered from a depth of 3800 m in the North Atlantic) attained a size of only 8 mm in 98 years of life time (Turekian et al. 1975). Although there are some exceptionally large bivalves in the deep sea, most do not exceed 5 mm in the adult size (Allen 1979). In some deep-sea organism, reduction in biomass is also observed instead of size or volume reduction. Oliver (1979) has shown that abyssal species of the bivalve genus *Limopsis* have extremely small visceral mass relative to the size of the mantle cavity and gills. In these species there is a reduction of the biomass/volume ratio that may be attributed to low metabolic rates.

8.2.2 Biochemical Adaptations

The deep-sea animals have oriented their enzyme flexibility to cope with the extreme pressure and temperature prevailing there. Flexibility suggests structural labiality, i.e. the very features of enzyme structure that provides superfast efficiency of their activity, and this flexibility allows enzymes to undergo rapid and reversible changes in shape (conformation) during the catalytic sequence. It has been observed that the extreme temperature and pressure in the deep oceanic environment play important roles in altering the enzyme conformation. A study conducted in this context shows that a pressure of

Table 8.2 Common morphological adaptations of deep-sea animals

Morphological adaptation	Description	Example
Elongation of appendages and sense organs	Deep-sea animals often increase their effective size through development of long, thin appendages or processes such as walking legs, sense organs or spines. Certain abyssal pycnogonids with a leg span of 60 cm have only 5 cm body length. In certain benthic isopods (*Munopsis latifrons*), the walking legs are 116 mm long, more than seven times longer than their body length	*Colossendeis colossea* and *Munopsis latifrons*
Extreme flatness	Deep-sea organisms sometimes increase effective size at the expense of their three-dimensional volume by flattening or expanding to occupy a greater amount of two-dimensional space. Organisms living on fine-grained soft substrates increase their body surface area by flattening that helps them to minimize the potential for sinking into the substrate	*Serolis* sp.
Mineralized lattice construction	Deep-sea animals sometimes increase their size through construction of a mineralized skeleton. In extreme cold water, this cannot be accomplished with calcium carbonate, but a large size can be attained using silica lattices	Hexactinellid sponges
Stalk elongation	Many deep-sea animals specially the passive suspension-feeding organisms achieve large size through stalk elongation. This helps the organism to place itself in the proper current regime so that maximum water can pass through a given cross-sectional area and provide nourishment to the animal	Hexactinellid sponges
Agglutination	Several groups of marine organisms dwelling in the deep sea utilize prefabricated building blocks by removing particles from sediment or aquatic compartment and finally incorporate them in their body skeletons. Many agglutinated foraminifera are able to achieve large size with this type of construction	*Lena neglecta*
Big bag construction	Large body size of deep-sea creatures is often attained by assuming the form of a gelatinous or mucus bag that is filled primarily with water. Certain holothurians of the deep sea attain large size through use of gut cavity as a fluid internal skeleton (Barnes et al. 1976)	*Peniagone diaphana*

340 atm inhibits the velocity of LDH reaction of a deep-sea fish by only 5 %, while for shallow living fishes, inhibition is much greater averaging 17 %. This proves the unique ability of the deep-sea faunal enzymes to retain catalytic ability even under high pressure. However, this is not a generalized view for all the deep-sea organisms, and many studies are still needed to understand the modifications of enzymes in this hostile environment.

Bioluminescence is another notable adaptation of the deep-sea fauna which is mainly used for luring the prey or for sex identification. In case of Lantern fish, the downward portion of the female shines and the male shines upwards. The deep-sea angler fish dwelling in perpetual darkness uses the property of self-luminescence for sexual identification. In fact it is very difficult to distinguish between the male and the female in this case as the male fish, no more than 10 cm in length, is attached with the female fish of approximate length of 3.6–4.8 m. In this case, the light is possessed only by the female and is situated at the end of a movable rod attached to the front part of the head of the female fish. The property of bioluminescence is also used for defence purposes. Some deep water squid species release a bioluminescence fluid that clouds the water with light, confusing predators by distracting their attention. The opossum shrimp (belonging to Mysidae) release a substance on being chased by prey that bursts into a cloud of miniature light particles. The sudden burst of light frightens and confuses potential predators, allowing the shrimp to make an escape.

Bioluminescence occurs when a protein called luciferin is combined with oxygen in the presence of an enzyme called luciferase and ATP. During the series of reactions that follows, the chemical energy of ATP is converted into light energy. This process is efficient producing almost 100 % light and very negligible heat energy. This characteristic is especially common in animals dwelling between 300 and 2400 m depth. Some deep-sea fauna like certain species of squid, crustacean and fish have their own luminescent organ, while others harbour bioluminescent bacteria in species-specific locations.

8.2.3 Vent Communities of the Deep Sea

In addition to existence of life in the deep waters of the ocean, the floors are important habitats for biotic community. In 1977, oceanographers discovered a unique community off the Galapagos Islands along volcanic ridges in the ocean floor. Since this initial discovery, marine scientists have discovered several other vent communities in different regions of the world including off the coast of Oregon, the west of Florida, the Gulf of Mexico and in the central Gulf of California. These communities are usually located around the deep-sea hydrothermal vents that bring superheated water and nutrients from beneath the earth's crust. The submarine hydrothermal vents are associated with the volcanically active zone at the crest of the mid-oceanic ridge system. The base of the oceanic crust is extremely hot (>1000 °C), and its upper boundary is in contact with sediments, due to which the elevation of temperature of the ambient media occurs. Hydrothermal activity accounting for water temperature elevations ranging from a few degrees above ambient to 350–400 °C is now known to occur at seafloor spreading centres worldwide.

On the basis of temperature, three main types of hydrothermal vents have been recognized which are:

(a) **Diffuse vents:** emitting low temperature, clear waters up to 30 °C.
(b) **White smoker vents:** releasing milky fluids with temperature varying from 200 to 330 °C.
(c) **Black smoker vents:** discharging jets of water blackened by sulphide precipitates at temperatures between 300 and 400 °C. Gaill and Hunt (1991) described a fourth kind of smoker, **the basal mound variety** characterized by sulphide-cemented biogenic tubes. The most common type of basal mound smoker is found on the East Pacific rise.

On the basis of geological setting, hydrothermal vents can be classified into two main types (Tivey and Johnson 1989): (1) those at sedimented spreading centres and (2) those at unsedimented spreading centres.

Despite the extremes of temperature and pressure, these self-contained communities are some of the most productive in the sea and stand as the proof of dependence of biotic community on nonconventional energy source. Chemosynthesis forms the foundation of such community and chemosynthetic bacteria occupy the first tropic level. Few common fauna of the hydrothermal vent community are listed here:

(1) Giant white clams—*Calyptogena magnifica*
(2) Mussels—*Bathymodiolus thermophilus*
(3) Tube worm—*Riftia pachyptila*
(4) Galatheid crab—*Munidopsis* sp.
(5) Enteropneust worm—*Saxipendium coronatum*
(6) Brachyuran crab—*Bythograea thermydron*
(7) Turrid gastropod—*Phymorphynchus* sp.

The species of vent communities share certain common features as listed here:

1. Most species have planktotrophic larva and have non-planktotrophic development and hence low dispersion.
2. Long clams, mussels, tube worms or a large variety of other animals depend on bacteria that oxidize the hydrogen sulphide gas dissolved in the spring waters as a source of energy to chemosynthetically produce food for the community.
3. Ocean water sulphur, Mn and Fe support the sulphur bacteria of the vent community, which depend on metabolic systems that utilize sulphur instead of oxygen.
4. Ecosystem of the vent communities are unique because it exists in areas where the average depth is 2500 m (8200 ft) and the temperature is above 350 °C inside the vent and about 250 °C in the surrounding waters. The vent community changes its place or character when the vent ceases to emit mineral oxides and sulphides.

8.2.4 Properties of Vent Community

The physico-chemical variables in hydrothermal vent regions are extremely drastic. There are few megafauna, endemic in nature, low diversity and unique community arrangement with the chemosynthetic bacteria at the base of the trophic level. The hydrothermal vent has a discrete ecosystem and is very efficient to sustain high pressure at a depth of 2500 m and 350 °C temperature. These vent communities settle only at these places due to mineral-rich (sulphide) waters. The animals living here do not have gut and exchange the materials with the surrounding environment of the vent.

Karl et al. (1984) measured production rates at one vent of 19 μgC/h. The primary consumers are the various animals that filter out the bacteria from the water, graze the bacterial film from the rocks or are locked in a symbiotic association with the bacteria. The various adaptations in this very peculiar ecosystem may be grouped under three broad headings.

(1) **Interspecific relationship:** The most surprising events at the vents are the symbiotic relationship established between bacteria and a variety of metazoans, including vestimentiferan tube worms, vesicomyid clams, mytilids and other invertebrates. Symbiont-assisted animals that retain functional digestive organs have been the subject of interest. Of particular interest are the polychaetes that exhibit several feeding types.

(2) **Metabolism:** In general sulphide levels in the low micromolar range effectively inhibit aerobic respiration and lead to death of organisms when individuals are exposed to the sulphide for sufficiently long periods of time.

(3) **Chemosynthesis:** The ecosystem in hydrothermal vent is governed by geothermal energy as photosynthesis fails to occur in this zone due to complete absence of sunlight. The reaction of seawater with crustal rocks at high temperature produces inorganic compounds (e.g. H_2S, S, $S_2O_3{}^{-2}$, $NH_4{}^+$, Fe^{2+}, $NO_2{}^-$, Mn^{+2}) that discharge from the hot springs and provide energy for the free-living and/or symbiotic chemosynthetic bacteria, which form the base of the food chain in these specialized habitats. By oxidizing sulphides especially H_2S as well as other reduced substrates such as H_2, Fe^{2+} and Mn^{+2} released from vents or colder seeps, the microbes obtain energy to synthesize organic compounds from carbon dioxide in seawater, the process being referred to as autotrophy. In so doing, the bacteria support copious populations of specifically adapted invertebrates living in close proximity to the vents. More than 250 strains of free-living bacteria have been isolated from deep-sea thermal environments. Among these microbial populations, the most common are S-, CH_4-, H_2-, Fe- and Mn-oxidizing bacteria.

Brain Churners

1. What factors control the survival, abundance and type of organisms in the intertidal zone?
2. How competition or pressures of predators limit the abundance of organisms?
3. Why tide pool is a congenial habitat for marine and estuarine organisms?
4. What are the different methods by which rocky shore organisms get attached to the hard substrata?
5. How are oysters adapted to thrive on hard substrata in the seashore?
6. Name few organisms of the muddy substratum. How they adapt to the soft bottom?
7. How rocky shore organisms adapt themselves to desiccation?
8. How the mangroves regulate excess salt in their tissues?
9. Give some examples of detritivores. Do they have any role in bioturbation?
10. Why do the deep-sea animals construct their bodies with silica lattices instead of calcium carbonate?

Annexure 8A: Signals of Salinity Fluctuation in *Nypa fruticans* Seedling Growth, a Case Study from the Lower Gangetic Plain

1. Introduction

The mangrove community is greatly affected by climate change-induced salinity fluctuation, dilution by precipitation and run-off, temperature oscillation, atmospheric carbon dioxide concentration, etc. Of all the outcomes from changes in the atmosphere's composition and alterations to land surfaces, relative sea level rise may be the greatest threat (Field 1995; Lovelock and Ellison 2007) as the event increases the soil and aquatic salinity often beyond tolerance of the sensitive mangrove species. Excessive saline conditions retard seed germination and impede growth and development of mangroves. Indian Sundarbans, the famous mangrove chunk of the tropics, is gradually losing *Heritiera fomes* (commonly known as Sundari) owing to increase of salinity in the central sector of the delta complex around the Matla River (Mitra et al. 2009a). Reports of alteration of growth in mangroves due to difference in salinity between western and central sectors of Indian Sundarbans are available (Mitra et al. 2004). However, no study has yet been carried on the effect of salinity fluctuation on the photosynthetic pigments and carotenoid level of *Nypa fruticans* under culture conditions from this part of the Indian subcontinent. The effects of salinity on mangroves have been studied in relation to antioxidative enzymes (Takemura et al. 2000; Parida et al. 2004b), leaf structure, rates of transpiration, stomatal conductance and rates of photosynthesis (Santiago et al. 2000; Parida et al. 2004a) and changes in chloroplast structure and function (Parida et al. 2003). Tanaka et al. (2000) reported that Na^+/H^+ antiport catalysed exchange of Na^+ for H^+ across the vacuolar membrane of the cells of *Bruguiera sexangula* offered tolerance to ionic stress imposed by NaCl and this mechanism was important for cellular salinity adjustments. Also, the mechanism of acclimation to salt in mangroves was suggested to be linked to the changes in the vacuolar size in *B. sexangula* (Hotta et al. 2000). Further, one of the biochemical mechanisms by which mangroves counter the high osmolarity of salt was accumulation of compatible solutes (Takemura et al. 2000).

In this paper, we present the effect of salinity on pigments in *Nypa fruticans* under hydroponic culture with an aim to obtain insights into the changes in chlorophyll and carotenoid level with salt acclimation. Such study is important from the point of sea level rise and subsequent saline water intrusion into the islands of Indian Sundarbans as the lower Gangetic delta complex is extremely vulnerable to climate change-related effects owing to its location below the mean sea level and experiencing a sea level rise of 3.14 mm/year. Moreover unlike other mangrove species, *Nypa fruticans* prefer extremely low saline condition and hence can act as signature of climate change-induced sea level rise and subsequent intrusion of saline water into the islands.

2. Materials and Methods

2.1 Plant Materials and Culture Conditions

Seeds of *Nypa fruticans* were collected from Sundarbans mangrove system of India. Seedlings were raised in the laboratory condition by diluting the water collected from the lower stretch of Sundarbans adjacent to Bay of Bengal in the southern part (salinity = 30 psu) under photosynthetically active radiation (PAR) of 1220–-1236 μmol m^{-2} s^{-1}. Two-month-old healthy seedlings were selected for hydroponic culture in Hoagland's nutrient medium (pH = 5.8–6.0). The preliminary experiments were carried out in the selected species at five different salinities (2 psu, 5 psu, 10 psu, 15 psu and 20 psu) in order to determine the optimum range of salinities in context to photosynthetic pigments and carotenoid. The cultures were aerated continuously with an air bubbler. The hydroponic cultures were maintained in a culture room under a 14 h photoperiod at PAR of 300 μmol m^{-2} s^{-1}, 26 ± 3 °C and 80 % RH. The culture medium was changed every 7 days. Leaves were harvested at 7-, 14-, 21- and 30-day intervals to measure the pigment concentrations.

2.2 Extraction and Estimation of Pigments

We followed the method of Porra et al. (1989) for chlorophyll estimation and Arnon (1949) for carotenoid estimation in the leaves of *Nypa fruticans*. The procedures for pigment estimation are shown here in flow charts.

Estimation of chlorophyll *a* and chlorophyll *b*

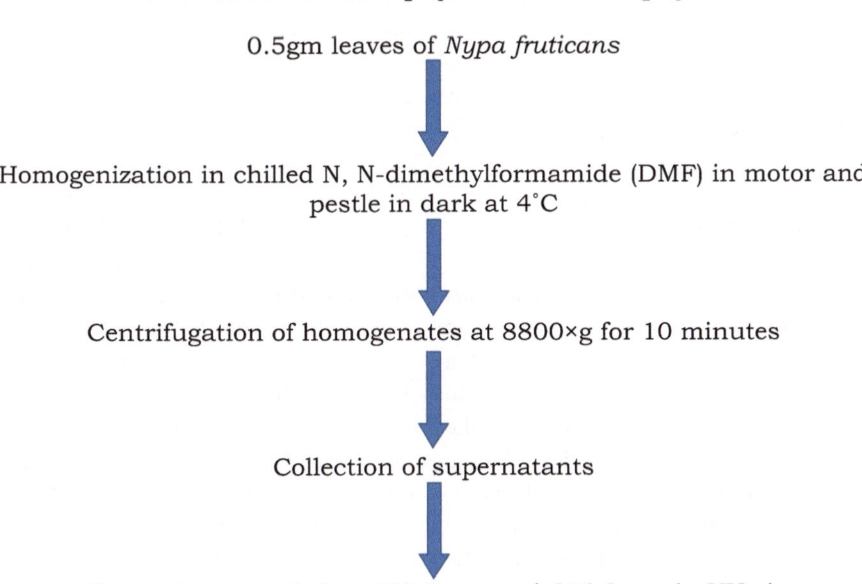

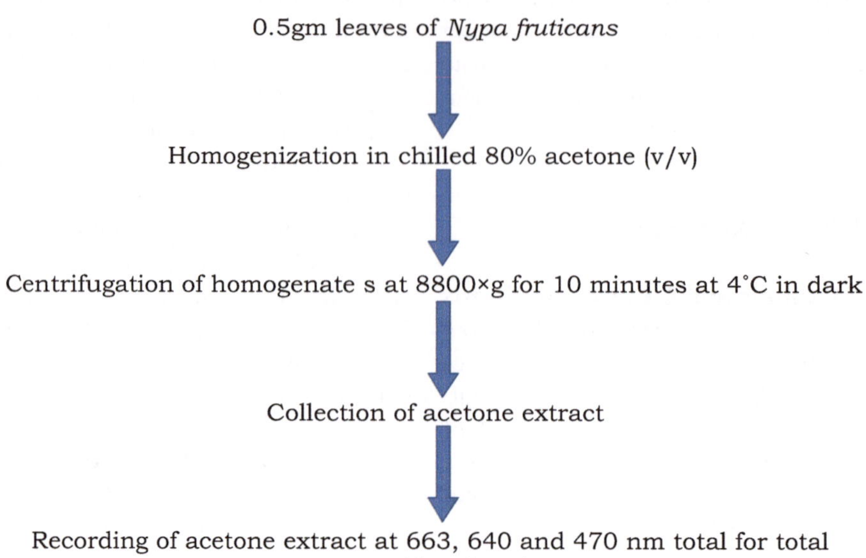

2.3 Statistical Analysis

Statistical analysis of the results was carried out according to Duncan's multiple range tests. Data were also subjected to analysis of correlation coefficient (r) in order to evaluate the interrelationship between salinity and selected pigments following the method of Sokal and Rohlf (1995).

3. Results and Discussion

All the collected seedlings of *Nypa fruticans* tolerate maximum salinity up to 15 psu and could be maintained for more than 30 days. On exposure to salinity of 20 psu, the leaves began to fall off after a week or so, and thus all the experiments were done up to 30 days in the salinity level 2, 5, 10 and 15 psu treated plants. The unhealthy conditions of the experimental seedlings of *Nypa fruticans* at 20 psu may be attributed to their ambient salinity in the western sector of deltaic Sundarbans region from where they were collected which usually ranges between 2 psu and 10 psu (Mitra 2000). Such a low saline belt in the western part of Indian Sundarbans may be attributed to freshwater supply from the Ganga–Bhagirathi channel, which originates from the Gangotri glacier of the Himalayan range.

The concentrations of chlorophyll and carotenoid pigments decreased significantly with the increase in salinity. The total chlorophyll expressed, on unit fresh wt. basis, decreased by 63.44 %, 73.33 %, 63.39 % and 63.89 % at 7-, 14-, 21- and 30-day intervals, respectively. The Chl *a:b* ratio in the plant, however, remained almost constant for the species and varied only marginally during the period under observation. In our experiments with differential salinity exposure, the Chl *a:b* ratio yielded a value of 3.00–3.41. It, thus, appears that high salinity did not affect Chl *a:b* ratio even though the total chlorophyll content decreased at high salt concentration. A similar trend in carotenoid content, expressed in fresh wt. basis, was observed. The pigment decreased by 36.84 % at the end of 7 days, 33.33 % at the end of 14 days, 33.33 % at the end of 21 days and 27.78 % at the end of 30 days. The decrease of the selected pigments with aquatic salinity is statistically significant.

The decrease in chlorophyll content at higher salinity might possibly due to changes in the lipid protein ratio of pigment–protein complexes or increased chlorophyllase activity (Iyengar and Reddy 1996). Our results agree with several reports of decrease content of chlorophyll and carotenoids by salinity as reported in a number of glycophytes (Gadallah 1999; Agastian and Kingsley 2000). As the Chl *a:b* ratio remained unaffected at high saline condition in the selected species, it appears that the light-harvesting complexes (LHCs) of thylakoid membranes are little altered by salt exposure.

The adverse impact of salinity on leaf chlorophyll of mangrove species significantly affects the rate of photosynthesis. Various studies have shown that a number of mangrove species grow best at salinities between 4 psu and 15 psu (Connor 1969; Clough 1985; Downton 1982; Burchett et al. 1984; Clough 1984). Till date there have been few studies on the effect of salinity on photosynthetic gas exchange in mangroves. Clough (1985) stated in his communication that the rate of light-saturated photosynthesis decreases with increasing salinity of ambient media, attributing this to co-limitation of assimilation rate by stomatal conductance and photosynthetic capacity in response to differences in water status induced by the various salinity treatments. Thus, on the evidences available so far, it is most likely that salinity exerts its effect on photosynthesis mainly through changes in leaf water status and this study confirms that the photosynthetic process may be affected at high saline condition due to decrease in chl *a* and *b* concentrations in mangroves. The present study is different from several works as the salinity of water has been altered naturally (through rainwater dilution) keeping all the constituent salts of brackish water constant unlike several previous studies where the plants were exposed to different NaCl concentrations (Mishra and Das 2003; Netondo et al. 2004) that are not the real image of ambient seawater.

Our results show that *Nypa fruticans* of Indian Sundarbans region can easily be propagated under low salinity conditions. At 15 psu, the plants become acclimated to salt after 1–2 weeks of exposure, but at 20 psu the seedlings could hardly adapt.

Indian Sundarbans and its adjacent estuaries at the apex of Bay of Bengal are one of the less studied regions of the world ocean in context to impact of rising salinity fluctuation on mangrove floral community, although the region sustains the 5th largest mangrove chunk in the world (2120 km^2 in the Indian part and 4500 km^2 in the Bangladesh part). The present study is extremely important from the point of view of rising salinity in the central sector of Indian Sundarbans over a period of two decades (Mitra et al. 2009b) due to complete obstruction of the freshwater supply of Ganga-Bhagirathi-Hooghly River as a result of heavy siltation since the late fifteenth century (Chaudhuri and Choudhury 1994) and rising sea level (Hazra et al. 2002) at the rate of 3.14 mm/year, which is higher than the global average sea level rise of 2.12 mm/year. The pigments, being the key machinery in regulating the growth and survival of the mangroves, require an optimum salinity range between 4 and 15 psu (Downton 1982; Burchett et al. 1984) for proper functioning. *Nypa fruticans*, the freshwater-loving mangrove species, prefers an optimum salinity between 2 and 5 psu (Mitra et al. 2004). It appears that the growth of the species would be better if freshwater of the western sector of Indian Sundarbans is channelized to the central sector.

Acknowledgments

The financial assistance from the Clean Blue Planet Consultancy Services, Kolkata, is gratefully acknowledged.

Table 8B.1 Effects of different salinities on pigment concentrations in *Nypa fruticans*

Duration of treatment (d)	Salinity (psu)	Chl *a*	Chl *b*	Total chl	Chl *a:b*	Carotenoid
7	2	0.83	0.29	1.12	2.86	0.21
	5	0.76	0.30	1.06	2.53	0.18
	10	0.65	0.25	0.90	2.60	0.16
	15	0.60	0.24	0.84	2.50	0.15
	20	0.41	0.16	0.57	2.56	0.09
14	2	0.79	0.31	1.10	2.55	0.19
	5	0.75	0.29	1.04	2.59	0.18
	10	0.67	0.26	0.93	2.58	0.13
	15	0.53	0.21	0.74	2.52	0.10
	20	–	–	–	–	–
21	2	0.73	0.30	1.03	2.43	0.16
	5	0.68	0.27	0.95	2.52	0.14
	10	0.51	0.20	0.71	2.55	0.12
	15	0.50	0.19	0.69	2.63	0.11
	20	–	–	–	–	–
30	2	0.66	0.26	0.92	2.54	0.18
	5	0.52	0.20	0.72	2.60	0.13
	10	0.48	0.19	0.67	2.53	0.12
	15	0.47	0.18	0.65	2.61	0.06
	20	–	–	–	–	–

Units of all pigments are mg.gm^{-1} fresh weight; different letters besides figures indicate statistically different means as at $p \leq 0.01$

Table 8B.2 Interrelationships between salinity and selected pigments in *Nypa fruticans*

Combination	'*r*' value	'*p*' value
Salinity × Chl *a*	−0.74913	<0.01
Salinity × Chl *b*	−0.74615	<0.01
Salinity × total Chl	−0.75187	<0.01
Salinity × carotenoid	−0.78281	<0.01

References

Allen, J. A. (1979). The adaptations and radiation of deep-sea bivalves. *Sarsia, 64*, 19–27.

Ansell, A. D., & Trueman, E. R. (1973). The energy cost of migration of the bivalve *Donax* on tropical sand beaches. *Marine Behavior and Physiology, 2*, 21–32.

Azocar, A., Rada, F., & Orozco, A. (1992). Relaciones hidricas e intercambio de gases en dos especies de mangle, con mecanismos contrastantes de regulacion de la salinidad interna. *Ectropicos, 5*(2), 11–19.

Balsamo, R. A., & Thomson, W. W. (1995). Salt effects on membranes of the hypodermis and mesophyll cells of *Avicennia germinans* (Avicenniaceae): A freeze-fracture study. *American Journal of Botany, 82*(4), 435–440.

Barnes, A. T., Queten, L. B., Childress, J. J., & Pawson, D. L. (1976). Deep-sea macroplanktonic sea cucumbers: Suspended sediment feeders captured from deep submergence vehicle. *Science, 194*, 1083–1108.

Boaden, P. J. S. (1989). Meiofauna and the origins of the Metazoa. *Zoological Journal of Linnaeus Society, 96*, 217–227.

Boaden, P. J. S., & Platt, H. M. (1971). Daily migration patterns in an intertidal meiobenthic community. *Thalassia Jugoslavica, 7*, 1–12.

Drennan, P. M., Berjak, P., & Pammenter, N. W. (1992). Ion gradients and adenosine triphosphatase localization in the salt glands of *Avicennia marina* (Forssk.) Vierh. *South African Journal of Botany, 58*(6), 486–490.

Fahn, A. (1979). *Secretory tissues in plants*. London: Academic.

Fenchel, T. M., & Redl, R. J. (1970). The sulfide system: A new biotic community underneath the oxidized layer of marine sand bottoms. *Marine Biology, 7*, 255–268.

Gaill, F., & Hunt, S. (1991). The biology of annelid worms from high temperature hydrothermal vent Remons. *Review of Aquatic Science, 4*(2–3), 107–137.

Hurd, C. L., Galvin, R. S., Norton, T. A., & Dring, M. J. (1993). Production of hyaline hairs by intertidal species of *Fucus* (Fucales) and their role in phosphate uptake. *Journal of Phycology, 29*, 160–165.

Karl, D. M., Knauer, G. A., Martin, J. H., & Ward, B. B. (1984). Bacterial chemolithotrophy in the ocean is associated with sinking particles. *Nature, 309*, 54–56.

Meyers, M. B., Fossing, H., & Powell, E. N. (1987). Micro-distribution of interstitial meiofauna, oxygen and sulfide gradients, and the tubes of macro-infauna. *Marine Ecology Progress Series, 35*, 223–241.

Naylor, E. (1985). Tidally rhythmic behaviour of marine animals. *Symposium Society of experimental Biology, 39*, 63–93.

Nott, J. A. (1973). Settlement of the larvae of *Spirorbis spirorbis* L. *Journal of Marine Biology Association, UK, 53*, 437–453.

Odum, E. P. (1971). *Fundamentals of ecology* (3rd ed.). Philadelphia: W.B. Saunders Company. 574 pp.

Oliver, P. G. (1979). Adaptations of some deep-sea suspension feeding bivalves (*Limopsis* and *Bathyarca*). *Sarsia, 64*, 33–36.

Reise, K., & Ax, P. (1979). A meiofaunal 'thiobios' limited to the anaerobic sulfide system of marine sand does not exist. *Marine Biology, 54*, 225–237.

Sanders, H. L. (1977). Evolutionary ecology and the deep-sea benthos. In C. E. Goulden (Ed.), *The changing scenes in natural sciences 1776–1976* (pp. 223–243). Philadelphia: Academy of Natural Sciences Special Publication.

Schmitz, K., & Srivastava, L. M. (1980). Long distance transport in *Macrocystis integrifolia*. III. Movement of THO. *Plant Physiology, 66*, 66–69.

Scholander, P. F., Hammel, H. T., Hemmingsen, E. A., & Cray, W. (1962). Salt balance in mangroves. *Plant Physiology, 37*, 722–729.

Smith, K. L., Jr., & Teal, J. M. (1973). Temperature and pressure effects on respiration of thecosomatous pteropods. *Deep Sea Research, 20*, 853–858.

Tivey, M. A., & Johnson, H. P. (1989). High-resolution geophysical studies of oceanic hydrothermal systems. *CRC Critical Review of Aquatic Science, 1*, 473–496.

Tomlinson, P. B. (1986). *The botany of mangroves.* Cambridge: Cambridge University Press. 413 pp.

Turekian, K. K., Cochran, J. K., Kharkar, D. P., Cerrato, R. M., Vaisnys, J. R., Sanders, H. L., Grassle, J. F., & Allen, J. A. (1975). Slow growth rate of a deep-sea clam determined by 228Ra chronology. *Proceedings of National Academy of Science of the United States of America, 72*, 2829–2832.

Ugolini, A., Scapini, F., & Pardi, L. (1986). Interaction between solar orientation and landscape visibility in *Talitrus saiwror* Montagu (Crustacea-Amphipoda). *Marine Biology, 90*, 449–460.

Walter, H. (1961). Salinity problems in the acid zones. The adaptations of plants to saline soils. *Arid Zone Research, 14*, 65–68.

Williams, J. A. (1983). The endogenous locomotor activity rhythm of four supralittoral peracarid crustaceans. *Journal of the Marine Biological Association U.K., 63*, 481–492.

Annexures References

Agastian, P., & Kingsley, S. J. (2000). Effect of salinity on photosynthesis and biochemical characteristics in mulberry genotypes. *Photosynthetica, 38*, 287–290.

Arnon, D. I. (1949). Copper enzymes in isolated chloroplasts, polyphenol oxidase in *Beta vulgaris. Plant Physiology, 24*, 1–15.

Burchett, M. D., Field, C. D., & Pulkownik, A. (1984). Salinity, growth and root respiration in the grey mangrove *Avicennia marina. Physiologia Plantarum, 60*, 113–118.

Chaudhuri, A. B., & Choudhury, A. (1994). *Mangroves of the Sundarbans, India* (Vol. I). Published by IUCN.

Clough, B. F. (1984). Growth and salt balance of the mangroves *Avicennia marin*a (Forsk.) Vierh, and *Rhizoplhora slylosa* griff. in relation to salinity. *Australian Journal of Plant Physiology, 11*, 419–430.

Clough, B. F. (1985). Effect of nutrient supply on photosynthesis in mangroves. In: The mangroves. Proceedings of the National Symposium on Biology, Utililization and Conservation of Mangroves. Shivaji University, Kolhapur, pp. 80–88.

Connor, D. J. (1969). Growth of grey mangrove *(Avicennia marina)* in nutrient culture. *Biotropica, 1*, 36–40.

Downton, W. J. S. (1982). Growth and osmotic relabor. s 01 the mangrove *Avicennia marina,* as influenced by salinity. *Australian Journal of Plant Physiology, 9*, 519–528.

Field, C. (1995). Impacts of expected climate change on mangroves. *Hydrobiologia, 295*, 75–81.

Gadallah, M. A. A. (1999). Effects of proline and glycinebetaine on *Vicia faba* response to salt stress. *Biology Plant, 42*, 249–257.

Hazra, S., Ghosh, T., Dasgupta, R., & Sen, G. (2002). Sea level and associated changes in Sundarbans. *Science and Culture, 68*, 309–321.

Hotta, M., Nemoto, S., & Mimura, T. (2000). Re-evaluation of role of vacuole during salt adaptation in higher plant cells. *Plant Cell Physiology, 41*, 79.

Iyengar, E. R. R., & Reddy, M. P. (1996). Photosynthesis in high salt-tolerant plants. In M. Pesserkali (Ed.), *Handbook of photosynthesis* (pp. 56–65). Baten Rose: Marshal Dekar.

Lovelock, C. E., & Ellison, J. C. (2007). Vulnerability of mangroves and tidal wetlands of the Great Barrier Reef to climate change. In J. E. Johnson & P. A. Marshall (Eds.), *Climate change and the Great Barrier Reef: A vulnerability assessment* (pp. 237–269). Australia: Great Barrier Reef Marine Park Authority and Australian Greenhouse Office.

Mishra, S., & Das, A. B. (2003). Effect of NaCl on leaf salt secretion and antioxidative enzyme level in roots of a mangrove, Aegiceras corniculatum. *Indian Journal of Experimental Biology, 41*, 160–166.

Mitra, A. (2000). The north-west coast of the Bay of Bengal and deltaic Sundarbans. *Seas at the Millennium: An Environmental Evaluation, U.K., 2*, 160.

Mitra, A., Banerjee, K., & Bhattacharyya, D. P. (2004). *The other face of mangroves.* Department of Environment, Government of West Bengal Publication Kolkata (India).

Mitra, A., Gangopadhyay, A., Dube, A., Schmidt, C. K., & Banerjee, K. (2009a). Observed changes in water mass properties in the Indian Sundarbans (Northwestern Bay of Bengal) during 1980–2007. *Current Science, 97*, 1445–1452.

Mitra, A., Zaman, S., Chakraborty, R., Halder, P., & Banerjee, K. (2009b). Abiotic indicators of climate change in Indian Sundarbans. *Indian Science Cruiser, 23*(2), 53–56.

Netondo, G. W., Onyango, J. C., & Beck, E. (2004). Sorghum and salinity: I. Response of growth, water relations, and ion accumulation to NaCl salinity. *Crop Science, 44*, 797–805.

Parida, A. K., Das, A. B., & Mittra, B. (2003). Effects of NaCl stress on the structure, pigment complex composition and photosynthetic activity of mangrove Bruguiera parviflora chloroplasts. *Photosynthetica, 41*, 191–200.

Parida, A. K., Das, A. B., & Mittra, B. (2004a). Effects of salt on growth, ion accumulation, photosynthesis and leaf anatomy of the mangrove *Bruguiera parviflora. Trees-Structural Function, 18*, 167–174.

Parida, A. K., Das, A. B., & Mohanty, P. (2004b). Defense potentials to NaCl in a mangrove, *Bruguiera parviflora*: Differential changes of isoforms of some antioxidative enzymes. *Journal of Plant Physiology, 161*, 531–542.

Porra, R. J., Thompson, W. A., & Kriendemann, P. E. (1989). Determination of accurate extinction coefficients and simultaneous equations for assaying chlorophyll a and b extracted with four different solvents: Verification of the concentration of chlorophyll standards by atomic absorption spectroscopy. *Biochim Biophysics Acta, 975*, 384–394.

Santiago, L. S., Lau, T. S., Melcher, P. J., Steele, O. C., & Goldstein, G. (2000). Morphological and physiological responses of Hawaiian *Hibiscus tiliaceus* population to light and salinity. *International Journal of Plant Science, 161*, 99–106.

Sokal, R. R., & Rohlf, F. J. (1995). *Biometry: The principles and practice of statistics in biological research* (3rd ed., pp. 321–356). New York: W. H. Freeman and Company.

Takemura, T., Hanagata, N., Sugihara, K., Baba, S., Karube, I., & Dubinsky, Z. (2000). Physiological and biochemical responses to salt stress in the mangrove, *Bruguiera gymnorrhiza. Aquatic Botany, 68*, 15–28.

Tanaka, Y., Fukuda, A., Nakamura, A., Yamada, A., & Saito, T. (2000). Molecular cloning and characterization of mangrove Na+/H+ antiporter cDNA. *Plant Cell Physiol, 41*, 27.

9 Hot Spots of Marine and Estuarine Ecosystems

Contents

9.1 Mangroves

Mangroves are a special type of vegetation that thrive in sediments that have a high salt concentration and are usually surrounded by seawater or estuarine water. As a result of these environmental conditions, they must conserve water and they exhibit adaptation similar to those found in salt marsh plants. These adaptations help to reduce the loss of water by evaporation from the leaves. In mangroves, the epidermis of the leaf is usually covered with a thick cuticle, and the stomata are sunken in nature that are usually confined to the undersurface. Another important adaptation in mangroves involves the germination of seeds. Unlike in most of the plant, the embryo in most mangroves germinates while the seed is still attached to parent plant.

The area occupied by mangroves in the planet Earth is roughly around 170,000 km^2. The ecosystem services provided by mangroves preferably to coastal population and land dwellers drive the economic wheel of the region. According to policy researchers, mangroves have tremendous social and ecological value. The annual economic value of mangroves, estimated by the cost of the products and services they provide, has been estimated to be $200,000–$900,000 per hectare (Wells et al. 2006).

In 2005, Badola and Hussain worked upon the function of mangroves in protecting the Bhitarkanika region in the state of Odisha,

A. Mitra, S. Zaman, *Basics of Marine and Estuarine Ecology*,
DOI 10.1007/978-81-322-2707-6_9

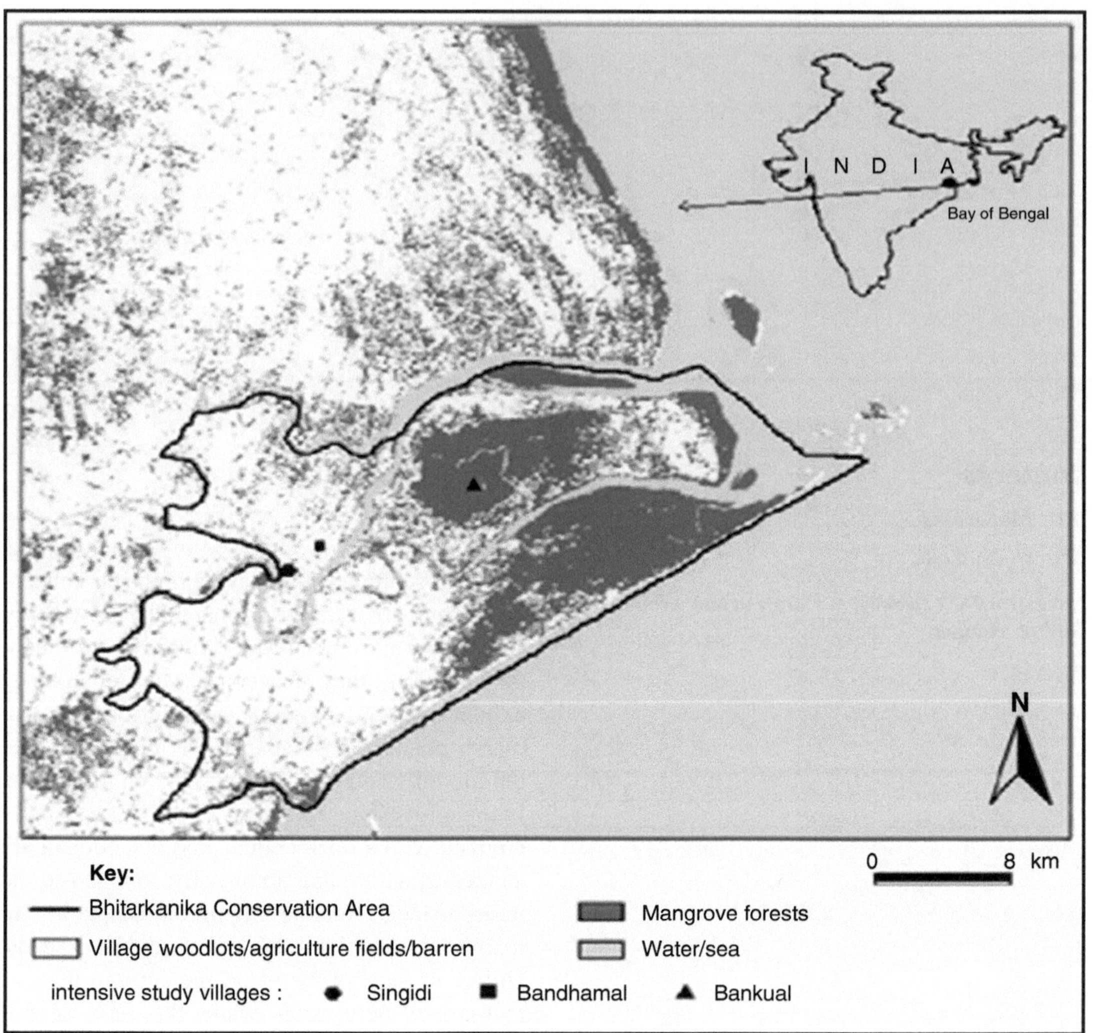

Fig. 9.1 Kendrapara and Bhadrak districts of Orissa, India, with the location of Bhitarkanika Conservation Area and the extent of mangrove forests

India. The second largest mangrove forest of the Indian subcontinent is the mangrove ecosystem of the Bhitarkanika (Fig. 9.1). Maximum biodiversity of mangrove flora and fauna is associated with this mangrove forest and its adjacent coastal areas. The dominant tree species *Sonneratia apetala*, *Heritiera fomes*, *Heritiera littoralis* and other *Avicennia* species make this mangrove forest different from others.

The report of Badola and Hussain (2005) did not aim to assess the biodiversity of the mangroves, but its primary objective was to measure the economic losses attributed to the 1999 supercyclone relative to the prevailing socio-economic conditions of the study villages. It evaluated the extent of damage caused in areas that were under the umbrella of mangrove forests and areas that were not in the wake of this supercyclone. In 1971, an embankment was created along the entire Orissa coast to prevent seawater intrusion into reclaimed paddy fields. Therefore, the report also studied the effectiveness of such artificial structures in providing storm protection, as opposed to mangrove forests.

Hence, the following three situations were identified: (1) a village in the shadow of mangroves; (2) a village not in the shadow of mangroves and with no embankment; and (3) a village not in the shadow of mangroves but with an embankment on the seaward side. Bankual village was in the shadow of a mangrove forest, Singidi village was neither in the shadow of mangroves nor protected by an embankment from storm surge and Bandhamal village was not in the shadow of mangroves but had a seaward side embankment. The report indicated that the intensity of the impact of the 1999 cyclone on these villages should have been fairly uniform, as all the three selected villages were equidistant from the seashore and had similar aspects. The two villages outside mangrove cover were located close to each other, but both were far from the mangrove forest in order to eliminate any effect of mangrove forest presence.

Services provided by the Bhitarkanika mangrove ecosystem in India and estimated cyclone damage avoided in the three selected villages, taking the supercyclone of 1999 as a reference point, were evaluated by assessing the socio-economic status of the villages; the cyclone damage to houses, livestock, fisheries, trees and other assets owned by the people; and the level and duration of flooding. Eleven variables were used to compare damage in the villages (Table 9.1). Attitude surveys were carried out in 10 % of the households in 35 villages located in the Bhitarkanika Conservation Area to assess local people's perceptions regarding the storm protection function of mangroves and their attitudes towards mangrove forests in general. In the mangrove-protected village, variables had either the lowest values for adverse factors (such as damage to houses) or the highest values for positive factors (such as crop yield). The loss incurred per household was greatest (US $153.74) in the village that was not sheltered by mangroves but had an embankment, followed by the village that was neither in the shadow of mangroves nor the embankment (US$44.02) and the village that was protected by mangrove forests (US$33.31). The local people were aware of and appreciated the functions performed by the mangrove forests in protecting their lives and property from cyclones and were willing to cooperate with the forest department with regard to mangrove restoration.

Although only indicative, the report shows that the damage attributed to the cyclone was more extensive in the village further away from the mangrove shadow. The embankments constructed in 1971, after a previous cyclone, to prevent the intrusion of saltwater into agricultural fields and villages were ineffective during the high storm surge; in fact, they acted as a barrier to run off when the water was receding. The embankments suffered a number of breaches that resulted in the flooding of villages such as Bandhamal, which was surrounded on all sides by the embankment. Singidi village, with no mangrove cover and no embankment, suffered the highest level of field inundation; however,

Table 9.1 Basic description and mean values of the variables (per household) in the three study villages in Bhitarkanika Conservation Area, India (US$1.00 = INR 45, August 2004)

		Villages		
Variable	Description	Singidi	Bankual	Bandhamal
DR	Damage to residences (0–19 scale)	9.40	5.34	10.44
PTD	Tree damage (%)	21.0	3.3	15.5
DPP	Damage to other personal properties (INR)	108.11	0.00	2375.00
DL	Damage to livestock in money terms (INR)	54.05	127.63	1044.37
FP	Flooding in premises (m)	0.34	0.29	0.58
FF	Flooding in fields (m)	1.99	1.09	1.39
WLF	Waterlogging in fields	9.46	5.63	12.87
CR	Cost of repair and reconstruction	996.97	682.86	973.21
Y99	Yield for the year 1999 (kg ha^{-1})	531	1479.5	335.9
LFS	Loss of fish	310.81	69.74	260.94
		1983.3	61,454.13	6918.62
TML	Total quantifiable variables (INR)	1983.3	61,454.13	6918.62

the seawater receded quickly, resulting in less damage to agricultural crops. Bankual village, which was in the shadow of mangrove forest and had minimal embankment around it, suffered the least. Although this study is not conclusive, the lack of breaches in the embankment closer to the forest is indicative of the protection provided by mangroves to the embankment. In areas far from the forest, several breaches in the embankment were observed. Water levels were higher and the flooding was of longer duration in Bandhamal.

Extensive *Casuarina* plantations established as a storm protection measure along the Orissa coast were ineffective in preventing damage; rather, they caused destruction to olive ridley sea turtle (*Lepidochelys olivacea*) nesting beaches. The cyclone uprooted almost all the trees in the immediate vicinity of the coast and caused much damage to trees several kilometres inland. However, mangrove forests and trees in the shadow of mangrove forests remained intact. The report contends that the vulnerability of many coastal communities to cyclones is heightened by the removal of mangroves for development, agriculture and habitation purposes (Valiela et al. 2001). Mangrove forests are natural buffers against storm surges and protect tropical shores from erosion by tides and currents. Ecological functions such as storm protection may be very important components in the total economic value of a wetland and may constitute almost 80 % of the estimated value. These major benefits are often the principal reasons for restoring mangrove forests along much of the low-lying deltaic coasts. In the aforementioned study in Orissa, there was a 20–30 % reduction in repair and maintenance costs of sea dyke systems due to the presence of mangroves in front of the dyke. The report realized that the artificial sea defences were not only expensive to build and repair, but they were also, in many cases, ineffective.

The mangrove ecosystem exchanges materials and energy with the adjacent estuarine water (Fig. 9.2), and these indirect benefits of mangroves have been pointed by several researchers (Ong 1993; Kristensen et al. 2008).

Mangroves are noted for their property to cope with the sea level rise through two basic routes, namely, (i) by an increase in soil surface elevation and/or (ii) by colonizing more landward areas. Mangroves accrete sediment and soil particles that lead to the rise of substratum compared to adjacent sea level. When the height of a mudflat reaches a height above mean sea level suitable for mangroves, and providing mangrove propagules (i.e. seeds) is available, then mangroves are expected to colonize such an area (Fig. 9.4).

Once mangroves have established, they have the potential to change the environment. They retard the velocity of water flows and reduce wave energy, which allow further deposition of sediments, and through the growth of subsurface roots, they may increase the soil volume. Both processes can further increase the height of the soil surface. If a time comes when soil inputs and losses approximately balance such that the soil surface height (i.e. the **surface elevation**) remains relatively stable (e.g. Fig. 9.3), then mangroves may remain as the climax vegetation for many years (sometimes thousands of years, e.g. in Twin Cays, Belize). If the height of the soil surface continues to increase due to soil inputs exceeding soil losses, then the soil surface height may continue to rise until it reaches the upper limit for mangroves to survive; ultimately, terrestrial vegetation may outcompete mangroves.

The difference in height between the current soil surface height within a mangrove forest and the maximum soil surface height that can be achieved with mangroves present (limited either by the balance of soil inputs and losses or by mangrove vegetation being outcompeted by terrestrial vegetation) is referred to as the mangrove **accommodation space** (Fig. 9.5). More generally, the term 'accommodation space' describes the available space for soil expansion or growth, both vertically and laterally, given the current position of the soil surface, the tidal frame and erosive forces. Over a particular stretch of coast, an accommodation volume may also be defined as the volume of space above the substrate that could be filled with sediment and allow

Net primary production: (Tg C y^{-1})

Litter fall: 68 ± 20

Wood production: 67 ± 40

Fine root production: 82 ± 57

Total: 218 ± 72

Fate of mangrove production (in Tg C y^{-1})

Burial: 18.4(°)

CO_2 efflux: 42 ± 31

POC export: 21 ± 22

DOC export: 24 ± 21

unaccounted 112 ± 85

Fig. 9.2 Fate of mangrove productivity

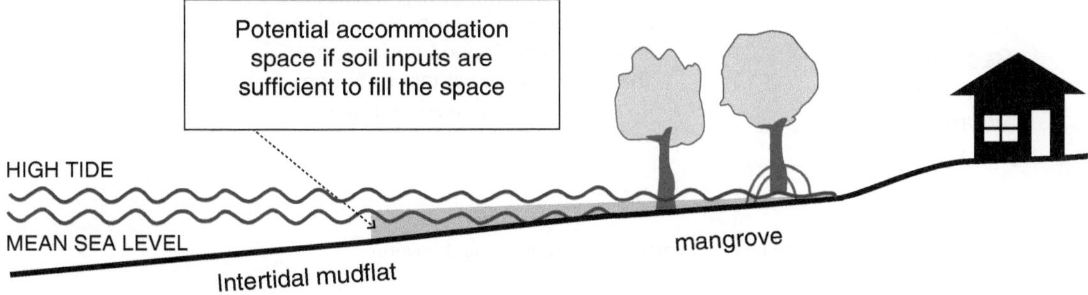

Fig. 9.3 Increase in soil surface height and sea level rise resulting in the relative height of the mangrove surface remaining constant within the tidal range

mangroves to grow there; this allows for a 'lateral accommodation space', meaning seaward areas where mangroves could live if sediment filled the space (limited also by bathymetry and wave conditions eroding sediment, these factors limit the seaward edge of the accommodation space shown in Fig. 9.6). The accommodation concept is widely used in geology (Schlager 1993; Miall 1996); in relation to coastal ecosystems, it has been applied more frequently to coral reef systems (Pomar 2001; Kennedy and Woodroffe 2002; Montaggioni 2005) but only occasionally in relation to salt marshes (French 2006) and mangroves (Spencer and Möller 2013).

When sea level rises or land subsides, the volume of accommodation space increases (Fig. 9.3), as the difference in height between

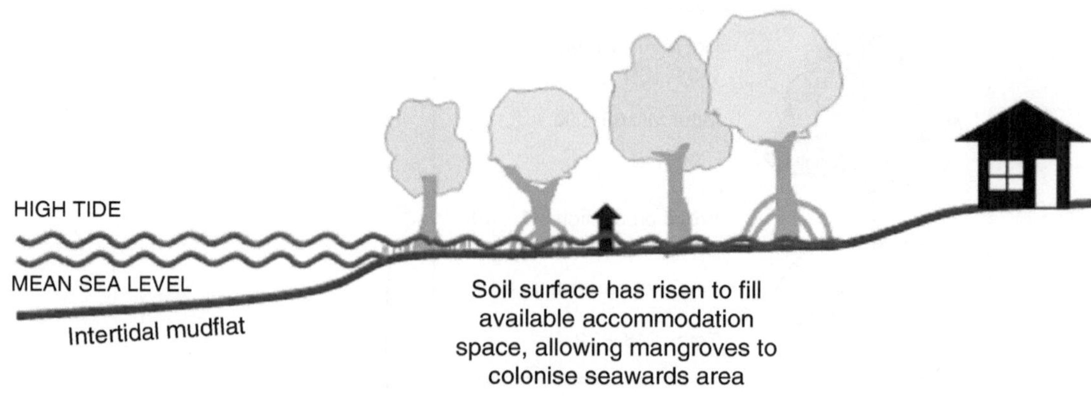

Fig. 9.4 Soil surface rise to fill up the available space allowing mangroves to colonize seawards area

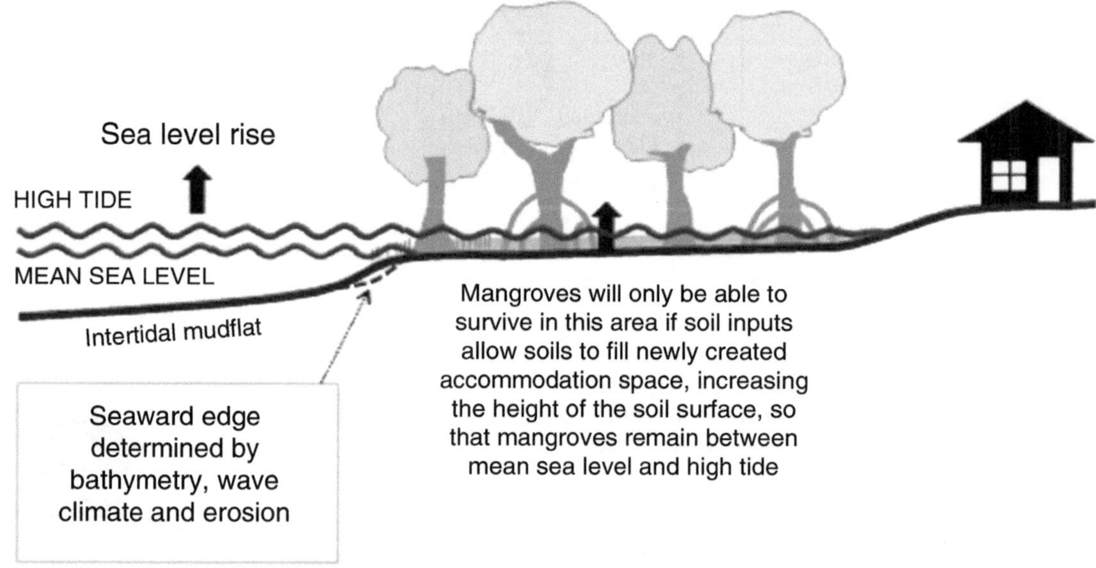

Fig. 9.5 Seaward edge determined by bathymetry, wave climate and erosion

the height of the substrate and mean sea level has increased. This volume can now be filled with soil if soil inputs are high enough, allowing the soil surface to rise until the newly created accommodation space has been filled. Soil inputs include organic or inorganic sediments and subsurface roots. The increase in height of the mangrove soil surface can result in mangroves remaining in their preferred part of the tidal frame, i.e. between mean sea level and high tide. Without such an increase in soil surface height, the mangrove surface could end up below mean sea level, creating stress on mangrove trees and probably resulting in their death. If the change in soil surface height exactly matches the change in sea level, this results in the relative height of the mangrove surface remaining constant within the tidal range (Figs. 9.4 and 9.5).

A number of processes may result in changes in the mangrove surface elevation, and these are illustrated in Fig. 9.6 (lower part).

These processes may be divided into surface processes and subsurface processes. For the purposes of this report, the **soil surface** refers to the interface between the soil and the air

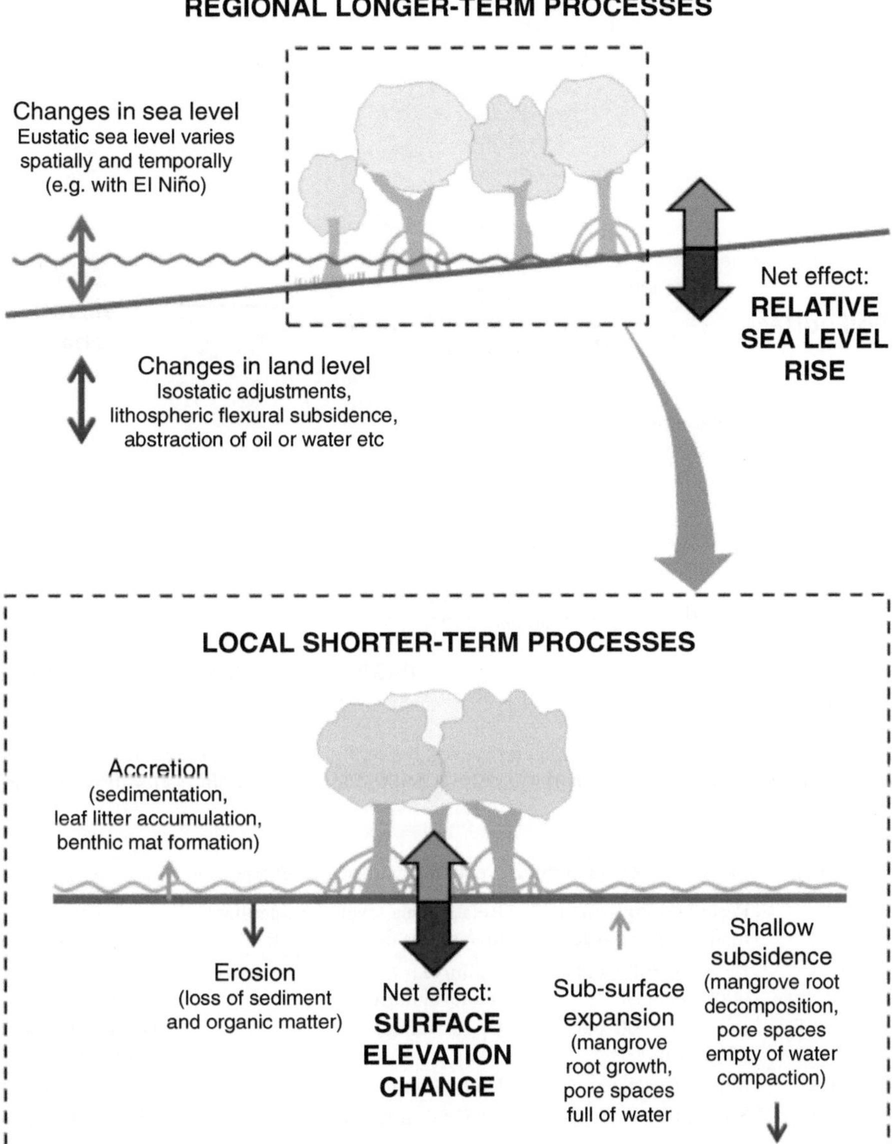

Fig. 9.6 Regional and local processes affecting the elevation of the mangrove surface relative to local mean sea level

(or water, when the tide covers the soil) (Fig. 9.7). **Surface processes** refer to those processes which occur at or above the mangrove soil surface, including sedimentation (the deposition of material onto the surface of the soil), accretion (the binding of this material in place) and erosion (the loss of surface material). **Subsurface processes** refer to processes that occur below the soil surface but above the basement or consolidated layer (Fig. 9.7); these include growth and decomposition of roots, swelling and shrinkage of soils related to water content and compaction, compression and rebound of soils due to changes in the weight of material above.

When subsurface processes result in a change in volume of the soil, this is called **subsurface expansion** or **shallow subsidence**. Shallow subsidence refers to the loss of elevation caused by

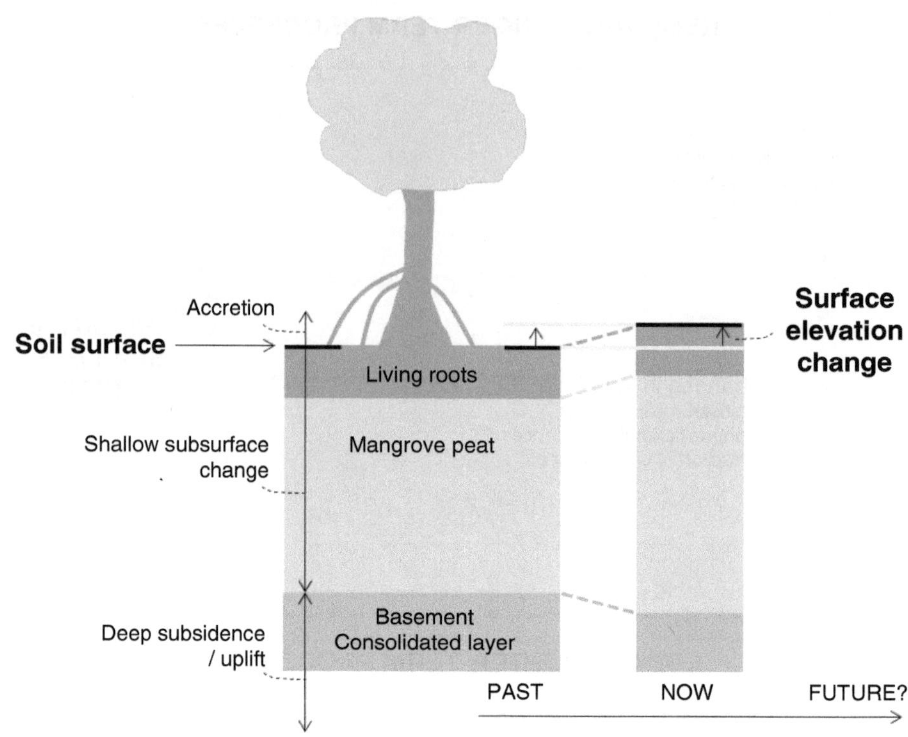

Fig. 9.7 Schematic diagram of a mangrove tree and the soil beneath it, showing where accretion, shallow subsurface change and deep subsidence/uplift occur in the profile and illustrating how surface elevation change may occur over time

Table 9.2 Locations and periods where mangroves kept pace with sea level rise

Location	Period during which mangroves persisted	Relative sea level rise rate that mangroves kept pace with	References
South Alligator River, Australia	Between 8000 and 6000 years BP	6 mm/year (12 m rise in relative sea level during this period)	Woodroffe (1990), Ellison (2009)
Mary River, Australia	Between 6500 and 4000 years BP	Up to 10 mm/year	Woodroffe and Mulrennan (1993), Ellison (2009)
Twin Cays, Belize	Since 7600 years BP	Up to 3 mm/year	McKee et al. (2007)
Hungry Bay, Bermuda	Since 2000 years BP	0.85–1.1 mm/year	Ellison (1993, 2009)
Fanga'uta Lagoon, Tonga	Between 7000 and 5500 years BP	1.2 mm/year	Ellison (2009)
Kosrae, Federated States of Micronesia	Since 2000 years BP	1–2 mm/year	Ellison (2008)

these subsurface processes, which act above the bedrock or consolidated layer; it is called *shallow* to distinguish it from *deep* **subsidence**, caused by longer-term geological processes (Cahoon et al. 1995), which are accounted for in relative sea level rise rates.

There are plenty of examples around the world, which show that mangroves have adapted with the sea level rise in different times and in different regions (Table 9.2).

Mangroves are also noted to increase productivity of adjacent water through litter fall. The mangrove litters (Fig. 9.6) are decomposed by bacteria and provide nutrients to the adjacent water bodies. Because of the abundance of the nutrients, the mangrove ecosystems act as the breeding and nursery ground for the fishes. Mangroves are also noted for their carbon sequestration capacity. Mangroves respond well to high carbon dioxide by exhibiting greater accumulation of biomass with increasing carbon dioxide level. For example, *Rhizophora mangle* under high carbon dioxide conditions which was double than normal for 1 year showed greater accumulation of biomass (Farnsworth et al. 1996). About 55–85 % of the above-ground biomass (shoot) is held in the woody trunk and branches. In contrast, 15–17 % is held in aerial roots. The belowground biomass (root) is high compared to other forest types. The ratio between above-ground biomass and below-ground biomass is about 2.5:1, while it is 4 in upland forests. This relatively heavy investment in root biomass helps to ensure stability in the soft substrates of mangrove environment. Carbon sequestration potential of the mangroves is 50 times greater than other tropical forests. This is because of high levels of belowground biomass and also considerable storage of organic carbon in mangrove sediment soils. Mangrove forests usually create thick, organically rich sediments as their substrata. Most of the substrata in the tropics except under deltaic environments consist of mangrove peat which mainly derives from mangrove roots. This shows that mangrove forests have great belowground productivity and play a significant role in carbon sequestration not only above ground but also below ground. *Rhizophora* forest has higher belowground carbon sequestration ability than other types of mangrove forest and displays its greatest ability during the early stage of the forest. Carbon burial rate of *Rhizophora* forest was faster at 580 g/cm/year than those of other types of mangrove forest (Fujimoto 2000). Estimation of the potential of a forest in sequestrating carbon involves calculating the total biomass per hectare and then applying appropriate conversion factors to get the carbon equivalents. Ong et al. (1995) estimated the amount of carbon sequestered in a 20-year-old stand of *Rhizophora apiculata* mangrove forest as 7.14 tons of carbon per hectare per year. The rate of carbon sequestered in mangrove mud is estimated to be around 1.5 tC/ha/year. The upper layers of mangrove sediments have high carbon content (a conservative estimate is 10 %). Each hectare of mangrove sediment would then contain some 700 tons of carbon per metre depth (Ong 1993). The monetary value of the carbon sequestered by the forest is calculated by using an international price per unit amount of carbon reduced (e.g. $150 per ton of carbon in Norway). The mangrove habitat is proved to be efficient in carbon sequestration, 2.4-fold as high as salt marshes, 5.2-fold as high as seagrasses and 50 times as high as tropical forest. The organic carbon concentrations in vegetated marine sediments exceed by 2–10 folds those in shelf/deltaic sediments (Duarte et al. 2005). Thus, mangrove sediment has a significant role in carbon sequestration. Preliminary estimates indicate that the total above-ground biomass for the world's mangrove forests may be over 3700 trillion grams of carbon (Tg) and that sequestration of organic matter directly into mangrove sediments is in the range of 14–17 Tg of carbon per year. Organic-rich soils ranged from 0.5 m to more than 3 m in depth and accounted for 49–98 % of carbon storage in the mangrove systems. Thus, mangroves are capable of accumulating and storing carbon in the soil in large quantities. The mangroves may have an important role to play in global carbon budgets and in the process of mitigating climate change. Mangrove deforestation generates emissions of 0.02–0.12 picograms of carbon per year—as much as around 10 % of emissions from deforestation globally, despite accounting for just 0.7 % of tropical forest area. Thus, failing to preserve mangrove forests could cause considerable carbon emissions and thus global warming (Spalding et al. 2010).

Mangrove extracts are used in indigenous medicine; for example, *Bruguiera* species (leaves) are used for reducing blood pressures and *Excoecaria agallocha* for the treatment of leprosy and epilepsy. Roots and stems of *Derris trifoliata* are used for narcotizing fishes, whereas *Acanthus ilicifolius* is used in the treatment of rheumatic disorders. Seeds of *Xylocarpus* species have antidiarrhoeal properties and *Avicennia* species have tonic effect, whereas *Ceriops* produces haemostatic activity. Barks of *Rhizophora* species have astringent, antidiarrhoea and antiemetic activities. Tender leaves of *Acrostichum* are used as a vegetable and a beverage is prepared from the fruits of *Sonneratia* spp. Extracts from mangroves seem to have a potential for human, animal and plant pathogens and for the treatment of incurable viral diseases like AIDS (Kathiresan 2000).

Apart from the medicinal value, Indian researchers are successful in preparing food products from mangrove fruits (Pramanick et al. 2014) or fish feed from mangrove litter and mangrove-associated species like seaweeds (Mondal et al. 2014). Today, the belief of mangrove ecosystem as wasteland or marshland has been totally changed. Mangrove ecosystem is now considered as one of the most economically valuable ecosystems of the world. In general, the economic value has been calculated between US $2000 and 9000 per hectare per year (Table 9.3).

The global mangroves are presently under severe anthropogenic stress, but how long this fragile and taxonomically rich ecosystem will withstand this anthropogenic threat is a million dollar question. The more important question is that how the mangrove will withstand the temperature rise and salinity rise, which are inevitable signatures of climate change occurring throughout the globe?

We conducted a thorough study in ten sampling stations in Indian Sundarbans and observed a pronounced decadal variation of true mangrove floral species, both in terms of total numbers (Fig. 9.8) and diversity (Fig. 9.9), but no study has yet been conducted on the fate of mangrove

Table 9.3 Country-wise economic value of mangroves (on the basis of service)

Country	Service	Valuation	References
Indonesia	Traditional use	3000 (half of income among the poorest households)	Ruitenbeek (1992)
Thailand	Traditional use	230–1200	Christensen (1982), Sathirathai (1998)
Southern parts of Thailand	Traditional use	1500 per household (a quarter of per capita GDP)	Sathirathai (1998)
In southern Thailand	Coastline protection and stabilization services	3000	Sathirathai (1998)
	Carbon sequestration	100	Sathirathai (1998)
Koh Kong province in Cambodia	Local level uses and indirect values	500–1600	Ban (1997)
Rekawa, Sri Lanka	Coastal protection from storm and fisheries values	1000	Gunawardena and Rowan (2005)
Sri Lanka	Storm protection	800,000	Batagoda (2003)
Irian Jaya	Erosion control service	600 per household per year	Ruitenbeek (1992)
South of Vietnam	Protection against extreme weather events	500,000	Tri et al. (1998)
Southeast Thailand	Ecosystem function	10,000	Panapitukkul et al. (1998)
India (North Malabar)	Total monetary value	10,960	Khaleel (2008)
	Forestry and fisheries benefits	500–2500	Dixon (1989)
	Disturbance regulation	1839	Costanza et al. (1997)
	Waste treatment	6696	Costanza et al. (1997)
	Habitat/refugee	169	Costanza et al. (1997)
Global mangroves	Food production	466	Costanza et al. (1997)
	Raw materials	162	Costanza et al. (1997)
	Recreation	658	Costanza et al. (1997)
	Total benefits	3294	Costanza et al. (1997)
	Total economic value	9990	Costanza et al. (1997)

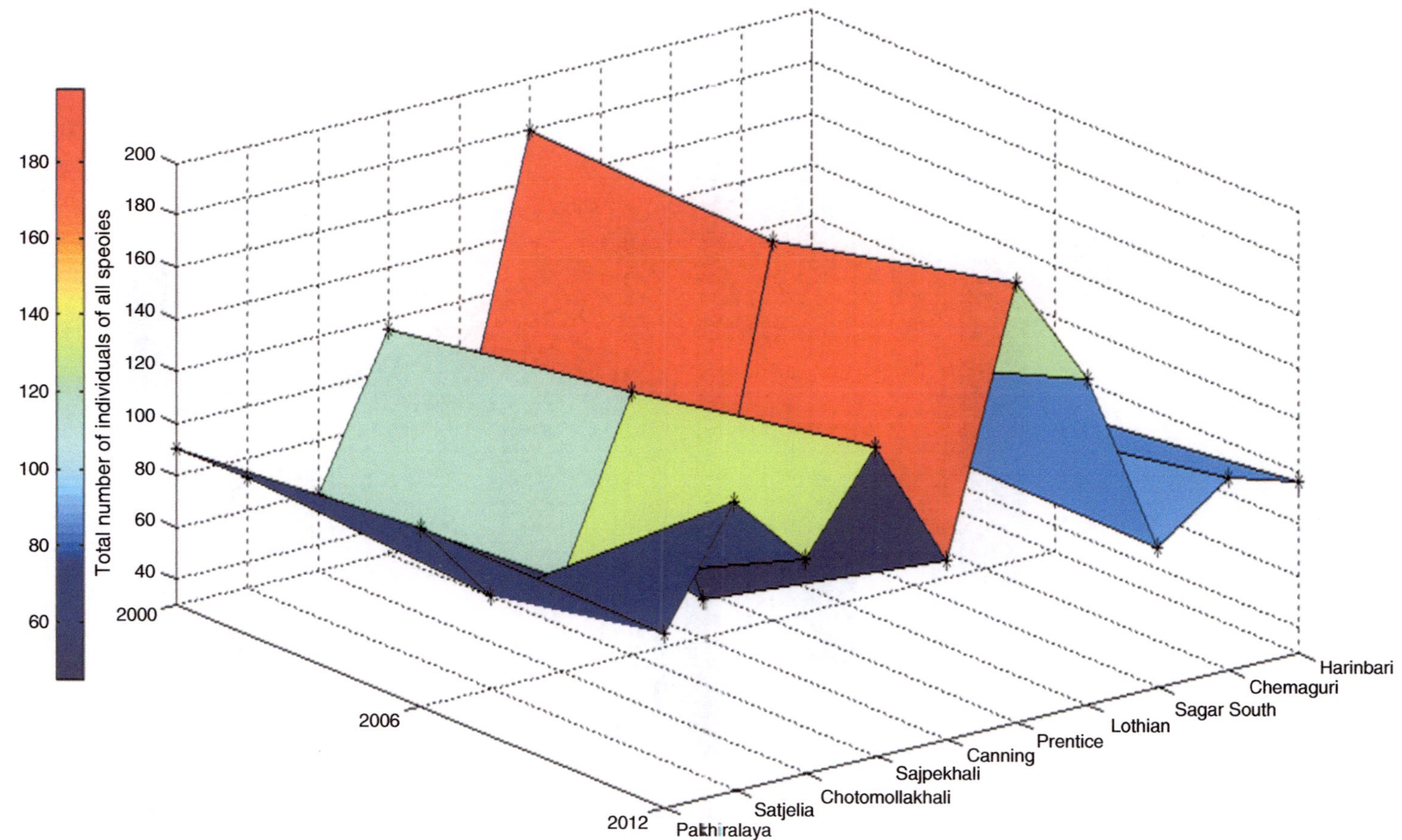

Fig. 9.8 Spatio-temporal variation of a total number of individuals of true mangrove floral species (N)

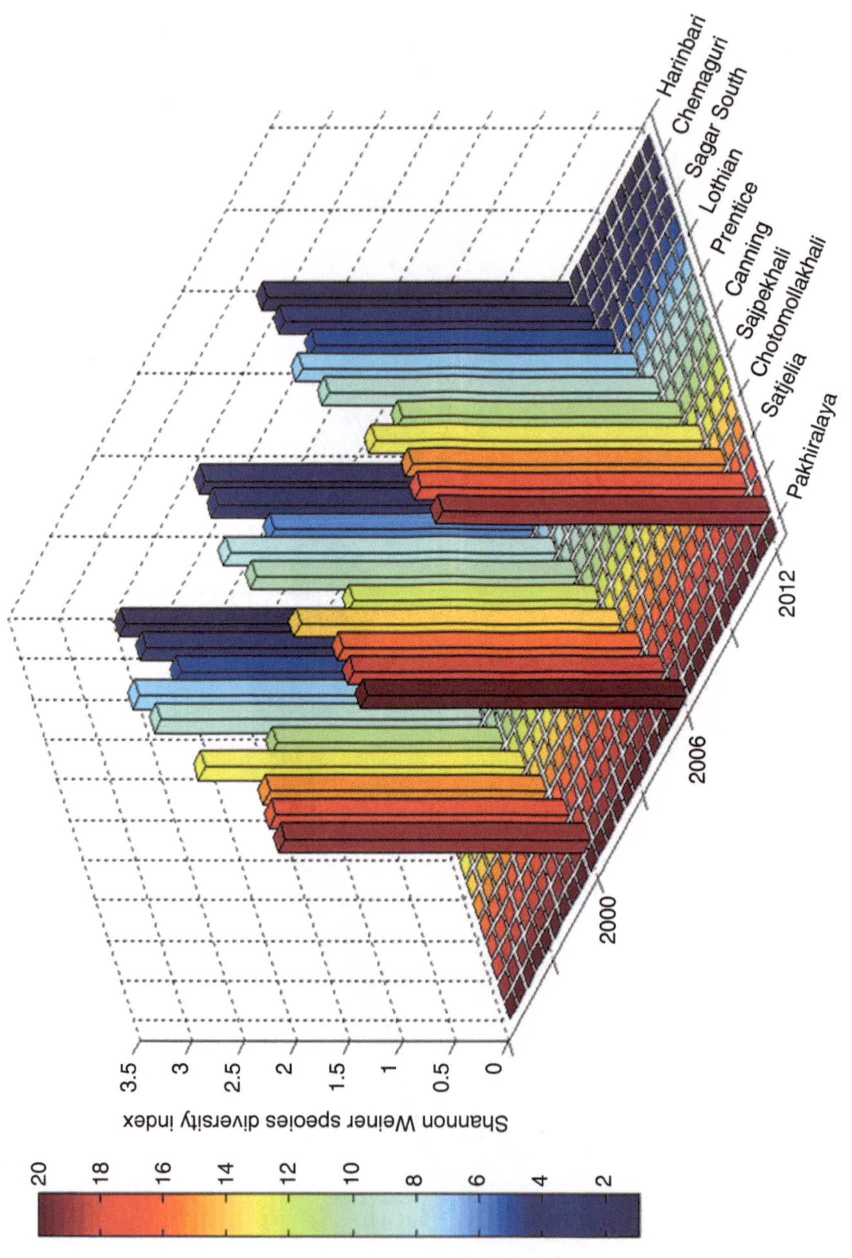

Fig. 9.9 Spatio-temporal variation of Shannon-Wiener species diversity index of true mangrove floral species

biomass in relation to rising salinity (in terms of predictive model), although it is a fact that salinity causes stunted growth of mangroves. In countries like India (Indian Sundarbans), few mangrove species like *Heritiera fomes* are already at the verge of extinction because of rising salinity.

Agricultural land conversion destroyed 17,179 ha of mangroves in India during 1975–2005. A further 7554 ha was lost due to shrimp cultivation. Over the last 30 years, some 7500 ha in Bangladesh has become submerged by rising seas. If this trend continues then not much days are left when mangroves will be some museum specimens beside dinosaur. Unfortunately, the issues are addressed in seminars and workshops with pomp and splendour, but implementation ultimately ends in smoke and we eagerly wait for the next seminar even after listening the warning/death bell for this very delicate ecosystem of the planet Earth.

9.2 Coral Reefs

Coral reefs, the rainforest of the ocean, occupy only 0.2 % of the total ocean basin but support not less than 25 % species of the ocean. The reefs sustain the most luxuriant and complex of all benthic communities. They are often regarded as the oases of the ocean. The Great Barrier Reef, the largest coral reef of the planet Earth, stretches more than 2000 km from New Guinea southward along the East Coast of Australia. The reefs are constructed and promoted by corals, which are colonial animals and whose individual members are called polyps (Fig. 9.10).

A coral polyp looks like a tiny sea anemone possessing stinging tentacles, but unlike the anemone, a coral polyp extracts calcium carbonate from the water and builds within its tissues a skeletal cup of calcareous material. Large numbers of polyps grow together in colonies of delicately branched forms or rounded masses. Corals are very slowly growing organisms; some species grow less than 1 cm in a year while others add up to 5 cm each year. The same coral may be found in different shapes and sizes, depending on the depth and wave action of the area. Within the tissues of the coral polyps, single-cell dinoflagellate algae called **zooxanthellae** are present. Polyps and zooxanthellae have symbiotic relationship in which the coral provides the algal cells with a protected environment, carbon dioxide and nutrients (like nitrate and phosphate), and the algal cells by the process of photosynthesis return oxygen, remove wastes and produce

Fig. 9.10 Coral polyps

carbon compounds, which help to nourish the coral. Some coral species receive as much as 60 % of their nutrition from the algae. Zooxanthellae also cause the coral to produce more calcium carbonate and increase the growth of their calcareous skeleton. The polyps feed actively at night by extending their tentacles for catching the zooplankton, but during the day, their tentacles are contracted, exposing their outer layer of cells containing zooxanthellae to the sunlight. The reef-building corals of tropical regions have specialized requirements like warm, clear, shallow, clean water and a firm substratum to which they can attach. The growth of the corals is restricted to tropical waters between 30°N and 30°S where the temperature usually ranges between 20 °C and 25 °C. Waters at depths greater than 50–100 m are too cold for considerable secretion of calcium carbonate. Perhaps because of this reason, most of the Caribbean corals are found in the upper 50 m of lighted zone, whereas Indian and Pacific corals are found at depths of 150 m in the more transparent water of these oceans.

Coral reefs are ecologically important ecosystems with extremely rich taxonomic diversity (Table 9.4). They sustain complex assemblages of many different types of plants and animals having intense competition for space and food. Globally, coral reefs comprise about 1 % of total continental shelf area or slightly more than that (Groombridge and Jenkins 2002), but they are among the most diverse of the world's ecosystems. It is estimated that about 793 species of scleractinian corals are present worldwide (Veron 2000) of which 719 occur in the Indo-West Pacific region. Very few species of any taxa are found in both the Indo-Pacific coral reef area and Caribbean reef area (Burke et al. 2001). It has been reported that the coral species richness of Indian Ocean Rim countries is very similar to those in the Pacific Ocean with both of these regions having significantly greater coral diversity than the Atlantic. Countries with large coral diversities such as Indonesia (443), Australia (428) and Thailand (238) border both the Indian and Pacific Oceans, thus increasing the apparent diversity of both.

Table 9.4 Species richness of coral reef-associated fauna

Taxon	Number of species	Reference
General Indo-Pacific reef faunas		
Fishes	>2200	Sale (1980)
Corals	390	Sebens (1994)
Nudibranchs	>500	Gosliner (1993)
Estimates of faunas of single reefs		
Fishes	860	Sale et al. (1994)
Molluscs	150–500	Sorokin (1993)
Crustaceans	100–250	Sorokin (1993)
Polychaetes	100–200	Sorokin (1993)
Echinoderms	50–100	Sorokin (1993)
Sponges	>50	Sorokin (1993)

Indonesia is blessed to have 75,000 km^2 of coral reef, which are distributed throughout the archipelago. Throughout the Indonesian coastal water, fringing reefs are most common and are present around most small- and medium-sized islands particularly in the eastern sector of Indonesia. Researchers documented that there are 590 species of hard corals (Suharsono 2004), 2010 species of soft corals (Moosa 1999) and 350 species of gorgonians (Moosa 1999). The coral reef ecosystems in Indonesia sustain diverse categories of marine algae. In the eastern part, 782 species of seaweeds have been documented out of which 179 species are under Chlorophyceae, 134 species are under Phaeophyceae and 452 are under Rhodophyceae (van Bosse 1928). Seaweeds are collected or cultured on reef flats. In Indonesia, the coral reef is facing severe threats. It has been estimated that 70 % of the reefs are either fair or richly damaged and only 6.69 % are in excellent condition (Suharsono 2004). Reports available on Indonesian coastal reef highlight that Indonesia has approximately 18,110 islands which are partly or wholly surrounded by coral reefs. These coral reefs are taxonomically rich natural display ground sustaining a wide spectrum of biological species.

The contribution of the Indian Ocean to this diversity is significant with six other countries spread from Seychelles in the west to India in the north to Malaysia in the east, all having over

200 species of coral recorded (Keesing and Irvine 2005).

Coral reef is the housing complex of a wide spectrum of fauna. It has been documented that as many as 300 animal species may live together on a single reef. The giant clam (*Tridacna* sp.) measuring up to a metre in length and weighing over 150 kg is an important species of reef ecosystem, but overharvesting and poaching have greatly reduced their numbers. These clams also possess zooxanthellae in large numbers in the colourful tissues that line the edges of the shell. The bubblegum corals are another surprise to the reef community. They are octocorals belonging to the phylum Cnidaria, which also includes stony corals and jellyfish. Despite forming enormous tree-like structures, the trunk and branches of bubblegum corals are not corneous or calcareous like most branching octocorals. Instead, the branches are accumulation and near fusion of microscopic sclerites made of calcite (Fig. 9.11).

Crabs, moray eels, colourful reef fish, poisonous stonefish, spiny sea urchins, seahorses, shrimps, lobsters, sponges and various life forms are found in the coral reef ecosystem. Global pattern of coral reef fish diversity has been examined in detail by McAllister et al. (1994). The biodiversity of coral reef fishes covers a wide array of species. Reef fishes show patterns of distribution and diversity at all scales ranging from global to regional to local. At the largest scale, diversity appears to be principally controlled by the interaction of tectonic and geomorphological events with evolution and dispersion processes. At a regional level, water parameter like turbidity and current strength, perhaps requiring differential adaptation at the larval stage, may control the species distribution pattern. At a local level, diversity is influenced by the habitat characteristics such as depth, heterogeneity and complexity. Because of the vagaries of larval distribution and recruitment, the assemblages to be found at a site are quite variable, although the extent of these variations probably differs among species and areas (Ormond and Callum 1997).

Fig. 9.11 Bubblegum corals

In Indian subcontinent, coral reefs are concentrated in regions like Gulf of Mannar, Andaman and Nicobar Islands, Lakshadweep Islands, etc. Coral reefs occur in and around 20 islands between Tuticorin and Pamban in the Gulf of Mannar and Palk Bay. The Gulf of Mannar is located in Tamil Nadu, in the southeast coast of India (Fig. 9.12). Coral reefs in this zone have developed around a chain of 21 uninhabited islands in four groups (Table 9.5) that lie along the 140 km coastal stretch between Rameshwaram and Tuticorin, at an average distance of 9 km from the mainland.

We conducted an intensive survey during 2013 in the Gulf of Mannar region and documented 36 genera of corals under different families like Pocilloporidae, Acroporidae, Fungiidae, Poritidae, Faviidae, Pectiniidae, etc. (vide **Annexure 9A** for the checklist).

Coral reefs are also the sites of microbial diversity. A wide spectrum of free-living nitrogen fixing bacteria, nitrate-reducing bacteria, ammonifying bacteria, inorganic phosphate-solubilizing bacteria, phosphate-producing bacteria, filamentous fungi and yeast were isolated from the coral mucus samples of the Gulf of Mannar. The positive correlation observed between the content of inorganic phosphate with solubilizing bacteria and phosphatase-producing bacteria indicates that the release of inorganic phosphate is mostly governed by bacterial action in the coral reef environment (Kannan 2004).

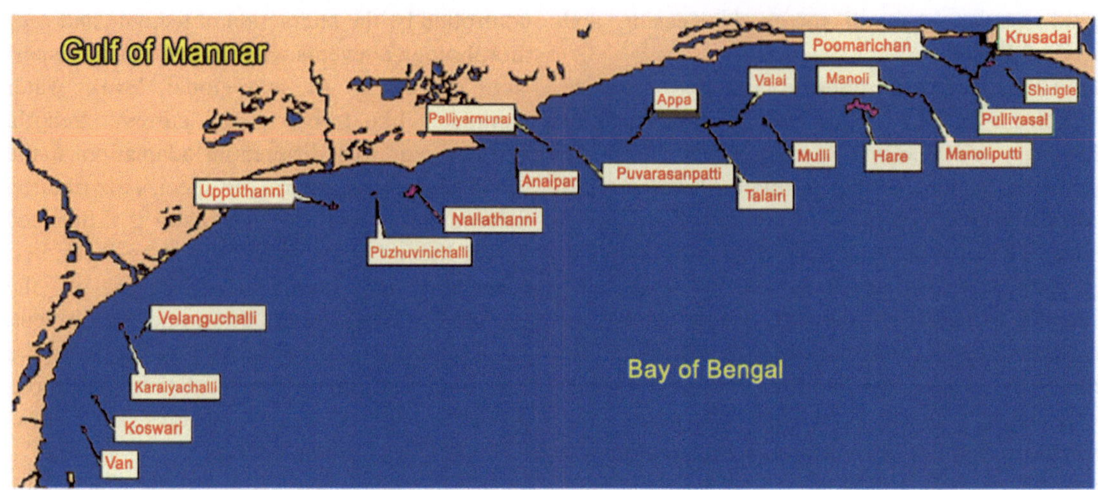

Fig. 9.12 Map of the Gulf of Mannar showing the location of the 21 islands

Table 9.5 Islands in the Gulf of Mannar and type of coral reef

Islands	Reef type
Mandapam group	
Shingle	Fringing reef that extends down to 2 m depth
Krusadai	Fringing reef extends down to 3 m, with patch reefs extending towards the eastern side
Pullivasal	Fringing reefs to 2.3 m depth with patch reefs extending to the northern side
Poomarichan	Fringing reef extends down to 2 m depth with patch reefs to the northeast
Manoliputti	Fringing reefs extend down to 2.2 m depth
Manoli	Fringing reefs extend to 2.2 m with patch reefs to the north
Hare	Fringing reef extends to 2.2 m depth, two patch reefs to the northwest 3 m deep
Keezhakarai group	
Mulli	Fringing reef extends to 3.5 m with patch reefs to the southeast and south at 2.9 and 3.2 m
Valai	Limited fringing reefs extend to 2.9 m
Thalaiyari	Fringing reefs extending to 2 m depth
Appa	Fringing reef extends to 3.2 m with patch reefs to the southeast and northwest side at 3.5 m
Poovarasanpatti	Discontinuous patch reefs distributed up to 2.5 m depth
Valimunai	Fringing reef to 2.5 m depth, patch reef to the southeast at 3.4 m depth
Anaipar	Fringing reef extends to 2.8 m
Vembar group	
Nallathanni	Fringing reef extending to a depth of 3 m and small patch reefs to the south side down to 3.9 m depth
Pulivinichalli	Fringing type that extends up to a depth of 2.5 m and patch reefs on the south side at 3.2 m depth
Upputhanni	Fringing reef that extends to 2.8 m depth with patch reefs to the south and west at 3 and 3.5 m depths

The coral reef community has diverse ecological and economic importance. Productivity of these reefs is high due to their symbiotic association with zooxanthellae and the efficient recycling of nutrients. These corals check erosion and are closely associated with diverse organisms such as brachyuran crabs (e.g. *Trapezia* spp., *Tetralia* sp., *Cymo* spp., *Domecia glabra*), gastropods (e.g. *Coralliophila violacea*, *Ouoyola* sp.) and fishes (e.g. *Abudefduf* spp., *Dascyllus* spp.). Turtles consume seagrasses in the reefs. In addition, coral reefs are nursery grounds for many fishes and habitat for rare species. Corals grow very slowly and its gradual damage will not only affect the biodiversity of the area but will also pose an adverse impact on the recruitment and yield of fisheries.

Though corals are economically important from the point of view of biodiversity, they have been exploited for commercial purposes. Their aesthetic value makes them costly souvenirs for tourists. Coralline boulders (mainly *Porites* sp.) have been used as cheap substitute for construction of houses. Coral reefs host many organisms that are of commercial importance and many of them are found on the accessible reef flats. The algae present on the coral reef system are often utilized for the extraction of agar and algin.

Holothurians, being member of the reef community, have been exploited in India for the production of *bêche-de-mer*, an industry introduced by the Chinese more than a thousand year ago (Hornell 1917). The product containing about 35–50 % digestible proteins imparts a flavour to soup. Though there is an internal market for *bêche-de-mer*, foreign exchange equivalent to one crore rupees is earned every year (James and James 1994). This industry has been mainly restricted to Gulf of Mannar and the Palk Bay and has been facing a crisis due to declining stocks. To mitigate the crisis, new areas are sought to be exploited. The crisis appears to arise mainly because of indiscriminate exploitation, lured by foreign exchange. Some echinoids, e.g. *Tripneustes gratilla*, are potentially economically important as their highly nutritious gonads are considered a delicacy in European, American and Far Eastern Asian countries. Ornamental fishes of reefs are important because of their high market value.

Extraction of bioactive substances is another important economic benefit obtained from coral reef organisms. The pufferfishes *Tetraodon* spp. contain tetrodotoxin, which is a neurotoxin. Many of the compounds extracted from coral reef-associated species, useful to the pharmaceutical industry, have been identified (Kennish 1989), and some of them are being commercially marketed. Using biotechnological techniques, these compounds can be mass-produced in the laboratory without exploiting the natural population. Agarolytic bacteria, of potential commercial importance, have been isolated from Agatti lagoon. Coral reef organisms are economically very important because many of them are the sources of potential drugs and to alleviate human ailments, whose research is under progress. Recent reports from Europe suggest that coral skeleton can be used as a bone material substitute in orthosurgery.

9.2.1 Types of Coral Reefs

On the basis of the structure and their relationship to the underlying geologic features, coral reefs can be subdivided into three categories (Fig. 9.13a, b, c). They are fringing reefs, barrier reefs and atolls.

1. **Fringing reefs**: These are found very close to or near the newly formed volcanic islands as well as surrounding the continental landmasses. They are located directly offshore and project outward to the sea, e.g. the Caribbean reef.
2. **Barrier reefs**: These reefs act as barrier separated by a lagoon from the landmass, with which they are associated. Those surrounding volcanic islands are formed from subsiding islands with fringing reefs. The lagoon forms between the island and the reef. The largest barrier reef is the Great Barrier Reef of Australia, which runs over 2000 km along the north-eastern coast of Australia to New Guinea. It is so large that it is visible from space shuttle. The second largest barrier reef is located in the Caribbean Sea off the coast of Belize.
3. **Atolls**: These reefs are somewhat circular in shape with a centrally located lagoon. According to Darwin, they are formed in areas where barrier reefs subside below sea level, whereas according to Alexander Agassiz, they are formed on the top of the cones of submerged volcanoes, leaving only the reef around a central lagoon. But in any of the case, the lagoon remains connected to the open sea by breaks in the reef. Over time, the

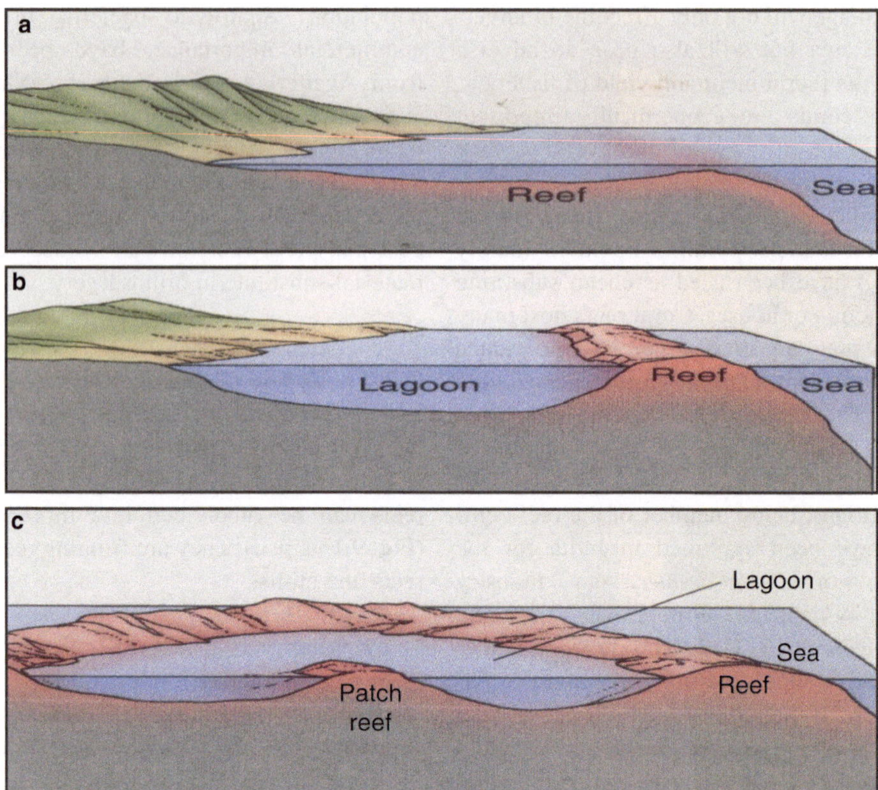

Fig. 9.13 (**a**) Fringing reef. (**b**) Barrier reef. (**c**) Atoll

reefs may become eroded or exposed to wind action or waves. These physical processes sometimes give rise to formation of islands surrounding the lagoons, which is perhaps the depositional feature of sand carried by the wind or wave action. More than 300 atolls are present in and around the Pacific and Indian Ocean, whereas 10 atolls are present in the Atlantic Ocean as the water is too cool or turbid.

Most atolls have northeast–southwest orientation with an island on the east, a broad well-developed reef on the west and a lagoon in between. Reefs of all atolls are widest on the southwest side.

Low-lying islands distributed in the Indian and South Pacific Oceans with a circular ring of reefs encircling a lagoon are known as an atoll. Other coral reef formations in Indian waters such as fringing reefs occur in the Palk Bay, Gulf of Kutch and Andaman and Nicobar Islands, while coral patches have been recorded at several locations along the coast.

In addition to these three main reef types, there are several special reef arrangements. These are **table reefs** and **patch reefs**. Table reefs are small reefs found in the open ocean that have no central islands or lagoons. Patch reefs occur numerously in patches located in lagoons associated with atolls and barrier reefs.

The **Lakshadweep archipelago** (8°N–12°13′ N, 71°E–74°E), the smallest union territory of India, is located on the 2500 km Laccadive-Chagos ridge, presumed to be the hot spot trace resulting from the northward migration of the Indian plate (Morgan 1981). The islands of this archipelago are located about 220–440 km from the mainland city of Kochi (Cochin) in Kerala. The taxonomic survey of the Lakshadweep revealed the presence of more than 600 species that includes more than 100 few records

Table 9.6 Synopsis of number of species recorded during surveys at Lakshadweep

Faunal/floral groups	No. of species recorded	No. of new records
Corals	96	29
Seagrasses/seaweeds	86	17
Anomuran crabs	10	3
Brachyuran crabs	81	30
Gastropods	155	45
Bivalves	24	7
Sea stars	13	2
Brittlestars	6	1
Sea cucumbers	23	–
Sea urchins	15	–
Fishes	120	1
Total	**629**	**135**

Table 9.7 Diversity of marine organisms in the Lakshadweep

	No. of species		
Faunal/floral groups	Reported[a]	New Records[b]	Total
Corals	104	29	133
Seagrasses/seaweeds/mangroves	181	17	198
Sponges	91	–	91
Polychaetes	69	–	69
Sipunculids	17	–	17
Echiurids	7	–	7
Decapods	156	33	189
Molluscs	425	52	477
Echinoderms	88	3	91
Fishes	601	1	602
Turtles	4	–	4
Sea snakes	5	–	5
Birds	101	–	101
Marine mammals	8	–	8
Total	**1857**	**135**	**1992**

[a]*Sources*: Appukuttan et al. (1989), Deshmukh (1991), Ghosh (1991), Haldar (1991), Jones and Kumaran (1980), Kaliaperumal et al. (1989), Lal Mohan (1989), Lal Mohan et al. (1989), Mathew et al. (1991), Misra and Chakraborty (1991), Pillai and Jasmine (1989), Rao (1991), Subba Rao and Subba Rao (1991), Rao et al. (1989), Sankarankutty (1961), Thomas (1989), Untawale et al. (1983), Untawale and Jagtap (1984)

[b]New records

(Table 9.6). Species belonging to other groups such as mangroves, sponges, soft corals, zoanthids, polychaetes, nemerteans, sipunculids, echiurids, other crustaceans, crinoids, ascidians, etc., also constitute some of the dominant members. The large number of new records seems to suggest that many species are not yet reported. A compilation of species reported to date, including new records observed during the present surveys, is given in Table 9.7. The table excludes the micro/meiobenthic and planktonic species. It can thus be seemed that the Lakshadweep coral reefs are important reservoirs of marine biodiversity in Indian territorial waters (Rodrigues 1996).

Assessment of biodiversity of flora and fauna of the coral reef environment of the **Great Nicobar Island** showed the presence of

150 species of phytoplankton, 80 species of seaweeds, 9 species of seagrasses, 109 species of zooplankton, 87 species of crustaceans (including 55 species of brachyuran crabs, 18 species of hermit crabs, 3 species of lobsters and 9 species of shrimps and 2 species of stomatopods), 119 species of molluscs (including 30 species of bivalves, 83 species of gastropods and 6 species of cephalopods), 49 species of echinoderms (including 9 species of starfish, 8 species of brittlestars, 11 species of sea urchins and sand dollars and 21 species of sea cucumbers) and 258 species of fishes (belonging to 14 genera). Besides, 39 species of coral-boring organisms were also identified from the coral reef area of the Great Nicobar Island. Many of the floral (3 species of seagrasses, 12 species of mangroves) and faunal (35 species of echinoderms, 22 species of benthic polychaetes, 144 species of fishes) species identified from here were found to be new distributional records for the Great Nicobar Island (Kannan 2004). It has also been reported that the reef area is the feeding centre for green turtle and resort for *Dugong dugon*.

At **Gulf of Mannar**, studies on plankton recorded 122 species of phytoplankton that include 94 species of diatoms, 15 species of dinoflagellates, 8 species of blue-greens, 3 species of greens and 2 species of others. There were 98 species of zooplankton with 51 species of copepods; 8 species of tintinnids; 6 species of polychaetes; 4 species of chaetognaths; 3 species of cladocerans; 3 species of thaliacea; 9 species of decapods; 2 species of hydrozoa; 1 species each in foraminifera, mysidaceae, phytoflagellates and pteropod larvae; and 8 juveniles. The phytoplankton population density ranged between 11,000 and 82,500 cells per litre, and the species diversity and population density were found to be more in the reef environs than the coastal waters. Meiofaunal density was higher in the lagoon zone followed by reef flat and seaward slope, with nematodes as the major component besides the Foraminiferans, Turbellarians, Polychaetes, Harpacticoida, Amphipods, Gastrotrics and Ostracods. A study on brachyuran crabs recorded 105 species belonging to 56 genera and 16 families. It was found that there is close association between corals and brachyuran crabs, and the species diversity of these crabs increases with the increase in the interbranchial space in the corals. Molluscs were represented by 60 species of gastropods and 20 species of bivalves. Coral reef fishes represent 129 species belonging to 73 genera and 34 families that are mostly caught for marine aquaria (Kannan 2004).

9.2.2 Coral Reef Characteristics

All three types of reef share several common characteristics. On the seaward side, the reef rises from the lower depths of the ocean to a level just at or just below the surface of the water. This portion of the reef is called **reef front** or **fore reef** (Fig. 9.14). The slope of the reef front can be either gentle or quite steep. In some cases, the reef front forms a vertical wall referred to as a **drop-off**. The reef front does not generally form a solid sea wall instead; finger-like projections of the reef protrude seaward. This arrangement, called a spur and groove formation, disperses wave energy and prevents damage to the reef and its inhabitants. Between some of these projections (spurs) lie sand-filled pockets (grooves) that allow sediments to be channelled down and away from the living coral surface and provide a habitat for many species of burrowing organisms.

The highest point on the reef is called the reef crest, and the area opposite the reef front levels off and is referred to as a reef flat or back reef. This area exhibits a high degree of variability. The flat may be short or several hundred metres long. It may be shallow or cut through by channels several metres deep. The bottom of the flat may consist of rock, sand, coral rubble or some combination of these. Seagrass beds are commonly found in the reef flat area. The reef flat of fringing reefs ends at the shoreline. The reef flat of atolls and barrier reefs descends into the lagoon.

Different areas of a reef support different species of coral as well as other organisms. Coral

Fig. 9.14 Fore reef and back reef

populations on the reef front are usually found at depths of 10–60 m (33–200 ft). Massive, dome-shaped brain corals (*Diploria*) and columnar pillar corals (*Dendrogyra*) are found on intermediate slopes. Below this region, species that form platelike formations, like lettuce leaf and elephant ear coral (*Pectinia*, *Pavona* and *Agaracia species*), predominate. Higher up on the reef, where wave stress is greatest, branching species of coral are found. Wave stress is one of the most important factors in determining what species of coral and other organisms can occupy the reef crest. In the Caribbean, this upper area is the habitat of elkhorn coral (*Acropora palmata*). This coral's heavy, spreading branches project towards sea, where they break the force of incoming waves.

In more protected areas behind the reef front, the deeper and less turbulent water supports more delicate species of coral. In the Caribbean, this is frequently a staghorn coral (*Acropora cervicornis*). In the Indo-Pacific region, species such as staghorn, finger (*Stylopora*), cluster (*Pocillopora*) and lace corals (*Pocillopora damicornis*) are prominent. Farther from the reef front in shallow, calmer water are small species of coral such as rose (*Meandrina* and *Manicina*), flower (*Mussa* and *Eusmilia*) and star (*Montastraea*). Species of this type found associated with both Caribbean and Indo-Pacific reefs. On some reefs, there are signs of zonation as certain species of coral gradually appear and disappear along the reef.

There are approximately 20 genera of coral represented on Caribbean reefs, and most Caribbean reefs are dominated by 10 species of hard coral and the hydrozoan coral *Millepora*, also known as fire or stinging coral. By contrast, there are more than 80 genera of coral in the Indo-Pacific, and the number of species is equally diverse. For instance, in the Caribbean, there are only three species of the genus *Acropora*, whereas there are at least 150 species of this genus in the Indo-Pacific. The Great Barrier Reef of Australia is inhabited by over 200 species in the genus *Acropora*. Cooler water temperatures and higher turbidity are the primary reasons that there are fewer coral species in Caribbean waters. The Indo-Pacific region is also geologically older and covers a larger area than the Caribbean; therefore, more species are to be expected.

9.2.3 Reef Productivity

The water that surrounds and bathes coral reefs is not very rich in nutrients. The presence of suspended material and organisms in productive waters reduces the amount of light that can penetrate the water. As a result, important wavelengths of visible light disappear rapidly below the surface, and the water appears green. Since corals require clear water so that there will be enough light to support their photosynthetic symbionts, it is not surprising to find that coral reefs develop only in water that contains minimal amounts of plankton and is quite blue. Although the water that bathes the coral reefs is nutrient poor, the reef itself is one of the most productive of all marine ecosystems.

The key to the high productivity of coral reefs is the symbiotic relationship between zooxanthellae and many of the reef's inhabitants, especially the hard coral polyps. Other reef animals, such as giant clams and their relatives (*Tridacnidae*), also support an enormous number of symbiotic zooxanthellae. Symbiotic zooxanthellae can be as much as three times more productive than the equivalent amount of phytoplankton. Coral animals and their symbionts are so intimately connected that they behave as a single organism. As previously mentioned, zooxanthellae provide the corals with nutrients essential for reef building. A large portion of the zooxanthellae's tremendous photosynthetic output is channelled back to their host, and much of the oxygen produced by the photosynthesis is consumed by the coral polyps. Zooxanthellae manufacture a variety of amino acids, sugars and other organic compounds that are absorbed directly into their host's tissues. In turn, carbon dioxide and ammonia, wastes of the coral animal's metabolism, are perfect nutrients for zooxanthellae. The symbionts absorb these materials directly from the animal's tissues so efficiently that several species of coral release virtually no nitrogen wastes to the surrounding water. Some compounds may be passed back and forth between polyp and zooxanthellae several times, each time being altered or combined with other molecules. It is not clear what regulates the exchange of nutrients, considering the fact that many of the materials taken by the coral could just as easily serve as growth factors for the zooxanthellae. It appears that coral polyps somehow stimulate the zooxanthellae to release the nutrients, but the nature of this process is not clear.

Coral animals (Fig. 9.15) actively poison their algal symbionts to receive maximum exposure to sunlight. During the day, the polyp's tissue layer containing the zooxanthellae (Fig. 9.15) is spread out like a carpet to catch as much sunlight as possible. At night, when photosynthesis is not carried out, the tissue containing the zooxanthellae is withdrawn, and the coral polyp extends its feeding tentacles and preys upon phytoplankton. In shallow water, where sunlight is abundant, coral tissues contain several layers of zooxanthellae, packing in as many zooxanthellae per unit volume as possible. The colonies formed by these species are generally quite branched, allowing maximum numbers of zooxanthellae to be exposed to sunlight. In deeper water, where sunlight is less available, coral colonies tend to be flatter, forming large horizontal tables over which polyps can spread their tissues containing the zooxanthellae.

When coral polyps are stressed, they expel the zooxanthellae from their tissues, a phenomenon known as bleaching. Lacking zooxanthellae that give coral its characteristic yellow-brown colour, the coral colonies appear white, like the bleached specimens in souvenir shops, thus the term bleaching. Without symbiotic zooxanthellae, coral polyps will cease to grow and will die within a few months. It appears that slight increase in water temperature (heat stress) is one of the most important causes of bleaching. During the 1980s, marine biologists recorded widespread coral bleaching throughout the world, a process that is unfortunately continuing today. Some researchers think that a combination of global warming due to increased carbon dioxide in the atmosphere and El Niño events is the primary culprit. Whatever the cause of bleaching, if environmental conditions return to a more normal range, the coral polyps will collect new zooxanthellae and continue to grow and thrive.

In addition to deriving nutrients from the symbiotic zooxanthellae, some coral polyps use sticky strands or nets of mucus to trap bacteria, plankton and detritus on which they feed. Soft

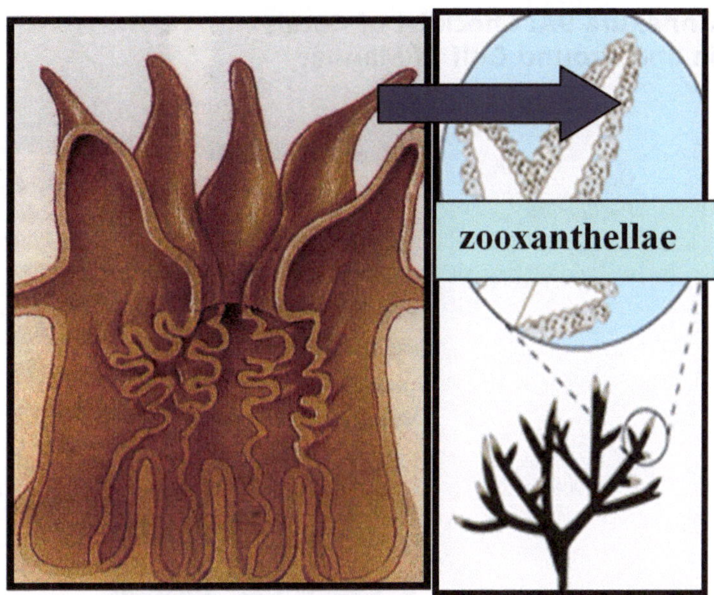

Fig. 9.15 Coral animal with zooxanthellae

corals frequently lack zooxanthellae and probably employ methods such as absorption and mucous nets to gather food. The coral mucus is an important source of food for many reef organisms. Other coral species that lack zooxanthellae are covered with numerous microscopic projections that probably allow them to absorb scarce, dissolved nutrients from the seawater.

Many organisms, including crabs, shrimp, molluscs and fishes, depend on the abundant coral mucus for energy-rich triglycerides and fatty acids. Some species of benthic shrimp can live almost entirely on a diet of coral mucus. The corals themselves provide food for some echinoderms, molluscs, crabs and fishes.

Coral reefs are highly visible, 'charismatic' and metabolically active benthic ecosystems. They contain two primary carbon (C) reservoirs found in other marine ecosystems: organic matter and calcium carbonate. They are recognized to be among the most rapid producers, per unit area, of both organic carbon (the transformation of inorganic carbon by photosynthetic organisms called 'primary production') and skeletal calcium carbonate ($CaCO_3$)—precipitated through calcification process).

Brain Churners

1. Why mangrove forests are known as 'natural buffers'?
2. How can mangroves cope with sea level rise?
3. State few ecosystem services of mangroves.
4. What is referred to as the 'mangrove accommodation space'?
5. What is known as carbon burial rate? How is the monetary value of carbon sequestration calculated?
6. Why is the growth of the corals restricted to tropical waters between 30°N and 30°S?
7. State few uses of corals. What are the three major types of coral reefs?
8. What is the key to the high productivity of coral reefs? Name two hard corals and two soft corals.
9. What is coral bleaching and how does it occur?
10. What is '0.6 rule' in estimating coral reefs as 'carbon source or sink'?

Annexure 9A: Checklist of Corals in and Around Gulf of Mannar

Sl. no.	Genus	Representative	Species documented
1.	*Pocillopora*	 *Pocillopora damicornis*	*Pocillopora damicornis Pocillopora verrucosa* *Pocillopora eydouxi*
2.	*Madracis*	 *Madracis kirbyi*	*Madracis interjecta* *Madracis kirbyi*
3.	*Acropora*	 *Acropora hemprichii*	*Acropora formosa* *Acropora intermedia* *Acropora corymbosa* *Acropora nobilis* *Acropora humilis* *Acropora valida* *Acropora hemprichii* *Acropora hyacinthus* *Acropora stoddarti* *Acropora millepora* *Acropora diversa* *Acropora brevicollis* *Acropora cytherea* *Acropora hebes* *Acropora echinata* *Acropora nasuta*

(continued)

Sl. no.	Genus	Representative	Species documented
4.	*Montipora*	 *Montipora aequituberculata*	*Montipora subtilis* *Montipora digitata* *Montipora divaricata* *Montipora venosa* *Montipora spumosa* *Montipora tuberculosa* *Montipora monasteriata* *Montipora jonesi* *Montipora granulosa* *Montipora exserta* *Montipora turgescens* *Montipora manauliensis* *Montipora verrucosa* *Montipora hispida* *Montipora foliosa* *Montipora verrilli* *Montipora aequituberculata*
5.	*Astreopora*	 *Astreopora myriophthalma*	*Astreopora myriophthalma*
6.	*Pavona*	 *Pavona varians*	*Pavona duerdeni* *Pavona varians* *Pavona decussata* *Pavona divaricata*
7.	*Pachyseris*	 *Pachyseris rugosa*	*Pachyseris rugosa*

(continued)

Sl. no.	Genus	Representative	Species documented
8.	*Siderastrea*	 *Siderastrea savignyana*	*Siderastrea savignyana*
9.	*Pseudosiderastrea*	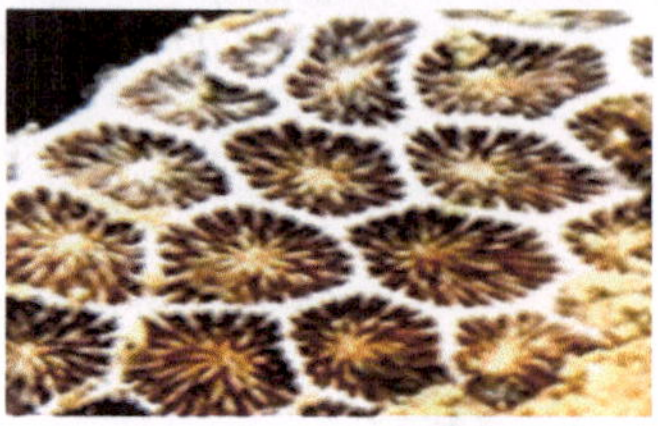 *Pseudosiderastrea tayami*	*Pseudosiderastrea tayami*
10.	*Coscinaraea*	 *Coscinaraea monile*	*Coscinaraea monile*
11.	*Psammocora*	 *Psammocora contigua*	*Psammocora contigua*

(continued)

Sl. no.	Genus	Representative	Species documented
12.	*Cycloseris*	*Cycloseris cyclolites*	*Cycloseris cyclolites*
13.	*Goniopora*	*Goniopora planulata*	*Goniopora stokesi* *Goniopora planulata* *Goniopora minor*
14.	*Porites*	*Porites lichen*	*Porites solida* *Porites mannarensis* *Porites lutea* *Porites lichen* *Porites exserta* *Porites compressa* *Porites complanata* *Porites nodifera*
15.	*Favia*	*Favia speciosa*	*Favia stelligera* *Favia pallida* *Favia speciosa* *Favia favus* *Favia valenciennesi* *Favia matthai*
16.	*Favites*	*Favites halicora*	*Favites abdita* *Favites halicora* *Favites pentagona* *Favites melicerum* *Favites complanata* *Favites flexuosa*

(continued)

Sl. no.	Genus	Representative	Species documented
17.	*Goniastrea*	 *Goniastrea pectinata*	*Goniastrea pectinata* *Goniastrea retiformis*
18.	*Platygyra*	 *Platygyra sinensis*	*Platygyra daedalea* *Platygyra sinensis* *Platygyra lamellina*
19.	*Leptoria*	 *Leptoria phrygia*	*Leptoria phrygia*
20.	*Hydnophora*	 *Hydnophora microconos*	*Hydnophora microconos Hydnophora exesa*
21.	*Leptastrea*	 *Leptastrea transversa*	*Leptastrea transversa Leptastrea purpurea*

(continued)

Sl. no.	Genus	Representative	Species documented
22.	*Cyphastrea*	 *Cyphastrea microphthalma*	*Cyphastrea serailia* *Cyphastrea microphthalma* *Cyphastrea japonica*
23.	*Echinopora*	 *Echinopora lamellosa*	*Echinopora lamellosa*
24.	*Plesiastrea*	 *Plesiastrea versipora*	*Plesiastrea versipora*
25.	*Galaxea*	 *Galaxea fascicularis*	*Galaxea fascicularis* *Galaxea astreata*
26.	*Merulina*	 *Merulina ampliata*	*Merulina ampliata*

(continued)

Sl. no.	Genus	Representative	Species documented
27.	*Acanthastrea*	 *Acanthastrea echinata*	*Acanthastrea echinata*
28.	*Lobophyllia*	 *Lobophyllia corymbosa*	*Lobophyllia corymbosa*
29.	*Symphyllia*	 *Symphyllia radians*	*Symphyllia radians* *Symphyllia recta*
30.	*Mycedium*	 *Mycedium elephantotus*	*Mycedium elephantotus*
31.	*Polycyathus*	 *Polycyathus verrilli*	*Polycyathus verrilli*

(continued)

Sl. no.	Genus	Representative	Species documented
32.	*Heterocyathus*	 *Heterocyathus aequicostatus*	*Heterocyathus aequicostatus*
33.	*Heteropsammia*	 *Heteropsammia* sp.	*Heteropsammia michelini*
34.	*Tubastrea*	 *Tubastrea aurea*	*Tubastrea aurea*
35.	*Dendrophyllia*	 *Dendrophyllia coarctata*	*Dendrophyllia coarctata* *Dendrophyllia indica*
36.	*Turbinaria*	 *Turbinaria peltata*	*Turbinaria crater* *Turbinaria peltata* *Turbinaria mesenterina*

In addition to the common coral species stated in Annexure 9A, some new species have been added up in the list like *Acropora hebes*, *Acropora echinata*, *Acropora nasuta*, *Acropora abrolhosensis*, *Montipora aequituberculata*, *Montipora* sp. *Novo and Goniopora* sp. (Patterson Edward et al. 2007).

References

Appukuttan, K. K., Chellam, A., Ramadoss, K., Victor, A. C. C., & Meiyappan, M. M. (1989). Molluscan resources. *CMFRI Bulletin, 43*, 77–92.

Badola, R., & Hussain, S. A. (2005). Valuing ecosystem functions: An empirical study on the storm protection function of Bhitarkanika mangrove ecosystem, India. *Environmental Conservation, 32*(1), 85–92.

Ban, C. (1997). *The economic valuation of tropical forest land use options*. Singapore: The Economy and Environment Programme for Southeast Asia (EPSEA).

Batagoda, B. M. S. (2003). *The economic valuation of alternative uses of mangrove forests in Sri Lanka*. The Hague: UNEP/Global programme of Action for the protection of the Marine Environment from Land-based Activities.

Burke, L., Kura, Y., Kassem, K., Revenga, C., Spalding, M., & McAllister, D. (2001). *Pilot analysis of global ecosystems: Coastal ecosystems*. Washington, DC: World Resources Institute.

Cahoon, D. R., Reed, D. J., & Day, J. W. (1995). Estimating shallow subsidence in microtidal salt marshes of the southeastern United States – Kaye and Barghoorn revisited. *Marine Geology, 128*(1–2), 1–9.

Christensen, B. (1982). *Management and utilization of mangroves in Asia and the Pacific* (FAO environment paper No. 3). Rome: United Nations Food and Agricultural Organization.

Costanza, R., D'Arge, R., De Groot, R., Farber, S., Grasso, M., Hannon, B., Linnberg, K., Naeema, S., O'Neill, R. V., Parvelo, J., Raskin, R. G., Sutton, P., & Van den Belt, M. (1997). The value of the world's ecosystem services and natural capital. *Nature, 387*, 253–260.

Deshmukh, V. D. (1991). Utilisation of paste shrimp Acetes: A review. *Marine Fisheries Information Service Technical and Extension Series, 110*, 7–8.

Dixon, J. A. (1989). The value of mangrove ecosystems. *Tropical Coastal Area Management Newsletter, 4*, 5–8.

Duarte, C. M., Borum, J., Short, F. T., & Walker, D. I. (2005). Seagrass ecosystems: Their global status and prospects. In N. V. C. Polunin (Ed.), *Aquatic ecosystems: Trends and global prospects*. Cambridge: Cambridge University Press.

Ellison, J. C. (1993). Mangrove retreat with rising sea-level, Bermuda. *Estuarine, Coastal and Shelf Science, 37*, 75–87.

Ellison, J. C. (2008). Long-term retrospection on mangrove development using sediment cores and pollen analysis: A review. *Aquatic Botany, 89*, 93–104.

Ellison, J. C. (2009). Geomorphology and sedimentology of mangroves. In G. Perillo, E. Wolanski, D. Cahoon, & M. Brinson (Eds.), *Coastal wetlands: An integrated ecosystem approach* (pp. 565–591). Amsterdam: Elsevier.

Farnsworth, E. J., Ellison, A. M., & Gong, W. K. (1996). Elevated CO_2 alters anatomy, physiology, growth, and reproduction of red mangrove (*Rhizophora mangle* L.). *Oecologia, 108*(4), 599–609.

French, J. (2006). Tidal marsh sedimentation and resilience to environmental change: Exploratory modelling of tidal, sea-level and sediment supply forcing in predominantly allochthonous systems. *Marine Geology, 235*(1–4), 119–136.

Fujimoto, K. (2000). Belowground carbon sequestration of mangrove forests in the Asia-Pacific region. In *Proceedings of Asia-Pacific cooperation on research for conservation of Mangroves*, Okinawa, Japan (pp. 87–96), PMid:10680662.

Ghosh, A. K. (1991). Fauna of Lakshadweep: An overview. In A. K. Ghosh (Ed.), *Fauna of Lakshadweep* (pp. 1–4). Chennai: Ed. Director, Zoological Survey of India.

Gosliner, T. M. (1993). Biodiversity of tropical opisthobranch gastropod faunas. *Proceedings of 7th International Coral Reef Symposium, 2*, 702–709.

Groombridge, B., & Jenkins, M. D. (2002). *World atlas of biodiversity*. Berkeley: Prepared by the UNEP World Conservation Monitoring Centre, University of California Press.

Gunawardena, M., & Rowan, J. S. (2005). Economic valuation of a mangrove ecosystem threatened by shrimp aquaculture in Sri Lanka. *Environmental Management, 36*, 535–536.

Haldar, B. P. (1991). Sipunculans of the Indian coast. *Memoirs of the Zoological Survey of India, 17*, 1–169.

Hornell, J. (1917). The Indian Beche-de-mer industry, its history and recent revival. *Madras Fisheries Bulletin, 11*(4), 119–150.

James, P. S. B. R., & James, D. B. (1994). Resources, exploitation, conservation and management of holothurians management of *beche-de-mer* industry in India. *Bulletin of Central Marine Fisheries Research Institute, 46*, 17–22.

Jones, S., & Kumaran, M. (1980). *Fishes of the Laccadive Archipelago*. Trivandrum: Nature Conservation and Aquatic Sciences Service. 760 pp.

Kaliaperumal, N., Kaladharan, P., & Kalimauthu, S. (1989). Seaweed and seagrass resources. *CMFRI Bulletin, 43*, 162–175.

Kannan, P. (2004). *Studies on the biology and incidental catch of sea turtles in selected centres along the Indian coast*. Chennai: University of Madras.

Kathiresan, K. (2000). A review of studies on Pichavaram mangrove, southeast India. *Hydrobiologia, 430*, 185–205.

Keesing, J., & Irvine, T. (2005). Coastal biodiversity in the Indian Ocean: The known, the unknown and the unknowable. *Indian Journal of Marine Sciences, 34* (1), 11–26.

Kennedy, D. M., & Woodroffe, C. D. (2002). Fringing reef growth and morphology: A review. *Earth-Science Reviews, 57*(3–4), 255–277.

Kennish, M. J. (Ed.). (1989). *Practical handbook of marine science*. Boca Raton: CRC Press.

Khaleel, K. M. (2008). *Management strategies for the mangrove wetlands of North Malabar* (p. 73). Bangalore: Institute of Wood Science and Technology. Abstracts.

Kristensen, E., Bouillon, S., Dittmar, T., & Marchand, C. (2008). Organic carbon dynamics in mangrove ecosystems: A review. *Aquatic Botany, 89*, 201–219.

Lal Mohan, R. S. (1989). Turtle resources. *Bulletin of Central Marine Fisheries Research Institute, 43*, 71–76.

Lal Mohan, R. S., James, D. B., & Kalimuthu, S. (1989). Mariculture potentials. *CMFRI Bulletin Marine Living Resources of the Union Territory of Lakshadweep an Indicative Survey with Suggestions for Development, 43*, 243–247.

Mathew, G., Thulasidas, K., & Venugopal, K. M. (1991). On the first record of the deep sea shark Centrophorus granulosus (Bloch and Schneider) from Indian seas. *Marine Fisheries Information Service Technical and Extension Series, 113*, 22–23.

McAllister, D. E., Schueler, F. W., Roberts, C. M., & Hawkins, J. P. (1994). Mapping and GIS analysis of the global distribution of coral reef fishes on an equal-area grid. In R. I. Miller (Ed.), *Mapping the diversity of nature* (pp. 155–175). London: Chapman & Hall.

McKee, K. L., Cahoon, D. R., & Feller, I. C. (2007). Caribbean mangroves adjust to rising sea level through biotic controls on change in soil elevation. *Global Ecology and Biogeography, 16*(5), 545–556.

Miall, A. D. (1996). *The geology of fluvial deposits: Sedimentary facies, basin analysis and petroleum geology*. Berlin: Springer. 582 pp.

Misra, A., & Chakraborty, R. K. (1991). Polychaetes from Lakshadweep. Fauna of Lakshadweep. Zoological Survey India. *State Fauna Series, 2*, 137–165.

Mondal, K., Bhattacharyya, S. B., & Mitra, A. (2014). Marine algae *Enteromorpha intestinalis* acts a potential growth promoter in prawn feed. *World Journal of Pharmaceutical Research, 3*(5), 764–775. ISSN 2277-7105.

Montaggioni, L. F. (2005). History of Indo-Pacific coral reef systems since the last glaciation: Development patterns and controlling factors. *Earth-Science Reviews, 71*(1–2), 1–75.

Moosa, M. K. (1999). The extent of knowledge about marine biodiversity in Indonesia. In *Proceedings of the international symposium on integrated coastal and marine resource management* (pp. 126–153). Malang.

Morgan, W. J. (1981). Hotspot tracks and the opening of the Atlantic and Indian oceans. In C. Emiliani (Ed.), *The sea* (The oceanic lithosphere, Vol. 7, pp. 443–487). New York: Wiley-Interscience.

Ong, J. E. (1993). Mangroves – A carbon source and sink. *Chemosphere, 27*, 1097–1107.

Ong, J. E., Gong, W. K., & Clough, B. F. (1995). Structure and productivity of a 20-year-old stand of *Rhizophora apiculata* Bl. mangrove forest. *Journal of Biogeography, 22*, 417–424.

Ormond, R. F. G., & Callum, M. (1997). The biodiversity of coral reef fishes. In R. F. G. Ormond, J. D. Gage, & M. V. Angel (Eds.), *Marine biodiversity – Pattern and processes* (pp. 216–257). Cambridge: Cambridge University Press.

Panapitukkul, N., Duarte, C. M., Thampanya, U., Kheowvongsri, P., Srichai, N., Geertz-Hansen, O., Terrados, J., & Boromthanarath, S. (1998). Mangrove colonization: Mangrove progression over the growing Pak Phanang (SE Thailand) mudflat. *Estuarine, Coastal and Shelf Science, 47*, 51–61.

Patterson Edward, J. K., Mathews, G., Patterson, J., Wilhelmsson, D., Tamelander, J., & Linden, O. (2007). *Coral reefs of the Gulf of Mannar, Southeastern India – Distribution, diversity and status* (SDMRI special research publication No. 12) (113 p.).

Pillai, C. S., G., & Jasmine, S. (1989). The coral fauna Lakshadweep. *CMFRI Bulletin, 43*, 179–195.

Pomar, L. (2001). Ecological control of sedimentary accommodation: Evolution from a carbonate ramp to rimmed shelf, Upper Miocene, Balearic Islands. *Palaeogeography, Palaeoclimatology, Palaeoecology, 175*(1–4), 249–272.

Pramanick, P., Zaman, S., Bera, D., Raha, A. K., & Mitra, A. (2014). Mangrove fruit products: A search for alternative livelihood for island dwellers of Gangetic Delta. *International Journal for Pharmaceutical Research Scholars, 3*(1), 131–137.

Rao, C. V. S. (1991). Marine crocodile landed. *CMFRI Newsletter* No. 51, January–March 1991 (p. 8).

Rao, R. R., Molinari, R. L., & Festa, J. F. (1989). Evolution of the climatological near surface thermal structure of the tropical Indian Ocean 1. Description of mean monthly mixed layer depths at surface temperatures, surface current, and surface meteorological fields. *Journal of Geophysical Research, 94*, 10801–10815.

Rodrigues, C. L. (1996). *Taxonomic and ecological survey of the Lakshadweep for Perumal Marine Park*. Goa: Project completion report, Department of Marine Sciences and Marine Biotechnology, Goa University. 46 pp.

Ruitenbeek, H. J. (1992). *Mangrove management: An economic analysis of management options with a focus on Bintuni Bay, Irian Jaya* (EMDI environmental reports No. 8). Jakarta/Halifax: Environmental

Management Development in Indonesia Project (EMDI).

Sale, F. P. (1980). The ecology of fishes in coral reefs. *Oceanography and Marine Biology: An Annual Review, 18*, 367–421.

Sale, P. F., Guy, J. A., & Steel, W. J. (1994). Ecological structure of assemblages of coral reef fishes on isolated patch reefs. *Oecologia, 98*, 83–99.

Sankarankutty, C. (1961). On some decapoda Brachyura from the Andaman and Nicobar Islands. 1. Families Portunidae, Ocypodidae, Grapsidae and Mictyridae. *Journal of the Marine Biology Association of India, 3*, 101–119.

Sathirathai, S. (1998). *Economic valuation of mangroves and the roles of local communities in the conservation of natural resources: Case study of Surat Thani, South of Thailand.* Singapore: Economy and Environment Program for Southeast Asia.

Schlager, W. (1993). Accommodation and supply – A dual control on stratigraphic sequences. *Sedimentary Geology, 86*, 111–136.

Sebens, K. P. (1994). Biodiversity of coral reefs: What are we losing and Why? *American Zoology, 34*, 115–133.

Sorokin, Y. I. (1993). *Coral reef ecology*. Berlin: Springer. 465 pp.

Spalding, M., Kainuma, M., & Collins, L. (2010). *World atlas of mangroves* (p. 319). London/Washington DC: Earthscan.

Spencer, T., & Möller, I. (2013). Mangrove systems. In J. F. Shroder (Ed.), *Treatise on geomorphology* (Vol. 10, pp. 360–391). San Diego: Academic Press.

Subba Rao, K. V., & Subba Rao, N. V. (1991). *Mollusca state fauna series* (Vol. 2). Calcutta: Fauna of Lakshadweep ZSI.

Suharsono. (2004). *Poster presentation at the tenth international coral reef symposium*, Okinawa.

Thomas. (1989). In Van Soest, R. W. M., Boury-Esnault, N., Hooper, J. N. A., Rützler, K., de Voogd, N. J., Alvarez de Glasby, B., et al. (2014) *World Porifera database*. Accessed through: World Register of Marine Species at http://www.marinespecies.org/aphia.php?p=taxdetails&id=233966 on August 1, 2014.

Tri, N. H., Adger, W. N., & Kelly, M. (1998). Natural resource management in mitigating climate impacts: The example of mangrove restoration in Vietnam. *Global Environmental Change, 8*, 49–61.

Untawale, A. G., & Jagtap, T. G. (1984). Marine microphytes of Minicoy (Lakshadweep) coral Atoll of the Arabian Sea. *Aquatic Botany, 19*, 97–103.

Untawale, A. G., Dhargalkar, V. K., & Agadi, V. V. (1983). *List of marine algae from India: Goa*. Goa: Records of the National Institute of Oceanography.

van Bosse, A. W. (1928). Rhodophycea, Gigartinales et. Rhodomeniales. Liste des Algues du Siboga Expedition. *Siboga Expedition Monography, 59d*, 1–41.

Veron, J. E. N. (2000). *Corals of the world* (Vols. 1–3, 1382 pp.). Townsville: Australian Institute of Marine Science.

Wells, S., Ravilous, C., & Corcoran, E. (2006). *In the front line: Shoreline protection and other ecosystem services from mangroves and coral reefs*. Cambridge: United Nations Environment Programme World Conservation Monitoring Centre. 33 pp.

Woodroffe, C. D. (1990). The impact of sea-level rise on mangrove shorelines. *Progress in Physical Geography, 14*(4), 483–520.

Woodroffe, C. D., & Mulrennan, M. E. (1993). *Geomorphology of the lower Mary River Plains, Northern Territory*. Darwin: North Australia Research Unit. 152 pp.

Threats to Marine and Estuarine Ecosystems

10

Contents

Oceans and estuaries comprise about 70 % of the Earth's land mass and provide more potential biodiversity for novel products and services than any other ecosystem in the world. Forty percent of the world inhabitants are living along the coastline of the oceans. Marine lives ranging from bacteria to higher forms of lives are sources of micro- and macromolecules required for the advancement of civilization. Despite the vast socio-economic potentials of oceans and estuaries, they are treated as the final receptacle of all wastes generated from anthropogenic activities. In addition, natural hazards like supercyclone, volcanic activities, tectonic movements, etc. are threats of megascale on the marine and estuarine ecosystems. Climate change-related effects like salinity fluctuation, acidification and extreme weather events greatly damage the salt water domain of the planet Earth. People all over the world face the reality of climate variability in different mode and magnitude. More than 530,000 people died as a direct result of almost 15,000 extreme weather events, and losses of more than USD 2.5 trillion (in PPP) occurred from 1993 to 2012 globally. The list of natural disaster is vast, and the magnitude of damage caused by such disasters is difficult to assess with extreme accuracy.

The marine and estuarine ecosystems mostly face four major threats: (i) climate change-related threats, which include alteration of aquatic temperature, salinity, pH and patterns of

A. Mitra, S. Zaman, *Basics of Marine and Estuarine Ecology*,
DOI 10.1007/978-81-322-2707-6_10

water movement (including currents, eddies and fronts); (ii) overfishing, which includes the by-catch problems both from commercial fishing, recreational fishing, illegal fishing and unregulated fishing; (iii) habitat damage largely caused by fishing gear (particularly during bottom trawling), destruction of coral reefs, mangroves, seagrass beds, salt marsh grass beds, etc. for coastal development programmes that include a wide range of activities like development of industrial units, tourism, aquaculture farms, saltpans, etc.; and (iv) pollution including nutrients, plastic litter, heavy metals, microbial load, persistent organic pollutants (POP), etc. All these threats are mixed in nature, e.g. natural threats like tsunami causes hypersaline environment due to the intrusion of salt water in areas where large number of saltpans exists. It is thus difficult to segregate the natural and human-induced threats on marine and estuarine ecosystems as both of them go hand in hand. Often the threats are synergistic in nature. For the convenience of the readers, we have discussed the threats under two major heads, viz. natural and anthropogenic threats.

10.1 Natural Threats

The natural factors that affect the abiotic and biotic components of marine and estuarine ecosystems may be broadly divided into six heads.

(A) Alteration of the Earth's orbit
(B) Natural oscillation of atmospheric carbon dioxide
(C) Volcanic activities
(D) Variations in solar output
(E) Plate tectonics
(F) Natural disasters and extreme weather events

10.1.1 Alteration of the Earth's Orbit

In astronomy, the **Earth's orbit** is the motion of the Earth around the sun, from an average distance of 149.59787 million kilometres (93 million miles) away. A complete orbit of the Earth around the sun occurs every 365.256363004 days (1 sidereal year). This motion gives an apparent movement of the sun with respect to the stars at a rate of about 1°/day (or a sun or moon diameter every 12 h) eastwards, as seen from the Earth. On average, it takes 24 h—a solar day—for the Earth to complete a full rotation about its axis relative to the sun so that the sun returns to the meridian. The orbital speed of the Earth around the sun averages about 30 km/s (108,000 km/h, or 67,108 mph), which is fast enough to cover the planet's diameter (about 12,700 km, or 7900 miles) in 7 min, and the distance to the moon of 384,000 km (239,000 miles) in 4 h.

There are several factors that regulate the alteration of the Earth's orbit. The first cyclical variation, known as eccentricity, controls the shape of the Earth's orbit around the sun. The orbit gradually changes from being elliptical to being nearly circular and then back to elliptical in a period of about 100,000 years. The greater the eccentricity of the orbit (i.e. the more elliptical it is), the greater is the variation in solar energy received at the top of the atmosphere between the Earth's closest (perihelion) and farthest (aphelion) approach to the sun. Currently, the Earth is experiencing a period of low eccentricity. The difference in the Earth's distance from the sun between perihelion and aphelion (which is only about 3 %) is responsible for approximately a 7 % variation in the amount of solar energy received at the top of the atmosphere. When the difference in this distance is at its maximum (9 %), the difference in solar energy received is about 20 %.

The second cyclical variation results from the fact that, as the Earth rotates on its polar axis, it wobbles like a spinning top changing the orbital timing of the equinoxes and solstices (Fig. 10.1). This effect is known as the precession of the equinox. The precession of the equinox has a cycle of approximately 26,000 years. According to illustration (A), the Earth is closer to the sun in January (perihelion) and farther away in July (aphelion) at the present time. Because of precession, the reverse will be true in 13,000 years, and the Earth will then be closer to the sun in July. This means, of course, that if everything else remains constant, 13,000 years from now, seasonal variations in the Northern Hemisphere

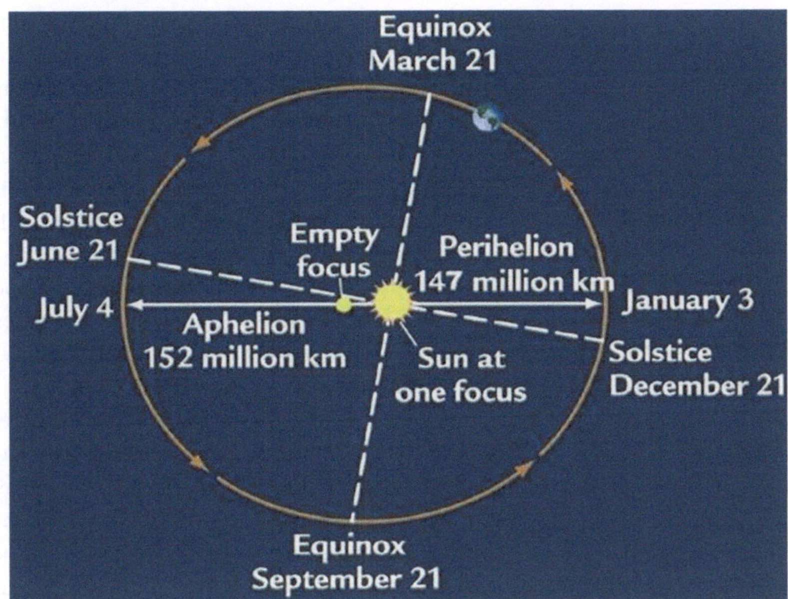

Fig. 10.1 Modification of the timing of aphelion and perihelion over time

should be greater than at present (colder winters and warmer summers) because of the closer proximity of the Earth to the sun.

The third cyclical variation is related to the changes in the tilt (obliquity) of the Earth's axis of rotation over a 41,000 years period. During the 41,000 year cycle, the tilt can deviate from approximately 22.5–24.5°. At the present time, the tilt of the Earth's axis is 23.5°. When the tilt is small, there is less climatic variation between the summer and winter seasons in the middle and high latitudes. Winters tend to be milder and summers cooler. Warmer winters allow for more snow to fall in the high latitude regions. When the atmosphere is warmer, it has a greater ability to hold water vapour, and therefore, more snow is produced at areas of frontal or orographic uplift. Cooler summers cause snow and ice to accumulate on the Earth's surface because less of this frozen water is melted. Thus, the net effect of a smaller tilt would be more extensive formation of glaciers in the polar latitudes.

Periods of a larger tilt result in greater seasonal climatic variation in the middle and high latitudes. At these times, winters tend to be colder and summers warmer. Colder winters produce more snow because of lower atmospheric temperatures. As a result, more snow and ice accumulates on the ground surface. Moreover, the warmer summers produced by the larger tilt provide additional energy to melt and evaporate the snow that fall and accumulate during the winter months. In conclusion, glaciers in the polar regions should be generally receding, with other contributing factors constant, during this part of the obliquity cycle.

The alteration of temperature and incoming solar radiation has profound influence on the distribution pattern and species diversity of coastal vegetation. Increased surface temperature is expected to affect mangroves (Field 1995; Ellison 2000) by:

1. Changing species composition
2. Changing phenological patterns (e.g. timing of flowering and fruiting)
3. Increasing mangrove productivity where temperature does not exceed an upper threshold
4. Expanding mangrove ranges to higher latitudes where range is limited by temperature, but is not limited by other factors, including a supply of propagules and suitable physiographic conditions

Since 1880, the Earth has warmed 0.6–0.8 °C, and it is projected to warm 2–6 °C by 2100 mostly due to human activity (Houghton et al. 2001). Mangroves are not expected to be adversely impacted by the projected increases in sea temperature (Field 1995). Most mangroves produce maximal shoot density when mean air temperature rises to 25 °C and stop producing leaves when the mean air temperature drops below 15 °C (Hutchings and Saenger 1987). At temperatures above 25 °C, some species show a declining leaf formation rate (Saenger and Moverly 1985). Temperatures above 35 °C have led to thermal stress affecting mangrove root structures and the establishment of mangrove seedlings (UNESCO 1992). At leaf temperatures of 38–40 °C, almost no photosynthesis occurs (Clough et al. 1982; Andrews et al. 1984). Some scientists have suggested that mangroves will move polewards with increasing air temperatures (UNEP 1994; Field 1995; Ellison 2005). Although it is possible that some species of mangroves will migrate to higher latitudes where such range extension is limited by temperature, Woodroffe and Grindrod (1991) and Snedaker (1995) suggest that extreme cold events are more likely to limit mangrove expansion into higher latitudes.

Temperature controls the rate of fundamental biochemical processes in organisms, and consequently, changes in the environmental temperature can influence population-, species- and community-level processes. In the marine environment, temperature can alter the number and diversity of adult species in a certain area by changing larval development time. Nevertheless, to date, the influence of temperature on larval duration has only been studied for specific species, and therefore, the generality of the temperature dependence of larval duration remains untested. Knowing the larval dispersal distance, which is believed to be influenced by the duration of the larval period, is a critical component for managing commercially important or invasive species. Recently, American researchers have studied the effect of temperature on larval development using data from 72 marine species, including cod, herring, American lobster, horseshoe crabs and clams. The researchers first used a multilevel model to estimate parameters that describe the influence of temperature on the development of marine larvae. Then, they used the results to develop a model that predicts the effect of temperature on dispersal and survival. The researchers found that the distance larvae travelled varied with ocean temperature. They observed that larvae from the same species travel far less in warmer waters than in colder waters. In particular, mean dispersal distance differs greatly (20 vs. 225 km) as temperature varies from tropical conditions (30 °C) to cold temperate waters (5 °C).

The authors argue that this is due to the fact that larvae in cold waters develop more slowly and drift further before beginning their next development stage because colder temperatures cause sluggish metabolisms. For endangered species, the survival of some animals may depend on whether offspring from parents in one protected area can get to another area where they are safe from harvest. Consequently, in warmer waters, marine protected areas may need to be closer together than in colder water, since in warmer water dispersal distances tend to be shorter. Moreover, they found that the predictions made by the developed model hold true for virtually all marine animals with a larval life cycle. In the context of global warming and associated changes in ocean temperatures, it is important to understand how fish populations are affected by changes in temperature. The current study provides new insights in this regard that may be useful in marine ecology and conservation. The model developed by the authors could represent a useful tool for studying larval movement with knowledge about ocean temperature. It may make it possible to predict what large-scale changes in ocean temperature may mean for adult populations (O'Connor et al. 2006).

10.1.2 Natural Oscillation of Atmospheric Carbon Dioxide

In prehistoric times, during the Permian period, in the Palaeozoic *Era*, the concentration of carbon dioxide dropped below 210 ppmV. The Permian Period was the phase of divergence and evolution and appearance of species. Dinosaurs prospered and predominated over all the other orders of vertebrates. Coniferous plants first appeared in the Permian period. The change of atmospheric temperature at the time of the Permian period was around 10 °C. Compared to the Permian period, the current change of global temperature is only 0.52 °C, while the concentration of atmospheric carbon dioxide is 385 ppmV. If the global temperature is dependent on carbon dioxide, then the change of temperature at present would be around 10 °C or higher, as it was during the Permian period.

From the Early Triassic to the Middle Cretaceous, the concentration of atmospheric carbon dioxide was similar to its current density. From the Late Cretaceous to the Early Miocene, the concentration climbed above 210 ppmV. During the Holocene period, the concentration has oscillated from 210 ppmV to 385 ppmV.

It is possible that the concentration of atmospheric carbon dioxide will increase normally in the course of the next 50 million years to 1050 ppmV or 2500 ppmV.

Researchers observed that the concentration of atmospheric carbon dioxide increases several centuries after glaciations. Perhaps this is due to the fact that most plants perish at sub-zero temperatures, and plants are organisms that capture carbon dioxide from their surroundings to make food through the process of photosynthesis.

Scientists have also observed that the concentration of atmospheric carbon dioxide increases during periods of warming. However, an increase in temperature always precedes an increase in carbon dioxide, which generally occurs decades or centuries after any change of temperature. Researchers have not observed an increase in the concentration of carbon dioxide to have preceded a period of warming. This latter phenomenon occurs because when oceans absorb more heat from an increase in the amount of direct solar irradiance incident upon the Earth's surface, they release more carbon dioxide molecules into the atmosphere. Nevertheless, most drastic increases in carbon dioxide concentration occur decades or centuries after the oceans have warmed up. For example, the present increase of atmospheric carbon dioxide was caused by an extraordinary increase in solar activity in 1998 which warmed up the El Niño South Atlantic Oceanic Oscillation (Fig. 10.2).

Several research works on climate change, however, have confirmed the findings that the temperature variations are closely correlated to the concentration of carbon dioxide in the atmosphere and variations in solar radiation received by the planet as controlled by the Milankovitch cycles. Measurements indicated that atmospheric carbon dioxide levels were about 30 % lower during colder glacial periods. It was also theorized that the oceans were a major storehouse of carbon dioxide and that they controlled the movement of this gas to and from the atmosphere depending on the temperature. Carbon dioxide is released from the oceans when global temperatures become warmer and diffuses into the ocean when temperatures are cooler. Initial alterations in global temperature were triggered by changes in received solar radiation by the Earth through the Milankovitch cycles. The increase in carbon dioxide then amplified the global warming by enhancing the greenhouse effect. The direct relationship between carbon dioxide concentration and temperature is shown in Fig. 10.3.

The effect of enhanced carbon dioxide on blue carbon community is poorly understood and sometimes contradictory, and there is a paucity of research in this area. A direct effect of elevated atmospheric carbon dioxide levels may be the increased productivity of some mangrove species (Field 1995; Ball et al. 1997; Komiyama et al. 2007). Mangrove metabolic responses to increased atmospheric carbon dioxide levels are likely to be increased growth rates (Farnsworth et al. 1996) and more efficient regulation of water loss (UNEP 1994). For some mangrove species, the response to elevated carbon dioxide may be sufficient to induce substantial change of vegetation along natural salinity and aridity gradients. Ball et al. (1997) showed that doubled carbon dioxide had little effect on mangrove growth rates in hypersaline areas, and this may combine with reduced rainfall to create some stress. The greatest effect may be under low salinity

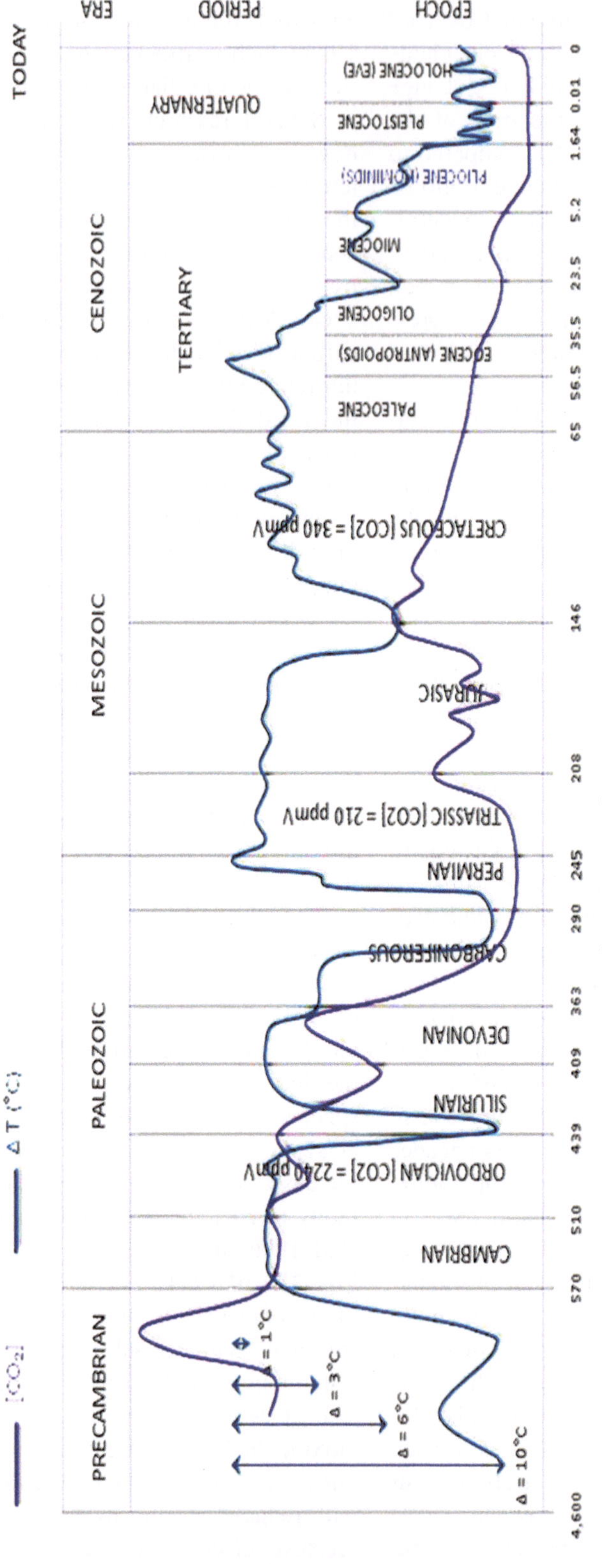

Fig. 10.2 Geological timescale: concentration of carbon dioxide and temperature fluctuation (*Source*: 1. Analysis of the temperature oscillations in Geological Eras by Dr. C.R. Scotese ©2002; 2001; 2. Ruddiman, W.F. (2001). Earth's Climate: past and future. W.H., Freeman and Sons, New York, NY; 3. Mark Pagani et al. (2005). Marked decline in Atmospheric carbon dioxide concentrations during the Paleocene. Science, 309(5734): 600–603. Conclusion and interpretation by Nasif Nahle ©(2005, 2007). Corrected on 07 July 2008 (CO_2: Ordovician Period))

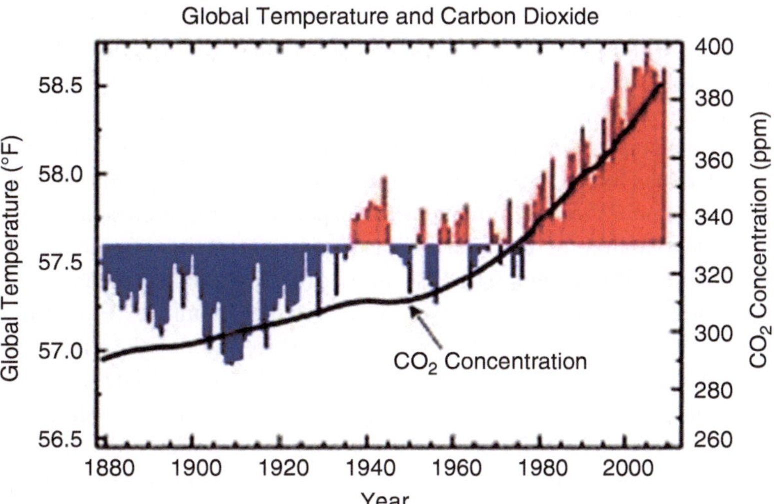

Fig 10.3 Global average temperature and carbon dioxide concentrations

conditions. Elevated carbon dioxide conditions may enhance the growth of mangroves when carbon gain is limited by evaporative demand at the leaves, but not when it is limited by salinity at the roots. There is no evidence that elevated carbon dioxide will increase the range of salinities in which mangrove species can grow.

The increase of atmospheric carbon dioxide and its subsequent dissolution in ocean and estuarine waters is, however, not favourable for molluscan communities as ocean acidification may dissolve their shell. The calcification rates of the edible mussel (*Mytilus edulis*) and Pacific oyster (*Crassostrea gigas*) decline linearly with increasing carbon dioxide levels (Gazeau et al. 2007). Squids are especially sensitive to ocean acidification because it directly impacts their blood oxygen transport and respiration (Portner et al. 2005). Sea urchins raised in lower-pH waters show evidence for inhibited growth due to their inability to maintain internal acid–base balance (Kurihara and Shirayama 2004). Scientists have also seen a reduced ability of marine algae and animals to produce protective carbonate shells (Gattuso et al. 1998; Langdon et al. 2000; Riebesell et al. 2000). These organisms are important food sources for many marine species. One such example is a pteropod, a free-swimming molluscan species that is eaten by organisms ranging in size from tiny zooplankton to whales. In particular, pteropods are an important source for North Pacific juvenile salmon. Mackerel, pollock, herring and cod are also known to feed on pteropods. Many species of other marine calcifers, such as coccolithophores (microscopic algae), foraminifera (microscopic protozoans), coralline algae (benthic algae), echinoderms (sea urchins and starfish) and molluscs (snails, clams and squid), also exhibit a general decline in their ability to produce shells with decreasing pH (Kleypas et al. 2006). A very recent study conducted by the present authors exhibit significant inverse relationship between aquatic pH and shell weight of *Saccostrea cucullata* in Indian Sundarbans estuarine water. Thus, in conclusion, it can be advocated that an increase of atmospheric carbon dioxide may in one way cause a positive impact on coastal vegetation but, on the other hand, may pose an adverse impact on shell-forming organisms by triggering the process of ocean acidification.

10.1.3 Volcanic Activities

Volcanic activity frequently occurs at the boundaries of the Earth's tectonic plates which are a series of large blocks moving between each other. The movement of these plates plays a significant role in the type of volcano formed, which influences its shape. Volcanoes also occur in the deep sea (suboceanic volcanoes), which affect the deep-sea organisms particularly the organisms constituting the vent communities.

Volcanoes affect the marine and estuarine flora and fauna through increment of suspended particulate matter or by altering the chemical composition of seawater. The case study of Barren Island volcano can be an eye opener for the readers. After the first eruption noticed in April 1991 began with hot gases and strong ash emissions with activity continuing through October, another eruption was noticed in December 1994 and January–March 1995, where thick clouds of pale brownish gas, dark ash particles and white steam were observed. Very high values of sulphate ion concentrations in rainwater were observed from January to April in 1995 when predominant winds are northeasterly. The volcanic emission of smoke was also reported during January 2000. The pH values showed a decreasing and sulphate and nitrate concentrations exhibited an increasing tendency from the year 1995 onwards. As the volcanic eruption at Barren Island during 1991 almost coincided with the major volcanic eruption of Mt. Pinatubo in June 1991 in the Philippines and Mt. Pinatubo's eruption was of very large magnitude, it is very difficult to differentiate individual contributions during the 1991 episode, and hence, it may be considered as a combined effect of both the eruptions. The sulphate and nitrate values of the aquatic phase increased abruptly after the eruption along with a lowering of pH.

According to the researchers, increase of suspended solid decreases the transparency of the aquatic phase thereby reducing the penetration of solar radiation. This has high probability to reduce the photosynthetic activity of phytoplankton and other marine producers. The members of higher trophic level may also get affected if the producer communities face crisis or extinction.

The volcanic ash also poses an adverse impact on the coral communities through the process like burial and smothering. Researchers have also documented sudden increase of phytoplankton species after the incidence of volcanoes, which may be attributed to increase of iron concentration in the ambient aquatic phase due to eruptions. The role of iron in the increase of phytoplankton has been observed by several researches in recent times (Zaman et al. 2013; Mitra and Zaman 2014). The case study of Pagan Island volcano is also a relevant example in this context (Fig. 10.4). However, little is known about the effect of moderate amounts of small particulate ash deposits on reef communities. Volcanic ash contains a diversity of chemical compounds that can induce nutrient enrichments triggering changes in benthic composition. Two independently collected data sets on the marine benthos of the pristine and remote reefs around Pagan Island, Northern Mariana Islands, reveal a sudden critical transition to cyanobacteria-dominated communities in 2009–2010, which coincides with a period of continuous volcanic ash eruptions.

Pagan's volcanic rocks mainly consist of silicon dioxide (SiO_2, 50.5 %), aluminium oxide (Al_2O_3, 17.4 %) and iron oxides (Fe_2O_3 and FeO, 11.2 %). The latter compounds are the most likely candidates to stimulate cyanobacterial growth as iron requirements of cyanobacteria are greater than other algal groups because of their high PS I to PS II ratio and the presence of nitrogenase in diazotrophs. The iron limitation of cyanobacteria in oligotrophic coral habitats has been attributed to their evolutionary origin in an anoxic ocean with high bioavailability of iron. When volcanic ash dissolves in seawater, bioactive trace metals like iron become immediately available to primary producers, and experimental studies have demonstrated that this increase in iron concentration can initiate cyanobacterial blooms.

10.1.4 Variations in Solar Output

The presence/absence and number of sunspots is one of the important reasons that regulate the magnitude of solar radiation. Sunspots have magnetic fields of strength up to 3000 times as

Fig. 10.4 Emission of gas and ash at Pagan Island

great as the average magnetic field of either the sun or the Earth. Astronomers believe the cause of sunspots is attributed to this fact. According to a standard explanation, the strong magnetic fields of the sun have the shape of tubes just below the solar surface at the beginning of the sunspot cycle. These tubes lie perpendicular to the sun's equator. The sun rotates faster at its equator than at its poles, and so the tubes are stretched out in the east to west direction. Kinks then develop in the magnetic tubes and push through the solar surface. A pair of sunspots appears wherever a kink penetrates, because the kink both leaves and re-enters the surface. The number and size of sunspots show cyclical patterns, reaching a maximum about every 11, 90 and 180 years. The decrease in solar energy observed in the early 1980s corresponds to a period of maximum sunspot activity based on the 11-year cycle. In addition, measurements made with a solar telescope from 1976 to 1980 showed that during this period, as the number and size of sunspots increased, the sun's surface cooled by about 6 °C. Apparently, the sunspots prevented some of the sun's energy from leaving its surface. However, these findings tend to contradict observations made on longer timescales.

During periods of maximum sunspot activity, the sun's magnetic field is strong. When sunspot activity is low, the sun's magnetic field weakens. The magnetic field of the sun also reverses every 22 years, during a sunspot minimum. Some researchers believe that the periodic droughts on the Great Plains of the United States are in some way correlated with this 22-year cycle.

Temperature, a direct function of solar output, determines the performance of coastal vegetation, and indeed all organisms, at the fundamental levels of enzymatic processes and metabolic function. Seaweeds have evolved biochemical and physiological adaptations, including variation in the identity and concentration of proteins and the properties of cell membranes, that enable them to optimize their performance with respect to the temperatures they encounter (Eggert 2012). Although seaweeds are generally well adapted to their thermal environment, they nevertheless experience temperatures in nature—

particularly during periods of environmental change—that are sufficiently high or low to result in disruptive stress in the form of cellular and subcellular damage (reviewed in Davison and Pearson 1996; Eggert 2012). Such damage and any reallocation of resources for protection and repair can slow growth, delay development and lead to mortality (Davison and Pearson 1996). In response, seaweeds can produce heat shock proteins that repair or remove damaged proteins (e.g. Vayda and Yuan 1994; Lewis et al. 2001). However, protein thermal physiology is not well understood in macroalgae (Eggert 2012) and the upregulation of heat shock protein production is only one of many transcriptional changes that occur in seaweeds during periods of thermal stress (Collen et al. 2007; Kim et al. 2011). Relevant genomic, transcriptomic and proteomic studies are only just beginning to scratch the surface, and most links from gene expression to organismal performance are far from well established. As a result of non-stressful conditions at intermediate temperatures and stress at the extremes, the relationship between temperature and most subcellular, tissue-level or whole-organism processes is described by a hump-shaped thermal performance curve. From colder to warmer, these curves generally rise exponentially as rates of biochemical reactions increase, peak at some optimum temperature and then fall rapidly as the biological components of the system become less efficient or damaged (Kordas et al. 2011; Eggert 2012). When properly parameterized across the full-temperature tolerance range of a species, thermal performance curves have the potential to predict the physiological effects of any given warming or cooling scenario (barring any further acclimatization, adaptation or context-dependent surprises). The effect of a small increase in thallus temperature will be beneficial when the initial temperature is cooler than optimal and detrimental when it is warmer than optimal, and the precise change in performance can be predicted from the starting and ending temperature values along the curve. Unfortunately, the shapes of thermal performance curves and the positions of their optima are poorly described in most seaweed. Although many physiological and ecological studies have linked seaweed performance to temperature, a substantial fraction of these studies do not investigate enough temperatures across a wide enough range to characterize the underlying, nonlinear relationship between the two. Furthermore, various physiological parameters within an organism differ in the shape and optimum temperature of their thermal performance curves, which limits our ability to use an easily measured parameter (e.g. photosynthesis) as a proxy for parameters that may be more ecologically relevant (e.g. growth and reproduction). Indeed, growth rates do tend to peak at lower temperatures than photosynthetic rates (Eggert 2012), presumably because metabolic rates increase faster than photosynthetic rates at higher temperatures. Much remains to be learned regarding the thermal dependence of the key physiological processes that control growth, reproduction and survival across the full range of temperatures experienced by an individual in its lifetime.

A study conducted by Indian researchers (Banerjee et al. 2013) exhibit that the biochemical composition of seaweed species in lower Gangetic delta region is greatly influenced by temperature, particularly the carbohydrate content exhibited significant positive correlation with ambient aquatic temperature.

10.1.5 Plate Tectonics

Plate tectonics governs the topography and motions of the surface of the Earth, and the loss of heat from the Earth's interior, but appears to be found uniquely on the Earth in the Solar System.

The movement of the Earth's crust is the basis of the theory of plate tectonics. In this theory, the lithosphere is viewed as a series of rigid plates that are separated by the earthquake belts of the world—that is, the trenches, ridges and faults. There are seven major lithospheric plates: the Pacific, Eurasian, African, Australian, North American, South American and Antarctic plates.

Each plate is composed of continental and/or oceanic crust (Fig. 10.5).

At the mid-oceanic ridges, where plate boundaries move apart as new lithosphere is formed, divergent plate boundaries occur. Convergent plate boundaries occur at trenches, where plates move towards each other and old lithosphere is destroyed. The plates move past each other at regions known as faults, which represent breaks in the Earth's crust where one plate can move past the other.

A transform fault is a special kind of fault that is found in sections of the mid-ocean ridge. Each side of a transform fault is formed by a different plate, and these plates are moving away from each other in opposite directions. The fault zone produced by this movement is quite active and is the site of frequent earthquakes. The motion of the plates along these faults produces a nearly continuous line of cliffs with sharp vertical drops, known as escarpments. In these regions, there are sudden changes in the ocean depth. Regions where the lithosphere splits, separates and moves apart as new crust is formed are called rift zones. The mid-ocean ridge and rise systems represent the major rift zones at this time. It is generally thought that rifting occurs when rising magma causes enough tension to stretch the overlying crust, creating a sunken rift zone. As this process continues, the rift spreads and the fault deepens and cracks, allowing the magma to seep through and eventually form a ridge. When this happens to the lithosphere under the continents, a sunken rift zone can occur. As the process continues, the fault gets thinner and deeper and can eventually fill with seawater. The Red Sea is an example of this type of formation.

As a plate moves away from the rift zone, it cools and thickens. At the rift, thinning of the crustal plate and increased flow of magma into the rift causes the land masses to separate. A low-lying region of oceanic basalt is formed as well.

Coastal drainages and strandlines are sensitive to land-level changes by tectonic movements and deformations. In the case of slow deformation as during the inter-seismic intervals, the coastlines and the drainage systems adjust slowly to the changes. The adjustments, however, may be abrupt in response to the fast, almost instantaneous (in geological perspective) coseismic land-level changes associated with morphogenic earthquakes (i.e. the earthquakes that produce recognizable surface deformation). Coseismic subsidence induces landward shift of the coastline resulting in drowning of the coasts and consequent distress, whereas uplift induces seaward migration of the coastline, exposing stretches of the seafloor. Moreover, the land-level changes lead to a change in the configuration of the intertidal zones (tidal swamps/mudflats/wetlands/estuaries), which may also shrink (Fig. 10.6) or expand due to ground uplift and subsidence, respectively.

The case study of North Andaman is very relevant in this context where the dimension of blue carbon reservoir has undergone a change due to tectonic movement. The western coastline of North Andaman is indented by several tidal marshes/swamps, bays, estuaries and mudflats. Vast stretches of intertidal wetlands covered by luxuriant mangrove forests (i.e. mangrove swamps) and traversed by a network of tidal creeks occur along the western coast.

During the event of tsunami on 26 December 2004, the estimated 0.3–1.5 m uplift and its easterly tilt in the North Andaman Island have caused many changes in the coastal geomorphic pattern. The strandline has moved seawards exposing vast stretches of seabed and the fringing coral reefs. Upper reaches of the tidal creeks and streams that flow from the hilly interior of the island to the sea have dried up, leaving dry streambeds with stagnant pools of water. The boundary of the tidal zones has migrated towards the central drainage channels, converting the upper intertidal swamps to subaerial coastal uplands (Fig. 10.7). It was documented that after 2 years, the vegetation of the supratidal zone died, and there was mass mortality in blue carbon reservoir due to the non-availability of tidal water.

There are also instances of complete blockage of freshwater supply due to plate tilting that not only caused massive alteration of coastal and

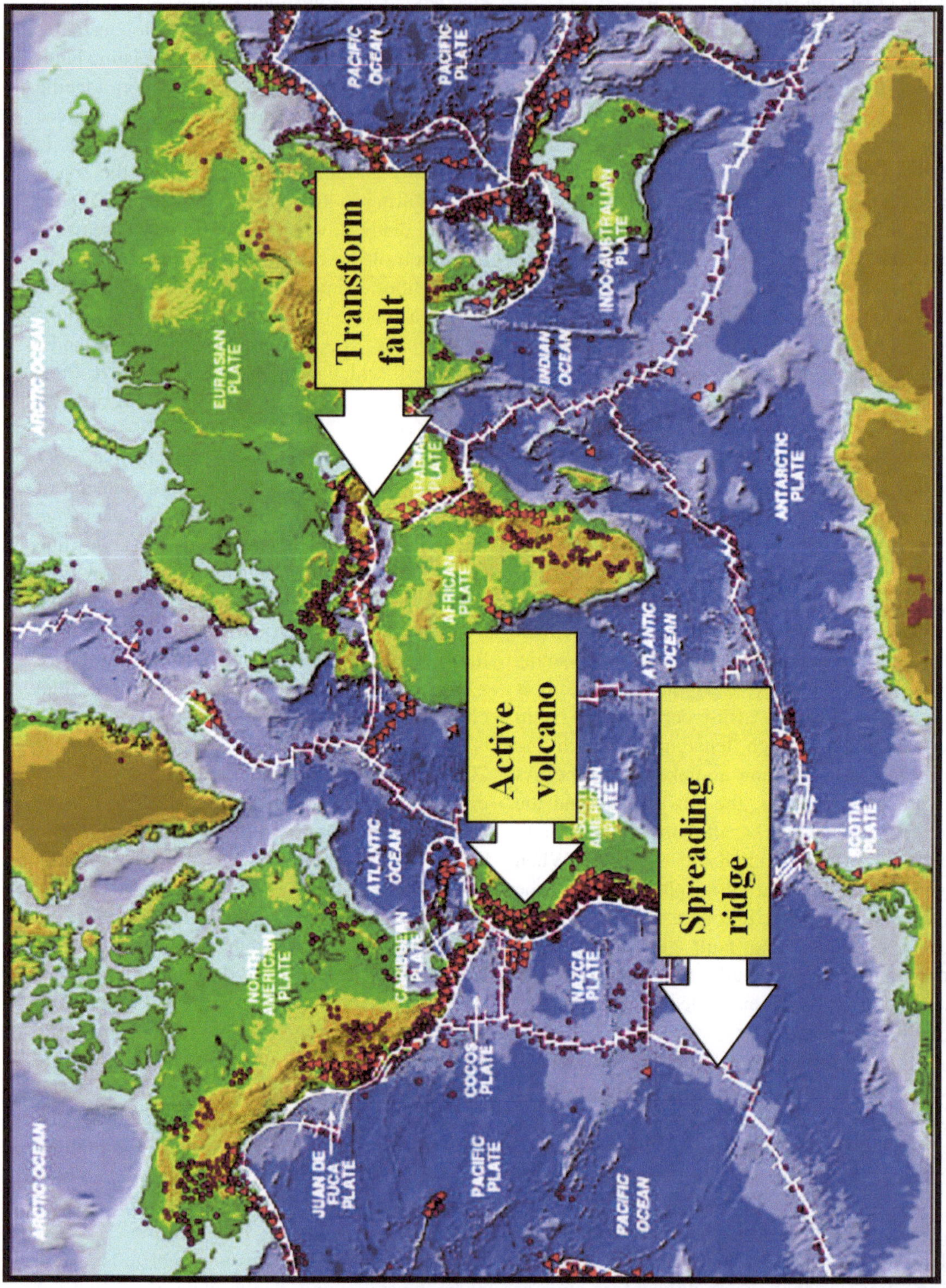

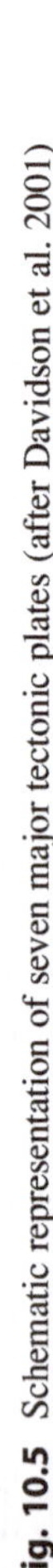

Fig. 10.5 Schematic representation of seven major tectonic plates (after Davidson et al. 2001)

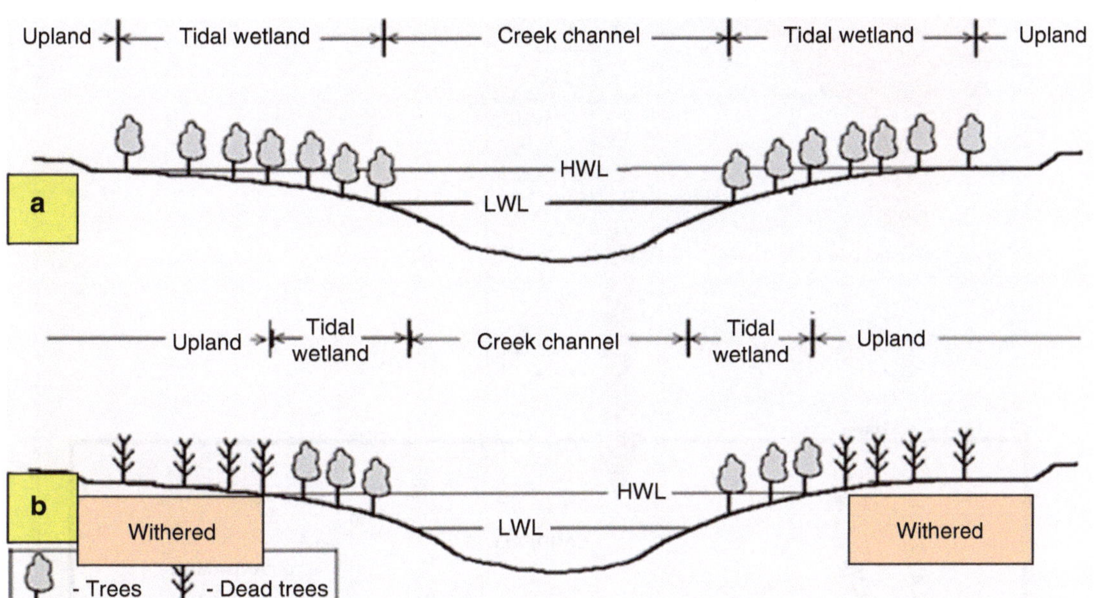

Fig. 10.6 Rise and fall of relative sea level due to ground uplift. (**a**) Ground uplift and consequent fall of sea level. (**b**) Distribution after the earthquake and coseismic uplift. The plants in the intertidal zone that have been shifted above the high tide level slowly wither away because of the unavailability of seawater

estuarine floral composition, but also resulted in stunted growth of some mangrove species. Neotectonic movements in the Bengal Basin between the twelfth and fifteenth century AD resulted in an easterly tilt of the deltaic complex (Chaudhuri and Choudury 1994). During the sixteenth century, the river Ganga changed its course to shift eastwards and join the Brahmaputra (Deb 1956; Blasco 1975; Snedaker 1991). Later, in the mid-eighteenth century, the combined Ganga (now called Padma) and Brahmaputra again tilted eastwards to empty into the river Meghna (Snedaker 1991). This continuing tectonic activity greatly influenced the hydrology of the deltaic region because of changes in the sedimentation patterns and the reduction in freshwater inflows. Most rivers (distributaries) other than the Hooghly, which contributed to the formation of the Ganga Delta (from west to east: Muriganga, Saptamukhi, Thakuran, Matla, Gosaba and Bidya), have lost original connections with the Ganga because of siltation, and their estuarine character is now maintained by the monsoonal runoff (Cole and Vaidyaraman 1966) and tidal actions (Mitra et al. 2009, 2011). The aquatic salinity of these rivers has increased that resulted in the stunted growth of the mangroves. The readers can consult Annexure 10A for further information.

10.1.6 Natural Disasters and Extreme Weather Events

Extreme weather describes weather phenomena that are at the extremes of the historical distribution, especially severe or unseasonal weather. The meteorological or statistical definition of extreme weather events is the events at the extremes (or edges) of the complete range of weather experienced in the past. Defined in this way, extreme weather events include, but are not limited to, *severe* events like heat waves or intense rainfall. For example, the warmest day of winter can be described as a weather extreme.

Whenever there is an extreme weather event, such as a sudden temperature fall, heat waves, flood or drought, people usually point towards global warming. Unfortunately, there is no straightforward logic to support this pointing.

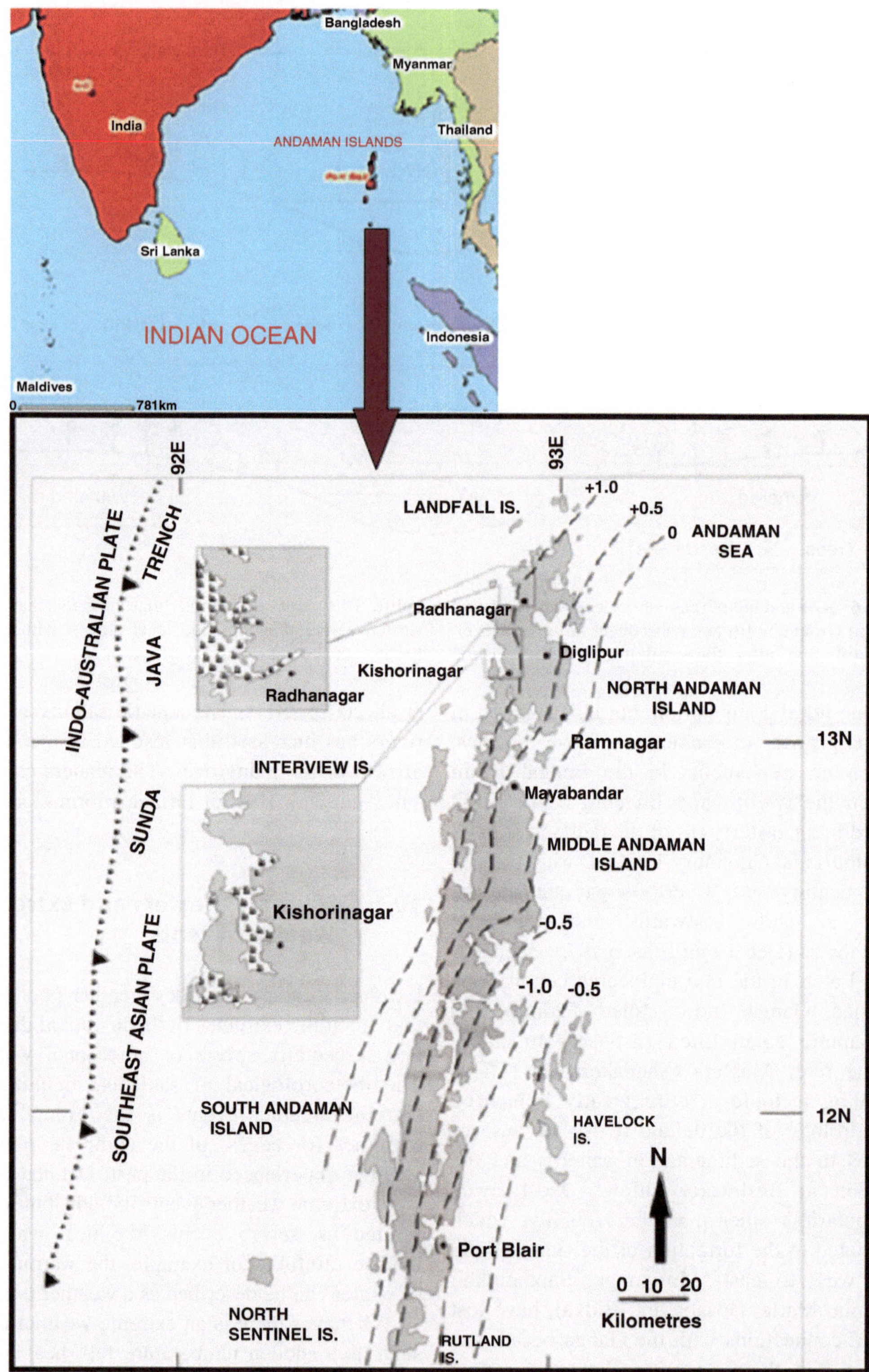

Fig. 10.7 Map showing the location of the Sunda–Java trench and distribution of coseismic vertical ground movement in the Andaman region, caused by the 26 December 2004 megathrust earthquake. The dashed lines represent the contours of vertical ground movement. Values assigned to the contour lines give the estimated vertical offset in metres. The '+' and '–' signs indicate ground uplift and subsidence, respectively. The '0' value contour represents the neutral line

Weather is highly variable and extreme weather events have always happened. Detecting trends takes time, particularly when observational records are rare or even missing in certain regions. An increase in extreme weather is expected with global warming because rising temperatures affect weather parameters in several ways. Changes in the frequency of extreme events coinciding with global warming have already been observed, and there is increasing evidence that some of these changes are caused by the impacts of human activities on the climate.

The extreme weather events influence the biotic communities of the oceans, seas, coastal regions, bays, estuaries and river mouths in various ways. The Extreme Weather Response Program (launched in 1 June 2008) focused on how floods and cyclones in the summer of 2010–2011 affected coral reefs, seagrass, dugong and marine turtles and islands in the Great Barrier Reef Marine Park in Australia. The effects of extreme weather events are region/location specific and cannot be generalized for all parts of the globe, but the present programme in the Great Barrier Reef may be considered as a road map in this domain of climate change.

10.1.6.1 Coral Reefs

Flood plumes and cyclone Yasi together posed adverse impacts on the coral reefs in the Great Barrier Reef Marine Park. Cyclone damage was both severe and widespread. Approximately 6 % of the reef area in the Marine Park suffered severe damage, with broken corals reported across an area exceeding 89,000 km^2. The most severe damage was confined to the area between Cairns and Townsville, sparing the key tourism areas off Cairns and Airlie Beach from major impacts. In contrast, the impacts of flood plumes were generally confined to shallow areas of inshore coral reefs near major rivers, such as those in the Keppel Bay region.

10.1.6.2 Seagrass

Seagrass can be vulnerable to the effects of reduced light during long periods of exposure to flood plumes. Scientists are still evaluating the impacts on seagrass meadows; however, early indications suggest significant seagrass loss in some areas between Cairns and Hervey Bay. Initial inspections of deep sites off Townsville using autonomous underwater vehicle found there has been damage to deep-water seagrass meadows in areas affected by cyclone Yasi.

10.1.6.3 Dugong and Turtle

Declines in seagrass from extreme weather events have serious implications for turtles and dugong as this is their main food source. Impacts on populations can take many months or years to fully eventuate. However, the number of dead dugong and turtle reported along the Queensland coast in the first 7 months after the extreme weather events exceeds any previous full year of records since the reporting programme began in 1996.

10.1.6.4 Islands

Islands are very exposed to the destructive forces of cyclones, while major floods can substantially increase the risk of debris and pests arriving on islands.

Aerial photographic surveys of islands have revealed a number of cays have disappeared or altered shape and size following the recent cyclones. Several new rubble cays have appeared. Changes in islands and cays are being mapped and analyzed using a geographic information system.

10.2 Anthropogenic Threats

Anthropogenic threats on marine and estuarine waters, sediment or on the biodiversity are consequences of human activities. Such activities cover a wide range of subjects like industrialization, urbanization, tourism, aquaculture, deforestation to meet the daily needs, destructive fishing activities, etc. The degree and types of threats are, however, site and time specific, e.g. the type of threats in Indian Sundarbans (like loss of ichthyoplankton due to the wild harvest of tiger prawn seeds or the cutting of mangroves for fuel) is not similar to that of Goa coast (where tourism is a major activity) or South Indian coasts (where salt production through development of mass-scale salt pans makes the surrounding coastal environment highly hypersaline). Again honey or mangrove fruit collection by the island dwellers of Sundarbans is regulated by season (time factor).

The anthropogenic threats on marine and estuarine ecosystem may be broadly divided into six broad heads:

(A) Pollution
(B) Aquaculture
(C) Unplanned tourism
(D) Introduction of alien species
(E) Development of coastal structures
(F) Negative fishing

10.2.1 Pollution

The magnitude and threat of marine pollution becomes more and more acute everyday with the growing use of sea for commerce, the increasing size and variety of cargo ships and tankers and also the use of seabed for mineral extraction. The other reasons may be the river runoff from the land carrying large amount of pollutants, atmospheric deposition, oil spills and other wastes. Nuclear weapon testing has added another dark chapter to the health and biodiversity of oceans.

Marine pollution has been defined as 'the introduction by human, directly or indirectly of substances or energy into the marine environment (including estuaries) resulting in such deleterious effects as harm to living resources, hazards to human health, hindrance to marine activities including fishing, impairment of quality for use of sea water and reduction of amenities'. Thus, marine pollution is the harmful entry of chemicals, energy or particles into the ocean compartment. A big problem is that many toxins adhere to tiny particles, which are taken up by plankton and benthic organisms, most of which are filter feeders in nature. This results in bioaccumulation and biomagnification, while going up the ocean food chain.

The problems of global marine pollution are examined in terms of both short-term, acute, local effects and long-term, chronic, cumulative, worldwide effects of pollutants in the world's oceans. The present pollution problems are most serious in coastal waters, which constitute only 10 % of the area of the oceans yet, together with the upwelling areas of the world, produce 99 % of the world fish catch. However, the long-term consequences of persistent, cumulative substances pose the greatest concern. The critical marine pollutants can be conveniently classified into five categories: (1) metals, (2) synthetic chemicals, (3) petroleum hydrocarbons, (4) radionuclides and (5) solid wastes. Pollutants may enter the sea through rivers, atmospheric transport, ocean outfalls, ocean dumping, ships and such marine activities as mining. Atmospheric transport of pollutants and entry into the sea through fallout, rainout and washout has been considered an important rapid route for certain pollutants from man's activities on land to the world oceans. This particularly applies to the radionuclides and the polyhalogenated hydrocarbons, e.g. DDT and PCBs, but may also be important to worldwide dispersion of metals and petroleum hydrocarbons.

Problems of marine pollution from ships, especially by oil, and from ocean dumping are rapidly coming under control through national legislation and international conventions, such as the International Convention for the Prevention of Pollution from Ships (1973) and the International Convention for the Prevention of Marine Pollution by Dumping of Wastes and Other Matter (1972). Control of atmospheric testing of nuclear weapons among the major nuclear powers, under the Treaty Banning Nuclear Weapons Tests in the Atmosphere, in Outer Space and Underwater, 1963, has largely stopped entry of radionuclides into the sea from fallout. There are stringent controls on nuclear power reactors, and other peaceful uses of atomic energy, so that in the absence of accidents, there is minimal entry of radionuclides into the sea from these sources.

Many scientific conferences on the marine environment are identifying the major pollution problems, examining the critical issues and determining ways in which a meaningful baseline survey can be conducted and a useful monitoring programme established. All the United Nations Specialized Agencies concerned with the marine environment continue to review certain pollution problems within their areas of responsibility. The MCO /FAO/UNESCO/ WMO/WHO/IAEA/United Nations/UNEP (1) Joint Group of Experts on the Scientific Aspects of Marine Pollution (GESAMP 2001) examines some of the problems on an interdisciplinary basis in annual meetings and in intersessional working groups. It advises member agencies and/or member governments, through the agencies, on marine pollution matters. The IOC/WMO IGOSS Pilot Project on Marine Pollution (Petroleum) Monitoring commenced in January 1975, with a review of the first year's results in May and June 1976, and another review is planned for an indefinite date after 1977. ICG for the Global Investigation of Pollution in the Marine Environment (GIPME) met for its third and last session at UNESCO, IOC/INF (1976) Paris, in July 1975 and prepared a comprehensive plan for the global investigation of pollution in the marine environment and baseline study guidelines. The working committee for GIPME, which replaced ICG, met for its first session in Hamburg, from 18 to 22 October 1976. Progress is being made on scientific data exchange in marine pollution. Training courses and workshops are providing a means of educating technical people in developing countries on techniques of measuring and assessing marine pollution and in initiating regional projects investigating problems of marine pollution.

Different substances find their entry in the marine and estuarine environment in three basic forms, namely, solid, liquid and gas. The major categories of waste that enter into the ocean are listed in Table 10.1.

The pollutants discharged from these sources pose considerable effects on the community, species, cellular and even genetic levels. The magnitude of hazard ranges from moderate to extreme depending on the nature of the chemicals. A list of major marine pollutants and their effects are presented in Table 10.2.

Table 10.1 Potential sources of marine and estuarine pollution

S. No.	Major sources of pollution
1	Domestic sewage
2	Sewage sludge
3	Industrial wastes
4	Solid wastes
5	Shipboard wastes
6	Aquacultural farms
7	Pesticides and fertilizers
8	Offshore oil exploration and production wastes
9	Oil spills
10	Radioactive wastes
11	Heat: thermal pollution from power plants
12	Fly ash from thermal power plants
13	Continental runoff
14	Antifouling paints
15	Barges and other metallic structures
16	Ocean mining
17	Precipitation of air borne pollutants
18	Oil from tanker cleaning and deballasting
19	Weathering of the Earth's crust
20	Volcanic eruptions
21	Natural submarine oil seeps
22	Dredge spoils
23	Military wastes (weapon testing, explosives, chemical warfare agents etc.)
24	Tourism and recreational activities

On the basis of origin, the wastes discharged in the marine and estuarine compartments may be classified into point and non-point sources. The effects produced by such wastes are discussed here.

10.2.1.1 Effect from Point Sources

Domestic Wastes

The domestic sewage is an untreated sewage that comes from coastal cities and towns through long pipelines, which would carry the waste away from recreational beaches. Wastes with high organic substances can lead to deoxygenation condition (effect at the ecosystem level). Again decomposition of these wastes can release large amounts of plant nutrients, phosphates, nitrates

Table 10.2 A ranking of major marine pollutants (Johnston 1976)

Pollutant	Harm to living resources	Hazard to human health	Hindrance to maritime activities	Reduction of amenity
Domestic sewage	Important	Important	Slight	Important
Pesticides	Important	Significant	Negligible	Negligible
Inorganic trade wastes	Mainly slight	Slight except Hg	Negligible	Negligible to slight
Radioactive wastes	Negligible	Significant	Negligible	Negligible
Petroleum	Significant	Uncertain	Significant	Important
Organic chemicals	Highly variable	Some significant	Negligible to slight	Variable
Organic trade wastes	Some Important/ significant	Negligible	Negligible to slight	Significant to Important
Military wastes	Uncertain	Uncertain	Negligible	Negligible
Waste heat	Significant	Negligible	Negligible	Negligible
Detergents	Uncertain	Negligible	Important	Important
Solid objects	Slight	Negligible	Uncertain	Significant
Inert solids and dredged spoil	Significant	Negligible	Uncertain	Significant

Notes: 'Important'—requires restrictive or preventive measures
'Slight'—caution required further study needed
'Negligible'—no restrictive action needed
'Uncertain'—special situations apart, no restrictive action needed

and silicates, which lead to massive algal bloom (effect at the community level since alteration of diversity indices occur). These also cause harm to the fish stock. As pointed out earlier in the beginning of this chapter, there exists lot of "noise" in any signal and these noises are the results of interactions between natural and man-made factors. A relevant case study in this context is the nitrate, phosphate and silicate levels in the Ganga–Bhagirathi and Hooghly estuarine system in the northeast coast of Indian subcontinent. Located adjacent to the city of Kolkata, the nutrient level is considerably high in the aquatic phase, but the abrupt increase in the nutrient levels during premonsoon 2009 (Figs. 10.8, 10.9 and 10.10) may be attributed to AILA (a supercyclone that occurred during 22–26 May 2009). This supercyclone not only devastated the lower Gangetic delta region but contributed huge nutrient load in the estuarine water through river bank erosion, huge surface runoff and churning of the water mass.

Due to rapid industrialization, synthetic detergents also come to marine environment, which contain up to 30 % phosphate in the form of polyphosphates that degrade into orthophosphate. Sewage, which contains ammonia and nitrate, also triggers the process of eutrophication. Deoxygenation caused by algal bloom can result in fish mortality (effect at the species level and community level).

As the disposal of sludge on the land becomes difficult, it is disposed into the ocean, and this method was practiced by several countries including UK, where in London City alone, five million tons of sludge/year from the treatment plants are dumped in the Thames Estuary. Table 10.3 shows the discharge of metals at the Thames Estuary.

The picture of discharging wastes into the sea has become almost uniform throughout the globe, irrespective of developed or developing nations. Table 10.4 shows the discharge of organic load directly to the Mediterranean Sea.

Heavy Metals

Most of the industries situated along the coasts, river and estuaries discharge untreated or insufficiently treated wastewater. Almost every industrial process involving water is a potential source of heavy metals in river and estuaries, and their traces are always found in marine organisms. Large volumes of liquid wastes are also produced in electroplating processes. The effect of heavy

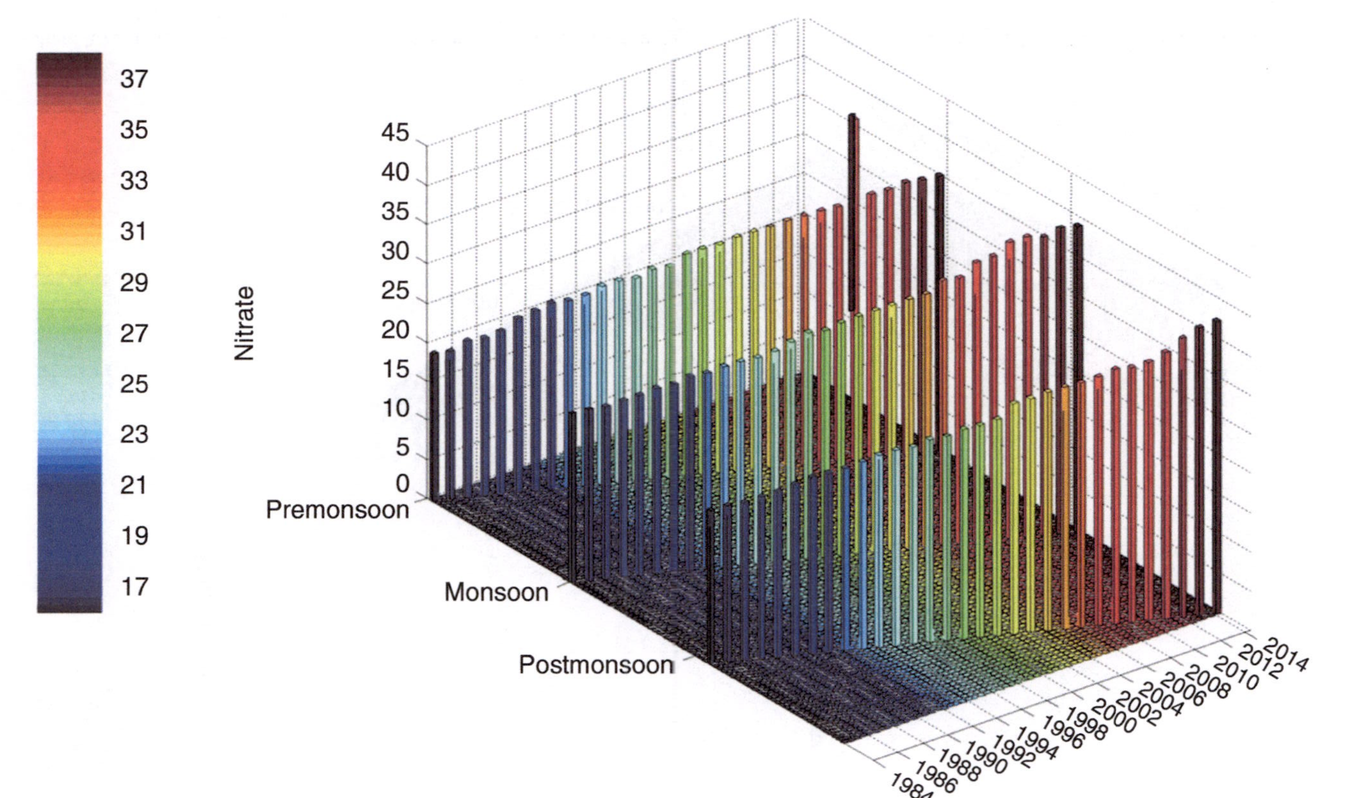

Fig. 10.8 Nitrate level in the Ganga–Bhagirathi and Hooghly estuarine water (in μgatl^{-1}) during three decades (1984–2014). Note the abrupt rise in level in premonsoon 2009 (synergistic effect of AILA and local level pollution)

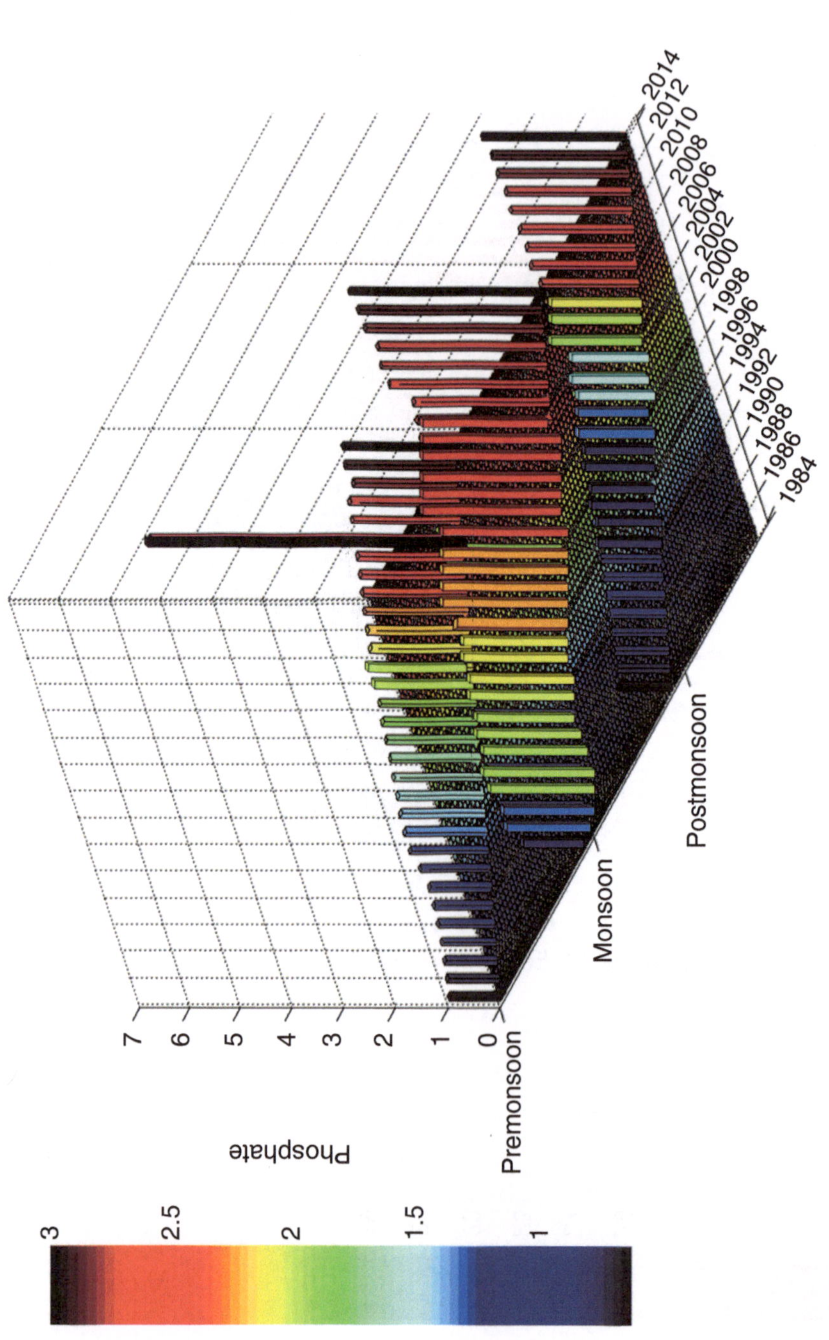

Fig. 10.9 Phosphate level in the Ganga–Bhagirathi and Hooghly estuarine water (in $\mu gatl^{-1}$) during three decades (1984–2014). Note the abrupt rise in level in premonsoon 2009 (synergistic effect of AILA and local level pollution)

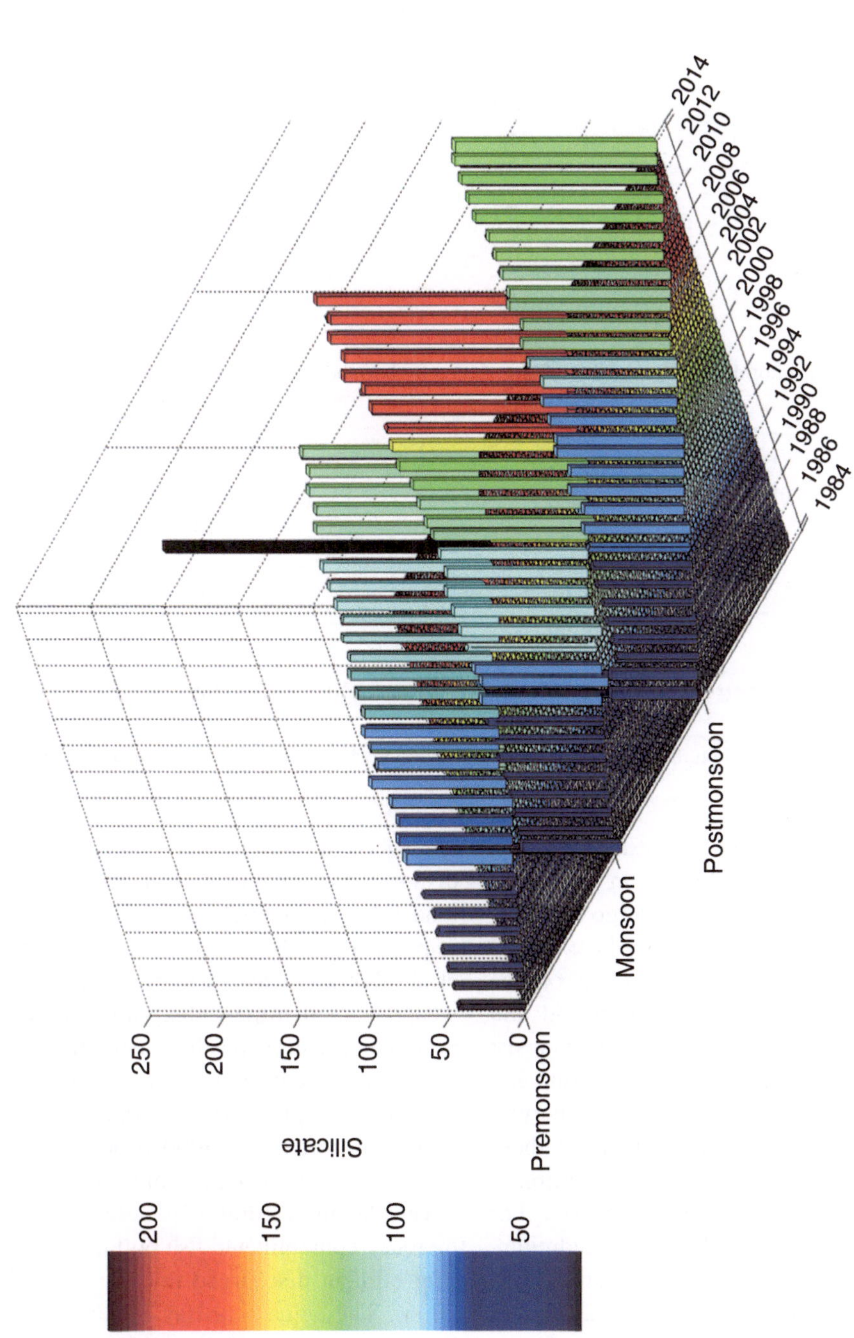

Fig. 10.10 Silicate level in the Ganga–Bhagirathi and Hooghly estuarine water (in μgatl^{-1}) during three decades (1984–2014). Note the abrupt rise in level in premonsoon 2009 (synergistic effect of AILA and local level pollution)

Table 10.3 Thames Estuary: inputs of contaminants (tonnes/annum)

	Sewage sludge	River	Sewage outfalls	Direct industrial	Total
Cadmium	2.3	1.8	5.2	0.6	9.9
Chromium	35.4	18.1	20.5	1	75
Copper	75.9	26.3	26.2	4.6	133
Lead	96	34	16.3	3.9	150.2
Mercury	0.7	0.2	0.8	0.1	1.8
Nickel	11.6	34	37.9	3.8	87.3
Zinc	345.0	60.3	131.3	20.2	556.8
Arsenic	0.5	2.5	0.2	0.1	3.3
Lindane (kg)		60.0	8	5	73
DDT (kg)		100	6	1	107
Average flow (M 1/d)		7511	2693	489	10,693

Table 10.4 Organic load of domestic sewage discharged directly into the sea or through rivers

	BOD_5[a] (tons/year)	BOD_5/km coastline (tons/year)
Spain	130,000	60
Northwestern Basin	360,000	336
Italy	400,000	61
Yugoslavia	17,800	27
Malta	8000	67
Greece	100,000	36
Turkey	100,000	36
Cyprus	9600	15
Lebanon	31,250	149
Israel	32,000	145

[a]The BOD_5 has been calculated on the assumption that the organic load of domestic sewage equals 20–25 kg/person/year

metals on marine organisms varies depending on their concentrations of pollutants in the tissue systems. In the body, they may undergo biotransformation and metabolism and may be excreted without the risk of toxicity depending on the chemical characteristics of the compound and the dose. However, some of the pollutants resist chemical and biological transformation and accumulate in the tissues, including the nerves, to cause toxicity. The adverse effects of these pollutants on the nerves give rise to neurotoxicity. Some heavy metals are neurotoxic. For instance, lead, mercury, nickel, zinc, cadmium, chromium and manganese (Stewart 1975; Klaassen 1995; Plaa 2001; Wagman 2000) are highly neurotoxic in nature. The most common sources of heavy metal pollution are industrial and mining activities, petroleum exploration, exploitation, processing and effluent management, atmospheric fallout and sewage disposal. Natural phenomena such as earthquake, landslides, tornadoes and cyclones have been implicated (Nemerow 1986; Nathaniel et al. 2000). Nuclear reactor accidents and solid weapons sometimes discharge heavy metal pollutants, which constitute potential dangers to the environment. For instance, the Chernobyl (Ukraine, April 1986) radioactive metal pollutants left about 4000 dead and 60,000 at the risk of dying from thyroid cancer. The Jintsu River, Japan, victims died after eating methyl mercury-contaminated fish polluted by an industrial effluent discharged into the river (Maduka 2006). The Minamata Bay in Japan has caused human losses in 1950s and 1960s which killed more than 100 persons in the area through the intake of fish and shellfish contaminated with methyl mercury discharged from a chemical industry. Consumption of oysters with high concentration of cadmium caused another endemic disease named "itai-itai" in a community in Japan.

Oil Spill

Oil is one of the world's main sources of energy, and it is transported by ship across oceans and by pipelines across land. This results in accidents while transferring oil to vessels, when transporting oil and when pipelines break, as well as when drilling for oil. While massive and catastrophic oil spills receive most of the attention, smaller and chronic oil spills and seeps occur almost at regular intervals. These spills contaminate coasts and estuaries, and they can cause harm to humans. Oil can refer to many different materials, including crude oil, refined petroleum products (such as gasoline or diesel fuel) or by-products, ships bunkers, oily refuse or oil mixed in waste. The contribution of oil to the marine environment arises from the disposal of waste oil from domestic and industrial automotive sources and from industrial operations (e.g. lubrication, hydraulic and cutting oils, etc). Most of the oil comes from unauthorized sources such as illegal cleaning of oil tankers and residue dumping. For example, only 5 out of 14 harbours in the Mediterranean have the facilities for receiving oil residues from tankers. Thus, in most cases, the residues are dumped into the sea. It is estimated that in this particular dumping area, 300,000 tons of oil is released annually. Oil is also released into the environment from natural geologic seeps on the seafloor, as along the California coastline. Overall production of petroleum products rose from about 500 million tons in 1950 to over 2500 million tons by the mid-1990s, resulting in massive transport and associated oil spills.

Oil is a mixture of hydrocarbon compounds. Crude oil is a mixture of gas, naphtha, kerosene, light gas and residuals, which have different adverse effects on the health of marine biota. The fate, behaviour and environmental effects of spilled oil can vary, depending upon the type and amount of material spilled. In general, lighter refined petroleum products such as diesel and gasoline are more likely to mix in the water column and are more toxic to marine life, but tend to evaporate more quickly and do not persist long in the environment. Heavier crude or fuel oil, while of less immediate toxicity, can remain on the water surface or stranded on the shoreline for a much longer period. A heavy oil spill across the shore blankets rock pools and prevents gas exchange and eliminates light as well as directly leaching toxins into the water, and can also become mixed deeply into pebble, shingle or sandy beaches, where it may remain for months or even years. Fishes often escape from toxicity by swimming away from the polluted area, but the benthic organisms suffer large numbers of mortalities as the oil settles with the sediments. Coral reefs often get damaged both by oil deposition and by the water-soluble chemicals contained in oil. Seabirds are adversely affected as the oil penetrates and opens up the structure of their plumage, so they become chilled, lose the ability to fly and lose their buoyancy in water. They are thus unable to feed normally, but ingest the oil as they attempt to preen. This condition finally draws these avian fauna towards death.

Oil in the seas also affects human health. Short-term public health impacts from oil spills include accidents suffered by those on damaged tankers or those involved in the cleanup and illnesses caused by toxic fumes or by eating contaminated fish or shellfish. However, there are other less obvious public health impacts, including losses and disruptions of commercial and recreational fisheries, seaweed culture units, boating and a variety of other uses of affected water. The concern involves the ingestion of carcinogenic compounds, especially the polycyclic aromatic hydrocarbons present in petroleum.

The various components of oil have varied effects on the biotic community. Insoluble petroleum fractions can be very damaging as they may coat the organisms and thereby cause suffocation or they may cause tainting of edible species. The toxicity due to oil is caused by the soluble fractions. Aliphatic hydrocarbons having low molecular weight, which are readily soluble in seawater, produce narcosis and anaesthesia at low concentration. At higher concentration, they can cause cell damage and death, particularly to the larval forms of marine life. Low-boiling-point aromatic hydrocarbons have been found to be the most toxic fractions and to be the primary cause of organism mortality. Certain heterocyclic compounds are also toxic at very high concentrations such as 3,4-benzopyrene, and other polycyclic aromatic compounds are known to be carcinogenic.

Susceptibility to oil toxicity has been best correlated with habitat. It has been found that pelagic fish and shrimp are most sensitive to crude oil or No. 2 fuel oil, and the intertidal animals (fish, crabs, starfish and molluscs) are the most tolerant.

The degree of initial impact by oil on marine and estuarine communities and the expected recovery rate are summarized in Table 10.5.

Effects of Oil Spills on Marine Biota

Birds die from oil spills if their feathers are covered with oil. Animals may die because they get hypothermia, causing their body temperature to be really low. Oil may also cause the death of an animal by entering the animal's lungs or liver. Oil also can kill an animal by blinding it. The animal will not be able to see and be aware of their predators. Oil spills sometimes are the reason for animals becoming endangered.

Seabirds

Seabirds are strongly affected by oil spills. A seabird may get covered with the oil. The thick black oil is too heavy for the birds to fly and so they attempt to clean themselves. The bird then eats the oil to clean its feathers and poisons itself.

Killer Whales

Oil spills are one of the many ways killer whales have become endangered. The oil may be eaten or may enter the whale's blowhole. A blowhole is a hole to help them breath. Plugging of the blowhole by oil will inhibit the entry of atmospheric air inside the whale's system. This may result in the death of the animal.

Plankton

Plankton are drifting community and are fully dependent on exchange of gases for their survival. Oil, being lighter than the seawater, floats on the surface and cuts off the diffusion of atmospheric gases into the water. This results in the shortage of oxygen resulting in the death of the planktonic forms. The layer of oil also prevents the solar radiation to reach the seawater due to which the photosynthetic process in phytoplankton is also affected.

The entire spectrum of biological effects that might occur due to oil pollution may be of the following types:

- Lethal toxic effects
- Sublethal effects that disrupt physiological or behavioural activities but do not cause immediate mortality
- Uptake of the oil or certain fractions of it by organisms causing tainting or in some cases carcinomas
- Ingestion by organisms which leads oil to enter the food web
- Direct smothering and suffocation or interference with movements to obtain food or escape from predators as a result of being coated by oil

Table 10.5 Effects of oil on marine populations and communities

Community or population type	Expected degree of initial impact	Expected recovery rate
Plankton	Light to moderate	Fast to moderate
Benthic communities: 1. Rocky intertidal 2. Sandy or muddy intertidal 3. Subtidal, offshore	Light Moderate Heavy	Fast Moderate Slow
Fish	Light to moderate	Fast to moderate
Birds	Heavy	Slow
Marine mammals	Light	Slow, if population seriously affected

- Alteration to the chemical and physical properties of marine habitat
- Mortalities caused by indiscriminate use of detergents to dispose the oil

Hazardous Materials

The hazardous materials mainly include artificially produced radioactive substances arising from nuclear detonations or waste products from nuclear plants. The release of these materials into the ocean system results in a widespread contamination.

Radionuclides reaching estuarine areas can remain in solution or in suspension, precipitate and settle at the bottom or be taken by plants and animals. The radionuclides which enter estuaries as waste from nuclear reactors are biologically active elements, and therefore, the ones that would be transferred through food webs are manganese, iron and zinc.

Thermal Pollution

Thermal pollution produces alteration of temperature in natural water bodies caused by human influence. The main cause of thermal pollution is the use of water as a coolant, especially in nuclear power plants and fossil fuel. Water used as a coolant is returned to the natural environment at a higher temperature. Increases in water temperature can alter aquatic organisms by (a) decreasing oxygen supply; (b) killing fish juveniles/ichthyoplankton, which are vulnerable to small increases in temperature; and (c) affecting ecosystem composition. Primary producers are affected by thermal pollution because higher water temperature often accelerates plant growth rates, resulting in a shorter lifespan and species overpopulation. This can cause an algal bloom that reduces the oxygen levels in the water. The higher plant density leads to an increased plant respiration rate because the reduced light intensity decreases photosynthesis. Tropical estuaries are nursery grounds for penaeid shrimps, lobsters, crabs, clupeid fish and sciaenid fish that constitute vast food supplies and high-value fisheries. The added heat lowers the dissolved oxygen content and may cause serious problems for the plants and animals living there. In extreme cases, fish lethality may occur. Thermal pollution is considered to be the main reason of ciguatera poisoning. Ciguatera poisoning causes severe discomfort in victims and can be fatal to man, which occurs throughout shallow tropical seas. Ciguatera cycle may begin with colonization of newly cleaned surfaces by pioneering blue-green algae, which prefers seawater of high temperature for growth.

Persistent Organic Pollutants (POPs)

Persistent organic pollutants (**POP**s) are organic compounds that are resistant to environmental degradation through chemical, biological and photolytic processes. Due to this nature, they persist in the environment, bioaccumulate in human and animal tissue, biomagnify in food chains and have potential significant impacts on human health and the environment. Thus, these groups of chemicals are not materially broken down over a reasonable period of time and are usually measured in decades or more. The POPs of most concern are those that build up in the environment or are bioaccumulated and/or biomagnified in the food chain.

POPs are often halogenated and characterized by low water solubility and high lipid solubility, leading to their bioaccumulation in fatty tissues. They are also semi-volatile, enabling them to move long distances in the atmosphere before deposition occurs. The Stockholm Convention, a global treaty to protect human health and the environment from POPs, describes POPs as 'chemicals that remain intact in the environment for long periods, become widely distributed geographically, accumulate in the fatty tissue of living organisms and are toxic to humans and wildlife. POPs circulate globally and can cause damage wherever they travel'. The twelve POPs have been identified by the Convention and are commonly known as the *dirty dozen*. They are DDT, aldrin, dieldrin, endrin, chlordane, heptachlor, mirex, toxaphene, hexachlorobenzene (HCB), polychlorinated biphenyls (PCBs), dioxins and furans. Of these, nine are organochlorine pesticides. PCBs are compounds with varied industrial applications, while HCB,

PCBs, dioxins and furans are produced unintentionally, industrially as well as otherwise.

The effects of POP on marine and estuarine organisms have been well documented. If analyzed for tissue or environmental samples, some POPs will almost always be found. As is the case with many environmental pollutants, it is most difficult to establish causality of illness or disease that is directly attributable to exposure to a specific persistent organic pollutant or group of POPs. This difficulty is further underscored by the fact that POPs rarely occur as single compounds and individual field studies are frequently insufficient to provide compelling evidence of cause and effect in their own right. More to the point, however, is the fact that the significant lipophilicity of these compounds means that POPs are likely to accumulate, persist and bioconcentrate and could, thus, achieve toxicologically relevant concentrations even though discrete exposure may appear limited.

Experimentally, POPs have been associated with significant environmental impact in a wide range of species and at virtually all trophic levels. While acute effects of POP intoxication have been well documented, adverse effects associated with chronic low-level exposure in the environment are of particular concern. Noteworthy in this context is the long biological half-life of POPs in biological organisms thereby facilitating accumulation of seemingly small unit concentrations over extended periods of time. For some POPs, there is some experimental evidence that such cumulative low-level exposures may be associated with chronic nonlethal effects including potential immunotoxicity, dermal effects, impairment of reproductive performance and frank carcinogenicity.

Several authors have reported immunotoxicity in association with exposure to different POPs. Investigators have demonstrated immune dysfunction as a plausible cause for increased mortality among marine mammals and have also demonstrated that consumption of persistent organic pollutant-contaminated diets in seals may lead to vitamin and thyroid deficiencies and concomitant susceptibility to microbial infections and reproductive disorders. Investigators have also noted that immunodeficiency has been induced in a variety of wildlife species by a number of prevalent POPs, including TCDDs, PCBs, chlordane, HCB, toxaphene and DDT.

Exposure to POPs has been correlated with population declines in a number of marine mammals including the common seal, the harbour porpoise, bottle-nosed dolphins and beluga whales from the St. Lawrence River. More notably, a clear cause and effect relationship has been established between reproductive failure in mink and exposure to some POPs.

The scientific literature has demonstrated a direct cause and effect relationship in mink and ferrets between PCB exposure and immune dysfunction, reproductive failure, increased kit mortality, deformations and adult mortality. Similarly, investigators have also demonstrated a convincing correlation between environmental concentrations of PCBs and dioxins with reduced viability of larvae in several species of fish. Noteworthy as well is the report suggesting significant reproductive impairment in a number of Great Lakes species described as top-level predators dependent on the Great Lakes aquatic food chain. Supporting this is the observation that wildlife, including stranded carcasses of St. Lawrence beluga whales, with reported high incidence of tumours, has contained significantly elevated concentrations of PCBs mirex, chlordane and toxaphene. A 100 % incidence of thyroid lesions in coho, pink and chinook salmon sampled in the Great Lakes over the last two decades has also been reported to be associated with increased body burdens of POPs (Tables 10.6, 10.7, 10.8, 10.9, 10.10, 10.11, and 10.12).

Miscellaneous

Apart from above pollutants, there are also some other minor pollutants that have considerable deleterious effects. Organic wastes with very high BOD levels coming from industries like palm oil and pulp mill could harm marine life through direct toxicity, oxygen depletion, turbidity, colouring of the water, etc. Kraft mill effluents pose a threat to fish because of methyl mercaptan; hydrogen sulphide toxicity and sulphite pulp mill wastes cause reduction of spawning and settling of oysters. Phenol, which is very toxic to aquatic

Table 10.6 Toxicity of chlordane to selected aquatic organisms

Species	Size/age	Temp (°C)	96-h LC_{50} (μg/L)	Reference
Penaeus duorarum (pink shrimp)	50–65 mm	28.4	0.4	Parrish et al. (1976)
Cancer magister (Dungeness crab)	Zoeal	13	1.3	Cladwell (1977)
Lepomis macrochirus (bluegill)	38–44	25	22	Henderson et al. (1959)
Pimephales promelas (fathead minnow)	38–84 mm	25	52	Henderson et al. (1959)
Salmo gairdneri (rainbow trout)	0.9 g	13	7.8	Cope (1965)

Table 10.7 Toxicity of DDT and its derivatives to selected marine organisms[a]

Organism	Compound	Temp (°C)	96-h LC50 (μg/l)	Reference
Crangon septemspinosa (shrimp)	DDT	20	0.4	McLeese and Metcalfe (1980)
Mysidopsis bahia (mysid shrimp)	DDT	25	0.45 (0.39–0.52)	Mayer (1987)
Cyprinodon variegatus (sheepshead minnow)	DDT	15	2.0 (48-h)	Mayer (1987)
Morone saxatilis (striped bass)	DDT	17	0.53 (0.38–0.84)	Korn and Earnest (1974)
	DDE	17	2.5 (1.6–4.0)	Korn and Earnest (1974)

[a]Taken from WHO (1989)

Table 10.8 Acute toxicity of dieldrin to selected aquatic organisms[a]

Species	Developmental stage, body weight or length	Temperature (°C)	96 h LC50 (μg/L)	Reference
Daphnia magna			330[b]	Anderson (1960)
Crangon septemspinosa (sand shrimp)	0.25 g 2.6 cm	20	7	Eisler (1969)
Pteronarcys californica (stonefly)	Naiad, 3–3.5	15.5	0.5	Sanders and Cope (1968), Johnson and Finley (1980)
Salmo gairdneri (rainbow trout)	1.4 g	13	12	Johnson and Finley (1980)
Pimephales promelas (fathead minnow)	0.6 g	18	3.8	Johnson and Finley (1980)
Lepomis macrochirus (bluegill)	1.3 g	18	3.1	Johnson and Finley (1980)

[a]Taken from WHO (1989)
[b]48-h LC50

Table 10.9 Toxicity of endrin to selected marine organisms[a]

Species	Developmental stage, body weight or length	Temperature (°C)	96 h LC50 (μg/L)	Reference
Penaeus duorarum (pink shrimp)	Adult	17	0.037 μg/L	Mayer (1987)
Pagurus longicarpus (hermit crab)	–	–	1.2 μg/L	Eisler (1970a)
Cyprinodon variegatus (sheepshead minnow)	Adult	18	0.38 μg/L	Mayer (1987)
Mugil cephalus (striped mullet)	83 mm	20	0.3 μg/L	Eisler (1970b)

[a]Taken from WHO (1989)

Table 10.10 Toxicity of heptachlor to selected aquatic organisms

Species	Developmental stage, body weight or length	Temperature (°C)	96 h LC50 (μg/L)	Reference
Pteronarcys californica (stonefly)	Technical (72 %)	15.5	0.9–1.1	Sanders and Cope (1968)
Penaeus duorarum (pink shrimp)	Technical (72 %)	27.5–30	0.11	Schimmel et al. (1976a)
Pimephales promelas (fathead minnow)	Technical (72 %)	25	130	Henderson et al. (1959)
Lepomis macrochirus (bluegill)	Technical (72 %)	25	26	Henderson et al. (1959)
Salmo gairdneri (rainbow trout)	Technical (72 %)	25	7.0	Macek et al. (1969)

Table 10.11 Toxicity of PCB mixtures to select aquatic organisms[a]

Species	Developmental stage, body weight or length	Temperature (°C)	96 h LC50 (μg/L)	Reference
Gammarus fasciatus (scud)	Mature	21	0.052	Mayer and Ellersieck (1986)
Ischnura verticalis (damselfly)	Late instar	15	0.4	Mayer and Ellersieck (1986)
Salmo gairdneri (rainbow trout)	1.8 g	17	>0.23	Mayer and Ellersieck (1986)
Pimephales promelas (fathead minnow)	Fry	24	0.008	Nebeker et al. (1974)
Lepomis macrochirus (bluegill)	0.8 g	18	0.69	Mayer and Ellersieck (1986)

[a]Taken from WHO (1993)

Table 10.12 Toxicity of toxaphene mixtures to select aquatic organisms[a]

Species	Developmental stage, body weight or length	Temperature (°C)	96 h LC50 (μg/L)	Reference
Penaeus duorarum (pink shrimp)	Nauplii	–	2.2	Schimmel et al. (1977)
Lepomis macrochirus (bluegill)	0.6–1.7 g	18	21(14–30)	Macek and McAllister (1970)
Pimephales promelas (fathead minnow)	0.6–1.5 g	12.7	3.2 (2.8–3.7)	Macek et al. (1969)
Salmo gairdneri (rainbow trout)		18.3	1.8	Cope (1965)

[a]Taken from WHO (1984)

animals, comes out from petrochemical industries and mix with water to produce unpleasant taste and smell in fish and shellfish.

The accumulation of phenols increases in the presence of detergents that impair respiration and salt balance. The by-products from the production of vinyl chloride such as chlorinated aliphatic hydrocarbons also act as a pollutant of the marine environment.

10.2.1.2 Effect from Non-point Sources

Non-point source pollution occurs as water moves across the land or through the ground and picks up natural and human-made pollutants, which can then be deposited in lakes, rivers, wetlands, coastal waters and even groundwater. The water that carries non-point source pollution may originate from natural processes, such as rainfall or snowmelt, or from human activities, such as crop irrigation or lawn maintenance. The most common non-point source pollutants are sediment, nutrients, microorganisms and toxics. Sediment can degrade water quality by contaminating drinking water supplies or silting in spawning grounds for fish and other aquatic species. Nutrients, microorganisms and other toxic substances can be hazardous to human health and aquatic life. In areas where crops are grown or in areas with landscaping (including grassy areas of residential lawns and city parks), irrigation and rainfall can carry soil, pesticides, fertilizers, herbicides and insecticides to surface water and groundwater. Bacteria, microorganisms and nutrients (nitrogen and phosphorus) are common non-point source pollutants from agricultural livestock areas and residential pet wastes.

The main non-point sources are:

- Soil erosion from farmland fields as well as construction sites
- Fertilizer runoff from both rural and urban areas
- Pesticide runoff from both rural and urban areas
- Animal waste management
- Paint, oil, anti-freeze and other contaminants poured directly into storm drains
- Activates near a water source that can easily be contaminated without care
- Illegal dumpsites
- Failing septic systems

The problem of pollution from non-point sources is very complex. Most of the non-point sources are carried by surface runoff, which enters the surface water in a diffuse manner and at intermittent intervals that are mostly related to the occurrence of meteorological events.

In the United States, problems associated with non-point pollution are as follows:

(a) Out of the four billion tons of sediments delivered into streams and rivers, almost half of the amount originates from approximately 170 million hectares of land.
(b) Strip mining, which affects approximately 150,000 ha of lands annually, results in the discharge of millions of tons of sediments and high acidity into receiving water.
(c) Non-point sources contribute roughly 80 % of the total nitrogen and more than 50 % of the phosphorus load into receiving water.
(d) Agricultural runoff may contain large quantities of toxic metals, pesticides and other organic chemicals.
(e) Large amounts of decomposable organics originate as a part of soil loss from non-point source that may form objectionable mud deposits in the surface water.

For faecal and total coliform counts, non-point sources account for over 98 %.

10.2.1.3 Effects from Solid Waste Disposal

Many of the larger cities and metropolitan areas in the world are on or near coastlines. The over-increasing amount of solid wastes generated from these places have greatly increased the pressure to dispose them off to the sea. Pollution of the ocean by solid waste is caused by the oceanic dumping of wastes and by the pollutants that reach the ocean through surface runoff. Thickened sewage sludge; industrial solid wastes such as spent clays, catalysts, sludge and drilling mud; and refuse and explosive materials from military and chemical warfare agents have been dumped for years and continue to be disposed of at designated and approved locations in many countries. Two basic methods are used for chemical waste disposal in the ocean. Wastes are either transported to sea aboard tank barges, or disposed of in sealed weighted barrels, which will ultimately leak. Refuse dumping methods are two types—loose and bale dumping. Baling of refuse is more efficient, neat and ecologically sound than loose dumping, but it is expensive. Loose dumping is difficult to handle and requires more storage space, and it floats. Sometimes, incinerator residues are dumped into the ocean instead of spreading in landfill. The radioactive wastes are also dumped into the ocean. Certain chemicals if released untreated, e.g. cyanides, mercury and polychlorinated biphenyls, are highly toxic, and exposure can lead to disease or death of organisms inhabiting the ecosystem.

10.2.2 Aquaculture

Aquaculture has become an important source of seaweed, shellfish and fish, especially for human food, and production is likely to continue to expand well into the next century. Mariculture has both direct and indirect impacts on biodiversity through the consumption of natural resources and the production of wastes. Natural resources such as land, water, seed and feed are required, consumption varying with intensity of production. Wastes, comprising uneaten food, faecal and urinary products, chemicals, pathogens and feral animals, are released into the environment, quantities also being largely dependent upon production methods.

World population continues to increase faster than global food supply (Anon, 1993). While 99 % of food comes from terrestrial agriculture (Pimentel et al. 1994), this disguises the fact that in many, especially developing, countries the bulk of animal protein comes from fish and other aquatic products. Aquatic foods have until recently been derived almost exclusively from capture fisheries sources. In recent years, however, aquaculture has been playing an increasingly important role. Mariculture can be defined as the farming of the marine environment. A large and increasing range of plants and animals is being farmed in the sea. Unlike inland water aquaculture, mariculture involves the culture of plants and invertebrates not only for food but also for decoration (shells and pearls) and chemicals (alginates). Systems and methods for the most commonly grown tropical and temperate species are summarized in Table 10.13. The terms 'intensive', 'semi-intensive' and 'extensive' are used here with regard to inputs of foods. In extensive aquaculture, the farmed organism is reliant on the environment for food or nutrients, while in semi-intensive farming, natural food is supplemented with additions of fertilizer and/or food, the latter usually being derived from agricultural by-products such as animal manures and rice bran. In intensive mariculture, all, or almost all, of the nutritional requirements are supplied by the farmer, and diets are largely fishmeal based. There is also a correlation between intensity of production, as defined here, and energy consumption.

It is an economic activity that transforms natural resources through inputs of capital and labour into products valued by society. In so doing, wastes are inevitably produced. The impact of aquaculture on the environment and on biodiversity thus arises from these three processes: the consumption of resources, the aquaculture processes itself and the production of wastes (Beveridge et al. 1994).

Statistics produced by the FAO show that world aquaculture production is currently around 25 million tonnes (FAO 1996), equivalent to 20 % of world fisheries (capture + culture) production by weight and around twice this by value. Production from the marine environment accounts for around 51 % of aquaculture production by weight (53 % by value) and is growing by some 5 % per annum. While only 4 % of farmed fish production comes from the sea, all farmed macroalgae, almost all farmed mollusks and more than 90 % of farmed crustaceans are produced in the marine environment. The fastest growing sectors of mariculture are in high-market-value products such as shrimp and fish, production of the former having doubled over the past 5 years. By contrast, farmed production of aquatic plants and molluscs has increased only slowly.

Aquaculture, which is essentially cage or pond culture, will have an impact on the carrying capacity of the coastal marine environment. The impacts are several folds and not fully appreciated. Eutrophication is a major issue. In general, from the fish feed, 85 % of the phosphorus (P), 80–88 % carbon (C) and 52–95 % of nitrogen (N) are lost to the environment (Wu 1995). Nearly 53 % of P, 23 % C and 21 % N end up in the sediment. Specifically, in the shrimp culture, such as those on the East Coast of India, which are run on Thailand design, 24 % N and 13 % P were incorporated, and the rest is exported to the environment (Briggs and Funge-Smith 1994).

Aquaculture preferably the shrimp culture is practiced in mangrove-dominated zone, because of the availability of brackish water, congenial nutrients and environmental conditions. For centuries, mangrove systems have contributed significantly to the wellbeing of coastal

Table 10.13 Summary of the principal rearing systems and methods employed in tropical and temperate mariculture

Group	Species	System	Method
Tropical			
Macroalgae	*Laminaria japonica* *Undaria pinnatifida* *Porphyra tenida* *Eucheuma spp.*	Beds (stake and line rafts)	Extensive
Molluscs	*Crassotrea spp.* *Mytilus spp.* *Pecten yessoensis* *Venerupis japonica* *Solen spp.*	Suspended (rafts, long lines)	Extensive
Crustaceans	*Scylla serrata* *Penaeus spp.*	Ponds	Semi-intensive Intensive
Finfish	*Chanos chanos* *Mugil spp.* *Epinephelus spp.* Serranidae Pagrus major *Seriola quinqueradiata*	Land based (ponds) and Water based (cages)	Intensive
Temperate			
Macroalgae	*Gracilaria*	Beds, rafts	Extensive
Molluscs	*Ostrea edulis* *Crassotrea spp.* *Mytilus spp.* *Tapes spp.*	Bottom (tressels, trays) and suspended (rafts, long lines)	Extensive
Finfish	*Salmo salar* *Oncorhynchus spp.* *Dicentrarchus labrax*	Water based (cages) and land based (tanks)	Intensive

communities through their provision of a wide array of goods (resource function) and services (regulatory function). These products come from forestry (wood used for fuel, construction and fishing poles, forage for livestock, honey, medicines, wax, etc.) and the higher-valued fish, crustaceans and molluscs from fisheries. A positive correlation between mangrove area and shrimp/fish catches has been documented for the Philippines, Malaysia, Indonesia and Australia (Primavera 1995, 1998). Mangrove ecosystem services include coastal protection provided by a buffer zone during typhoons and storm surges, reduction of shoreline and riverbank erosion, flood control, nutrient recycling and habitat for wildlife. However, such a valuable compartment of nature was cleared in mass scale to promote shrimp culture. Although conversion to salt beds, agriculture, settlements and overexploitation by coastal dwellers have caused mangrove decline, aquaculture remains the major causative factor at least in Southeast Asia (Primavera 1995, 1997). The high rates (25–80 %) of mangrove loss in the region over the last three decades (Low et al. 1994) have coincided with the Shrimp Fever of the 1980s (Primavera 1997, 1998). In Vietnam, a total of 102,000 ha of mangroves have been cleared for shrimp farming from 1983 to 1987 (Tuan 1997). Shrimp farms in Thailand accounted for 32 % or 65,150 ha of the total 203,600 ha of mangrove area destroyed between 1961 and 1993 (Menasveta 1997). For example, in Ao Ko Nok, Chanthaburi province, the increase in prawn ponds from zero in 1975 to 1836 ha in 1991 saw a parallel decline in mangrove forests by 1428 ha in the same period (Raine 1994). Although Southeast Asia has the greatest area of shrimp ponds totalling close to a million ha, mangrove–shrimp pond conversion is also widespread in Latin America. Seventy-five percent of around 100,000 ha of shrimp ponds established in Ecuador by 1995 and more than a third of the total 11,515 ha of shrimp farms in Honduras were developed in mangroves (Skladany and Harris 1995; DeWalt et al. 1996).

In the Indian subcontinent, almost a similar picture was witnessed in all the maritime states during the 1980s. The fast development of the shrimp sector required the conversion of flat, coastal lands to shrimp ponds. Part of the shrimp pond construction took place in mangroves, and shrimp aquaculture has been an important cause of the conversion of mangroves in India in the last decade (Lakshmana Rao et al. 1994; Holmgren 1994; Alagarswami 1995; Krishnamoorthy 1995; James 1999). A recent survey by the aquaculture sector found that about 5 % of the shrimp aquaculture farms in India have been constructed in former mangrove areas (ADB/NACA 1998) (Table 10.14). Mangrove conversion has been undertaken by both small-scale extensive farms and by larger-scale semi-intensive and intensive farms (Vivekanandan et al. 1997; ADB/NACA 1998).

In order to determine the proportion of the destruction of mangroves by aquaculture in relation to other factors having an impact on mangrove ecosystems, the Andhra Pradesh Remote Sensing Application Centre carried out a case study for the Godavari delta in Andhra Pradesh. The results of the classification of the images are presented in Table 10.15.

From the remote sensing images, it is apparent that in the Godavari delta areas, about 14 % of the aquaculture farms have been constructed on mangrove lands. Shrimp aquaculture is responsible for about 80 % of the conversion of mangrove land. Shrimp ponds are often located in sparse mangrove forests (see maps). The decrease in the area of sparse mangrove cover is partly reversed by the conversion of dense into sparse mangroves, probably through fuelwood collection and grazing.

The rate of conversion of mangroves into shrimp ponds increased in the period 1997–1999, suggesting that shrimp pond construction started in fallow and croplands but then encroached on mangroves in the absence of suitable fallow land. Policy regulations banning the conversion of mangroves to shrimp ponds and the protected status of the Godavari forest (Figs. 10.11 and 10.12) have not been able to prevent the conversion of mangroves into shrimp ponds.

Table 10.14 Prior land use of shrimp farms (%)

Production system	Mangroves	Intertidal wetland	Rice farming land	Other, including fallow land
Traditional and extensive	3	20	32	45
Semi-intensive	7	8	5	80
Total[a]	5	14	18	63

Source: ADB/NACA (1998)
[a]966 farms, with a total surface of 3560 ha

Table 10.15 The impact of shrimp aquaculture in the Godavari delta (ha)

Land use	Land use area	Converted to shrimp farms				
	1989	1997	1999	1987–1997	1997–1999	1989–1999
Crop land				4543	2324	6903
Fallow land				3149	1327	4497
Dense mangrove	16,586	15,987	15,318	433	471	1137
Sparse mangrove	4530	3786	3,199	604	666	1030
Total mangroves	21,116	19,773	18,517	1037	1137	2167
Other				2281	1493	3714
Aquafarms	2006	13,032	19,239			
Total				*17,281*	*6251*	*17,281*

Source: Remote Sensing Images from the Andhra Pradesh Remote Sensing Application Centre (1999)

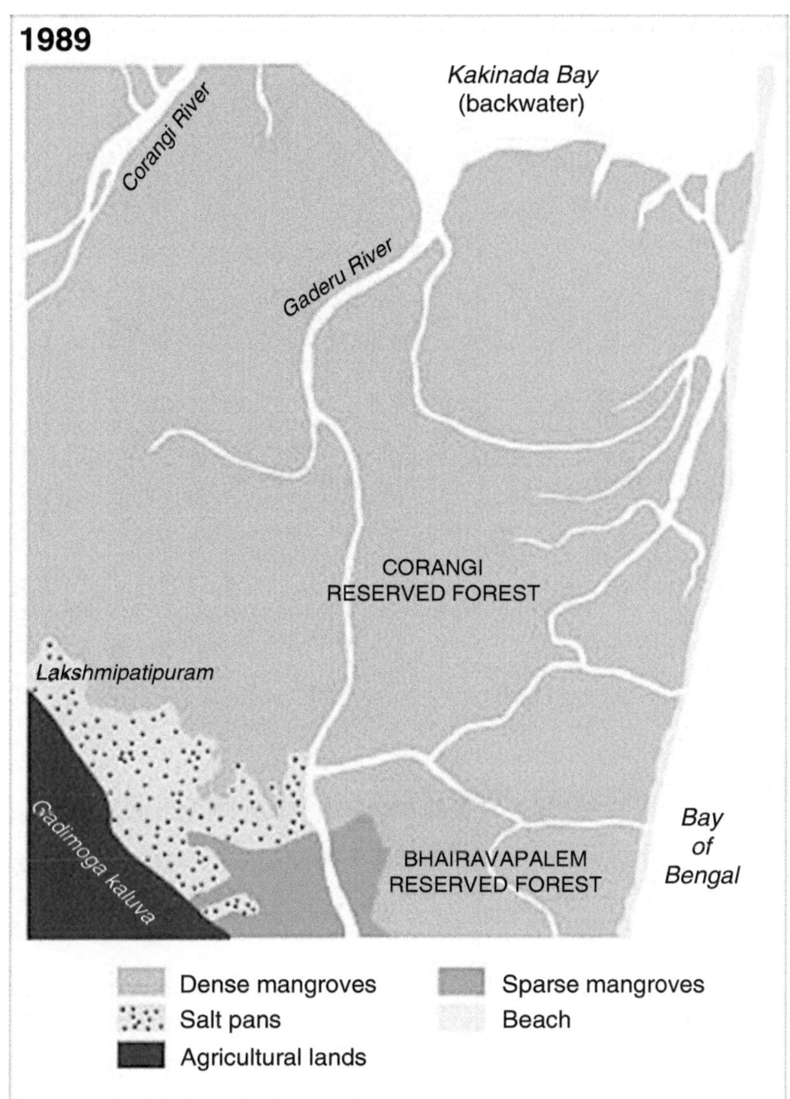

Fig. 10.11 Land use and land cover in the Godavari delta in 1989

10.2.3 Unplanned Tourism

There are greatly increasing stresses on coasts caused by tourism even in Antarctica and the Arctic. The most serious threats are those of habitat destruction. Mangroves are often removed, wetland areas filled in and estuaries reclaimed to make way for tourist complexes without there being any evaluation of the benefits of the intact systems or the ecosystem services connected with the system (Wells 2006). Once built, the resort may lead to effects on adjacent habitats through sewage discharge and other threats and ultimately to the loss of habitats and their resources. Establishment of hotels on coral reefs is becoming popular and often leads to the destruction of the habitat.

Several tourist units and hotels have been constructed without giving any importance to the endemic biodiversity in developing countries. A relevant example in this context is the developmental activities in the Goa coast. Goa is India's smallest state by area and the fourth smallest by population. Located in West India in the region known as the Konkan, it is bounded by the state of Maharashtra to the north and by

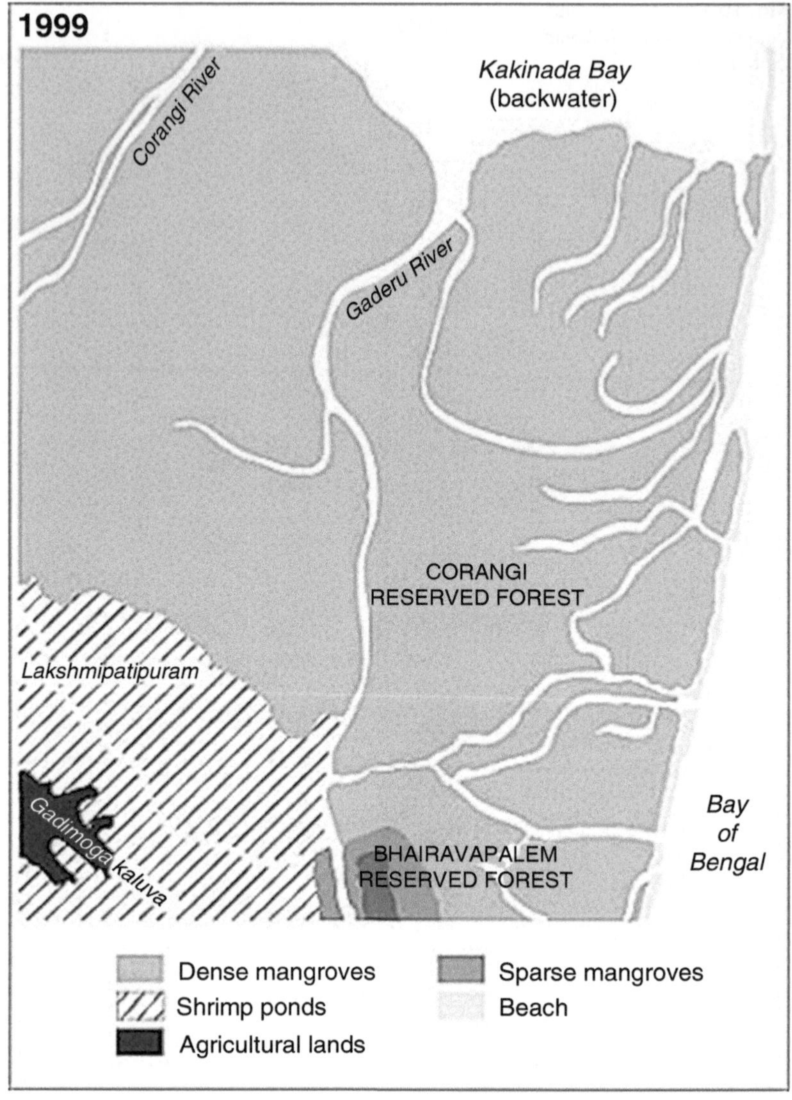

Fig. 10.12 Land use and land cover in the Godavari delta in 1999

Karnataka to the east and south, while the Arabian Sea forms its western coast. Goa is India's richest state with a GDP per capita two and a half times that of the country as a whole. This maritime state encompasses an area of 3702 km^2 (1429 sq mi) and lies between the latitudes 14°53′54″N and 15°40′00″N and longitudes 73°40′33″E and 74°20′13″E. Goa has a coastline of 101 km (63 mi). In Goa, several coastal areas have changed from virtual wilderness in 1970s to haphazardly developed stretches, full of concrete buildings and related structures in less than 20 years. The Baga–Candolim coast is a classic example of frenzied development (Mascarenhas 1997). Several shore fronts have been designed and built in such a manner that they bear little resemblance to the coast that formerly existed. Many coastal communities thus experienced a dramatic growth during this period with constructions of high-rise buildings, resorts and residential dwellings that mushroomed almost all over the coastal zone of Goa. During the last two decades, the advent of tourism, population increase coupled with building activity and modern societal demands has resulted in large-scale changes in the geological and ecological setup and has indelibly altered ecosystems, land use patterns and the coastal zone landscape

Fig. 10.13 Hard concrete structures adjacent to the HTL on the Goa coast (15°29′5.0″N and 73°47′40.4″E) (Photograph taken on 04.04.2013 by Dr. Subhadra Devi Gadi (zoologist))

(Mascarenhas 1997). Figures 10.13 and 10.14 show some developmental concrete constructions almost adjacent to the high tide line (HTL), which can accelerate the rate and magnitude of erosion subsequently leading to destruction of coastal vegetation.

In addition to the concrete structures, some additional threats of tourism include the left-away plastic bottles, cans, etc. by the tourists or enjoying in the estuarine system with lighted vessels and boats (Fig. 10.15), thereby posing an adverse impact on the endemic wildlife associated with the estuaries.

10.2.4 Introduction of Alien Species

About half of the crude oil produced per year is transported through the sea. After unloading of the cargo of oil from a tanker, it carries seawater as ballast. It is a general practice to fill several tanks (25–30 % of the total capacity of the tanker) with seawater to ballast the ship for the voyage back to the loading terminal. These ballast waters are discharged prior to filling the tanks again with new oil. Alien species are mostly introduced into a system mainly through ballast discharge.

The ctenophore *Mnemiopsis leidyi* was imported from the United States East Coast to the Black Sea, probably in ballast water. The introduction of the ctenophore species caused a catastrophic alteration in the whole trophic web and contributed to a huge reduction in stocks of commercial fisheries (GESAMP 1995). Other concerns covered by GESAMP are the transport of species of algae that may cause toxic blooms in new areas and other introductions which have led to dramatic effects at regional levels. Alterations in biodiversity are also highly likely although this is poorly documented.

10.2.5 Development of Coastal Structures

Coastal zones face a variety of pressures. Coastal development results in the infilling of lagoons and reclamation of coastline. Historically, the Dutch have fought a long battle with the North

Fig. 10.14 A hotel under construction on the Goa coast (15°29′5.1″N and 73°47′40.9″E) [Photograph taken on 04.04.2013 by Ms. Ankita Mitra (environmentalist)]

Fig. 10.15 Lighted vessels in the estuarine water adjacent to mangrove forests

Sea and by an extensive system of dykes have extended their landmass and turned the semi-enclosed marine Zuiderzee into the freshwater IJsselmeer. However, most coastal development is piecemeal and insidious and hence difficult to effectively regulate. The cumulative effects of such developments within a particular estuary can be depressing. For instance, about 2500 ha of mudflats in the Firth of Forth, Scotland, have disappeared over the last 200 years through a series of individually unspectacular schemes, leading to a reduction in the estuary's fish biomass by 50 % (McLusky et al. 1991). This scale of loss of intertidal zone has occurred or is forecast for many UK areas (Prater 1981) and must apply to most of the world's industrialized estuaries. Development, whether by dykes or coastal defences, seaside promenades, residential 'marinas' or dock complexes, tends to shorten the foreshore and reduces the extent of mudflats, upper-shore creek and salt marsh systems.

Sometimes coastal structures become indispensable to promote the rapid pace of industrialization and urbanization in the coastal zone, which is now a common scene in developed and developing countries. This can cause dramatic change in the coastal landscape. Examples of such alterations include the construction of sea walls, breakwaters, revetments, groins and jetties. Each of these structures, aimed at protecting coastal property, refracts energy away from the shore and can cause erosion and increase currents that can harm seagrasses.

10.2.6 Negative Fishing

Overfishing is one of the major anthropogenic threats on the living resources of the marine and estuarine ecosystems. It may be defined as the level of fishing which puts at-risk values endorsed either by the fishery management agency, by the nation in whose water fishing takes place or within widely accepted international agreements. A point of critical importance in this regard successfully needs traditional stock sustainability criteria (e.g. fishing a stock at maximum sustainable yield) which may well be considerably higher than a level of fishing intensity which meets the criteria designed to protect marine biodiversity.

Despite the fact that most fisheries resources are now within the jurisdiction of coastal states, nearly all the world's fish resources are overexploited (FAO 1991). Between 1988 and 1990, the marine fish catch declined in nine key fishing areas and especially of Peru, pelagic fish of Japan, of the Northeast coast of the United States and in European seas. The consequences of heavy fishing pressure on commercial species are that the size distribution changes, and this leads to loss of genetic diversity, e.g. orange roughy (Elliott and Ward 1992). In many areas of the Northwest Atlantic, there have been dramatic changes in the composition of fish stocks as a consequence of fishing. Highly important commercial species have declined (e.g. herring and Arctic cod), and other less valuable species have increased, e.g. sand eels (Sherman and Alexander 1990) and sharks. Several studies show that changes in fish species composition have dramatic effects on other species dependent on fish such as seabirds and mammals (Monaghan 1992; Hamre 1994). Exploitation of fish resources can lead to local or regional species extinctions. The Blue Walleye (*Stizostedion vitreum glaucum*) was over-shed in Lake Erie and became locally extinct (Scott and Crossman 1973). The Coelacanth (*Latimeria chalumnae*), which lives in caves in the Cormora islands, has a total world population of under 500 individuals and is being harvested accidentally as a by-catch of fishing for other species (Mackenzie 1995) and is in real danger of becoming extinct. Local extinctions of fish can also occur where estuaries are made unfit for spawning. Trawling for bottom-living fish species is having a major effect on the habitat for species other than target species. It has been estimated that all of the seabed of the North Sea is trawled over at least twice per year and the gear is getting heavier over time (Sydow 1990). Trawls have destroyed long-lived species of molluscs and echinoderms in the North Sea. Since these species play important functional roles in biogeochemical cycling, the consequences may be far-reaching. There are plans to designate trawl-free areas where by comparison with trawled areas, effects of trawling can be assessed. Fishing using explosives on coral reefs (Lundin and Linde 1993) occurs globally in areas where reefs are not properly protected. The ensuing destruction of the reef habitat, which sustains not only the fish but all other species dependent on the reef, has catastrophic consequences for biodiversity. In the Philippines, in addition to dynamite fishing and fishing for the aquarium industry, there is a further serious problem of the widespread and increasing use of cyanide to obtain live fish for restaurants. Although the fish recover when placed in clean water, the cyanide has major effects on the reefs. It is not known what effects the loss of large numbers of reef fish will have on the reef system as a whole. There are relatively few quantitative data on local species extinctions.

Brain Churners

1. Alteration of salinity in an estuarine ecosystem is one of the major threats related to climate change and man-made construction of barrages. Explain the statement with a case study.
2. What are the major types of pollution in marine and estuarine ecosystems?
3. How does plate tectonics affect the biodiversity of marine ecosystem?
4. What factors regulate the alteration of the Earth's orbit?
5. Briefly discuss the role of temperature on the mangrove vegetation.
6. What is the probable reason behind the increase of carbon dioxide after the phase of glaciation?
7. What is acidification? Which phyla are expected to be harmed by the process?
8. How do the coastal structures damage the seagrass beds?
9. How are alien species introduced into the aquatic phase of the marine ecosystem?
10. How does bottom trawling affect the benthic and pelagic communities of marine ecosystem?

Annexure 10A: Impact of Salinity on Above-Ground Biomass and Stored Carbon in a Common Mangrove *Excoecaria Agallocha* Inhabiting Lower Gangetic Delta

1. Introduction

Mangroves are a taxonomically diverse group of salt-tolerant, mainly arboreal, flowering plants that grow primarily in tropical and subtropical regions (Ellison and Stoddart 1991). Salinity plays a crucial role in the growth and survival of mangroves. Based on the physiological studies, Bowman (1917) and Davis (1940) concluded that mangroves are not salt lovers, rather salt tolerant. However, excessive saline conditions retard seed germination and impede growth and development of mangroves. Indian Sundarbans, the famous mangrove chunk of the tropics, is gradually losing few mangroves species (like *Heritiera fomes, Nypa fruticans,* etc.) owing to the increase of salinity in the central sector of the deltaic complex around the Matla River. Reports on the adverse impact of salinity on growth of mangroves in Indian Sundarbans are available (Mitra et al. 2004). However, no study has yet been carried out to investigate the effect of salinity on the carbon content of mangroves from this part of the Indian subcontinent.

The present study aims to establish a baseline data set of stored carbon in the AGB of *Excoecaria agallocha*, a dominant mangrove species of Indian Sundarbans. The species thrives luxuriantly in a wide range of salinity (4 psu–28 psu), and hence, an attempt was also made to find the AGB and carbon content in above-ground structures (stem, branches and leaves) of the species with respect to ambient aquatic salinity.

2. Materials and Methods

2.1 Study Area

The mighty river Ganga emerges from the Himalayas and flows down to the Bay of Bengal covering a distance of 2525 km. At the apex of Bay of Bengal, a delta has been formed which is recognized as one of the most diversified and productive ecosystems of the tropics and is referred to as Indian Sundarbans. The deltaic complex has a Biosphere Reserve area of 9630 sq. km and houses 102 islands. The western sector of the deltaic lobe receives the snowmelt water of mighty Himalayan glaciers after being regulated through several barrages on the way. The central sector, on the other hand, is fully deprived from such supply due to heavy siltation and clogging of the Bidyadhari channel in the late fifteenth century (Chaudhuri and Choudhury 1994). Such variation causes sharp difference in salinity between the two sectors (Mitra et al. 2009). Two sampling sites were selected each in the western and central sectors of this lower Gangetic delta (Fig. 10A.1). The station in the western part lies at the confluence of the river Hooghly (a continuation of Ganga–Bhagirathi system) and the Bay of Bengal. The site is locally known as Sagar South (21°31′4.68″N latitude and 88°01′47.28″E longitude). In the central sector, the sampling station was selected at Canning (22°18′37.44″N latitude and 88°40′36.84″E longitude), near to tide-fed Matla River. The study was undertaken in both of these sectors during low tide period through three seasons, viz. premonsoon (March), monsoon (September) and postmonsoon (December) for 5 consecutive years (2005–2010).

In each sector, a plot size of 10 m × 10 m was selected, and the average readings were documented from 15 such plots. The mean relative density of *Excoecaria agallocha* was evaluated for relative abundance of the species.

2.2 Above-Ground Stem Biomass Estimation

The stem volume for each tree of the species in every plot was estimated using the Newton's formula (Husch et al. 1982) as per the expression $V = h/6\ (A_b + 4A_m + A_t)$, where V is the volume (in m^3), h the height measured with laser beam (BOSCH DLE 70 Professional model) and A_b, A_m and A_t are the areas of the selected tree at base, middle and top, respectively. Specific gravity (G) of the wood was estimated, taking the stem cores from 5 to 10 cm depth with a motorized corer, which was further converted into stem biomass (B_S) as per the expression $B_S = GV$. The stem biomass of individual tree was finally multiplied with the number of trees of the species in 15 selected plots in both western and central Indian Sundarbans.

2.3 Above-Ground Branch Biomass Estimation

The total number of branches irrespective of size was counted on each of the sample trees. These branches were categorized on the basis of basal diameter into three groups, viz. <6 cm, 6–10 cm and >10 cm. Dry weight of two branches from each size group was recorded separately using the equation of Chidumaya (1990). Total branch biomass (dry weight) of individual tree was determined after drying at 80 ± 5 °C as per the expression $B_{db} = n_1bw_1 + n_2bw_2 + n_3bw_3 = \Sigma\ n_ibw_i$, where B_{db} is the dry branch biomass per tree, n_i the number of branches in the *i*th branch group, b_{wi} the average weight of branches in the *i*th group and $i = 1, 2, 3, \ldots n$ are the branch groups. The branch biomass of individual tree was finally multiplied with the number of trees of the species in all the 15 plots for each station.

2.4 Above-Ground Leaf Biomass Estimation

Leaves from nine branches (three of each size group) of individual trees were plucked, weighed and oven dried separately to a constant weight at 80 ± 5 °C. Three trees per plot were considered for estimation. The leaf biomass was then estimated by multiplying the average biomass of the leaves per branch with the number of branches in a single tree and the average number of trees per plot as per the expression

$L_{db} = n_1Lw_1N_1 + n_2Lw_2N_2 + \ldots\ldots\ldots n_iLw_iN_i$, where, L_{db} is the dry leaf biomass of selected mangrove species per plot, $n_1 \ldots n_i$ are the number of branches of each tree of the species, $Lw_1 \ldots\ldots Lw_i$ are the average dry weight of leaves removed from the branches and $N_1 \ldots\ldots N_i$ are the number of trees of the species in the plots.

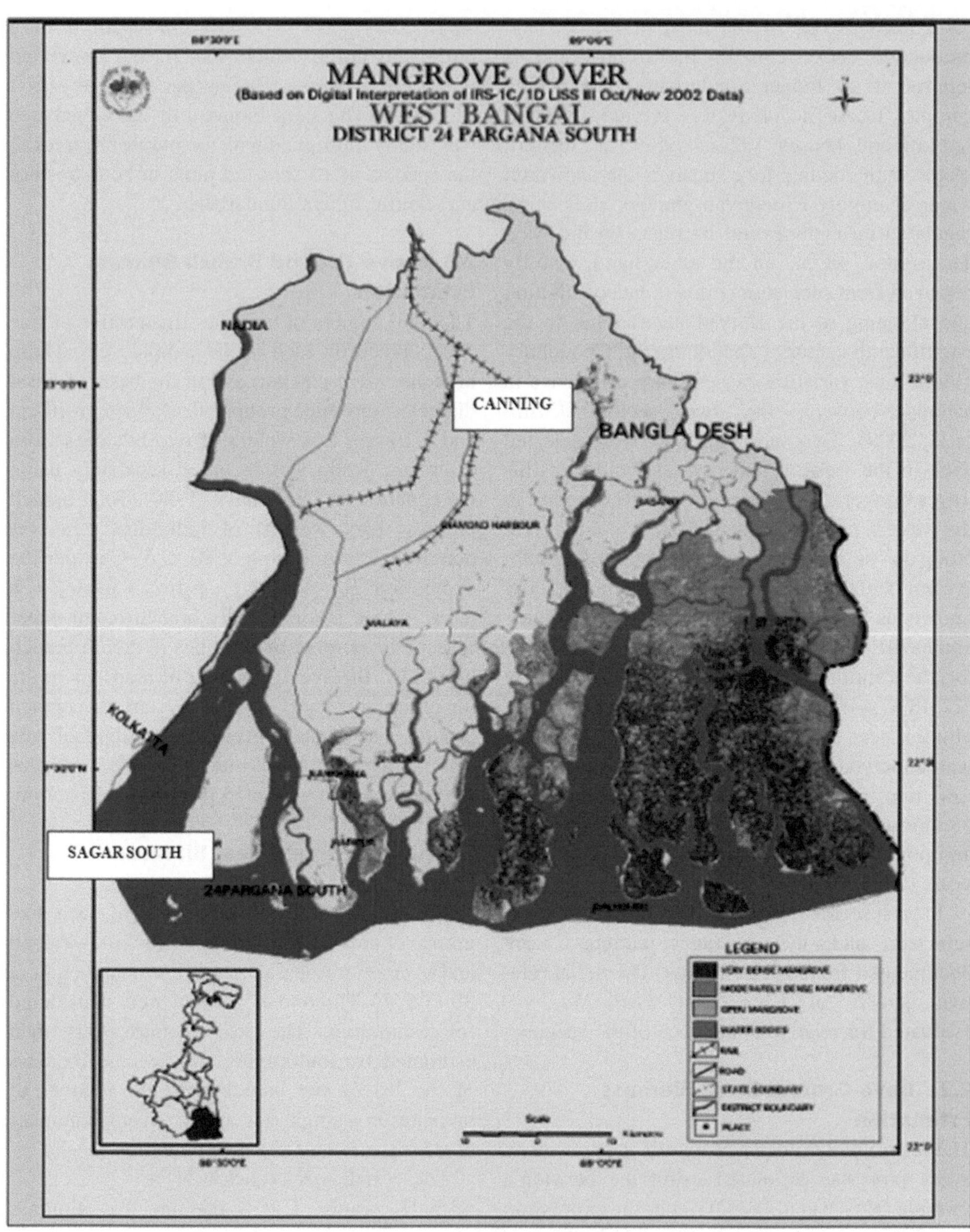

Fig. 10A.1 Location of sampling stations in the western and central sectors of Indian Sundarbans

2.5 Carbon Estimation

Direct estimation of percent carbon was done by a CHN analyzer. For this, a portion of fresh sample of stem, branch and leaf from 30 trees (two trees/plot) of the species (covering all the 15 plots) was collected. The vegetative parts were oven dried separately at 70 °C and ground to pass through a 0.5 mm screen (1.0 mm screen for leaves). The carbon content (in %) was finally analyzed on a *Vario MACRO* **elementar CHN analyzer.**

2.6 Salinity

The surface water salinity was recorded by means of an optical refractometer (Atago, Japan) in the field and cross-checked in the laboratory by employing Mohr–Knudsen method. The correction factor was found out by titrating the silver nitrate solution against standard seawater (IAPO Standard Seawater Service Charlottenlund, Slot Denmark, chlorinity = 19.376 ‰). Our method was applied to estimate the salinity of standard seawater procured from NIO, and a standard deviation of 0.02 % was obtained for salinity. The average accuracy for salinity (in connection to our triplicate sampling) is ± 0.28 psu.

2.7 Statistical Analysis

Scatter plots, allometric equations and correlations were computed with a sample size of 240 for each sector to observe the interrelationships between AGB, DBH, stem, branch and leaf biomass along with stored carbon in these above-ground structures. Analysis of variance (ANOVA) was performed to assess whether biomass and carbon content varied significantly between sites, years and seasons; possibilities less than 0.01 ($p < 0.01$) were considered statistically significant. All statistical calculations were performed with SPSS 9.0 for Windows.

3. Results

3.1 Relative Abundance

Nine species of mangroves were documented in the selected plots in the western sector, but in the central sector, only six species were recorded. The mean relative abundance of *Excoecaria agallocha* was 18.75 % and 25.81 % in the western and central sectors, respectively. In both the sectors, the trees are ~12 years old, but high salinity in the central sector probably stunted the growth of the species.

3.2 Above-Ground Biomass

The stem, branch, leaf and AGB of the mangrove species increased with age. The increment was, however, not uniform in both the sectors as revealed from the trend line equations (Figs. 10A.2, 10A.3, 10A.4 and 10A.5). We observed significant variation in the rate of AGB increase between sites ($p < 0.01$). It was 0.63 t/ha/month and 0.75 t/ha/month in the western and central sectors, respectively. The yearly variation of AGB was also significant ($p < 0.01$), but the seasonal variation was not pronounced. It is

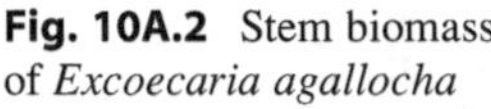

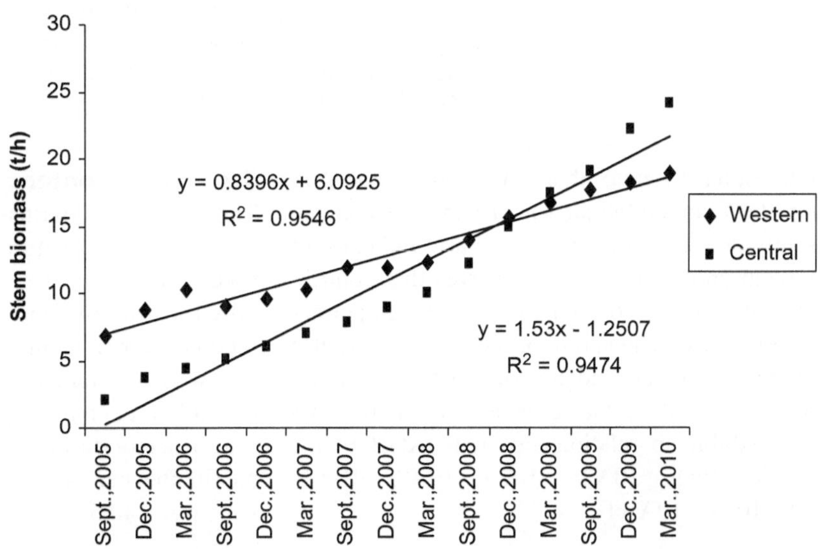

Fig. 10A.2 Stem biomass of *Excoecaria agallocha*

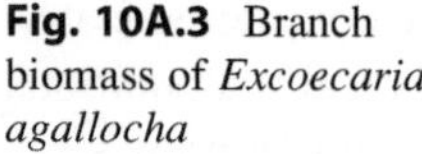

Fig. 10A.3 Branch biomass of *Excoecaria agallocha*

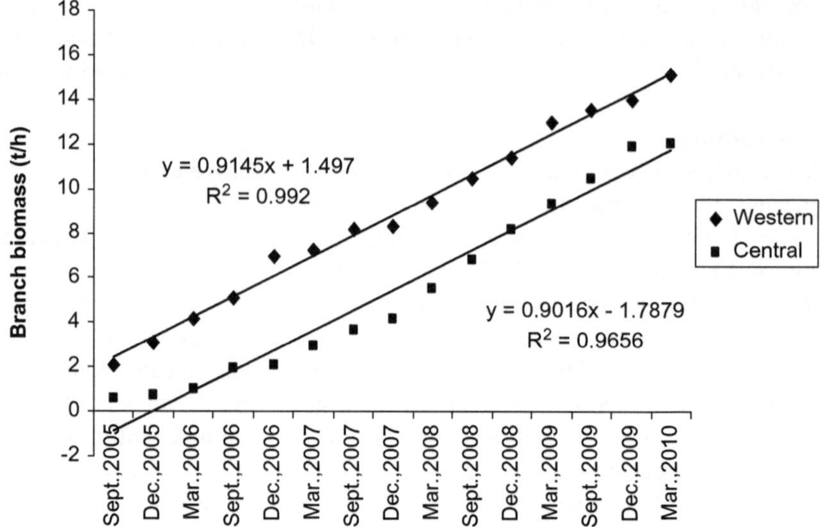

Fig. 10A.4 Leaf biomass of *Excoecaria agallocha*

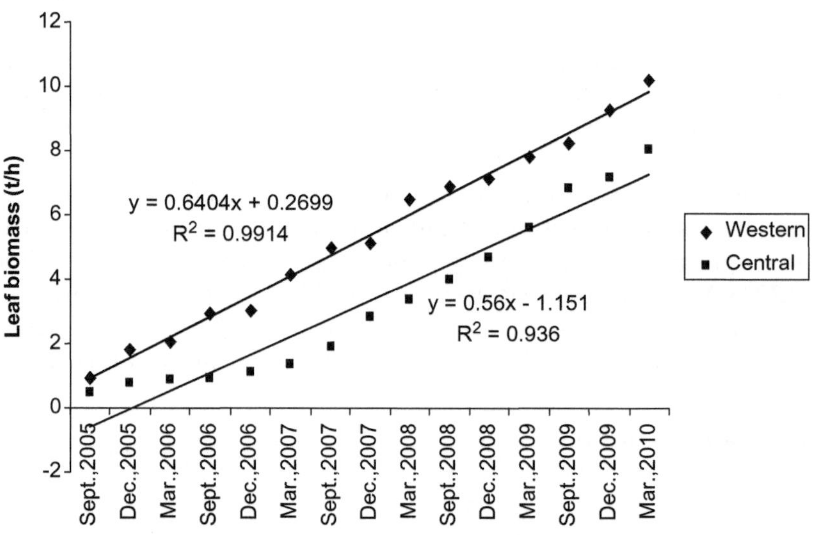

interesting to note that AGB of *Excoecaria agallocha* in the Indian Sundarbans is accounted solely due to stem, which is a basic indicator of growth unlike branches and leaves that contribute substantially to litter fall and less to permanent biomass. The nature of the scatter plots also confirm strong dependency of AGB on stem biomass and DBH unlike branch and leaf biomass that exhibit no relationships with AGB of the species (Figs. 10A.6, 10A.7, 10A.8, 10A.9, 10A.10 and 10A.11).

3.3 Carbon Content

The seasonal variations of stored carbon in the above-ground structures of the species for five successive years are shown in Figs. 10A.12, 10A.13, 10A.14, and 10A.15. In both of the sectors, carbon content was highest in stems, followed by branches and leaves. In the stem, the carbon content ranged from 0.81 t/ha (in the central sector during September 2005) to 10.13 t/ha (in the central sector during March 2010), which are 40.5 % and 42.0 % of the biomass,

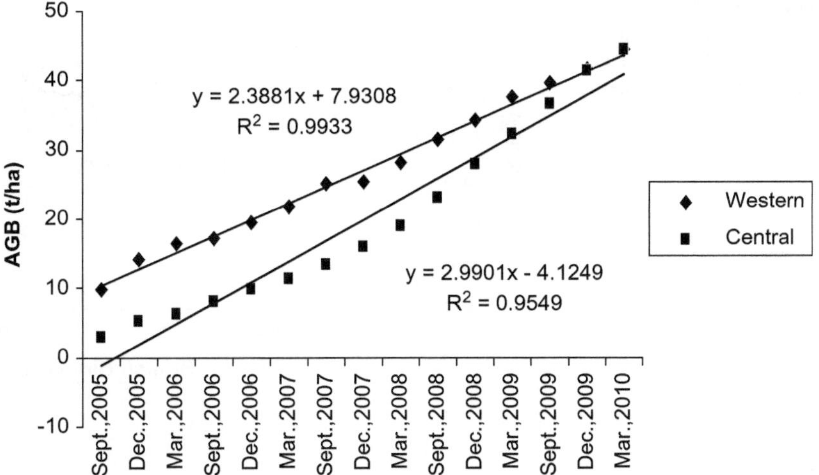

Fig. 10A.5 AGB of *Excoecaria agallocha*

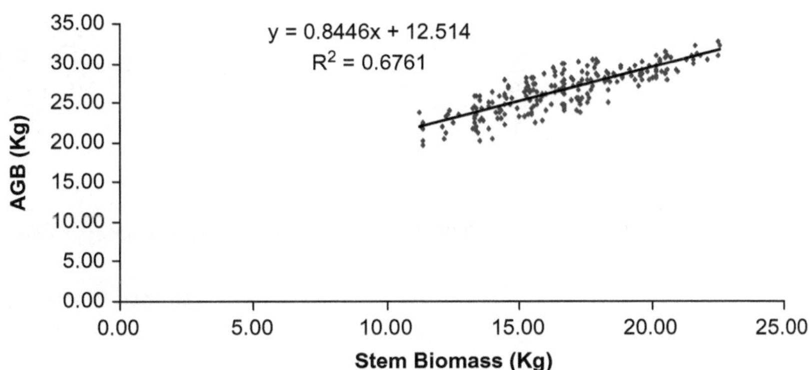

Fig. 10A.6 Relationship between stem biomass and AGB of *Excoecaria agallocha* in the western sector

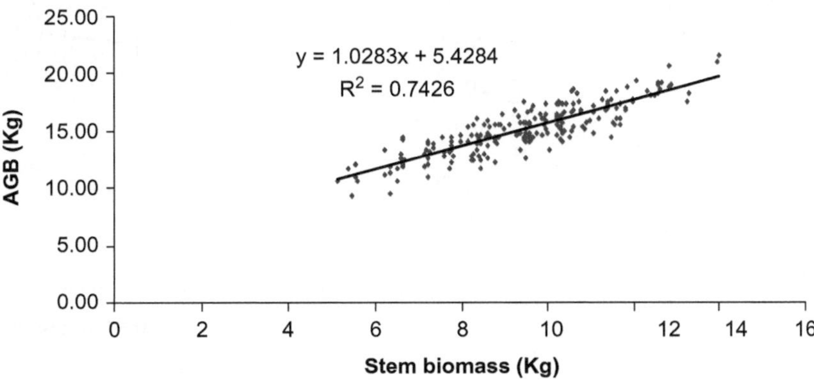

Fig. 10A.7 Relationship between stem biomass and AGB of *Excoecaria agallocha* in the central sector

respectively. The sequestration rates of carbon in the stem of the western and central sectors are significantly different ($p < 0.01$) with values of 0.10 t/ha/month and 0.17 t/ha/month, respectively. In the branch, the range of stored carbon was 0.22 t/ha (39.2 % of the branch biomass in the central sector during September 2005) to 6.40 t/ha (42.2 % of the branch biomass

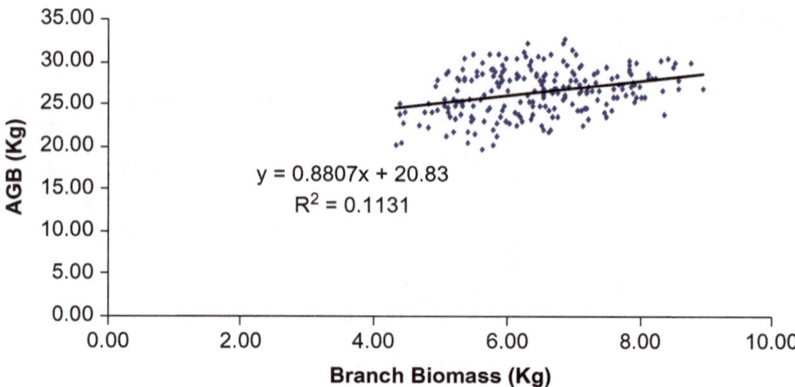

Fig. 10A.8 Relationship between branch biomass and AGB of *Excoecaria agallocha* in the western sector

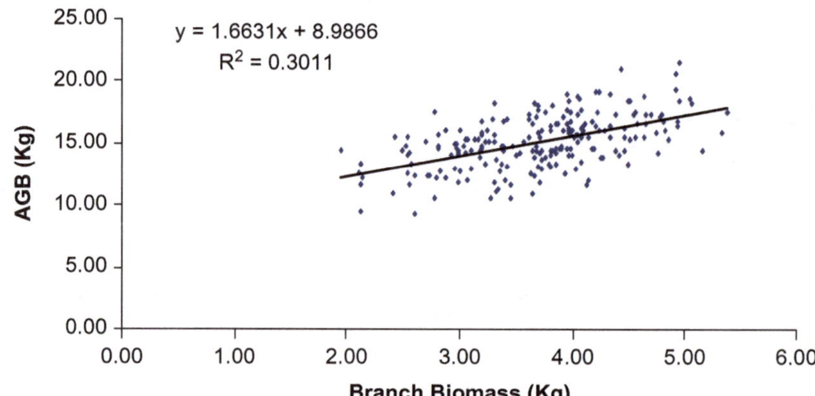

Fig. 10A.9 Relationship between branch biomass and AGB of *Excoecaria agallocha* in the central sector

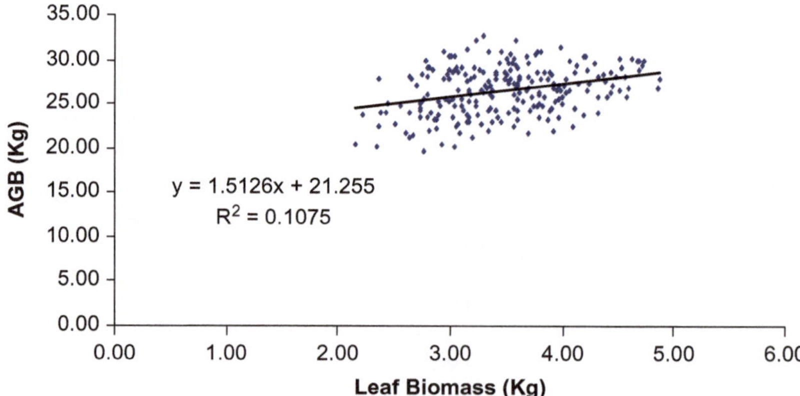

Fig. 10A.10 Relationship between leaf biomass and AGB of *Excoecaria agallocha* in western sector

in the western sector during March 2010). The branch sequestered 0.10 t/ha/month and 0.09 t/ha/month in the western and central sectors, respectively. In the leaf, minimum carbon content (0.22 t/ha which is equivalent to 43.1 % of leaf biomass) was observed in the central sector in September 2005 and the maximum value (4.74 t/ha which is equivalent to 46.4 % of leaf biomass) was recorded in the western sector in March 2010. The sequestration rates are 0.08 t/

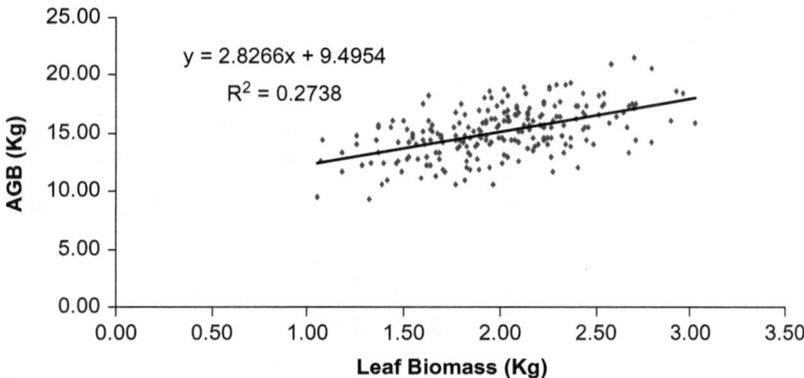

Fig. 10A.11 Relationship between leaf biomass and AGB of *Excoecaria agallocha* in the central sector

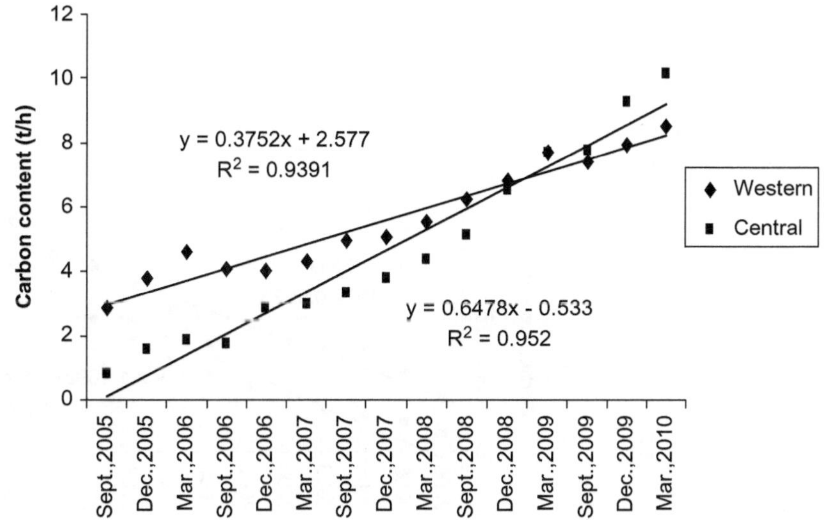

Fig. 10A.12 Carbon content in *Excoecaria agallocha* stem

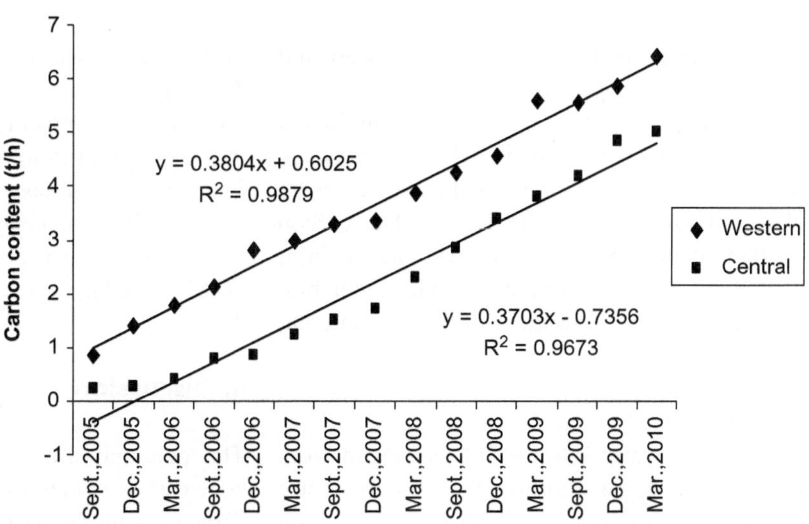

Fig. 10A.13 Carbon content in *Excoecaria agallocha* branch

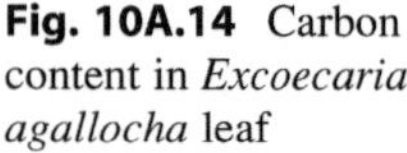

Fig. 10A.14 Carbon content in *Excoecaria agallocha* leaf

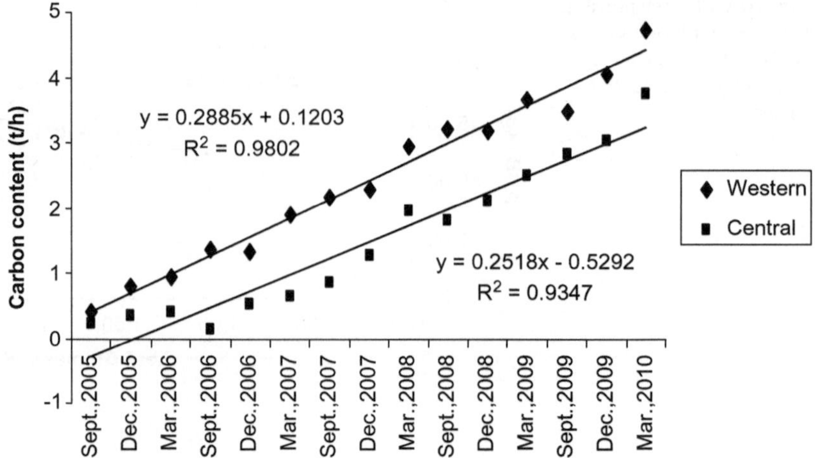

Fig. 10A.15 Carbon content in AGB of *Excoecaria agallocha*

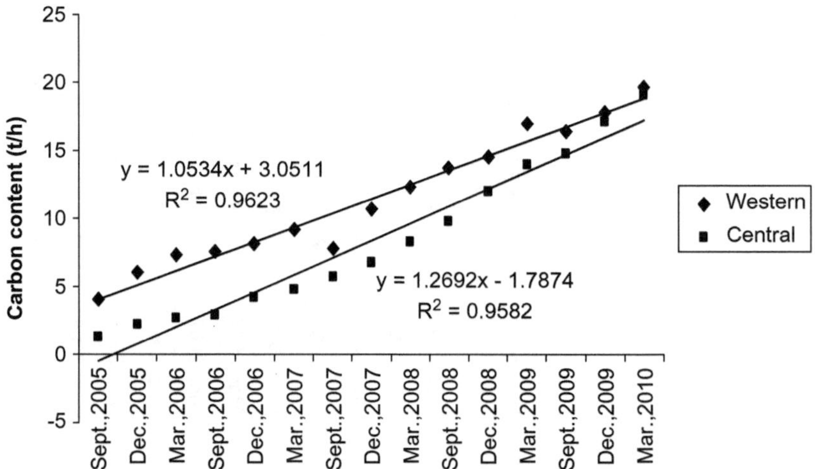

ha/month and 0.06 t/ha/month in the western and central sectors, respectively.

ANOVA results confirmed significant differences in stored carbon of the stem between the sites ($p < 0.01$), but no differences were observed for branches and leaves. The carbon content in the above-ground structures exhibit significant positive correlations with stem biomass and its DBH, but not with branch and leaf biomass.

3.4 Salinity

The surface water salinity values ranged from 8.66 psu (at Sagar South in the western sector during 2010 monsoon) to 26.59 psu (at Canning in the central sector during 2008 premonsoon). The salinity values varied as per the order premonsoon > postmonsoon > monsoon, and the seasonal variation is significant ($p < 0.01$). The salinity values were significantly higher ($p < 0.01$) in the central sector compared to the western sector irrespective of seasons and year (Table 10A.1).

4. Discussion

The potential impact of mangrove on coastal zone carbon dynamics has been a topic of intense debate during the past decades. Despite the large

Table 10A.1 Seasonal variation of surface water salinity (in psu) in the selected stations during 2005–2010

	2005		2006		2007		2008		2009		2010	
Season	A	B	A	B	A	B	A	B	A	B	A	B
Premonsoon	26.99	24.99	26.10	26.50	25.12	26.00	29.11	26.59	24.04	26.08	23.58	25.95
Monsoon	9.16	10.44	9.02	9.65	9.30	9.98	8.76	9.90	9.08	10.02	8.66	10.13
Postmonsoon	22.32	23.10	21.67	23.10	21.80	23.88	20.73	24.06	21.04	24.32	19.88	25.31

A Sagar South (western sector), *B* Canning (central sector)

Table 10A.2 List of AGB in few mixed mangrove forests

Region	Location	Condition or age	AGB (t/ha)	Reference
Sri Lanka	8°15′N latitude and 79°50′E longitude	Fringe forest	172.0	Amarasinghe and Balasubramaniam (1992)
Sri Lanka	8°15′N latitude and 79°50′E longitude	Riverine forest	57.0	Amarasinghe and Balasubramaniam (1992)
Thailand (Trat Eastern)	12°12′N latitude and 102°33′E longitude	Secondary forest	142.2	Poungparn (2003)
Western Indian Sundarbans (Sagar South)	88°01′47.28″N latitude and 21°31′4.68″E longitude	~18 years	19.64	This study
Around central Indian Sundarbans (Canning)	88°40′36.84″N latitude and 22°18′37.44″E longitude	~18 years	19.08	This study

number of case studies dealing with various aspects of organic matter cycling in mangrove systems (Kristensen et al. 2008), there is a very limited consensus on the carbon sequestering potential of mangroves. It has been opined by several workers that the carbon sequestration in this unique producer community is a function of biomass production capacity, which in turn depends upon interaction between edaphic, climate and topographic factors of an area (Chaudhuri and Choudhury 1994; Mitra and Banerjee 2005). Hence, results obtained at one place may not be applicable to another. We therefore attempted to establish allometric equations for *Excoecaria agallocha* of the Indian Sundarbans relating its DBH, stem biomass, branch biomass, leaf biomass, AGB and stored carbon. The nature of the scatter plots indicates significant positive correlations between AGB, stem biomass, DBH and stored carbon in both of the sectors. The AGB and stored carbon do not exhibit any dependency on branch and leaf biomass of the species, Sengupta et al. (2013). This indicates the sole contribution of stem biomass and DBH to AGB and carbon stored in the above-ground structures.

Mangroves, in general, prefer brackish water environment, and in extreme saline condition, stunted growth is observed (Mitra et al. 2004). The present study, however, presents a different picture and reveals the adaptation of *Excoecaria agallocha* in the high-saline central sector. The relatively higher growth rate of the above-ground structures of the species in the central sector (0.75 t/ha/month) compared to the western part (0.63 t/ha/month) confirms its tolerance to salinity. A critical analysis of biochemical mechanisms may throw light on the adaptation of *Excoecaria agallocha* in the high-saline environment of the central Indian Sundarbans.

The carbon content and sequestration rate of above-ground structures are also higher in the central sector as a direct function of above-ground biomass. During our study period, the average surface water salinity in the central Indian Sundarbans were relatively higher (26.22 psu during premonsoon, 10.02 psu during monsoon and 23.70 during postmonsoon) compared to the western part (25.59 psu during premonsoon, 9.00 psu during monsoon and 21.31 during postmonsoon). This could not retard the carbon sequestration of the species (by the total AGB) as evidenced from the stored carbon and sequestration rate in the central sector (0.32 t/ha/month) compared to the western sector (0.28 t/ha/month). The results of our study have been compared with the AGB of few mixed mangrove forests (Table 10A.2) to evaluate the potential of Indian Sundarbans mangrove as carbon sink. The values of the present study are less when compared with other regions, but efficient adaptation of the species in high-saline zone has multiplied the importance of the species as the present geographical locale is vulnerable to climate change-induced salinity rise owing to its location below the mean sea level and experiencing a sea level rise of 3.14 mm/year as compared to global average of 2.5 mm/year.

References

ADB/NACA. (1998). *Aquaculture sustainability and the environment. Report on a regional study and workshop on aquaculture sustainability and the environment.* Bangkok: Asian Development Bank and Network of Aquaculture Centers in Asia and the Pacific.

Alagarswami, K. (1995). India country case study. In *Regional study and workshop on the environmental assessment and management of aquaculture development* (TCP/RAS/2253). NACA environment and aquaculture development series no. 1. Bangkok: Network of Aquaculture Centers in Asia and in the Pacific.

Amarasinghe, M. D., & Balasubramaniam, S. (1992). Net primary productivity of two mangrove forest stands on the northwestern coast of Sri Lanka, status, primary productivity and taxonomy of mangroves, seagrasses and algae. *Hydrobiologia, 247*(1), 37–47.

Anderson, B. G. (1960). The toxicity of organic insecticides to *Daphnia*. In C. M. Tarzwell (Ed.), *Biological problems in water pollution* (Proceedings of the 2nd seminar), Cincinnati, Ohio. Robert, A. Taft sanitary and engineering center (pp. 94–95) (Technical report W60-3). Cited in WHO (1989).

Andhra Pradesh Remote Sensing Application Centre (APRSAC). (1999). *Environmental management and monitoring of shrimp culture project, East Godavari District, Andhra Pradesh – land use/land cover*. Hyderabad, India.

Andrews, T. J., Clough, B. F., & Muller, G. J. (1984). Photosynthetic gas exchange properties and carbon isotope ratios of some mangroves in North Queensland. In H. J. Teas (Ed.), *Physiology and management of mangroves, tasks for vegetation science* (Vol. 9, pp. 15–23). The Hague: Dr. W. Junk.

Ball, M. C., Cochrane, M. J., & Rawason, H. M. (1997). Growth and water use of the mangroves *Rhizophora apiculata* and *R. stylosa* in response to salinity and humidity under ambient and elevated concentration of atmospheric CO_2. *Plant Cell Environment, 20*, 1158–1166.

Banerjee, K., Sengupta, K., Raha, A., & Mitra, A. (2013). Salinity based allometric equations for biomass estimation of Sundarban mangroves. *Biomass and Bioenergy (Elsevier), 56*, 382–391.

Beveridge, M. C. M., Ross, L. G., & Kelly, L. A. (1994). Aquaculture and biodiversity. *Ambio, 23*, 497–502.

Blasco, F. (1975). The mangroves of India. *Institut Français Pondichéry Travaux Section Scientifique et Technologie, 14*, 1–175.

Briggs, M. R. P., & Funge-Smith, S. J. (1994). A nutrient budget of some intensive marine shrimp ponds in Thailand. *Aquaculture Fishery Management, 5*, 89–811.

Chaudhuri, A. B., & Choudhury, A. (1994). *Mangroves of the Sundarbans*. IUCN Wetlands Programme (247 pp.). Bangkok: IUCN

Cladwell, R. S. (1977). Biological effects of pesticides on the dungeness crab. *Ecological Research Services*, EPA-600/3: 77–131.

Clough, B. F., Andrews, T. J., & Cowan, I. R. (1982). Physiological processes in mangroves. In B. F. Clough (Ed.), *Mangrove ecosystems in Australia: Structure, function and management* (pp. 193–210). Canberra: Australian National University Press.

Cole, C. V., & Vaidyaraman, P. P. (1966). *Salinity distribution and effect of freshwater flows in the Hooghly River* (Proceedings tenth conference on coastal engineering, Tokyo, pp. 1312–1434). New York: American Society of Civil Engineers.

Collen, J., Guisle-Marsollier, I., Leger, J. J., & Boyen, C. (2007). Response of the transcriptome of the intertidal red seaweed *Chondrus crispus* to controlled and natural stresses. *New Phytology, 176*, 45–55.

Cope, O. B. (1965). *Sport fishery investigations*. Washington, DC: US Department of the Interior, Fish and Wildlife Service (Circular 226).

Davidson, J. P., Reed, W. E., & Paul, M. D. (2001). *Exploring earth: An introduction to physical geology* (2nd ed.). Upper Saddle River: Prentice Hall.

Davison, I. R., & Pearson, G. A. (1996). Stress tolerance in intertidal seaweeds. *Journal of Phycology, 32*, 197–211.

Deb, S. C. (1956). Paleoclimatology and geophysics of the gangetic delta. *Geographical Reviews India, 18*, 11–18.

DeWalt, B. R., Vergne, P., & Hardin, M. (1996). Shrimp aquaculture development and the environment: People, mangroves and fisheries on the Gulf of Fonseca, Honduras. *World Development, 24*, 1193–1208.

Eggert, A. (2012). Seaweed responses to temperature. In C. Wiencke & K. Bischof (Eds.), *Seaweed biology* (pp. 47–66). Berlin: Springer.

Eisler, R. (1969). Acute toxicities of insecticides to Marine Decapod Crustaceans. *Crustaceana, 16*(3), 302–310.

Eisler, R. (1970a). Latent effects of insecticide intoxication to marine molluscs. *Hydrobiologia, 36*(3–4), 345–352.

Eisler, R. (1970b). Acute toxicities of organochlorine and organophosphorus insecticides to estuarine fish. *Technical Paper Bouchure on Sport Fishery and Wildlife, 46*, 1–12.

Elliott, N. G., & Ward, R. D. (1992). Enzyme variation in orange roughy, *Hoplostethus atlanticus* (Teleostei: Trachichthyidae), from southern Australian and New Zealand waters. *Australian Journal of Marine Freshwater Research, 43*, 1561–71.

Ellison, J. (2000). How South Pacific mangroves may respond to predicted climate change and sea level rise. In A. Gillespie & W. Burns (Eds.), *Climate change in the south pacific: Impacts and responses in Australia, New Zealand, and small Islands states* (pp. 289–301). Dordrecht: Kluwer Academic (Chapter 15).

Ellison, J. C. (2005). Impacts on mangrove ecosystems. In *Proceedings of the conference on the great greenhouse gamble: A conference on the impacts of climate change on biodiversity and natural resource management*, Sydney.

FAO. (1991). *Environment and the sustainability of fisheries*. Rome: FAO.

FAO. (1996). *Feasibility of freshwater prawn hatcheries in Sindh Province*. Technical Cooperation Programme (FI: TCP/PAK/4559) Technical report. Rome: FAO, 73 p.

FAO. (2003). *Status and trends in mangrove area extent worldwide*. Paris: Food and Agriculture Organization of the United Nations, Forest Resources Division.

Farnsworth, E. J., Ellison, A. M., & Gong, W. K. (1996). Elevated CO_2 alters anatomy, physiology, growth and reproduction of red mangrove (*Rhizophora mangle* L.). *Oecologia, 108*, 599–609.

Field, C. (1995). Impacts of expected climate change on mangroves. *Hydrobiologia, 295*, 75–81.

Gattuso, J. P., Frankignoulle, M., Bourge, I., Romaine, S., & Buddemeier, R. W. (1998). Effect of calcium carbonate saturation of seawater on coral calcification. *Global and Planetary Change, 18*, 37–46.

Gazeau, F., Quiblier, C., Jansen, J. M., Gattuso, J.-P., Middelburg, J. J., & Heip, C. H. R. (2007). Impact of elevated CO_2 on shellfish calcification. *Geophysical Research Letters, 34*, L07603. doi:10.1029/2006GL028554.

GESAMP. (1995). The invasion of the ctenophore *Mnemiopsis leidyi* in the Black Sea. *Report Study GESAMP, 58* (In press).

GESAMP (IMO/FAO/UNESCO-IOC/WMO/WHO/IAEA/UN/UNEP Joint Group of Experts on the Scientific Aspects of Marine Environmental Protection). (2001). Planning and management for sustainable coastal aquaculture development. *Report Study GESAMP, 68*: 90.

Hamre, J. (1994). Biodiversity and exploitation of the main fish stocks in the Norwegian, Barents Sea ecosystem. *Biodiversity Conservation, 3*, 473–92.

Harvey, S. (1975). Heavy metals. In L. S. Goodman & A. Gilman (Eds.), *Pharmacological basis of therapeutics* (5th ed., pp. 924–941). New York: Macmillan.

Henderson, C., Pickering, Q. H., & Tarzwell, C. M. (1959). Relative toxicity of ten chlorinated hydrocarbon insecticide to four species of fish. *Transactions of the American Fisheries Society, 88*, 23.

Holmgren, S. (Ed.). (1994). *An environmental assessment of the Bay of Bengal region* (Swedish Centre for Coastal Development and Management of Aquatic Resources, March 1994). Madras: Bay of Bengal Programme.

Houghton, J., Ding, Y., Griggs, D., Noguer, M., Van Der Linden, P., Dai, X., Maskell, K., & Johnson, C. (Eds.). (2001). *Climate change: The scientific basis. Published for the intergovernmental panel on climate change*. Cambridge: Cambridge University Press. 881 pp.

Hutchings, P., & Saenger, P. (1987). *Ecology of mangroves*. Queensland (Australia): University of Queensland Press.

International Convention for the Prevention of Pollution from Ships (MARPOL) Adoption: 1973 (Convention), 1978 (1978 Protocol), 1997 (Protocol – Annex VI); Entry into force: 2 October 1983 (Annexes I and II).

International Convention on the Prevention of Marine Pollution by Dumping of Wastes and Other Matter Adoption: 13 November 1972; Entry into force: 30 August 1975; 1996 Protocol: Adoption: 7 November 1996; Entry into force: 24 March 2006.

International Co-ordination Group (ICG) for the Global Investigation of Pollution in the Marine Environment (GIPME). Report, Unesco-IOC Rept. IOC/GIPME-111/3, 9 p. + IV Annexes. International Co-ordination Group (ICG), 1st Session London Bloomsbury Center Hotel, 2–6 Apr 1973.

IOC/INF-263 rev., entitled "A comprehensive plan for a Global Investigation of Pollution in the Marine Environment (GIPME)", and document IOC/INF-276, entitled "Report of the IOC/ICES Working Group on Baseline Study Guidelines", as amended by the IOC Executive Council at its seventh session (Bergen, 21–26 June 1976).

IUCN. (1989). *The impact of climatic change and sea level rise on ecosystems*. Report for the Commonwealth Secretariat, London.

James, P. S. B. R. (1999). *Shrimp farming development in India – an overview of environmental, socio-economic, legal and other implications*. Published on line by Aquaculture Magazine, December 1999. www.ioa.com/ãquamag

Johnson, W. W., & Finley, M. T. (1980). *Handbook of acute toxicity of chemicals to fish and aquatic invertebrates* (Resource publication, Vol. 137). Washington, D.C.: U.S. Department of the Interior, Fish and Wildlife Service.

Johnston, R. (1976). *Marine pollution*. London: Academic Press. Science, 729 pages. ISBN 0123876508, 9780123876508.

Kim, E., Park, H. S., Jung, Y., Choi, D. W., Jeong, W. J., Hwang, M. S., Park, E. J., & Gong, Y. G. (2011). Identification of the high-temperature response genes from *Porphyra seriata* (Rhodophyta) expression sequence tags and enhancement of heat tolerance of *Chlamydomonas (*Chlorophyta) by expression of the Porphyra HTR2 gene. *Journal of Phycology, 47*, 821–8.

Klaassen, C. D. (1995). Heavy metals. In J. Hardman & L. Limbird (Eds.), *Goodman and Gilman's pharmacological basis of therapeutics* (9th ed., pp. 1049–73). New York: McGrw Hill.

Kleypas, J. A., Feely, R. A., Fabry, V. J., Langdon, C., Sabine, C. L., & Robbins, L. L. (2006). *Impacts of ocean acidification on coral reefs and other marine calcifers. A guide to future research*. Report of a workshop held 18–20 April 2005, St. Petersburg, FL, sponsored by NSF, NOAA, and the U.S. Geological Survey, 88 pp.

Komiyama, A., Ong, J. E., & Poungparn, S. (2007). Allometry, biomass and productivity of mangrove forests: a review. *Aquatic Botany*. doi:10.1016/j.aquabot.2007.12.006, this issue.

Kordas, R. L., Harley, C. D. G., & O'Connor, M. I. (2011). Community ecology in a warming world: The influence of temperature on interspecific interactions in marine systems. *Journal of Experimental Marine Biology Ecology, 400*, 218–26.

Korn, S., & Earnest, R. (1974). Acute toxicity of twenty insecticides to striped bass, Morone saxatilis. *California Fish Game, 60*, 128.

Krishnamoorthy. (1995). *Remote sensing of mangrove forests in Tamil Nadu Coast, India*. Faculty of Civil Engineering, Anna University, Madras (thesis).

Kurihara, K., & Shirayama, Y. (2004). Impacts of increased atmospheric CO_2 on sea urchin early development. *Marine Ecological Progress Series, 274*, 161–169.

Lakshmana Rao, M. V., Mahapatra, K., & Subba Rao, D. V. (1994). The Coastal Zone of Orissa, Bay of Bengal: Threats and prospects for Sustainable Development. *Coastal Zone, 1*: 304–319, Coastal zone Canada '94: Cooperation in the coastal zone International conference, Coastal zone Canada '94: Cooperation in the coastal zone. ISBN: 1896246001. ISSN: 0821–1302.

Langdon, C., Takahashi, T., Marubini, F., Atkinson, M., Sweeney, C., Aceves, H., Barnett, H., Chipman, D., & Goddard, J. (2000). Effect of calcium carbonate saturation state on the calcification rate of an experimental coral reef. *Global Biogeochemical Cycles, 14*, 639–654.

Lewis, S., Donkin, M. E., & Depledge, M. H. (2001). Hsp70 expression in *Enteromorpha intestinalis* (Chlorophyta) exposed to environmental stressors. *Aquatic Toxicology, 51*(277), 91.

Low, J. K. Y., Arshad, A., & Lim, K. H. (1994). Mangroves as a habitat for endangered species and biodiversity conservation. In C. Wilkinson, S. Sudara, & C. L. Ming (Eds.), *Proceedings of the third ASEAN – Australia symposium on living coastal resources: Status reviews, Chulalongkorn University, Bangkok, May 16–20 1994*. Townsville: Australian Institute of Marine Science.

Lundin, C. G., & Linde, Â. O. (1993). Coastal ecosystem: Attempts to manage a threatened resource. *Ambio, 22*, 468–73.

Macek, K. J., & McAllister, W. A. (1970). Insecticide susceptibility of some common fish family representatives. *Transactions of the American Fisheries Society, 99*, 20–27.

Macek, K. J., Hutchinson, C., & Cope, O. B. (1969). The effects of temperature on the susceptibility of bluegills and rainbow trout to selected pesticides. *Bulletin Environmental Contamination Toxicology, 4*, 174–183.

Mackenzie, D. (1995). End of the line for the living fossil. *New Science, 1978*, 14–5.

Maduka, H. C. C. (2006). Water pollution and Man's health. *The Internet Journal of Gastroenterology, 4* (1), 1–11.

Mascarenhas, B. (1997). The order and size of entry into international markets. *Journal of Business Venturing, 12*(4), 287–99.

Mayer, F. L. (1987). *Acute toxicity handbook of chemicals to estuarine organisms* (United States Environmental Protection Agency). Gulf Breeze: Environmental Research Laboratory.

Mayer, F. L., Jr., & Ellersieck, M. R. (1986). *Manual of acute toxicity: Interpretation and data base for4l0chemicals and66 species of fresh water animals*. Washington, DC: US Department of the Interior Fish and Wildlife Service (Resource publication 160).

McLeese, D. W., & Metcalfe, C. D. (1980). Toxicities of eight organochlorine compounds In: Sediment and in seawater to *Crangon septemspinosa. Bulletin Environmental Contamination Toxicology, 25*, 921–928.

McLusky, D. S., Bryant, D. M., & Elliot, M. (1991). The impact of reclamation on benthic production and prey availability in the Forth Estuary, eastern Scotland. In J. Pethick (Ed.), *The changing coastline*. Dordrecht: Kluwer.

Menasveta, P. (1997). Mangrove destruction and shrimp culture systems. *World Aquaculture, 28*(4), 36–42.

Mitra, A., & Zaman, S. (2014). *Carbon sequestration by Coastal Floral Community*, published by The Energy and Resources Institute (TERI). India: TERI Press (Copyright The Energy and Resources Institute 2014; ISBN- 978-81-7993-551-4).

Mitra, A., Banerjee, K., Sengupta, K., Gangopadhyay, A. (2009). Pulse of climate change in Indian sundarbans: A myth or reality? *National Academy of Science Letters, 32*: 1–2.

Mitra, A., Mondal, K., & Banerjee, K. (2011). Spatial and tidal variations of physico–chemical parameters in the lower gangetic delta region, West Bengal, India. *Journal of Spatial Hydrology, American Spatial Hydrology Union Spring, 11*(1), 52–69.

Monaghan, P. (1992). Seabirds and sand eels: The conflict between exploitation and conservation in the northern North Sea. *Biodiversity Conservation, 1*, 98–111.

Nahle, N. (2005) updated 2008. Cycles of global climate change. *Biology Cabinet Journal Online Article* No. 295; www.nipcc.ch/datei/1290446298.

Nahle, N. (2007). Heat stored by greenhouse gases. *Biology cabinet*. Cycles of global climate change. *Biology Cabinet Journal Online*. Article no. 295. http://www.biocab.org/Climate_Geologic_Timescale.html and http://www.biocab.org/Carbon_Dioxide_Geological_Timescale.html. Accessed April 27, 2007.

Nathaniel, I. T., Salami, A. T., & Olajuyigbe, A. C. (2000). Environmental features of Nigerian economic exclusive zone (EEZ): Ibino and Bonny as case study. *African Journal and Environmental Studies, 1*(1), 9–17.

Nebeker, A. V., Puglisi, F. A., & DeFoe, D. L. (1974). Effect of polychlorinated biphenyl compounds on survival and reproduction of the fathead minnow and flag fish. *Transactions of the American Fisheries Society, 103*, 562–568.

Nemerow, N. L. (1986). *Streams, lakes, estuary and ocean pollution* (pp. 381–398). New York: Van Nostrand Reinhold.

O'Connor, M. I., John, F., Bruno, S., Gaines, D., Benjamin, S., Halpern, S., et al. (2006). Temperature control of larval dispersal and the implications for marine ecology, evolution, and conservation, PNAS early edition. In H. James (Ed.), Albuquerque, NM:

Brown, University of New Mexico (Approved 13 Nov 2006, received for review April 27, 2006).

Pagani, M., Zachos, J. C., Freeman, K. H., Tipple, B., & Bohaty, S. (2005). Marked decline in atmospheric carbon dioxide concentrations during the Paleocene. *Science, 309*(5734), 600–603.

Parrish, P. R., Schimmel, S. C., Hansen, D. J., Patrick, J. M., & Forester, J. (1976). Chlordane: Effect on several estuarine organisms. *Journal of Toxicology and Environmental Health, 1*, 485–494.

Pimentel, D., Giampietro, M., & Bukkens, S. G. F. (1994). Optimum population for US and Latin America. In *Proceedings of the first world optimum population congress* (pp. 19–21). Cambridge: Cambridge University, August 9–11, 1993. Better World, the Magazine of the Optimum Population Trust (OPT), No. 3 (January 1994).

Plaa, G. L. (2001). Introduction to toxicology of occupation and environment. In B. Katzung (Ed.), *Basic and clinical pharmacology* (8th ed., pp. 851, 987–1022). New York: Laing Medical Books/McGraw Hill.

Portner, H. O., Langenbunch, M., & Michaelidis, B. (2005). Synergistic effects of temperature extremes, hypoxia, and increases in CO_2 on marine animals, From Earth history to global change. *Journal of Geophysical Research, 110*(C09S10), 1–15. doi:10.1029/2004JC002561.

Poungparn, S. (2003). *Common allometric relationships for estimating the biomass of mangrove forests*. Doctoral dissertation, Gifu University, Gifu.

Prater, A. J. (1981). *Estuary birds of Britain and Ireland, Poyser, T&D Poyser*. Berkhamstead: Poyser, T&D Poyser, Calton. 440pp.

Primavera, J. H. (1995). Mangroves and brackish water pond culture in the Philippines. *Hydrobiologia, 295*, 303–309.

Primavera, J. (1997). Socio-economic impacts of shrimp culture. *Aquaculture Resource, 28*, 815–827.

Primavera, J. H. (1998). Tropical shrimp farming and its sustainability. In S. de Silva (Ed.), *Tropical mariculture* (pp. 257–289). London: Academic.

Raine, R. M. (1994). Current land use and changes in land use over time in the coastal zone of Chanthaburi Province, Thailand. *Biological Conservation, 67*, 201–204.

Ramsar Secretariat. (2001). *Wetland values and functions: Climate change mitigation*. Switzerland: Gland.

Riebesell, U., Zondervan, I., Rost, B., Tortell, P. D., Zeebe, R. E., & Morel, F. M. M. (2000). Reduced calcification of marine plankton in response to increased atmosphere CO_2. *Nature, 407*, 364–367.

Ruddiman, W. F. (2001). *Earth's climate: Past and future*. New York: W. H. Freeman & Sons.

Saenger, P., & Moverly, J. (1985). Vegetative phenology of mangroves along the Queensland coastline. *Proceedings of the Ecological Society of Australia, 13*, 257–265.

Sanders, H. O., & Cope, O. B. (1968). The relative toxicities of several pesticides to naiads of three species of stoneflies. *Limnology Oceanography, 13*, 112–117.

Schimmel, S. C., Patric, J. M., Jr., & Forester, J. (1976). Heptachlor: Toxicity to and uptake by several estuarine organisms. *Journal of Toxicology and Environmental Health, 1*, 955–965.

Schimmel, S. C., Patrick, J. M., & Forester, J. (1977). Uptake and toxicity of toxaphene in several estuarine organisms. *Archive Environmental Contamination Toxicology, 5*, 353–367.

Scotese, C. R. (2001). *Atlas of earth history*. PALEOMAP Project, Arlington, 52 p.

Scott, W. B., & Crossman, E. J. (1973). Freshwater fishes of Canada. *Bulletin of Fishery Research Board (Canada), 752*, 996.

Sengupta, K., Roy Chowdhury, M., Bhattacharya, S. B., Raha, A., Zaman, S., & Mitra, A. (2013). Spatial variation of stored carbon in Avicennia alba of Indian Sundarbans. *Discovery Natura, 3*(8), 19–24.

Serebrennikov, M. G., & Pervozvansky, A. A. (1965). *Revealing hidden periodicities* (p. 244p). Moscow: Nauka. in Russian.

Sherman, K., & Alexander, L. (1990). *Large marine ecosystems: Patterns processes and yields* (Vol. 111, p. 493). Washington, DC: American Association of Advanced Science Selection Symposium.

Skladany, M., & Harris, C. K. (1995). On global pond: International development and commodity chains in the shrimp industry. In P. McMichael (Ed.), *Food and agrarian orders in the world economy* (pp. 171–189). Connecticut: Greenwood Press.

Smith, J., Schellinhuber, H., & Mirza, M. (2001). Chapter 19. Vulnerability to climate change and reasons for concern: A synthesis. In J. McCarthy, O. Canziani, N. Leary, D. Dokken, & K. White (Eds.), *Climate Change 2001: Impacts, adaptation, and vulnerability* (Published for the Intergovernmental Panel on Climate Change, pp. 913–967). Cambridge: Cambridge University Press.

Snedaker, C. (1991). Notes on the Sundarbans with emphasis on geology, hydrology and forestry. In J. Seidensticker, R. Kurin, & A. K. Townsend (Eds.), *The commons in South Asia: Societal pressures and environmental integrity in the sundarbans*. Washington, DC: The International Center, Smithsonian Institution.

Snedaker, S. (1995). Mangroves and climate change in the Florida and Caribbean region: Scenarios and hypotheses. *Hydrobiologia, 295*, 43–49.

Stewart, H. (1975). Heavy metals. In L. S. Goodman & A. Gilman (Eds.), *Pharmacological basis of therapeutics* (5th ed., pp. 924–941). New York: Macmillan.

Sydow, J. S. (1990). Cruise report experiments on the interaction fishing gear (beam trawl)-benthos with R.V. Mitra. *BEON Rapport, 8*, 1–57.

Tuan, M. S. (1997). Building up the strategy for mangrove management in Vietnam. In P. N. Hong, N. Ishwaran, H. T. San, N. H. Tri, & M. S. Tuan (Eds.), *Proceedings of Ecotone V, Community participation in conservation, sustainable use and rehabilitation of mangroves in Southeast Asia* (pp. 244–255). Vietnam: UNESCO, Japanese Man and the Biosphere National Committee and Mangrove Ecosystem Research Centre.

UNEP. (1994). *Assessment and monitoring of climate change impacts on mangrove ecosystems* (UNEP regional seas reports and studies No. 154). Nairobi: Regional Seas Programme.

UNESCO. (1992). Coastal systems studies and sustainable development. In *Proceedings of the COMAR interregional scientific conference* (276 pp.). Paris: UNESCO, May 21–25, 1992.

Valiela, I., Bowen, J., & York, J. (2001). Mangrove forests: One of the world's threatened major tropical environments. *Bioscience, 51*, 807–815.

Vayda, M. E., & Yuan, M. L. (1994). The heat-shock response of an Antarctic alga is evident at 5°C. *Plant Molecular Biology, 24*, 229–33.

Vivekanandan, V., Muralidharan, C. M., & Subba Rao, M. (1997). *A study on the marine fisheries of Andhra Pradesh*. As supported by BILANCE, 1997 (draft report).

Wagman, R. J. (2000). The environment and health. In *The new complete medical and health encyclopedia* (Vol. 3, pp. 956–964). Ferguson: Chicago. unesdoc.unesco.org/images/0003/000304/030469eb.pdf.

Wells, S. (2006). *Establishing national and regional systems of MPAs-a review of progress with lessons learned*. Marine: UNEP World Conservation Monitoring Centre, UNEP Regional Seas Programme, ICRAN, IUCN/WCPA.

WHO. (1989). *Environmental health criteria No. 91: Aldrin and dieldrin*. Geneva: World Health Organization. 335pp.

WHO. (1993). *Guidelines for drinking-water quality* (2nd ed.). Geneva: WHO. ISBN 92 4 154460.

WHO (World Health Organization). (1984). *Camphechlor* (Environmental health criteria, Vol. 45, pp. 1–66). Geneva: WHO.

Woodroffe, C. D., & Grindrod, J. (1991). Mangrove biogeography: The role of quaternary environmental and sea-level change. *Journal of Biogeography, 18*, 479–492.

Wu, R. S. S. (1995). The environmental impact of marine fish culture: Towards a sustainable future. *Marine Pollution Bulletin, 31*, 159–166.

Zaman, S., Raha, A. K., & Mitra, A. (2013). Signals of salinity fluctuation in *Nypa fruticans* seedling growth: A case study from lower Gangetic plain. *Indian Journal of Forestry, 36*(3), 327–331.

Annex A: References

Bowman. (1917). *Mangrove regeneration and management*. In Hoque, A. K. F. (1995). Mimeograph.

Chaudhuri, A. B., & Choudhury, A. (1994). *Mangroves of the Sundarbans*. India, IUCN- The World Conservation Union, 1. The IUCN Wetlands Programme. Bangkok: IUCN. 247 pp.

Chidumaya, E. N. (1990). Above ground woody biomass structure and productivity in a Zambezian woodland. *Forest Ecology and Management, 36*, 33–46.

Davis, J. H. (1940). The ecology and geological role of mangroves in Florida. *Paper Tortugas Laboratory, 32*, 303–412.

Ellison, J. C., & Stoddart, D. R. (1991). Mangrove ecosystem collapse during predicted sea-level rise: Holocene analogues and implications. *Journal of Coastal Research, 7*, 151–165.

Husch, B., Miller, C. J., & Beers, T. W. (1982). *Forest mensuration*. New York: Ronald Press.

Kristensen, E., Bouillon, S., Dittmar, T., & Marchand, C. (2008). Organic matter dynamics in mangrove ecosystems. *Aquatic Botany*. doi:10.1016/JAqua.Bot.200715.05 (in press).

Mitra, A., & Banerjee, K. (2005). *Living resources of the sea: Focus Indian Sundarbans*, Published by WWF India. In Col S. R. Banerjee (Ed.), Canning Field Office, 24 Parganas (S), W.B. 96 pp.

Mitra, A., Banerjee, K., & Bhattacharyya, D. P. (2004). *In the other face of mangroves*. India: Published by Department of Environment, Government of West Bengal.

Mitra, A., Banerjee, K., Sengupta, K., Gangopadhyay, A. (2009). Pulse of climate change in Indian sundarbans: A myth or reality? *National Academy of Science Letters, 32*: 1–2.

Conservation of Marine and Estuarine Resources 11

Contents

11.1 Conservation or Preservation?

Protecting the environment of the planet Earth is the prime goal of mankind. The domain of protection of natural resources encompasses two major verticals, namely, conservation and preservation. Conservation is the sustainable use and management of natural resources which include flora, fauna, air, water, earth deposits, etc. Conservation of natural resources primarily focuses on the needs and interests of human beings, for example, biological, economic, cultural and aesthetic values these resources possess. The mangrove ecosystem, for example, contains a wide spectrum of flora and fauna, which provides timber, fuel wood, honey, wax, fishes, medicinal ingredients, etc. The seed of conservation germinates for developing a better future. Preservation, on the contrary, is an attempt to maintain the existing condition of the environment of a particular ecosystem or habitat that is mostly in a wilderness condition. The essence of preservation spreads from the fact that mankind is encroaching the natural habitats at such a rate that many untouched landscapes are now getting sacrificed for industrial development, urban development, farming, tourism, aquaculture, etc. Strong lobbies of preservationists support protection of natural reservoir with priority and give less importance to the ecosystem services of natural resources (flora, fauna, etc.) to mankind.

The original version of this chapter was revised. An erratum to this chapter can be found at https://doi.org/10.1007/978-81-322-2707-6_13

A. Mitra, S. Zaman, *Basics of Marine and Estuarine Ecology*,
DOI 10.1007/978-81-322-2707-6_11

A comparative account of conservation and preservation is given in Table 11.1.

The vertical of conservation emphasizes on alternative livelihood schemes through which human encroachment can be reduced and natural reservoir be preserved. We have strong inclination to recommend the term conservation as the ecosystem services of flora, fauna and other organisms need to be explored and rationally utilized to mitigate the needs of the growing population (that is considered by us as the root cause of all evils in this planet). The generation of alternative livelihood schemes also falls under this category.

Whether the term is preservation or conservation is not the primary question, but protecting the natural resources from the natural and anthropogenic threats is the main goal of both. In context to marine and estuarine ecosystems, it is essential to develop a concrete plan for ensuring protection to the flora and fauna thriving in the system. For achieving this primary goal, programmes and planning with activities must be developed with thrust area and priority (Table 11.2).

11.2 Tiers for Conservation

Conservation of marine and estuarine ecosystems requires a concrete understanding of the different tiers of biodiversity. The taxonomic diversity is higher in benthic rather than pelagic system, although considerable uncertainty exists in this statement. Also, if one compares the coastal biodiversity with the open ocean, the former will be much ahead compared to the latter. A snapshot of different tiers of biodiversity, namely, habitat diversity, community and ecosystem diversity, functional diversity, population diversity, phyletic diversity, species diversity and genetic diversity, is presented here.

11.2.1 Habitat Diversity

The most frequently used quantitative measure of biodiversity is for a given area rather than for a given biological community. In ecological terms, physical areas and the biotic components they contain are referred to as habitats. Habitat diversity is a more useful term than that of ecosystem diversity since habitats are easy to envisage (e.g. a mangrove forest, a coral reef, an estuary, etc.). Furthermore, habitats often have clear boundaries. Habitats have been termed the template for ecology (Southwood 1977). There are strong relationships between sampling scale and the processes that influence diversity (Huston 1994). At small scales, all species are presumed to interact with each other and to be competing for similar limiting resources. Ecologists have called this within habitat (or alpha) diversity (Fisher et al. 1943; Whittaker 1960, 1967). At slightly larger scales, habitat and/or community boundaries are crossed, and sampling covers more than one habitat or community. This scale has been called between-habitat (or beta) diversity (Whittaker 1960, 1975, 1977). At an even larger scale (regional scale) where evolutionary rather than ecological processes operate, the pattern has been called gamma diversity or, more recently, 'landscape diversity' (Whittaker 1960; Cody 1986). Landscape diversity can be defined as the

Table 11.1 Comparative account preservation and conservation

Issue	Preservation	Conservation
Objectives	Landscapes without humans	Flora, fauna, microbes and food webs—members of all tiers of ecological pyramids
Justifications	Aesthetic	Intellectual interest; present and future utility preferably for the benefit of mankind
Target	National parks, sanctuary, etc.	Hot spots of biodiversity and even natural resources
Obstacles	Encroachment, overexploitation, etc.	Encroachment, overexploitation, shrinkage of livelihood, lack of proper management of man–natural resource conflict, etc.

Table 11.2 Planning, components and domain of investigation with priority for conserving marine and estuarine ecosystems

S. no.	Planning	Components	Domain of investigation	Priority
1	Integrated landscape development	1. Biodiversity assessment 2. Identification and scaling of threats 3. Identification of degraded and fragile ecosystems 4. Zonation of the operational area into core, buffer and tourism/urbanization/ industrial (activity) zones 5. Identification of ecorestoration plans befitted to the landscape	Ecosystem	A
2	Assessment of the impact of climate change on the marine and estuarine ecosystems	1. Assessment and prediction of the impact of climate change on flora, fauna and energy flow through different tiers of trophic levels 2. Assessment and prediction of impact of climate change on the community structure of flora and fauna 3. Identification of keystone species 4. Assessment and prediction of climate change on the socio-economic and demographic profile of coastal communities 5. Preparation of a long-term conservation plan with special emphasis on disaster management	Ecosystem	A
3	Ecosystem services	1. Identification of knowledge, innovations and practices of indigenous and local communities 2. Assessment of impacts of developmental activities on ecosystem goods and services 3. Evaluation of negative and positive impacts using tools of market and non-market valuation 4. Economic analysis of conservation–development scenarios using multiple-criteria assessment and cost–benefit analysis 5. Ranking of scenarios and identification of options for conservation and sustainable development	Ecosystem	A
4	Biodiversity census	1. Assessment and inventorization of flora and fauna 2. Assessment and scaling of community health of flora and fauna 3. Identification of threatened species 4. Short-term and long-term conservation plan involving the local population (coastal population, island dwellers, etc.)	Ecosystem	A

(continued)

Table 11.2 (continued)

S. no.	Planning	Components	Domain of investigation	Priority
5	Assessment of anthropogenic threats	1. Listing the industries around the landscape and finding out the effluent release 2. Analysis of samples of contaminated soil and water 3. Laboratory analysis of affected plants/animals to determine the concentration of toxicants (bioaccumulation) 4. Identification of indicator organisms	Ecosystem	A
6	Assessment of coastal geomorphology and shoreline configuration	1. Collection of available data on various aspects of coastal geomorphology and analysis in GIS domain to record the changes 2. Identification of drivers which might have changed the geomorphology and developed ecological modelling for expected future changes 3. Quantification and determination of importance of sediment and freshwater flow	Ecosystem	A
7	Assessment of dilution factor in the estuarine and coastal zones	1. Collection of data on freshwater flow, barrage discharge, land run-off and sediment influx 2. Recording and documentation of tidal amplitude 3. Regular monitoring of salinity of sample station (under investigation) and that of source water (particularly for estuarine ecosystem)	Ecosystem	A
8	Assessment of intertidal mudflats/sandy shore/rocky shore and their ecological significance	1. Assessment of biodiversity in the intertidal mudflats/sandy shore/rocky shore along with their adaptations 2. Assessment of avifaunal diversity and migratory species 3. Texture and composition of the intertidal zone	Habitat	A
9	Assessment of mangroves/ seagrass/salt marsh grass/coral reefs	1. Identification and marking of monitoring plots for biomass study on regular basis 2. Identification of interlinkage between the habitats, e.g. mangroves, salt marsh grasses or coral reefs and mangroves, etc (if any)	Habitat	A
10	Coastal lagoon ecology and biodiversity	1. Developmental history and origin of lagoon (through satellite imagery) 2. Flora, fauna and keystone species assessment 3. Primary and secondary productivity of the lagoon 4. Livelihood associated with the lagoon water and other resources	Habitat	A
11	Threatened animals	1. Identification and assessment of status of critical habitats of threatened species 2. Long-term planning for ecorestoration	Habitat	A

(continued)

Table 11.2 (continued)

S. no.	Planning	Components	Domain of investigation	Priority
12	Monitoring of primary production	1. Preparation of phytoplankton species inventory 2. Tidal, diurnal and seasonal variations in primary productivity 3. Monitoring phytopigment level on a regular basis during high tide and low tide	Species	A
13	Monitoring of secondary production	1. Documentation of zooplankton 2. Documentation of nekton (considering both osteichthyes and chondrichthyes) 3. Inventorization of benthic organisms 4. Application of DNA barcoding as a tool for biodiversity assessment	Species	A
14	Invasive species	1. Monitoring of ballast discharge 2. Monitoring ecology of invasive species 3. Monitoring the interaction between endemic and alien species 4. Long-term planning to manage invasive species particularly those which intrude through ballast discharge	Species	B
15	Natural calamities	1. Collection of secondary data on magnitude and occurrence of natural calamities 2. Long term planning for disaster management 3. Monitoring of resources (flora, fauna, groundwater, intertidal zone, etc.) before and after the natural calamities through satellite imagery and ground truthing 4. Development of bio-shield to reduce the impact of natural calamities 5. Development of reclamation models for the affected sites 6. Development of species recovery plans particularly for the threatened species	Ecosystem and production sectors	B
16	Livelihood	1. Identification of alternative livelihood schemes befitted to the area of investigation 2. Identification of proper beneficiaries associated with the livelihood schemes 3. Monitoring the viability of schemes through cost–benefit analysis and market demand 4. Establishment of local and regional marketing linkage 5. Attempt to develop international market through quality production and subsequent certification	Socio-economic	A

(continued)

Table 11.2 (continued)

S. no.	Planning	Components	Domain of investigation	Priority
17	Tourism	1. Assessment of existing tourism units 2. Monitoring the effluents released from tourism units and prediction of the level of ingredients of these effluents (viz. nitrate, phosphate, etc.) through statistical tools/programmes 3. Assessment of tourist carrying capacity and footfalls 4. Development of ecotourism involving the local people 5. Training the local people on flora, fauna and other endemic species on a regular basis so that they can provide service as eco-guide 6. Development of ecofriendly waste management system (bioremediation)	Socio-economic	B
18	Good management practice	1. Collection of successful case studies from district level/state level/country level/international level 2. Group activity through formation of homogeneous group 3. Capacity building in the sphere of aquaculture, agriculture, horticulture, etc. 4. Value addition in the field of aquaculture, agriculture, horticulture, etc.	Socio-economic	C
19	Production and development sectors	1. Rapid and long-term EIA studies for any developmental activities in the coastal and estuarine ecosystems 2. Impact studies and habitat degradation 3. Monitoring ecosystem services and deletion of natural resources	Socio-economic and developmental activities	B
20	Policy-level research	1. Development of a national policy on the coastal and marine environment 2. The efficacy of international, national and state policies and legal instruments in resource management in the coastal and marine environment needs to be assessed. The readers may consult Annexure 11A to have an idea about the existing legal framework to protect the marine/coastal/estuarine resources 3. Mechanism of networking various governmental, non-governmental and community-based institutions in the participatory management of coastal and marine resources 4. Identification of various components that need to be part of a national- or state-level policy for restoration and rehabilitation of communities affected by conservation initiatives, natural disasters, etc.	Policy research	C

Note: *A* high priority, *B* moderate priority, *C* low priority

mosaic of habitats over larger scales of often hundreds of kilometres. Franklin (1993) discussed landscape diversity in relation to biodiversity conservation. Ray (1991) called the marine equivalents seascapes. Much attention has been given to ways of conserving landscape diversity on land. Clearly, a given habitat can be maintained, but landscape diversity can be reduced if the mosaic of habitats is altered. It is clearly important, therefore, to specify what scale (and hence type of diversity) is being studied. In an important recent paper, Tuomisto et al. (1995) have shown from an analysis of satellite images followed by extensive ground truthing that beta diversity has been greatly underestimated in tropical rainforests. Since between-habitat (beta) diversity has been underestimated, then the landscape diversity will also be underestimated. The conservation value of different areas primarily depends on a sound estimate of between-habitat and landscape diversity. This is a topic that must be emphasized thorough consideration and discussion in any future conservation strategy. Within coastal areas, there are a wide variety of habitats with known high species diversity such as seagrass beds (McRoy and Lloyd 1981), coastal sedimentary habitats (Gray 1994), mangal (MacNae 1968; Walsh 1974) and coral reefs (Loya 1972; Huston 1985; Sheppard 1980). Ray and Gregg (1991) critically analysed the coastal wetland areas of Virginia and the Carolinas, USA. They concluded that there are large differences in the proportions of salt and freshwater marshes, forest/scrub–shrub and tidal-at areas which lead to differences in biodiversity between the two areas. In Indian Sundarbans also, difference in salt concentrations in the estuarine water causes significant differences in mangrove floral diversity. *Excoecaria agallocha* and *Avicennia* spp. are common in high-saline to moderately saline areas (15–28 psu), whereas *Heritiera fomes* and *Sonneratia apetala* are found in low saline with a salinity around 5 psu. Ray (1991) classified marine habitats into 20 categories as a basis for characterizing coastal areas. Coral reefs are also highly variable with large differences between the reef top, reef crest and reef slope both in coral and associated species, and each component is probably best considered as between-habitat biodiversity. Hard rocky surfaces have a rich encrusting flora and fauna, for example, in clumps of mussels, Suchanek (1992) found over 300 species in Washington, USA.

11.2.2 Community and Ecosystem Diversity

Biodiversity can also be considered at levels other than that of taxonomic organization, for example, at the level of the community or/and ecosystem. In fact, when biodiversity is measured quantitatively, it is usually as the number of species or the value of a diversity index for a given community or area of habitat. A great ecological debate started in the 1930s on whether or not species occurred in distinct groups which could be classified as communities. Today, the generally accepted view is that species are distributed along environmental gradients in approximately lognormal abundance patterns (Mills 1969). However, interactions between species (predator–prey relationship, commensalism, symbiotic relationship and competitive relationship) lead to there being co-occurring groups of species under given environmental conditions. Thus, communities are convenient groupings of species which merge gradually into other groupings unless there are sharp boundaries in environmental conditions. Recently, another term has found favour, assemblage, which is a more neutral term and does not imply the tight interspecies organization that is implied in the term 'community', with its anthropomorphic connotations. The diversity of a community (or assemblage) is often measured. In the Biodiversity Convention, an ecosystem is defined as 'A dynamic complex of plant, animal and microorganism communities and their non-living environment interacting as a functional unit'. Terms such as 'estuarine ecosystem' or 'coral reef ecosystem' are used commonly. Yet the boundaries of such systems are loosely defined and are especially difficult to demarcate in the sea since the fluxes of energy and material within and exported from a system are rarely known. It is perhaps significant that in the Research Agenda for Biodiversity (Solbrig 1991), no mention is made of ecosystem diversity. Huston (1994) in his book used the terms 'community' and 'ecosystem' interchangeably. However, while the distinction between community and ecosystem may be helpful, in some ways the implication that communities and ecosystems can be studied as separate entities is wrong. No ecological system, whether individual, population or community, can be studied in isolation from the environment in which it exists. In this context, we have analysed the macro-benthic molluscan community diversity and structure in four regions of the East Coast of India (distributed in four maritime states of the country), namely, West Bengal (Shankarpur coast), Odisha (Bahuda estuary), Andhra Pradesh (East Godavari estuary) and Puducherry (Kalapet coast), during the 2014 summer. Data from 50 quadrates (10 quadrates selected at random from each coordinate) were collected (Fig. 11.1) to ensure quality to the population data, and these data sets were finally used to enumerate the community structure of macro-benthic molluscs in the study area through computation of the Shannon–Wiener species diversity index (1949), as per the expression:

Species diversity index $(\overline{H}) =$

$$-\sum_{i=1}^{s} P_i \log_e P_i \quad or, \ (\overline{H}) = -\sum_{i=1}^{s} n_i/N \log_e n_i/N$$

where:

P_i = Importance probability for each species
n_i = Importance value for each species
N = Total of importance values

The macro-benthic molluscan community inhabiting the intertidal zone of four major estuaries/coasts in the eastern part of India showed the presence of 17 dominant species. In the Shankarpur area of West Bengal, six bivalve species and four gastropod species were recorded during the study period. In the Bahuda estuary of Odisha, the number of bivalve species was nine, and the number of gastropod species was eight. In the East Godavari mudflats in the state of Andhra Pradesh, we documented eight bivalve species and eight gastropod species. In the Kalapet coast of Puducherry, the number of bivalve and gastropod species was nine and seven, respectively. The community structure of the molluscan community also exhibited pronounced spatial variation (Table 11.3).

Question arises at this point regarding significant spatial variation of the macro-benthic molluscan community structure. The only logical

Fig. 11.1 Population density study of macro-benthic molluscan species

Table 11.3 Mean population density of macro-benthic molluscan species (of 50 quadrates at an interval of 500 m selected at random) in the East Coast of India

Species	Shankarpur coast	Bahuda estuary	East Godavari estuary	Kalapet coast
 *Nerita articulata*	2	6.6	3.2	2
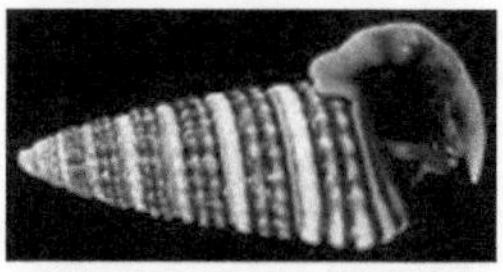 *Cerithidea cingulata*	13.4	10	6.4	13.4
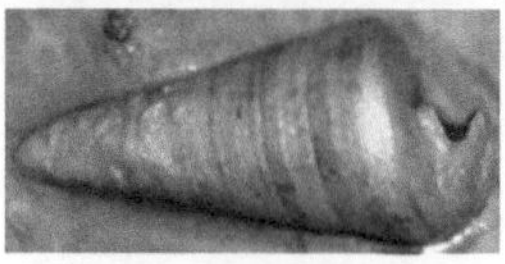 *Telescopium telescopium*	11.8	9.8	8	6
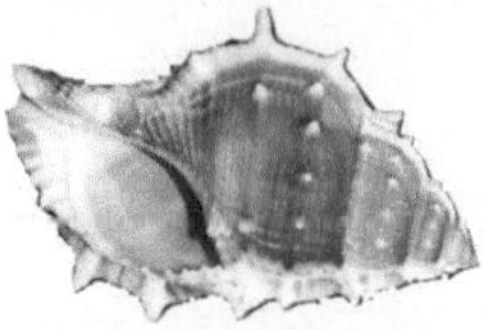 *Bursa rana*	10.4	7.6	1.2	5.6
 *Nassarius* spp.	13.2	8.8	6.4	3
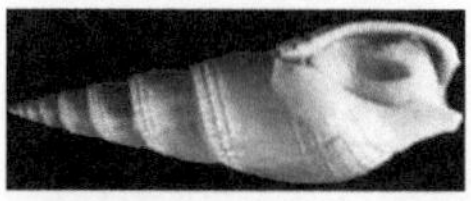 *Bullia vittata*	0	6.4	3	3.6
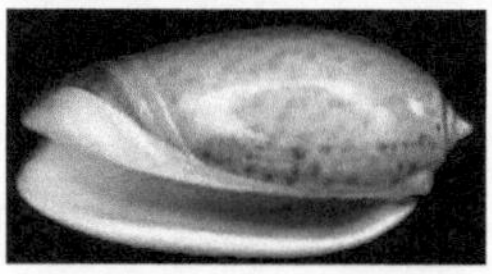 *Oliva oliva*	0	7.8	4.6	0
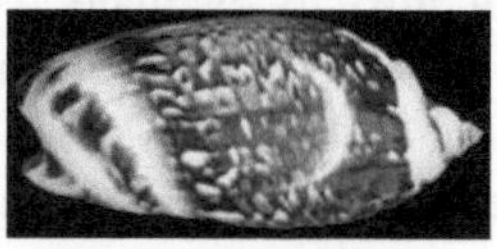 *Olivancillaria gibbosa*	0	7.8	6.8	4.6

(continued)

Table 11.3 (continued)

Species	Shankarpur coast	Bahuda estuary	East Godavari estuary	Kalapet coast
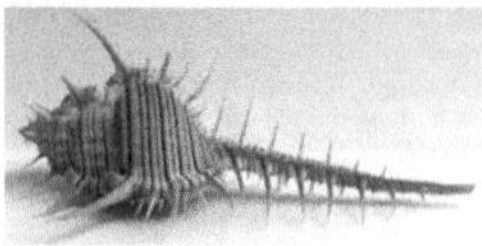 *Murex troscheli*	13.6	6.8	5.4	4
 *Anadara* sp.	4.4	3.4	1.8	3.8
 *Perna viridis*	0	4.2	2.2	4
 *Crassostrea cuttackensis*	3.4	3.4	8.6	4.4
 *Saccostrea cucullata*	14.6	6.2	7.4	13.6
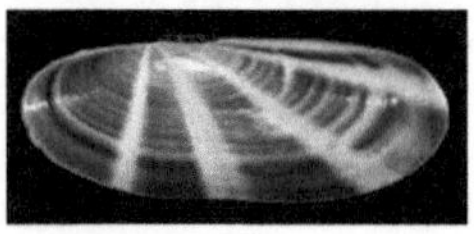 *Siliqua radiata*	0	4.4	0	6

(continued)

answer is the effect of environment. Thus, no community can be studied without evaluating the ambient environment. The present study was therefore undertaken not only to prepare a checklist of the most common bivalves and gastropods in the major estuaries/coasts of the eastern part of India but also to evaluate the health of the community in context to environmental stress. For this we used the Shannon–Wiener species diversity index as

Table 11.3 (continued)

Species	Shankarpur coast	Bahuda estuary	East Godavari estuary	Kalapet coast
Donax scortum	8	8.8	2.6	12.4
Sunetta scripta	0	8	2	8.4
Meretrix meretrix	12.8	5.6	4.6	15.6
N	107.6	115.6	74.2	110.4
H	2.278	2.787	2.638	2.599

proxy and observed maximum value of the index in the Bahuda estuary of Odisha. This coast is relatively free of any type of human interference/ anthropogenic pressure. The East Godavari estuary in the state of Andhra Pradesh and Kalapet coast in Puducherry experience considerable anthropogenic stress particularly due to intense fishing activities and tourism-related pressure. Shrimp culture and salt pan also fall within the domain of anthropogenic activities in these two sites. The moderate value of Shannon–Wiener index in the East Godavari estuary and Kalapet coast is an indicator of such stress related to human activities in these areas. In the state of West Bengal, the Shankarpur coast exhibited the lowest value of Shannon–Wiener index, which may be attributed to high pollution load in the environment due to presence of shrimp farms, tourism units and a major fish landing station. Several boat and trawler repairing units are also concentrated along the coast of Shankarpur due to which the area experiences stress of a high degree. The frequent movements of cargo vessels and oil tankers in the adjacent water of Shankarpur coast have made the area much more vulnerable to stress factor (Fig. 11.2).

The value of Shannon–Wiener index has several ecological explanations. As Margalef (1968) stated, 'the ecologists find in any measure of diversity an expression of the possibilities of constructing feedback system'. Higher diversity then signifies longer food chains and more cases of symbiosis (mutualism, commensalisms, etc.) and greater probabilities for negative feedback control, which reduces the drastic oscillation and hence increases stability.

The value of Shannon–Wiener index is a unique indicator of environmental stress. As the sensitive species gradually shift or get eliminated from the habitat with the increase of magnitude

Fig. 11.2 Magnification of stress (as indicated by lowest Shannon–Wiener index value in Table 11.3) in the Shankarpur area of West Bengal by frequent movements of vessels and trawlers

of environmental stress, therefore the species diversity index has been claimed as an effective statistics for predicting the change in environment (Wilhm and Dorris 1968; Cairns and Dickson 1971).

A scale of pollution in terms of species diversity (3.0–4.5 slight, 2.0–3.0 light, 1.0–2.0 moderate and 0.0–1.0 heavy pollution) has been described by Staub et al. (1970). In a Shannon–Wiener legislation, the aquatic environment of soil and water is divided as good when H > 4; good quality is 4–3, moderate quality 3–2, poor quality 2–1 and very poor quality <1. The Shannon–Wiener index in the present study is within the range of 2.114–2.863 which indicates less environmental stress on the macrobenthic molluscan species inhabiting the East Coast of India. The distribution of species (in terms of population density) becomes more dissimilar as the environmental stress increases and accordingly species diversity decreases with poor water quality. A community dominated by relatively few species indicates that the environment is under stress (Plafkin et al. 1989). Considering the two primary variables, namely, (i) number of species and (ii) uniformity/non-uniformity, among the population of different species in the present macro-benthic molluscan community, we conclude that the selected estuaries/coasts in the eastern part of the Indian subcontinent experience moderate environmental stress, the magnitude being in the order Shankarpur coast (in West Bengal) > Kalapet coast (in Puducherry) > East Godavari estuary (in Andhra Pradesh) > Bahuda estuary (in Odisha).

A three-class stress scale (2.000–2.300, 2.300–2.600, 2.600–2.900) constructed on the basis of mean values of Shannon–Wiener index (1949) thus reflects the most congenial environment for the growth and survival of molluscan biodiversity in the Bahuda estuary of Odisha. On the contrary, in the Shankarpur coast of West Bengal, the situation is worst as per the range and classification of our constructed stress scale (Fig. 11.3).

11.2.3 Functional Diversity

Functional diversity is the range of functions that are performed by organisms in a system. The species within a habitat or community can be divided into different functional types such as feeding guilds or plant growth forms or into functionally similar taxa such as suspension feeders or deposit feeders. Functionally similar species may be from quite different taxonomic entities. One of the major current topics of debate is that of functional redundancy (Di Castri and Younes 1990; Walker 1991) where it is suggested that there are more species present in communities than are needed for efficient biogeochemical and trophic functions. Recent data, however, show that this is not the case and the higher the number of species in a community, the greater the efficiency of biogeochemical processes (Naeem et al. 1994; Tilman and Downing 1994). Such experiments, however, have not been done in the marine environment. Steele (1991) defines functional diversity in a different and idiosyncratic way as 'the variety of different responses to environmental change, especially the diverse space and time scales with which organisms react to each other and to the environment'. Steele's main point is that marine organisms are closely linked to physical processes at decadal scales whereas on land undisturbed systems change at scales of centuries to millennia. The data of decadal variation of true mangrove floral count and diversity in Indian Sundarbans is a befitting example in this context, where salinity is playing a major role in regulating the growth and diversity of species. The present study was conducted in the Indian Sundarban mangrove ecosystem during November 2000, 2006 and 2012 in ten sampling sites of Indian Sundarbans (Fig. 11.4). Several field trips were made to select the field stations (Figs. 11.5, 11.6, 11.7, and 11.8) and study the distribution of mangrove vegetation, geographic nature of the riverine system, salinity, environmental quality and anthropogenic pressure. While selecting the sampling sites, we focused on two major issues, namely, natural threats (like salinity fluctuation) and human-induced factors (like industrialization, urbanization, tourism, aquacultural practice, etc.). These criteria have been selected to observe the factors regulating the survival/degradation of mangroves in and around the selected sampling stations.

Quadrates of 10 m × 10 m were laid randomly up to 500 m from LTL, and data from each one were recorded from 15 such quadrates. Plant materials collected during the sampling were identified and confirmed from the Botanical Survey of India, Kolkata.

Relative abundance of the species was estimated as per the expression: **RA** = abundance of a particular species/sum of the

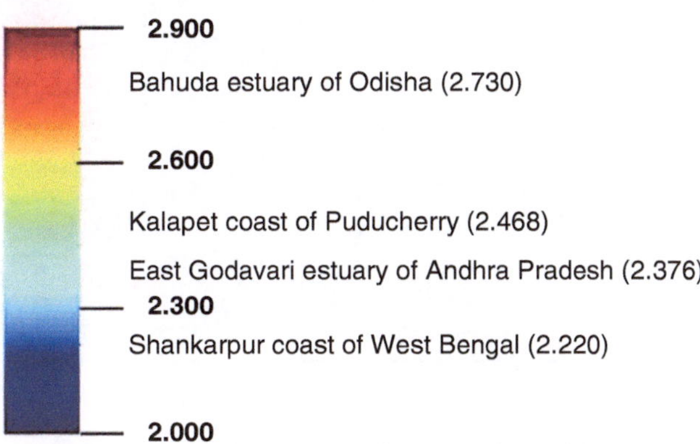

Fig. 11.3 A three-class stress scale to indicate the health of estuary/coastal zone

abundance of all species × 100, where **RA** represents relative abundance.

The Shannon–Wiener index for diversity (H) was calculated based on the abundance value of plant species in different categories as per the expression stated earlier (see Sect. 11.2.2). It is to be noted that the Shannon–Wiener index was calculated on the basis more than equal to 40 % abundance of the species common to all the ten selected sampling stations. To assess whether the total number of individuals of all the true mangrove species (N) and Shannon–Wiener species diversity index (H) vary significantly among sites and years (6-year interval data sets), analysis of variance (ANOVA) was performed; possibilities less than 0.05 ($p < 0.05$) were considered statistically significant.

A total of 25 mangrove species (*Acanthus ilicifolius*, *Acrostichum aureum*, *Aegiceras corniculatum*, *Aegialitis rotundifolia*, *Avicennia alba*, *Avicennia marina*, *Avicennia officinalis*, *Bruguiera cylindrica*, *Bruguiera gymnorrhiza*, *Bruguiera hexangula*, *Bruguiera parviflora*, *Ceriops decandra*, *Ceriops tagal*, *Excoecaria agallocha*, *Heritiera fomes*, *Kandelia candel*, *Lumnitzera racemosa*, *Nypa fruticans*, *Phoenix paludosa*, *Rhizophora apiculata*, *Rhizophora mucronata*, *Sonneratia apetala*, *Sonneratia caseolaris*, *Xylocarpus granatum*, *Xylocarpus mekongensis*) were recorded from the selected sites.

The total number of individuals of all the documented species per unit area, which represents the dense/sparse growth of the trees, exhibits pronounced variation due to which the

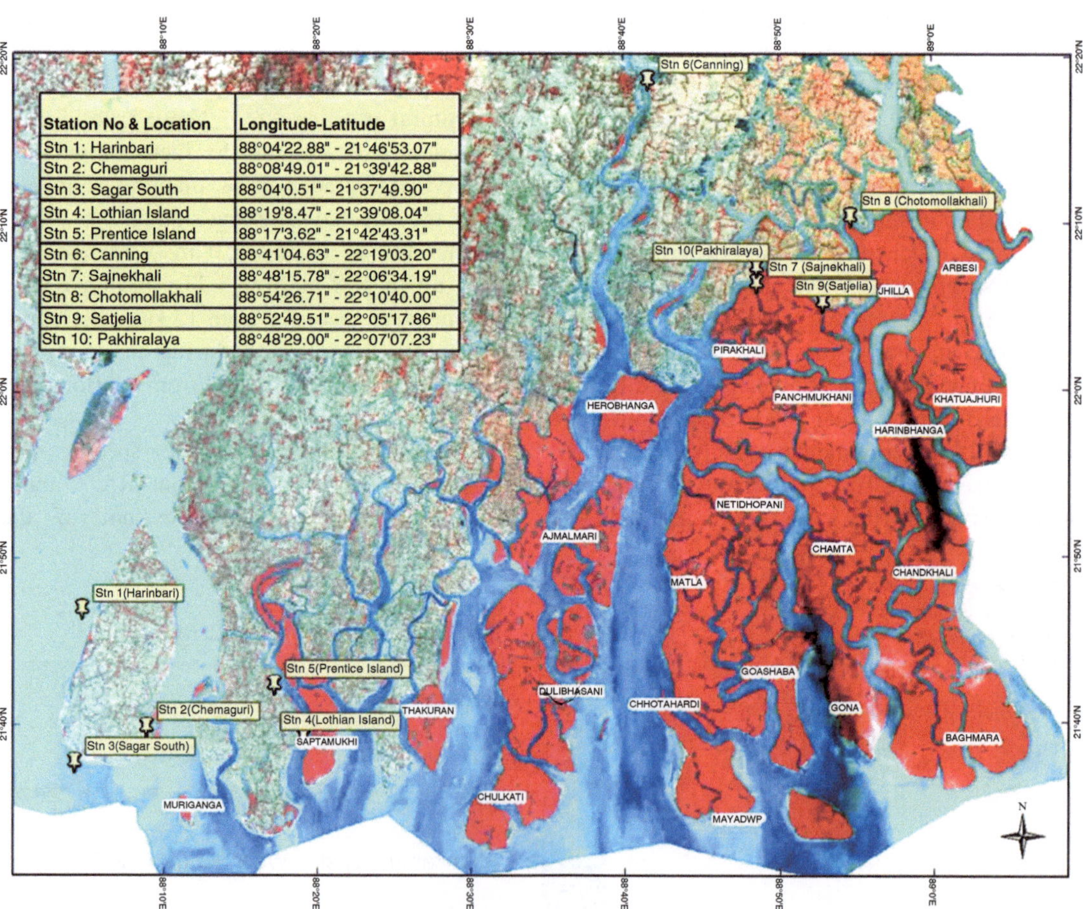

Station No & Location	Longitude-Latitude
Stn 1: Harinbari	88°04'22.88" - 21°46'53.07"
Stn 2: Chemaguri	88°08'49.01" - 21°39'42.88"
Stn 3: Sagar South	88°04'0.51" - 21°37'49.90"
Stn 4: Lothian Island	88°19'8.47" - 21°39'08.04"
Stn 5: Prentice Island	88°17'3.62" - 21°42'43.31"
Stn 6: Canning	88°41'04.63" - 22°19'03.20"
Stn 7: Sajnekhali	88°48'15.78" - 22°06'34.19"
Stn 8: Chotomollakhali	88°54'26.71" - 22°10'40.00"
Stn 9: Satjelia	88°52'49.51" - 22°05'17.86"
Stn 10: Pakhiralaya	88°48'29.00" - 22°07'07.23"

Fig. 11.4 Location of sampling stations in Indian Sundarbans

Fig. 11.5 A view of the field site in the Indian Sundarbans

Fig. 11.6 Mangroves thrive on the intertidal mudflat, and their growth and metabolism are greatly regulated by salinity

Shannon–Wiener species diversity is affected. Highest numbers of individuals (N) are observed in the quadrate of Lothian Island, a pristine forest patch almost with no human intervention, whereas lowest value is observed in Canning, the sampling station with maximum anthropogenic stress (Fig. 11.9).

Fig. 11.7 Field survey on the intertidal mudflat of the study area

Fig. 11.8 Researchers laying quadrates of dimension 10 m × 10 m in the mangrove forest for evaluating relative abundance (RA) of the true mangrove floral species

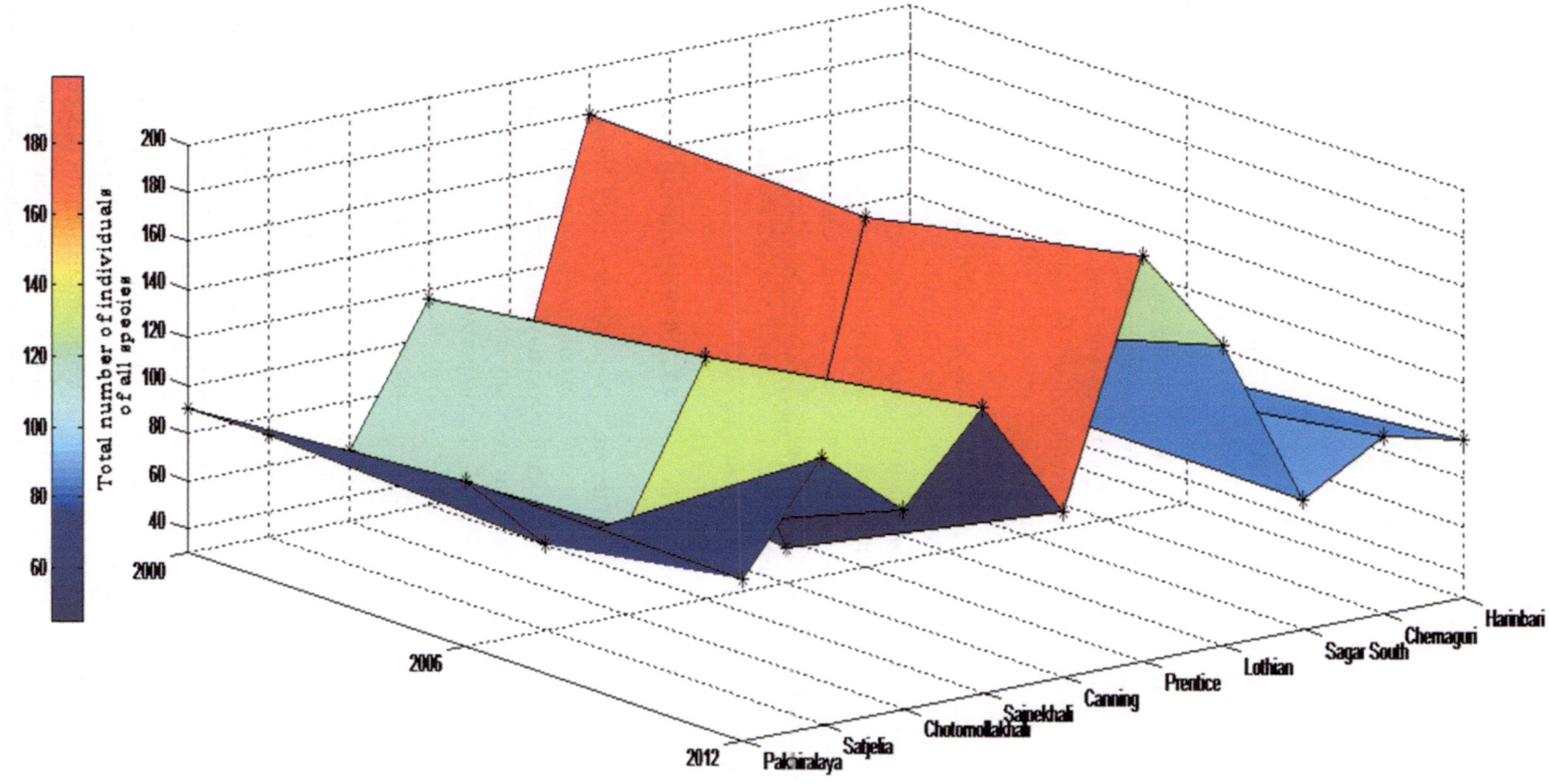

Fig. 11.9 Spatio-temporal variation of total number of individuals of all species (N)

The results of the Shannon–Wiener index are shown in Table 11.4 and Fig. 11.10. The spatial order of the index is Lothian Island (3.052, 3.083 and 3.174 in 2000, 2006 and 2012, respectively) > Prentice Island (3.043, 3.033 and 3.13 in 2000, 2006 and 2012, respectively) > Sajnekhali (3.035, 3.027 and 3.091 in 2000, 2006 and 2012, respectively) > Pakhiralaya (2.883, 3.008 and 3.074 in 2000, 2006 and 2012, respectively) > Satjelia (2.765, 2.928 and 3.07 in 2000, 2006 and 2012, respectively) > Chotomollakhali (2.622, 2.828 and 2.948 in 2000, 2006 and 2012, respectively) > Harinbari (2.562, 2.701 and 2.88 in 2000, 2006 and 2012, respectively) > Chemaguri (2.563, 2.784 and 2.924 in 2000, 2006 and 2012, respectively) > Sagar South (2.44, 2.449 and 2.837 in 2000, 2006 and 2012, respectively) > Canning (2.145, 2.319 and 2.636 in 2000, 2006 and 2012, respectively).

The variation in the value of the index reflects (i) the degree of stress (both natural and anthropogenic) and (ii) conditions of the ambient environment (in terms of hydrological parameters and soil quality). Greater value of the index represents a more congenial environment which usually occurs due to the survival of more number of species or even distribution of the number of individuals among different species in the quadrate.

The true mangrove floral diversity values are relatively lower in the sampling stations of central Indian Sundarbans (Sajnekhali, Chotomollakhali and Pakhiralaya). This may be attributed to the synergistic effects of both hypersalinity and human intrusion in these sampling stations, except Sajnekhali, which is a protected reserve forest under the West Bengal forest department. The hypersaline water in the central Indian Sundarbans is the effect of Bidyadhari siltation since the late fifteenth century (Chaudhuri and Choudhury 1994) due to which the mangroves are less diverse and stunted in this zone.

The overall investigation thus pinpoints the hypersalinity and human intrusion as the major threats to mangrove floral diversity of Indian Sundarbans and advocates for an ecorestoration-oriented management plan that (i) encompasses provision of alternative livelihood to reduce the exploitation of mangrove resources, (ii) increases the dilution factor of the estuarine water through periodic dredging of the silted Bidyadhari river and (iii) is freshening the central Indian Sundarbans through construction of rainwater harvesting ponds and plantation of mangrove associate species in mass scale (as ground cover) like *Suaeda maritima*, *Salicornia* sp., etc. (Fig. 11.11), that are potential absorbers of salt from the ambient media.

Table 11.4 Shannon–Wiener species diversity index at selected stations of Indian Sundarbans

Station	2000	2006	2012
Harinbari	2.562	2.701	2.88
Chemaguri	2.563	2.784	2.924
Sagar South	2.44	2.449	2.837
Lothian	3.052	3.083	3.174
Prentice	3.043	3.033	3.13
Canning	2.145	2.319	2.636
Sajnekhali	3.035	3.027	3.091
Chotomollakhali	2.622	2.828	2.948
Satjelia	2.765	2.928	3.07
Pakhiralaya	2.883	3.008	3.074

11.2.4 Population Diversity

A population may be defined as a group of organisms belonging to the same species and occupying a particular space. A population, like any other level of organization, has a number of important group properties like density, birth rate, death rate, growth rate, dispersion, etc., not shared by the adjacent levels (the organism on the one hand and the community on the other).

The term population diversity is definitely a new entrant in the field of biodiversity. However, the diversity of population between two or more sites is a common feature of biodiversity-related work. It has been observed by researchers that population of a species is a function of suitability of habitat in terms of nutrients, light availability and other natural resources in addition to the predation pressure. In this context, the variation of molluscan population can be considered as an example. The authors observed that even in the

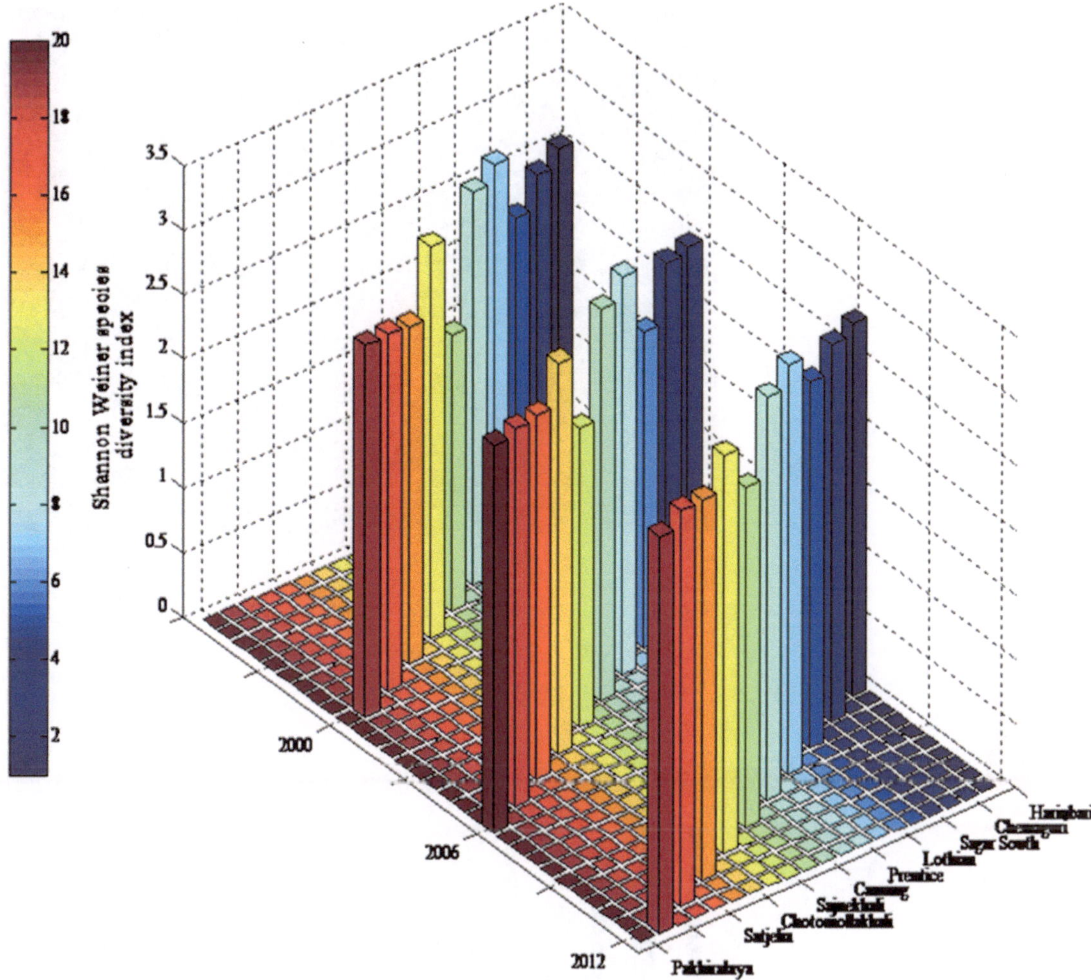

Fig. 11.10 Spatio-temporal variation of Shannon–Wiener species diversity index (H)

same area, the population of molluscs fluctuates significantly (Figs. 11.12 and 11.13). This may extend our discussion to niche preference of species.

In the coastal floral community, variation in relative abundance of species is a function of edaphic factors, climatic factors and availability of resources like nutrients, light, temperature, salinity, etc. In mangrove forests of coastal zone, river mouths or deltaic system, a significant variation in population is observed, e.g. in the lower Gangetic region, the abundance of *Sonneratia apetala* in the western sector and abundance of *Avicennia* spp. in the entire deltaic complex are the result of spatial variation of salinity (Mitra et al. 2011; Banerjee et al. 2013; Sengupta et al. 2013; Raha et al. 2014). The primary cause behind this variation in population is attributed to the adaptation of the species to factors like salinity (primarily). *S. apetala*, *Heritiera fomes* and *Nypa fruticans* can tolerate low-saline environment (between 2 psu and 7 psu), whereas *Avicennia* spp. can survive and grow luxuriantly even at a salinity around 25 psu. *Rhizophora* sp. can again adapt to extremely high-saline condition. Thus, salinity is a limiting factor governing the distribution and relative abundance (RA) of the species.

The functional diversity also encompasses the grouping of species on the basis of functionally

Fig. 11.11 *Suaeda maritima* in the intertidal mudflats of Sundarban mangroves

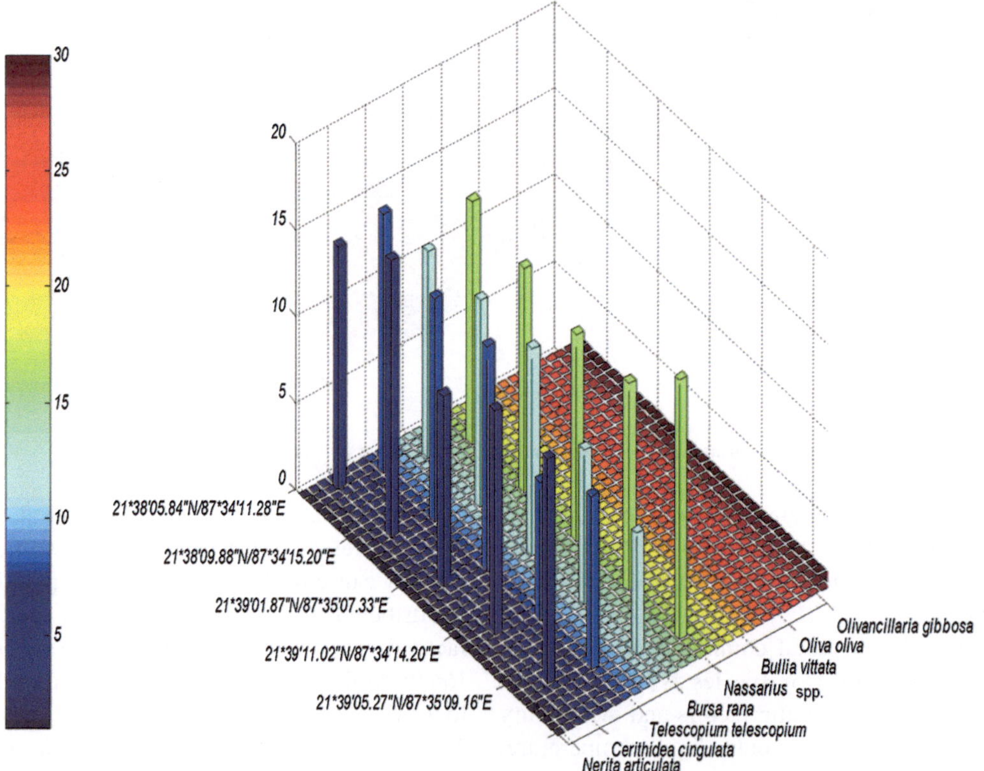

Fig. 11.12 Significant variation of population of gastropod species in five different quadrates in the same area in Shankarpur coast (a site in the East Coast of India)

similar taxa like suspension feeders or detritivores. Although such classification is still in its infancy, but in some mangrove ecosystem, documentation and subsequent grouping of species are frequently done. In mangrove-dominated Indian Sundarbans, the major filter feeders are the oysters (Fig. 11.14), and the detritivores are the gastropods (Fig. 11.15).

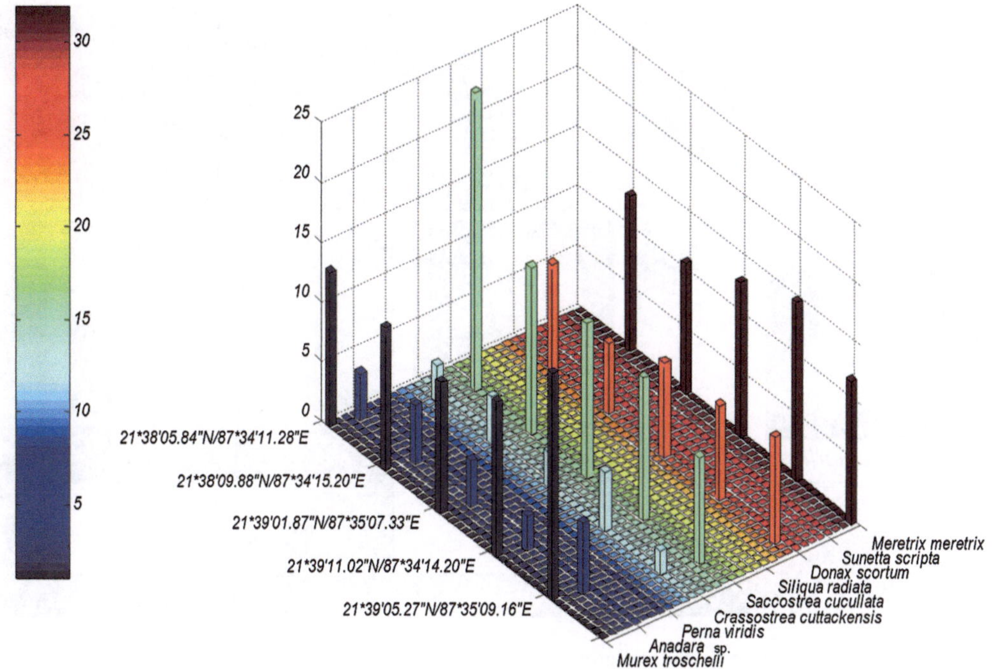

Fig. 11.13 Significant variation of population of bivalve species in five different quadrates in the same area in Shankarpur coast (a site in the East Coast of India)

11.2.5 Phyletic Diversity

In the marine and estuarine ecosystems, there are more animal phyla than on land. Most phyla are encompassed within the benthic compartment. Even within the benthic environment, diversity varies with depth as seen in the case of epifaunal diversity in the north-east Indian shelf (Table 11.5) (Ganesh and Raman 2007).

A similar picture was also observed in the infaunal community, where different depths witness variation in diversity of infaunal species (Table 11.6) (Ganesh and Raman 2007).

The phyletic diversity in the marine ecosystem is very pronounced in the microbial communities. The synthesis of 9.6 million bacterial V6-ribosomal RNA amplicons collected from marine microbial samples exhibit that the pelagic and benthic communities are significantly different at all taxonomic levels. Even in the pelagic compartment, there is significant variation in phyletic diversity between surface and deep water, coastal and open ocean and anoxic and oxic waters. This variation may be attributed to availability of oxygen, light and nutrients (from adjacent land masses, mangrove litter and other coastal vegetation decomposition). The difference in physical mixing also regulates the distribution pattern of marine microbes as benthic communities exhibit more diversity with increasing distance than pelagic communities. The higher phyletic diversity of marine microbial community may be the result of substantially higher densities of bacterial population in sediments, greater temporal stability of habitat (compared to that of the pelagic compartment), higher niche diversity and resource partitioning in the benthic realm.

11.2.6 Species Diversity

The most common usage of diversity is the number of species found in a given area, species diversity. Most ecologists would regard a community comprising of 50 individuals of species A and 50 of species B as more diverse than a community comprising 99 individuals of species

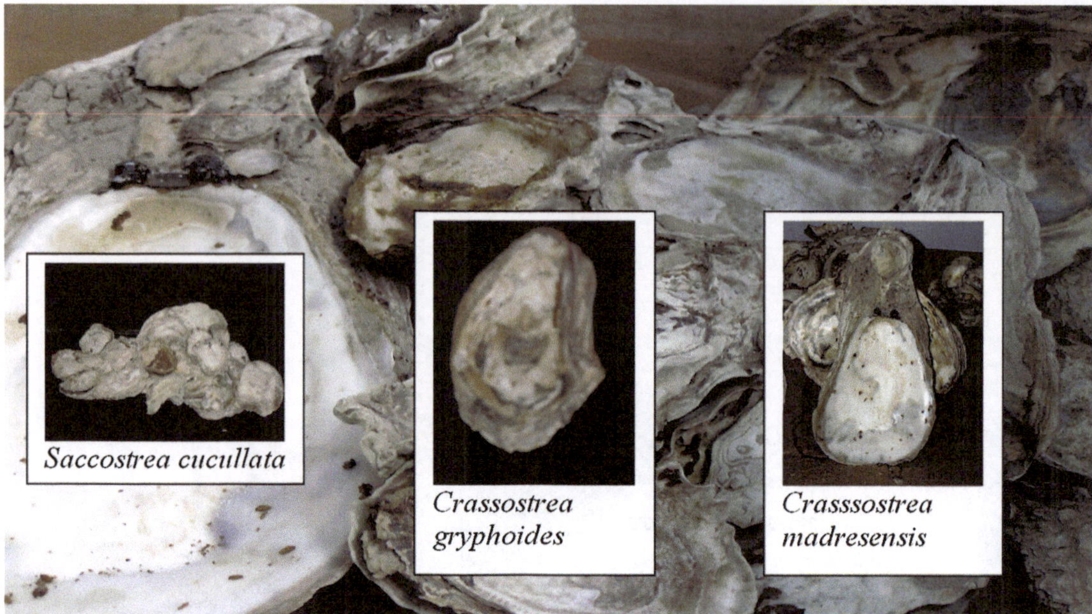

Fig. 11.14 Filter-feeding bivalves in the mangrove ecosystem of Indian Sundarbans

Fig. 11.15 Detritivores (gastropods) in the mangrove ecosystem of Indian Sundarbans

Table 11.5 Distribution of important epifaunal species (ind. haul^{-1}) at different depths on the north-east Indian shelf (Ganesh and Raman 2007)

Species	30–50 m	51–75 m	76–100 m	>100 m
Virgularia sp.	–	–	0.75 ± 0.31 (0–2)	–
Panthalis sp.	2.00 ± 0.92 (1–17)	0.33 ± 0.33 (0–1)	1.5 ± 0.98 (0–8)	–
Eunice indica	6.80 ± 3.48 (1–18)	–	–	–
Diopatra neapolitana	5.74 ± 2.56 (0–35)	–	7.63 ± 5.95 (1–48)	–
Maldanidae sp. 1	1.32 ± 0.74 (1–14)	0.33 ± 0.33 (0–1)	–	–
Maldanidae sp. 2	–	–	2.67 ± 0.88 (0–4)	–
Flabelligera sp.	–	0.33 ± 0.33 (0–1)	0.50 ± 0.38 (1–3)	–
Terebellidae	0.95 ± 0.53 (1–10)	–	0.25 ± 0.25 (1–2)	–
Pista sp.	0.89 ± 0.79 (1–15)	–	–	–
Oratosquilla sp.	–	0.67 ± 0.67 (1–2)	0.63 ± 0.26 (12)	–
Parapenaeus sp.	–	–	1.25 ± 0.86 (1–7)	–
Polychelid lobsters	–	–	3.63 ± 3.48 (0–28)	–
Raninoides serratifrons	0.58 ± 0.27 (1–5)	1.00 ± 0.58 (1–2)	–	–
Calappa lophos	0.21 ± 0.16 (1–3)	–	–	–
Charybdis sp.	6.89 ± 3.70 (1–70)	–	0.38 ± 0.18 (0–1)	–
Liagore sp.	0.16 ± 0.12 (1–2)	1 ± 0 (1)	–	–
Majidae	–	–	1.00 ± 0.68 (1–5)	–
Tibia delicatula	–	–	0.50 ± 0.19 (0–1)	6.00 ± 1.00 (5–7)
Natica vitellus	–	0.33 ± 0.33 (0–1)	0.75 ± 0.41 (1–3)	–
Nassarius variegatus	1.05 ± 0.52 (1–7)	–	–	–
Gemmula speciosa	0.21 ± 0.12 (1–2)	0.33 ± 0.33 (0–1)	–	–
Lophiotoma indica	1	2.00 ± 1.00 (0–3)	–	–
Conus inscriptus	0.89 ± 0.45 (1–8)	0.67 ± 0.33 (0–1)	–	–
Terebra collumelaris	0.32 ± 0.15 (1–2)	–	–	–
Amygdalum watsoni	0.05 ± 0.05 (0–1)	–	–	19.50 ± 9.50 (10–29)
Amphioplus depressus	2.32 ± 0.66 (1–12)	–	1.25 ± 0.37 (1–3)	–
Clypeaster rarispinus	0.84 ± 0.41 (1–7)	0.33 ± 0.33 (0–1)	1.25 ± 1.25 (1–10)	–
Echinodiscus auritus	0.21 ± 0.16 (1–3)	–	–	–
Sipunculida	0.84 ± 0.53 (1–10)	–	–	–
Pleuronectidae	–	0.67 ± 0.67 (1–2)	0.38 ± 0.26 (1–2)	0.50 ± 0.50 (0–1)
Overall abundance	51 ± 10 (6–160)	9 ± 4 (3–15)	37 ± 14 (5–102)	28 ± 13 (15–40)

Note: Data presented as mean ± SD (range). –: not found in this depth range

A and 1 individual of B. Thus, in addition to the number of species in a given area, diversity indices have been proposed that take into account the distribution of individuals among species (Magurran 1988). The number of species currently described on Earth is between 1.4 and 1.7 million (Stork 1988), but the Global Biodiversity Assessment suggests a conservative estimate of

Table 11.6 Distribution of important infaunal species (ind. m^{-2}) at different depths on the north-east Indian shelf (Ganesh and Raman 2007)

Species	30–75 m	76–100 m	>100 m
Chloeia rosea	39.7 ± 17.2 (10–640)	36.2 ± 11.0 (10–130)	1.7 ± 1.7 (10–20)
Ancistrosyllis parva	13.0 ± 2.7 (10–60)	73.1 ± 30.8 (10–390)	4.2 ± 2.3 (10–20)
Nereis sp.	10.5 ± 3.8 (10–100)	6.2 ± 2.1 (0–20)	0.8 ± 0.8 (0–10)
Nephtys sp.	45.4 ± 6.9 (10–210)	125.4 ± 312.0 (0–420)	1.7 ± 1.1 (0–10)
Lumbrineris sp. 1	14.6 ± 2.9 (10–60)	3.1 ± 1.3 (0–10)	–
Prionospio sp. 2	9.7 ± 3.9 (10–130)	7.7 ± 2.3 (10–20)	86.7 ± 43.5 (10–530)
Spionidae sp. 1	17.0 ± 5.8 (10–200)	0.7 ± 0.7 (0–10)	0.8 ± 0.8 (0–10)
Spionidae sp. 3	7.3 ± 2.6 (10–60)	49.2 ± 17.3 (0–200)	
Magelona sp.	19.2 ± 4.9 (0–140)	13.9 ± 7.7 (10–100)	–
Cirratulidae sp. 1	43 ± 8.6 (10–250)	53.1 ± 15.5 (10–150)	4.2 ± 3.4 (10–40)
Cossura coasta	10.8 ± 2.4 (10–70)	44.6 ± 18.4 (10–230)	174.2 ± 69.1 (10–700)
Mediomastus sp.	19.2 ± 8.2 (10–260)	2.3 ± 1.7 (10–20)	–
Notomastus sp.	16.8 ± 4.3 (10–100)	4.6 ± 4.6 (10–60)	–
Ampelisca sp.	211.1 ± 43.1 (10–1390)	–	–
Maera sp.	25.7 ± 5.5 (10–140)	1.5 ± 1.5 (10–20)	–
Corophium sp.	1.4 ± 1.4 (10–50)	22.3 ± 5.0 (10–70)	70.8 ± 49.0 (10–560)
Nemertea	10.5 ± 1.9 (10–40)	6.2 ± 3.1 (10–30)	8.3 ± 3.2 (10–40)
Overall abundance	808 ± 97 (230–2460)	586 ± 82 (290–1140)	379 ± 124 (30–1280)

Note: Data presented as mean ± SE (range). –: not found in this depth range

1.75 million (Heywood and Watson 1995). However, this figure does not include microbial species. Little is known about microbial diversity in general. New genetic techniques will change this. For example, Giovannoni et al. (1990) using ribosomal RNA techniques found a completely novel group of bacteria in the Sargasso Sea. On land, there are more species known than in the sea. This is due largely to the extraordinary diversity of beetles (Coleoptera); 400,000 species are described (Heywood and Watson 1995). Recently, in a highly controversial paper, Grassle and Maciolek (1992) have suggested that there may be ten million undescribed species in the deep sea. Briggs (1991) and May (1992) disagree with the methods used, and May suggests that a more realistic estimate may be around 500,000 undescribed deep-sea species. Nevertheless, even this lower figure would be a substantial increase in the approximate figure of 300,000 known marine species (Grassle 1991). Over geological time there has been a large change in the ratios of orders of families to genera to species (Briggs 1994). A rapid increase occurred in higher taxa (orders and families) until the Ordovician when diversity levelled off. In the Permian, some 50 % of marine families became extinct (Raup 1979; Sepkoski 1979, 1984, 1991). The number of species has increased enormously in recent geological time more than doubling compared to those present 100 million years ago (Signor 1994). Most of the marine species diversity is benthic rather than pelagic (Angel 1993). This is a consequence of the fact that the marine fauna originated in benthic sediments. The pelagic realm has an enormous volume compared with the inhabitable part of the benthic realm. Yet there are only 3500–4500 species of phytoplankton (Sournia and Chretiennot-Dinet 1991) compared with the 250,000 species of flowering plants on land. Angel estimates that there are probably only 1200 oceanic fish species against 13,000 coastal species. In the pelagic realm, diversity is higher in coastal rather than oceanic areas (Angel 1993), and, therefore, efforts should be concentrated in coastal areas. Another highly important aspect of species diversity is endemism (i.e. the species occurring in a restricted locality). The Antarctic has a higher degree of endemism than the Arctic. In the Red Sea 90 % of some groups of fishes are endemic. Overall, however, only 17 % of Red

Sea fishes are endemic (Shepherd 1994). In a survey of 799 pantropical fish species, Roberts et al. (in Shepherd 1994) showed that 17 % occupied only one grid square (223 × 223 km). In the Indian Ocean, of the 482 coral species recorded, 27 % occur only at one site (Shepherd 1994), and of the 1200 species of echinoderms found at 16 sites, 47 % occurred at only one site (Clark and Rowe 1971). High degrees of endemicity pose problems for development of conservation strategies. Because of their extreme non-elastic nature, they need the variables (environmental parameters needed for conservation) within a very narrow range. However, species with a wide range of tolerance to environmental parameters can be easily conserved even in places far away from their original habitat. The example of *Ipomoea pes-caprae* is very pertinent here. *Ipomoea pes-caprae* commonly known as goat's foot is a trailing herb found abundantly in the sand-dominated intertidal mudflats of the coastal region of India. The herb has long stems, sometimes twining and reaching up to 30 m in length. The leaves are simple, ovate, quadrangular and rounded, measuring 2.5 cm by 10 cm with slender petioles that can be as long as 17 cm (Fig. 11.16).

A unique experiment was conducted during 2014 with this coastal herb under the joint research collaboration between KPO of Tata Steel (Odisha) and Techno India University (Salt Lake, Kolkata) in terms of its conservation. Artificial sandy environment was created in the Tata Steel KPO complex of Odisha (where laterite soil is predominant) to grow this coastal vegetation, and the result was cent percent successful. The species not only survived and exhibited growth (Fig. 11.17), but the leaf chlorophyll (total) was almost similar when compared to that collected from the Sundarban estuarine region.

This experiment is an eye-opener of species-level conservation through creation of artificial environment and is now widely used in Environment Management Plan (EMP) of any developmental activities.

11.2.7 Genetic Diversity

The most basic level of biological diversity is that found within a species and is known as genetic diversity. Genetic diversity encompasses the variation among individuals within a population in their genetic make-up and the genetic variation among populations (Begon et al. 1990). Each species consists of one or more populations of individuals. A population is usually defined as a group of individuals that can interbreed and, if sexually reproducing, can interchange genetic material. Different populations tend to diverge genetically due to their having limited genetic mixing or mutations, natural selection, genetic drift and the accumulation of selectively neutral mutations. Thus, there are genetic differences both among individuals and among populations. Populations with higher genetic diversity are more likely to have some individuals that can withstand environmental change and thereby pass on their genes to the next generation (Nevo et al. 1987). On an evolutionary timescale, (over many generations) genetic diversity is higher in species which characterize unstable, stressed environments when compared with counterparts from more stable environments (Nevo et al. 1984). However, on an ecological timescale (few generations), stress reduces genetic diversity. Gillespie and Guttman (1988) showed that long-term exposure to contaminants decreased genetic diversity, and the remaining population was more vulnerable to extinction. Alberte et al. (1994) have shown that stressed eelgrass has lower genetic diversity than non-stressed populations. Commercial fishing, concentrating on specific size ranges, has significantly altered the genetic composition of populations (Elliott and Ward 1992). In general, marine species have higher genetic diversity than freshwater and terrestrial species. In a comparative study of fish, Ward et al. (1994) showed that average heterozygosity was similar in marine and freshwater species subpopulations, but was considerably less in freshwater species. High genetic diversity is found in marine algae (Wood 1989), and *Pinctada margaritifera* is an exploited

Fig. 11.16 *Ipomoea pes-caprae*—a trailing herb found abundantly in the Indian coasts

Fig. 11.17 Growth of *Ipomoea pes-caprae* in the artificial substratum of KPO of Tata Steel (Odisha) complex

tropical bivalve (Durand and Blanc 1988). Elliott and Ward (1992) found that a minimum of only 200 migrants per year were enough to maintain the genetic diversity of the orange roughy (*Hoplostethus atlanticus*) which suggests that marine populations probably exchange between 10 and 100 times more migrants per generation than freshwater species. Not all marine populations have high numbers, and Scudder (1989) argues that for marginal populations the best way to maintain genetic (and species) diversity is by 'marginal habitat conservation'. This is an alternative strategy to the conservation of high-biodiversity 'hot spots' advocated by some. Much work has been done on the genetics of species used in aquaculture: on clams (Bushek and Allen 1989), Manila clam (Mattoccia et al. 1991), oysters (Blanc and Jaziri 1990; Hedgecock and Sly 1990; Hedgecock et al. 1991; Jaziri et al. 1987; Sly and Hedgecock 1989), penaeid shrimps (Qiu 1991; Benzie et al. 1992), salmonids (Gall et al. 1992) and the orange roughy (Elliott and Ward 1992). Doyle et al. (1991) have reviewed genetic aspects of aquaculture and conclude that current practices lead to reductions in genetic diversity and maintenance of many breeds and meta-populations of marine species is needed (see also reviews by Cataudella and Crosetti 1993; Blanc and Bonhomme 1987). Grassle et al. (1991) argues that a considerable proportion of the genetic diversity of the planet is probably found in deep-sea organisms and recommends genetic studies of hydrothermal vent fauna which are naturally tolerant of high concentrations of toxic elements produced by the vents.

Brain Churners

1. 'A' is an island where developmental activities occurred in the early 1990s. 'B' is another island in the same geographical locale, which is yet uninhabited. State with reason which island needs to be preserved.
2. How does alternative livelihood regulate the conservation approach in a fragile ecosystem?
3. State few alternative livelihoods that can upgrade the economic profile of coastal population.
4. Which of the two terms 'preservation' and 'conservation' should be given priority in environmental protection? Explain your answer with two reasons.
5. State four steps that must be taken to conserve *Heritiera fomes* (an endangered true mangrove flora) in an estuarine system?
6. How can an island be reclaimed from the effects of sea level rise?
7. Would you like to coin the term 'ecosystem gift' instead of 'ecosystem service'? Justify your answer with proper logic.
8. Would you encourage the participation of local people in the process of conservation? Why?
9. What is ecorestoration? Do you expect the Shannon–Wiener index value to change after successful ecorestoration?
10. Lagoons '1' and '2' have Shannon–Wiener index values 3.025 and 4.123, respectively. Which environment is more congenial and why?

Annexure 11A: International Legal Framework to Safeguard Marine and Estuarine Ecosystems

1. 1982 UNCLOS

The 1982 United Nations Convention on the Law of the Sea (UNCLOS) was adopted in 1982 and enters into force in 1994. UNCLOS established the types of maritime zones that can be claimed by coastal states, as well as the rights, jurisdiction and obligations of coastal states in the various zones. It gave coastal states the right to claim the territorial sea of 12 nautical miles and an exclusive economic zone of 200 nautical miles. The territorial sea is under the sovereignty of the coastal states, but such sovereignty is subject to the right of ships of all states to pass through the territorial sea. The coastal states have the right to pass laws and regulations to protect and preserve the marine and coastal environment within its territorial sea, subject to restrictions on its right to regulate ships exercising passage rights through its waters. The 200-nautical-mile EEZ is not under the sovereignty of the coastal states. It is a 'specific legal regime' in which the coastal state has the right to explore and exploit the living and nonliving resources. The coastal state has jurisdiction to pass laws and regulations only in so far as they are connected to its right to explore and exploit the living and nonliving resources. Other states have the right to exercise traditional high seas freedoms in the EEZ such as freedom of navigation and freedom of overflight. The right of coastal states to pass laws to regulate pollution from ships of other states in the EEZ is severely restricted.

UNCLOS has several important provisions in context to conservation of marine and coastal habitats and ecosystems. First, it provides that states have a general obligation to protect and preserve the marine environment. Second, it provides that states have an obligation to take all measures as necessary to prevent pollution of the marine environment from any source, using for this purpose the best practicable means at their disposal and in accordance with their capabilities. Third, it provides in Article 194 (5) that such measures 'Shall include those necessary to protect and preserve rare or fragile ecosystem as well as the habitat of depleted, threatened or endangered species and other forms of marine life'.

2. Agenda 21, Chapter 17 (1992)

Chapter 17 of Agenda 21, the programme of action adopted by states at the 1992 Earth Summit in Rio de Janeiro, was a significant advance in several respects. Firstly, it made clear that currently the management of the marine and coastal resources has not always proved capable of achieving sustainable development and that coastal resources and the coastal environment are being rapidly degraded and eroded in many parts of the world. Secondly, it suggested an approach for establishing a situation of governance or integrated management. It provided that each coastal state should consider establishing, or where necessary strengthening, appropriate coordinating mechanisms, for integrated management and sustainable development of coastal and marine areas and their resources, at both the local and national levels. It also stated that states should cooperate, as appropriate, in the preparation of national guidelines for integrated coastal zone management and development, drawing on existing experience. Thirdly, it provided that coastal states undertake measures to maintain biological diversity and productivity of marine species and habitats under national jurisdiction. These measures might include surveys of marine biodiversity, inventories of endangered and management of protected areas and support of scientific research and dissemination of its results.

3. 1992 Biodiversity Convention

The 1992 Convention on Biological Diversity (Biodiversity Convention) was adopted at the Earth Summit in Rio de Janeiro in 1992. The objectives of the Biodiversity Convention are the conservation of biological diversity, the sustainable use of its components and the fair and equitable sharing of the benefits arising out of the utilization of genetic resources. The Biodiversity Convention is the first global, legally binding comprehensive agreement to address all aspects of biological diversity: genetic resources, species and ecosystems. It establishes a proactive, holistic, cross-sectoral and ecosystem-based approach to conserve and sustain using their biological resources.

The Biodiversity Convention establishes a framework of genera, obligations that Parties are obliged to elaborate in more detail at the national level. Parties are required to develop and implement comprehensive national biodiversity strategies and action plans. Parties are also required to identify research and monitor their biodiversity, to establish protected areas, to regulate or manage activities with significant adverse effects on biodiversity and to conduct assessments of the biodiversity impacts of proposed projects. The convention also requires Parties to take special measures to protect customary required uses and the knowledge, innovation and practices of local and indigenous communities.

4. The Jakarta Mandate (1995)

The Biodiversity Convention does not contain any specific provisions on the conservation of marine and coastal biodiversity. At their second Conference of Parties (COP) in Jakarta in 1995, the Parties agreed on a programme of action for implementing the Biodiversity Convention with respect to marine and coastal biodiversity. This programme is referred to as the Jakarta Mandate of Marine and Coastal Biodiversity or 'The Jakarta Mandate'.

The Jakarta Mandate recommended a 'checklist' of actions that Parties should take to fulfil their obligations under the Biodiversity Convention in marine and coastal environments. The recommendation cover, in particular, five thematic areas: integrated marine and coastal area management, mariculture, alien species, living marine resources and marine protected areas. The two areas that are of the most relevance to the conservation of marine and coastal habitats are integrated marine and coastal area management (IAM) and marine protected areas (MPAs).

5. 1971 Ramsar Convention on Wetlands

The 1971 Convention on Wetlands is of international importance, particularly for waterfowl. The convention was conducted in Ramsar, Iran, on 2 February 1971. It is popularly referred to as the Ramsar Convention. The Ramsar Convention entered into force in 1975. More than 1000 wetlands have been designated for inclusion in the List of Wetlands of International Importance (Ramsar Sites), covering some 73 million hectares.

The main decision-making body of the Ramsar Convention is the Conference of the Parties, which is comprised of delegates from all the member states. UNESCO serves as Depositary for the Convention, but its administration has been entrusted to the secretariat known as the 'Ramsar Bureau', which is housed in the headquarters of IUCN (1980), the World Conservation Union in Gland, Switzerland. The Ramsar Secretariat (2001) acts under the authority of the Conference of the Parties and the Standing Committee of the Convention.

6. 1972 World Heritage Convention

The Convention concerning the protection of the world cultural and natural heritage is the UNESCO Convention that was adopted by the General Conference of UNESCO in 1972. It is popularly known as the World Heritage Convention. This Convention covers both cultural heritage and natural heritage sites of outstanding value. Sites which can be designated as natural heritage sites under the Convention include sites with outstanding physical, biological and geological features, habitats of threatened plants or animal species and areas of value on scientific or aesthetic rounds or from the point of view of conservation. The criteria for selecting natural heritage sites require that they be of sufficient size and contain the necessary elements to ensure the integrity of ongoing ecological and biological processes. For example, the criteria may require that of coral reef designation should include seagrass, mangrove or other adjacent ecosystems. An international committee of government representatives selects the sites with the consent of the states concerned.

The Convention establishes a fund for conservation of the cultural and natural sites on the list of the World Heritage Sites. The fund can be used for various kinds of aid and technical cooperation, including the conduct of expert studies to plan conservation measures, the training of local specialists in conservation or renovation measures, the training of local specialists in conservation or renovation techniques and the supplying of equipment for the protection of the natural park.

7. UNESCO Man and the Biosphere Programme

The Man and the Biosphere (MAB) Programme is not based on a legally binding treaty or convention. It is an interdisciplinary programme of research and training intended to develop the basis, within the natural and the social sciences, for the rational use and conservation of the resources between people and the environment. The overall programme is guided by the MBA International Coordinating Council consisting of 34 member states elected by the UNESCO General Conference. Programme activities are conducted in more than 100 countries under the direction of their MBA National Committees or focal points.

Global Programme of Action for the Protection of the Marine Environment from Land-Based Activities, 1995 (GPA)

The Global Programme of Action for the Protection of the Marine Environment from Land-

Based Activities (GPA) was adopted at a global intergovernmental conference in Washington in 1995. The United Nations Environment Programme (UNEP) was tasked to lead the coordination effort and to establish a GPA Coordination Office. The comprehensive, multisectoral approach of the GPA reflects the desire of governments to strengthen the collaboration and coordination of all agencies with mandates relevant to the impact of land-based activities on the marine environment.

The GPA is designed to be a source of conceptual and practical guidance to be drawn upon by national and/or regional authorities for devising and implementing sustained action to prevent, reduce, control and/or eliminate degradation from land-based activities. One of the issues addressed by the GPA is physical alterations and destruction of habitat. The objectives of the GPA on this issue are to:

(a) Safeguard the ecosystem and maintain the integrity and biological diversity of habitats which are of major socio-economic and ecological interest through integrated management of coastal areas.
(b) Where practicable, restore marine and coastal habitats that have been adversely affected by anthropogenic activities.

8. FAO Code of Conduct on Responsible Fisheries

The Code of Conduct for Responsible Fisheries was unanimously adopted on 31 October 1995 by the FAO Conference. The code provides a necessary framework for national and international efforts to ensure sustainable exploitation of aquatic living resources in harmony with the environment and in a manner that is consistent with applicable global instruments. It is global in scope and is directed towards members and non-members of FAO; fishing entities' subregional and global organizations, whether governmental or non-governmental; and all persons concerned with the conservation of fishery resources and management and development of fisheries; it provides principles and standards applicable to the conservation, management and development of all fisheries.

The General Principles stated in the Code include two which are of particular relevance to the conservation of marine and coastal ecosystems and habitats.

All critical fishery habitats in marine and freshwater ecosystems such as wetland, mangroves, reefs, lagoons, nurseries and spawning areas should be protected and rehabilitated as far as possible and where necessary (FAO 2003).

Particular effort should be made to protect such habitats from destruction, degradation, pollution and other significant activities that threaten the health and viability of the fishery resources.

9. Convention of International Maritime Organization

The Convention of International Maritime Organization (IMO) gives coastal states an opportunity to establish regimes in certain vulnerable coastal areas to reduce the oil pollution from ships. Under the IMO conventions, there are two types of areas that can be established where special rules will apply to ships. Both types are established by the IMO after the submission of scientific and technical evidence by the coastal state.

The first is 'Special Areas' under the International Convention on the Prevention of Pollution from Ships, or MARPOL 73/78. A Special Area under MARPOL 73/78 is defined as follows:

> A sea area where, for recognized technical reasons in relation to its oceanographic and ecological condition and to the particular character of its traffic, the adoption of special mandatory methods for the prevention of sea pollution by oil, noxious liquid substances, or garbage, as applicable, is required.

Special Areas provide with a higher level of protection from such discharges from ships than other areas of the sea. Special Areas can be established by the IMO; it can be shown by scientific and technical evidence that they are vulnerable to pollution by discharges from ships and that special mandatory measure of protection is required. Several areas, such as the Black Sea, Mediterranean Sea, Baltic Sea and the Wider Caribbean Region, have been designated by the IMO as Special Areas.

The second areas where special IMO rules can apply to shops are Particularly Sensitive Sea Areas, or PSSAs. A PSSA is 'an area that needs special protection through IMO because of its significance for recognized ecological or socio-economic or scientific reasons and which may be vulnerable to damage by international maritime activities'. If an area is designated as a PSSA, 'associated protective measures' may be ordered by the IMO. Associated protective measures are international rules and standards that are within the purview of the IMO which regulate international maritime activities for the protection of the area at risk. Examples of such measures are discharge restrictions, routing measures, operational criteria and prohibited activities. Such measures must be specially tailored to meet the needs of the particular PSSA. To date, two areas have been designated by the IMO as PSSAs. They are the Great Barrier Reef Marine Park in Australia and the Sabana-Camagüey Archipelago in Cuba.

Special Areas and PSSAs are important tools for marine protected areas that are located beyond 12 nautical miles from the coast in the EEZ of the coastal states.

References

Alberte, R. S., Suba, G. K., Procaccini, G., Zimmerman, R. C., & Fain, S. R. (1994). Assessment of genetic diversity of seagrass populations using DNA fingerprinting: Implications for population stability and management. *Proceedings of National Academy Science of the United States of America, 91*, 1049–1053.

Angel, M. V. (1993). Biodiversity of the pelagic ocean. *Conservation Biology, 7*, 760–772.

Anon. (1993). *World Population Prospects: The 1992 revision*. New York: United Nations. ASEAN-Australia Marine Science Project (1992 J ASEAN Marine Science).

Banerjee, K., Sengupta, K., Raha, A., & Mitra., A. (2013). Salinity based allometric equations for biomass estimation of Sundarban mangroves. *Biomass & Bioenergy*, (ELSEVIER), 56, 382–391.

Begon, M., Harper, J. L., & Townsend, C. R. (1990). *Ecology, individuals, population and communities* (Reprint ed. 1995, pp. 1–929). Boston: Blackwell Science.

Benzie, J. A. H., Frusher, S., & Ballment, E. (1992). Geographical variation in allozyme frequencies of populations of *Penaeus monodon* (Crustacea: Decapoda) in Australia. *Australian Journal of Marine Freshwater Research, 43*, 715–725.

Blanc, F., & Bonhomme, F. (1987). Generic polymorphism in natural molluscs populations of aquacultural interest. In K. Tiews (Ed.), *Selection, hybridization and genetic engineering in aquaculture* (Vol. 1, pp. 59–78). Rome: FAO European Inland Fisheries Advisory Committee, 18/19.

Blanc, F., & Jaziri, H. (1990). Variation of allozymic polymorphism in *Ostrea angasi* and *O. edulis*. In T. Gjedrem (Ed.), *Genetics in aquaculture* (pp. 331–332). International Maritime Organization, 01 Jan 1997 – Marine Pollution – 24 pages.

Briggs, J. C. (1991). Global species diversity. *Journal of Natural History, 25*, 1403–1406.

Briggs, J. C. (1994). Species diversity: Land and sea compared. *System Biology, 43*, 130–135. Marine biodiversity, 169.

Bushek, D., & Allen, S. K. (1989). Effective population size for shellfish brood stock management: Conflicts between theory and practice. *Journal of Shellfish Research, 8*, 446–447.

Cairns, J., & Dickson, K. L. (1971). A simple method for the biological assessment of the effects of waste discharges on aquatic bottom-dwelling organisms. *Journal of Water Pollution Control Federation, 43* (5), 755–772.

Cataudella, S., & Crosetti, D. (1993). Aquaculture and conservation of genetic diversity. In R. S. V. Pullin, H. Rosenthal, & J. L. Maclean (Eds.), *Environment and aquaculture in developing countries* (Vol. 31, pp. 60–73). Manila: International Cent for Living Aquatic Resources Management (ICLARM).

Chaudhuri, A. B., & Choudhury, A. (1994). *Mangroves of the Sundarbans*. Vol. 1: India. Bangkok: International Union for Conservation of Nature and Natural Resources (IUCN).

Clark, A. M., & Rowe, F. W. E. (1971). *Monograph of shallow water Indo-West Pacific echinoderms* (p. 238). London: British Museum (Natural History).

Cody, M. L. (1986). Diversity, rarity, and conservation in Mediterranean-climate regions. In M. E. SouleÂ (Ed.), *Conservation biology: The science of scarcity and diversity* (pp. 122–152). Sunderland: Sinauer.

Di Castri, F., & Younes, T. (1990). Ecosystem function of biological diversity. *Biology International Special Issue, 22*, 1–20.

Doyle, R. W., Shackell, N. L., Basiao, Z., Urawain, S., Matricia, T., & Talbot, A. J. (1991). Selective diversification of aquaculture stocks: A proposal for economically sustainable genetic conservation. In N. Billington & P. D. N. Herbert (Eds.), *Ecological and genetic implications of fish introductions* (pp. 148–154). Ottawa: Department of Fisheries and Oceans/Canadian Journal of Fisheries and Aquatic Sciences.

Durand, P., & Blanc, F. (1988). Genetic diversity in a tropical marine bivalve: *Pinctada margaritifera* (Linne, 1758). *Bulletin of Social Zoology, France, 113*, 293–304.

Elliott, N. G., & Ward, R. D. (1992). Enzyme variation in orange roughy, *Hoplostethus atlanticus* (Teleostei: Trachichthyidae), from southern Australian and New Zealand waters. *Australian Journal of Marine Freshwater Research, 43*, 1561–1571.

Fisher, R. A., Corbet, A. S., & Williams, C. B. (1943). The relationship between the number of species and the number of individuals in a random sample of an animal population. *Journal of Animal Ecology, 12*, 42–58.

Franklin, J. (1993). Preserving biodiversity: Species, ecosystems or landscapes? *Ecology Application, 3*, 202–205.

Gall, G. A. E., Bartley, D., Bentley, B., Brodziak, J., Gomulkiewicz, R., & Mangel, M. (1992). Geographic variation in population genetic structure of Chinook salmon from California and Oregon. *Fishery Bulletin, 90*, 77–100.

Ganesh, T., & Raman, A. V. (2007). Macrobenthic community structure of the northeast Indian shelf, Bay of Bengal. *Marine Ecology Progress Series, 341*, 59–73.

Gillespie, R. B., & Guttman, S. I. (1988). Relationships between genetic structure of electrophoretically-determined allozymes in fish populations and exposure to contaminants. In L. Meserve (Ed.), *97th annual meeting: The Ohio academy of science* (Vol. 88, p. 46). Aquatic Sciences and Fisheries Abstracts, Volume 19.

Giovannoni, S. J., Britschgi, T. B., Moyer, C. L., & Field, K. G. (1990). Genetic diversity in Sargasso Sea bacterioplankton. *Nature, 345*, 60–63.

Grassle, J. F. (1991). Deep-sea benthic biodiversity. *Bioscience, 41*, 464–469.

Grassle, J. F., & Maciolek, N. J. (1992). Deep-sea species richness: Regional and local diversity estimates from quantitative bottom samples. *American Nature, 139*, 313–341.

Grassle, J. F., Lasserre, P., McIntyre, A. D., & Ray, G. C. (1991). Marine biodiversity and ecosystem function. *Biology International, 23*, 1–19.

Gray, J. S. (1994). Is the deep sea really so diverse? Species diversity from the Norwegian continental shelf. *Marine Ecology Progress Series, 112*, 205–209.

Hedgecock, D., & Sly, F. (1990). Genetic drift and effective population sizes of hatchery-propagated stocks of the Pacific oyster, *Crassostrea gigas. Aquaculture, 88*, 21–38.

Hedgecock, D., Nelson, K., & Banks, M. A. (1991). Does variance in reproductive success limit effective population sizes of marine organisms? A proposed test in the Dabob Bay population of Pacific oysters, using enzymatic amplification of mitochondrial DNA. *Journal of Shellfish Research, 10*, 234–237.

Heywood, V. H., & Watson, R. T. (Eds.). (1995). *Global biodiversity assessment* (p. 1140). Cambridge: Cambridge University Press.

Huston, M. A. (1985). Patterns of species diversity in relation to depth at Discovery Bay, Jamaica. *Bulletin of Marine Science, 37*, 928–935.

Huston, M. (1994). *Biological diversity: The coexistence of species in changing landscapes*. Cambridge: Cambridge University Press.

IUCN. (1980). *World conservation strategy: Living resource conservation for sustainable development.* Gland: IUCN-UNEP-WWF. 44 pp.

Jaziri, H., Durand, P., & Pichot, P. (1987). Genetic diversity between and within populations of the European Oyster, *Ostrea edulis*. Selection, Hybridization and Genetic Engineering. In K.Tiews, (Ed.), *Aquaculture* (Vol. 1, pp. 177–187). Rome: FAO European Inland Fisheries Advisory Committee, 18/19.

Loya, Y. (1972). Community structure and species diversity of hermatypic corals at Eilat, Red Sea. *Marine Biology, 13*, 100–123.

MacNae, W. (1968). A general account of the fauna and flora of mangrove swamps and forests of the Indo-West-Pacific region. *Advance Marine Biology, 6*, 73–270.

Magurran, A. E. (1988). *Ecological diversity and its measurement*. Croom Helm: Princeton University Press.

Margalef, R. (1968). *Perspectives in ecological theory*. Chicago: University of Chicago Press. 111 p.

Mattoccia, M., Cesaroni, D., & Matarazzo, P. (1991). Genetic changes in the Manila clam, *Tapes philippinarum*: Temporal and spatial variation. In N. Depauw, & J. Joyce (Eds.), *Aquaculture and the Environment* (pp. 210–211). Aquatic Sciences and Fisheries Abstracts, Volume 24.

May, R. M. (1992). Bottoms up for the oceans. *Nature, 357*, 278–279.

McRoy, P. C., & Lloyd, D. S. (1981). Comparative function and stability of macrophyte-based ecosystems. In A. R. Longhurst (Ed.), *Analysis of marine ecosystems* (pp. 473–490). New York: Academic.

Mills, E. L. (1969). The community concept in marine Zoology, with comments on continua and instability in some marine communities: A review. *Journal Fishery Research Board, Canada, 26*, 1415–1428.

Mitra, A., Sengupta, K., & Banerjee, K. (2011). Standing biomass and carbon storage of above-ground structures in dominant mangrove trees in the Sundarbans. *Forest*

Ecology and Management, 261(7), 1325–1335. doi:10.1016/j.foreco.2011.01.012. Elsevier.

Naeem, S., Thompson, L. J., Lawler, S. P., Lawton, J. H., & Woodfin, R. M. (1994). Declining biodiversity can alter the performance of ecosystems. *Nature, 368*, 734–737.

Nevo, E., Beiles, A., & Ben-Shlomo, R. (1984). The evolutionary significance of genetic diversity: Ecological, demographic and life history correlates. In G. S. Mani (Ed.), *Evolutionary dynamics of genetic diversity* (Lecture notes in biomathematics, Vol. 53, pp. 13–213). Berlin/Heidelberg: Springer.

Nevo, E., Noy, R., & Lavie, B. (1987). Levels of genetic diversity and resistance to pollution in marine organisms. *FAO/UNEP meeting on the effects of pollution on marine ecosystems* (pp. 175–182). Blanes, October 7–11, 1985.

Plafkin, J. L., Barber, M. T., Poter, K. D., Gross, S. K., & Highes, R. M. (1989). *Rapid bioassessment protocol for use in stream and rivers: Benthic macro invertebrates and fish.* U. S. Environmental Protection Agency, Office of Water, Washington, DC.

Qiu, Y. (1991). Utility of protein and mtDNA markers in penaeid shrimp: A review. *Aquaculture Environment, 14*, 270.

Raha, A. K., Mishra, A., Das, S., Zaman, S., Ghatak, S., Bhattacharjee, S., Raha, S., & Mitra, A. (2014). Time Series Analysis of forest and tree cover of West Bengal from 1988 to 2010, using RS/GIS, for monitoring afforestation programmes. *The Journal of Ecology (Photon), 108*, 255–265.

Raup, D. M. (1979). Size of the Permo-Triassic bottleneck and its evolutionary implications. *Science, 206*, 217–218.

Ray, G. C. (1991). Coastal-zone biodiversity patterns. *Bioscience, 41*, 490–498.

Ray, G. C., & Grassle, J. F. (1991). Marine biological diversity. *Bioscience, 41*, 453–461.

Ray, G. C., & Gregg, W. P., Jr. (1991). Establishing biosphere reserves for coastal barrier ecosystems. *Bioscience, 41*, 301–309.

Scudder, G. G. E. (1989). The adaptive significance of marginal populations: A general perspective. In C. D. Levings, L. B. Holtby, & M. A. Henderson (Eds.), *Proceedings of the national workshop on effects of habitat alteration on Salmonid stocks* (pp. 180–185). Ottawa: Department of Fisheries and Oceans/Canadian Special Publication of Fisheries and Aquatic Sciences.

Sengupta, K., Roy Chowdhury, M., Bhattacharyya, S. B., Raha, A., Zaman, S., & Mitra, A. (2013). Spatial variation of stored carbon in *Avicennia alba* of Indian Sundarbans. *Discovery Nature, 3*(8), 19–24. ISSN: 2319–5703.

Sepkoski, J. J. J. (1979). A kinetic model of Phanerozoic taxonomic diversity. II. Early Phanerozoic families and multiple equilibria. *Palaeobiology, 5*, 222–251.

Sepkoski, J. J. J. (1984). A kinetic model of Phanerozoic taxonomic diversity. III. Post-Palaeozoic families and mass extinctions. *Palaeobiology, 10*, 246–267.

Sepkoski, J. J. J. (1991). Biodiversity and Conservation: Causes and consequences of biodiversity loss 2 : pollution, climate change and unsustainable exploitation, Volume 3. In E. C. Dudley (Ed.), *The unity of evolutionary biology* (pp. 210–236). Poraand: Diosorides Press.

Shannon, C. E., & Weiner, W. (1949). A mathematical theory of communication. *Bell System Technical Journal, 27*, 379–423.

Shepherd, P. A. (1994). A review of plant communities of derelict land in the city of Nottingham, England and their value for nature conservation. *Memorabilia Zoologica, 49*, 129–137.

Sheppard, C. R. C. (1980). Coral cover, zonation and diversity on reef slopes of Chagos Atolls, and population structures of major species. *Marine Ecology Progress Series, 2*, 193–205.

Signor, P. W. (1994). Biodiversity in geological time. *American Zoology, 34*, 23–32.

Sly, F., & Hedgecock, D. (1989). Genetic drift and effective population sizes in commercial stocks of the Pacific Oyster, *Crassostrea gigas*, on the U.S. West Coast. *Journal of Shellfish Research, 8*, 324–334.

Solbrig, O. T. (Ed.). (1991). *IUBS/SCOPE/UNESCO from genes to ecosystems: A research agenda for biodiversity.* Cambridge: IUBS.

Sournia, A., & Chretiennot-Dinet, G. (1991). Marine phytoplankton: How many species in the world ocean? *Journal of Plankton Research, 13*, 1093–1039.

Southwood, T. R. E. (1977). Habitat, the template for ecological strategies? *Journal of Animal Ecology, 30*, 1–8.

Staub, R. H., & Hass, I. J. (1970). The effect of industrial effluents of Memphis and Shelby country on primary plankton production. *Bioscience, 20*, 905–912.

Steele, J. H. (1991). Marine functional diversity. *Bioscience, 41*, 470–474.

Stork, N. (1988). Insect diversity: Facts, fiction and speculation. *Biology Journal Linnaeus Society, 35*, 321–337.

Suchanek, T. H. (1992). Extreme biodiversity in the marine environment: Mussel bed communities of *Mytilus californianus. Northwest Environnent Journal, 8*, 150–152.

Tilman, D., & Downing, J. A. (1994). Biodiversity and stability in grasslands. *Nature, 367*, 363–365.

Tuomisto, H., Ruokolainen, K., Kalliola, R., Liina, A., Danjoy, W., & Rodriguez, Z. (1995). Dissecting Amazonian biodiversity. *Science, 269*, 63–66.

Walker, B. H. (1991). Biodiversity and ecological redundancy. *Conservation Biology, 6*, 18–23.

Walsh, G. E. (1974). In R. J. Reimold, & W. H. Queen (Eds.), *Ecology of halophytes* (pp. 51–174). New York: Academic.

Ward, R. D., Woodwark, M., & Skibinski, D. O. F. (1994). A comparison of genetic diversity levels in marine, freshwater, and anadromous fishes. *Journal of Fishery Biology, 44*, 213–232.

Whittaker, R. H. (1960). Vegetation of the Siskiyou Mountains, Oregon and California. *Ecology Monograph, 30*, 279–338.

Whittaker, R. H. (1967). Gradient analysis of vegetation. *Biology Review, 42*, 207–264.
Whittaker, R. H. (1975). *Communities and ecosystems* (2nd ed.). New York: Macmillan.
Whittaker, R. H. (1977). Evolution of species diversity in land communities. *Evolution Biology, 10*, 1–67.
Wilhm, J. L., & Dorris, T. C. (1968). Species diversity of benthic macro-invertebrates in a stream receiving domestic and oil refinery effluents. *Amererican Middle Nature, 76*, 427–449.
Wood, A. M. (1989). Population genetic studies of algae. *Journal of Phycology, 25*, 11.

Instruments and Methods 12

Contents

12.1 Instruments

The marine and estuarine compartments are the storehouses of vast resources, but instrumentation sector is a vital wing not only to monitor the magnitude and variation of these resources but also to harness them in a cost-effective way. Research vessels from different countries are constantly monitoring the oceans and generating data on temperature, salinity, pH, dissolved oxy gen (DO), nutrients chlorophyll and several other parameters. Many of these research vessels have sophisticated laboratories inside, where analysis of water sediment and other biological samples are carried out.

The modern research vessel (RV) is equipped with precision navigational instruments that communicate with satellites and computers (Fig. 12.1).

RVs are often dedicated to undertake research activities in physical oceanography, chemical oceanography, geological oceanography and biological oceanography and are thus equipped with specific types of scientific instruments. Some subfields of the various disciplines are hydrology, marine chemistry, marine geophysics and geology of the ocean, hydroacoustics, hydrography, meteorology and marine biology. Also included in the scientific fleet are vessels for telemetry of space vehicles, which are often used to collect meteorological information. The largest fleets of RVs belong to the Russia and the

A. Mitra, S. Zaman, *Basics of Marine and Estuarine Ecology*,
DOI 10.1007/978-81-322-2707-6_12

Fig. 12.1 A view of a modern research vessel (RV)

United States. Vessels displacing 2000–4000 tonnes are the common types of ships. The present-day oceanographic research activities are totally dependent on scientifically equipped RVs, which have a high degree of seaworthiness, good speed, manoeuvrability and stability. These vessels possess devices for 'over-the-tide studies, such as winches, cranes, booms and folding platforms. Removable (container) laboratories are also important components of modern RVs'. Descriptions of few RVs are presented in Table 12.1.

An ideal research vessel is reliable, manoeuvrable, stable at sea and has comfortable living and working spaces. The RV Revelle (Fig. 12.2) is a typical large research vessel (Scripps Institution of Oceanography 2009; scilib.ucsd.edu/sio/annual/Annual_report_2010_ pdf). It was built in 1996. Its overall length is 277 ft and its displacement is 3180 tonnes.

Icebreakers are important components of modern research vessel. Most of the icebreakers used for research have dual purposes, including as supply and rescue ships. Icebreakers have double hulls and rounded bows, and the ice is broken by running the ship up onto the ice.

These sophisticated RVs are equipped with several instruments meant for carrying out researches in several fields of oceanography.

A remarkable revolution has occurred in the field of marine instrumentation. Instead of conventional reversing thermometer, scientists and researchers are today using CTD probes for simultaneous monitoring of conductivity, temperature and depth. These data are finally transmitted for further processing. Gas-analysis probes, intended to measure the concentration of dissolved gases in water column, have also been developed. Electromagnetic and acoustic devices now make current measurements in the ocean.

Modern research vessels are also equipped with meteorological observation devices, temperature sensors, current accelerometers, echo sounders, hydro-locators, etc. They use towed

Table 12.1 Description of few RVs engaged in oceanographic research

Name	Displacement	Speed and range	Overview
Jean Charcot (built in 1965)	2200 tonnes	Speed—15 knots Range—10,000 miles	Length—74.5 m Beam—14.1 m Draught—5.0 m
Professor Bogorov (built in 1975)	1677 tonnes	Speed—13.5 knots	Length—68.8 m Beam—12.4 m Draught—4.2 m
Academician Mstislav Keldysh (built in 1981)	6339 tonnes	Speed—15.7 knots Range—20,000 miles	Length—122 m Beam—17.8 m Draught—5.8 m
Vadim Popov (between 1986 and 1988, six ships of the Vadim Popov class were constructed)	929 tonnes	Range—5500 miles	Length—49.9 m Beam—10.0 m Draught—3.6 m
Admiral Vladimirsky (hydrographic vessels of this type were built between 1970 and 1977)	9100 tonnes	Speed—20 knots Range—23,000 miles	Length—147 m Beam—18.6 m Draught—6.3 m
Geolog Primor'ya (built in 1984)	791 tonnes	Speed—9 knots Range—1320 miles	Length—35.1 m Beam—18.2 m Draught—3.3 m
Meteor (built in 1986)	4000 tonnes	Speed—15 knots Range—10,000 miles	Length—97.5 m Draught—5.3 m

magnetometers, heat-flow (geothermal) probes, radioactivity sensors, wave meters and underwater television cameras and even carry submersible vehicles. However, students and scholars pursuing oceanographic studies at the college/university level use several common instruments as listed in Tables 12.2, 12.3, 12.4 and 12.5.

12.1.1 Instruments Used for Physical Oceanographic Study

Physical oceanography is a branch of oceanography, which focuses on the dynamic property of oceans, seas, bays and estuaries like waves, currents, tides, etc. The air–water interaction is

Fig. 12.2 A view of RV Roger Revelle

Table 12.2 Common instruments in the sphere of physical oceanography

Instrument	General description
Bathythermograph	The bathythermograph (BT) is a device used for recording graphically the in situ water temperature relative to the depth at which the measurements are made. It is capable of measuring temperature with an accuracy of ± 0.05 °C. It is particularly useful for oceanographic research and can be towed from a ship while moving at speeds of up to about 18 knots, or it may be used to obtain a temperature-depth profile at a specific point. Thus, the instrument can be used for measuring temperature and temperature gradients, locating thermocline, identifying thermal regimes of different water masses as well as providing supplemental data for water current studies
Tide gauge	It is used to determine the tidal amplitude. The earliest version of the modern tidal gauge is perhaps the Aime's tidal gauge. In 1838 this tide gauge was first used in the port of Algiers. In course of time, Crouzet Marine Oceanology Corporation developed autonomous ultrasonic tide recorder from a prototype developed by the Studies and Research Department of France Electric. This instrument was mounted on the bottom and emitted sound waves that reflected off the water surface. With the change of the water level, this instrument could record the apparent changes in depth
Piezometers	Liquids posses the property of compressibility and this property is measured by piezometers
Viscometer	It is used to measure the viscosity of a liquid. The instrument works by measuring the force, which oppose the rotation of a disc or a cylinder that is immersed in the liquid
Current meter	This device is used to know the magnitude and direction of the current in the marine and estuarine waters
Secchi disc	It is a device which is very commonly used to get an indication of transparency of natural waterbodies. It is a heavy circular plate made up of metal, plastic or wood and normally of approximately 30 cm in diameter. The upper surface is painted plain white or white and black alternate band with the bottom painted black. It is fastened to a rope through a ring located at the centre of its upper surface. The disc is submerged vertically into the water for observation and is viewed from above to determine the depth at which it just disappears

also encompassed in the vertical of physical oceanography. The most productive areas of planet Earth are located in these regions of upwelling, which occur largely on western coasts, as evidenced by the large fisheries in such regions. Upwelling produced by the Peru Current creates one of the richest fisheries in the world. In addition, this upwelling supports large populations of seabirds that deposit countless tons of nitrate- and phosphate-rich guano on

Table 12.3 Common instruments in the sphere of geological oceanography

Instrument	General description
Echo sounder	This instrument works on the principle of reflection of sound wave. The instrument has great application in studying the depth and bottom features of the ocean floor. The transducer in echo sounder emits signals directed perpendicularly along the track of the vessel. The nature of return of these signals after hitting the ocean floor determines the bottom relief (bathymetry)
Autonomous sampler	This device collects a sample of bottom sediments of about 0.15 m in area. The sample weighs about 30 kg. The time of submergence and return to the surface is about 1.5 m/s
Geological coring device	Geological coring devices are used to collect the sediment samples from the seafloor. These are further analyzed to know the mineral resources, metal content, sediment texture, composition, etc. In general vibrating corer, piston corer and gravity corer are used to draw samples from the ocean floor
Proton magnetometer	This instrument is used to monitor the vector of geomagnetic induction. There are measurement intervals to know the vector of geomagnetic induction and normally the intervals are of 1, 3, 5, 20 and 60 s
Marine multichannel geothermal complex	This is used to measure the geothermal gradients (heat flow) in the bottom sediments. Measurements are taken with regard to the temperature of the near-bottom layer of water and heat conductivity of the ocean flow
Manned submersibles	These are equipped with modern instrumentations and are the perfect tools for underwater research, capable of researches for opening opportunities for undersea technical work. The first manned submersibles are usually considered to be the bathyspheres and bathyscaphs. The Swiss scientist A. Piccerd designed the bathyscaphs in 1953–1954 and in 1960 dived to the bottom of the deepest part of the ocean, the Mariana Trench

coastal islands, and thus the ocean water gets naturally fertilized. Such phenomenon of upwelling and distribution of nutrients is also addressed through the lens of physical oceanography.

The knowledge of physical oceanography helps to get a basic idea on the circulation pattern of the ocean and weather condition and predict the impact of global warming, El Niño, etc., on the marine and estuarine environment. Physical oceanography also gives idea on the depth profile of the ocean, which is important for the movement of submarine. The reservoir under the land is depleting very fast, owing to man's need in terms of energy. Coal has already come on the verge of depletion, and days are not very far when the coal-powered electrical plants will have to stop their production journey. Under this situation, tidal energy, ocean thermal energy or the wind energy may be the source of future power, whose magnitude and application is dependent on the knowledge base of physical oceanography. The in-depth study and researches on physical oceanography are carried out with some basic instruments whose names and short description are given in Table 12.2.

12.1.2 Instruments Used for Geological Oceanographic Study

Geological oceanography, sometimes called marine geology, covers a vast range of time and space scales since it considers processes that occur in minutes over millimetre-scale distances as well as processes that effect entire ocean basins and continents over time spans of millions of years. The phenomena like continental drift, formation of trench due to subduction, seafloor spreading, formation of manganese nodules on the ocean floor, sedimentation, formation of islands, shoreline changes due to waves, currents and tidal actions, etc., are all encompassed under the branch of Geological Oceanography.

The tectonic movement is an important aspect of geophysics through which knowledge of natural calamities like tsunamis can be gained. Thus,

Table 12.4 Common instruments in the sphere of chemical oceanography

Instrument	General description
Nephelometric turbidity meter	The instrument is used to determine the turbidity of the water and is based on the Tyndall effect of scattering of light. The nephelometric method is applicable to the measurement of turbidity in all water types over a practical working range of about 0.05 to 1000 FTU. In Hach model 2100A turbidimeter, a strong light beam is transmitted upwards through a turbidity optical tube containing the sample. The amount of light reflected at 90° angle to the incident light beam by the suspended particulate matter is directly proportional to the turbidity present and is received by a photomultiplier tube. This diffracted is transformed into an electrical signal, which is measured on the instrument meter graduated in turbidity units
Colorimeter and spectrophotometer	Colorimetric and spectrophotometric methods have been extensively used in the analysis of nutrients, because of their speed, simplicity and accuracy. Nutrients such as nitrogen—or phosphorus—containing substances can be detected in microgram per litre range without any separation or pre-concentration. These methods are based on absorption of visible or ultraviolet light by molecules, and the intensity of absorption is proportional to the concentration of the absorbing species in the solution. When the absorption of light occurs in the visible range (approximately 400-700 nm) and can be monitored visually or by simple devices such as colorimeters or filters and photometers, the technique is referred to as colorimetry Spectrophotometry differs from colorimetry in that a narrower band of wavelength for absorption is used, and the absorption of a substance is measured over a wider wavelength range that includes the ultraviolet region of the spectrum (250–700 nm)
Atomic absorption spectrophotometer (AAS)	Atomic absorption spectrophotometry is the most widely used technique for the determination of trace metals in waters, sediment and biological samples. In this technique, atoms absorb radiation only at discrete wavelengths characteristic of the absorbing species. Thus, radiation from the source, produced from a vapour of the metal of interest, is absorbed at a discrete wavelength(s) by atoms of that element in the atomizer. As a result, the radiation beam intensity is attenuated by an amount that is proportional to the concentration of the element of interest in the atomizer. In AAS, the atomizer is usually a flame—or furnace-type electrothermal device. Radiation of a characteristic wavelength is usually produced by a hollow cathode or electrodeless discharge lamp. A monochromator placed after the atomizer is used to isolate the desired wavelength from the absorbing and non-absorbing lines
Gas chromatograph (GC)	Gas chromatography is an analytical technique for separating compounds on the basis of their volatility. It provides both qualitative and quantitative information for individual compounds present in a sample. Compounds move through a GC column in the form of gases mainly because of two reasons: either the compounds are normally gases or they are heated and vaporized into a gaseous state. The components of GC are injector (hollow heated glass-lined cylinder, where the sample is introduced), carrier gas (usually helium, hydrogen or nitrogen, which is the mobile phase and moves the sample through the column), capillary GC column (coated with stationary phase like large molecular weight polysiloxane, polyethylene glycol or polyester polymers of 0.1 to 2.5 μm film thickness), GC oven, electron capture detector (ECD) and flame ionization detector (FID)
High-performance liquid chromatography (HPLC)	High-performance liquid chromatography is an analytical technology for quantifying and analyzing mixtures of chemical compounds. It is used to determine the amount of a chemical compound within a mixture

(continued)

Table 12.4 (continued)

Instrument	General description
	of other chemicals. For this purpose, the sample is dissolved in a solvent (like water or alcohol). A detector measures response changes between the solvent itself and the solvent and sample when passing through it. Thus, in HPLC the mobile phase is a solvent, which is pumped under high pressure through a column, and the stationary phase is a finely divided solid held inside the column
Inductively coupled plasma atomic emission spectrometer (ICP-AES)	This sophisticated instrument allows multielemental analysis and is fast revolutionizing the conventional flame emission method of seawater analysis in terms of trace metals. In this case, the sample is introduced into plasma via an ultrasonic or pneumatic nebulizer chamber system similar to that used in the conventional AAS. In the plasma, the sample is subjected to temperatures of about 7000–10,000 K, and electrons of the atoms of elements contained in the sample undergo a transition to excited state by absorbing thermal energy from the plasma source that are then re-emitted at a characteristic frequency (atomic spectra) as they return to lower energy state. Thus, the elements present may be determined by resolving individual spectral lines of the element of interest and choosing and measuring the intensity of one or more of the carefully selected lines for each element
Neutron activation analyser (NAA)	Neutron activation analyser is an extremely sensitive, precise and accurate method of trace elements determination. Because of its simultaneous multielemental capability, this technique can give a wealth of elemental information even in small amounts of sample, e.g. aerosols, lunar materials, special precious rare minerals, etc. AA is based on measurement of the intensity of radioactivity of an element produced by the bombardment of the element with nuclear projectiles. When the sample is bombarded with neutrons, radioactive nuclei are invariably formed due to their interaction. The compound spectrum produced by the mixture of radionuclides is resolved either instrumentally (INAA), using high-resolution g-spectrometer with high purity Germanium detector (HPGe) system or by carrying out chemical separation of the radionuclides. Induced radioactivities are measured and compared with that of the standard, which is irradiated along with sample under the same conditions to evaluate the amount of element in the sample
Gas chromatograph mass spectrophotometer (GC-MS)	Foods and beverages contain several types of aromatic compounds, some naturally present in the raw materials and some are formed during processing. GC-MS is extensively used for the analysis of these compounds which include esters, fatty acids, alcohols, aldehydes, terpenes, etc. It is also used to detect and estimate contaminants from spoilage or adulteration which may be harmful and which are often controlled by governmental agencies, for example, pesticides

geological oceanography has become a vital component in present-day disaster management sector.

The seafloor is covered with sediments, which are unique reservoirs of minerals. Basically the mineral resources of the ocean may be divided into four main categories, namely, (1) elements dissolved in seawater, (2) minerals recoverable from the underlying bed rock, (3) minerals found in ocean bottom and (4) minerals within marine sediments like oil and gas. The availability and extraction of variable resources in different zones of the marine and estuarine compartment come under purview of geological oceanography. The common instruments used in the field of geological oceanography are given in Table 12.3.

Table 12.5 Common instruments in the sphere of biological oceanography

Instrument	General description
Nansen bottles	Nansen bottles are used to collect non-sterile water samples specially phytoplankton from subsurface levels. The reversing bottle is made of brass and plated inside with tin or silver or coated with a special lacquer. It is fitted with a drain cock and an air vent to facilitate draining the trapped water samples. The reversing water samplers are available in different capacities (750–1250 ml)
Niskin sampler	Niskin samplers are used to collect sterile water. This involves a presterilized plastic bag, which fits onto a winged sampler. The wings are locked together, fitted with a plastic bag and positioned on a line. At the predetermined depth, a messenger operates a guillotine, which opens a hose into the bag and simultaneously releases the wings of the sampler. These wings spring apart, drawing water into the bag. The bag may be retrieved for subsequent examination
SCUBA equipment	The SCUBA equipment is used to collect the epibiotic and the benthic communities. The epibiotic communities are collected by immersing inert structures, e.g. glass slides into the sea for predefined periods
Grab sampler	This device is used to grab material from the upper layers of seafloor sediment for studying the habit and habitat of benthic community

12.1.3 Instruments Used for Chemical Oceanographic Study

The knowledge of chemical oceanography is essential not only to know the mineral formation and their deposition in the ocean but also to know the phytoplankton status of the seawater; since these tiny free-floating floral components are constituted of carbon, nitrogen, phosphate and silicate, which they absorb from the ambient seawater.

Marine corrosion, settlements of biofoulars on submersible objects, pollution due to industrialization and urbanization, ocean biogeochemical cycle and even global warming are all studied under the banner of chemical oceanography. The knowledge of this branch of oceanography is extended to extract useful substances like magnesium, sodium, bromine, etc., from seawater. In the sphere of chemical oceanography, the common instruments (Table 12.4) are used mainly to analyze water and sediment samples, but nowadays with increasing thrust on climate change, analysis of stored carbon in soil, coastal vegetation and even molluscan tissue has become extremely important. The stored carbon is monitored through CHN analyzer (Fig. 12.3). A study conducted in 24 stations in deltaic Sundarbans (Indian part) during 2012 was completely based on the use of CHN analyzer. The study report is presented as Annexure 12A for the convenience of the readers. In this study, the bulk density and carbon density of the soil samples were also analyzed to assess the role of soil in the mangrove-dominated lower Gangetic delta as the store house carbon.

12.1.4 Instruments Used for Biological Oceanographic Study

Biological oceanography, sometimes called marine biology, is the study of life in the sea ranging from viruses to whales. It has several subtopics like census of marine/estuarine/coastal biodiversity, adaptation of flora and fauna to marine and estuarine ecosystems, energy flow from primary producers to consumers, decomposition process by microbes, biodiversity in different depths of the ocean and living resources with their ecosystem services.

Fish form a rich source of food and nutrition to man. Fishing has long been the economic activity of variable importance in all the latitudes. It is regarded as the oldest occupation of man, and now all the great maritime nations have well-developed fishing industry.

Fisheries have played a part in the destiny of nations. The rise and downfall of nations have been the result of the prosperity and decay of the industries connected with fisheries. The Hanseatic League from the small beginnings of the

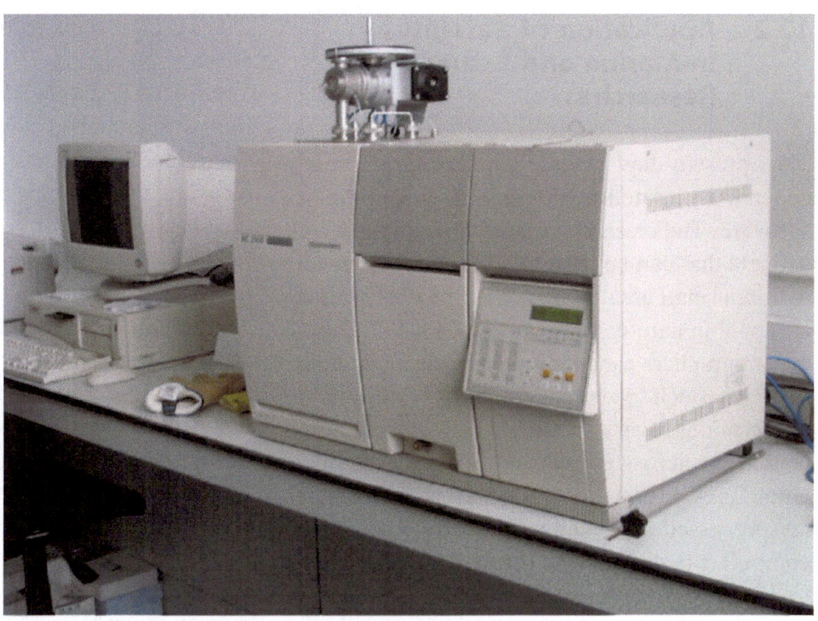

Fig. 12.3 A view of CHN analyzer

herring fisheries in Scania rose to a power of great magnitude. The rise of the sea power of the Dutch in seventeenth century came as the result of development of fisheries in the North Sea. The Tudors in England feared a loss of sea power with the decline of fisheries. The progress of fisheries also resulted in the development of mercantile marines and the royal navy. The location and extent of commercial fisheries in comparison to the great extent of oceans is very restricted, but the fish as an inexhaustible source has the incredible capacity for multiplying. The wanton exploitation and dwindling local supplies has, however, affected the commercial fishing in some areas. The world's important fisheries are of three general types: (1) inshore, (2) pelagic and (3) demersal. The general distribution and abundance of fish population depend upon the spawning habits of fish and the presence of food and congenial water quality.

Apart from fishing, the study of biological oceanography also helps to culture edible oyster, pearl oyster, shrimp, seaweeds and even phytoplankton, which are extremely important for sustaining modern aquaculture industry. The future world is waiting on the optimistic researches related to Marine Biotechnology, as several strains of microbes, flora and fauna have been identified as the potential sources of bioactive substances. Marine phytoplankton have been identified to be a rich source of astaxanthin, which is a powerful, bioactive antioxidant that has demonstrated efficacy in animal or human models for addressing several heath problems like muscular degeneration, Alzheimer's (Markesbery and Carney 1999; Behl 1999) and Parkinson's diseases (Ebadi et al. 1996), cardiovascular diseases, stroke and several types of cancer.

Recent researches have shown that astaxanthin acts like a vitamin for salmon. Nowadays, astaxanthin is used in the feed of farm-raised salmon to impart their flesh the same pink colour of wild salmon and also in poultry feed to augment the yellow colour of egg yolk. It is expected that the branch of biological oceanography will handover bio-gems to mankind through the corridors of biotechnology, bioinformatics and genetic engineering. The branch of biological oceanography needs some basic instruments (Table 12.5) for its expansion. However, apart from the instruments stated in Table 12.5, several programmes/softwares are used to compute community structure of flora and fauna inhabiting marine and estuarine ecosystems.

12.2 Application of Satellites in Marine and Estuarine Researches

The present-day oceanographers are greatly dependent on satellite observations and computer software. The research vessels are equipped with gadgets that can be used to study the parameters within a small area. In this case, the observations are local in nature. With the use of satellites, the oceanographers can study the oceans as a global system. Each satellite records millions of observations everyday as it follows changing conditions across the world's oceans. The huge amounts of information collected by the satellites are processed and manipulated by sophisticated computers to develop the databank in connection to seashore configuration, currents, waves, pigment level, biodiversity, sea ice, storms and even the seafloor. The continuity, global coverage and high temporal and spatial resolution of satellite data make it an important tool for monitoring and characterizing marine ecosystems. The potential fishing zone (PFZ) is detected through satellites. It is very important to note in this context that satellites do not observe fish stocks directly, but measurements such as sea-surface temperature (SST), sea-surface height (SSH), ocean colour, ocean winds and sea ice characterize critical habitat that influences marine resources. Most of the spatial features that are important to marine ecosystems like ocean fronts, eddies, convergence zones, river plumes and coastal regions cannot be adequately resolved without satellite data. Chlorophyll is the only biological component of the marine ecosystem accessible to remote sensing (*via* ocean colour) and as such provides a key metric for evaluating the health and productivity of marine ecosystems on a global scale. Long-term ocean colour satellite monitoring provides an important tool for better understanding of the marine processes, ecology and the coastal environmental changes (Tang and Kawamura 2001). Modern oceanographic vessels are, therefore, linked to satellites via computers, allowing scientists to use immediate data to plan their sampling programmes while at sea.

SEASAT, a specialized oceanographic satellite, was launched in June 1978, but remained operational only until October. Its radar could measure the distance between the satellite and the sea surface with an accuracy of about 5 cm (2 in.), allowing the measurements of wave heights. In addition to this, the radar of SEASAT could monitor temperatures, wind speeds, sea ice cover, currents and phytopigment level of the aquatic system. In the ocean, the physical, chemical and biological processes are linked in an intimate manner (Tang et al. 2002). Oceanic features such as chlorophyll concentration, current boundaries, sea-surface temperature, ocean fronts and eddy, suspended matter and dissolved organic matter (DOM), etc., influence the ocean dynamics and its interaction with the atmosphere. To study this intricate interrelationship in a more detailed way, several global sensors with their spectral characteristics are presently in use (Table 12.6).

The satellite data (after being calibrated with ground reality) are also used to generate maps of nutrient load in different segments of the estuary. We conducted a nutrient-level monitoring during June 2014 in the Hooghly estuary in the north-east coast of India (Fig. 12.4) and generated separate maps each for nitrate (Fig. 12.5), phosphate (Fig. 12.6) and silicate (Fig. 12.7). This study was later correlated with the phytopigment level (Fig. 12.8) in the estuarine water procured from the ocean colour monitor (OCM) of the Indian Remote Sensing Satellite IRS-P4, which is optimally designed for the estimation of chlorophyll in coastal and oceanic waters.

Another interesting area of studying and monitoring marine and estuarine fauna is biologging. The vertical of biologging, i.e. the deployment of recording and transmitting tags on animals to study their movements, behaviour, physiology and habitat usage, has rapidly expanded over the past decade, because of advances in miniaturization of electronic tags (Bograd et al. 2010). Fishes are tagged to monitor their migratory

Table 12.6 Ocean colour satellites and sensors

Sensor	Agency	Satellite	Launch date	Swath (km)	Resolution (m)	No. of bands	Spectral coverage (nm)	Orbit
MERIS	ESA (Europe)	ENVISAT (Europe)	01/03/02	1150	300/1200	15	412–1050	Polar
MMRS	CONAE (Argentina)	SAC-C (Argentina)	21/11/00	360	175	5	480–1700	Polar
MODIS-Aqua	NASA (USA)	Aqua (EOS-PM1)	04/05/02	2330	1000	36	405–14,385	Polar
MODIS-Terra	NASA (USA)	Terra (USA)	18/12/99	2330	1000	36	405–14,385	Polar
OCM	ISRO (India)	IRS-P4 (India)	26/05/99	1420	350	8	402–885	Polar
OSMI	KARI (Korea)	KOMPSAT (Korea)	20/12/99	800	850	6	400–900	Polar
PARASOL	CNES (France)	Myriade series	18/12/04	2100	6000	9	443–1020	Polar
SeaWiFS	NASA (USA)	OrbView-2 (USA)	01/08/97	2806	1100	8	402–885	Polar

routes and preferred breeding and feeding regions (Fig. 12.9).

Satellite remote sensing (SRS) oceanographic data coupled with tracking data are now widely accepted and used to identify the niches, breeding grounds, nesting sites and travelling patterns of marine animals. The data generated through SRS provide both the meso- and larger-scale oceanographic context for each available animal position and time. The types of SRS data most commonly used with animal tracking include SST, surface Chl *a* and geostrophic currents. Before linking tracking and SRS data, it is preferable to estimate the most likely tracks using a state–space modelling approach (Patterson et al. 2008). In addition, improved tag position data is obtained by including satellite-derived SST in the estimation process (Nielsen et al. 2006). A recently developed alternative modelling approach validated with GPS data consists of bootstrapping random walks generated from the probability distributions of animal locations and trajectories for the geolocation of tagged animals (Tremblay et al. 2009). The method provides a flexible framework for including remotely sensed data sets and has the advantage of being easier to implement than state–space models.

SSTs are the most commonly SRS data used in combination with tagging data. These can be analyzed to determine whether an animal uses mesoscale features, including temperature fronts and cyclonic eddies, and to characterize its habitat regarding preferred SSTs (Polovina et al. 2000; Kobayashi et al. 2008). For loggerhead sea turtles (*Caretta caretta*), preferred habitat north of Hawaii constitutes a temperature and chlorophyll front delineated by a SST of 18 °C. Daily maps of probable turtle habitat, defined by a narrow band around the 18 °C SST isotherm, are distributed to longline fishers to help them avoid the area and reduce turtle bycatch (Howell et al. 2008).

SRS chlorophyll data often serve as a valuable proxy for water mass boundaries and may identify upwelling associated with mesoscale features. The range of surface chlorophyll values used by an animal may help characterize its habitats (Polovina et al. 2000; Kobayashi et al. 2008). For example, by combining turtle tracking with SeaWiFS chlorophyll data, Polovina et al. (2001) characterized and described interannual changes in the position and dynamics of a North Pacific basin-wide chlorophyll front, the transition zone chlorophyll front (TZCF), which has proven to be an

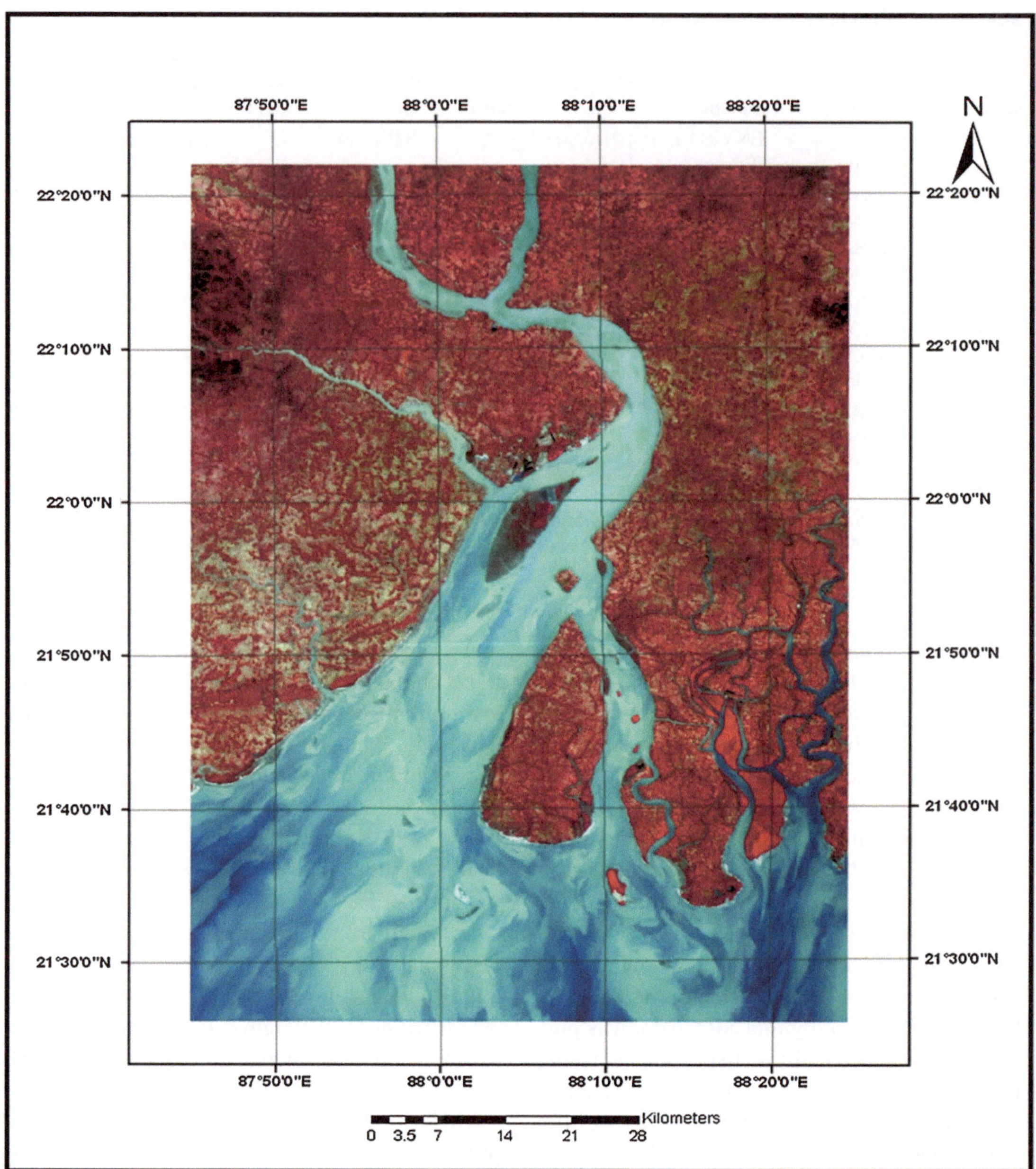

Fig. 12.4 Hooghly estuary in the north-east coast of India

important migration and forage habitat for a variety of species.

Brain Churners

1. How does the vertical of physical oceanography help in exploring alternative energy from oceans and estuaries?
2. What is the basic working principle of viscometer?
3. According to you, which colour should be used to paint the upper surface of the Secchi disc—white or alternate band of white and black? Justify your answer with reason.
4. What are the different categories of mineral resources of the ocean?
5. What substances are used as mobile and stationary phases in HPLC?
6. How is the compound spectrum produced by radionuclide mixture resolved in neutron activation analyzer?
7. Write the difference between pelagic and demersal fishes with some examples.
8. State the role of carotenoid pigments in finfish and shellfish species.
9. Distinguish between stenohaline and euryhaline species. Which type of species (between stenohaline and euryhaline) will be excluded if the dilution factor of the estuarine system abruptly increases due to barrage discharge?
10. How can the migration of horseshoe crab in mangrove swamps be monitored through updated technology?

Annexure 12A: Study on Soil Organic Carbon in the Intertidal Mudflats of Indian Sundarbans

1. Introduction

Human activities have led to considerable emissions of greenhouse gases (Murako 2004). In particular, for the period from 1980 to 1989, carbon dioxide emission from fossil-fuel burning and tropical deforestation amounted to 7.1 billion tons of carbon being released a year (Table 12A.1) (IPCC 1994). Increase in atmospheric carbon dioxide concentration can account for about half of the carbon dioxide emission for this period (Siegenthaler and Sarmiento 1993). This has led to study the capacity of carbon sequestration in forests and other terrestrial and wetland ecosystems.

Most of the studies so far available are related to forest ecosystems and crops, and there is not enough information on carbon sequestration potential of wetland soil. Wetlands provide several important ecosystem services, among which soil carbon sequestration is most crucial particularly in the backdrop of rising carbon dioxide in the present century. Wetlands cover about 5 % of the terrestrial surface and are important carbon sinks containing 40 % of SOC at global level (Mitsch and Gosselink 2000). Estuarine wetlands have a capacity of carbon sequestration per unit area of approximately one order of magnitude greater than other systems of wetlands (Cerón-Bretón et al. 2010) and store carbon with a minimum emission of greenhouse gases due to inhibition of methanogenesis because of sulphate (Bridgham et al. 2006). The reservoirs of SOC, however, can act as sources or sinks of atmospheric carbon dioxide, depending on land use practices, climate, texture and topography

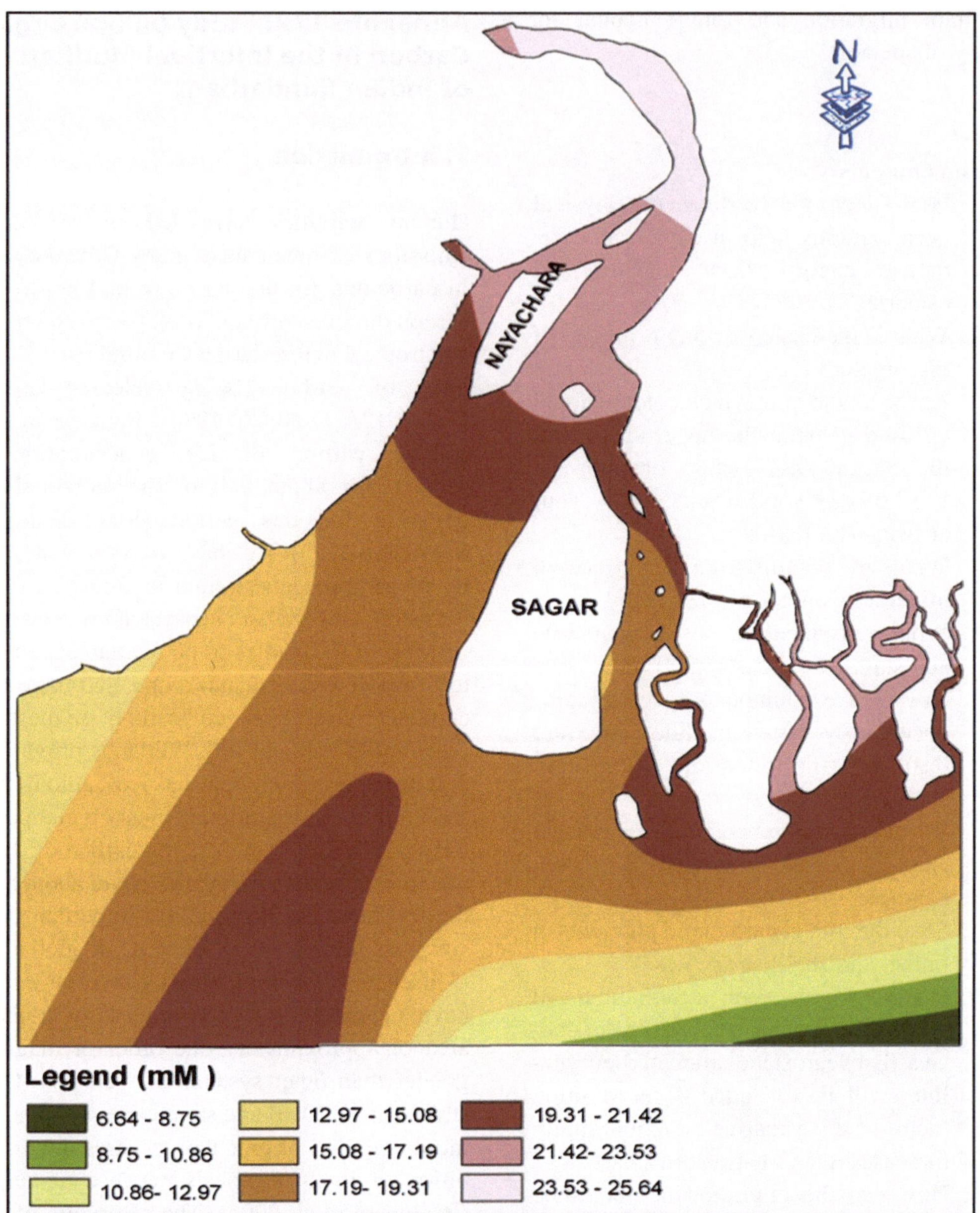

Fig. 12.5 Nitrate level in the Hooghly estuary in the north-east coast of India

(Vesterdal et al. 2002; Zinn et al. 2005; Homann et al. 2004; Shukla and Lal 2005).

Vertical patterns of SOC can contribute as an input or as an independent validation for biogeochemical models and thus provide valuable information for examining the responses of terrestrial ecosystems to global change (Jobb'agy and Jackson, 2000; Wang et al. 2004; Mi et al. 2008). A large number of biogeochemical models, however, do not contain explicit algorithms of belowground ecosystem structure and function (Jackson et al. 2000). Most of the studies primarily focused on the topsoil carbon stock, and carbon dynamics in deeper soil layers and driving factors behind vertical distributions of soil organic carbon remain poorly understood

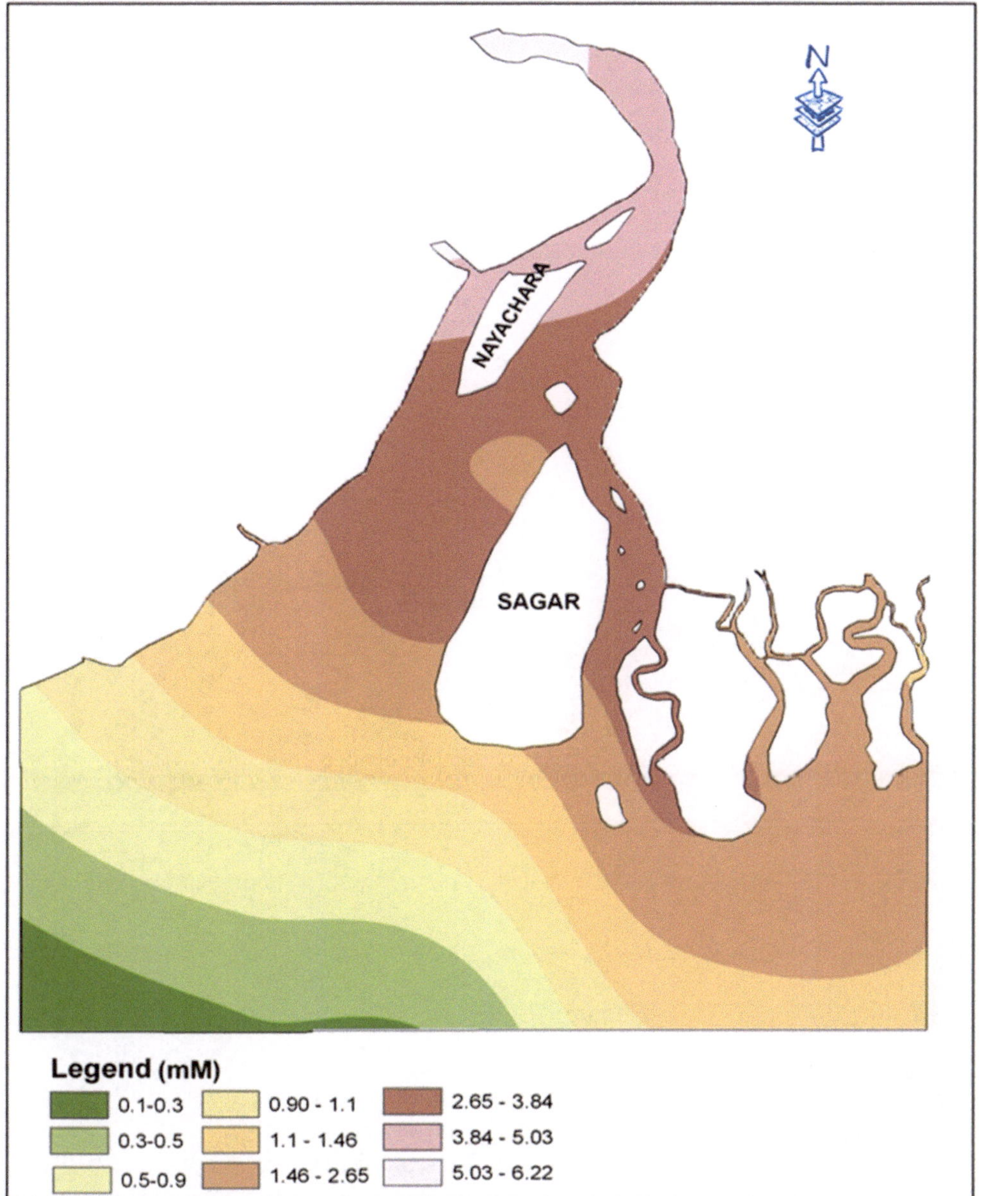

Fig. 12.6 Phosphate level in the Hooghly estuary in the north-east coast of India

(Jobb'agy and Jackson, 2000; Gill et al. 1999; Meersmans et al. 2009). Thus, improved knowledge of distributions and determinants of SOC across different soil depth is essential to determine whether carbon in deep soil layers will react to global change and accelerate the increase in atmospheric carbon dioxide concentration (Meersmans et al. 2009; Fontaine et al. 2007).

With this background, the present study was undertaken to estimate the SOC in four different depths in the mangrove-dominated Indian Sundarbans that sustains some 34 true mangrove species and some 62 mangrove associate species (Mitra 2000). This deltaic lobe together with Bangladesh Sundarbans constitutes the world's largest brackish water wetland. Hence, it is

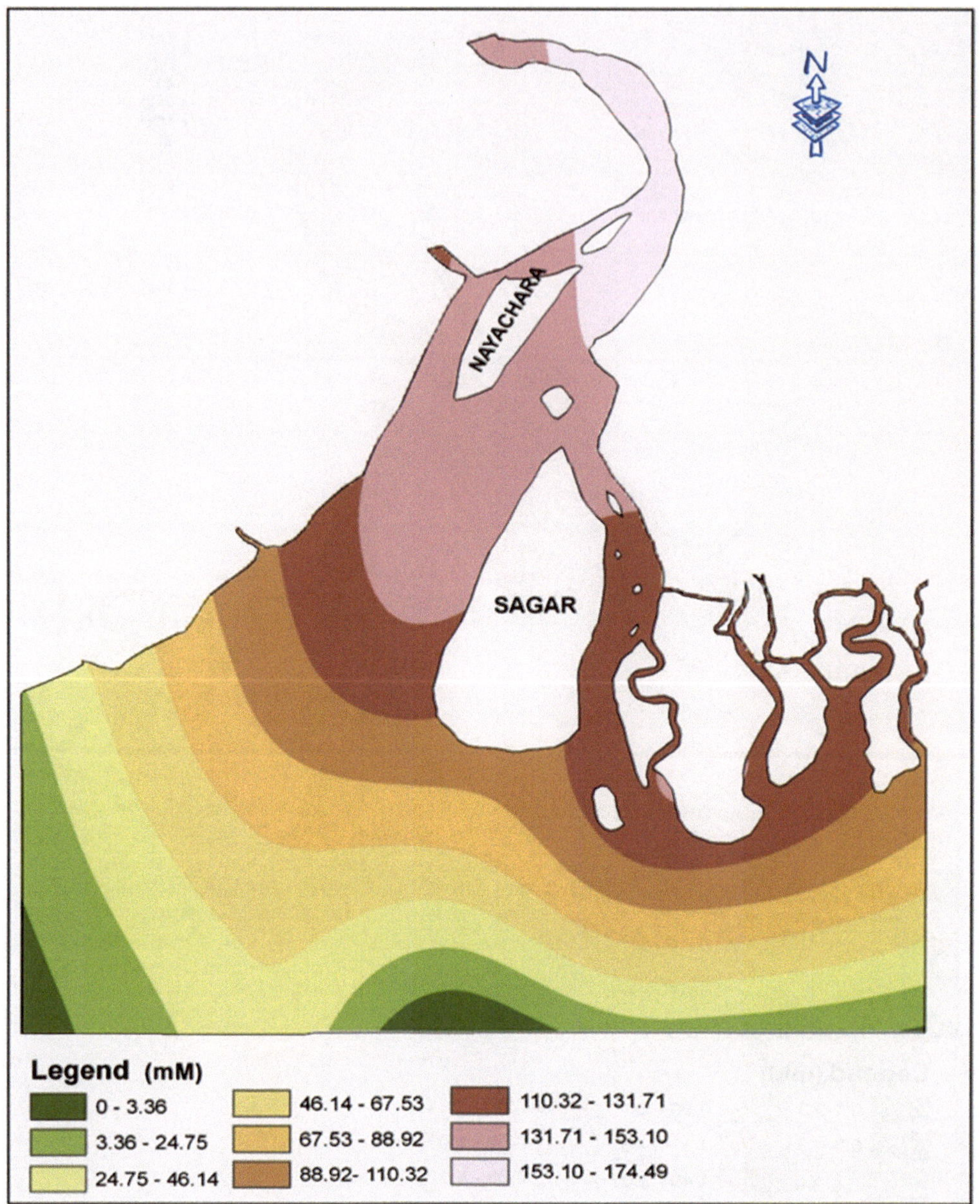

Fig. 12.7 Silicate level in the Hooghly estuary in the north-east coast of India

essential to establish a baseline data of soil carbon pool of this mangrove ecosystem. In this study, we used our unpublished data of SOC and bulk density to evaluate the spatial variations of OCD in the intertidal mudflats of western and eastern Indian Sundarbans that are markedly different with respect to anthropogenic activities and mangrove vegetation.

2. Materials and Methods

2.1 The Study Area

The Sundarban mangrove ecosystem covering about one million ha in the deltaic complex of the rivers Ganga, Brahmaputra and Meghna is shared between Bangladesh (62 %) and India (38 %) and is the world's largest coastal wetland.

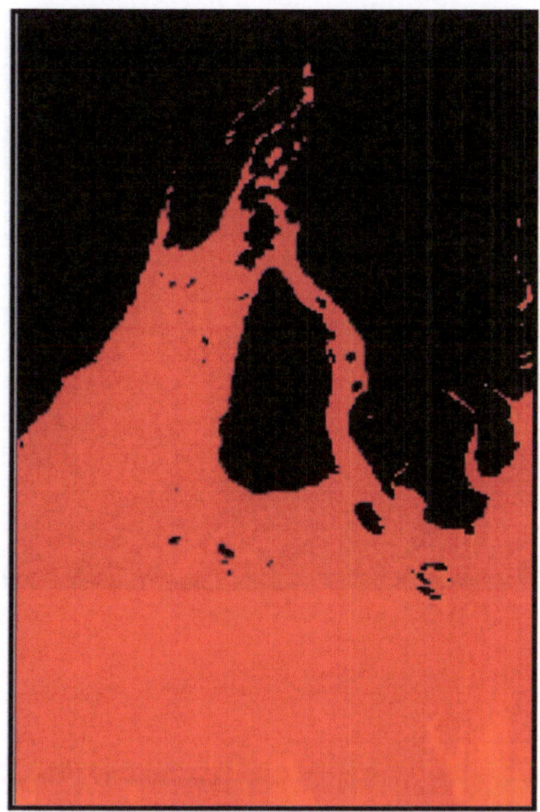

Fig. 12.8 Phytopigment level in the Hooghly estuarine water

Enormous load of sediments carried by the rivers contribute to its expansion and dynamics.

The Indian Sundarbans (between 21°13′N and 22°40′ N latitude and 88°03′E and 89°07′E longitude) is bordered by Bangladesh in the east, the Hooghly River (a continuation of the river Ganga) in the west, the Dampier and Hodges line in the north and the Bay of Bengal in the south. The important morphotypes of deltaic Sundarbans include beaches, mudflats, coastal dunes, sand flats, estuaries, creeks, inlets and mangrove swamps (Chaudhuri and Choudhury 1994). The temperature is moderate due to its proximity to the Bay of Bengal in the south. Average annual maximum temperature is around 35 °C. The summer (premonsoon) extends from the mid-March to mid-June and the winter (postmonsoon) from mid-November to February. The monsoon usually sets in around the mid of June and lasts up to the mid of October. Rough weather with frequent cyclonic depressions occurs during mid-March to mid-September. Average annual rainfall is 1920 mm. Average humidity is about 82 % and is more or less uniform throughout the year. This unique ecosystem is also the home ground of Royal Bengal tiger (*Panthera tigris tigris*). The deltaic complex sustains 102 islands, 48 of which are inhabited. The ecosystem is extremely prone to erosion, accretion, tidal surges and several natural disasters, which directly affect the top soil and the subsequent carbon density. The average tidal amplitude is around 3.0 m.

We conducted survey at 24 stations in the Indian Sundarban region in February, 2012. Station selection was primarily based considering the blocks in Indian Sundarbans.

2.2 Sampling

Table 12A.2 and Fig. 12A.1 represent our study site in which sampling plots of 10 m × 5 m were considered for each station. Care was taken to collect the samples within the same distance from the estuarine edge, tidal creeks and the same microtopography. Under such conditions, spatial variability of external parameters such as tidal amplitude and frequency of inundation (Ovalle et al. 1990) inputs of material from the adjacent bay/estuary and soil granulometry and salinity (Lacerda et al. 1993; Tanizaki 1994) are minimal.

10 cores were collected from the selected plots in each station by inserting PVC core of known volume into the soil to a maximum depth of 0.40 m during low tide condition. Each core was sliced in 0.10 m layers up to 0.40 m depth. The uppermost 0.01 m, which frequently includes debris and freshly fallen litter, was not used in this study. Each core section was placed in aluminium foil and packed in ice for transport. In the laboratory, the collected samples were carefully sieved and homogenized to remove roots and other plant and animal debris prior to oven-drying to constant weight at 105 °C for bulk density determination considering the volume of the PVC core. SOC of the collected samples ($n = 10$) from each plot was analyzed by standard method (Walkley and Black 1934), and the

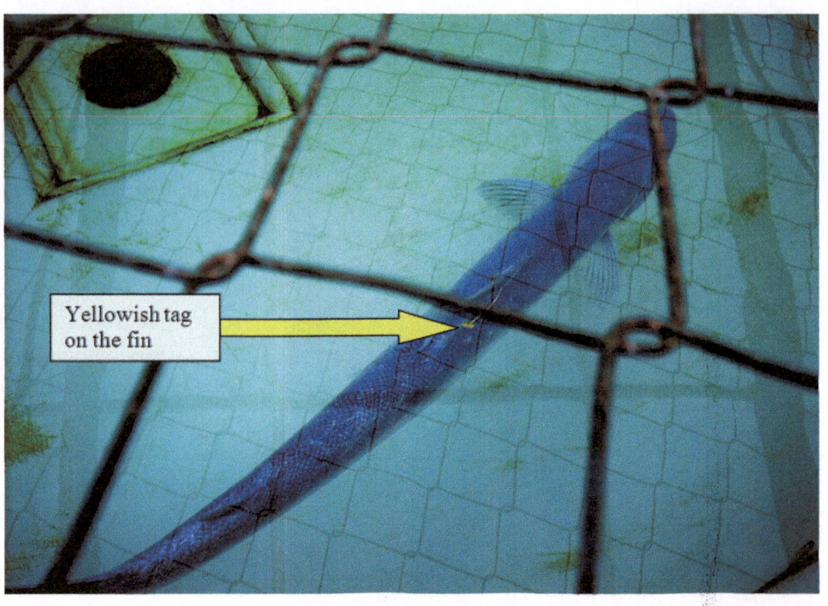

Fig. 12.9 Fishes are tagged to monitor their migratory routes and preferred breeding and feeding regions

Table 12A.1 Anthropogenic carbon fluxes, 1980 to 1989

Carbon dioxide sources	GtC/year
Fossil fuel burning, cement production	5.5 ± 0.5
Changes in tropical land use	1.6 ± 1.0
Total anthropogenic emission	7.1 ± 1.1
Partitioning among reservoirs	3.2 ± 0.2
Storage in the atmosphere	2.0 ± 0.8
Oceanic uptake	0.5 ± 0.5
Uptake by Northern Hemisphere forest regrowth	1.4 ± 1.5
Additional terrestrial sinks: CO_2 fertilization, nitrogen fertilization, climatic effects	

Source: IPCC (1994)

mean value was considered for determination of OCD in (kg/m^2) as per the expression:

OCD = % SOC × bulk density (BD) × soil depth

3. Results and Discussion

3.1 Soil Organic Carbon (SOC)

The organic carbon in soil differs significantly between stations. The spatial trend of SOC follows the order Stn. 1 (1.41 %) > Stn. 2 (1.28 %) > Stn. 23 (1.27 %) > Stn. 3 (1.23 %) > Stn. 8 (1.06 %) > Stn. 9 (1.04 %) > Stn. 4 (1.02 %) = Stn. 13 (1.02 %) > Stn. 10 (1.01 %) > Stn. 12 (0.99 %) > Stn. 11 (0.98 %) > Stn. 14 (0.96 %) > Stn. 5 (0.88 %) > Stn. 15 (0.83 %) > Stn. 6 (0.80 %) = Stn. 16 (0.80 %) > Stn. 19 (0.75 %) > Stn. 7 (0.74 %) = Stn. 22 (0.74 %) = Stn. 24 (0.74 %) > Stn. 17 (0.73 %) > Stn. 18 (0.72 %) . Stn. 20 (0.68 %) > Stn. 21 (0.66 %) (Annexure 12A.1 and Fig. 12A.2).

The significant spatial variation of SOC ($p < 0.001$) may be attributed to a large extent by mangrove diversity, anthropogenic activity, accretion and erosion processes (Table 12A.3 and Fig. 12A.3).

The relatively low SOC at Sagar South (Stn. 4) is due to its location at sea front where wave action and tidal amplitude is maximum (~3.5 m mean amplitude). This station experiences the freshwater discharge from the Farakka Barrage (located in the upstream zone), which is about 40,000 cusec/day. This huge quantum of

Table 12A.2 Sampling stations in Western and Eastern Indian Sundarbans

Sl. No.	Sampling station	Latitude	Longitude
1	Muriganga	21°38′25.86″N	88°08′53.55″ E
2	Saptamukhi	21°36′02.49″N	88°23′47.18″E
3	Thakuran	21°49′43.17″N	88°33′21.57″E
4	Herobhanga	21°59′34.32″N	88°41′46.52″E
5	Ajmalmari	21°51′34.72″N	88°39′00.68″E
6	Dhulibasani	21°47′06.62″N	88°33′48.20″E
7	Chulkathi	21°41′53.62″N	88°34′10.31″E
8	Arbesi	22°11′43.14″N	89°01′09.04″E
9	Jhilla	22°09′51.53″N	88°57′57.07″E
10	Pirkhali	22°06′00.97″N	88°51′06.04″E
11	Panchmukhani	21°59′41.58″N	88°54′14.71″E
12	Harinbhanga	21°57′17.85″N	88°59′33.24″E
13	Khatuajhuri	22°03′06.55″N	89°01′05.33″E
14	Chamta	21°53′18.56″N	88°57′11.40″E
15	Matla	21°53′15.30″N	88°44′08.74″E
16	Chandkhali	21°51′13.59″N	89°00′44.68″E
17	Goashaba	21°43′50.64″N	88°46′41.44″E
18	Gona	21°41′15.44″N	88°54′31.09″E
19	Chhotahardi	21°44′42.24″N	88°44′17.79″E
20	Baghmara	21°39′04.45″N	89°04′40.59″E
21	Mayadwip	21°35′50.23″N	88°47′09.95″E
22	Jambu Island	21°35′42.03″N	88°10′22.76″E
23	Lothian	21°38′21.20″N	88°20′29.32″E
24	Sagar Island	21°38′51.55″N	88°02′20.97″E

freshwater discharge through the Hooghly channel also causes erosion of the Sagar Island. Continuous erosion of the southern part of this island may be the reason behind minimum retention of organic matter in the intertidal zone (Fig. 12A.4).

The variation of SOC in the Indian Sundarban is thus regulated through an intricate interaction of biological, physical and anthropogenic activities.

The factors governing variation of below-ground carbon storage in mangrove soils is difficult to pinpoint (Bouillon et al. 2009; Alongi 2008) as it is not a simple function of measured flux rates, but also integrates thousands of years of variable deposition, transformation and erosion dynamics associated with fluctuating sea levels and episodic disturbances (Chmura et al. 2003). The mean value of SOC shows a decrease with depth. Similar trend is also observed by several researchers. The organic carbon levels under *Rhizophora mangle* soil were 2.80 %, 2.70 % and 2.70 % in the 0.01–0.05 m, 0.05–0.10 m and 0.10–0.15 m depth respectively (Lacerda et al. 1995). Similar decrease of SOC with depth was also observed under *Avicennia* soil (Lacerda et al. 1995). Report of decreasing mangrove SOC below 1 m was also documented in several mangrove ecosystems (Donato et al., 2011).

3.2 Bulk Density

The bulk density of mangrove soil is attributable to the relative proportion of sand, silt and clay and more specifically to the specific gravity of solid organic and inorganic particles and porosity of the soil. The compactness of mangrove soil increases with depth both in western and eastern Indian Sundarbans due to which the bulk density exhibits higher values with depths in all the stations. Basically the bulk density in the present study area is regulated by sediment texture and deposition/erosion which is the effect of current pattern, tidal amplitude and wind action. The order of bulk density variation is Stn.

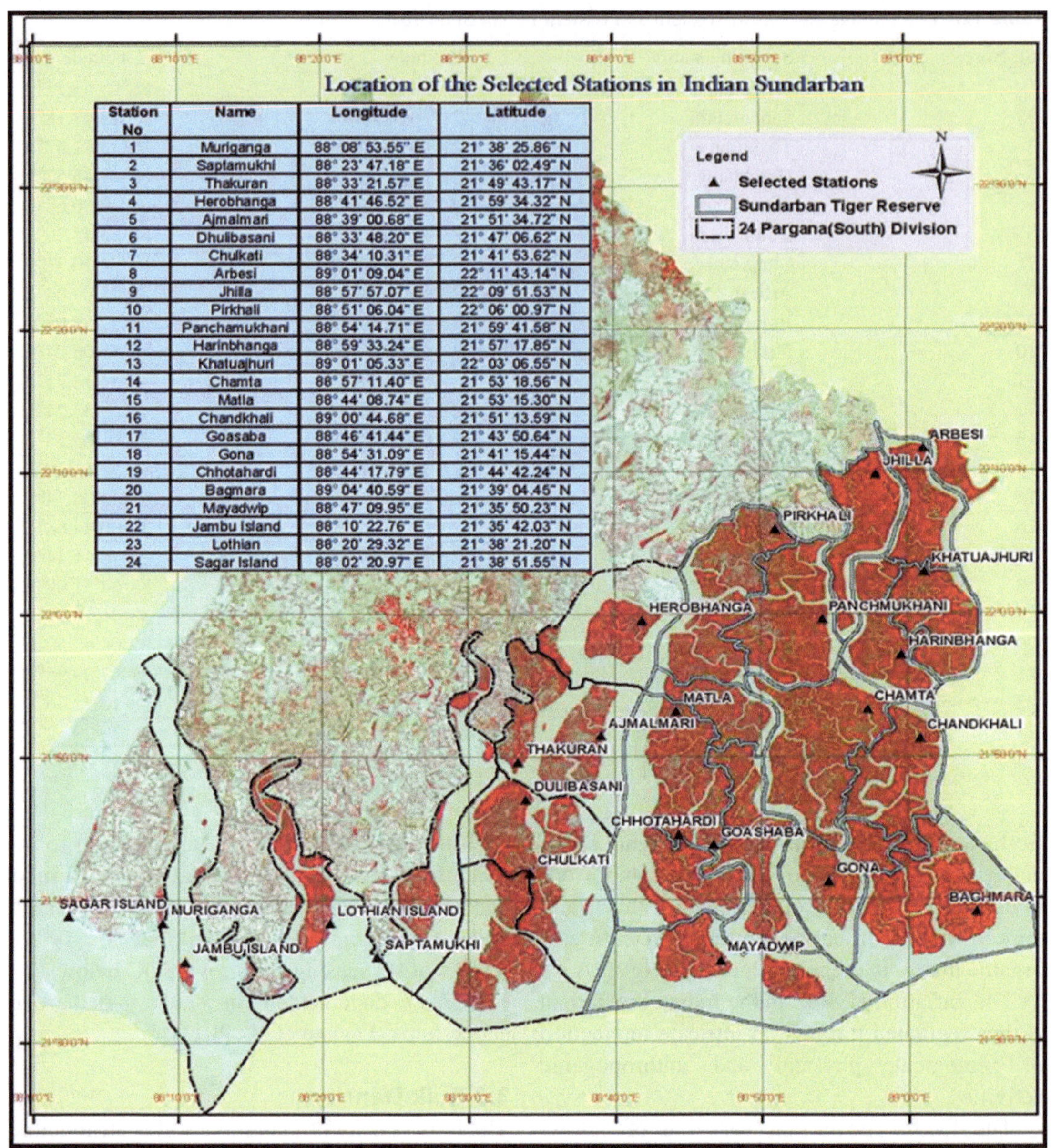

Station No	Name	Longitude	Latitude
1	Muriganga	88° 08' 53.55" E	21° 38' 25.86" N
2	Saptamukhi	88° 23' 47.18" E	21° 36' 02.49" N
3	Thakuran	88° 33' 21.57" E	21° 49' 43.17" N
4	Herobhanga	88° 41' 46.52" E	21° 59' 34.32" N
5	Ajmalmari	88° 39' 00.68" E	21° 51' 34.72" N
6	Dhulibasani	88° 33' 48.20" E	21° 47' 06.62" N
7	Chulkati	88° 34' 10.31" E	21° 41' 53.62" N
8	Arbesi	89° 01' 09.04" E	22° 11' 43.14" N
9	Jhilla	88° 57' 57.07" E	22° 09' 51.53" N
10	Pirkhali	88° 51' 06.04" E	22° 06' 00.97" N
11	Panchamukhani	88° 54' 14.71" E	21° 59' 41.58" N
12	Harinbhanga	88° 59' 33.24" E	21° 57' 17.85" N
13	Khatuajhuri	89° 01' 05.33" E	22° 03' 06.55" N
14	Chamta	88° 57' 11.40" E	21° 53' 18.56" N
15	Matla	88° 44' 08.74" E	21° 53' 15.30" N
16	Chandkhali	89° 00' 44.68" E	21° 51' 13.59" N
17	Goasaba	88° 46' 41.44" E	21° 43' 50.64" N
18	Gona	88° 54' 31.09" E	21° 41' 15.44" N
19	Chhotahardi	88° 44' 17.79" E	21° 44' 42.24" N
20	Bagmara	89° 04' 40.59" E	21° 39' 04.45" N
21	Mayadwip	88° 47' 09.95" E	21° 35' 50.23" N
22	Jambu Island	88° 10' 22.76" E	21° 35' 42.03" N
23	Lothian	88° 20' 29.32" E	21° 38' 21.20" N
24	Sagar Island	88° 02' 20.97" E	21° 38' 51.55" N

Fig. 12A.1 Location of sampling stations

24 (1.44 gm/cc) > Stn. 22 (1.38 gm/cc) > Stn. 6 (1.35 gm/cc) = Stn. 11 (1.35 gm/cc) > Stn. 9 (1.34 gm/cc) = Stn. 13 (1.34 gm/cc) = Stn. 23 (1.34 gm/cc) > Stn. 1 (1.33 gm/cc) = Stn. 5 (1.33 gm/cc) = Stn. 20 (1.33 gm/cc) > Stn. 4 (1.32 gm/cc) = Stn. 8 (1.32 gm/cc) = Stn. 10 (1.32 %) > Stn. 2 (1.31 gm/cc) = Stn. 18 (1.31 gm/cc) = Stn. 19 (1.31 gm/cc) = Stn. 21 (1.31 gm/cc) > Stn. 3 (1.30 gm/cc) = Stn. 7 (1.30 gm/cc) = Stn. 12 (1.30 gm/cc) = Stn. 15 (1.30 gm/cc) = Stn. 16 (1.30 gm/cc) = Stn. 17 (1.30 gm/cc) > Stn. 14 (1.29 gm/cc) (Figs. 12A.5 and 12A.6). The significant spatial variations of bulk density ($p < 0.001$) as shown in Annexure 12A.2 are thus regulated by geo-physical processes.

Annexure 12A.1 SOC (in %) in 24 blocks of Indian Sundarbans

Station no.	Name	1–10 cm	10–20 cm	20–30 cm	30–40 cm
1	Muriganga	1.46	1.44	1.38	1.36
2	Saptamukhi	1.31	1.29	1.27	1.25
3	Thakuran	1.29	1.25	1.19	1.19
4	Herobhanga	1.05	1.02	1.01	0.99
5	Ajmalmari	0.91	0.89	0.88	0.85
6	Dhulibasani	0.83	0.8	0.79	0.79
7	Chulkathi	0.75	0.75	0.74	0.71
8	Arbesi	1.09	1.05	1.05	1.04
9	Jhilla	1.07	1.04	1.04	1.02
10	Pirkhali	1.04	1.01	0.99	0.98
11	Panchmukhani	1.02	1.00	0.98	0.92
12	Harinbhanga	1.07	1.03	0.98	0.89
13	Khatuajhuri	1.09	1.05	1.00	0.93
14	Chamta	1.05	0.99	0.92	0.86
15	Matla	0.86	0.84	0.82	0.81
16	Chandkhali	0.82	0.81	0.79	0.76
17	Goashaba	0.77	0.74	0.73	0.69
18	Gona	0.75	0.73	0.73	0.67
19	Chhotahardi	0.78	0.75	0.74	0.71
20	Baghmara	0.73	0.7	0.68	0.62
21	Mayadwip	0.71	0.69	0.65	0.60
22	Jambu Island	0.80	0.79	0.69	0.66
23	Lothian	1.31	1.31	1.26	1.18
24	Sagar Island	0.81	0.78	0.71	0.66

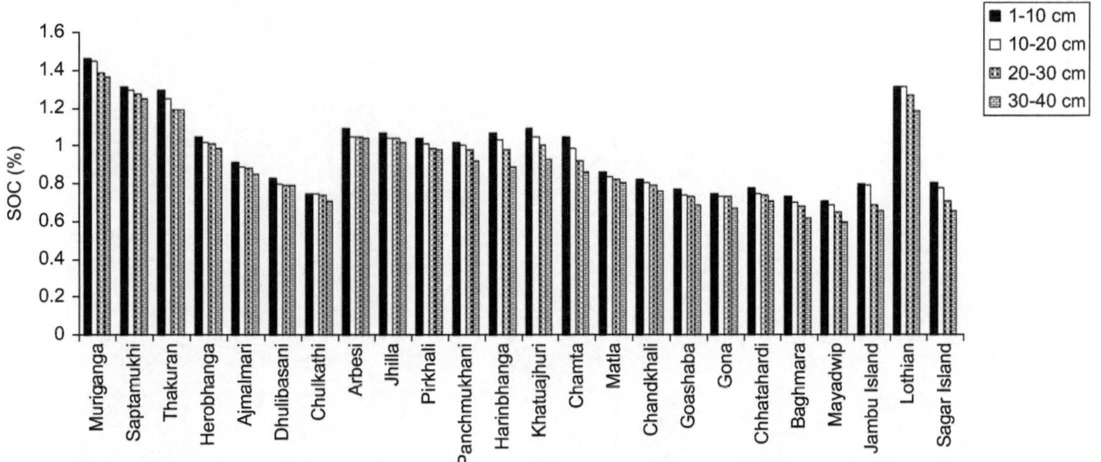

Fig. 12A.2 Spatial variation of mean soil organic carbon (SOC) in Indian Sundarbans

3.3 Organic Carbon Density (OCD)

OCD being a direct function of SOC and bulk density exhibits almost similar spatial variation to that of SOC. The OCD differs significantly between stations ($p < 0.001$). The spatial trend of OCD is in the order Stn.1 (1.875 kg/m^2) > Stn.23 (1.697 kg/m^2) > Stn.2 (1.680 kg/m^2) > Stn.3 (1.595 kg/m^2) > Stn.8 (1.398 kg/m^2) >

Table 12A.3 ANOVA for spatial variation of SOC, BD and OCD

	Source of variation	SS	DF	MS	F_{OBS}	P VALUE	F_{CRIT}
SOC	Between stations	4.0783	23	0.177317	337.127	6.94E-62	1.686
	Between depth	0.115408	3	0.038469	73.140	2.18E-21	2.737
BD	Between stations	0.099166	23	0.004312	12.801	8.66E-17	1.686
	Between depth	0.043586	3	0.014529	43.139	8.13E-16	2.737
OCD	Between stations	7.093164	23	0.308398	339.927	5.23E-62	1.686
	Between depth	0.071925	3	0.023975	26.426	1.71E-11	2.737

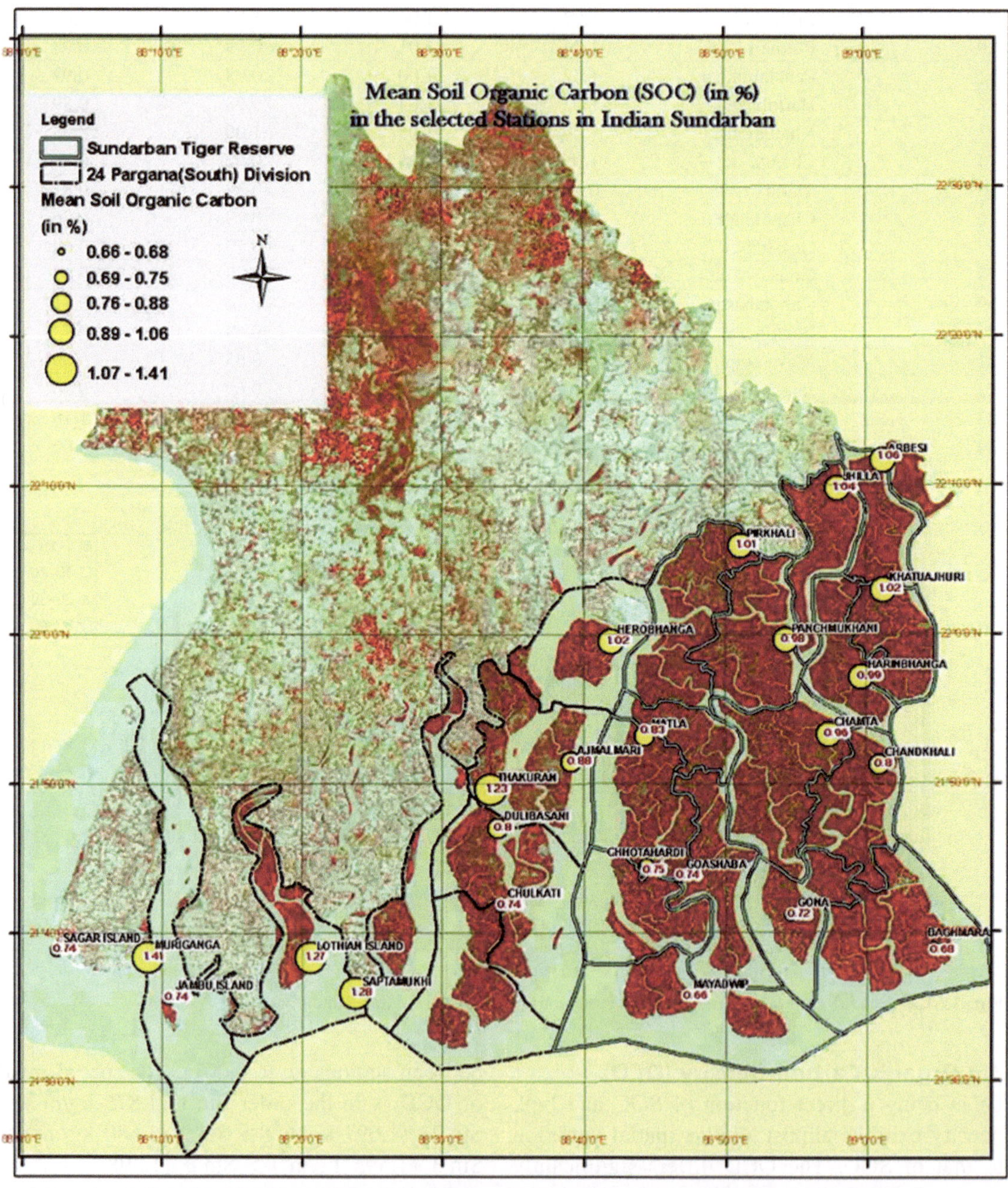

Fig. 12A.3 Map showing spatial variation of mean soil organic carbon (in %) in Indian Sundarbans

Fig. 12A.4 Shoreline changes of Sagar Island confirming the process of erosion in the Southern part

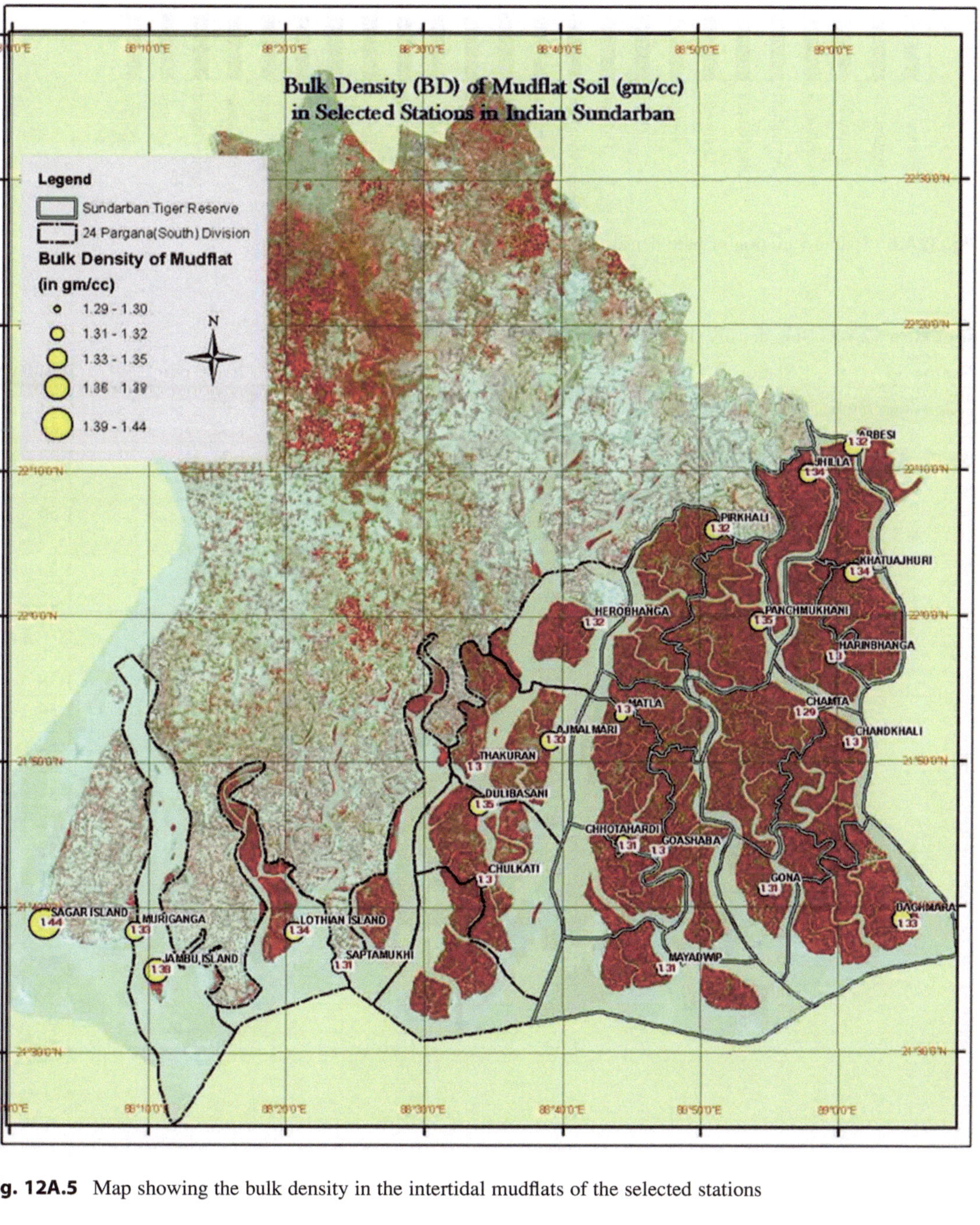

Fig. 12A.5 Map showing the bulk density in the intertidal mudflats of the selected stations

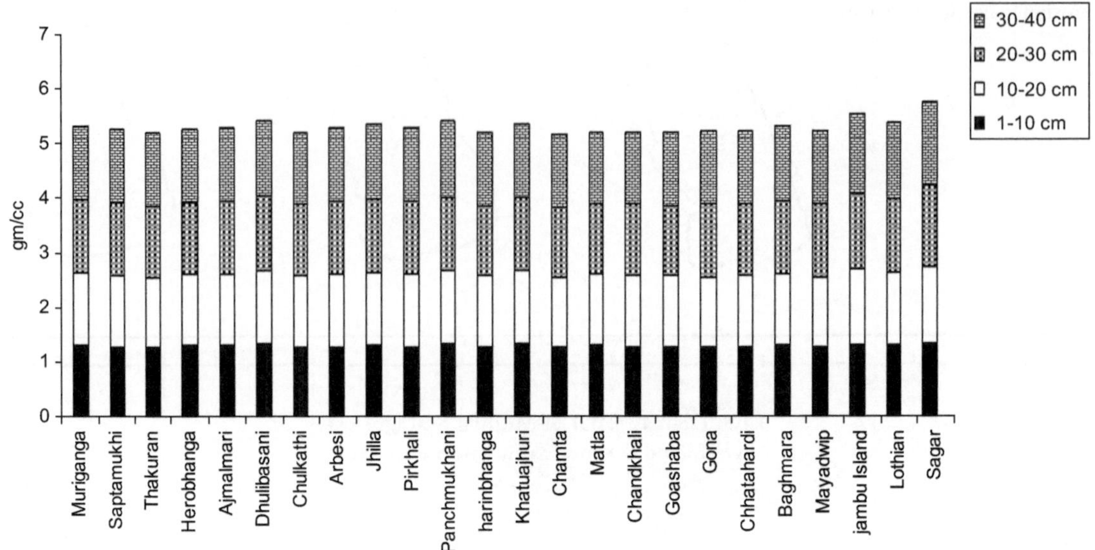

Fig. 12A.6 Spatial variation of bulk density

Annexure 12A.2 Bulk density (in gm/cc) of mudflat soil

Station no.	Station	1–10 cm	10–20 cm	20–30 cm	30–40 cm
1	Muriganga	1.32	1.33	1.33	1.34
2	Saptamukhi	1.28	1.31	1.32	1.34
3	Thakuran	1.27	1.29	1.31	1.32
4	Herobhanga	1.30	1.31	1.32	1.33
5	Ajmalmari	1.31	1.31	1.33	1.35
6	Dhulibasani	1.34	1.34	1.35	1.37
7	Chulkathi	1.28	1.30	1.30	1.33
8	Arbesi	1.29	1.31	1.34	1.35
9	Jhilla	1.31	1.33	1.33	1.37
10	Pirkhali	1.28	1.32	1.34	1.35
11	Panchmukhani	1.33	1.34	1.35	1.38
12	Harinbhanga	1.28	1.29	1.30	1.34
13	Khatuajhuri	1.33	1.34	1.35	1.33
14	Chamta	1.26	1.28	1.29	1.32
15	Matla	1.30	1.3	1.30	1.31
16	Chandkhali	1.28	1.3	1.31	1.32
17	Goashaba	1.28	1.29	1.29	1.32
18	Gona	1.26	1.29	1.33	1.34
19	Chhotahardi	1.29	1.3	1.31	1.33
20	Baghmara	1.30	1.32	1.34	1.37
21	Mayadwip	1.27	1.29	1.32	1.35
22	Jambu Island	1.32	1.38	1.39	1.44
23	Lothian	1.30	1.33	1.36	1.38
24	Sagar	1.33	1.41	1.50	1.52

Annexure 12A.3 Mean OCD (in kg/m^2) of Sundarban soil

Station	1–10 cm	10–20 cm	20–30 cm	30–40 cm
Muriganga	1.927	1.915	1.835	1.822
Saptamukhi	1.677	1.690	1.676	1.675
Thakuran	1.638	1.613	1.559	1.571
Herobhanga	1.365	1.336	1.333	1.317
Ajmalmari	1.192	1.166	1.170	1.148
Dhulibasani	1.112	1.072	1.067	1.082
Chulkathi	0.960	0.975	0.962	0.944
Arbesi	1.406	1.376	1.407	1.404
Jhilla	1.402	1.383	1.383	1.397
Pirkhali	1.331	1.333	1.327	1.323
Panchmukhani	1.357	1.340	1.323	1.270
Harinbhanga	1.370	1.329	1.274	1.193
Khatuajhuri	1.450	1.407	1.350	1.237
Chamta	1.323	1.267	1.187	1.135
Matla	1.118	1.092	1.066	1.061
Chandkhali	1.050	1.053	1.035	1.003
Goashaba	0.986	0.955	0.942	0.911
Gona	0.945	0.942	0.971	0.898
Chhotahardi	1.006	0.975	0.969	0.944
Baghmara	0.949	0.924	0.911	0.849
Mayadwip	0.902	0.890	0.858	0.810
Jambu Island	1.056	1.090	0.959	0.950
Lothian	1.703	1.742	1.714	1.628
Sagar Island	1.077	1.100	1.065	1.003

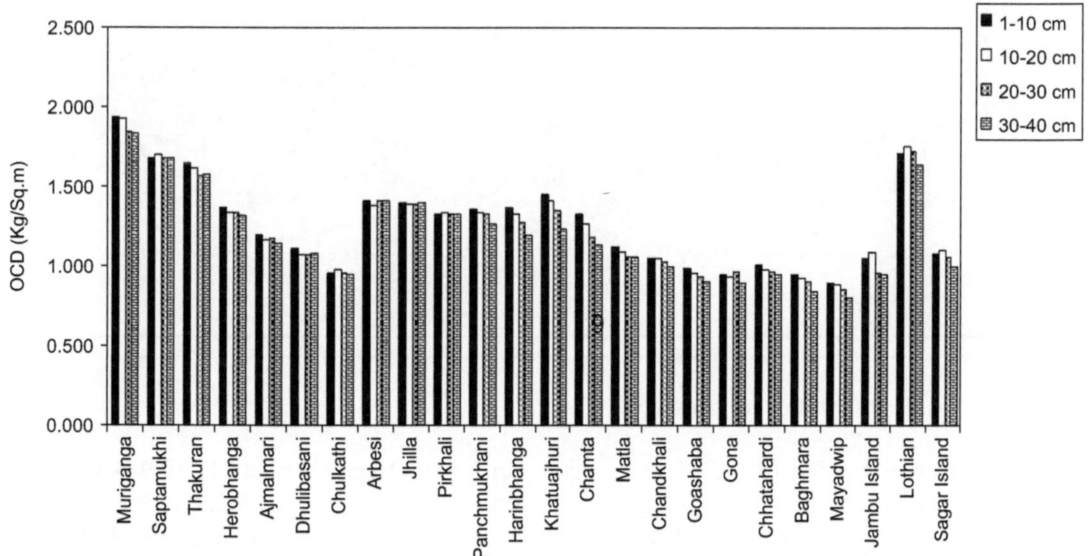

Fig. 12A.7 Spatial variation of OCD

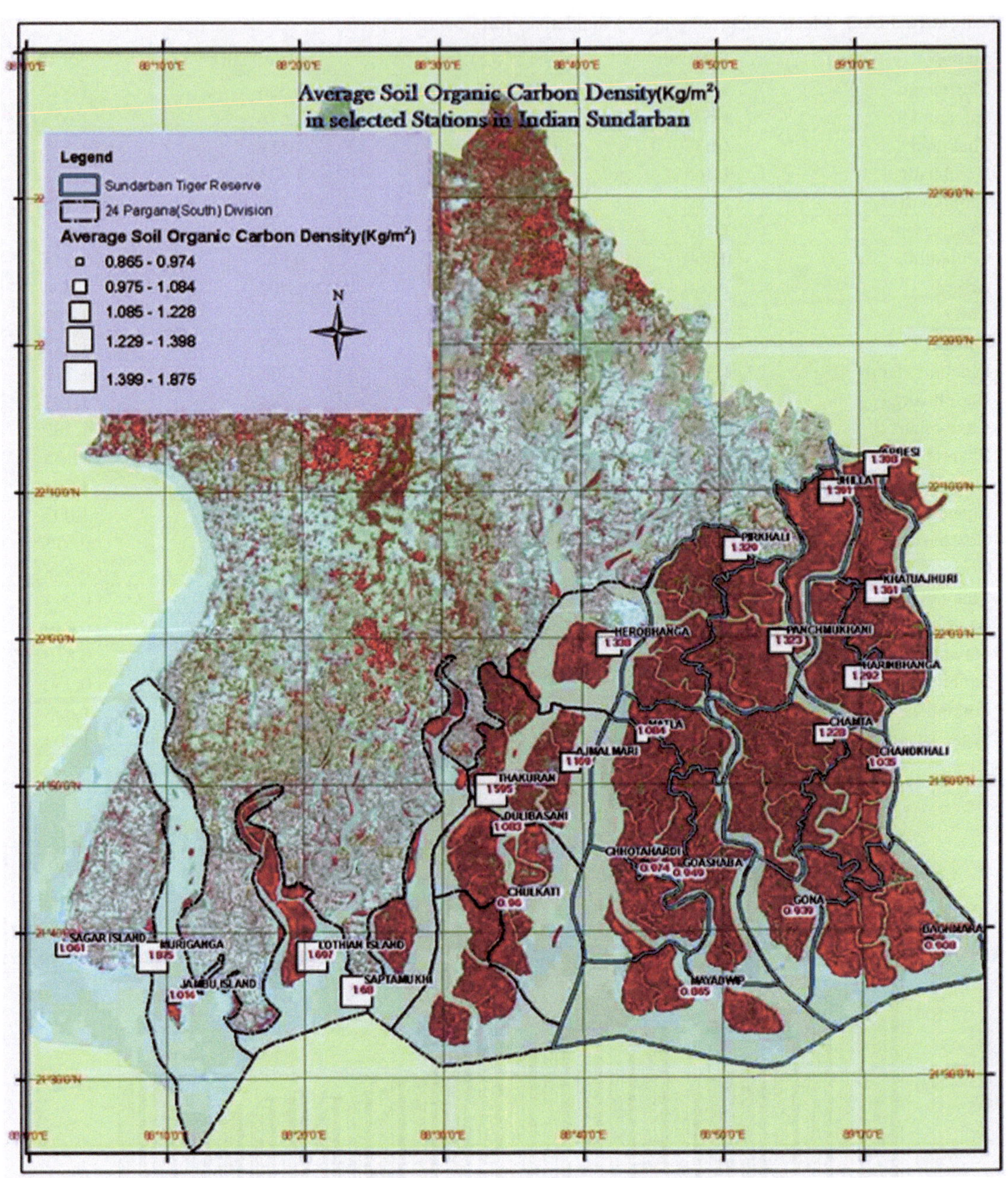

Fig. 12A.8 Map showing OCD in the study area

Stn.9 (1.391 kg/m^2) > Stn.13 (1.361 kg/m^2) > Stn.4 (1.338 kg/m^2) > Stn.10 (1.329 kg/m^2) > Stn.11 (1.322 kg/m^2) > Stn.12 (1.291 kg/m^2) > Stn.14 (1.228 kg/m^2) > Stn.5 (1.169 kg/m^2) > Stn.15 (1.084 kg/m^2) > Stn.6 (1.083 kg/m^2) > Stn.24 (1.061 kg/m^2) > Stn.16 (1.035 kg/m^2) > Stn.22 (1.014 kg/m^2) > Stn.19 (0.974 kg/m^2) > Stn.7 (0.960 kg/m^2) > Stn.17 (0.948 kg/m^2) > Stn.18 (0.939 kg/m^2) > Stn.20 (0.908 kg/m^2) > Stn.21 (0.865 kg/m^2) (Annexure 12A.3 and Figs. 12A.7 and 12A.8).

We compared our carbon density data (ranging from 0.865 kg/m^2 to 1.875 kg/m^2) with several global reports published between 2004 and

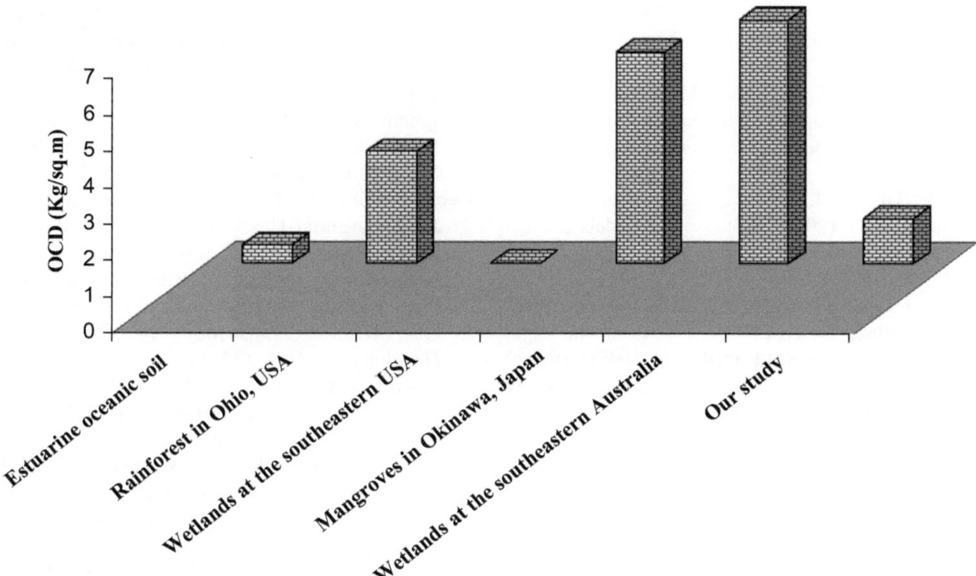

Fig. 12A.9 Global figures of OCD compared to our study in February 2012

2011. OCD of 3.03 kg/m^2, 0.033 kg/m^2, 5.73 kg/m^2, 6.61 kg/m^2 and 0.38 kg/m^2 was observed in rainforests of Ohio, USA (Bernal and Mitsch 2008); wetlands at the southeastern United States (Brevik and Homburg 2004); mangroves in Okinawa, Japan (Khan et al. 2007); wetlands at the southeastern Australia (Howe et al. 2009) and estuarine oceanic soil (Donato et al. 2011) respectively (Fig. 12A.9).

The present study is significant from the point that the area has not yet witnessed the light of documentation of soil carbon content, although above-ground mangrove biomass (AGMB) and carbon storage have been studied by several workers (Mitra et al. 2010, 2011). A thorough study has been done on the whole-ecosystem C storage in mangroves across a broad tract of the Indo-Pacific region, the geographic core of mangrove area (40 % globally) and diversity and the study sites comprised wide variation in stand composition and stature spanning 30° of latitude (8°S–22° N), 73° of longitude (90°–163°E) and including eastern Micronesia (Kosrae); western Micronesia (Yap and Palau); Sulawesi, Java, Borneo (Indonesia); and the Sundarban (Ganges-Brahmaputra Delta, Bangladesh) (Donato et al. 2011). The study, however, left out the lower Gangetic region sustaining the Indian Sundarban. The present approach is thus an attempt to fill this gap area and establish a baseline data of SOC and OCD in the mangrove-dominated Indian part of Sundarban delta.

References

Behl, C. (1999). Alzheimer's disease and oxidative stress: Implications for novel therapeutic approaches. *Progressive Neurobiology, 57*, 301.

Bograd, S. J., Block, B. A., Costa, D. P., & Godley, B. J. (2010). Biologging technologies: New tools for conservation. Introduction. *Endangered Species Research, 10*, 1–7.

Ebadi, M., Srinivasan, S. K., & Baxi, M. D. (1996). Oxidative stress and antioxidant therapy in Parkinson's disease. *Progressive Neurobiology, 48*, 1.

Howell, E. A., Kobayashi, D. R., Parker, D. M., Balazs, G. H., & Polovina, J. J. (2008). TurtleWatch: A tool to aid in the bycatch reduction of loggerhead turtles *Caretta caretta* in the Hawaii-based pelagic longline fishery. *Endangered Species Research, 5*, 267–278.

Kobayashi, D. R., Polovina, J. J., Parker, D. M., Kamezaki, N., Cheng, I.-J., Uchida, I., et al. (2008). Pelagic habitat characterization of loggerhead sea turtles, *Caretta caretta*, in the North Pacific Ocean (1997–2006): Insights from satellite tag tracking and remotely sensed data. *Journal of Experimental Marine Biology and Ecology, 356*, 96–114.

Markesbery, W. R., & Carney, J. M. (1999). Oxidative alterations in Alzheimer's disease. *Brain Pathology, 9*, 133.

Nielsen, A., Bigelow, K. A., Musyl, M. K., & Sibert, J. R. (2006). Improving light-based geolocation by including sea surface temperature. *Fisheries Oceanography, 15*, 314–325.

Patterson, T., Thomas, L., Wilcox, X., Ovaskainen, O., & Matthiopoulos, J. (2008). State–space models of individual animal movement. *Trends in Ecology & Evolution, 23*, 87–94.

Polovina, J. J., Kobayashi, D. R., Parker, D. M., Seki, M. P., & Balazs, G. H. (2000). Turtles on the edge: Movement of loggerhead turtles (*Caretta caretta*) along oceanic fronts, spanning longline fishing grounds in the central North Pacific, 1997–1998. *Fisheries Oceanography, 9*, 71–82.

Polovina, J. J., Howell, E., Kobayashi, D. R., & Seki, M. P. (2001). The transition zone chlorophyll front, a dynamic global feature defining migration and forage habitat for marine resources. *Progress in Oceanography, 49*, 469–483.

Tang, D. L., & Kawamura, H. (2001). Long-term time series satellite ocean color products on the Asian waters. In *Proceedings of the 11 th PAMS/JECSS workshop* (pp. 49–52). Seoul, South Korea: Hanrimwon Publishing (CD-ROM: O112-P03).

Tang, D. L., Kawamura, H., & Luis, A. J. (2002). Short-term variability of Phytoplankton blooms associated with a cold eddy in the northeastern Arabian Sea. *Remote Sensing of Environment, 81*, 82–89.

Tremblay, Y., Robinson, P. W., & Costa, D. P. (2009). A parsimonious approach to modeling animal movement data. *PLoS ONE, 4*, e4711.

Internet Reference

scilib.ucsd.edu/sio/annual/Annual_report_2010_pdf

Annexure References

Alongi, D. M. (2008). Mangrove forests: Resilience, protection from tsunamis, and responses to global climate change. *Estuarine, Coastal and Shelf Science, 76*, 1–13.

Anthropogenic carbon fluxes; 1980 to 1989. IPCC 1994

Bernal, B., & Mitsch, W. J. (2008). A comparison of soil carbon pools and profiles in wetlands in Costa Rica and Ohio. *Ecological Engineering, 34*, 311–323.

Bouillon S., Rivera-Monroy V. H., Twilley R. R., & Kairo J. G. (2009). Mangroves. In: D. Laffoley, & G. Grimsditch (Eds.), *The management of natural coastal carbon sinks*. International Union for the Conservation of Nature and Natural Resources, IUCN, Gland, Switzerland, pp. 53.

Brevik, E., & Homburg, J. (2004). A 500 year record of carbon sequestration from a coastal lagoon and wetland complex, Southern California, USA. *Catena, 57*, 221–232.

Bridgham, S. D., Megonigal, J. P., Keller, J. K., Bliss, N. B., & Trettin, C. (2006). The carbon balance of North American wetlands. *Wetlands, 26*, 889–916.

Cerón-Bretón, J. G., Cerón-Bretón, R. M., Guerra-Santos, J. J., Aguilar-Ucán, C., Montalvo-Romero, C., Vargas-Cáliz, C., et al. (2010). Effects of simulated tropospheric ozone on nutrients levels and photosynthetic pigments concentrations of three mangrove species. *WSEAS Transactions on Environment and Development, 6*(2), 133–143.

Chaudhuri, A. B., & Choudhury, A. (1994). *Mangroves of the Sundarban – India* (1st ed.). Bangkok: IUCN- The World Conservation Union.

Chmura, G. L., Anisfeld, S. C., Cahoon, D. R., & Lynch, J. C. (2003). Global carbon sequestration in tidal, saline wetland soils. *Global Biogeochemical Cycles, 17*, 1111.

Donato, D. C., Kauffman Brone, J., Murdiyarso, D., Sofyan, K., Melanie, S., & Markku, K. (2011). Mangroves amongst the most carbon-rich forests in the tropics. *Nature Geoscience, 4*, 293–297.

Fontaine, S., Barot, S., Barre, P., Bdioui, N., Mary, B., & Rumpel, C. (2007). Stability of organic carbon in deep soil layers controlled by fresh carbon supply. *Nature, 450*, 277–280.

Gill, R., Burke, I. C., Milchunas, D. G., & Lauenroth, W. K. (1999). Relationship between root biomass and soil organic matter pools in the short-grass steppe of Eastern Colorado. *Ecosystems, 2*, 226–236.

Homann, P. S., Remillard, S. M., Harmon, M. E., & Bormann, B. T. (2004). Carbon storage in coarse and fine fractions of Pacific Northwest old growth forest soils. *Soil Science Society of America Journal, 68*, 2023–2030.

Howe, A. J., Rodríguez, J. F., & Saco, P. M. (2009). Surface evolution and carbon sequestration in disturbed and undisturbed wetland soils of the Hunter estuary, southeast Australia. *Estuarine, Coastal and Shelf Science, 84*, 75–86.

Jackson, R. B., Schenk, H. J., Jobba'gy, E. G., Canadell, J., Colello, G. D., Dickinson, R. E., et al. (2000). Belowground consequences of vegetation change and their treatment in models. *Ecological Applications, 10*, 470–483.

Jobb'agy, E. G., & Jackson, R. B. (2000). The vertical distribution of soil organic carbon and its relation to climate and vegetation. *Ecological Applications, 10*, 423–436.

Khan, M. N. I., Suwa, R., & Hagihara, A. (2007). Carbon and nitrogen pools in a mangrove stand of Kandelia obovata (S., L.) Yong: Vertical distribution in the soil-vegetation systems. *Wetlands Ecology and Management, 15*, 141–153.

Lacerda, L. D., Carvalho, C. E. V., Tanizaki, K. F., Ovalle, A. R. C., & Rezende, C. E. (1993). The biogeochemistry and trace metals distribution of mangrove rhizospheres. *Biotropica, 25*, 251–256.

Lacerda, L. D., Ittekkot, V., & Patchineelam, S. R. (1995). Biochemistry of mangrove soil organic matter: A comparison between Rhizophora and Avicennia soils in South-eastern Brazil. *Estuarine, Coastal and Shelf Science, 40*, 713–720.

Meersmans, J., van Wesemael, B., De Ridder, F., Fallas Dotti, M., De Baets, S., & Van Molle, M. (2009). Changes in organic carbon distribution with depth in agricultural soils in Northern Belgium, 1960–2006. *Global Change Biology, 15*, 2739–2750.

Mi, N., Wang, S. Q., Liu, J. Y., Yu, G. R., Zhang, W. J., & Jobb'agy, E. G. (2008). Soil inorganic carbon storage pattern in China. *Global Change Biology, 14*, 2380–2387.

Mitra, A. (2000). The Northeast coast of the Bay of Bengal and deltaic Sundarban. In C. Sheppard (Ed.), *Seas at the Millennium – An environmental evaluation, Chapter 62* (pp. 143–157). Amsterdam: Elsevier Science.

Mitra, A., Banerjee, K., & Sengupta, K. (2010). The affect of salinity on the mangrove growth in the lower Gangetic plain. *Journal of Coastal Environment, 1*(1), 71–82.

Mitra, A., Sengupta, K., & Banerjee, K. (2011). Standing biomass and carbon storage of above-ground structures in dominant mangrove trees in the Sundarban. *Forest Ecology and Management, 261* (7), 1325–1335.

Mitsch, W. J., & Gosselink, J. G. (2000). *Wetlands* (p. 920). New York: Wiley.

Murako, D. (2004). Seaweed resources as a source of carbon fixation. *Bulletin Fisheries Research Agency, 1*, 59–63.

Ovalle, A. R. C., Rezende, C. E., Lacerda, L. D., & Silva, C. A. R. (1990). Factors affecting the hydrochemistry of a mangrove tidal creek, Sepetiba Bay, Brazil. *Estuarine, Coastal and Shelf Science, 31*, 639–650.

Shukla, M. K., & Lal, R. (2005). Erosional effects on soil organic carbon stock in an on-farm study on Alfisols in west central Ohio. *Soil and Tillage Research, 81*, 173–181.

Siegenthaler, U., & Sarmiento, J. L. (1993). Atmospheric carbon dioxide and the ocean. *Nature, 365*, 119–125.

Tanizaki, K. F. (1994). *Biogeoquímica de metais pesados na rizosfera de plantas de um manguezal do Rio de Janeiro*, M.Sc. thesis, Departamento de Geoquímica, Universidade Federal Fluminense, Niterói, p. 67.

Vesterdal, L., Ritter, E., & Gundersen, P. (2002). Change in soil organic carbon following afforestation of former arable land. *Forest Ecology and Management, 162*, 137–147.

Walkley, A., & Black, I. A. (1934). An examination of the Degtjareff method for determining soil organic matter and a proposed modification of the chromic acid titration method. *Soil Science, 37*, 29–38.

Wang, S. Q., Huang, M., Shao, X. M., Mickler, R. A., Li, K. R., & Ji, J. J. (2004). Vertical distribution of soil organic carbon in China. *Environmental Management, 33*(Suppl.1), S200–S209.

Zinn, Y. L., Lal, R., & Resck, D. V. S. (2005). Changes in soil organic carbon stocks under agriculture in Brazil. *Soil and Tillage Research, 84*, 28–40.

Erratum to: Conservation of Marine and Estuarine Resources

Abhijit Mitra and Sufia Zaman

Erratum to: Chapter 11 in: A. Mitra, S. Zaman, *Basics of Marine and Estuarine Ecology*, https://doi.org/10.1007/978-81-322-2707-6_11

The original version of the book was inadvertently published without citations to the original copyright holder for Tables 11.5 and 11.6.

The correct citation has been included as reference citation in text on page 439 and in the table caption on page 441 and 442 as follows:

On page 439:

In the marine and estuarine ecosystems, there are more animal phyla than on land. Most phyla are encompassed within the benthic compartment. Even within the benthic environment, diversity varies with depth as seen in the case of epifaunal diversity in the north-east Indian shelf (Table 11.5) (Ganesh and Raman 2007).

A similar picture was also observed in the infaunal community, where different depths witness variation in diversity of infaunal species (Table 11.6) (Ganesh and Raman 2007).

On Pages 441 and 442:

Table 11.5 Distribution of important epifaunal species (ind. haul^{-1}) at different depths on the north-east Indian shelf (Ganesh and Raman 2007)

Table 11.6 Distribution of important infaunal species (ind. m^{-2}) at different depths on the north-east Indian shelf (Ganesh and Raman 2007)

The updated online version of this chapter can be found at https://doi.org/10.1007/978-81-322-2707-6_11

A. Mitra, S. Zaman, *Basics of Marine and Estuarine Ecology*,
DOI 10.1007/978-81-322-2707-6_13